AF837188

Application of Solution Protein Chemistry to Biotechnology

Application of Solution Protein Chemistry to Biotechnology

Roger L. Lundblad

CRC Press
Taylor & Francis Group
Boca Raton London New York

CRC Press is an imprint of the
Taylor & Francis Group, an **informa** business

CRC Press
Taylor & Francis Group
6000 Broken Sound Parkway NW, Suite 300
Boca Raton, FL 33487-2742

© 2009 by Taylor & Francis Group, LLC
CRC Press is an imprint of Taylor & Francis Group, an Informa business

No claim to original U.S. Government works
Printed in the United States of America on acid-free paper
10 9 8 7 6 5 4 3 2 1

International Standard Book Number-13: 978-1-4200-7341-6 (Hardcover)

This book contains information obtained from authentic and highly regarded sources. Reasonable efforts have been made to publish reliable data and information, but the author and publisher cannot assume responsibility for the validity of all materials or the consequences of their use. The authors and publishers have attempted to trace the copyright holders of all material reproduced in this publication and apologize to copyright holders if permission to publish in this form has not been obtained. If any copyright material has not been acknowledged please write and let us know so we may rectify in any future reprint.

Except as permitted under U.S. Copyright Law, no part of this book may be reprinted, reproduced, transmitted, or utilized in any form by any electronic, mechanical, or other means, now known or hereafter invented, including photocopying, microfilming, and recording, or in any information storage or retrieval system, without written permission from the publishers.

For permission to photocopy or use material electronically from this work, please access www.copyright.com (http://www.copyright.com/) or contact the Copyright Clearance Center, Inc. (CCC), 222 Rosewood Drive, Danvers, MA 01923, 978-750-8400. CCC is a not-for-profit organization that provides licenses and registration for a variety of users. For organizations that have been granted a photocopy license by the CCC, a separate system of payment has been arranged.

Trademark Notice: Product or corporate names may be trademarks or registered trademarks, and are used only for identification and explanation without intent to infringe.

Library of Congress Cataloging-in-Publication Data

Lundblad, Roger L.
Application of solution protein chemistry to biotechnology / Roger L. Lundblad.
p. ; cm.
Includes bibliographical references and index.
ISBN 978-1-4200-7341-6 (hardcover : alk. paper)
1. Proteins--Biotechnology. 2. Proteins--Solubility. 3. Solution (Chemistry) I. Title.
[DNLM: 1. Proteins--chemistry. 2. Biotechnology--methods. QU 55 L962a 2009]

TP248.65.P76L86 2009
660.6'3--dc22 2009006362

Visit the Taylor & Francis Web site at
http://www.taylorandfrancis.com

and the CRC Press Web site at
http://www.crcpress.com

Contents

Preface

I have a broad view of biotechnology, ranging from the use of yeast in the preparation of baked goods, to the use of the browning reaction in cooking, to biological adhesives, to the high technology of cell culture preparation of protein therapeutics. Thus, this book has chapters on the use of classical protein chemistry in food; the protein and carbohydrate chemistry of adhesives, glues and sealants; protein chemistry and hydrogels; and the use of chemical modification to prepare protein therapeutics. This book intends to demonstrate current use and, as important, the development of the applications of technologies for use in biotechnology. There is considerable material that is not available in electronic format. I would emphasize that just because you cannot find it on the Internet doesn't mean it does not exist. This understanding may prevent someone from reinventing the wheel. As a caveat, I would note that giving something a new name does not constitute a discovery. It is hoped that investigators will find this approach useful.

I would like to once again thank Professor Bryce Plapp of the University of Iowa for his tolerance of the thermodynamically challenged and Professor Don Gabriel of the University of North Carolina at Chapel Hill for helpful discussions on adhesives, glues, and sealant. The author also thanks the Barbara Norwitz and Jill Jurgensen of Taylor & Francis for their hard work in bringing this work to fruition.

Roger L. Lundblad

Chapel Hill, North Carolina

The Author

Roger L. Lundblad is a native of San Francisco, California. He received his undergraduate education at Pacific Lutheran University and his Ph.D. in biochemistry at the University of Washington. After postdoctoral work in the laboratories of Stanford Moore and William Stein at The Rockefeller University, he joined the faculty of the University of North Carolina at Chapel Hill. He joined the Hyland Division of Baxter Healthcare in 1990. Currently Dr. Lundblad is an independent consultant and writer in biotechnology in Chapel Hill, North Carolina. He is an adjunct professor of pathology at the University of North Carolina at Chapel Hill and editor-in-chief of the Internet *Journal of Genomics and Proteomics*.

1 Introduction to the Solution Chemistry of Proteins

Proteins are polymers composed of different monomer units, so protein is considered a heteropolymer.[1–4] As with other polymers, proteins have functional and structural roles; for example, proteins can be turned into plastics (see Chapter 6). The solution chemistry of proteins includes the response of proteins to changes in solvent,[5–9] as well as the reactivity of functional groups on proteins. This chapter will focus on the latter subject, including discussion of the effect of solvent and microenvironment. Changes in environment such as solvent or temperature that affect protein conformation and the tools used to study shape changes in protein conformation are discussed in Chapter 2. Changes in solvent composition, such as changes in metal ion concentration, are usually reversible, whereas changes induced by changes in pH or temperature are frequently irreversible and are associated with protein denaturation (see Chapters 2, 5, and 6 for further discussion of protein denaturation). The chemical modification of a protein may or may not result in a conformational change (see Chapter 2).

The chemical modification of a protein can occur either from the addition of a specific reagent or reagents, resulting in the random or nonrandom chemical modification of different amino acid residues (e.g., modification of tyrosine and lysine residues by acylating reagents such as acetic anhydride), random or nonrandom modification of a single type of amino acid (e.g., the modification of several lysine residues without modification of other amino acid residues such as tyrosine as with pyridoxal phosphate), or the site-specific chemical modification of a single amino acid residue (most frequently a functional group in binding or catalysis). A variety of reagents are used to effect the modification of amino acid residues in proteins. A list of some of the more commonly used reagents is provided in Table 1.1.

Site-specific chemical modification is strictly defined as a process that yields a stoichiometrically altered protein with the quantitative covalent derivatization of a single unique amino acid residue, without either modification of any other amino acid residue or conformational change. In fact, this objective is rarely obtained with most reagents as there are several factors that confound this goal. First, few reagents are specific for the modification of a single functional group. Most reagents react with nucleophiles on proteins, and the nucleophilic character is, in part, dependent on the protonation state of the residue. The acid dissociation constants for "typical" amino acid function groups are presented in Table 1.2. The acid dissociation constant is dependent on the microenvironment surrounding the specific amino acid residues, and this issue is discussed in the following text.

TABLE 1.1
Reagents for the Chemical Modification of Proteins and Other Biological Macromolecules

Reagent	Specificity/Conditions[a]	Molecular Weight	References
Acetic anhydride	Lysine, α-amino groups, tyrosine hydroxyl; preferred reaction is at lysine; pH 8 or greater; reaction can be "driven" α-amino groups at pH less than 6.5. Avoid nucleophilic buffers such as Tris; hydrolysis of the reagent is an issue above pH 9.5. Acetic anhydride has been used for trace labeling in the study of protein conformation and, more recently, the deuterated derivative has been used in proteomics for differential isotope tagging.	102.1	1–5
N-acetylimidazole (1-acetylimidazole)	Tyrosine hydroxyl groups, lysine ε-amino groups, transient reaction at histidine; neutral pH.	110.1	6–10
BNPS-skatole [bromo-3-methyl-2-(2-nitro-phenylmercapto)-3*H*-indole; 2-(2′ nitrophenyl-sulfenyl)-3-methyl-3 bromoindolenine	Tryptophan; 50–70% acetic acid; associated with peptide bond cleavage.	363.3	11–15
Bromoacetamide (2-bromoacetamide)	Cysteine; reaction with active site histidine residues; also reaction with lysine, methionine and, possibly, carboxylic acids. Reaction at pH 5–9 but reaction with methionine at pH 3.0. Reaction rate below pH 7.5 is usually slow as the modification of cysteine requires thiolate anion (p*K*a for cysteine is 8.7). Reaction is slower than iodoacetamide. A neutral reagent.[b]	138	16–20
Bromoacetic acid (2-bromoacetic acid)	Reaction parameters similar to bromoacetamide except bromoacetic acid is a charged reagent at pH greater than 4 (p*K*a is 2.7 at 25°C). Amide and acid derivatives can show different reaction patterns.[c]	139	21–25
Bromoethylamine	Modification of sulfhydryl groups; conversion of cysteine to lysine analog (*S*-2-aminoethylcysteine); reaction with cysteine at alkaline pH (see bromoacetamide). Reaction is reasonably specific for cysteine, with possible modification at the amino-terminal α-amino group and histidine.	204.9 as HBr salt	26–30
N-bromosuccinimide	Modification of tryptophan with some oxidative side reactions; Ph 4–6.	178	31–35

TABLE 1.1 (CONTINUED)
Reagents for the Chemical Modification of Proteins and Other Biological Macromolecules

Reagent	Specificity/Conditions[a]	Molecular Weight	References
2,3-Butanedione (diacetyl)	Modification of arginine residues; reversible reaction with the product stabilized by the presence of borate; reaction at alkaline pH.	86.1	36–40
Citraconic anhydride	Reversible modification of lysine residues; modification at pH above 8.0 and reversed below pH 6.0.	112.1	41–45
Cyanate[d]	Carbamylation of α-amino groups in proteins at alkaline pH, with some preference toward the modification of N-terminal α-amino groups. Reaction also occurs at cysteine residues using 2-nitro-5-thiocyanatobenzoic acid.[d]	N/A	46–50
1,2-Cyclohexanedione	Modification of arginine in the presence of borate. At pH 7–9, the reaction is reversible; above pH 9, the reaction is irreversible with the formation of several products.	112.1	51–55
DCC (1,3-dicyclohexyl-carbodiimide)	Modification of carboxyl groups in proteins (activated carboxyl group, which is then modified with a nucleophile such as glycine methyl ester; solubility issues have made EDC a more attractive reagent); also used for synthesis of phosphate ester bonds and peptide bonds; pH less than 5, modification requires protonated carboxyl group. The reagent has been used at more alkaline pH values. The modification of tyrosine and cysteine has been reported. DCC is an inhibitor of the proton-translocating ATPase in mitochondria and has been extensively used to characterize that activity.	206.3	56–60
Diethylpyrocarbonate	Modification of histidine residues in proteins with transient modification of tyrosine; possible reaction at amino groups. Disubstitution of histidine results in ring opening.	162.1	61–65
EDC [1-ethyl-3(3-dimethylaminopropyl)-carbodimide]; *N*-(3-dimethylamino-propyl)-*N*′-ethylcarbo-diimide	Modification of carboxyl groups in proteins frequently with *N*-hydroxylsuccinimide. Used for zero-length cross-linking in proteins and for the coupling of proteins to matrices and for the preparation of protein conjugates.	155.2	66–70
Ellman's reagent (5,5′-dithio-*bis*-nitrobenzoic acid)	Modification and measurement of cysteine (sulfhydryl) groups in proteins.	396.4	71–75

TABLE 1.1 (CONTINUED)
Reagents for the Chemical Modification of Proteins and Other Biological Macromolecules

Reagent	Specificity/Conditions[a]	Molecular Weight	References
N-ethylmaleimide	Modification of sulfhydryl groups via Michael addition to the maleimide ring. There are some proteins that are distinguished by their modification with *N*-ethylmaleimide, including *N*-ethylmaleimide-sensitive fusion protein (NSF) and soluble *N*-ethylmaleimide-sensitive factor attachment protein receptors (SNARES).	125.1	76–80
Ethyleneimine (aziridine)	Earlier used as modification reagent for cysteine but has been largely supplanted by bromoethylamine. Aziridine (ethyleneimine) serves as a functional base for reagents used for the modification of cysteine in proteins.	43.1	81–85
Formaldehyde[e,f] (reductive methylation)	Formation of a Schiff base with a primary amine (e.g., ε-amino group of lysine), which is reduced with sodium borohydride or more often with sodium cyanoborohydride, resulting in the formation, in the case of lysine, of ε-methyllysine and ε-dimethyllysine. The modification is performed at alkaline pH. In a related reaction, the formylation of tryptophan occurs with formic acid under acidic conditions (HCOOH/HCl), which is reduced by base.[g]	30.0	86–90
Gold	Reaction with cysteine. This process is used to bind proteins to solid matrices. The reaction of gold with proteins provides some of the basis for the use of gold and gold compounds of the treatment of arthritis.	N/A	91–95
2-Hydroxy-5-nitrobenzyl bromide (Koshland's reagent)	Modification of tryptophan by alkylation of the indole ring. Reaction at acid pH (pH 2–6). Disubstitution can occur. This was one of the first "reporter groups."[h]	232.0	96–100
N-hydroxysuccinimide	Functional group for modification of amino groups in proteins, frequently with carboiimide for carboxyl modification and coupling in proteins.	115.1	101–105
2-Iminothiolane (Traut's reagent)	Placement of a sulfhydryl group by the modification of amino groups in proteins and other amino-containing materials.	137.6 as the HCl	105–110
Iodoacetamide	Reaction is faster than with bromo or chloro derivatives. Reaction characteristics similar to bromoacetamide; similar to bromoacetamide, iodoacetamide is neutral.	185	111–115

TABLE 1.1 (CONTINUED)
Reagents for the Chemical Modification of Proteins and Other Biological Macromolecules

Reagent	Specificity/Conditions[a]	Molecular Weight	References
Iodoacetic acid	See bromoacetic acid for reaction conditions. As with bromoacetic acid, reagent is charged at pH greater than 5 (p*K*a is 3.12 at 25°C).	186	116–120
2-Mercaptoethanol (β-mercaptoethanol); 2-thioethanol	Used to maintain proteins in reduced states; previous use was for the reduction of disulfide bonds but has been largely replaced by TCEP (see later text).	78.1	121–125
Mercuric chloride ($HgCl_2$)	Reaction with organic sulfhydryl groups such as cysteine. The chemistry of this reaction is poorly understood.[i] Mercuric chloride is used for the inhibition of membrane enzymes and is described as a specific inhibitor of aquaporin.[j]	271.5	126–130
Methyl acetimidate	Modification of amino groups. Imido esters are the functional groups for a number of cross-linking agents such as dimethylsuberimidate.[k] One of the more interesting imido esters is methyl picolinimidate.[l] Reaction at pH 8–10.	109.6 as HCl	131–135
Methyl methane-thiosulfonate	Methyl methanethiosulfonate is one of a group of alkyl methanethiosulfonate derivatives that reversibly modify cysteine residues in proteins.[m] Reaction occurs at slightly alkaline pH (pH 7.8).	126.2	136–140
Ninhydrin (1*H*-indene-1,2,3-trione monohydrate; 2,2-dihydroxy-1,3-indanedione)	Determination of amino groups; provides the basis for detection in amino acid analysis; modification of arginine residues.[n] Ninhydrin is also used for the detection of cyanide.	178.1	141–145
2-Nitrophenylsulfenyl chloride (*o*-nitrophenyl-sulfenyl chloride)	Modification of tryptophan residues in proteins; reaction occurs at acid pH; modified tryptophan can be converted to the 2-thioltryptophan derivatives. This modification has been used to purify tryptophan peptides from protein hydrolyzates. An analog, 2-(trifluoromethyl)-benzenesulfenyl chloride, has been developed for use in mass spectrometry.[o]	189.6	146–150
Performic acid (peroxyformic acid)	Cleavage of disulfide bonds; oxidation of cysteine and methionine in proteins with side reactions at tryptophan with other minor modifications.[p]	62	151–155

TABLE 1.1 (CONTINUED)
Reagents for the Chemical Modification of Proteins and Other Biological Macromolecules

Reagent	Specificity/Conditions[a]	Molecular Weight	References
Phenylglyoxal	Modification of arginine residues in proteins; reaction accelerated in the presence of bicarbonate buffers; reaction at alkaline pH. *p*-hydroxyphenylglyoxal and *p*-nitrophenylglyoxal are useful derivates.[q]	134.1 as hydrate	156–160
Sodium cyanoborohydride	Reducing agent for Schiff bases in proteins; reduces ketones, aldehydes, hydrazones, and enamines but does not reduce lactones, amides, or disulfide bonds. Used to "stabilize" Schiff base linkages between amine "probes" and aldehyde-based matrices.	62.8	161–165
Sodium sulfite	Oxidative sulfitolysis to cleave disulfide bonds; conversion of cysteine to *S*-sulfocysteine.	126	166–170
Sodium tetrathionate ($Na_2S_4O_6$)	Modification of cysteine to *S*-sulfoderivatives; a reversible modification. Also a mild oxidizing agent that can, via the *S*-sulfoderivative, drive the formation of disulfide bonds, thus serving a protein cross-linking reagent.[r]	306.3	171–175
TCEP (tris[2-carboxyethyl] phosphine)	Reduction of disulfide bonds in proteins; reagent of choice. The reagent is readily soluble in water and reduces disulfide bonds at low pH (3.0).	286.7 at HCl	176–180
Tetranitromethane (TNM)	Nitration of tyrosine residues in proteins with nitration and possible cross-linking; also reacts with sulfhydryl groups; possible reaction with indole ring of tryptophan. Reaction at alkaline pH; does introduce a "reporter group" in proteins. The nitrotyrosine function can be reduced to aminotyrosine with sodium dithionite (sodium hydrosulfite). The modification with peroxynitrite is a similar reaction.[s]	196	181–185
TNBS (trinitro-benzenesulfonic acid)	Modification of amino groups in proteins; used for the determination of amino groups. Reaction at alkaline pH; also reacts with sulfhydryl groups.	293.2	186–190
Vinyl pyridine	Modification of sulfhydryl groups with majority of use in protein structure analysis; both the 2-vinylpyridine and the 4-vinylpyridine are used.	105.1	191–195
Woodward's reagent K (*N*-ethyl-5-phenylisoxazolium-3-sulfonate)	Reagent used for the modification of carboxyl groups in proteins. Reaction at acidic pH.	253.3	196–200

TABLE 1.1 (CONTINUED)
Reagents for the Chemical Modification of Proteins and Other Biological Macromolecules

[a] Absolute specificity of modification cannot be guaranteed. Conditions are most common with the caveat that specific buffer effects are observed (see Buffers).

[b] Chaiken, I.M. and Smith, E.L., Reaction of chloroacetamide with the sulfhydryl group of papain, *J. Biol. Chem.* 244, 5087–5094, 1969; Chaiken, I.M. and Smith, E.L., Reaction of the sulfhydryl group of papain with chloroacetic acid, *J. Biol. Chem.* 244, 5095–5099, 1969.

[c] Gerwin, B.I., Properties of the single sulfhydryl group of streptococcal proteinase. A comparison of the rates of alkylation by chloroacetic acid and chloroacetamide, *J. Biol. Chem.* 242, 451–456, 1967.

[d] Early concern regarding the reaction of cyanate with protein was directed at urea. This was followed with work using potassium cyanate. More recent work has used 2-nitro-5-thiocyanatobenzoic acid for the modification of cysteine residues in proteins (Degani, Y. and Patchornik, A., Cyanylation of sulfhydryl groups by 2-nitro-5-thiocyanatobenzoic acid. High-yield modification and cleavage of peptides at cysteine residues, *Biochemistry* 13, 1–11, 1974; Price, N.C., Alternative products in the reaction of 2-nitro-5-thiocyanatobenzoic acid with thiol groups, *Biochem. J.* 159, 177–180, 1976; Wu, J. and Watson, J.T., Optimization of the cleavage reaction for cyanylated cysteinyl proteins for efficient and simplified mass mapping, *Anal. Biochem.* 258, 268–276, 1998).

[e] Other aldehydes such 4-hydroxy-2-nonenal, glyceraldehydes, and reducing sugars such as glucose also react in a similar manner. The reaction with reducing sugars and 4-hydroxy-2-nonenal is more complex; the reaction with glucose and reducing sugars is part of the Maillard reaction.

[f] Formaldehyde (paraformaldehyde) is used for tissue fixation, and the chemistry is complex with the formation of protein cross-links. Antigen retrieval techniques have been developed for immunohistocytochemistry (Chu, W.S., Furusato, B., Wong, K. et al., Ultrasound-accelerated formalin fixation of tissue improves morphology, antigen and mRNA preservation, *Mod. Pathol.* 18, 850–863, 2005; Namimatsu, S., Ghazizadeh, M., and Sugisaki, Y., Reversing the effects of formalin fixation with citraconic anhydride and heat: A universal antigen retrieval method, *J. Histochem. Cytochem.* 53, 3–11, 2005; Shi, S.R., Liu, C., Balgley, B.M. et al., Protein extraction from formalin-fixed, paraffin-embedded tissue sections: Quality evaluation by mass spectrometry, *J. Histochem. Cytochem.* 54, 739–743, 2006; Sompuram, S.R., Vani, K., Hafer, L.J., and Bogen, S.A., Antibodies immunoreactive with formalin-fixed tissue antigens recognize linear protein epitopes, *Am. J. Clin. Pathol.* 125, 82–90, 2006; Yamashita, S., Heat-induced antigen retrieval: Mechanisms and application to histochemistry, *Prog. Histochem. Cytochem.* 41, 141–200, 2007).

[g] Reviero, A., Coletti-Previero, M.A., and Cavadore, J.-C., A reversible chemical modification of the tryptophan residue, *Biochim. Biophys. Acta* 147, 453–461, 1967; Strosberg, A.D. and Kanarek, L., Immunochemical studies on hen's egg-white lysozyme. Effect of formylation of the tryptophan residues, *FEBS Lett.* 5, 324–326, 1969; Magous, R., Bali, J.P., Moroder, L., and Previero, A., Effect on N^{in}-formylation of the tryptophan residue on gastrin (HG-13) binding and on gastric acid secretion, *Eur. J. Pharmacol.* 77, 11–16, 1982.

[h] Burr, M. and Koshland, D.E., Jr., Use of "reporter groups" in structure-function studies of proteins, *Proc. Nat. Acad. Sci. USA* 52, 1017–1024, 1964.

TABLE 1.1 (CONTINUED)
Reagents for the Chemical Modification of Proteins and Other Biological Macromolecules

[i] Mercuric sulfide is the principle ore of mercury and exists in two forms: αHgS (red hexagonal), which is known as *cinnabar;* and βHgS (black amorphous), which is known *metacinnabar*. The solubility product of mercuric sulfide is approximately 10^{-52}. Mercury salts can exist as mercurous [mercury(I)] and mercuric [mercury(II)]. Mercurous exists as a polyanion Hg_2^{2+}, which can undergo disproportionation to form Hg and Hg^{2+}. Mercuric ion can form tight complexes with free sulfhydryl group and disulfide bonds. The chemistry and structure of these derivatives are still poorly understood. The possibility of an Hg cross-link between sulfhydryl groups is a possibility. The term *mercaptan* is derived from the phrase "mercury capture." (See *The Chemistry of Mercury*, Ed. C.A. McAuliffe, Macmillan, London, 1977; Roberts, H.L., Some general aspects of mercury chemistry, *Adv. Inorg. Chem. Radiochem.* 11, 309–339, 1968; Grant, G.J., Mercury: Inorganic and coordination chemistry, in *Encyclopedia of Inorganic Chemistry*, Ed. R.B. King, John Wiley., Chichester, U.K., Volume 4, 2136–2145, 1994).

[j] Martinez-Ballesta, M.C., Diaz, R., Martinez, V., and Carvajal, M., Different blocking effects of $HgCl_2$ and NaCl on aquaporins of pepper plants, *J. Plant Physiol.* 160, 1487–1492, 2003; Liu, K., Nagase, H., Huang, C., Purification and functional characterization of aquaporin-8, *Biol. Cell* 98, 153–161, 2006; Yang, B., Kim, J.K., and Verkman, A.S., Comparative efficacy of $HgCl_2$ with candidate aquaporin-1 inhibitors DMSO, gold, TEA^+ and acetazolamide, *FEBS Lett.* 580, 6679–6684, 2006.

[k] Coggins, J.R., Hooper, E.A., and Perham, R.N., Use of dimethyl suberimidate and novel periodate-cleavable bis(imido)esters to study the quaternary structure of the pyruvate dehydrogenase multienzyme complex of *Escherichia coli*, *Biochemistry* 15, 2527–2533, 1976.

[l] McKinley-McKee, J.S. and Morris, D.L., The lysines in liver alcohol dehydrogenase. Chemical modification with pyridoxal-5′-phosphate and methyl picolinimidate, *Eur. J. Biochem.* 28, 1–11, 1972; Fries, R.W., Bohlken, D.P., Blakley, R.T., and Plapp, B.V., Activation of liver alcohol dehydrogenase by imidoesters generated in solution, *Biochemistry* 14, 5233–5238, 1975; Shaw, A. and Marienetti, G.V., The effect of imidoesters, fluorodinitrobenzene and trinitrobenzenesulfonate on ion transport in human erythrocytes, *Chem. Phys. Lipids* 27, 329–335, 1980.

[m] The first derivative was methyl methanethiosulfonate (Smith, D.J., Maggio, E.T., and Kenyon, G.L., Simple alkanethiol groups for temporary blocking of sulfhydryl groups of enzymes, *Biochemistry* 14, 766–771, 1975). There are a number of alkyl derivatives in use including ionic derivatives such as ethylsulfonato methanethiosulfonate or 2-carboxyethyl methanethiosulfonate and ethyltrimethylammonium methanthiosulfonate.

[n] Takahashi, K., Specific modification of arginine residues in proteins with ninhydrin, *J. Biochem.* 80, 1173–1176, 1976; Chaplin, M.F., The use of ninhydrin as a reagent for the reversible modification of arginine residues in proteins, *Biochem. J.* 155, 457–459, 1976.

[o] Li, C., Gawandi, V., Protos, A. et al., A matrix-assisted laser desorption/ionization compatible reagent for tagging tryptophan residues, *Eur. J. Mass Spectrom.* 12, 213–221, 2006.

[p] Dai, J., Zhang, Y., Wang, J. et al., Identification of degradation products formed during performic oxidation of peptides and proteins by high-performance liquid chromatography with matrix-assisted laser desorption/ionization and tandem mass spectrometry, *Rapid Commun. Mass Spectrom.* 19, 1130–1138, 2005.

[q] Yamasaki, R.B., Shimer, D.A., and Feeney, R.E., Colorimetric determination of arginine residues in proteins by *p*-nitrophenylglyoxal, *Anal. Biochem.* 111, 220–226, 1981.

TABLE 1.1 (CONTINUED)
Reagents for the Chemical Modification of Proteins and Other Biological Macromolecules

[r] Parker, D.J. and Allison, W.S., The mechanism of inactivation of glyceraldehyde 3-phosphate dehydrogenase by tetrathionate, *o*-iodosobenzoate, and iodine monochloride, *J. Biol. Chem.* 244, 180–189, 1969; Chung, S.I., and Folk, J.E., Mechanism of the inactivation of guinea pig liver transglutaminase by tetrathionate, *J. Biol. Chem.* 245, 681–689, 1970; Silva, C.M. and Cidlowski, J.A., Direct evidence for intra- and intermolecular disulfide bond formation in the human glucocorticoid receptor. Inhibition of DNA binding and identification of a new receptor-associated protein, *J. Biol. Chem.* 264, 6638–6647, 1989; Kaufmann, S.H., Brunet, G, Talbot, B. et al., Association of poly(ADP-ribose) polymerase with the nuclear matrix: The role intermolecular disulfide bond formation, RNA retention, and cell type, *Exp. Cell Res.* 192, 524–535, 1991; Desnoyers, S., Kirkland, J.B., and Poirier, G.G., Association of poly(ADP-ribose) polymerase with nuclear subfractions catalyzed with sodium tetrathionate and hydrogen peroxide crosslinks, *Mol. Cell. Biochem.* 159, 155–161, 1996; Tramontano, F., di Meglio, S., and Quesada, P., Co-localization of poly(ADPR)polymerase 1 (PARP-1), poly(ADPR)polymerase 2 (PARP-2) and related proteins in rat testis nuclear matrix defined by chemical cross-linking, *J. Cell. Biochem.* 94, 58–66, 2005.

[s] Haddad, I.Y., Zhu, S., Ischiropoulos, H., and Matalon, S., Nitration of surfactant protein A results in decreased ability to aggregate lipids, *Am. J. Physiol.* 270, L281–L288, 1996; Greis, K.D., Zhu, S., and Matalon, S., Identification of nitration sites on surfactant protein A by tandem electrospray mass spectrometry, *Arch. Biochem. Biophys.* 335, 396–402, 1996; Petersson, A.B., Steen, H., Kalume, D.E. et al., Investigation of tyrosine nitration by mass spectrometry, *J. Mass Spectrom.* 36, 6160625, 2001; Lee, W.I. and Fung, H.L., Mechanism-based partial inactivation of glutathione *S*-transferase by nitroglycerin: tyrosine nitration vs. sulfhydryl oxidation, *Nitric Oxide* 8, 103–110, 2003; Batthyany, C., Souza, J.M., Duran, R. et al., Time course and site(s) of cytochrome c tyrosine nitration by peroxynitrite, *Biochemistry* 44, 8038–8046, 2003.

The environments of the various amino acid residues in a protein are not identical. As a result of this lack of homogeneity, a variety of surface polarities will surround the various functional groups. The physical and chemical properties of any given functional group will be strongly influenced by the nature (e.g., polarity) of the local microenvironment. For example, consider the effect of the addition of an organic solvent, ethyl alcohol, on the p*K*a of acetic acid. In 100% H_2O, acetic acid has a p*K*a of 4.70. The addition of 80% ethyl alcohol results in an increase of the p*K*a to 6.9. In 100% ethyl alcohol, the p*K*a of acetic acid is 10.3. The reader is directed to a study by García-Moreno and coworkers[10] for a listing of residues with marked changes in p*K*a values resulting from microenvironmental influences (see Table 1.3). The reader is also directed to the study by Hnízda[11] and coworkers on microenvironmental influences on the reactivity of lysine and histidine residues in proteins (lysozyme and human serum albumin). They concluded that any modification is an indication of surface accessibility, with other factors also contributing to reactivity. Considering the importance of this information, it is surprising that there are not more studies in this area. Some 70 years ago, Richardson[12] concluded that lowering the dielectric constant decreases the acidity (increases the p*K*a) of carboxylic acids

TABLE 1.2
Dissociation Constants for Ionizable Groups in Proteins[a]

Amino Acid	p*Ka*[b]
Aspartic acid–α-carboxyl	1.95
Aspartic acid–β-carboxyl	3.71
Glutamic acid–α-carboxyl	2.16
Glutamic acid–γ-carboxyl	4.15
γ-Carboxyglutamic	5.1[c]
C-terminal carboxyl	2.36
Lysine ε-amino	9.16
Histidine imidazole ring	6.04
Arginine–guanidine group	12.10
α-Amino group	9.68
Serine hydroxyl	13.60[d]
Threonine hydroxyl	≥ 14.0[e]
Tyrosine hydroxyl	10.10
Cysteine sulfhydryl	8.14

[a] Data is for free amino acids.

[b] Except as noted in the table, the data for Table 1.2 was adapted from *CRC Handbook of Chemistry and Physics*, 86th ed., Ed D. Lide, CRC Press, Boca Raton, FL, 2005–2006; See also Mooz, E.D., Data on the naturally occurring amino acids, in *Practical Handbook of Biochemistry and Molecular Biology*, G.D. Fasman, Ed., CRC Press, Boca Raton, Florida, 1989. Also see Dawson, Elliott, Elliott, and Jones, *Data for Biochemical Research*, Oxford University Press, Oxford, 1969.

[c] Inferred from pH dependence of calcium binding by bone protein; See Svärd, M., Drakenberg, T., Andersson, T., and Fernlung, P., Calcium binding to bone gamma-carboxyglutamic acid protein from calf studied by ^{43}Ca NMR, *Eur. J. Biochem.* 158, 372–378, 1986.

[d] Determined for *N*-acetylserineamide. See Bruice, T.C., Fife, T.H., Bruno, J.J., and Brandon, N.E., Hydroxyl group catalysis. II. The reactivity of the hydroxyl group of serine. The nucleophilicity of alcohols and the ease of hydrolysis of their acetyl esters as related to their p*Ka*, *Biochemistry* 1, 7–12, 1962.

[e] Assumed value based on increased alkyl chain length results in high p*Ka*, *cf.* MeOH p*Ka* = 15.5; EtOH p*Ka* = 15.9, Rochester, C.H., Acidity and inter- and intra-molecular H-bonds, in *The Chemistry of the Hydroxyl Group Part 1*, Ed. S. Patai, Interscience Publishers, London, 1971.

with little effect on the dissociation of protonated amino groups. These observations were confirmed by Duggan and Schmidt.[13] The increase in the p*Ka* of carboxyl groups in organic solvents has a favorable effect on transpeptidation reactions[14,15] where the carboxyl groups are required to be protonated.

Some modification reactions take advantage of differences in p*Ka* values in similar chemical groups. The difference in p*Ka* values between an α-amino group and

TABLE 1.3
Solvent Effects on Apparent p*K*a Values for Amino Acids and Related Compounds

	ΔpKa		
Functional Group	**86% EtOH**	**65% EtOH**	**20% Dioxane**
CH_3COOH	+2.24	+1.19	+0.37
Alanine-COOH	+1.79	+1,19	+0.23
Alanine-αNH_3^+	+0.13	+0.30	−0.05
Lys-COOH	+1.73	+1.05	+0.10
Lys-αNJ_3^+	+0.18	+0.05	−0.15
Lys-εNH_3^+	+0.14	0.00	−0.20
Arg-COOH	+1.12	+1.32	+0.11
Arg-αNH_3^+	−0.01	+0.36	−0.10
Arg-GuanidinoNH_3^+	+0.25	+1.52	+0.55

an ε-amino group makes it possible to selectively modify the α-amino group in a protein. Another example is the selective modification of the γ-carboxyl groups on a protein without modification of the α-carboxyl groups on a protein because the protonated form of the carboxylic acid is required for successful reaction.

Other factors that can influence the p*K*a of a functional group in a protein include hydrogen binding with an adjacent functional group, the direct electrostatic effect of the presence of a charged group in the immediate vicinity of a potential nucleophile, and direct steric effects on the availability of a given functional group.

There is another consideration that can in a sense be considered either a cause or consequence of microenvironmental polarity. This has to do with the environment immediately around the residue modified. These are the factors that can cause a "selective" increase (or decrease) in reagent concentration in the vicinity of a potentially reactive species. The most clearly understood example of this is the process of affinity labeling.[16] The modification of most functional groups in a protein by a chemical reagent is second-order with an observed linear relationship between reagent concentration and rate (actually, the rate is proportional to the square of the concentration of one reactant or to the product of the concentration of the two reactants). From a practical perspective, the relationship between reagent concentration and reaction rate is linear (Figure 1.1) with a second-order reaction. An affinity label shows saturation kinetics, where a point is reached at which an increase in reagent concentration does not increase reaction rate (Figure 1.1)

Another consideration is the partitioning of a reagent such as tetranitromethane between the aqueous environment, which is polar, and the interior of the protein, which is nonpolar. Tetranitromethane is an *organic* compound and, in principle, can react equally well with exposed and buried tyrosyl residues.[17] Skov and coworkers[18] modified horse heart cytochrome c with tetranitromethane (fourfold molar excess over tyrosine). Two of the four tyrosine residues, Y48 and Y67, were modified.

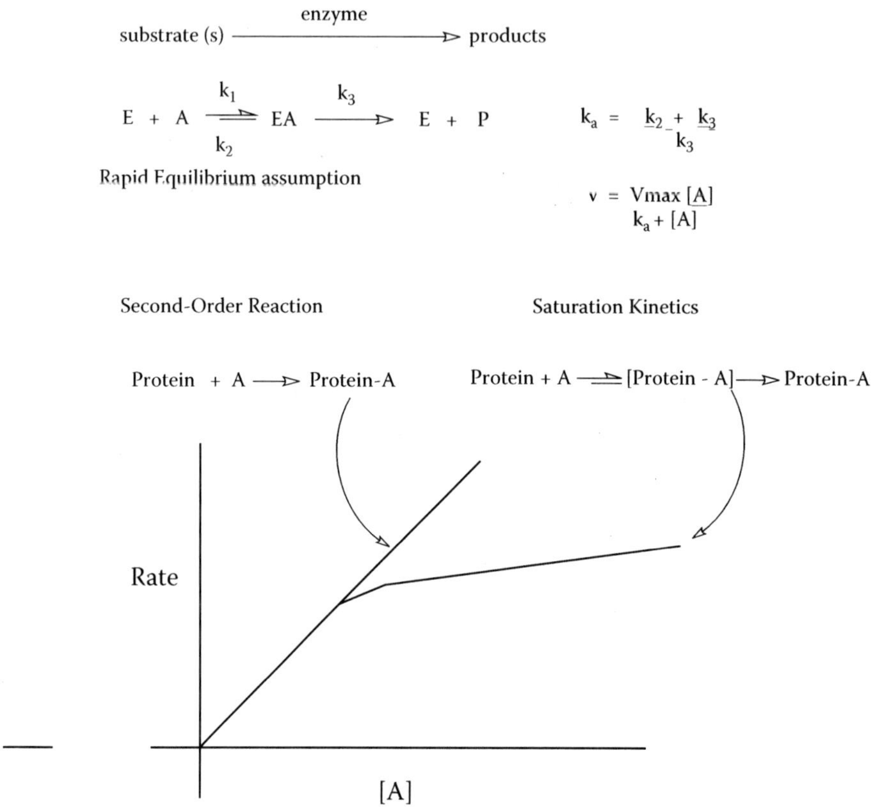

FIGURE 1.1 Saturation kinetics in the modification of proteins. The modification of a protein is usually a second-order reaction with the rate dependent on the concentration of both reactants. For practical purposes, the protein concentration is usually kept constant and the concentration of reagent is varied, when a straight line is obtained as shown in the figure indicated (pseudo first-order reaction). There are reactions in which the reagent binds to the protein prior to the modification reaction, resulting in saturation kinetics for the reaction as indicated in the figure. (See Shen, W.C. and Colman, R.F., Cyanate modification of essential lysyl residues of the diphosphopyridine nucleotide-specific isocitrate dehydrogenase of pig heart, *J. Biol. Chem.* 250, 2973–2978, 1975; Hummel, C.F., Gerber, B.R., and Carty, R.P., Chemical modification of ribonuclease A with 4-arsono-2-nitrofluorobenzene, *Int. J. Protein Res.* 24, 1–13, 1984; Huynh, Q.K., Mechanism of inactivation of *Escherichia coli* 5-enolpyruvoylshikimate-3-phosphate synthase by *o*-phthalaldehyde, *J. Biol. Chem.* 265, 6700–6704, 1990.) Although kinetically similar, these reactions are different from suicide substrates. (See Walsh, C.T., Suicide substrates, mechanism-based enzyme inactivators: recent developments, *Annu. Rev. Biochem.* 53, 493–535, 1984.)

These residues have reduced exposure to solvent. In more recent work,[19] Battghyány and coworkers observed that peroxynitrite readily modified Y97 and Y74l under more rigorous conditions; all four tyrosine residues were modified by peroxynitrite with dinitration and trinitration observed. Tyrosine 48 was the least susceptible to modification with peroxynitrite. These investigators also studied modification with tetranitromethane (tenfold molar excess over tyrosine) and obtained modification comparable to that obtained with peroxynitrite; Y67 was most susceptible to nitration with tetranitromethane. This is discussed later in greater detail in the consideration of tyrosine modification.

Establishing the stoichiometry of modification is a relatively straightforward process. First, the molar quantity of modified residue is established by analysis. This could be spectrophotometric, as, for example, with the trinitrophenylation of primary amino groups, the nitration of tyrosine with tetranitromethane, or the alkylation of tryptophan with 2-hydroxy-5-nitrobenzyl bromide or by amino acid analysis to determine either the loss of a residue as, for example, in photooxidation of histidine and the oxidation of the indole ring of tryptophan with *N*-bromosuccinimide or the appearance of a modified residue such as with *S*-carboxymethylcysteine or N^1-or N^3-carboxymethylhistidine. In the situation where spectral change or radiolabel incorporation is used to establish stoichiometry, analysis must be performed to determine that there is not a reaction with another amino acid. For example, the extent of oxidation of tryptophan by *N*-bromosuccinimide can be determined spectrophotometrically, but amino acid analysis or mass spectrometric analysis is required to determine if modification has also occurred with another amino acid such as histidine or methionine. It is clear that the evolution of mass spectrometry from an esoteric, specialized laboratory resource to a technique that is as common in the protein chemistry as amino acid analysis has provided another tool for the evaluation of protein structure after chemical modification.[20–32]

Site-specific chemical modification of a protein requires that the modification of one residue mole per mole of protein (or functional subunit) has occurred without modification of another amino acid (e.g., modification has only occurred with lysine and not with tyrosine). The reaction pattern of a given reagent with free amino acids or amino acid derivatives does not necessarily provide the basis for reaction with such amino acid residues in protein. Furthermore, the reaction pattern of a given reagent with one protein cannot necessarily be extrapolated to all proteins. The results of a chemical modification can be markedly affected by reaction conditions (e.g., pH, temperature, solvent and/or buffer used, degree of illumination, etc.). Establishment of stoichiometry does not necessarily mean that this modification has occurred at a unique residue (unique in terms of position in the linear peptide chain—not necessarily unique with respect to reactivity). It is, of course, useful if there is a change in biological activity (catalysis, substrate binding, ion binding, etc.) that occurs concomitant with the chemical modification. Ideally, one would like to establish a direct relationship (i.e., 0.5 mol/mol of protein with 50% activity modification; 1.0 mol/mol of protein with 100% activity modification). More frequently, there is the situation where there are several moles of a given residue modified per mole of protein, but there is reason to suspect stoichiometric chemical modification. In some of these situations it is possible to fractionate the protein into uniquely modified species. The separation of carboxy-methyl-His^{12}-pancreatic ribonuclease from carboxy-methyl-His^{119}-pancreatic ribonuclease is a classic example of this type of a situation.[33] It has also been possible to separate derivatives of lysozyme obtained from the modification of carboxyl groups.[34]

The determination of stoichiometry of modification from only the functional consequences of such modification is a far more difficult proposition. First, there must be a clear, unambiguous signal that can be effectively measured. In a situation where there are clearly multiple sites of reaction that can be distinguished by analytical techniques, the approach advanced by Ray and Koshland is useful.[35] This analysis is based on establishing a relationship between the rate of loss of biological activity and the rate of modification of a single residue. A similar approach has been advanced by Tsou[36–38] and is based on establishing a relationship between the number of residues modified and the change in biological activity. Horiike and McCormick[39] have explored the approach of relating changes in activity to extent of chemical modification. These investigators state that the original concepts which form the basis of this approach are sound, but that extrapolation from a plot of activity remaining versus residues modified is not necessarily sound. Such extrapolation is only valid if the "nonessential" residues react much slower (rate at least 10 times slower). Given a situation where all residues within a given group are equally reactive toward the reagent in question, the number of essential residues obtained from such a plot is correct only when the total number of residues is equal to the number of essential residues, which is in turn equal to 1.0. However, it is important to emphasize that this approach is useful when there is a difference in the rate of reaction of an essential residue or residues and all other residues in that class, as in the modification of histidyl residues with diethylpyrocarbonate in lactate dehydrogenase[40,41] and pyridoxamine-5′-phosphate oxidase.[42] The reader is referred to a review by Rakitzis[43] for a discussion of the kinetics of protein chemical modification. There has been continuing use of this approach during the 20 years since the publication of this article.[44–49] An example of the use of reaction rate is provided from the study of the modification of an aminopeptidase by diethylpyrocarbonate.[50] It was demonstrated that the reaction of the aminopeptidase with diethylpyrocarbonate resulted in the modification of histidine residues. A difference of the reactivity of the two histidine residues modified by diethylpyrocarbonate in the presence and absence of calcium ions permitted the identification of one of the two histidine residues as critical for the binding of calcium ions. Careful analysis of the effect of pH on the reaction rate in the presence and absence of calcium ions allowed the assignment of p*K*a value to the two residues. A similar problem is faced in the use of activity-based proteomics (see Chapter 8), where only qualitative observations are possible.

Chemical modification can be used for the selective fragmentation of proteins for determination of their primary structure, the preparation of large fragments for characterization by mass spectrometry, and the chemical synthesis of proteins. This includes reagents such as cyanogen bromide for the chemical cleavage of specific peptide bonds, citraconic anhydride for the reversible blocking of lysine residues to restrict tryptic cleavage to arginine residues, and the reversible blocking of arginine residues with 1,2-cyclohexanedione to restrict tryptic cleavage to lysine residues. Cyanogen bromide cleavage of proteins yields larger fragments than those obtained by tryptic digestion, facilitating structure assignments.[51,52] In a similar approach, modification of lysine residues with citraconic anhydride restricts tryptic cleavage to arginine residues which, as with the CNBr cleavage, provides larger fragments providing for more facile sequence coverage of the protein.[53] Cyanogen bromide fragments have been used in the engineering of semisynthetic proteins.[54] A 223-residue hybrid of *Streptomyces griseus* trypsin was synthesized by native chemical ligation[55] from a chemically synthesized amino-terminal fragment and a larger C-terminal fragment obtained by cyanogen bromide cleavage. Cyanogen bromide is also used for the cleavage of fusion proteins.[56–61]

There are a number of *in vivo* "chemical" modifications of proteins that can markedly influence function. Such *in vivo* chemical modifications include cotranslational[62,63,63a] and posttranslational reactions,[63b,64] including such reactions as glycosylation and carboxylation (Figure 1.2) as well as sulfation, methylation, and phosphorylation (Figure 1.3). There are also modifications that occur with reactive organic compounds such as 4-hydroxy-2-nonenal (Figure 1.4). It should be noted that the characterization of such reactions has benefited significantly from the excellent preceding work on the organic chemistry of proteins.

Oxidation of the "active-site" methionine residue in alpha-1-antitrypsin (alpha-1-antiprotease inhibitor) results in pulmonary damage.[65–67] Methionine is quite susceptible to oxidation, first, to the sulfoxide, a reversible reaction, and to the sulfone.[68,69] There is a spectrum of oxidizing agents, including free radicals such as hydroxyl radical,[70–72] organic peroxides,[73] and hypochlorites[74] (Figure 1.5). The oxidation of biological macromolecules is assumed to have unfavorable consequences but might also have a role in normal protein catabolism.[75–79]

Glycation is the term used to identify the reaction of reducing sugars with proteins (Figure 1.6). This involves the initial formation of a Schiff base followed by rearrangement in the Maillard reaction,[80,81] eventually resulting in advanced glycation end (AGE) products.[82] Reaction can occur at lysine and arginine residues with resulting cross-link formation.[83,84] Methyl glyoxal is an endogenous factor[85] that has proved useful in *in vitro* studies of the chemistry of glycation.[86,87]

Conformational change must be considered in the interpretation of the results of the site-specific modification of a protein.[88–90] It has been more frequent, though, that site-specific chemical modification has been used to assess conformational change in proteins.[91–100] Salhany and coworkers[98] used modification of a carboxyl group with Woodward's reagent K to study conformational effects in a membrane transport protein. D'Ambrosio et al.[99] used a combination of a chemical cross-linking, modification of lysine residues with acetic anhydride, and modification of tyrosine residues with tetranitromethane to study the dimeric structure of porcine aminoacylase 1. Mass spectrometry was used for analysis of the various chemical modification procedures. Li and Bigelow[100] modified tyrosine with tetranitromethane for use as a fluorescence resonance energy transfer (FRET) receptor for the study of the interaction between the transmembrane and cytosolic domains of phospholamban. The use of specific chemical modification to study changes in the environment surrounding amino acid residues in proteins has been studied over the past 30 years. The study of Kirtley and Koshland[101] provided the basis for the concept of using "reporter" groups to study changes in the microenvironment surrounding a site of modification. This study used 2-bromoacetamido-4-nitrophenol to modify a limited number of sulfhydryl groups in glyceraldehyde-3-phosphate dehydrogenase. The modified protein has a λ_{max} at 390 nm ($\varepsilon = 7100\ M^{-1}\ cm^{-1}$) between pH 7.0 and 7.6. The addition of the coenzyme NAD caused a marked change in the spectral properties (decrease in absorbance at

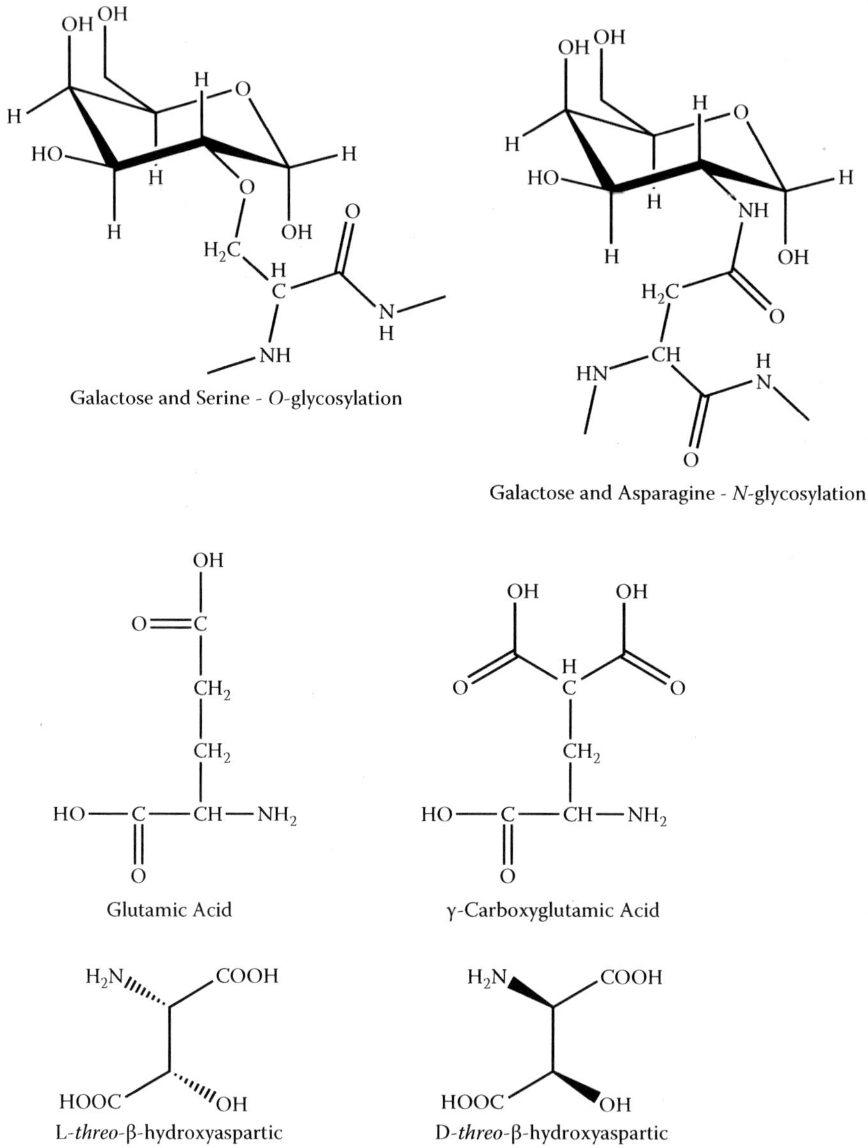

FIGURE 1.2 Some posttranslational modifications of proteins. (See Walsh, C.T., Garneau-Tsodikova, S., and Gatto, G.J., Jr., Protein posttranslational modifications: the chemistry of proteome diversification, *Andewandte Chem.* 44, 7342–7372, 2005.)

approximately 375 nm and increase in absorbance at approximately 420 nm) of the modified enzyme, which is consistent with a change in the microenvironment around the modified residue (increase in polarity of medium, which results in increased formation of the nitrophenolate ion). The reaction of 2-hydroxy-5-nitrobenzyl bromide with tryptophanyl residues to yield the 2-hydroxy-5-nitrobenzyl derivative[102] and

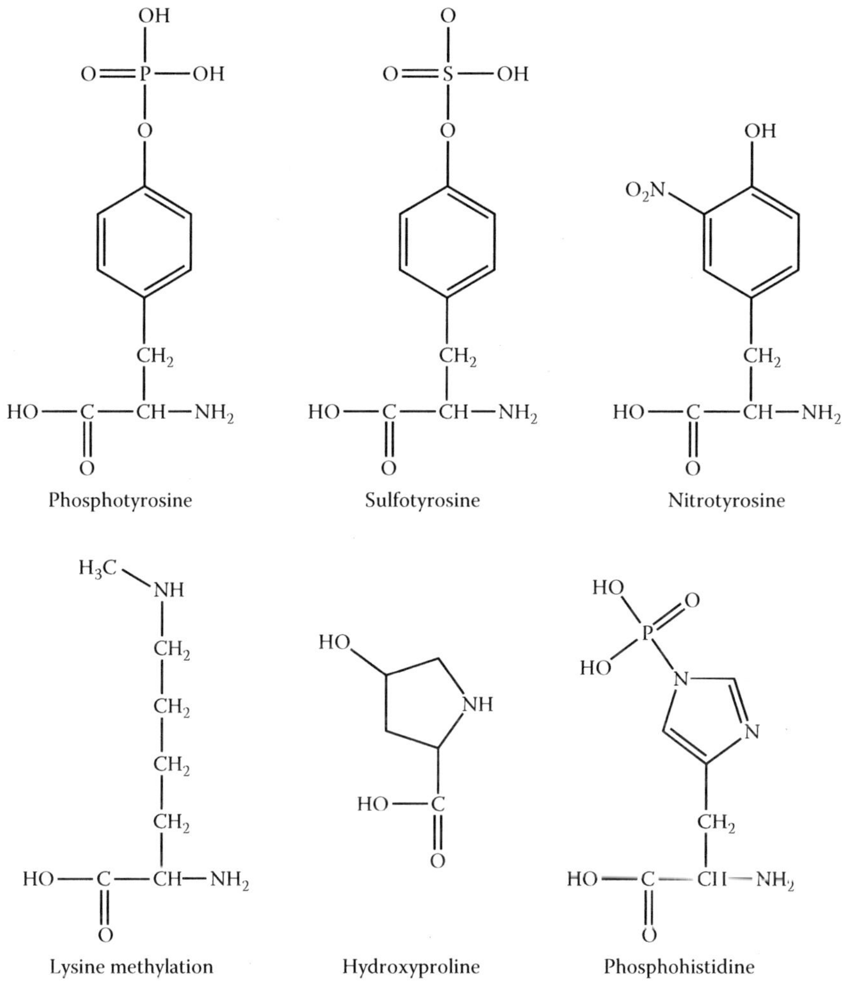

FIGURE 1.3 Examples of covalent modification of proteins that control protein function. The majority of these reactions are catalyzed by specific enzymes and occur in specific domains of the substrates. The best known are the various phosphorylation and methylation reactions.

the reaction of tetranitromethane with tyrosyl residues[103] to form the 3-nitrotyrosyl derivative was extensively used to study microenvironmental changes in the modified proteins.[104] Fluorescent probes have been used extensively in the study of protein conformation. The chemistry used for the covalent modification of proteins is described in detail in individual chapters. The majority of studies use either lysine or cysteine as a target residue for modification.[105] The insertion of cysteine into recombinant proteins[106] via oligonucleotide-directed mutagenesis[107,108] has provided an opportunity for the attachment of fluorescent probes to specific protein domains.[109] Fluorescent energy transfer (FRET) with covalently attached fluorescent probes are

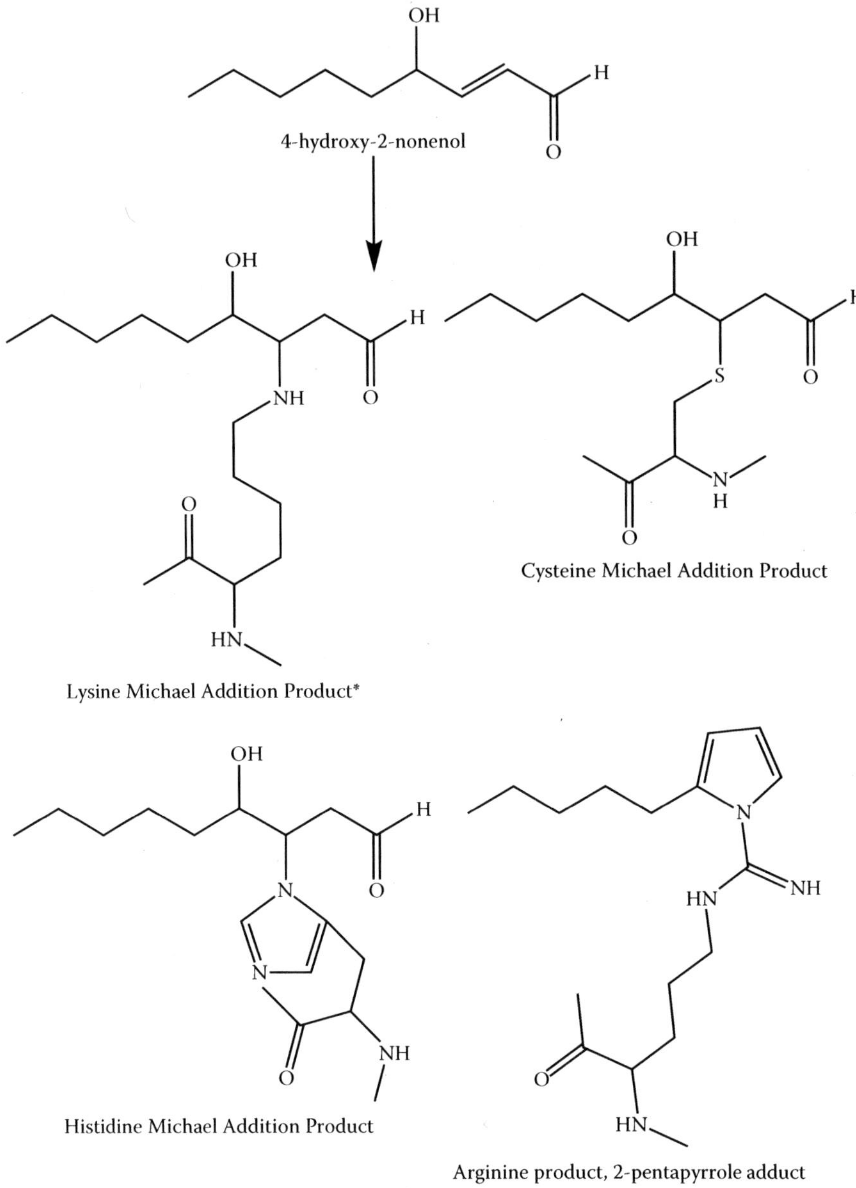

FIGURE 1.4 Modifications of proteins by 4-hydroxy-2-nonenal. (See Poli, G., Biasi, F., and Leonarduzzi, G., 4-Hydroxynonenal-protein adducts: A reliable biomarker of lipid oxidation in liver diseases, *Mol. Aspects Med.* 29, 67–71, 2008; Schneider, C., Porter, N.A., and Brash, A.R., Routes to 4-hydroxynonenal: Fundamental issues in the mechanisms of lipid peroxidation, *J. Biol. Chem.* 283, 15539–15543, 2008.)

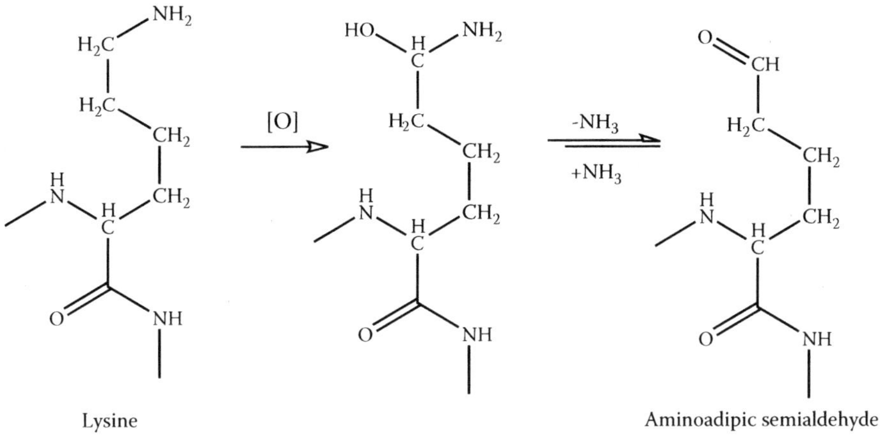

FIGURE 1.5 The oxidation of lysine residues in proteins to yield carbonyl groups. The oxidation of lysine is one of the several oxidation reactions which occur in proteins. The oxidation of cysteine is described in Figure 1.39 and the oxidation of methionine is described in Figure 1.42.

proving increasingly useful in the study of protein conformation.[100–113] Chapter 2 provides additional discussion of the use of chemical modification in the study of protein conformation.

The reaction of 1-dimethylaminonaphthalene-5-sulfonyl chloride (dansyl chloride) has been useful both in structural analysis and amino group modification with proteins. In one study,[116] dansyl chloride (in acetone) was added to a solution of trypsin in 0.1 *M* phosphate, pH 8.0. The reaction was terminated after 24 h at 25°C by acidification to pH 3.0 with 1.0 M HCl. Insoluble material was removed by centrifugation, and the supernatant fraction was placed in dialysis. These investigators reported modification of the amino terminal isoleucine and one lysine residue. The extent of modification was determined by absorbance at 336 nm ($\varepsilon_m = 3.4 \times 10^4$ M^{-1} cm^{-1}). Reaction of dansyl chloride with phosphoenolpyruvate carboxylase has been used to introduce a fluorescent probe into this protein at a single lysine residue.[117] Park and coworkers[118] used dansylation to improve sensitivity in peptide mass fingerprinting. Amoresano and coworkers[119] used danysl chloride for the analysis of nitrotyrosine residues in proteins after reduction of the nitrotyrosyl residues to aminotyrosine. Cirulli and coworkers[120] used dansyl chloride for the labeling of bacterial surface proteins prior to mass spectrometric analysis.

Modification of protein amino groups with isothiocyanate derivatives of various dyes has proved to be an effective means of introducing structural probes into proteins at specific sites.[121] Fluorescein isothiocyanate has been used to modify cytochrome P-450 (reaction performed in 30 mM Tris, pH 8.0 containing 0.1% Tween 80; 2 h at 0°C in the dark),[122] actin (2 mM borate, pH 8.5; 3 h at ambient temperature, then at 4°C for 16 h),[123] and ricin (pH 8.1, 6°C for 4 h).[124] The reader is directed to an elegant study on the effect of microenvironment on the fluorescence of arylaminophthalenesulfonates.[125]

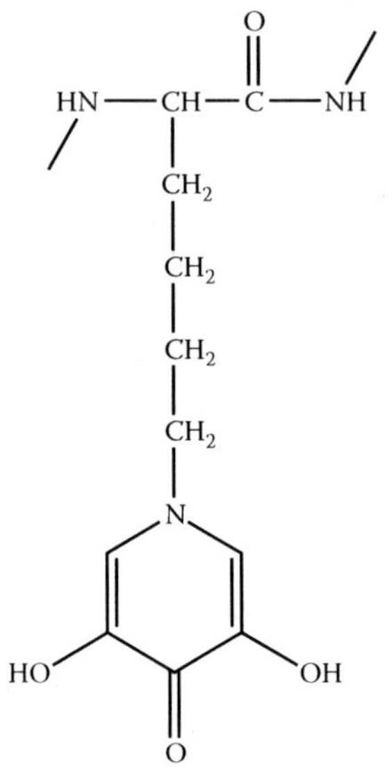

Trihydroxy-triosidine, derived from reaction with glyceraldehyde

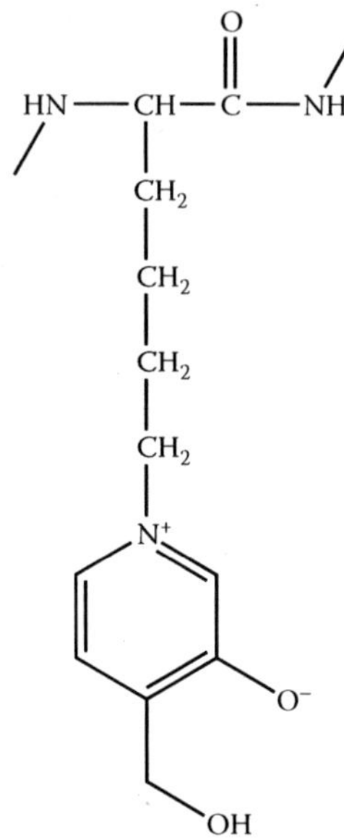

2-ammonium060[4-(hydroxymethyl)-3-oxidopyridium-1-yl-]hexanoate; derived from reaction with dehydroascorbic acid

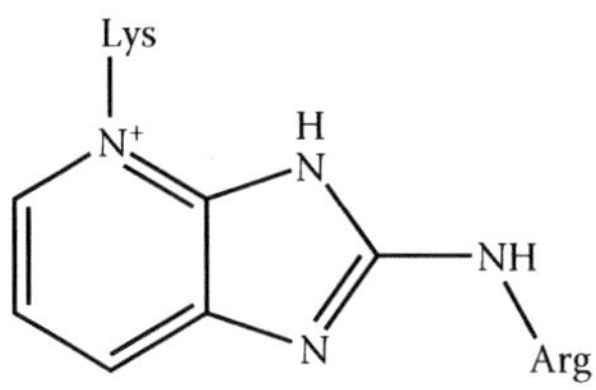

Pentosidine

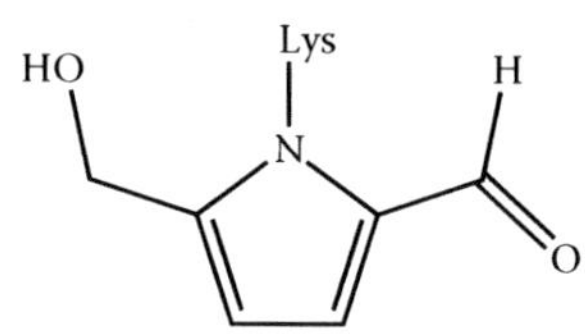

Pyrraline

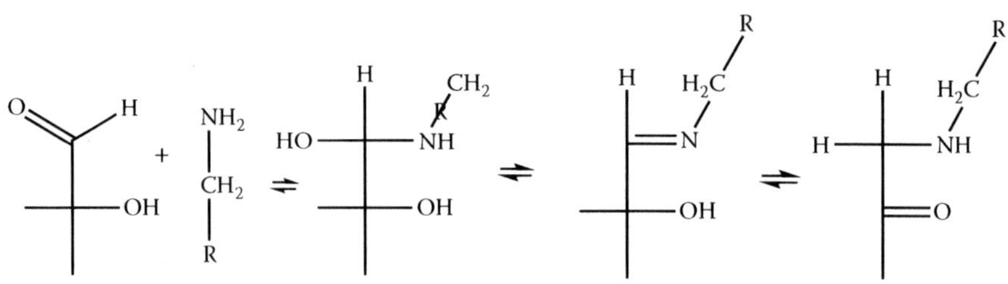

Schiff Base

Amadori Product

FIGURE 1.6 The modification of proteins by the process of glycation. This process can result in the formation of complex structures known as advanced glycation end products (AGE). (See Miyata, T., Taneda, S., Kawai, R. et al., Identification of pentosidine as a native structure for advanced glycation end products in β_2-microglobulin-containing amyloid fibrils in patients with dialysis-related amyloidosis, *Proc. Natl. Acad. Sci. USA* 93, 2353–2358, 1996.)

Electron paramagnetic resonance (EPR) is a technique that measures unpaired electrons.[126–131] Reorientation of the electron causes absorption of microwave energy when a sample is placed in a strong magnetic field.[132] EPR measures the fluid environment around a spin label and motion in that environment. The reader is directed to a recent discussion[133] of EPR . Intrinsic unpaired electrons are found in free radicals and some transition metal ions. However, most work in solution protein chemistry is performed with spin labels that are inserted into proteins and membranes. Nitroxide spin labels are very sensitive to the fluid environment. EPR has been used to probe protein function (Figure 1.7)[134–137] but has gained most interest for the study of membrane environments[127–131,126,138–146] (Figure 1.8). One early study used spin-labeled derivatives of diisopropylphosphorofluoridate to study the active site environment of trypsin.[147] Subsequent studies used various spin-labeled derivatives (piperidinyl nitroxide, pyrrolidinyl nitroxide, and pyrrolinyl nitroxide substituent groups) of phenylmethylsulfonyl fluoride to compare microenvironments surrounding the active sites in α-chymotrypsin and trypsin.[148,149] These reagents were used to study the active site of thrombin.[150,151] In subsequent studies,[152] Nienaber and Berliner obtained the crystal structures of thrombin modified at the active site serine with two substituted pyrrolidone nitroxide derivatives, 4-(2,2,5,5-tetramethylpyrrolidone-1-oxyl)-*p*-(fluorosulfonyl) benzamidine and 3-(2,2,5,5-tetrametnylpyrrolidone-1-oxyl)-*m*-(fluorosulfonyl) benzamidine. The crystal structures confirmed the earlier observations on the topography of the extended active site region of thrombin previously obtained by the electron spin resonance studies cited earlier. The preparation of spin-labeled pepsinogen has been reported.[153] This study used an *N*-hydroxysuccinimide ester derivative, 3-[[(2,5-dioxo-1-pyrrolidiny)oxyl] carbonyl]-2,5,-dihydro-2,2,5,5-tetramethyl-1H-pyrrolyl-1-oxy, to modify lysyl residues in pepsinogen. Coupling was accomplished at pH 7.0 (0.1 M sodium phosphate) for 7 h at 22°C, resulting in the derivatization of approximately three amino groups. Site-directed attachment of nitroxide spin labels to inserted cysteine residues has been used to study the conformation of *S*-adenosyl- methionine synthetase[154] and the mitochondrial oxoglutarate carrier.[155] The reader is directed to other recent studies of interest.[156–159] Voinov and coworkers[156] reported on the use of a thiol-specific nitroxide, methanethiosulfonic acid *S*-(1-oxyl-2,2,3,5,5,-pentamethylimidazolin-4-methyl) ester (Figure 1.3b) to study protein microenviroments; this derivative could be used with cysteine insertion to probe different regions of a protein, as has been done with other sulfhydryl-specific spin-labeled reagents.[157,158] The incorporation of spin-labeled amino acids into proteins under amber mutant technology has also been reported.[159]

AMINO GROUPS

Amino groups must be unprotonated to function as nucleophiles, so alkaline pH conditions are usually required. As shown in Table 1.1, the p*K*a for the epsilon amino group of lysine is 10.79 and 9.68 for an α-amino group, but there is a considerable range depending on the microenvironment of the functional group. For example, Schmidt and Westheimer[160] studied the effect of pH on the acylation of the amino group at the active site of acetoacetate decarboxylase by 2,4-dinitrophenyl propionate. These data suggest that the p*K*a for this amino group was 5.9, which is some 4 p*K* units less than

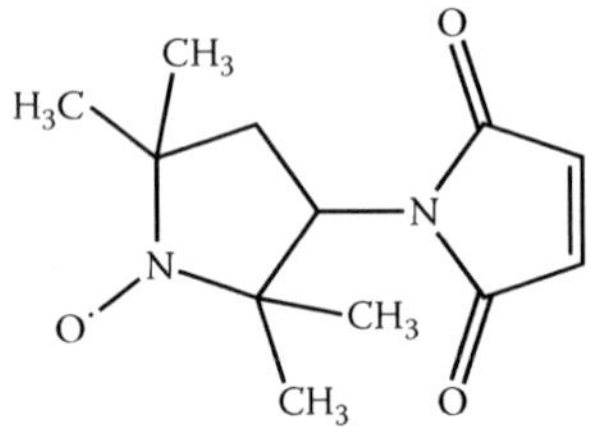

3-Maleimido-2,2,5,5-tetramethyl-1-pyrrolidinyloxyl (sulfhydryl)

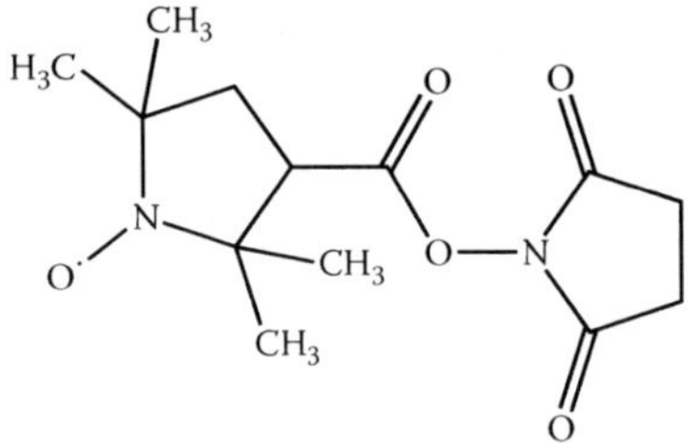

2,2,5,5-Tetramethyl-3-pyrrolin-1-oxyl-3-carboxylic acid *N*-hydroxysuccinimide (lysine)

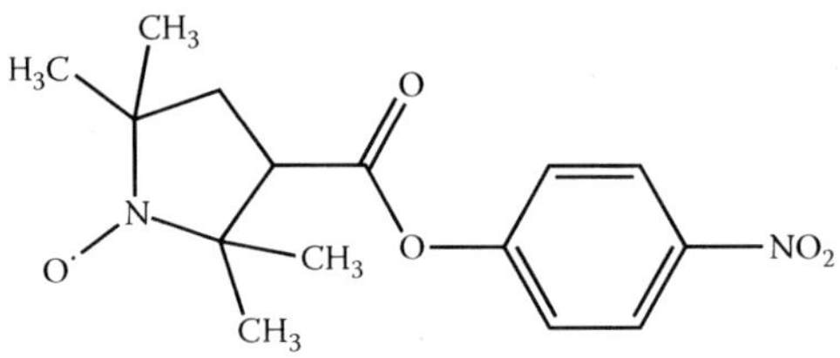

3-carboxy-2,2,5,5-tetramentyl-1-pyrrolidinyloxyl-*p*-nitrophenyl ester (serine at active site)

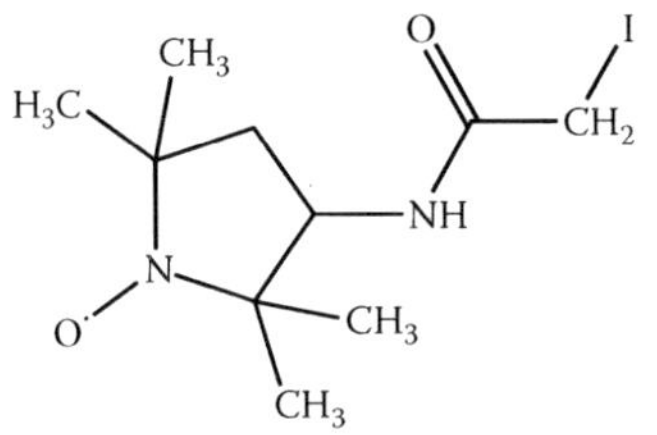

2,2,5,5-Tetramethyl-3-pyrrolin-1-oxyl-3-iodoacetmide

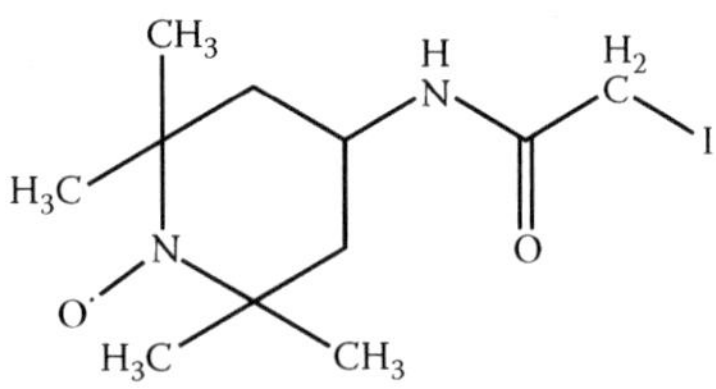

2,2,6,6-tetramethyl-1-oxylpiperidinyl-4-iodoacetamide

FIGURE 1.7 Some useful spin-labeled reagents that have been used for the site-specific modification of proteins. Spin label probes are generic, either tetramethylpyrrolidinyl or tetramethylpiperidinyl derivatives, with specificity of labeling provided by established chemistry. (See McConnell, H.M., Deal, W., and Ogata, R.T., Spin-labeled hemoglobin derivatives in solution, polycrystalline suspensions, and single crystals, *Biochemistry* 8, 2580–2585, 1969; Shimshick, E.J. and McConnell, H.M., Rotational correlation time of spin-labeled α-chymotrypsin, *Biochem. Biophys. Res. Commun.* 46, 321–327, 1972; Likhtenshtein, G.I. (tr. P.S. Shelnitz), *Spin Labeling Methods in Molecular Biology*, John Wiley & Sons, New York, 1976; Fajer, P.G., Electron spin resonance spectroscopy labeling in peptide and protein analysis, in *Encyclopedia of Analytical Chemistry*, ed., R.A. Meyers, Wiley, Chichester, U.K., 2000.)

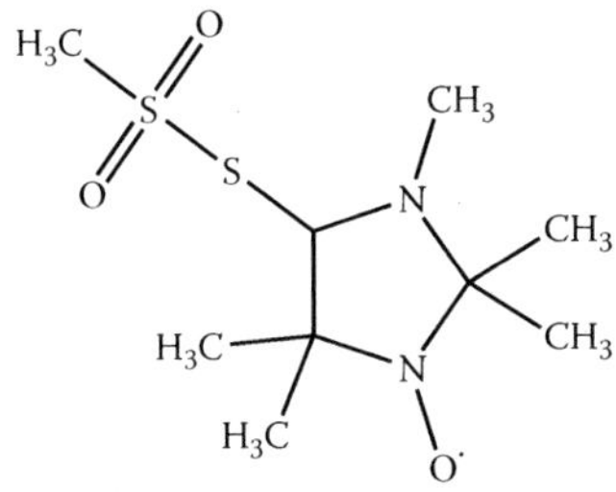

Methanethiosulfonic acid *S*-(1-oxyl-2,2,3,5,5-pentamethyl-imidazolidin-4-ylmethyl) ester (IMTSL)

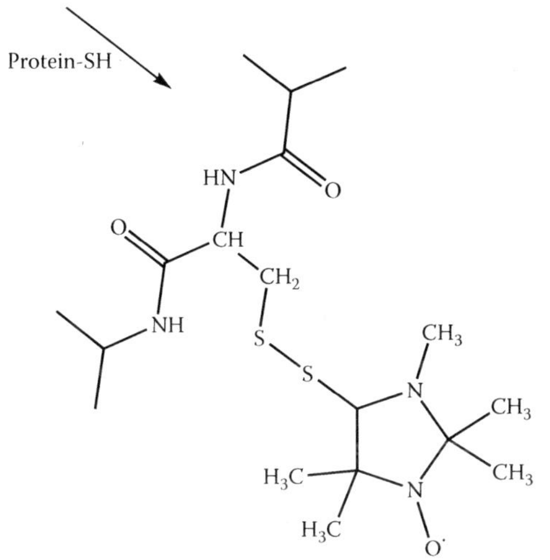

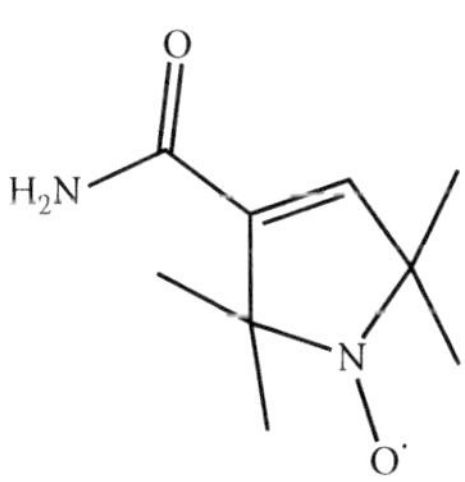

3-carbamoyl-2,2,5,5-tetramethyl-3-pyrroline-1-yloxyl (CTPO)

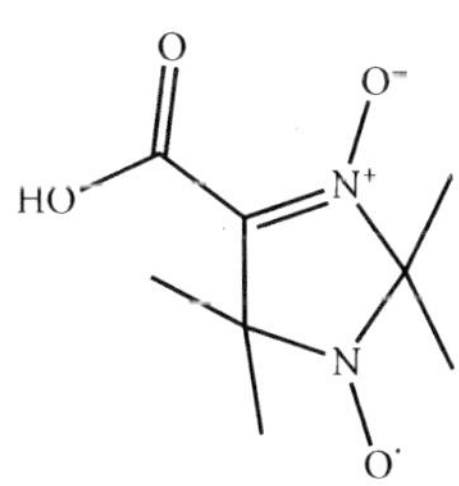

4-carboxy-2,2,5,5,-tetramethyl-3-imidazoline-3-oxide-1-oxyl

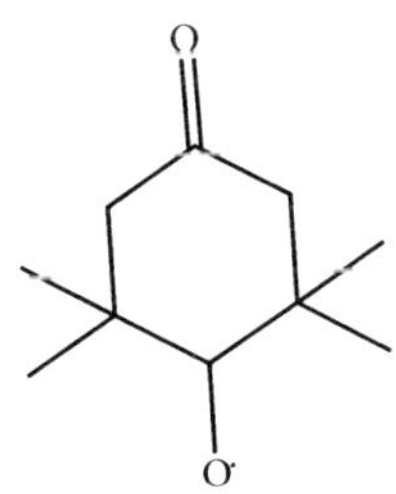

2,2,6,6-tetramethyl-4-piperidine-1-oxyl (TEMPONE)

FIGURE 1.8 Spin label reagents for membrane studies. (See Xu, Q., Kim, M., Ho, D. et al., Membrane hydrocarbon thickness modulates the dynamics of a membrane transport protein, *Biophys J.* (2008) doi:10.1529/biophysj.108.133629; Froncisz, W., Camenisch, T.G., Ratke, J.J. et al., Saturation recovery EPR and ELDOR at *W*-band for spin labels, *J.Magn. Reson.* (2008), doi:10.1016/j.jmr.05.008; *EPR Spectroscopy in Membrane Biophysics*, Ed. M.A. Hemminga and L.A. Berlinder, Springer, New York, 2007; Siminov, A.J., Ruuge, A., Resnikov, V.A. et al., Site-directed electrostatic measurements with a thiol-specific pH-sensitive nitroxide: Differentiating local p*K* and polarity effects by high-field EPR, *J. Am. Chem. Soc.*126, 8872–8873, 2004.)

that of an "ordinary" ε-amino group of lysine. Subsequently, Highbarger and coworkers[161] extended observations of this enzyme, confirming the suggestion that the low p*K*a of lysine[115] was caused by spatial proximity to lysine.[116] The presence of lysine residues with similarly low p*K*a values has been observed for other enzymes, including mitochondrial aspartate aminotransferase,[162] ac fructose-1,6-bisphosphatase,[163] phosphonoacetaldehye hydrolase,[164] sacrosine oxidase,[165] penicillin-binding protein 5 from *Escherichia coli*,[166] and mRNA cap-specific 2′-O-methyltransferase.[167] Lysine residues with low p*K*a values have also been observed for noncatalytic proteins such as cytochrome c and plastocyanin.[168] Matthews and coworkers[169] showed that a lysine residue replacing a "buried" methionine residues in T4 lysozyme had a p*K*a of 6.5. These workers suggest that burying an acidic group such as an aspartic acid results in an increased p*K*a, whereas burying a basic residue such as lysine lowers the p*K*a value. However, there are exceptions, as in staphylococcal nuclease, where Harms and coworkers[170] found that a buried lysine residue (Lys 38) had a normal p*K*a . These investigators suggest that this is an indication of the importance of local flexibility and water penetration as determinants of p*K*a values in proteins.

Acylation of amino groups (α-amino groups, ε-amino groups; amino sugars) in proteins with organic acid anhydrides[171,172] (Figure 1.9) is a facile modification procedure for the study of protein topology. Reaction can also occur at other nucleophilic functional groups, including sulfhydryl, phenolic hydroxyl, the imidazole ring of histidine, and at aspartic or glutamic groups via mixed acid anhydride formation. Most of these modifications are either exceedingly transient or labile under conditions (mild base) where *N*-acyl groups are stable; the formation of mixed anhydrides is rarely observed.

Izumi and coworkers[173] used reaction with acetic anhydride to study the topology of a cytochrome P450. These investigators observed that there are residues modified in the detergent-solubilized protein that were not available in the proteolipid form. This group subsequently used modification with acetic anhydride to study the protein surfaces of the cytochrome P450 17α and NADPH-cytochrome P450 reductase.[174] Gudiksen and coworkers[175] acetylated all 18 lysine residues in bovine carbonic anhydrase with acetic anhydride. There was no difference in the renaturation of the native and modified protein from sodium dodecyl sulfate. This suggests that the charged lysine residues do not have a role in the refolding of the enzyme. Miyazaki and Tsugita[176,177] used acetic anhydride and perfluoric acid as a C-terminal sequencing method based on the abilty of acetic anhydride to form a mixed anhydride with the C-terminal residue (Figure 1.10). Acylation of lactoferrin with acetic anhydride, which increased the net negative charge, decreased antimicrobial activity, whereas amidation, which decreased the net negative change, increased antibacterial activity.[178] Sanchez and coworkers[179] used modification with acetic anhydride or succinic anhydride to identify N-terminal peptides from *in situ* digestion of proteins in gels. Higashimoto[180] used acetic anhydride to identify areas of protein surface involved in the interaction between heme oxygenase-1 and NADPH-cytochrome p450 reductase. Calvete and coworkers[181] used a clever approach to identify the heparin-binding domain of bovine seminal plasma protein PDC-109. The PDC-109 protein was bound to heparin-agarose in 16.6 mM Tris–50 mM NaCl–1.6 mM EDTA–0.025% NaN_3, pH 7.4. After washing the column to remove protein not bound

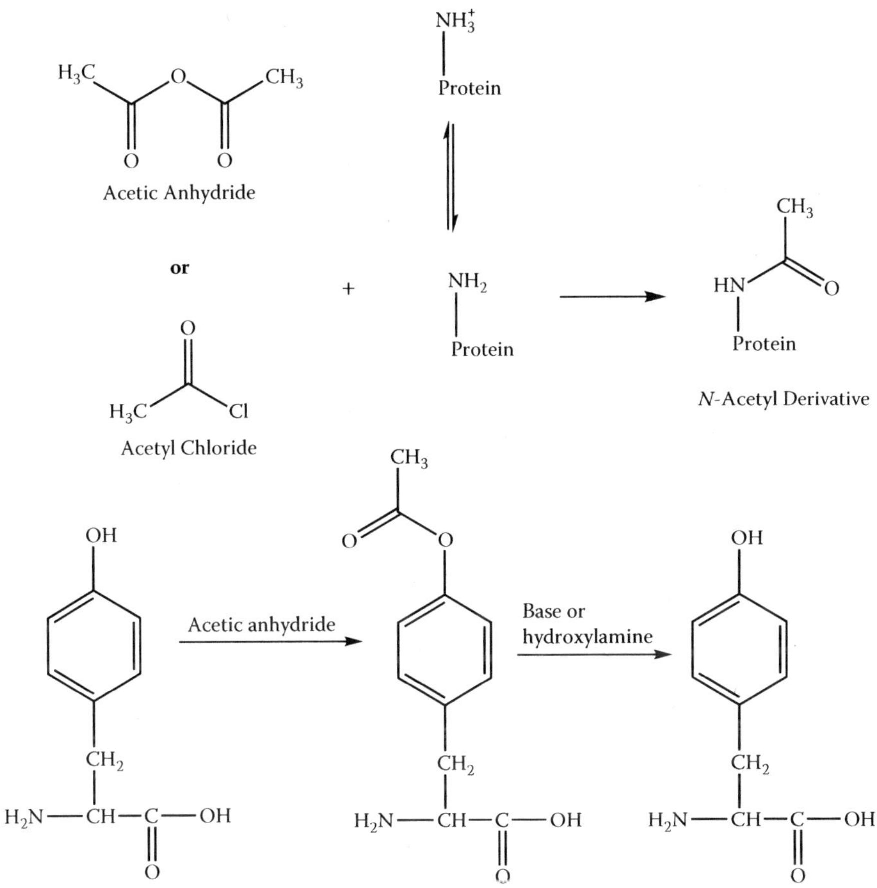

FIGURE 1.9 The reaction of acetic anhydride with amino and hydroxyl groups in peptides and proteins. The amide product formed with primary amines is stable, whereas the acetylation at hydroxyl groups and on the imidazole ring is reversible; the stability of such derivatives is dependent on the microenvironment.

to the matrix, the column was recycled at room temperature with the application buffer containing acetic anhydride (25- to 1600-fold molar excess over protein lysine). A similar experiment was performed with 1,2-cyclohexanedione to study arginine modification. Six basic residues were protected from modification by binding to the heparin matrix. Taralp and Kaplan[182] examined the reaction of acetic anhydride with lyophilized α-chymotrypsin in vacuo. α-Chymotrypsin was lyophilized from an unbuffered solution at pH 9.0 in one chamber in a reaction vessel. ^{3}H-Acetic anhydride was added to another compartment in the reaction vessel. The reaction vessel was evacuated and placed in an oven at 75°C. Several reaction vessels were used and removed at various time intervals for analysis. The proteins were then modified with ^{14}C-acetic anhydride, and the ratio of ^{3}H to ^{14}C was used to determine the extent of modification. Whereas complete modification of amino groups is achieved at pH 9.0 in aqueous solution, in the nonaqueous

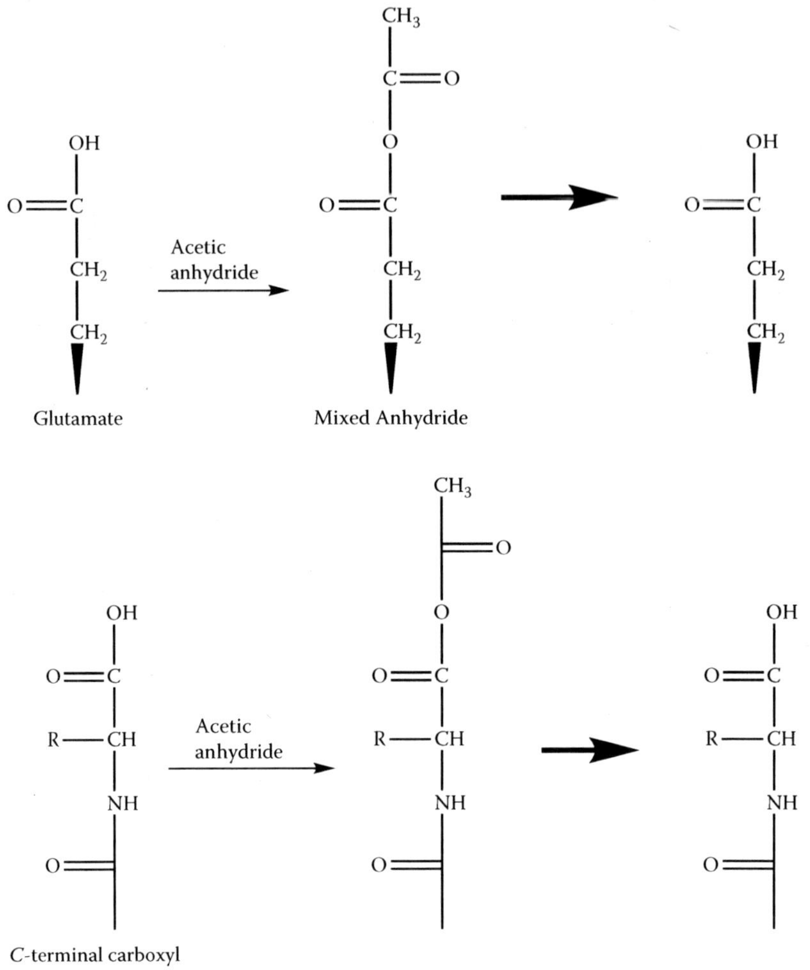

FIGURE 1.10 The formation of mixed anhydride in proteins. The mixed anhydride products are unstable and undergo rapid hydrolysis, but are stable under nonaqueous conditions (lyophilized proteins). (See Taralp, A. and Kaplan, H., Chemical modification of lyophilized proteins in nonaqueous environments, *J. Protein Chem.* 16, 183–193, 1997.)

system, only 25% of the ε-amino groups and 90% of the α-groups were modified. It also appeared that mixed anhydrides formed with carboxyl groups on the protein surface. Kaplan and coworkers subsequently reported on the modification of amino groups in lyophilized proteins with iodomethane.[183] The pH memory effect describes the correlation between solution pH prior to freeze-drying (lyophilization) and functional group reactivity in the lyophilized state.[184] The ionization state of a given functional group in solution is maintained in the lyophilized state.

In addition to acetic anhydride, acylation can be performed with a variety of acid anhydrides, including citraconic anhydride,[185–191] maleic anhydride,[192–197] succinic anhydride,[198–207] trimellitic anhydride,[208–212] cis-aconitic anhydride,[213–217] and various phthalic anhydride derivatives[218–221] (Figures 1.11 and 1.12). Modification with the dicarboxylic acid anhydrides such as succinic anhydride provide for charge reversal.[221–223] Dicarboxylic acid anhydrides are also used for the preparation of allergoids (see Chapter 9).

Competitive labeling (trace labeling; also see Chapter 2) is a technique for determining the ionization state or constant and intrinsic reactivity of individual amino groups in a protein.[224] The method is based on the hypothesis that the individual amino groups will compete for a trace amount of radiolabeled reagent (the reagent is selected on the basis of nonselective reactivity with amino groups; with most studies, acetic anhydride has been the reagent of choice). The extent of radiolabel incorporation into the protein at a given site will then be a function of the p*K*a, microenvironment, and inherent nucleophilicity of that particular amino group.[224] After the reaction with the radiolabeled reagent is complete, the protein is denatured, and complete modification at each amino group is achieved by the addition of an excess of unlabeled reagent. A reproducible digestion method (i.e., tryptic or chymotryptic hydrolysis) is used to obtain peptides from the completely modified protein. The peptides are separated by a chromatographic technique, and the extent of radiolabel at each site is determined. The extent of radiolabel incorporation at a given site is a

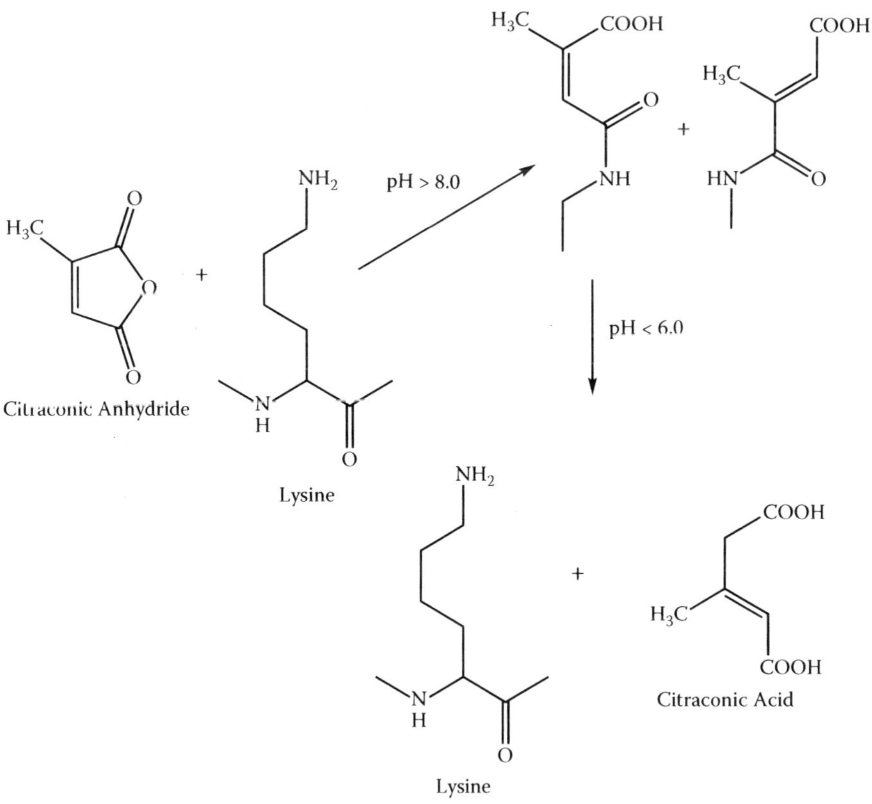

FIGURE 1.11 The reversible modification of lysine with citraconic anhydride.

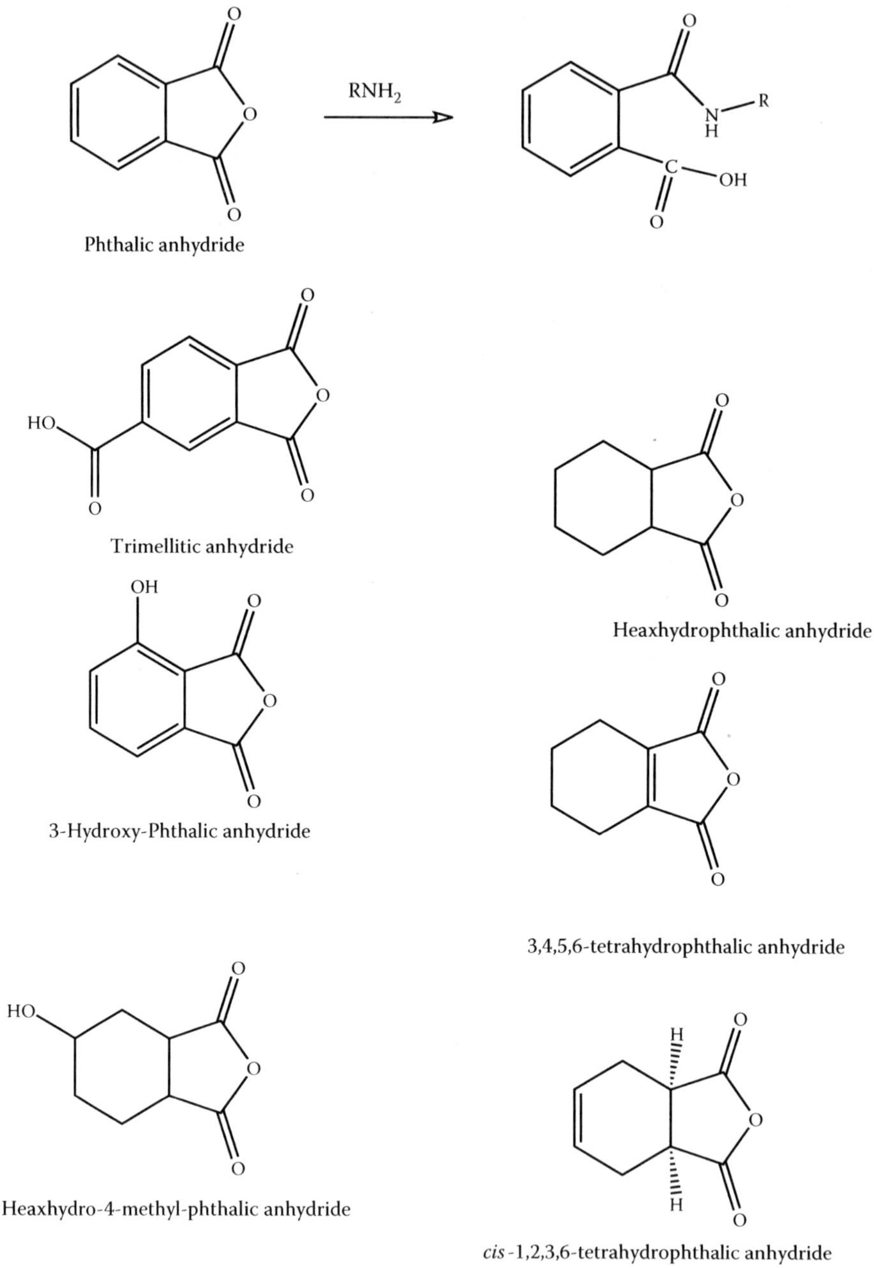

FIGURE 1.12 The modification of proteins with phthalic anhydride and various derivatives.

function of the reactivity of that individual amino group under the reaction conditions used at the radiolabel step. An alternative approach[225,226] involves a "trace" labeling step with tritiated acetic anhydride followed by complete modification with unlabeled acetic anhydride under denaturing conditions. This modified protein is then mixed with a preparation of the same protein that has been uniformly labeled with the ^{14}C-labeled acetic anhydride. Digestion and separation of peptide is performed by conventional techniques (see earlier text), and the extent of radiolabeling is determined. The ratio of ^{3}H to ^{14}C in peptides containing amino groups is an indication of functional group reactivity. This method is somewhat more sensitive than the original method. Reductive methylation has also been used.[227] Although this is a laborious technique, the data obtained are excellent and provide considerable insight into the solution structure of proteins. There has been a consistent use of this technique for the study of troponin-T,[228] troponin-C,[229] troponin-I,[230] calmodulin,[231–233] and tropomyosin.[234] In particular, studies[233,234] that have used this technique to assess conformational change in solution have been particularly rewarding. The use of mass spectrometry has greatly simplified the assay.

Lysine residues can be modified by reaction with α-ketoalkyl halides such as iodoacetic acid.[235] Acylation can occur at pH > 7.0, but the rate of reaction is much slower than reaction with cysteinyl residues. Both the mono- and disubstituted derivatives have been reported. The monosubstituted derivative migrates close to methionine on amino acid analysis, whereas the disubstituted derivative migrates near aspartic acid. Boja and Fales[236] noted the modification of lysine and protein α-amino groups with iodoacetamide. This modification poses problems for the subsequent analysis of samples by mass spectrometry and can be avoided by the inclusion of a thioether such as 2,2′-thiodiethanol. Nielsen and coworkers[237] observed disubstitution of lysine with iodoacetamide, yielding a derivative with the same atomic composition as diglycine, a marker for ubiquitinylation. The modification of lysine could be avoided by the use of chloroacetamide in place of iodoacetamide.

Both fluoronitrobenzene and fluorodinitrobenzene have been used in protein chemistry since Sanger and Tuppy's work on the structure of insulin.[238] Carty and Hirs[239] developed the use of 4-sulfonyl-2-nitrofluorobenzene for the modification of amino groups in pancreatic ribonuclease. This reagent also is more stable than, for example, fluorodinitrobenzene under alkaline conditions, permitting more accurate measurement at pH > 9.0. The lysine residue at position 41 is the site of major substitution which is a reflection of the lower p*K*a for the ε-amino group of this residue. Use of this compound did not present the solubility and reactivity problems posed by the fluoronitrobenzene compounds. It was possible to qualitatively determine the classes of amino groups in ribonuclease; these were the α-amino group, nine "normal" amino groups, and lysine 41. The reactivity of lysine 41 was influenced by neighboring functional groups. This effect was lost at pH > 11 or on thermal denaturation of the protein. Fluorodinitrobenzene is now used infrequently for protein chemistry but does see use as a hapten/sensitizer in immunology studies.[240–243]

Cyanate reacts with primary amine functions in proteins (Figure 1.13) and other biological polymers. Stark and coworkers[244] pursued the observation that ribonuclease was inactivated by urea in a time-dependent reaction. It was established that this inactivation was a reflection of the content of cyanate in the urea preparation.

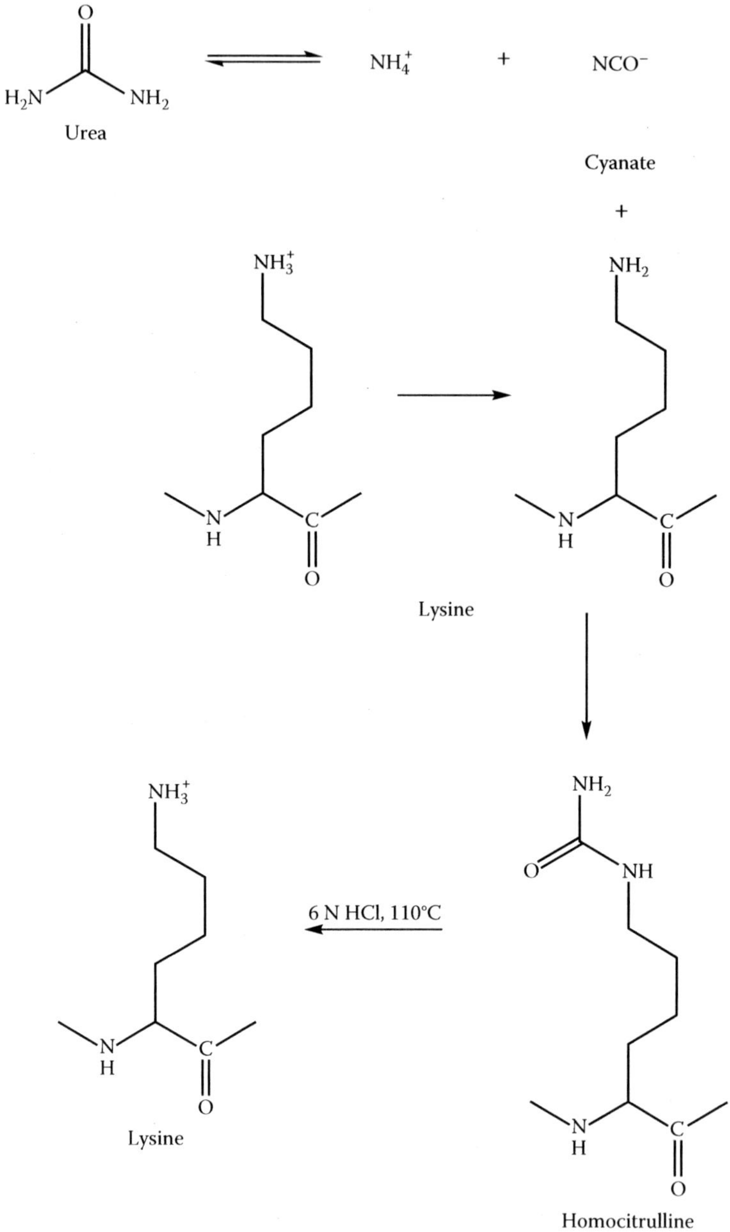

FIGURE 1.13 A scheme for the formation of cyanate from urea and subsequent carbamylation of lysine. (See Marier, J.R. and Rose, D., Determination of cyanate, and a study of its accumulation in aqueous solutions of urea, *Anal. Biochem.* 7, 304–314, 1964; Gerding, J.J., Koppers, A., Hagel, P., and Bloemendal, H., Cyanate formation in solutions of urea. II. Effect of urea on the eye lens protein-crystallin, *Biochim. Biophys. Acta* 243, 375–379, 1971.)

There have been a number of recent studies on the modification of proteins by cyanate derived from urea.[245–250] The possible modification of proteins during proteomic analysis by cyanate derived from urea has been of concern. However, there is some question of the importance of the dismutation of urea to form cyanate under normal conditions for the preparation of proteins for analysis.[251–254] Cyanate can react with other functional groups in protein.[255] The ε-amino group of lysine is the least reactive ($k = 2.0 \times 10^{-3}$ M^{-1} min^{-1}) compared to the α-amino group of glycylglycine ($k = 1.4 \times 10^{-1}$ $M^{-1}min^{-1}$). The carbamyl derivative of histidine is quite unstable as is the corresponding derivative of cysteine. The reaction of chymotrypsin with cyanate results in loss of catalytic activity associated with the carbamylation of the active-site serine residue.[256] Shen and Colman[257] observed saturation kinetics in the modification of a lysine residue in diphosphopyridine nucleotide-specific isocitrate dehydrogenase, suggesting the cyanate/isocyanic acid bound to the enzyme active site is an analog of carbon dioxide. Several different reagents, including cyanate, were used to study the role of lysine residues in bovine pancreatic deoxyribonuclease A.[258] Modification with cyanate was performed at 37°C in 1.0 M triethanolamine hydrochloride, pH 8.0. The extent of modification was determined by analysis for homocitrulline following acid hydrolysis. A time course of hydrolysis was utilized to provide for the accurate determination of homocitrulline, because this amino acid slowly decomposes to form lysine during acid hydrolysis.

The modification of lysine with imidoesters has the advantages that the charge of the lysine residue is maintained during the modification (Figure 1.14). Plapp and coworkers[258] examined the reaction of methyl picolinimidate with pancreatic deoxyribonuclease. Methyl picolinimidate (Figure 1.15) is an imidoester that reacts with the primary amino groups in proteins. The extent of modification of a protein by methyl picolinimidate can be determined by spectral analysis. Under these conditions, essentially all of the primary amino groups in deoxyribonuclease (nine lysine and one amino-terminal amino group) were modified, but there was no

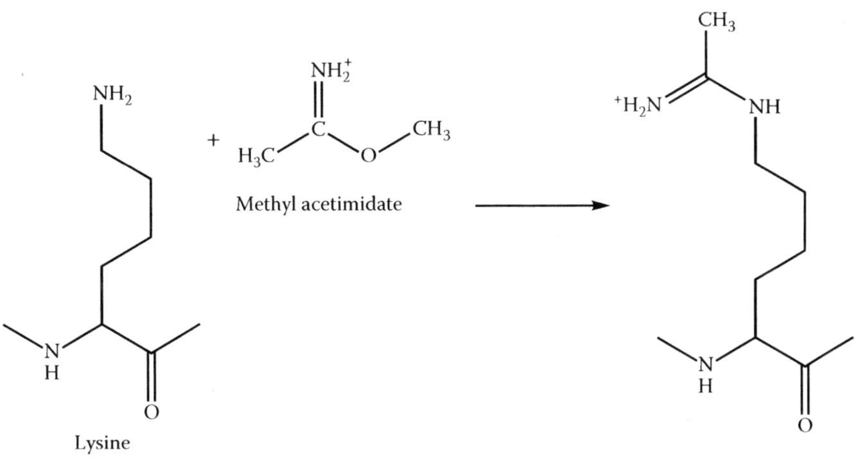

FIGURE 1.14 The reaction of methyl acetimidate with lysine.

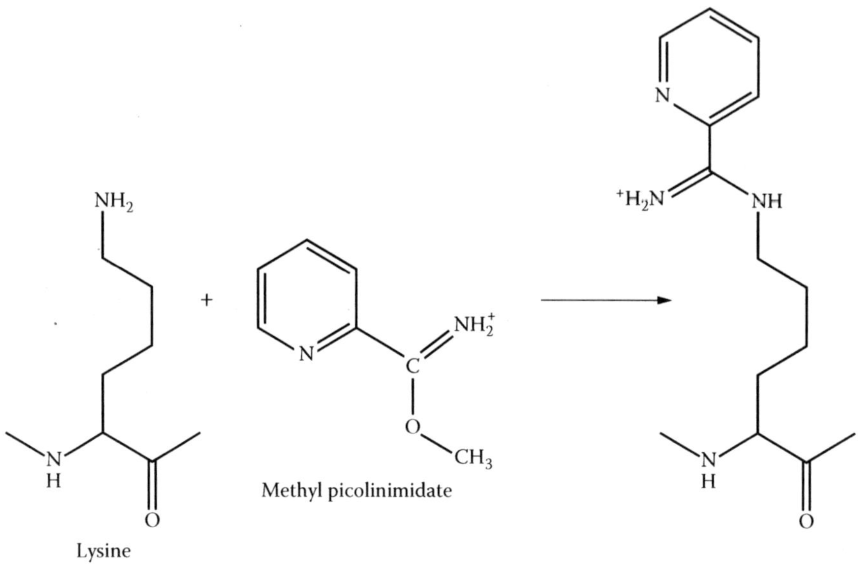

FIGURE 1.15 The reaction of methyl picolinimidate with lysine.

change in biological activity. Plapp has also studied the reaction of methyl picolinimidate with horse liver alcohol dehydrogenase.[259] This study was somewhat unique in that modification of the enzyme resulted in enhanced catalytic activity, reflecting more rapid dissociation of the enzyme–coenzyme complex. It should be noted that the derivatized lysine reverts to lysine (60% yield) under the normal conditions of acid hydrolysis.

Pyridoxal phosphate (Figure 1.16) is useful for the modification of lysine because of the selectivity of reaction, spectral properties of the modified residue, reversibility of reaction, and the establishment of stereochemistry by use of radiolabeled sodium borohydride (sodium borotritiide) to reduce the Schiff base initially formed on the reaction of pyridoxal phosphate with a primary amine. Pyridoxal phosphate will react with all primary amines (both ε-amino groups of lysine and the amino-terminal α-amino function) in a protein. In general, pyridoxal-5′-phosphate is far more reactive than pyridoxal because of intramolecular hemiacetal formation and the neighboring group effect of the phosphate moiety. Shapiro and coworkers investigated the reaction of pyridoxal phosphate with rabbit muscle aldolase.[260] The initial reaction produced a species with an absorbance maximum at 430 to 435 nm, reflecting the protonated Schiff base form of the pyridoxal phosphate–protein complex. After reduction with sodium borohydride, the absorbance maximum was at 325 nm, which is characteristic of the reduced Schiff base. This is a quite useful study, in that the difference in reactivity between pyridoxal and pyridoxal-5′-phosphate is demonstrated, as is the reversible nature of the initial complex.

Schnackerz and Noltmann[261] compared the reaction of pyridoxal-5′-phosphate and other aldehydes in reaction with rabbit muscle phosphoglucose at pH 8.0. Pyridoxal-5′-phosphate (0.19 mM) resulted in 82% inactivation, whereas the following results

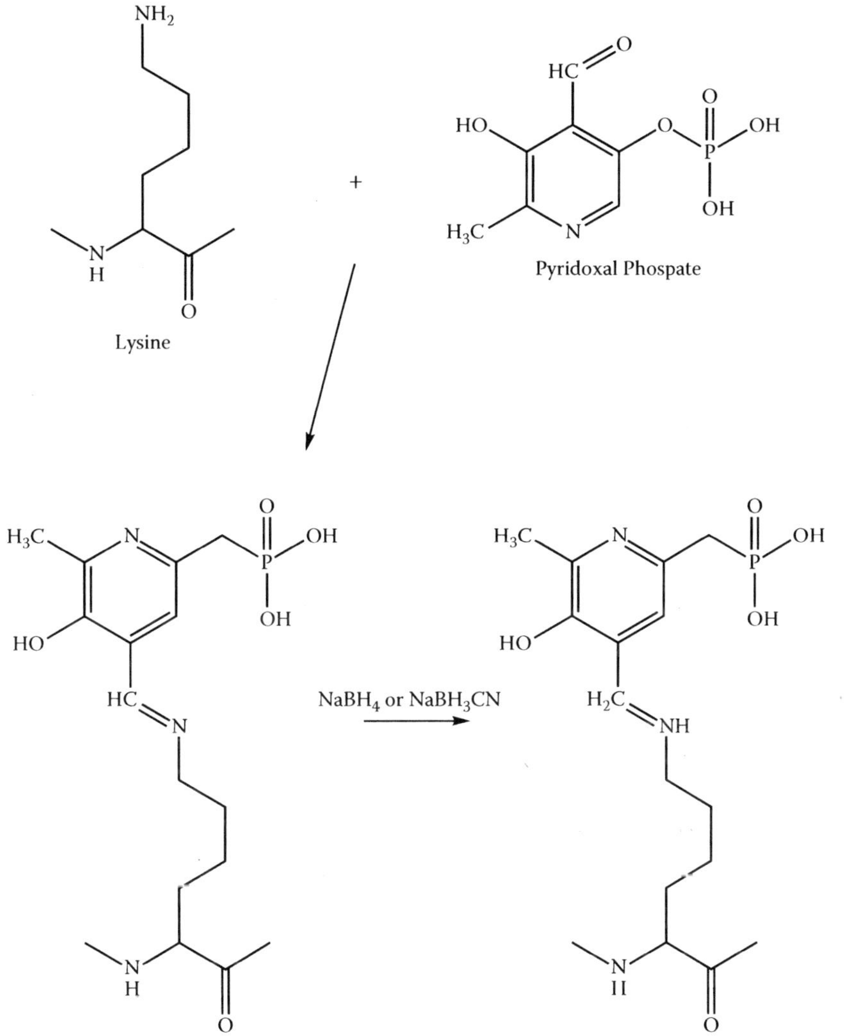

FIGURE 1.16 The reaction of pyridoxal phosphate with lysine and subsequent reduction of the Schiff base with borohydride.

were obtained with other aldehydes: pyridoxal (8.4 mM), 16% inactivation; acetaldehyde (75 mM), 75% inactivation; and acetone (75 mM), 31% inactivation. This last reaction is of interest as many investigators are unaware that acetone can react with amino groups in proteins. The reaction of acetone with primary amino groups has been known for some time[262] and is discussed in further detail later when discussing the topic of reductive alkylation. The importance of local environmental factors in the specificity of modification by pyridoxal phosphate is emphasized by Ohsawa and Gualerzi.[263] These investigators examined the modification of *Escherichia coli* initiation factor by pyridoxal phosphate in 0.020 M triethanolamine, 0.03 M KCl, pH 7.8.

In the course of the studies, it was observed that pyridoxal phosphate will not react with poly(AUG). These investigators also reported the preparation of N^6-pyridoxal lysine by reaction of pyridoxal phosphate with polylysine in 0.01 M sodium phosphate, pH 7.2 at 37°C followed by reduction with NaBH4. The reduction was terminated by the addition of acetic acid. Acid hydrolysis (6 N HCl, 110°C, 22 h) yielded N^6-pyridoxal-L-lysine. Bürger and Görisch[264] reported the inactivation of histidinol dehydrogenase upon reaction with pyridoxal phosphate in 0.02 M Tris, pH 7.6. This modification could be reversed by dialysis unless the putative Schiff base was stabilized by reduction with NaBH4 (*n*-octyl alcohol added to prevent foaming). These investigators used a $\Delta\varepsilon$ for ε-amino pyridoxal lysine of 1×10^4 $M^{-1}cm^{-1}$ at 325 nm.

The specificity of pyridoxal-5′-phosphate in protein modification is thought to be derived from electrostatic interactions via the phosphate group with positively charged groups (i.e., arginine) on the protein surface. A conceptually related compound is methyl acetyl phosphate. The reagent was originally developed as an affinity label for D-3-hydroxybutyrate dehydrogenase.[265] Manning and coworkers have examined the chemistry of the reaction of methyl acetyl phosphate with hemoglobin in some detail.[266,267] It appears to be an affinity label for the 2,3-diphosphoglycerate binding site.[266] More recent work suggests that this reagent may be a useful generic probe for anion-binding sites in proteins.[267] The use of methyl acetylphosphate as an affinity label has also been suggested by other investigators.[268]

The modification of primary amines in proteins by reductive alkylation (Figure 1.17) has the advantage that the basic charge properties of the modified residue are preserved. The early work on this modification has been reviewed by Means.[269] Both monosubstituted and disubstituted derivatives can be prepared depending on reaction conditions and the nature of the carbonyl compound. The introduction of sodium cyanoborohydride as a reducing agent for this reaction represented a real advance. Sodium cyanoborohydride is stable in aqueous solution at pH 7.0. Unlike sodium borohydride, which can reduce aldehydes and disulfide bonds, sodium cyanoborohydride only reduces the Schiff base formed in the initial process of reductive alkylation. Jentoft and Dearborn have studied the use of sodium cyanoborohydride in some detail.[270] In particular, the preparation of sodium cyanoborohydride is critical, and most, if not all, commercial preparations require recrystallization prior to use. The radiolabeling of proteins using ^{14}C-formaldehyde and sodium cyanoborohydride has been reported.[271] The modification was performed in 0.04 M phosphate, pH 7.0, at 25°C. The modification can be performed equally well at 0°C, but, as would be expected, it takes longer; there is no effect on the extent of the modification. In this regard, these authors estimated that the same extent of modification obtained in 1 h at 37°C could be achieved in 4 to 6 h at 25°C or 24 h at 0°C. Although the majority of experiments in this study were performed in phosphate buffer at pH 7.0, equivalent results can be obtained in Tris or HEPES buffer at pH 7.0. A greater extent of modification was observed with sodium cyanoborohydride at pH 7.0 than with sodium borohydride at pH 9.0. Reductive methylation with 3C-enriched formaldehyde has been used to introduce an NMR probe for the study of protein conformation.[272] A similar approach has been developed using deuterated acetone.[273]

The effect of carbonyl compounds of different size on the extent of reductive alkylation has been examined by Fretheim and coworkers.[274] The extent of modification

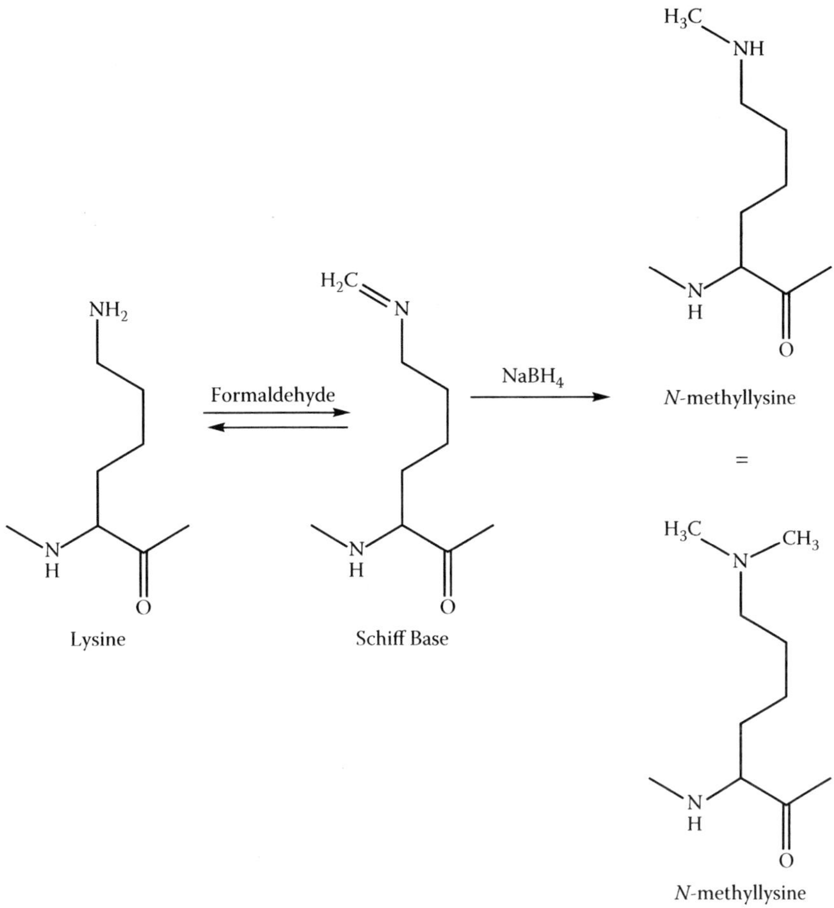

FIGURE 1.17 The reductive methylation of proteins.

is more a reflection of the type of alkylating agent and reaction conditions than an intrinsic property of the protein under study. For example, nearly 100% disubstitution can be obtained with formaldehyde and approximately 35% disubstitution with *n*-butanol, whereas only monosubstitution can be obtained with acetone, cyclopentanone, cyclohexanone, and benzaldehyde. Whereas most of the products of reductive alkylation retained solubility, the reaction products obtained with cyclohexanone and benzaldehyde tended to precipitate. Examination of the reductive alkylation of ovomucoid, lysozyme, and ovotransferrin with different aldehydes suggests that such modification occurs without major conformational change as judged by circular dichroism measurements.[275] The same study also examined the stability of the modified proteins by scanning differential calorimetry. The extensive modification of amino groups decreases thermal stability. The destabilization effect increases with increasing size (and hydrophobicity) of the modifying aldehyde.

The reaction of 2,4,6-trinitrobenzenesulfonic acid (TNBS) with amino groups (Figure 1.18) is used to study the function and reactivity of amino groups in proteins.[276–278] The modification of amino groups with TNBS is easy to monitor by spectral analysis. In the presence of an excess of sulfite, absorbance at 420 nm is the most sensitive index, having $\varepsilon = 2.0 \times 10^4\ M^{-1}cm^{-1}$; absorbance at 420 nm is dependent on the ability of the reaction product to form a complex with sulfite. Fields[279] suggests the recrystallization of TNBS from 2.0 M HCl prior to use. Modification reactions are usually performed in phosphate buffer (pH 6.0 to 9.0). Derivatives of α- and ε-amino groups have similar spectra; α-amino derivatives have a slightly higher

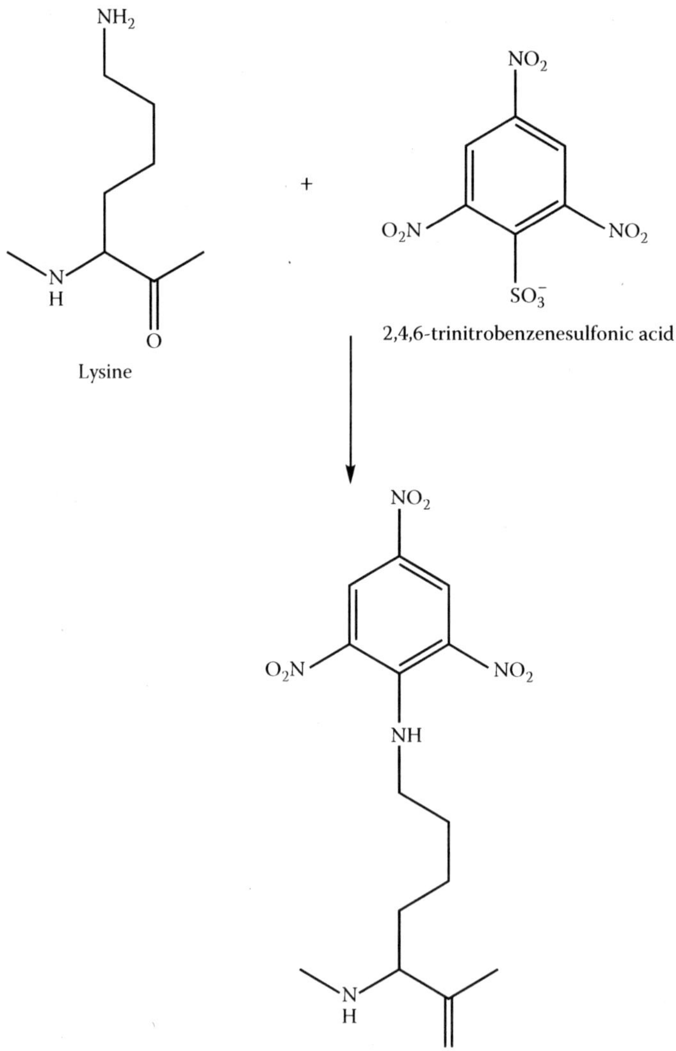

FIGURE 1.18 The reaction of trinitrobenzenesulfonic acid with lysine.

extinction coefficient at 420 nm ($\varepsilon = 2.20 \times 10^4\ M^{-1}cm^{-1}$) than ε-amino groups ($\varepsilon = 1.92 \times 10^4\ M^{-1}cm^{-1}$). Both of these derivatives have much higher extinction coefficients than the derivative obtained by reaction of TNBS with cysteinyl residues ($\varepsilon = 2.25 \times 10^3\ M^{-1}cm^{-1}$). The α-and ε-amino derivatives can be differentiated by their stability to acid or base hydrolysis. The α-amino derivatives are unstable to acid hydrolysis (8 h at 110°C) or base hydrolysis.[280] Cayot and Tainturier[281] carefully examined the conditions for the use of TNBS for the determination of amino groups in proteins. They observed that although raising the temperature of the reaction only slightly increases the rate of reaction with amino groups, the rate of the hydrolysis of the reagent is substantially increased. It is also observed that, as expected, accessibility of the functional groups is important for reaction with TNBS. They recommend reaction at pH 10 (0.1 M borate) with a 100-fold molar excess (to protein amino groups). Reaction is usually complete in 15 min and can be measured by absorbance at 420 nm. Frieden and coworkers have explored the reaction of trinitrobenzenesulfonic acid with bovine liver glutamate dehydrogenase.[282,283] In these studies, the modification was performed in 0.04 M potassium phosphate, pH 8.0. Under these reaction conditions, the cysteinyl residues were not modified. The preparative reactions were terminated by reaction with 2-mercaptoethanol. It is of interest that under certain conditions (with reduced coenzyme), glutamate dehydrogenase catalyzed the conversion of TNBS to trinitrobenzene.[284]

The reaction of TNBS with simple amines and hydroxide ions has been studied in some detail by Means and coworkers.[285] The reaction of TNBS with hydroxide is first order with respect to both trinitrobenzenesulfonate and hydroxide ions. Reaction with amines was considered in some detail. In general, reactivity of trinitrobenzenesulfonate with amines increases with increasing basicity except that secondary amines and *t*-alkylamines are comparatively unreactive. The specific binding of trinitrobenzenesulfonate to proteins must be considered in the study of the reaction of this compound with proteins. Only amines with a pKa > 8.7 follow a simple rate law. These investigators presented the following considerations regarding the reaction of trinitrobenzenesulfonic acid with proteins. Reactivity is a sensitive measure of the basicity of an amino group. Adjacent charged groups influence the rate of reaction with an increase observed with a positively charged group and a decrease with a negatively charged group. Proximity to surface hydrophobic regions that can bind TNBS can increase the observed reactivity of a particular amino group. It is possible to label components on the surface of membranes with TNBS, as the sulfonate moiety does not permit membrane penetration. The same is true for pyridoxal-5′-phosphate.

The use of trinitrobenzenesulfonate in the selective modification of membrane surface components has been explored by Salem and coworkers.[286] This study involved the modification of intact cells with the TNBS (dissolved in methyl alcohol) diluted to a 1% methanolic solution. As mentioned earlier, trinitrobenzenesulfonate does not pass across (or into) membranes, being more hydrophilic than, for example, fluorodinitrobenzene.

Haniu et al.[287] have examined the reaction of lysine residues in NAD(P)H:quinone reductase with TNBS as compared to the reaction of tyrosine residues with *p*-nitrobenzenesulfonyl fluoride. Isolation and characterization of the peptides containing the modified residues showed that the modified tyrosyl residues are in hydrophobic regions of the protein, whereas the modified lysine residues are in hydrophilic regions.

N-hydroxysuccinimide (NHS) ester derivatives were introduced by Anderson and coworkers[288] for the preparation of "active esters" of acyl peptides. *N*-hydroxysuccinimide-based reagents are reasonably specific for amino groups in proteins (Figure 1.19), but can react with hydroxyl functions and sulfhydryl functions,[289] yielding ester and thioester derivatives, respectively (Figure 1.20) and possible at imidazole rings. Any of these derivatives would be considerably less stable than the amide bond. The development of reagents based on *N*-hydroxysuccinimide chemistry can be challenging,[290] but the derivatives are quite useful.[291–295] *N*-hydroxysuccinimide chemistry is used for binding of proteins to matrices.[296–299] Takeda and coworkers[300] used a bifunctional reagent (Figure 1.21) that contained an NHS function and a benzylthioester function to prepare a DNA–protein hybrid. The Bolton–Hunter reagent[301] is based on *N*-hydroxysuccinimide chemistry.

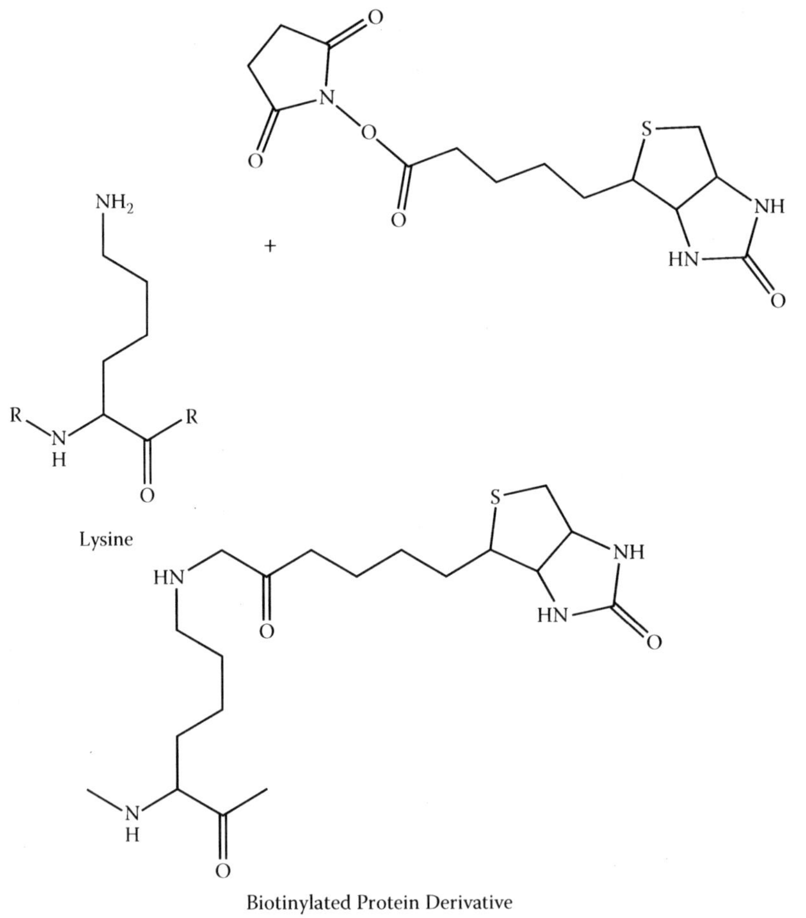

FIGURE 1.19 The reaction of *N*-hydroxysuccinimide ester with amino groups in proteins.

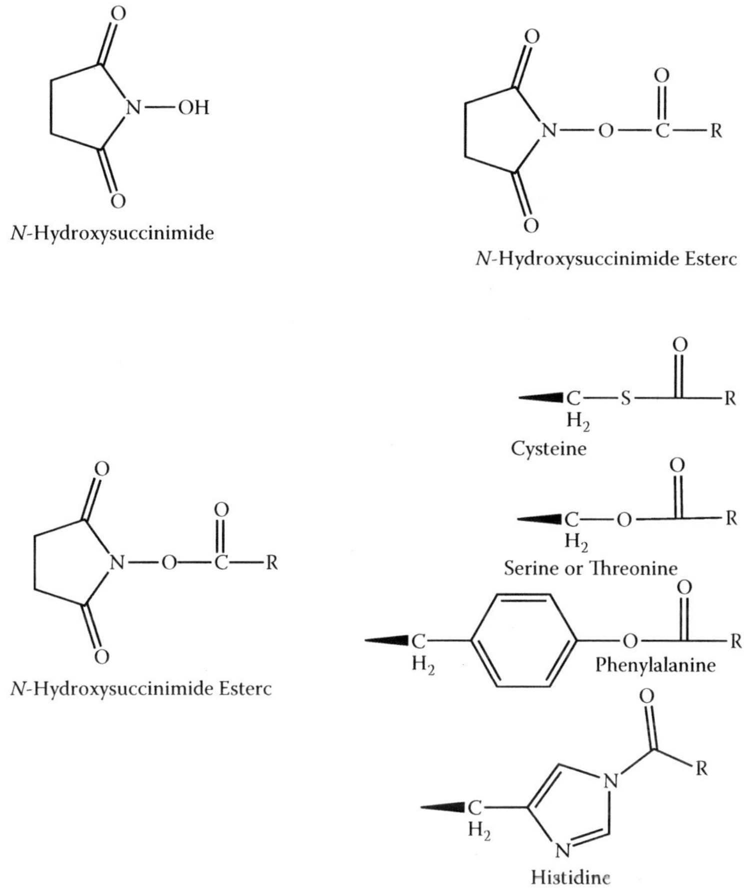

FIGURE 1.20 The reaction of *N*-hydroxysuccinide esters with hydroxyl, sulfhydryl, and imidazole groups in proteins.

Guanidination of lysine residues in proteins using *O*-methylisourea was reported by Greenstein in 1935.[302] Guanidation has been reported to stabilize proteins[303]; more recently, there has been interest[304] in guanidination in the cold stabilization of proteins. 2-*S*-thiuroniumethanesulfonate (Figure 1.22)[305] has been developed as an alternative to *O*-methylisourea. Improved detection by mass spectrometry of lysine-containing peptides is further improved by guanidination.[306]

TYROSINE

The chemical modification of tyrosine in proteins has proved to be useful in the study of protein topology as well as for the development of unique chemistry for binding to solid matrices. It is possible to modify tyrosyl residues in proteins under relatively mild conditions with reasonably high specificity with a variety of reagents, obtaining, in turn, a variety of derivatives as described in the following

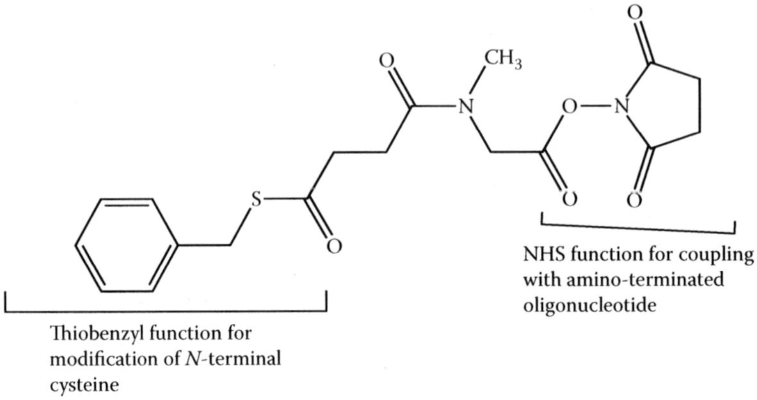

FIGURE 1.21 An *N*-hydroxysuccinimide thiobenzyl ester reagent for use in chemical ligation or cross-linking. (From Takeda, S., Tsukiji, S., Ueda, H., Covalent split-protein fragment–DNA hybrids generated through *N*-terminus-specific modification of proteins by oligonucleotides, *Org. Biomol. Chem.* 6, 2187–2194, 2008.)

text. *N*-acetylimidazole (NAI) acetylates the phenolic hydroxyl group in a reversible reaction. Iodination is used for the modification of tyrosine residues and provides for the production of radiolabeled derivatives. Tetranitromethane (TNM) nitrates tyrosyl residues to yield the 3-nitro derivative, which markedly lowers the p*K*a of the phenolic hydroxyl group. The 3-nitro function can be reduced under mild conditions to give the 3-amino derivatives, which can be subsequently modified with a

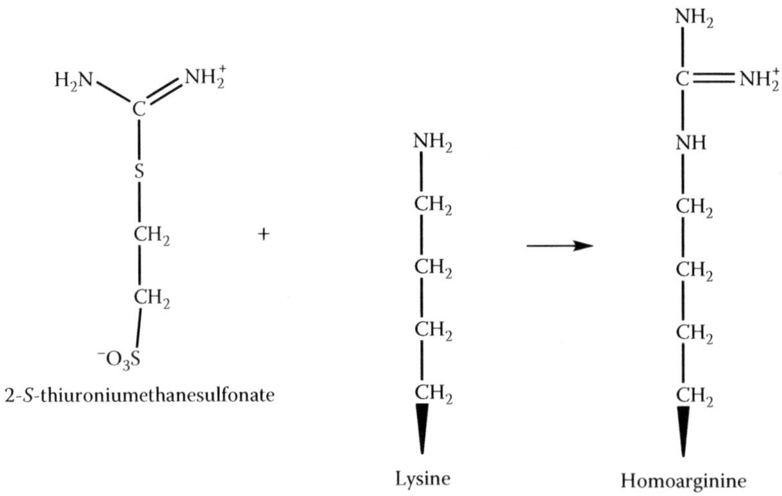

FIGURE 1.22 The formation of homoarginine in proteins from the modification of lysine with 2-*S*-thioroniumethanesulfonate. (From Hundle, B.S. and Richards, W.R., Use of a novel membrane-impermeable guandinating reagent, 2-*S*-[^{14}C]-thioroniumethanesulfonate, for the labeling of intracytoplasmic membrane proteins in *Rhodobacter sphaeroides*, *Biochemistry* 26, 4505–4511, 1987.)

variety of useful compounds such as dansyl chloride [5-(dimethylamino)-1-napthalenesulfonyl chloride) or biotinamidonexanoic acid sulfo-*N*-hydroxysuccinimide ester, sodium salt (sulfosuccinimidyl-6-biotinamidohexanoate). Reaction with TNM can also result in zero-length cross-linkage in proteins via the formation of dityrosine. Interest markedly increased with the discovery of the role of peroxynitrite in the modification of tyrosine.

Factors that influence the reactivity of tyrosyl residues in proteins are, at best, poorly understood. It would seem that the reactivity is influenced most by the ionization state of tyrosine, which is a function of the microenvironment around the residue. It is extremely useful for investigators to review early literature on the factors influencing tyrosine ionization in proteins.[306–315]

Iodination is used for the modification of tyrosyl residues in proteins.[316–318] Durr and coworkers[319] have observed that *O*-acetylation of tyrosine prevented iodination. In this work, the *O*-acetyl tyrosine was incorporated during the solid-phase synthesis of vasopression or oxytocin, allowing iodination at a second tyrosine residue. Radioisotope labeling of proteins with isotopes of iodine has been used extensively to study protein turnover (catabolism) and *in vivo* distribution.[320–323] The development of neoantigens secondary to iodination[324,325] can result in additional complications. Sohoel and coworkers[326] showed that iodination changes the subcutaneous absorption rate of an insulin derivative. The therapeutic use of the iodination of proteins and peptides is discussed in Chapter 9.

The preparation of *O*-acyl derivatives via the action of carboxylic acid anhydrides (i.e., acetic anhydride) have been used for some time, but it is very difficult to obtain selective modification of tyrosine as these reagents readily react with primary amines to form stable *N*-acyl derivatives.[327,328] It is, however, possible to obtain the selective modification of tyrosine with acetic anhydride (Figure 1.23) by reaction at mildly acidic pH (1.0 M acetate, pH 5.8 at 25°C), approximately 20,000-fold molar excess

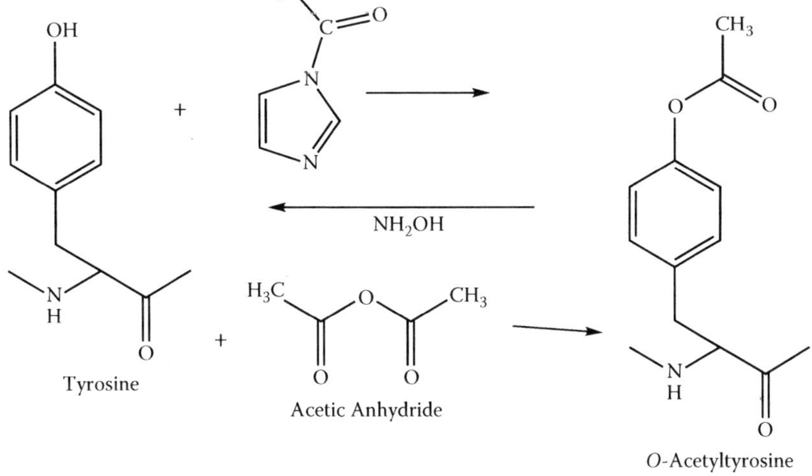

FIGURE 1.23 The *O*-acetylation of tyrosine by acetic anhydride or *N*-acetylimidazole.

of acetic anhydride (5.1×10^{-2} M acetic anhydride, 2.9×10^{-6} M enzyme).[329] Bernad and colleagues[330] have reported on an extensive study comparing the modification of lysyl and tyrosyl residues in lysozyme with dicarboxylic acid anhydrides. In 50 mM HEPES, 1.25 M NaCl, pH 8.2, amino groups (primarily lysine residues) were far more reactive than hydroxyl groups (including tyrosine, serine, and threonine).

N-acetylimidazole (NAI) was first used as a reagent for the modification of tyrosyl residues (Figure 1.23) in bovine pancreatic carboxypeptidase A.[331] This same group of investigators subsequently reported on the use of NAI in the determination of "free" tyrosyl residues in proteins[332] as opposed to "buried" residues. This has not necessarily proved to be the case.[333] A wide variety of buffers have been used for the study of the reaction of NAI but a high concentration of nucleophilic species such as Tris should be avoided because of reagent instability.[331] Likewise, although the modification occurs more rapidly at pH values more alkaline than 7.5, reagent and product (*O*-acetyl tyrosine) stability become a significant problem. *N*-acyl derivatives other than NAI have been prepared.[334–336]

El Kebbaj and Latruffe[337] examined the membrane penetration of NAI. NAI was similar to TNM in its ability to preferentially inactivate D-3-hydroxybutyrate dehydrogenase in "inside-out" membranes when compared to intact mitochondria. Evaluation of the partition of NAI in water–organic solvent systems showed an 81/19(%) distribution in an H_2O/hexane system and 31/69 in H_2O/1-octanol. Graves and coworkers have demonstrated that NAI will not acetylate 3-nitrotyrosine.[338] This study also examined the rates of acetylation of tyrosine and 3-fluorotyrosine as a function of pH between 7.5 and pH 9.5. The formation of the O-acetylated derivative was evaluated by HPLC analysis. As NAI is unstable at increasing pH, it was not possible to obtain reliable second-order rate constants; thus, the relative rates of reaction were reported. 3-Fluorotyrosine was acetylated more rapidly than tyrosine at pH 7.5, whereas the opposite was true at pH 9.5. It is suggested that the O-acetylation of tyrosine by NAI is facilitated by a microenvironment that promotes the ionization of the phenolic hydroxyl group, increasing nucleophilicity. There are several approaches to the determination of the extent of tyrosine modification by NAI. The amount of acetylhydroxamate produced by the reaction of hydroxylamine can be determined.[339] Second, *O*-acetylation of tyrosine produced a decrease in absorption at 278 nm ($\Delta\varepsilon = 1210\ M^{-1}\ cm^{-1}$).[332]

Use of TNM for the modification of tyrosyl residues in proteins was advanced over 50 years ago.[340] However, it was not until some two decades later that the studies of Vallee, Riordan, Sokolovsky, and Harell established the specificity and characteristics of the reaction of TNM (Figure 1.24) with proteins.[341,342] The modification proceeds optimally at mildly alkaline pH. The rate of modification of *N*-acetyltyrosine is twice as rapid at pH 8.0 as at pH 7.0; it is approximately ten times as rapid at pH 9.5 as at pH 7.0. Although the reaction of tetranitromethane with proteins is reasonably specific for tyrosine, oxidation of sulfhydryl groups has been reported, as has reaction with histidine, methionine, and tryptophan.[342–345] The reaction with sulfhydryl groups would seem to be the most common side reaction. Reaction of TNM with cysteine in proteins can result in disulfide bond formation and the formation of oxidation products such as sulfone and sulfenic acid derivatives. Reaction with TNM can also result in the covalent cross-linkage of tyrosyl residues, resulting in inter- and

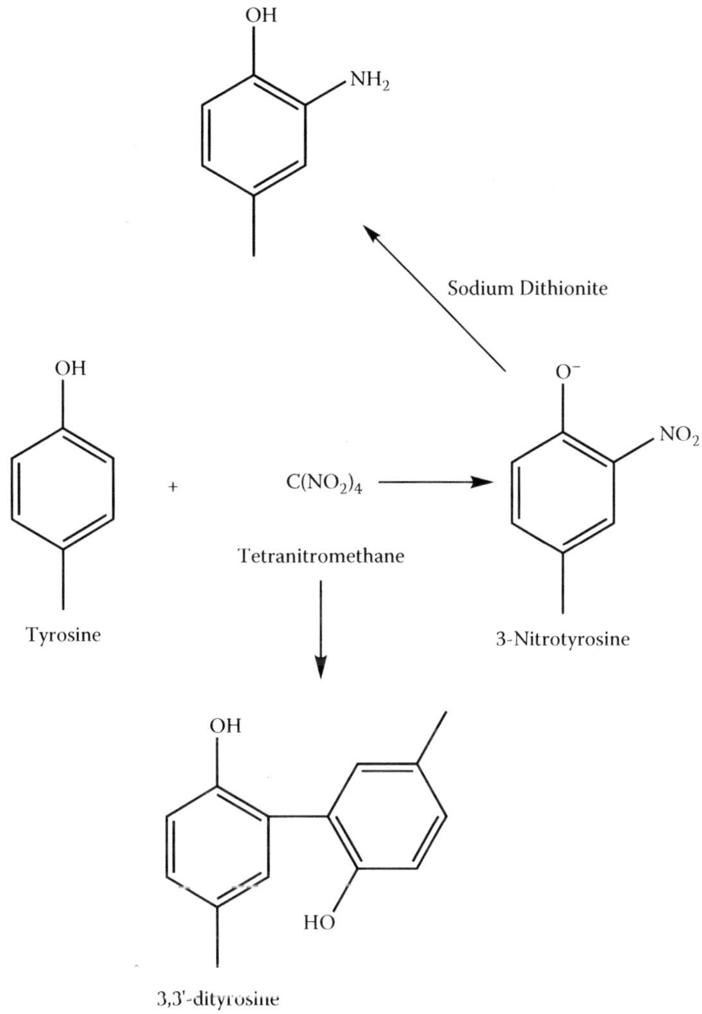

FIGURE 1.24 The modification of tyrosine by tetranitromethane.

intramolecular association of peptide chains.[346] Acidification of reaction mixtures tends to favor the cross-linkage reaction.[347] The extent of cross-linkage varies with the protein being studied. For example, reaction of pancreatic deoxyribonuclease with tetranitromethane results in extensive formation of dimer.[348]

The extent of modification of tyrosyl residues by tetranitromethane in proteins can be assessed by either spectrophotometric means or by amino acid analysis.[349,350] Mass spectrometry is proving to be of increasing value in the analysis of nitrotyrosine in proteins.[351–353] Mass spectrometry permitted the identification of a dinitrotyrosine secondary to reaction with TNM. Antibodies to 3-nitrotyrosine in proteins have been developed and are useful not only for the analysis of purified proteins[354–356] but also for the enrichment of nitrotyrosine-containing proteins,[354] *in situ* localization in cells,[357] and for identification on 2D gel electrophoretograms.[358] Reduction with sodium dithionite (see later text) improves the specificity for the detection of nitrotyrosine-containing proteins on Western Blots; nitrotyrosine-positive bands are eliminated on reduction, leaving the "false-positive spots."[358,359]

The 3-nitrotyrosine can be reduced to the corresponding amine (Figure 1.2) under relatively mild conditions ($Na_2S_2O_4$, 0.05 M Tris, pH 8.0).[360] The conversion of 3-nitrotyrosine to 3-aminotyrosine is associated with the loss of the absorption maximum at 428 nm and the change in the p*K*a of the phenolic hydroxyl group from approximately 7.0 to 10.0. On occasion, reduction of the nitro function in this manner reverses the modification of function observed on nitration. The resultant amine function can be subsequently modified.[361] Fox and colleagues[362] modified the single tyrosine residue in *Escherichia coli* acyl carrier protein with TNM, subsequently reduced the 3-nitrotyrosyl residue to 3-aminotyrosine with sodium dithionite, and modified the 3-aminotyrosine with dansyl chloride at pH 5.0 (50 mM sodium acetate; the low pH provides specificity of modification) to obtain dansyl acyl carrier protein, which was used for fluorescence anisotropy studies.[363]

Riordan and coworkers[343] suggested that the reactivity of tyrosyl residues in proteins with TNM was a measure of exposure to solvent. This concept of "free" and "buried" residues was introduced earlier by the same group in their studies on the reaction of tyrosine with *N*-acetylimidazole in proteins.[332] In this model, free tyrosyl residues are considered to be in direct contact with solvent and have p*K*a values between 9.5 and 10.5, whereas "buried" tyrosyl residues are relatively inaccessible to solvent and have p*K*a values above 10.5. The general applicability of this correlation between reactivity and solvent accessibility was challenged by Myers and Glazer[333] in studies on subtilisin. These investigators argued that tyrosyl residues in apolar locations are preferentially modified by TNM or *N*-acetylimidazole. It is not at all clear that reactivity correlates with solvent exposure or lack of solvent exposure.

A discussion of tyrosine nitration in proteins would be incomplete without a brief consideration of peroxynitrite-mediated nitration (Figure 1.25). Although

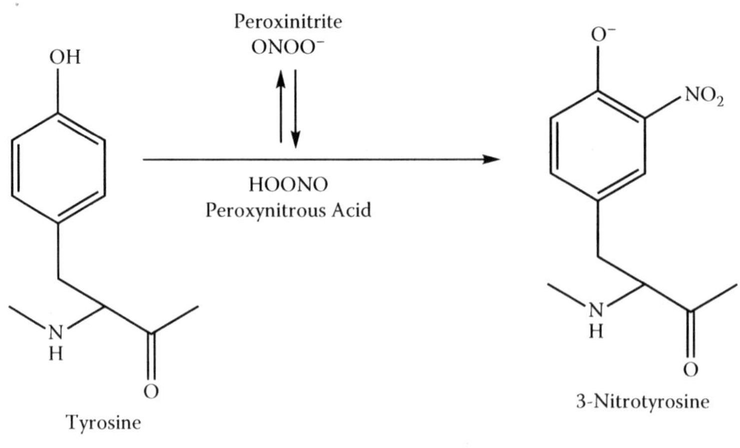

FIGURE 1.25 The modification of tyrosine with peroxynitrite.

peroxynitrite is not as convenient a reagent as TNM, there are facile methods for the synthesis and storage of the reagent.[364–366] There has been extensive work in this area, and it is fair to say that there are far more current studies on peroxynitrite than TNM. However, the work on peroxynitrite has been directed mostly toward physiological implications rather than site-specific chemical modification of proteins.[350,369–369]

Peroxynitrite is sensitive to microenvironmental factors and has been used to study membrane structure. Peroxynitrite, as TNM, can either nitrate tyrosine or oxidize tyrosine to dityrosine. Zheng and coworkers[370] used a hydrophobic probe, *N-t*-BOC-L-tyrosine-L-*t*-butyl ester to measure the modification of tyrosine in artificial membranes. The probe was preferentially nitrated, not oxidized, when inserted in a lipid bilayer. The use of this probe is the subject of a recent review.[371] A subsequent study used a series of 23-residue transmembrane peptides with single tyrosine residues at positions 4, 8, and 12 as a probe of a synthetic membrane.[372] The peptides were inserted into a multilamellar liposome composed of 1,2-dilauroyl-sn-glycero-3-phosphatidyl choline. Fluorescence spectra of the peptides incorporated into the membrane showed increasing derivatives with the highest value and the Tyr-4 with the least, suggesting the greatest penetration by residue 12. When peroxynitrite is generated *in situ*, nitration of the Tyr-12 derivatives is greater than that of the Tyr-8 derivative, which is in turn greater than that of the Tyr-4 peptide. The authors note that peroxynitrite is in equilibrium with peroxynitrous acid ($pKa = 6.8$). It is suggested that peroxynitrous acid diffuses into the membrane, where it undergoes hemolytic decomposition to form a nitric oxide radical, which then reacts with tyrosine to form nitrotyrosine. Shao and coworkers[373] showed that chlorination or nitration catalyzed by myeloperoxidase or peroxynitrite nitration of apolipoprotein A-I resulted in modification of Tyr-192. Other work showed that Tyr-192 was in a hydrophilic or exposed environment. Combination of apolipoprotein A-I with HDL-reduced modification suggests that exposure of tyrosine is important for modification with peroxynitrite. This is somewhat contrary to the previous observations and general suggestion that nitration is promoted by a hydrophobic environment.[374] There might be differences with some proteins but, in general, peroxynitrite and TNM demonstrate similar patterns in reacting with several proteins.[375]

Tyrosyl residues in proteins can also be modified by reaction with cyanuric fluoride (Figure 1.26).[376,377] The reaction proceeds at alkaline pH (9.1) via modification of the phenolic hydroxyl group with a change in the spectral properties of tyrosine. The phenolic hydroxyl groups must be ionized (phenoxide ion) for reaction with cyanuric fluoride. The modification of tyrosyl residues in elastase[378] and yeast hexokinase[379] with cyanuric fluoride has been reported. Modification of tyrosyl residues can occur as a side reaction with other residue-specific reagents such as 7-chloro-4-nitrobenzo-2-oxa-1,3-diazole (Figure 1.27) (7-chloro-4-nitrobenzofurazan; NBD-Cl; Nbf-Cl).[380,381] Modification of the phenolic hydroxyl group with 2,4-dinitrofluorobenzene has also been reported.[382] A novel reaction of PMSF with a tyrosyl residue in archaeon superoxide dismutase has been reported,[383] and it is noteworthy that Means and Wu reported the modification of a tyrosine residue in human serum albumin with diisopropylphosphorofluoridate.[384]

Diazonium salts readily couple with proteins (Figure 1.28) to form colored derivatives with interesting spectral properties.[385] Reaction with diazonium salts is

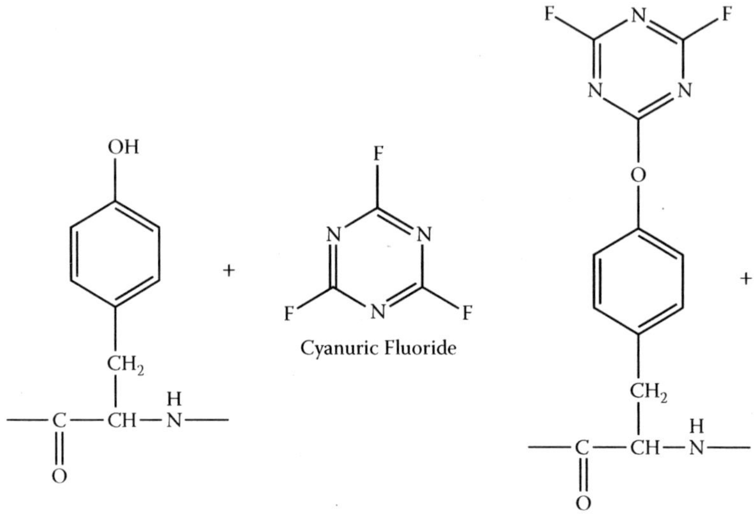

FIGURE 1.26 The modification of tyrosine with cyanuric fluoride.

accomplished at alkaline pH (pH 8 to 9, bicarbonate/carbonate or borate buffers). It is relatively difficult to obtain specific residue class modification with the aromatic diazonium salts, but tyrosine, lysine, and histidine are rapidly modified.[386,387] The reaction of chymotrypsinogen A with diazotized arsanilic acid has been investigated.[388] The reaction is terminated by the addition of a sufficient quantity of aqueous phenol (0.1 M) to react with excess reagent. The extent of the formation of monoazotyrosyl and monoazohistidyl derivatives is determined by spectral analysis.[386,387] The extent of reagent incorporation is determined by atomic absorption analysis for arsenic. Tyrosine (~1.0 mol/mol) and lysine (~4 mol/mol) were the only amino acid residues modified to any significant extent under these reaction conditions. The arsaniloazo functional group provides a spectral probe that can be used to study conformational change in proteins. In this particular study, there was a substantial change in the circular dichroism spectrum during the activation of the modified chymotrypsinogen preparation by trypsin.

Vallee and coworkers[389,390] reported on the reaction of bovine carboxypeptidase A crystals with diazotized *p*-arsanilic acid (conditions not specified) and obtained specific modification of Tyr.[248] Purification of the peptide containing the modified tyrosine residue was achieved by using antibody directed against the arsaniloazotyrosyl group. The antibodies were obtained from rabbits using arsaniloazovalbumin and arsaniloazobovine γ-globulin as antigen. The reaction of bovine carboxypeptidase A with diazotized 5-amino-1H-tetrazole has been reported.[291] Diazotized 5-amino-1H-tetrazole also specifically reacts with Tyr[248] in bovine carboxypeptidase A (in 0.67 M potassium bicarbonate/carbonate, 1.0 M NaCl, pH 8.8). A sevenfold molar excess of reagent was used, and the reaction was terminated after 30 min by the addition of Tris buffer. The extent of modification of tyrosine to tetrazolylazotyrosine is determined by absorbance at 483 nm (Figure 1.2) ($\varepsilon = 8.7 \times 10^3\ M^{-1}\ cm^{-1}$).

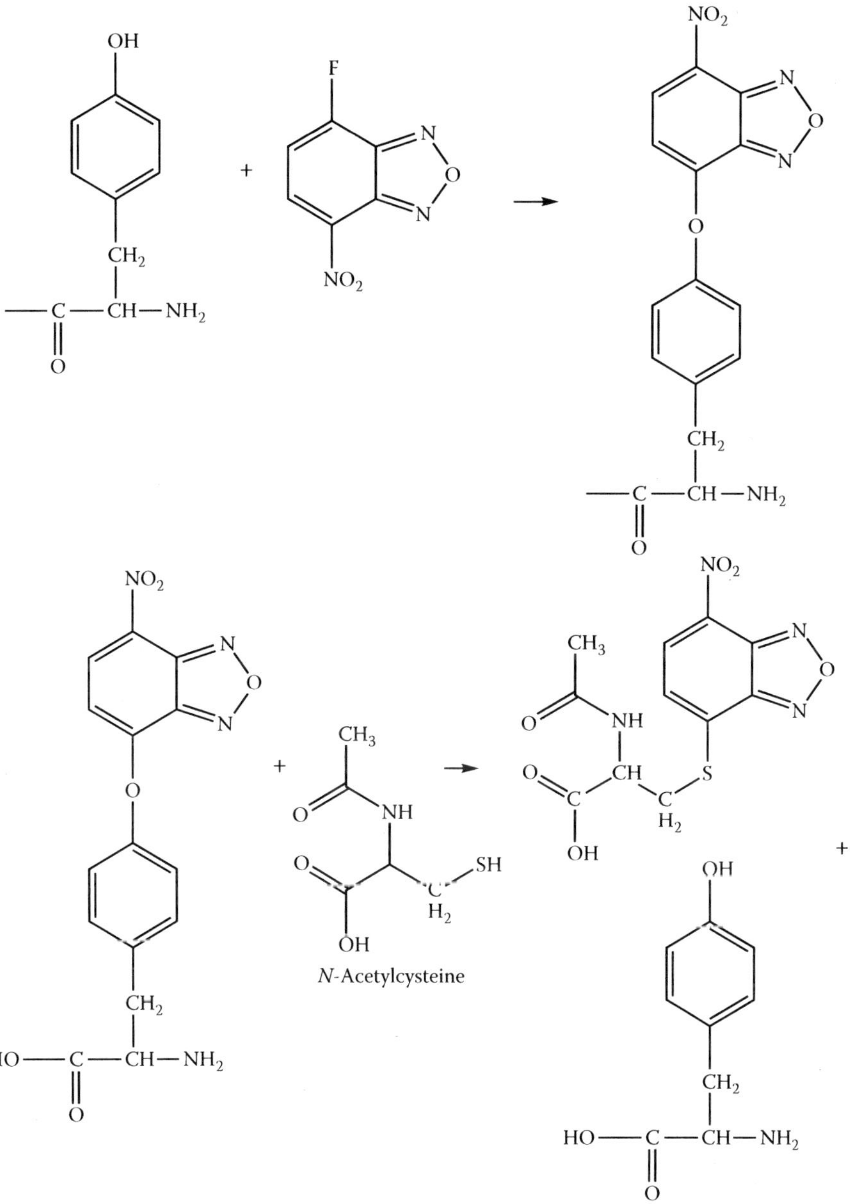

FIGURE 1.27 The modification of tyrosine with 4-fluoro-7-nitro-2,1,3-benzoxadiazole and reversal of the modification with *N*-acetylcysteine. (See Toyo'oka, T., Mantani, T., and Kato, M., Reversible labeling of tyrosine residues in peptide using 4-fluoro-7-2,1,3-benzoxadiazole and *N*-acetyl-L-cysteine, *Anal. Sci.* 19, 341–346, 2003.)

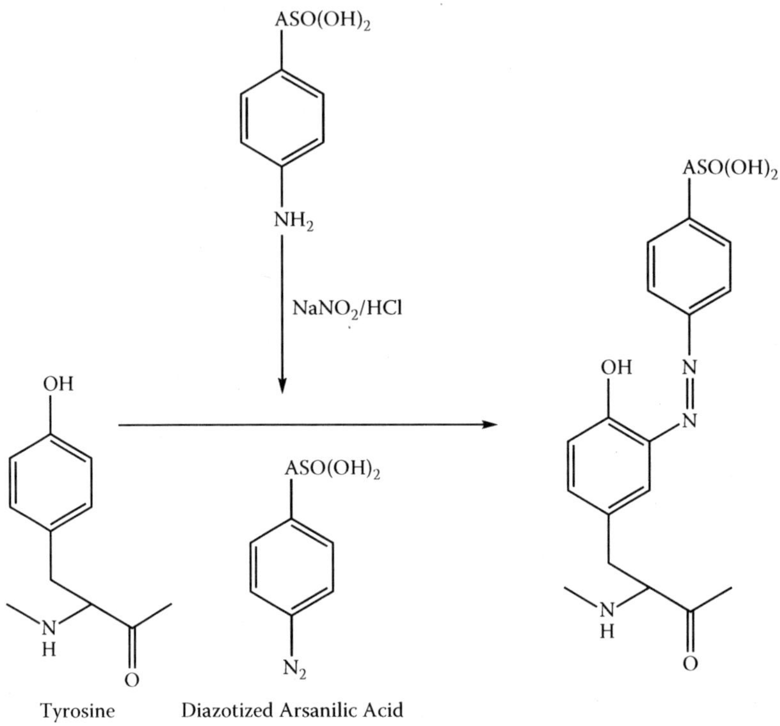

FIGURE 1.28 The modification of tyrosine with diazonium salts.

Modification of Tyr[248] in carboxypeptidase A by this reagent permits the subsequent modification of Tyr[198] by tetranitromethane.

p-Nitrobenzenesulfonyl fluoride (NBSF) is another environmentally sensitive reagent that can be used for the modification of tyrosyl residues in proteins. This reagent was developed by Liao and coworkers for the selective modification of tyrosyl residues of pancreatic DNase.[392] The modification reaction with NBSF can be performed in solvents (i.e., 0.1 M Tris-Cl, pH 8.0; 0.1 M *N*-ethylmorpholine acetate, pH 8.0) typically used for the modification of tyrosine residues by other reagents such as tetranitromethane or *N*-acetylimidazole. The rate of reagent hydrolysis is substantial and increases with increasing pH. In a recent study, NBSF has been used to characterize tyrosyl residues in NAD(P)H:quinone reductase.[287] Analysis of the product of the reaction showed that NBSF-modified tyrosyl residues were located in hydrophobic regions of the protein.

Cysteine is (usually) the most powerful nucleophile in a protein and, as a result, is frequently the easiest to selectively modify with a variety of reagents; it is also most frequently modified by reagents intended for other residues. Cysteine is the sulfur analog of serine, with the hydroxyl group replaced with a sulfhydryl group. The reader is directed to an excellent review of thiols[393,394] for a more thorough discussion of both aliphatic and aromatic thiols. The bond dissociation energy for sulfhydryl groups is substantially less than that of the corresponding alcohol function providing a basis for the increased acidity of sulfhydryl groups; for example, the p*K*a for ethanethiol is 10.6, whereas it is 18 for ethanol. As a consequence, whereas the reaction of cysteine with chloroacetate is slow (5.3×10^{-3} M^{-1} min^{-1}), reaction with serine is nonexistent under the same conditions; reaction of a cysteine residue at an enzyme active site (papain) is some 30,000 times faster (150 M^{-1} min^{-1}) than that of free cysteine at pH 6.0.[395]

Modification of cysteine residues proceeds via either a nucleophilic addition or displacement reaction with the thiolate anion as the nucleophile (Figure 1.29). The reaction with the α-keto-haloalkyl compounds such as iodoacetate is an example of a nucleophilic displacement reaction, whereas the reaction of maleimide is a nucleophilic addition to an olefin. This reaction is an example of a Michael reaction or Michael addition. In addition to the review cited earlier, there are other reviews on sulfhydryl chemistry[396–400] A variety of reagents are available for the modification of cysteine residues in proteins, as listed in Table 1.1.

Local environment has a profound effect on the reactivity of cysteine residues in proteins. It has been shown[401] that local electrostatic potential modulates reactivity of individual cysteine residue in rat brain tubulin. Rat brain tubulin dimer contains 20 cysteine residues: 12 residues in the α-subunit and 8 in the β-subunit. The rates of reaction of the cysteine residues in rat brain tubulin were determined with a variety of reagents in 0.3 M MES, pH 6.9, containing 1.0 mM EGTA and 1 mM $MgCl_2$ in the dark. The absence of light is critical because haloalkane compounds such as monobomobimane undergo photolysis. The reagents evaluated included syn-monobromobimane, *N*-ethylmaleimide, iodoacetamide, and [5- (2-iodoacetyl)amino) ethyl) amino) naphthalene-1-sulfonic acid] AEDANS. Approximately 50% of the 20 sulfhydryl groups react equivalently with all reagents. Reaction is slower with iodoacetamide than with *N*-ethylmaleimide, and a greater number of cysteine residues are modified with *N*-ethylmaleimide than with iodoacetamide. It is suggested that the difference in the rates of reaction is ascribed to the differences in the chemistry of the reaction of the two compounds with the thiolate ion with the reaction, with iodoacetamide being a nucleophilic displacement whereas the reaction with *N*-ethylmaleimide is an addition reaction. It was possible to identify a single highly reactive cysteine residue (347α) by reaction with chloroacetamide, which generally reacts with sulfhydryl groups more slowly than iodoacetamide.[402] The relative order of nucleofugacity of halide in displacement reactions is well known.[403] Horton and coworkers[404] have studied the modification of the active site cysteine in papain and free cysteine with chloroacetic acid, chloroaceamide, and *N*-ethylmaleimide (Table 1.4). The presence of a charged group can lower the p*K*a of cysteine. The presence of an aspartic acid residue or a lysine residue lowers the observed p*K*a of a cysteine residue from 8.1 to 7.1.[405] Isotope-coded iodoacetanilide (*N*-phenyl iodoacetamide) was used to determine individual cysteine p*K*a in thioredoxin[406]; the p*K*a for one cysteine residues was 6.5 and the other greater than 10.

The unique reactivity of cysteine has prompted investigators to use site-specific mutagenesis to place cysteine at particular points in a protein for the subsequent attachment of structural probes.[407–414] The major use of cysteine insertion/site-directed chemical modification[109] has been directed toward the elucidation of topology with emphasis on membrane proteins.[415–421]

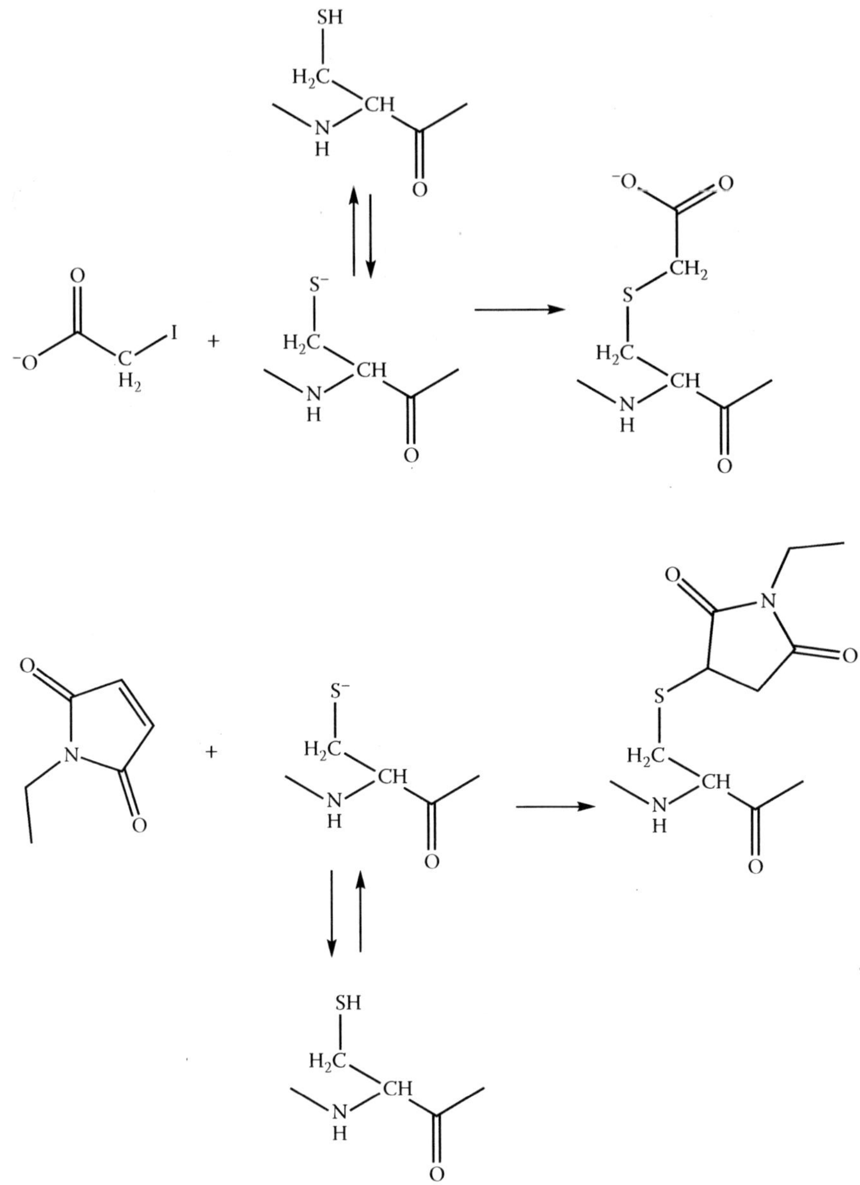

FIGURE 1.29 The reaction of cysteine with iodoacetate or *N*-ethylmaleimide. Also shown is the dissociation of a proton from the sulfhydryl group of cysteine to form the thiolate anion, which is the reactive species.

TABLE 1.4
Rate Constants for the Modification of Cysteine

	k (M^{-1} s^{-1})			
Reagent	**Cysteine**	**Glutathione**	**$HSCH_2CH_2OH$**	**Protein SH**
Chloroacetamide	0.00412[a]	0.00157[a]		0.144[a,b]
Chloroacetic acid	0.00057[a]	0.00108[a]		2.75[a,b]
N-Ethylmaleimide	593[a]	263[a]		2.55[a,b]
Iodoacetamide	0.6[c]			0.004[c,d]
N-phenyl iodoacetanilide	1.83[c]			0.028[c,d]
5,5′-Dithiobis(2-nitrobenzoic acid)				1.82[e]
5,5′-Dithiobis(2-nitrobenzoic acid)				3.37[f]
5,5′-Dithiobis(2-nitrobenzoic acid)				442[g]
5,5′-Dithiobis(2-nitrobenzoic acid)				40.2[h]
2-Aminoethyl-MTS[i]			0.00076[j]	
2-Hydroxyethyl-MTS[i]			0.000095[j]	
2-Sulfoethyl-MTS[i]			0.25[j]	
2-(Trimethylammonium)ethyl-MTS[i]			0.069[j]	
2-Bromoethanol	0.000011[k]			
Iodomethane	0.15[k]			
2-Bromoethylamine	0.0016[k]			
Bromoethane	0.00056[k]			
4-Bromoethylimidazole	0.0099[k]			
4-Chloromethylimidazole	0.51[k]			

[a] pH 6.5, 26°C (Evans, B.L.B., Effect of hydronitrobenzylation of tryptophan-177 on reactivity of active site cysteine-25 in papain, *Arch. Biochem. Biophys.* 206, 362–371, 1981).

[b] Papain (Evans, B.L.B., Effect of hydronitrobenzylation of tryptophan-177 on reactivity of active site cysteine-25 in papain, *Arch. Biochem. Biophys.* 206, 362–371, 1981).

[c] pH 7.0, 23°C (Nelson, K.J., Day, A.E., and Zeng, B.-B., Isotope-coded, iodoacetamide-based reagent to determine individual cysteine p*K*a values by matrix-assisted laser desorption/ionization time-of-flight mass spectrometry, *Anal. Biochem.* 175, 187–195, 2008).

[d] *Escherichia coli* thioredoxin (Nelson, K.J., Day, A.E., and Zeng, B.-B., Isotope-coded, iodoacetamide-based reagent to determine individual cysteine p*K*a values by matrix-assisted laser desorption/ionization time-of-flight mass spectrometry, *Anal. Biochem.* 175, 187–195, 2008).

[e] Bovine cardiac troponin C, pH 7.0 (Fuchs, F., Liou, Y.-M., and Grabarek, Z., The reactivity of sulfhydryl groups of bovine cardiac troponin C, *J. Biol. Chem.* 264, 20344–20349, 1989).

[f] Bovine cardiac troponin C, pH 7.0 with 2.1 mM $CaCl_2$ (Fuchs, F., Liou, Y.-M., and Grabarek, Z., The reactivity of sulfhydryl groups of bovine cardiac troponin C, *J. Biol. Chem.* 264, 20344–20349, 1989).

[g] Rabbit skeletal muscle troponin C, pH 7.0 (Fuchs, F., Liou, Y.-M., and Grabarek, Z., The reactivity of sulfhydryl groups of bovine cardiac troponin C, *J. Biol. Chem.* 264, 20344–20349, 1989).

[h] Rabbit skeletal muscle troponin C, pH 7.0 with 2.1 mM $CaCl_2$ (Fuchs, F., Liou, Y.-M., and Grabarek, Z., The reactivity of sulfhydryl groups of bovine cardiac troponin C, *J. Biol. Chem.* 264, 20344–20349, 1989).

[i] MTS, methanethiosulfonate.

[j] pH 7.0, 20°C (Karlin, A. and Akabas, M.H., Substituted-cysteine accessibility method, *Methods Enzymol.* 293, 123–145, 1998).

[k] pH 9.5 (Schindler, J.F. and Viola, R.E., Conversion of cysteinyl residues to unnatural amino acid analogues. Examination in a model system, *J. Protein Chem.* 15, 737–742, 1996).

Alkyl alkanethiosulfonates (e.g., methyl methanethiosulfonate) (Figure 1.30) have been extensively used in the past decade for the modification of cysteine residues in proteins. Methyl, ethyl, and trichloromethyl derivatives of methanethiosulfonate and propylpropanethiosulfonate were described by George Kenyon and colleagues in 1975.[422] These alkyl alkanethiosulfonates form mixed disulfides with sulfhydryl group of cysteine, with the release of sulfinic acid. The mixed disulfides are going to be of variable stability, more stable than thioesters but less stable than thioethers. The stability will be influenced by the nature of substituents on the alkyl function. The modification is highly specific for sulfhydryl groups and is easily reversed by mild reduced agents such as 2-mercaptoethanol or dithiothreitol. Reaction could occur at sites other than cysteine in proteins.[423] Such reactions would most likely occur at amino groups, and these modifications would not be reduced by reducing agents such as dithiothreitol or 2-mercaptoethanol. Although to the best of my knowledge there are no reports of the modification of amino groups in proteins with alkylthiosulfonates, there are modifications with alkylthiosulfonates that are only partially

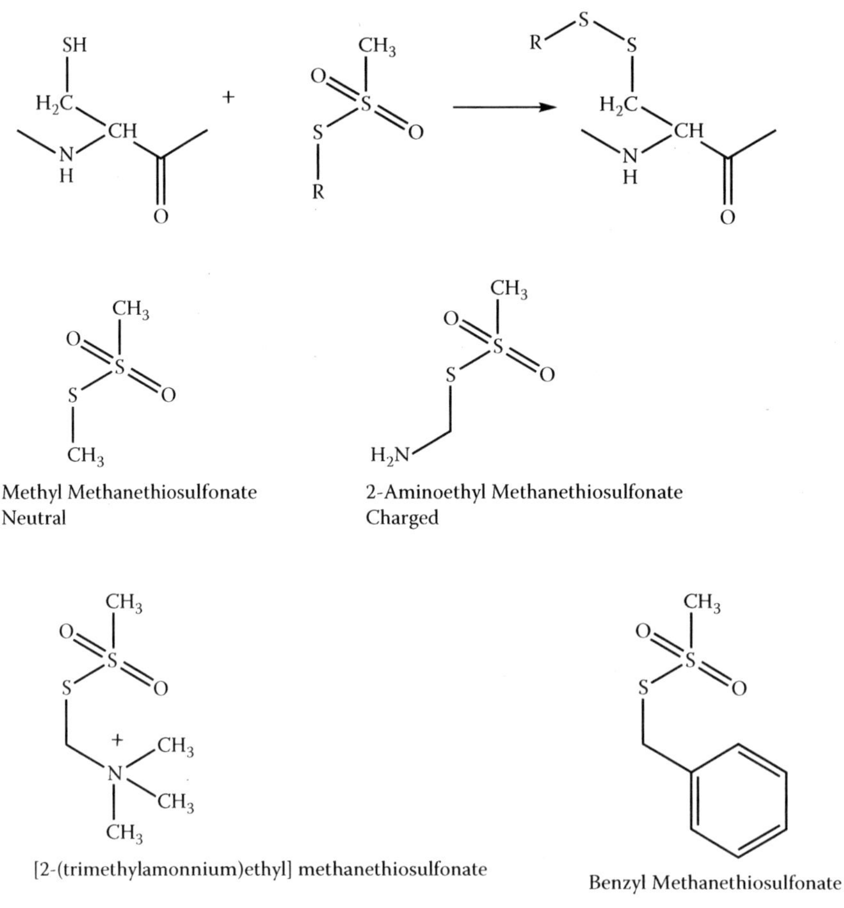

FIGURE 1.30 Some representative alkyl methane thiosulfonates.

reversed on reduction.[424] Modification of protein sulfhydryl groups is usually accomplished at pH 7–9, reflecting the importance of the thiolate anion as the reactive species.[425] In addition to pH dependence studies that demonstrate a clear preference for reaction with the thiolate anion, this paper presents a listing of the dependence of the second-order rate constant of the reaction on the p*K*a of the individual thiol. The reader is also directed to a study by Stauffer and Karlin[426] on the effect of ionic strength on the reaction of alkylthiosulfonates with simple and protein-bound thiols. The extent of modification is best determined by the incorporation of radiolabeled reagent. In later work,[427,428] Kenyon and Bruice extended the understanding of the use of this class of reagents as part of a more general review of the modification of cysteine residues. Most work has used alkyl derivatives of methanethiosulfonate and these reagents are frequently referred to as MTS reagents.[429,430] There is a wealth of information in the review by Karlin and Akabas[109] that should be carefully considered by those interested in the use of MTS reagents. MTS reagents do undergo base-catalyzed hydrolysis and therefore should be prepared in neutral media immediately before use. Some investigators do prepare stock solutions of reagent, which are stored at −20°C. Stability under these conditions has not been validated, and this practice is not recommended. Manipulation of the alkyl function does alter the membrane permeability of the various MTS reagents.[431]

Haloacetates such as chloroacetic acid and iodoacetic acid, the corresponding amides, and derivatives have been extremely useful reagents for the specific modification of cysteinyl residues. Haloacetates and haloacetamides react with cysteine via a S_N2 reaction mechanism to give the corresponding carboxymethyl or carboxamidomethyl derivatives (Figure 1.31). When a rapid reaction is desired, the iodine-containing compounds are generally used. Dahl and McKinley-McKee[432] have made a rather detailed study of the reaction of alkyl halides with thiols. It is emphasized that reactivity of alkyl halides not only depends on the halogen, but also on the nature of the alkyl groups. These investigators emphasized that the reactivity of an alkyl halide such as iodoacetate depends not only on the leaving potential of the halide substituent (I > Br>>> Cl; 130:90:1), but also on the nature of the alkyl group. The rate of reaction of 2-bromoethanol with the sulfhydryl group of L-cysteine (pH 9.0) is approximately 1000 times less than that observed with bromoacetic acid. The reactions are extremely pH dependent, emphasizing the importance of the thiolate anion in the reaction. Although these studies provide a useful framework, this is not always the situation. Iodide is highly polarizable and is a good leaving group, but salvation can be problematic. With nitrogen as the nucleophile, the order of reaction of α-haloacetic acids with 4-(*p*-nitrobenzyl) pyridine was I>Br>>Cl, consistent with the order of reaction with thiols as described by Dahl and McKinley-McKee; the order of reactivity with the histidine residues in RNase was Br>I>Cl and Br>I>Cl with DNase.

Reduction under denaturing conditions followed by alkylation with haloalkyl reagents or other alkylating agents is a common procedure in preparation of the protein for further analysis.[433–436] The success of analytical procedures such as two-dimensional electrophoresis[437–440] in proteomics depends on the effectiveness of this process. Alkylation prevents the formation of "spurious" spots during the subsequent electrophoretic analysis.[441,442] The reduction and alkylation step could be used in the

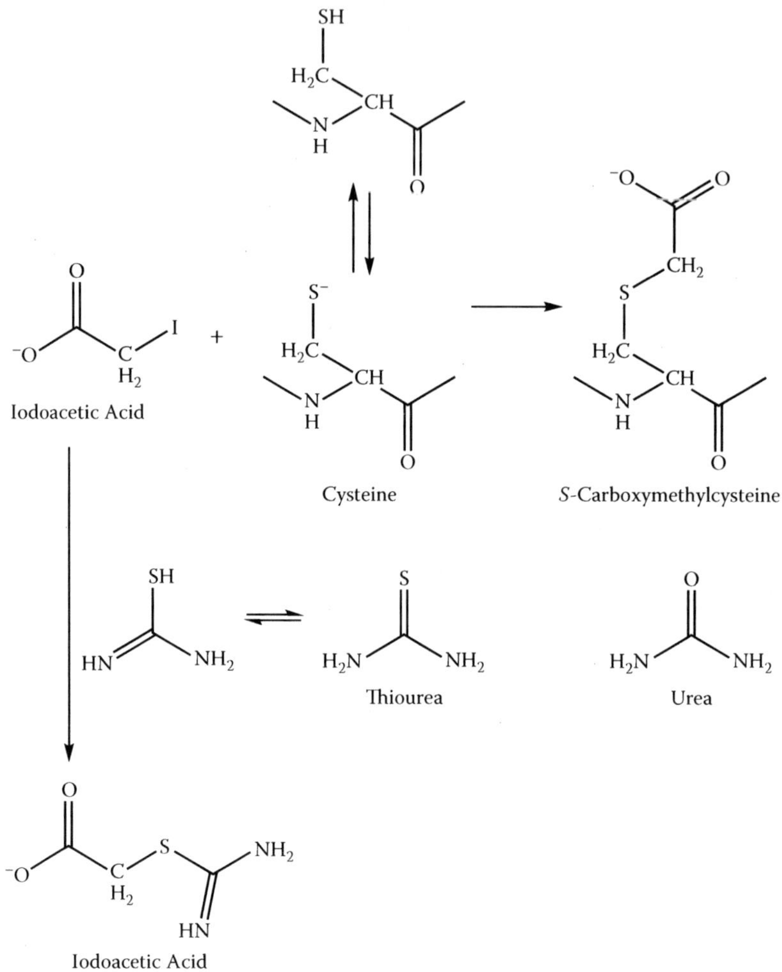

FIGURE 1.31 The alkylation of cysteine with iodoacetic acid to form *S*-carboxymethylcysteine and the reaction of the haloacetic acid with thiourea.

preparation of the sample prior to the isoelectric focusing step or between the isoelectric focusing step and the SDS/PAGE (polyacrylamide gel electrophoresis in the presence of sodium dodecyl sulfate) step.[443] In the latter, this would involve the *in situ* treatment of the immobilized pH gradient (IPG) strip prior to the SDS/PAGE step. It is probably best to alkylate prior to the isoelectric focusing step to avoid the issue of disulfide bond reformation ("disulfide scramble") during separation. Another consequence that can be avoided by alkylation is the beta-elimination of cysteine in the alkaline pH range forming dehydroalanine (see later text) and consequent unwanted peptide bond cleavage.[444]

The reduction and alkylation step is usually performed under denaturing conditions with an uncharged chaotropic agent such as urea. However, urea has the potential to create problems through a dismutation reaction that forms cyanate, which in turn can react with nucleophiles, including cysteine and lysine, in subject proteins, increasing heterogeneity (see later text). Recently, it was recommended that thiourea be included in the sample preparation "cocktail" (9.0 urea is the common concentration; the inclusion of 2 M thiourea is recommended) for the purpose of improving protein solubilization[445] prior to the isoelectric focusing step. However, the use of thiourea complicates the alkylation step because it reacts with iodoacetic acid or iodoacetamide[446] (Figure 1.31). Such inhibition is not observed with maleimides.[447] Given the operational advantages provided by thiourea in the preparation of the sample, it might be preferable to use maleimide reagents instead of iodoalkyl reagents for the blocking of sulfhydryl groups. It is noted that the rate of reaction of maleimides with the cysteine thiolate groups is more rapid than that for iodoalkyl compounds. This might become especially important with the introduction of a "reporter" group such as fluorescein.

In studies with wool proteins, carboxymethylation with iodoacetate was essentially complete after 10 min of reaction.[448] Side reactions with other amino acids occurred at longer times of reaction (Figure 1.32). Other investigators report that the alkylation process required a long period of time and was incomplete after 6 h of reaction.[449] Vinyl pyridine (Figure 1.33) is an alternative for the alkylation of cysteine residues in proteins,[450] and deuterated derivatives have been prepared[451] for use in the study of differential protein expression.

N-ethylmaleimide reacts with sulfhydryl groups in proteins (Figure 1.34) with considerable specificity.[452–454] This reaction is a Michael addition, which is a reaction between a nucleophile (thiolate anion) and an olefin (the maleimide ring). Bednar has examined the chemistry of the reaction of *N*-ethylmaleimide with cysteine and other thiols in some detail.[455] The reaction of NEM with simple thiols can be described by the Brønsted equation. The second-order rate constant for the reaction of NEM with the thiolate anion of 2-mercaptoethanol is 10^7 M^{-1} min^{-1}. This value is at least 5×10^{10} greater than for the reaction with the thiol. This study also reports data for the decomposition of NEM in several buffers and should be considered for the determination of truly accurate kinetic data. This reaction can be followed by the decrease in absorbance at 300 nm, the absorbance maximum of *N*-ethylmaleimide. The extinction coefficient of *N*-ethylmaleimide is 620 M^{-1} cm^{-1} at 302 nm.[452] The spectrophotometric assay is not sensitive, and the modification is usually monitored by the incorporation of radiolabeled reagent. The alkylation product (*S*-succinyl cysteine) is stable and can be determined by amino acid analysis following acid hydrolysis. Although the reagent is reasonably specific for cysteine, reaction with other nucleophiles must be considered.[456] A "diagonal" procedure for the isolation of cysteine-containing peptides modified with *N*-ethylmaleimide has been reported.[457] This procedure is based on the hydrolysis of the reaction product of cysteine and *N*-ethylmaleimide to cysteine-*S*-*N*-ethyl succinamic acid, generating a new negative charge.

Another example of a Michael-type addition reaction involving cysteine is the reaction with acrylamide (Figure 1.35). Originally considered to be an unwanted complication of the electrophoresis of reduced proteins in acrylamide gel systems, modification with acrylamide is now considered to be a useful site-specific modification of cysteine in proteins.[458–460]

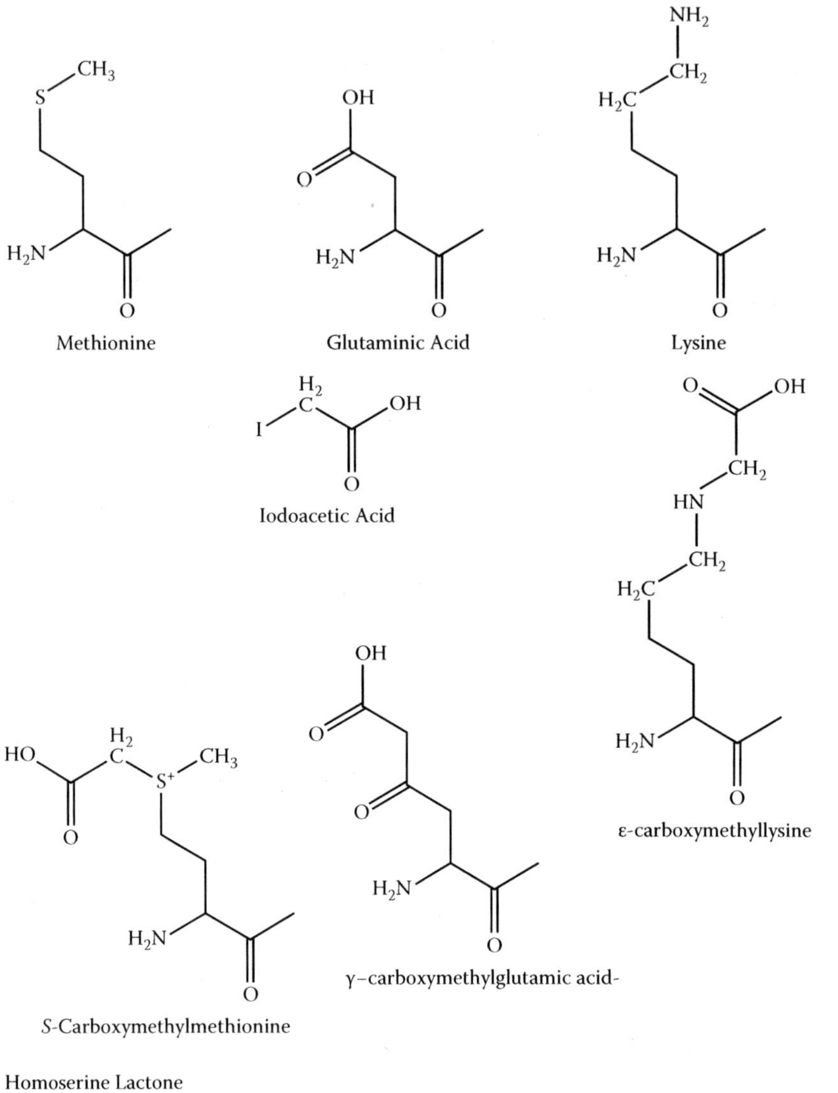

FIGURE 1.32 The alkylation of cysteine with iodoacetic acid to form *S*-carboxymethyl cysteine and the reaction of the haloacetic acid with thiourea.

Ellman developed 5,5′-dithiobis(2-nitrobenzoic acid) (DTNB; Ellman's reagent),[461] which today is one of the more popular reagents for the modification and determination of the sulfhydryl group (Figure 1.36). Reaction with sulfhydryl groups in proteins results in the release of 2-nitro-5-mercaptobenzoic acid, which has a molar extinction coefficient of 13,600 M^{-1} cm^{-1} at 410 nm. Riddles and coworkers[462,463] studied the chemistry of DTNB in alkaline solutions as well as the reaction of DTNB with thiols. They concluded that the rate of reaction of DTNB was dependent on the pH and the p*K*a of the thiol residue. The steric and electrostatic considerations that

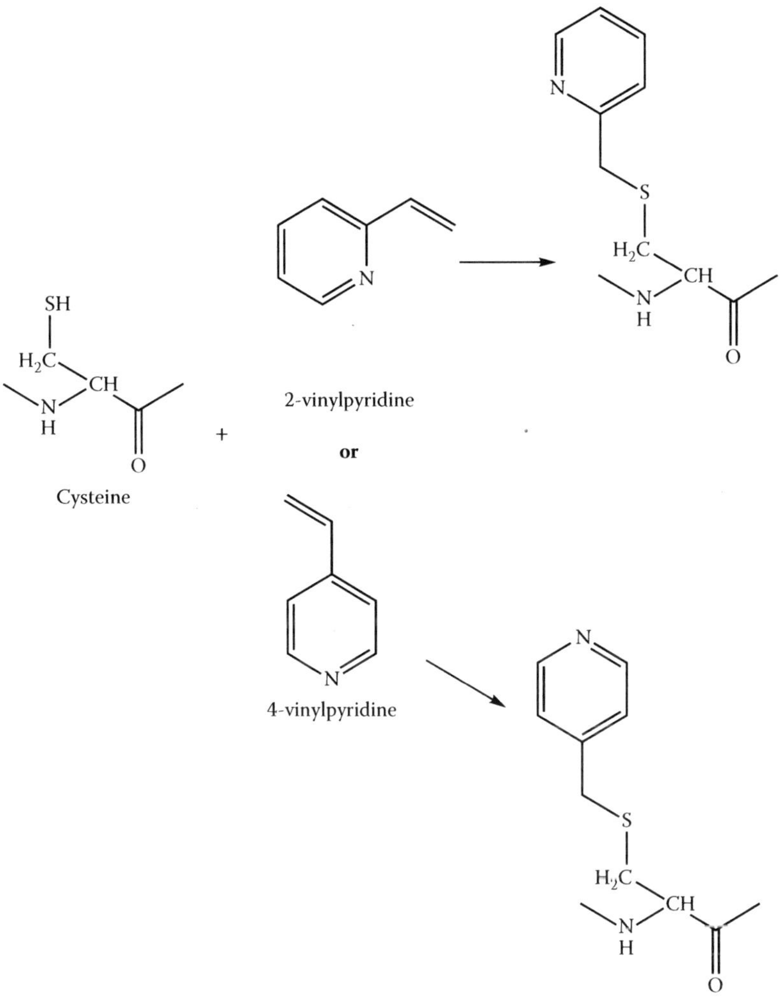

FIGURE 1.33 The reaction of vinyl pyridine with cysteine.

were discussed earlier are also applicable to this reaction. At pH 7.0 and 25°C, model thiols and some protein thiols react rapidly. There are examples of protein thiols that react more slowly ($t½ \geq 10$ min). There are several examples of the use of DTNB to study protein conformation.[464–467] 4,4′-Dithiodipyridine (Figure 1.37) is similar to 5,5′-dithiobis-(5-nitrobenzoic acid) in that a mixed disulfide is formed between a cysteinyl residue in the protein and the reagent with the concomitant release of pyridine-4-thione.[468–470] The reaction of 4,4′-dithiodipyridine with protein sulfhydryl groups can be followed by spectroscopy ($\varepsilon_{324\,nm} = 19{,}800\ M^{-1}\,cm^{-1}$). The reaction is readily reversed by the addition of a reducing agent such as dithiothreitol.

Kimura and coworkers[471] introduced methyl 3-nitro-2-pyridyl disulfide and methyl 2-pyridyl disulfide. Both of these reagents modify sulfhydryl groups forming the

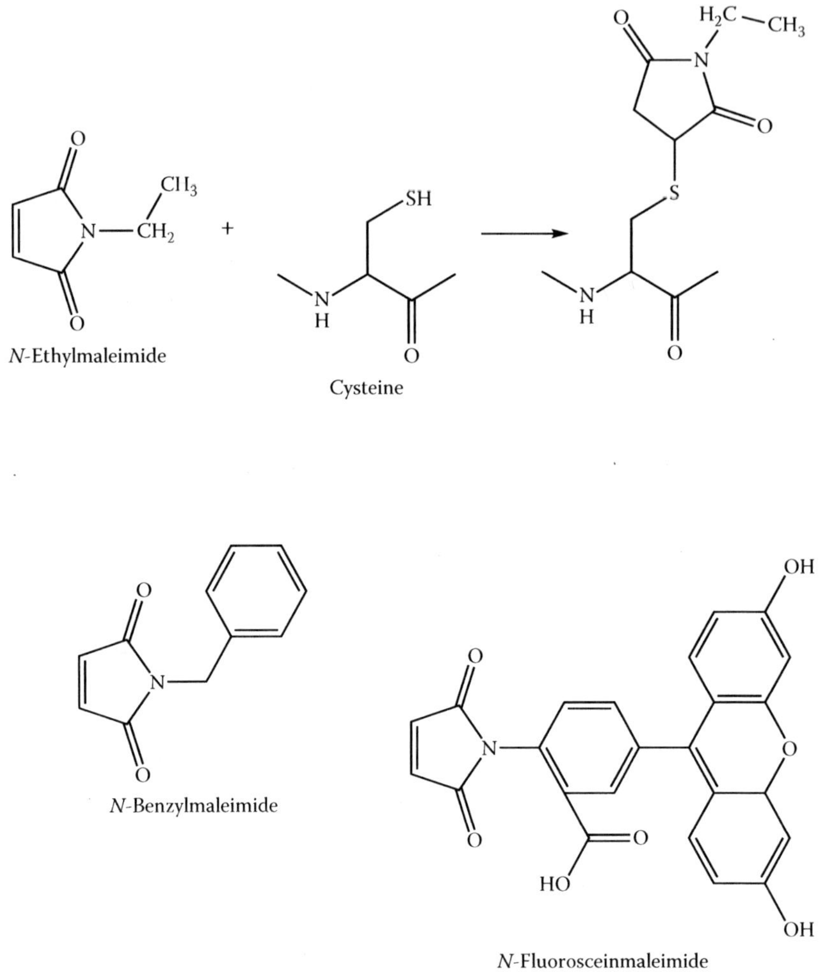

FIGURE 1.34 The reaction of *N*-ethylmaleimide with cysteine and some structures of several maleimide derivatives.

thiomethyl derivative. The spectrum of 3-nitro-2-pyridone is pH dependent. There is an isosbestic point at 310.4 nm that can be used to determine the extent of the reaction of methyl-3-nitro-2-pyridyl disulfide with sulfhydryl groups. The difference in spectrum obtained does not show the pH dependence of the nitropyridyl derivative. At 343 nm, the change in extinction coefficient is 7060 M^{-1} cm^{-1}. The extinction coefficient (7600 M^{-1} cm^{-1}) of the 2-thiopyridinone at 343 nm is relatively stable from pH 3 to 8.0. There is a marked decrease in absorbance above pH 8.0, reflecting the loss of a proton. Reaction with the sulfhydryl group in the protein clearly proceeds more rapidly at alkaline pH.

The use of *p*-hydroxymercuribenzoate (PCMB) for the modification of protein sulfhydryl groups dates at least to the work of Hellerman and coworkers.[472] The use

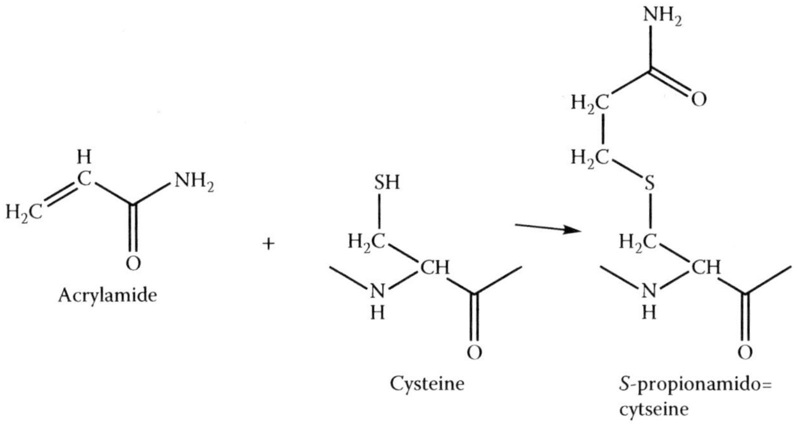

FIGURE 1.35 The reaction of acrylamide with cysteine. (See Brune, D.C., Alkylation of cysteine with acrylamide for protein sequence analysis, *Anal. Biochem.* 207, 285–290, 1992; Mineki, R., Taka, H., Fujimura, T. et al., *In situ* alkylation with acrylamide for identification of cysteinyl residues in proteins during one- and two-dimensional sodium dodecyl sulphate-polyacrylamide gel electrophoresis, *Proteomics* 2, 1672–1681, 2002.)

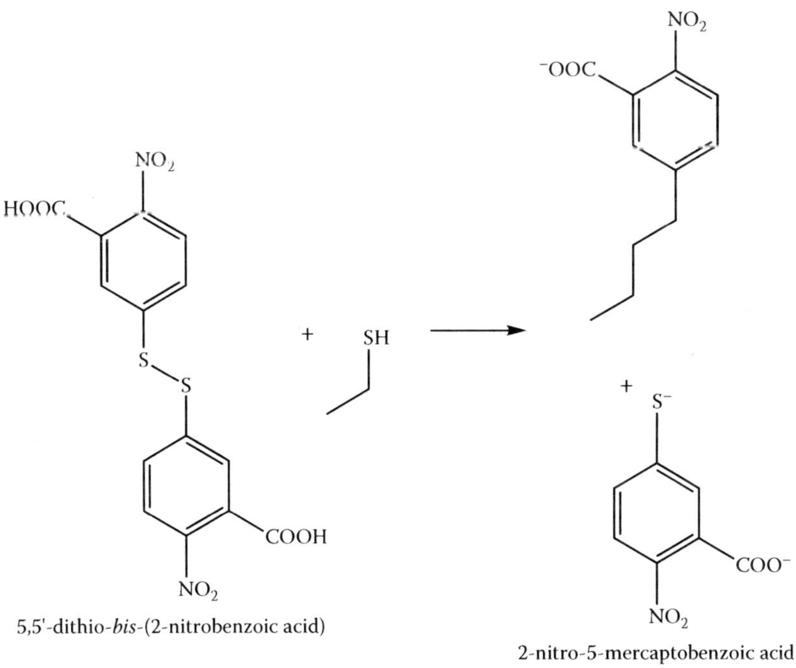

FIGURE 1.36 The reaction of 5,5′-dithio-*bis*-(2-nitrobenzoic acid) with cysteine.

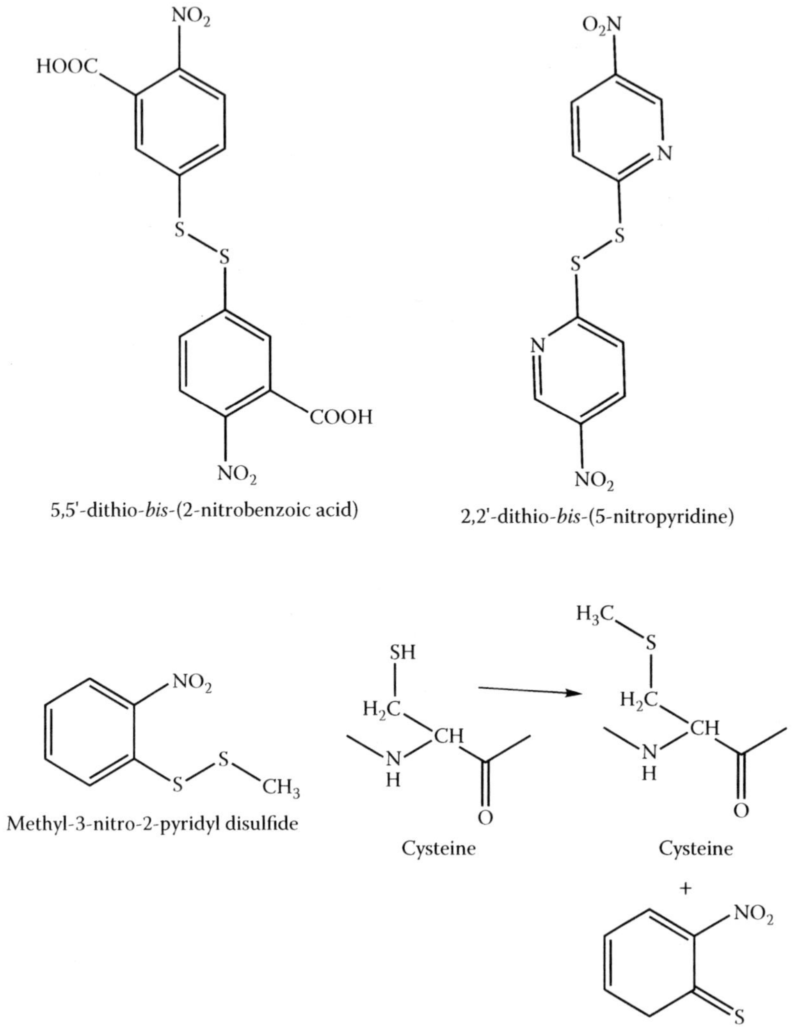

FIGURE 1.37 The structures of 5,5′-dithio-*bis*-(2-nitrobenzoid acid, 2,2′-dithio-*bis*-(5-nitropyridine), and methyl-3-nitro-2-pyridyl disulfide.

of the reagents was greatly advanced by Boyer,[473] who developed a spectrophotometric method for the determination of reaction stoichiometry. *p*-Chloromercuriphenylsulfonate was developed as a more hydrophilic reagent.[474] *p*-chloromercuribenzoate demonstrated membrane permeability, whereas *p*-chloromercuriphenylsulfonte does not penetrate the membrane.[475] The two reagents appear to have equivalent reactivity with proteins in solution, but can differ from other sulfhydryl reagents.[476] There appears to be greater use of the *p*-chloromercuribenzoate derivative.[477,478]

Serine can be changed to cysteine in proteins.[479–481] Neet and Koshland[479] converted the active site serine in subtilisin to cysteine. Subtilisin was modified with phenylmethylsulfonyl fluoride and then treated with potassium thiolacetate, resulting in the formation of cysteine. The resulting enzyme is active toward ester substrates but not protein substrates and is inhibited by PCMB. This observation was extended by Polgar and Bender.[480] Slade and coworkers[481] converted serine to cysteine in penicillin acylase. Serine and cysteine in proteins are converted to dehydroalanine by heating under alkaline conditions and can subsequently form adducts with cysteine and/or lysine via Michael addition reaction.[482] Less drastic methods have been developed for the formation of dehydroalanine from cysteine[483] and for subsequent use in protein ligation strategies.[484] Dehydroalanine derived from cysteine has recently been described as a common posttranslational modification in human serum albumin.[485]

Cysteine is also modified in a reversible reaction by cyanate[486–490] (Figure 1.38). The cyano derivative has been used for specific chemical cleavage (see following text) as a probe.[490] Cysteine is also subject to oxidation[491] by a variety of oxidizing agents, including hydrogen peroxide (Figure 1.39), performic acid, and copper ions.

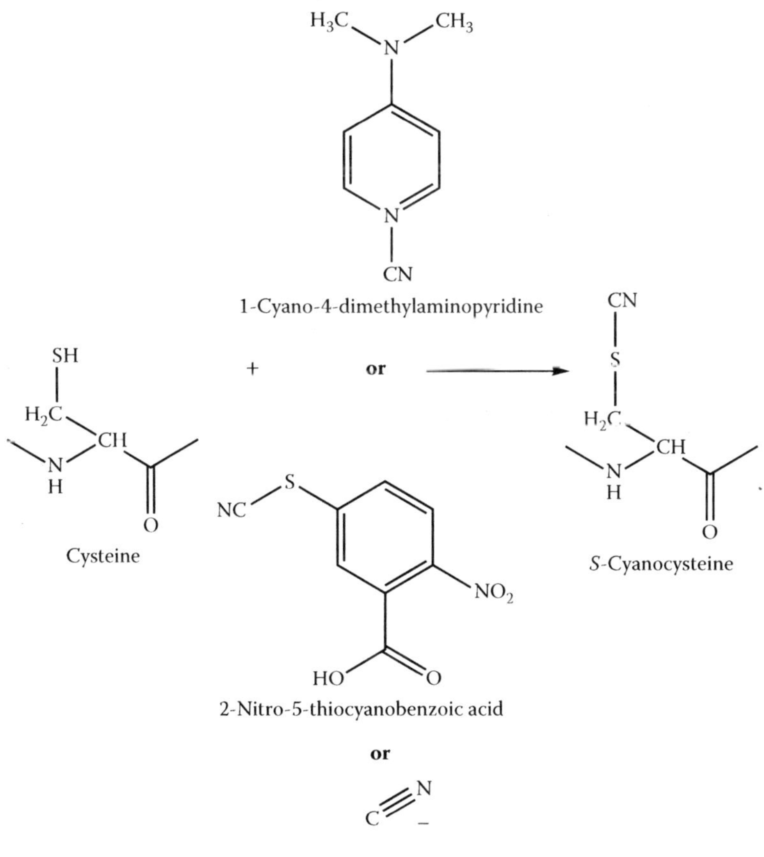

FIGURE 1.38 The cyanylation of cysteine.

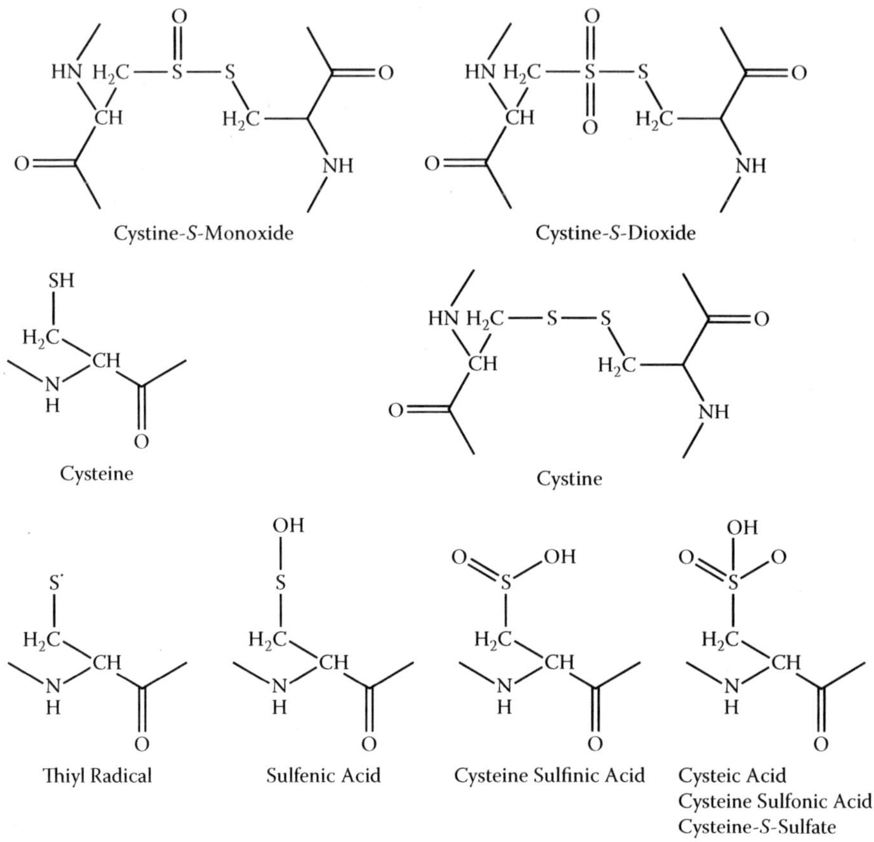

FIGURE 1.39 The oxidation of cysteine.

CYSTINE

Cystine residues are considered to be critical for maintaining the native structure of a protein.[492–497] Thiol-disulfide exchange refers to a reversible association process resulting either in cystine formation (disulfide cross-linking) or mixed disulfide formation. The formation of cysteine-glutathione is an example of a mixed disulfide.

Cleavage of cystine in proteins can be accomplished by oxidation or reduction. Gorin and Godwin[498] have reported that cystine can be quantitatively converted to cysteic acid by reaction with iodate in 0.1 to 1.0 M HCl. This reaction was complete in 15 to 30 min. After longer periods of reaction, the iodination of tyrosine residues occurred. Oxidation of cystine can be accomplished under more vigorous conditions with reagents such as performic acid[499] (Figure 1.40).

The reaction of sulfite with cystine yields the *S*-sulfo derivative of cystine and cysteine (Figure 1.41). The earlier literature and chemistry of this reaction has been reviewed by Cole.[500] The reaction proceeds optimally at alkaline pH (pH 9.0). An oxidizing agent such as cupric ions or *o*-iodosobenzoate can be included to ensure effective conversion (oxidative sulfitolysis) of all cystine residues to the corresponding

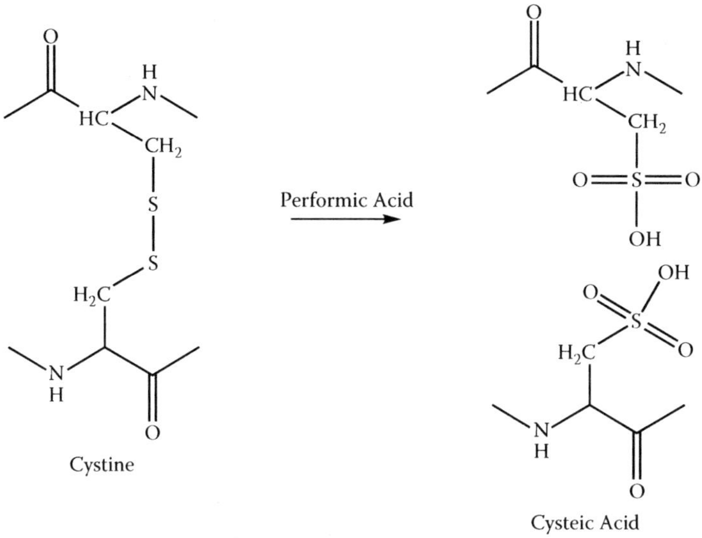

FIGURE 1.40 Performic acid oxidation of cystine.

S-sulfocysteine derivatives. Another approach involves the inclusion of sodium tetrathionate in the reaction to convert the cysteine to *S*-sulfocysteine.[501] An ingenious approach to the sulfitolysis reaction was developed by Scheraga and colleagues.[502] These investigators included 2-nitro-5-sulfothiobenzoate in the reaction, which resulted in the conversion of the cysteine to *S*-sulfocysteine with the concomitant formation of 2-nitro-5-thiobenzoate. 2-Nitro-5-thiobenzoate can be measured at 412 nm and is proportion to the cystine residues in the protein. The reaction is reversible, forming cysteine upon treatment with a thiol such as 2-mercaptoethanol or dithiothreitol.

This reaction has been adapted to the controlled reduction of disulfide bonds in proteins in the absence of denaturing agents.[503] Sulfitolysis in the presence of sodium tetrathionate has proved useful for the quantitative conversion of cystine to *S* sulfocysteine in the processing of biotherapeutics.[504–506]

Reduction of cystine can be accomplished with a mild reducing agent such as β-mercaptoethanol, dithiothreitol, or cysteine. Gorin and coworkers[507] have examined the rate of reaction of lysozyme with various thiols. At pH 10.0 (0.025 M borate), the relative rates of reaction were 2-mercaptoethanol, 0.2; dithiothreitol, 1.0; 3-mercaptopropionate, 0.4; and 2-aminoethanol, 0.01. The results with aminoethanethiol were somewhat surprising because the reaction (disulfide exchange) involves the thiolate anion, and 2-aminoethanethiol would be more extensively ionized than the other mercaptans. Dithiothreitol, as introduced by Cleland, is a useful reagent for the reduction of disulfide bonds in proteins.[508] Dithiothreitol and the isomeric form, dithioerythritol, are each capable of the quantitative reduction of disulfide bonds in proteins. The oxidized form of dithiothreitol has an absorbance maximum at 283 nm ($\Delta\varepsilon = 273$), which can be used to determine the extent of disulfide bond cleavage.[509] Phosphorothioate can be used to cleave disulfide bonds in proteins, forming the *S*-phosphorothioate derivative.[510, 511]

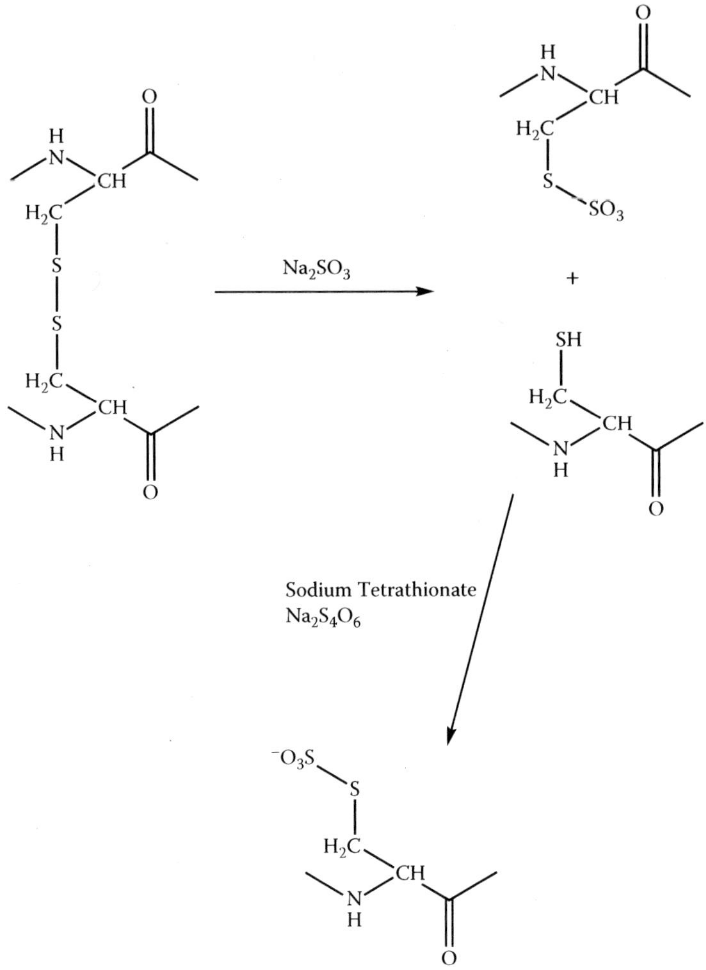

FIGURE 1.41 The oxidative sulfitolysis of cystine.

It is important to appreciate that the reduction of cystine to cysteine involves the addition of an electron to the sulfur, with the proton coming along for the ride (depends on pH). An exception would be direct hydride transfer. It has been demonstrated that disulfide bridges in α-lactalbumin are reduced by electron transfer from an excited tryptophan residue.[512,513] In studies with human α-lactalbumin,[513] ultraviolet light (270–290 nm, 1 mW/cm^2, 2–4 h, 50 mM HEPES-150 mM KCl, pH 7.8, 20°C) irradiation resulted in a 10 nm red shift of its tryptophan fluorescence emission spectrum (324–334 nm). Reaction of the irradiated protein with 5,5′-dithiobis(2-nitrobenzoic acid) demonstrated the presence of free sulfhydryl groups. The cleavage of a disulfide bond mediated by ultraviolet light has also been reported for bovine somatropin.[514] Lyophilized recombinant bovine somatotropin was photolyzed by ultraviolet light (305–410 nm, λ_{max} = 350 nm). The protein had been lyophilized from carbonate buffer, and the cake contained 6% moisture. Unlike the other examples of disulfide bond reduction mediated via tryptophan, the cysteine residues in photolyzed somatotropin appear to donate electrons back to tryptophan, leaving a pair of thiyl radicals, subsequently add oxygen to form the sulfonate. Disulfide bonds can also be cleaved by X-radiation (synchrotron radiation).[515] This study reported the reduction of a redox-active disulfide in a tryporedoxin. The radiation dose was less than that required to break a structural disulfide in lysozyme.[516]

The use of trivalent phosphorus nucleophiles to reduce organic disulfides has been known for some time.[517] Tri-*n*-butylphosphine will reduce *S*-sulfocysteine to cysteine[518] and will also reduce disulfide bonds in proteins. The extensive application of trialkylphosphines/triarylphosphines for the modification of cystine residues in proteins was hampered by insolubility of reagents such as tri-*n*-butylphosphine.[519] The synthesis of a water-soluble phosphine, tris(2-carboxyethyl)phosphine (TCEP), was a significant advance.[520,521] It is quite soluble in water (310 g/L). Dilute solutions (5 mM) are reasonably stable at acid pH values; at pH values above 7, the rate of conversion of the reagent to the oxide is significant. The reduction of disulfides proceeds very rapidly at pH 4.5 and below. Kinetic selectivity in the reduction of disulfides could be demonstrated. Gray[522] extended these early observations to permit the use of TCEP reduction to establish the position of disulfide bonds in proteins. Because the reduction is performed at low pH (stock solution of 20 mM TCEP in 0.17 M citrate, pH 3.0, is stable for weeks at 23°C; the reduction is performed in 0.1% trifluoroacetic acid with 1–10 μM TCEP), it is possible to obtain partially reduced peptides by HPLC separation. Alkylation of the free thiols in the isolated peptides with 4-vinylpyridine permitted subsequent structural analysis of the peptide and disulfide bond assignment.

Disulfide bonds are unstable at alkaline pH (pH ≥ 13.0).[89] This has been examined by Donovan in some detail.[523] With protein-bound cystine, there is change in the spectrum with an increase in absorbance at 300 nm. Florence[524] proposed that cleavage of disulfide bonds in proteins by base proceeds via β-elimination to form dehydroalanine and a persulfide intermediate that can decompose to form several products.

METHIONINE

The modification of methionine in a native protein is generally accomplished with considerable difficulty. It is possible to obtain highly selective oxidation with some reagents in certain proteins, and the results have been useful. Because the dissociation of a proton from sulfur is unnecessary to generate the nucleophile, relatively specific derivatization by alkylating agents can be accomplished at low pH. Whereas other residues such as cysteine and histidine are susceptible to alkylation, these residues are protonated and resist modification under acid conditions.

The oxidation of methionine (Figure 1.42) is the most carefully studied modification. Part of this interest stems from issues associated with the manufacture of recombinant proteins[525–530] and part from the increase of interest in biological oxidation.[531,532] The reader is directed to an excellent review by Vogt[533] for a

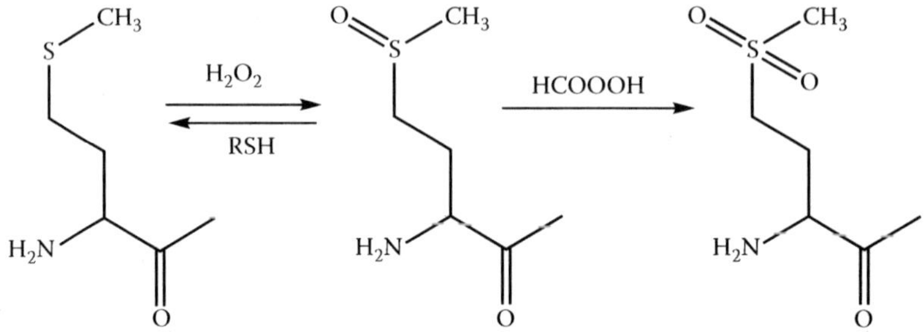

FIGURE 1.42 The oxidation of methionine. The reversible oxidation of methionine to methionine sulfoxide is shown as well as the subsequent further irreversible oxidation of methionine sulfoxide to methionine sulfone.

discussion of chemical and biological oxidation processes. It is useful to recognize that there is no clear division between chemical and biological processes as, for example, H_2O_2 is produced *in vivo*. It is possible to convert methionine sulfoxide to methionine under relatively mild conditions,[534] thus providing for the reversibility of the oxidative reactions described later. A systematic study has shown that of four reducing agents tested, mercaptoacetic acid, 2-mercaptoethanol, dithiothreitol, and *N*-methylmercaptoacetamide, the latter reagent, *N*-methylmercaptoacetamide, was the most effective. The reactions demonstrated little pH dependence, but did not proceed well at concentrations of acetic acid above 50% (v/v). Methionine sulfoxide reductases[535–537] catalyze the conversion of methionine sulfoxide to methionine. Conversion of methionine sulfoxide to methionine sulfone is essentially irreversible under common solvent conditions and requires more vigorous reagents such as performic acid.[538]

The oxidation of methionine is used for protein surface mapping.[539] Oxidation was performed in 2 mM sodium ascorbate–10 mM sodium phosphate–0.6 μM $NH_4Fe(SO_4)$–1.3 mM EDTA, pH 6.5. Oxidation was initiated by the addition of H_2O_2 to a final concentration of 0.3% and quenched by the addition of an equal volume of 2 M Tris, pH 5.0. Other reagents for the "selective" oxidation of methionine include chloramine T[540,541] and hydrogen peroxide.[542,543] The reaction of methionine with chloramine T can be followed spectrophotometrically.[542] It is noted that the oxidation of methionine is a possible side reaction of the treatment of proteins with *N*-bromosuccinimide.[544]

The development of *t*-butyl hydroperoxide by Keck[545] as a selective oxidizing agent for methionine in proteins represented a significant advance. Results obtained with native recombinant interferon and recombinant tissue-type plasminogen activator showed that this reagent was selective for the oxidation of exposed methionine residues in proteins. Tryptic peptides were separated by HPLC and analyzed by mass spectrometry. Two methionine residues were oxidized in recombinant interferon with *t*-butyl hydroperoxide, whereas all five residues were oxidized to a varying extent by H_2O_2 under the same reaction conditions. Three methionine residues were oxidized in native tissue-type plasminogen activator; all five residues were oxidized to varying degrees in the presence of 8.0 M urea. *t*-Butyl hydroperoxide has been successfully used for recombinant human leptin (100 mM sodium borate, pH 9.0, room temperature)[546] and recombinant human granulocyte colony-stimulating factor (25 mM sodium acetate, pH 4.5, 25°C) [547]

The reaction of iodoacetate with methionine (Figure 1.43) was reported by Gundlach and coworkers[548] in 1959. The reaction of iodoacetate with methionine does not appear to be pH dependent and proceeds much slower than the reaction with cysteine. The resulting sulfonium salt yields homoserine and homoserine lactone

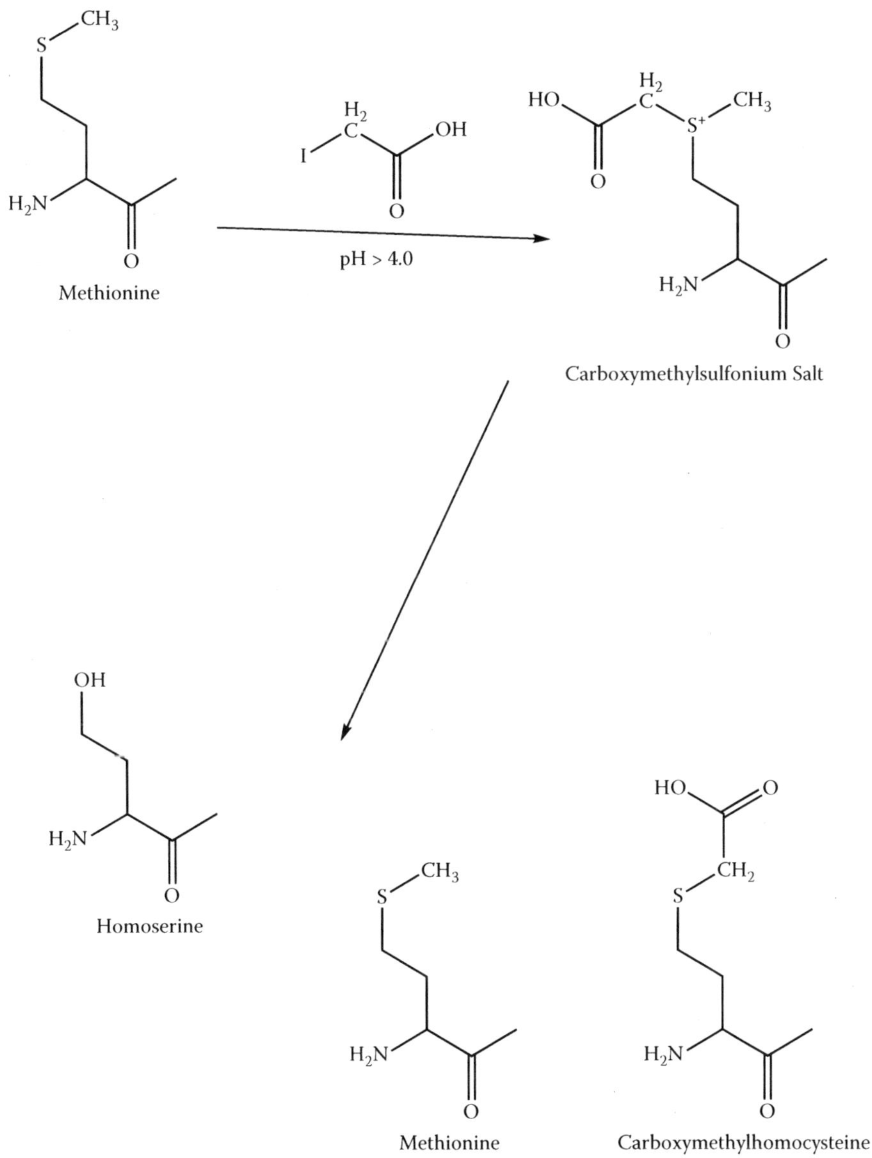

FIGURE 1.43 The reaction of methionine with iodoacetate.

when heated at 100°C at pH 6.5. On acid hydrolysis (6 *N* HCl, 110°C, 22 h), a mixture of methionine and *S*-carboxymethyl homocysteine together with a small amount of homoserine lactone was obtained (Figure 1.43). In general, methionine residues only react with the α-halo acids after the disruption of the secondary and tertiary structure of a protein.[549] Selectivity in the modification of methionine in proteins by α-halo acids can be achieved by performing the reaction at acid pH (pH 3.0 or less). The modification of methionine by ethyleneimine has been reported in a reaction producing a sulfonium salt derivative.[550] In the protein, four of six methionine residues were modified at pH 4.0, and all methionine residues were reactive at pH 3.2. Naider and Bohak[551] have reported that the sulfonium salt derivatives of methionine (e.g., *S*-carboxymethyl methionine, the reaction product of methionine and iodoacetic acid) can be converted to methionine by reaction with a suitable nucleophile. The reversible alkylation of methionine by iodoacetate in dehydroquinase has been reported by Kleanthous and coworkers.[552] In this reaction, iodoacetate behaves kinetically as an affinity label with a K_i of 30 μM and a k_{inact} of 0.014 min^{-1}, pH 7.0 (50 mM potassium phosphate). There is no reaction with iodoacetamide. Two methionine residues are modified during the reaction of dehydroquinase with iodoacetate. In a companion study, Kleanthous and Coggins[553] demonstrated that 2-mercaptoethanol treatment under alkaline conditions (0.5% ammonium bicarbonate, 37°C) could reverse modification at one of the two residues. If the modified protein is denatured, there is no reversal of modification at either residue. The results are interpreted in terms of the close proximity of a positive charge (i.e., lysine) to one of the two methionyl residues, which provides the basis for (1) the affinity labeling and (2) for the 2-mercaptoethanol-mediated reversal of modification. The ability to reverse the alkylation of methionine under relatively mild conditions as described previously has resulted in the development of a clever affinity approach to the purification of methionine peptides. Several groups[554–556] have reported the isolation of methionine peptide by reaction with bead containing a bromoacetyl function under acidic conditions (e.g., 25% acetic acid) and subsequent reducing agent under alkaline conditions as described earlier.

TRYPTOPHAN

The specific chemical modification of tryptophan in protein is one of the more challenging problems in protein chemistry. The solvent conditions for providing specificity of modification are, in general, somewhat harsh and there is the considerable possibility of either the concomitant or separate modification of other amino acid residues.

Treatment of tryptophan with hydrogen peroxide results in the oxidation of the indole ring.[557–560] Underberg and colleagues[561] have reviewed the methods for the qualitative and quantitative analysis of tryptophan oxidation in peptides and proteins, including UV spectroscopy, fluorescence, and HPLC analysis. HPLC analysis of tryptophan oxidation products has been described.[562] Detail is provided for the separation of kynureine, 5-hydroxytryptophan, tryptophan, and dioxindolealanine on a C_{18} column. The reader is directed to an excellent study by Mach and coworkers[563] for the extinction coefficients of tryptophan, tyrosine, and cystine in proteins.

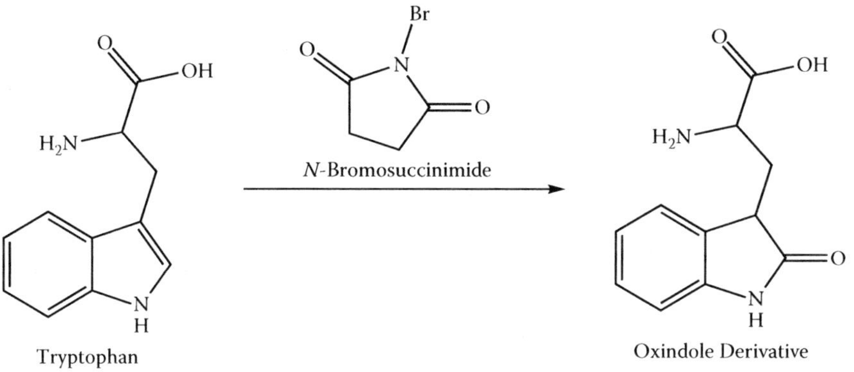

FIGURE 1.44 The reaction of *N*-bromosuccinimide with tryptophan.

The reaction of *N*-bromosuccinimide (NBS) with tryptophan (Figure 1.44) is a well-established method for the site-specific chemical modification of this residue in proteins.[564–566] In general, the modification reaction is performed in 0.1 M sodium acetate, pH 4 to 5. The *N*-bromosuccinimide should be recrystallized from water before use. The presence of halides such as chloride or bromide in the solvent must be avoided, because the addition of *N*-bromosuccinimide will oxidize these ions to the elemental form with disastrous and irreproducible effects on the proteins under study. In general, a twofold molar excess of *N*-bromosuccinimide per mole of tryptophan is necessary to achieve modification. Daniel and Trowbridge[567] found that (at pH 4.0) the reaction of *N*-bromosuccinimide with acetyl-L-tryptophan ethyl ester required 1.5 mol of *N*-bromosuccinimide per mole of the acetyl-L-tryptophan ethyl ester, whereas trypsinogen required 2.0 to 2.3 mol *N*-bromosuccinimide per mole of tryptophan oxidized, and trypsin required 1.5 to 2.0 mol *N*-bromosuccinimide per mole of tryptophan oxidized.

The reaction of *N*-bromosuccinimide with proteins can also result in the cleavage of peptide bonds at tryptophan, tyrosine, and histidine.[568] Feldhoff and Peters[569] have devised a procedure that has enhanced specificity for tryptophan. Their procedure uses 8.0 M urea–2.0 M acetic acid as the solvent with a 20-fold molar excess of *N*-bromosuccinimide. Their approach offers at least two advantages: first, the protein is denatured so that all residues should be equally available, and second, the *N*-bromosuccinimide reacts with urea to yield *N*-bromourea, a less severe oxidizing agent that should have increased specificity for tryptophanyl residues.

2-Hydroxy-5-nitrobenzyl bromide (Figure 1.45), frequently referred to as Koshland's reagent, was introduced by Koshland and coworkers.[570,571] Under appropriate reaction conditions (pH 4.0 or below), the reagent is highly specific for reaction with tryptophan. This reagent also has the advantage of being a "reporter" group, in the sense that the spectrum of the hydroxynitrobenzyl derivative is sensitive to changes in the microenvironment. This decrease observed in absorbance at 410 nm associated with an increase in absorbance at 320 nm upon the addition of dioxane is similar to that seen with acidification, and reflects the increase in the p*K*a of the phenolic hydroxyl group. The chemistry of the reaction of 2-hydroxy-5-nitrobenzyl

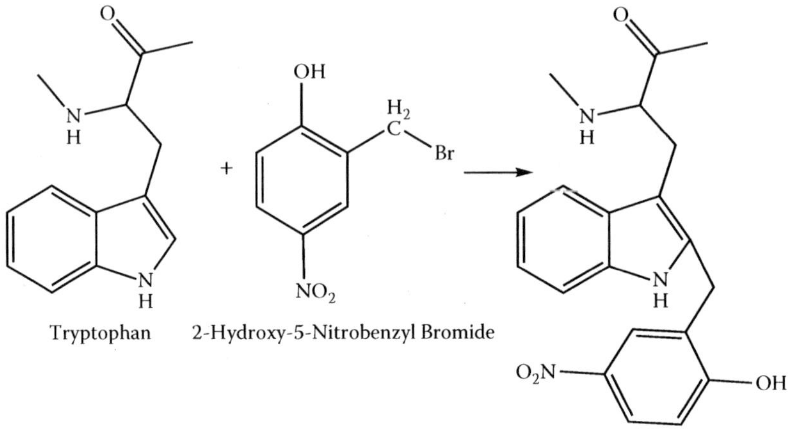

FIGURE 1.45 The reaction of tryptophan with 2-hydroxy-5-nitrobenzyl bromide.

bromide with tryptophan has been studied in some detail.[102,572,573] Disubstitution on the indole ring is a possibility and is usually seen as a sudden "break" in the plot of extent of modification versus reagent excess. MALDI-TOF mass spectrometry has been used in products derived from the modification of tryptophan in a model peptide (GEGKGWGEGK) with 2-hydroxy-5-nitrobenzyl bromide.[573] A total of five products were obtained that reflected various degrees of substitution at the single tryptophan residue.

Horton and Koshland[574] developed a clever approach for modification of hydrolytic enzymes. If 2-hydroxy-5-nitrobenzyl bromide is substituted at the phenolic hydroxyl, it is essentially unreactive, as originally shown for the methoxy derivative. Horton and Young[575] prepared 2-acetoxy-5-nitrobenzyl bromide. This derivative, similar to the methoxy derivative, is essentially unreactive. There is considerable structural identity between 2-acetoxy-5-nitrobenzyl bromide and *p*-nitrophenyl acetate, which is a nonspecific substrate for chymotrypsin. α-Chymotrypsin removes the acetyl group from 2-acetoxy-5-nitrobenzyl bromide, thus generating 2-hydroxy-5-nitrobenzyl bromide at the active site (Figure 1.46), which then either rapidly reacts with a neighboring nucleophile or undergoes hydrolysis. Reagents with reaction characteristics similar to 2-hydroxy-5-nitrobenzyl bromide are nitrophenylsulfenyl derivatives[576] (Figure 1.47). The reaction product resulting from the sulfenylation of lysozyme[577] with 2-nitrophenylsulfenyl chloride (40-fold molar excess), pH 3.5 (0.1 M sodium acetate), has spectral characteristics that can be used to determine the extent of reagent incorporation (at 365 nm $\varepsilon = 4 \times 10^3\ M^{-1}\ cm^{-1}$) . These reagents show considerable specificity for the modification of tryptophan at pH ≤ 4.0. Wilchek and Miron[578] have reported on the reaction of 2,4-dinitrophenylsulfenyl chloride with tryptophan in peptides and protein, and subsequent conversion of the modified tryptophan to 2-thiotryptophan by reaction with 2-mercaptoethanol at pH 8.0. The thiolysis of the modified tryptophan is responsible for changes in the spectral properties of the derivative. The characteristics of the modified tryptophan have resulted in the

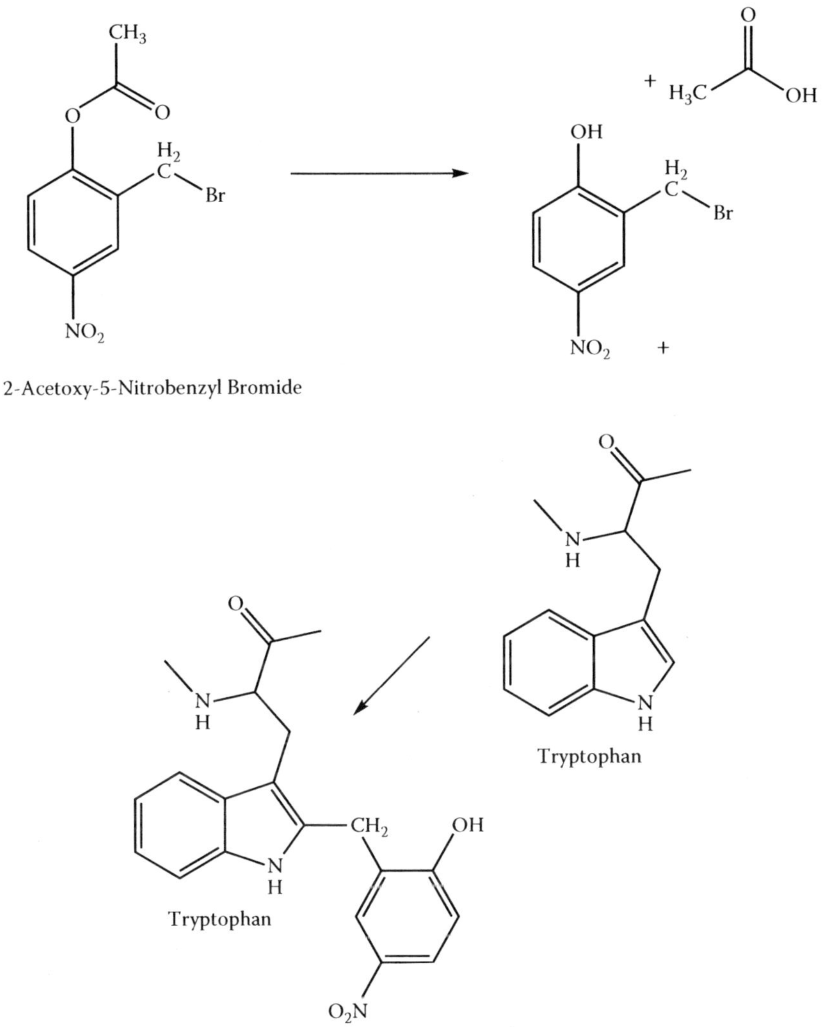

FIGURE 1.46 The hydrolysis of 2-acetoxy-5-nitrobenzyl bromide to release 2-hydroxy-5-nitrobenzyl bromide for subsequent reaction with tryptophan.

development of a facile purification scheme for peptides containing the modified tryptophan residues.[579,580] Nishimura and coworkers[581] have developed a heavy (^{13}C) form of 2-nitrobenzenesulfenyl chloride for the differential labeling of tryptophan residues in protein mixtures. The application of the isotope-coded affinity tag strategy[582] to tryptophanyl residues has significant advantage in that tryptophan is one of the least abundant residues in proteins. This chemistry has seen increased use in peptide mapping using mass spectrometry.[583–585]

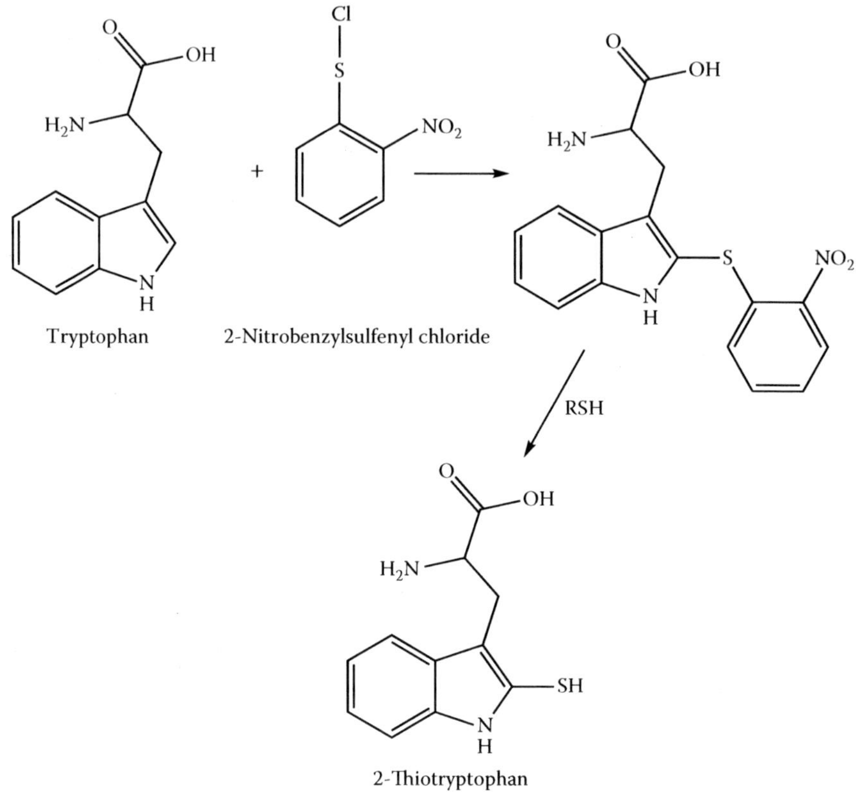

FIGURE 1.47 The reaction of tryptophan with 2-nitrophenylsulfenyl chloride.

ARGININE

Present approaches to the site-specific modification of arginyl residues in proteins used three reagents: phenylglyoxal (and derivatives such as *p*-hydroxyphenylglyoxal),[586,1] 2,3-butanedione,[587] and 1,2-cyclohexanedione.[588] A review of the literature from suggests that phenylglyoxal is the most extensively used reagent for the site-specific chemical modification of arginine in proteins. As with other site-specific chemical modifications of proteins, there has been increasing use of mass spectrometry to characterize the chemical modification of arginine in proteins.[589–593]

The use of phenylglyoxal (Figure 1.48) was developed by Takahashi[586] in 1968. The stoichiometry of the reaction involves the reaction of 2 mol of phenylglyoxal with 1 mol of arginine. Borders and coworkers[594] have reported the synthesis of a chromophoric derivative, 4-hydroxy-3-nitrophenylglyoxal. The adduct between 4-hydroxy-3-nitrophenylglyoxal and arginine absorbs light at 316 nm ($\varepsilon = 1.09 \times 10^4\ M^{-1}\ cm^{-1}$. The derivative is unstable to acid hydrolysis (6 N HCL, 110°C, 24 h) but can be stabilized by the inclusion of thioglycolic acid. This same group subsequently used this reagent to identify the reactive arginine in yeast Cu,Zn superoxide dismutase.[595] The reaction of 4-hydroxy-3-nitrophenylglyoxal (50 mM BICINE,

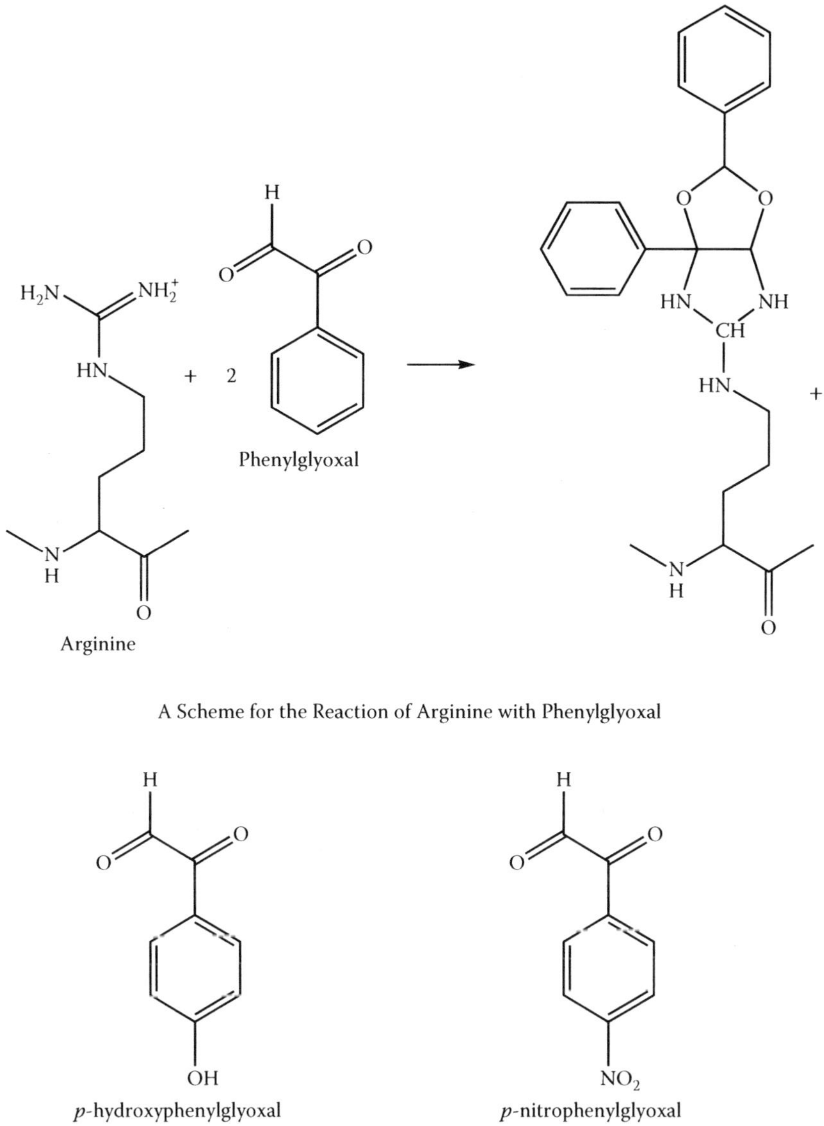

FIGURE 1.48 The reaction of phenylglyoxal with arginine in proteins.

100 mM $NaHCO_3$, pH 8.3) with yeast Cu,Zn superoxide dismutase is slower (0.57 M^{-1} min^{-1}) than that observed with phenylglyoxal (28. M^{-1} min^{-1}). A similar difference in the rate of reaction with the two reagents was observed with creatinine kinase.[596] *p*-Hydroxylphenylglyoxal was developed by Feeney and colleagues[597] for the detection of available arginine residues in proteins. As with phenylglyoxal, it reacts with arginine under mild conditions (pH 7–9, 25°C, 30–60 min). The concentration of the resulting adduct (2:1 stoichiometry) with arginine can be determined at 340 nm ($\varepsilon = 1.83 \times 10^4\ M^{-1}\ cm^{-1}$). The modification is slowly reversed under basic

conditions. The general characteristics of the reaction of *p*-hydroxyphenylglyoxal are similar to those described for phenylglyoxal. There have been several studies that have compared *p*-hydroxyphenylglyoxal and phenylglyoxal. It can be argued that *p*-hydroxyphenylglyoxal is more hydrophilic than phenylglyoxal. The most remarkable observation on differences between *p*-hydroxyphenylglyoxal and phenylglyoxal comes from studies by Ericksson and coworkers.[598] These investigators observed that treatment of mitochondria with phenylglyoxal (10 mM HEPES–250 mM sucrose–10 mM succinate, 100 μM EGTA–3 μM rotenone, pH 8.0) results in the closing of the permeability pore, whereas reaction with *p*-hydroxyphenylglyoxal results in pore opening. The reaction of arginine with phenylglyoxal is greatly accelerated in bicarbonate–carbonate buffer systems.[599] The reaction of methylglyoxal with arginine is also enhanced by bicarbonate, whereas a similar effect is not seen with either glyoxal or 2,3-butanedione. The molecular basis for this specific buffer effect is not clear at this time, nor is it known whether reaction with α-amino functional groups occurs at a different rate than with other solvent systems used for this modification of arginine with phenylglyoxal. Branlant and coworkers[600] have used *p*-carboxyphenylglyoxal in bicarbonate buffer at pH 8.0 to modify aldehyde reductase. A second-order rate constant of 26 M^{-1} min^{-1} was observed in 80 mM bicarbonate and 2.9 M^{-1} min^{-1} in 20 mM sodium phosphate, pH 7.0. Saturation kinetics was observed with this reagent under certain conditions. Eun[601] has examined the effect of borate on the reaction of arginine with phenylglyoxal and *p*-hydroxyphenylglyoxal. The base buffer of these studies was 0.1 M sodium pyrophosphate, pH 9.0. Spectroscopy was used to follow the rate of arginine modification. The rate of modification of either free arginine or *N*-acetyl-L-arginine with phenylglyoxal was 10 to 15 times higher than that of *p*-hydroxyphenylglyoxal in the base buffer system. The inclusion of sodium borate (10 to 50 mM) markedly increased the rate of the reaction (approximately 20-fold) of *p*-hydroxyphenylglyoxal with either arginine or *N*-acetyl-L-arginine, whereas there was only a slight enhancement of the phenylglyoxal reaction.

2,3-Butanedione (Figure 1.49) is the second well-characterized reagent for the selective modification of arginyl residues in proteins. There were problems with the specificity of the reaction[602] and the time required for modification until the observation of Riordan[603] that borate had a significant effect on the nature of the reaction of 2,3-butanedione with arginyl residues in proteins. Leitner and Linder[590] have developed an approach to the general labeling of guanidino groups in proteins via reaction with 2,3-butanedione in the presence of an arylboronic acid (e.g., phenylboronic acid) under alkaline conditions (pH 8–10). The sample is then subjected to electrospray ionization mass spectrometry without further processing.

The use of 1,2-cyclohexanedione under very basic conditions to modify arginyl residues was demonstrated in 1967.[604] However, it was not until Patthy and Smith[588] reported on the reaction of 1,2-cyclohexanedione in borate with arginyl residues in proteins that the use of this reagent became practical. At alkaline pH, reaction of 1,2-cyclohexanedione with arginine forms N^5-(4-oxo-1,3-diazaspiro[4,4]non-2-yliodene)-L-ornithine (CHD-arginine), a reaction that cannot be reversed. Between pH 7.0 and 9.0, a compound is formed from arginine and 1,2-cyclohexanedione, N^7-N^8-(1,2-dihydroxycyclohex-1,2-ylene)-L-arginine (DHCH-arginine) (Figure 1.50). This compound is stabilized by the presence of borate and is unstable in the presence of

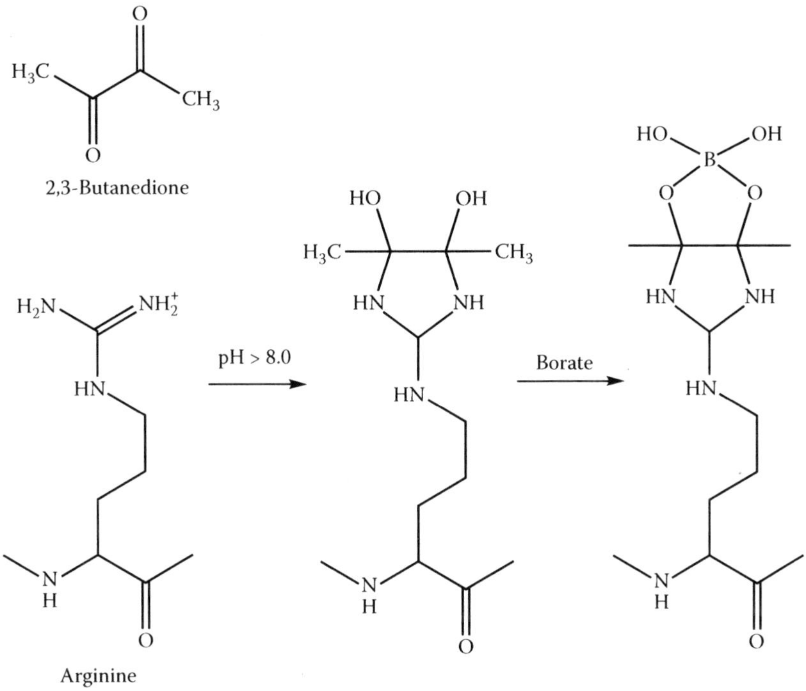

A Scheme for the Reaction of 2,3-Butanedione with Arginine

FIGURE 1.49 The reaction of 2,3-butanedione with arginine in proteins.

buffers such as Tris. This compound is readily converted back to free arginine in 0.5 M hydroxylamine, pH 7.0. Calvete and colleagues[605] used a novel approach to the modification of arginine residues in bovine seminal plasma protein PDC-109. The protein was bound to a heparin-agarose column and the 1,2-cyclohexanedione (in 16 mM Tris-50 mM NaCl–1.6 mM EDTA–0.025% NaN_3, pH 7.4) circulated through the column overnight at room temperature. The modified protein was eluted with 1.0 M NaCl. Residues shielded from modification were presumed to be the heparin-binding site.

HISTIDINE

Because many enzymes contain a histidine residue, which is critical for the catalytic process, the site-specific modification of this residue has been the subject of many studies. Most of these studies have been directed at the catalytic mechanism of enzymes and few at protein–protein interactions or substrate–cofactor binding. Thus, despite the importance of histidine, only a small number of reagents have been studied.

Histidine, methionine, and tryptophan are quite sensitive to photooxidation, whereas tyrosine, serine, and threonine are somewhat less sensitive.[606–608] Histidine residues are oxidized in the process of radiolytic protein footprinting.[609,610]

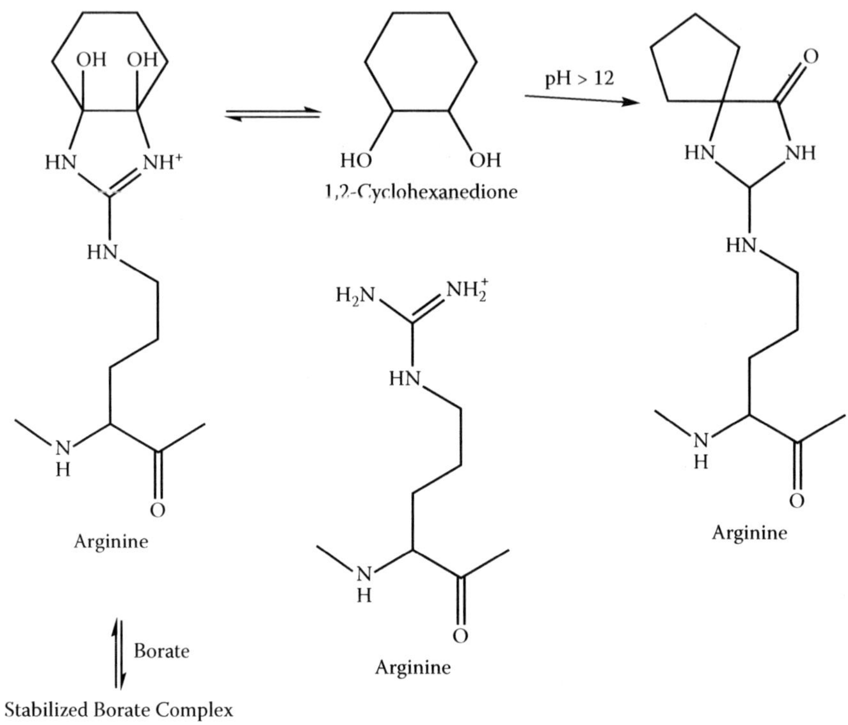

FIGURE 1.50 The reaction of 1,2-cyclohexanedione with arginine. (Adapted from Patthy, L. and Smith, E.L., Reversible modification of arginine residues. Application to sequence studies by restriction of tryptic hydrolysis in lysine residues. *J. Biol. Chem.* 557–564, 1975.)

Histidine residues can be modified by α-halo carboxylic acids and amides (i.e., bromoacetate and bromoacetamide) (Figure 1.51). The histidine residue must have enhanced nucleophilic character.[611–614] The chemistry of histidine alkylation with α-halo carboxylic acids and amides provides the basis for the development of peptide chloromethyl ketones for the affinity labeling of proteolytic enzymes.[615,616] Methyl *p*-nitrobenzenesulfonate (Figure 1.52) has been used to methylate histidine residues in ribosomal peptidyl transferase.[617] In these experiments, the ribosome preparation was modified by a 300-fold molar excess of methyl *p*-nitrobenzenesulfonate (from a stock solution dissolved in acetonitrile).

Diethyl pyrocarbonate is the most common reagent for the modification of histidine in proteins (Figure 1.53).[618–620] In the pH range from 5.5 to 7.5, diethylpyrocarbonate is reasonably specific for histidyl residues. Reaction of diethylpyrocarbonate with histidine residues at a moderate excess of diethylpyrocarbonate results in substitution at one of the nitrogen positions on the imidazole ring. This reaction is associated with an increase in absorbance at 240 nm ($\Delta\varepsilon = 3200\ M^{-1}\ cm^{-1}$). The modification is readily reversed at alkaline pH and, in particular, in the presence of nucleophiles such as hydroxylamine. Tris and other nucleophilic buffers can also reverse the modification and their use should be avoided with diethyl

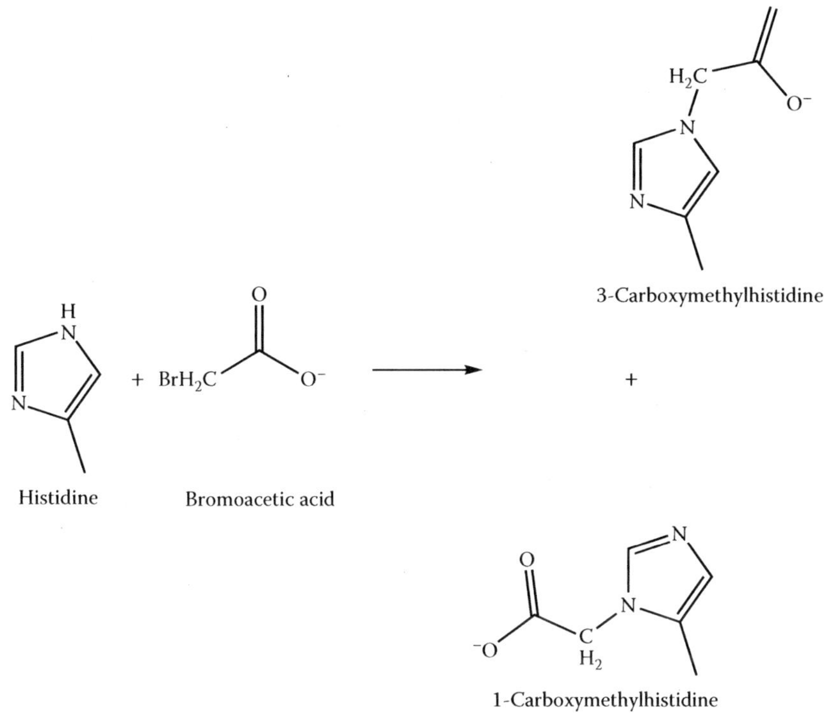

FIGURE 1.51 The reaction of histidine with 2-bromoacetic acid to form 1-carboxymethylhistidine and 3-carboxymethylhistidine.

pyrocarbonate. Generally, treatment with neutral hydroxylamine (0.1 to 1.0 M, pH 7.0) is used to regenerate histidine. Carboxyethylation at both the N_1 and N_3 positions results in a derivative with altered spectral properties compared to the monosubstituted derivative. This derivative does not regenerate histidine, and treatment with neutral hydroxylamine or base results in scission of the imidazole ring. Mass spectrometry is of increasing value in the analysis of the chemical modification of histidine in proteins, including carboxyethylated histidine.[621–623] It has been shown that there is good correlation between spectral measurements and mass spectrometry.[624] Diethylpyrocarbonate can also modify other nucleophiles such as cysteine, tyrosine, and primary amino groups. Modification at sulfhydryl residues, which is not well documented with protein-bound cysteine, can be determined by a decrease in free sulfhydryl groups. Reaction of tyrosine is easily assessed by a decrease in absorbance at 275 to 280 nm, similar to that observed on *O*-acetylation with *N*-acetylimidazole. This modification is reversed by neutral hydroxylamine. Reaction at primary amino groups (α-amino groups; ε-amino groups of lysine) results in a derivative that is stable to hydroxylamine. An elegant study[625] has examined the reaction of diethylpyrocarbonate with histidyl residues in cytochrome b_5. Using (NMR) spectroscopy with this well-characterized protein, it has been possible to identify factors influencing histidine modification with this reagent; three major factors include (1) the p*K*A of the individual histidine residue, (2) solvent exposure

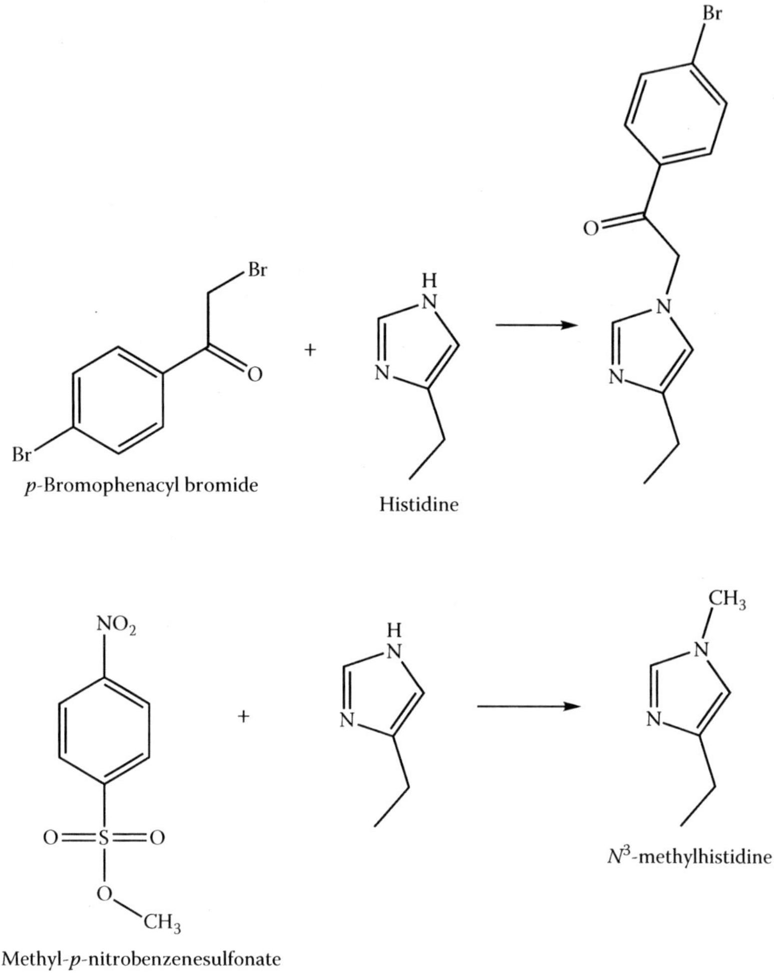

FIGURE 1.52 The *N*-methylation of histidine with methyl-*p*-nitrobenzene sulfonate.

of the residue, and (3) hydrogen bonding of the imidazolium ring. Furthermore, these investigators point out that tautamerization of the imidazolium ring leads to heterogeneity of modification, which in turn explains differences in the spectral properties of modified proteins.

CARBOXYL GROUPS

The use of carbodiimide-mediated modification[626,627] (Figure 1.54) is the most extensively used method for the modification of carboxyl groups in proteins. Carbodiimides react with protonated carboxyl groups, yielding an activated intermediate, most likely an acylisourea, which then reacts with a nucleophile such as an amine.[628] Carbodiimides are also used for zero-length cross-linking (Figure 1.55) of

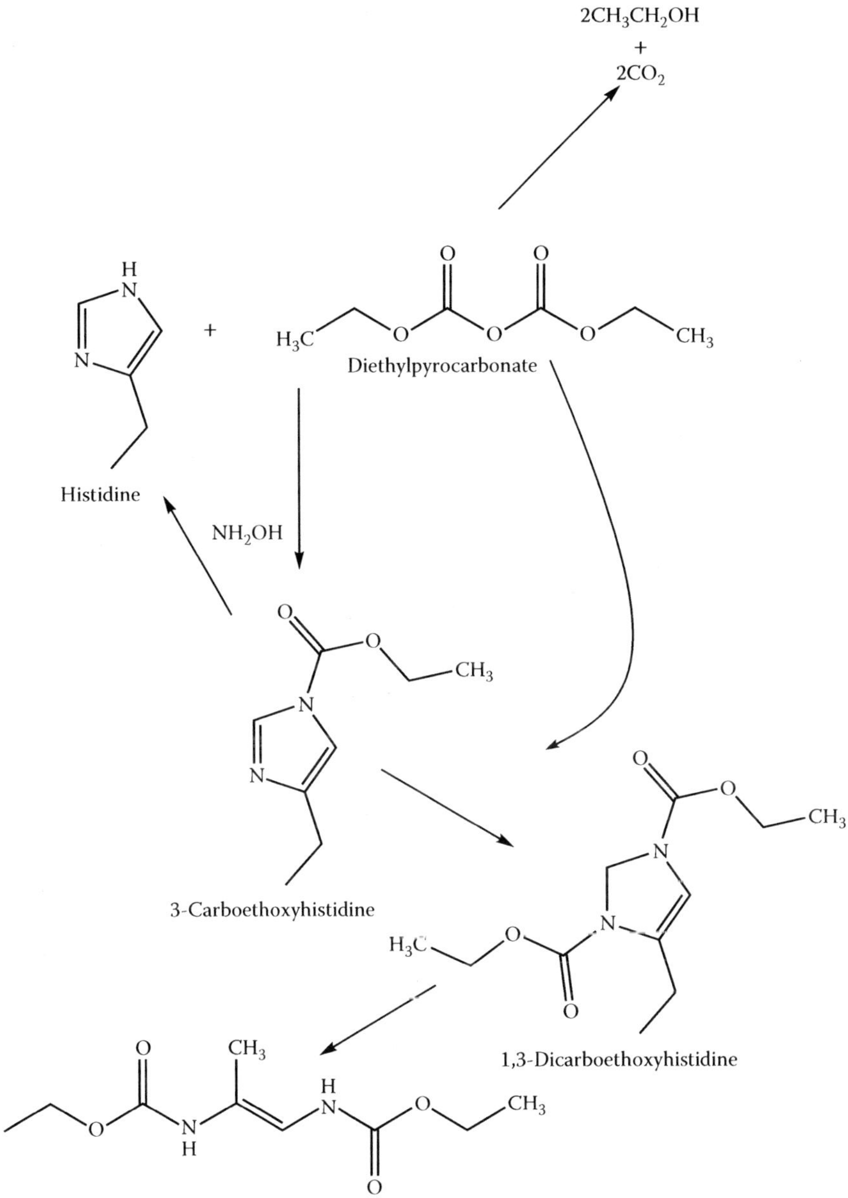

FIGURE 1.53 The chemistry of the reaction of diethylpyrocarbonate with histidine.

proteins between proximate lysine residues and carboxyl groups.[629,630] Zero-length cross-linking can be of value in the study of protein conformation (Chapter 2) and the preparation of hydrogels (Chapter 5).

Although water-insoluble carbodiimides such as *N,N′*-dicyclohexylcarbodiimide continue to be useful for the site-specific modification of carboxyl groups,[631,632] most

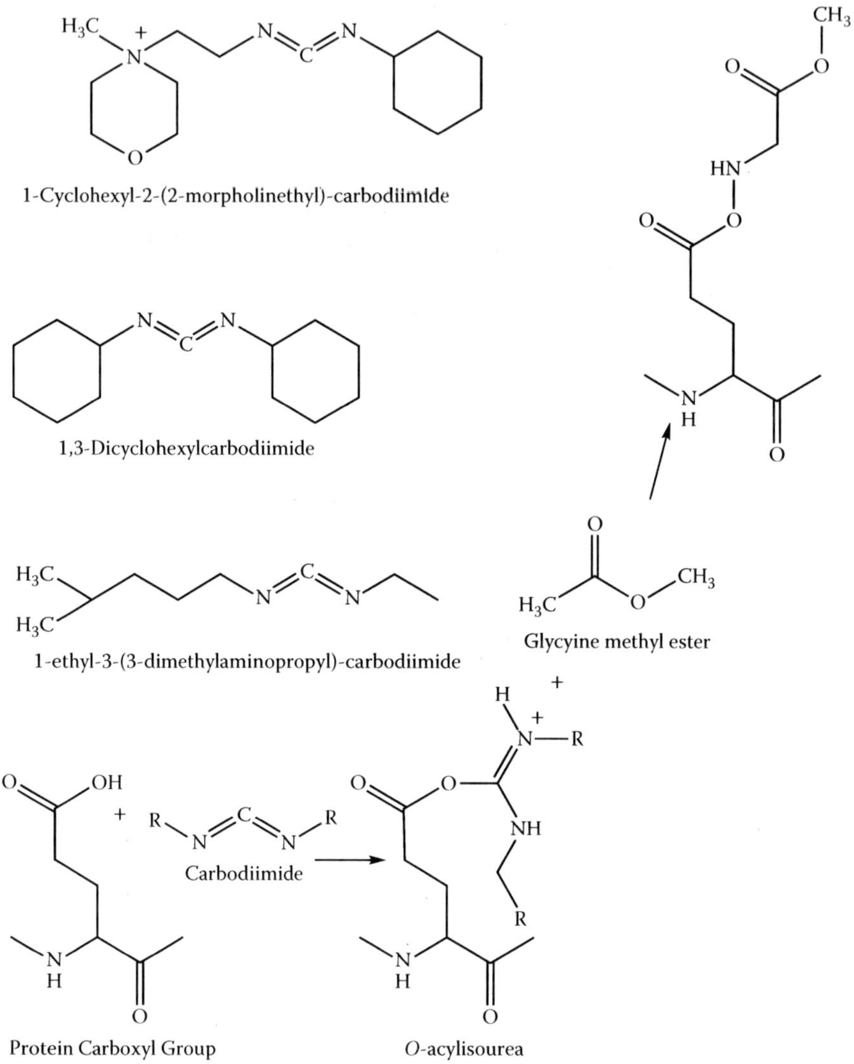

FIGURE 1.54 Structures of some commonly used carbodiimides and a scheme for the reaction of carbodiimides with carboxyl groups in proteins.

current work uses water-soluble carbodiimides such as 1-ethyl-3-(3-dimethylaminopropylcarbodiimide (*N*-ethyl-*N'*-(dimethylaminopropyl) carbodiimide; EDC). Water-soluble carbodiimides were developed by Sheehan and Hlavka.[633,634] It is of interest that the first application to proteins was the zero-length cross-linking of collagen.[634] This study used 1-ethyl-3-(2-morpholinyl-(4)-ethyl)carbodiimide metho-*p*-toluenesulfonate in unbuffered aqueous solution. Riehm and Scheraga[635] advanced the use of water-soluble carbodiimides for proteins in a study on the modification of ribonuclease with 1-cyclohexyl-3-(2-morpholinylethyl)carbodiimide at pH 4.5 with a pH-Stat. A number of different products were obtained that could be separated by

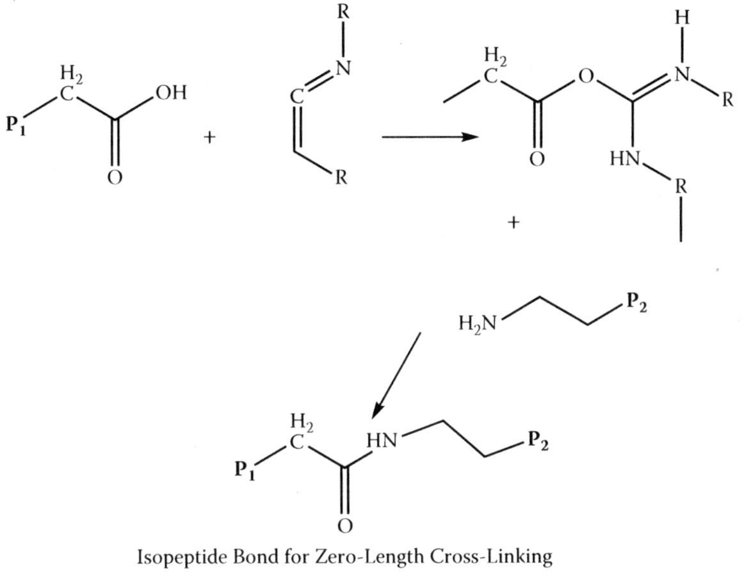

FIGURE 1.55 Zero-length cross-linking.

ion-exchange chromatography (BioRex® 70). These investigators also suggest that the mechanism between a carbodiimide and a carboxyl group lead to the formation of an unstable acylisourea, which can either decompose to an acylurea derivative or react with a nucleophile. Koshland and colleagues developed the use of a water-soluble carbodiimide in the presence of an excess of amino acid nucleophilic for the modification of carboxyl groups in proteins. The initial study by Hoare and Koshland used *N*-benzyl-*N′*-3-dimethylaminopropylcarbodiimide; subsequent studies used EDC and resulted in the development of a quantitative method for the measurement of carboxyl groups in proteins.[626,636] These initial studies also introduced the concept of using a unique nucleophile such as norleucine methyl ester, aminomethane-sulfonic acid, and norvaline. The possibility of a side reaction was discussed with reference to the possible modification of the phenolic hydroxyl of tyrosine to form the *O*-arylisourea. Border and colleagues[637] evaluated the stability of EDC in aqueous solution. EDC has a $t_{1/2}$ of 37 h (pH 7.0), 20 h (pH 6.0), and 3.9 h (pH 5.0) in 50 mM 2-(*N*-morpholino)ethanesulfonic acid at 25°C; in the presence of 100 mM glycine, the $t_{1/2}$ values were 15.8 h (pH 7.0), 6.7 h (pH 6.0), and 0.73 h (pH 5.0). The authors suggest that this supports the use of EDC at pH 6.0 or 7.0, but at pH 5.0 stability would be an issue. This study also reported a major decrease in the stability of EDC under the aforementioned solvent conditions in the presence of 10 mM phosphate or 10 mM ATP. Lei and coworkers[638] have reported kinetic studies on the hydrolysis of EDC in aqueous solution under acidic conditions. The conversion of EDC to the corresponding acylurea was measured by mass spectrometry and capillary electrophoresis. The rate of decomposition increased with increasing acidity. Mirsky and colleagues[639] used the decrease in absorbance at 214 nm to measure the stability of EDC. These investigators also report decreased stability with decreasing pH. The presence of citrate, acetate, or phosphate increased the rate of EDC decomposition. Shegal and Vijay[640] have optimized the conditions for EDC-mediated coupling of a carboxyl-containing compound to an amine matrix (Affi-Gel® 102). These investigators noted that the presence of *N*-hydroxysuccinimide greatly improved the coupling of butyric acid to the matrix. 1,2-Diaminoethane or diaminomethane can be coupled to aspartic acid residues to produce a trypsin-sensitive bond.[641] Lin and coworkers[642] used the water-soluble carbodiimide-mediated coupling of cystamine to protein carboxyl groups followed by the reduction of the coupled cystamine with dithiothreitol to give 2-aminothiol functional groups bound to protein carboxyl groups. Nitrotyrosine ethyl ester can be used as the modifying nucleophile with EDC,[643] providing a chromophore as well as a method for isolating the modified peptide by immunoaffinity chromatography.[644]

There are examples of carboxyl group modification with reagents expected to react far more effectively with other nucleophiles (Figure 1.56). An example of this is the reaction of iodoacetamide with ribonuclease T_1 to form the glycolic acid

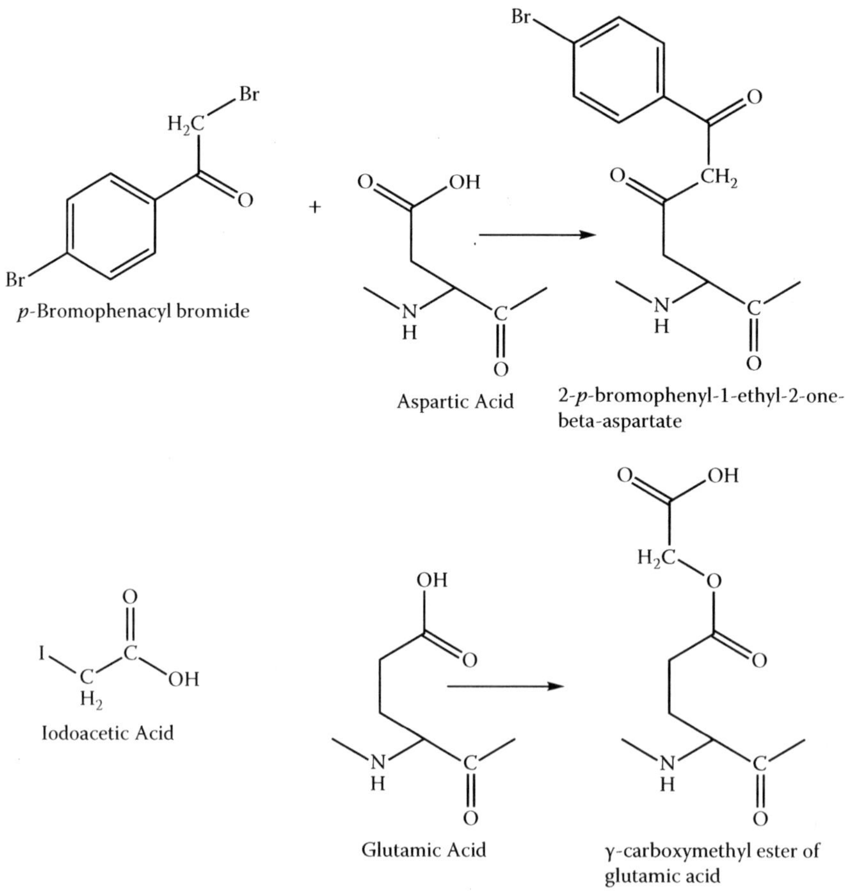

FIGURE 1.56 The modification of aspartic acid with *p*-bromophenacyl bromide.

derivative of the glutamic acid residue, as elegantly shown by Takahashi and coworkers.[645] Another example is the modification of a specific carboxyl group in pepsin by *p*-bromophenacyl bromide.[646,647]

Woodward and coworkers[648] developed *N*-ethyl-5-phenylisoxazolium-3′-sulfonate (Woodward's reagent K) (Figure 1.57) and various other *N*-alkyl-5-phenylisoxazolium fluoroborates as reagents for the "activation" of carboxyl groups for synthetic purposes. Anfinsen and coworkers[649] have studied the kinetics of the aqueous hydrolysis of this reagent and reaction with staphylococcal nuclease. This study demonstrated that Woodward's Reagent K is very unstable in aqueous solution above pH 3.0. Studies on the rate of enzyme inactivation by this reagent should be corrected for reagent hydrolysis to obtain accurate second-order rate constants. Bodlaender and coworkers[650]

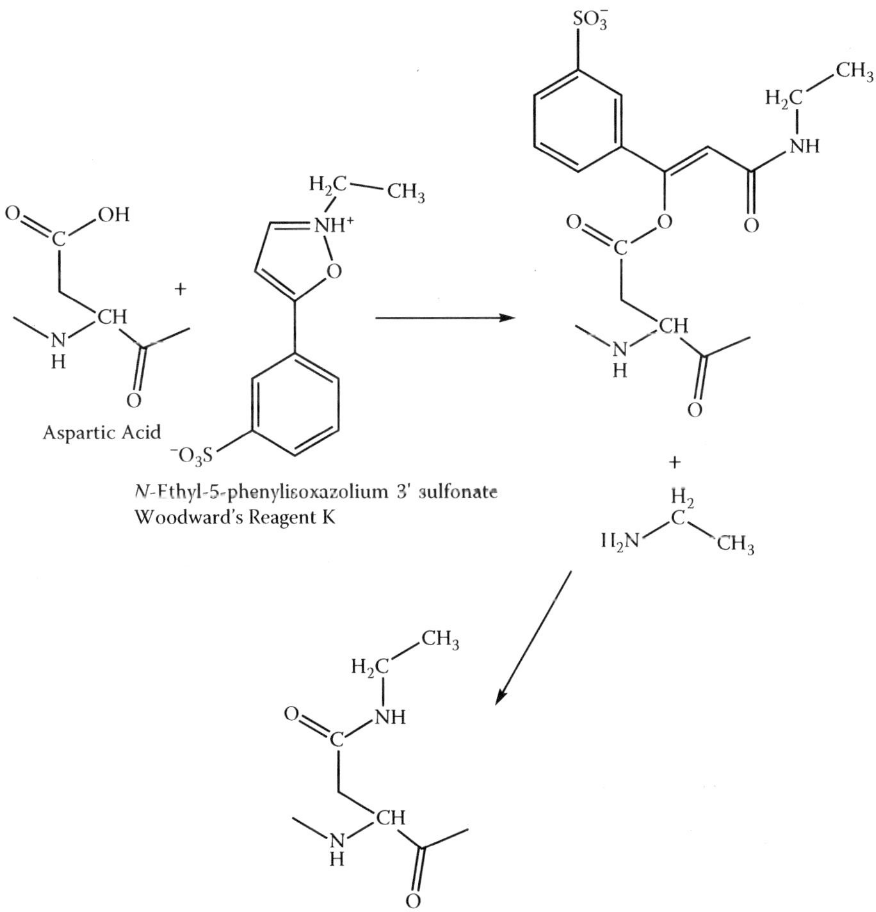

FIGURE 1.57 The modification of aspartic acid with Woodward's reagent K (*N*-ethyl-5-phenylisoxazolium-3′-sulfonate). The formation of the ketoketimine intermediate is shown with the subsequent reaction with a nucleophile (ethyl amine) to form a stable modified derivative.

used *N*-ethyl-5-phenylisoxazolium-3-sulfonate, the *N*-methyl and *N*-ethyl derivatives of 5-phenylisoxazolium fluoroborate, or *N*-methylbenzisoxazolium fluoroborate to activate carboxyl groups on trypsin for subsequent modification with methylamine or ethylamine.

CHEMICAL CLEAVAGE OF PEPTIDE CHAINS

Cleavage of methionine-containing peptide bonds with CNBr[651–655] is the most widely used method for specific chemical cleavage of peptide bonds. The reaction cleaves a peptide bond in which methionine contributes the carboxyl moiety. Methionine is converted into homoserine lactone and homoserine during this process with the *loss of methyl thiocyanate*. The reaction is reasonably quantitative although, as indicated in the following text, variable amounts of CNBr (CNBr) might be required. The methionine content of most proteins is low[656] enough that a reasonably small number of fragments are obtained, providing a distinct advantage in primary structure analysis.

The chemistry of this reaction is straightforward (Figure 1.58), involving the nucleophilic attack of the thioether sulfur on the carbon in CNBr followed by cyclization to form the iminolactone, which is hydrolyzed by water, resulting in cleavage of the peptide bond. At acid pH this reaction does not generally in and by itself affect any other amino acid with the exception of cysteine, which is converted to cysteic acid. Cleavage of peptide chains at methionine with CNBr proceeds best with a fully denatured protein in mild acid.[657, 658] The reaction proceeds effectively with a 20- to 100-fold molar excess of CNBr (added either as a solid to the protein or peptide dissolved in the solvent of choice). The molar ratio of CNBr to methionine residues should be established for each peptide and protein; it was necessary use a 3000-fold molar excess to cleave a particular methionine-serine peptide bond in pancreatic deoxyribonuclease.[659]

Robillard and coworkers[660,661] provided methods for the in-gel cleavage of integral membrane proteins for subsequent analysis by MALDI-TOF mass spectrometry. The CNBr cleavage reaction was performed in 70% TFA (one small CNBr crystal dissolved in 200–300 μL 70% trifluoroacetic acid was added to the gel slice) for 14 h in the dark at room temperature. The digested gel piece was sonicated for 5 min and then extracted twice with sonication with 30 μL 60% acetonitrile–1% TFA and concentrated in vacuo prior to analysis. This work has been extended by other investigators.[662,663]

Partial acid hydrolysis is the oldest of the various chemical approaches to the cleavage of specific peptide bonds. The general principle of partial acid hydrolysis is based on the use of dilute acid at a pH just adequate to maintain the β-carboxyl group of aspartic acid in the protonated form. Under these conditions, peptide bonds in which the carboxyl moiety is contributed by aspartic acid are cleaved 100-fold more rapidly than other peptide bonds. The use of 0.03 *N* HCl in vacuo at 105°C for 20 h has been found to be satisfactory for the partial acid hydrolysis of proteins. Hulmes and Pan[664] demonstrated that gas-phase trifluoroacetic acid preferentially cleaves peptide chains at the amino termini of threonine and serine (22°C, 1–15 days or 45°C for 2–3 days). The degradation of antiflammin-2 under acidic conditions has been studied.[665] The reaction was more rapid at pH 3.0 (37°C, sodium citrate) than at

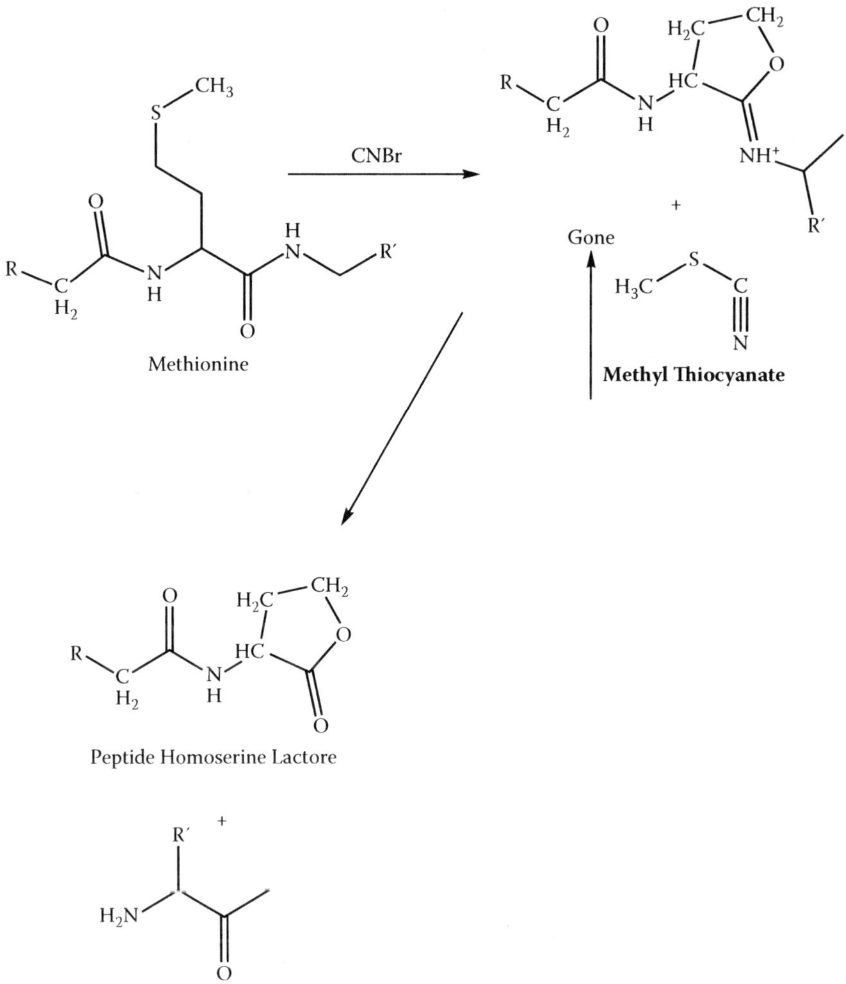

FIGURE 1.58 The cyanogen bromide cleavage of methionine peptide bonds with the formation of homoserine lactone and methyl isothiocyanate.

pH 2.37 (37°C, sodium phosphate), or pH 4.10 (37°C, sodium citrate). It is noted that partial acid hydrolysis is used more frequently for polysaccharide hydrolysis.[666]

S-cyanocysteine is obtained by reaction of cysteine or cystine with 2-nitro-5-thiocyano-benzoic acid (Figure 1.59). Cleavage of the *S*-cyanocysteine is achieved by incubation in 0.1 M sodium borate, 6 M guanidine, pH 9.0 at 37°C with the formation of 2-iminothiazolidine-4-carboxyl peptides. Lu and Gracy[667] used 2-nitro-5-thiocyanobenzoic acid to convert the cysteinyl residues in human placental glucosephosphate isomerase to *S*-cyanocysteine, followed by cleavage at the modified cysteine residues. Watson and colleagues[668] used cyanylation combined with mass spectrometric analysis to determine the disulfide structure of sillucin. These investigators used a combination of partial reduction and CN-induced cleavage. The peptide

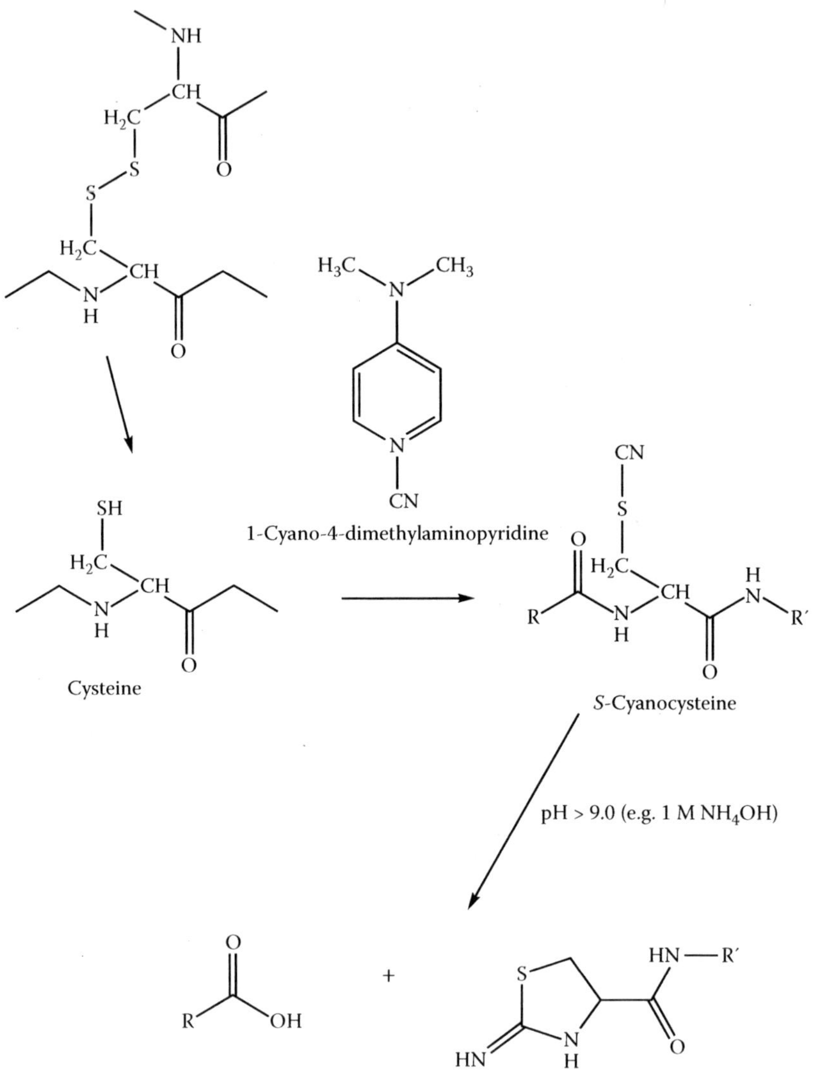

FIGURE 1.59 The cleavage of peptide bonds by cyanate. An excellent reference is Qi, J. et al., Determination of the disulfide structure of sillucin, a highly knotted, cysteine-rich peptide, by cyanylation/cleavage mass mapping. *Biochemistry* 40, 4531–4538, 2001.

was partially reduced with phosphine, and the resulting cysteine residues immediately cyanylated with 1-cyano-4-(dimethylamino)pyridinium tetrafluoroborate. The cyanylated peptides were isolated by HPLC and cleaved with aqueous ammonia. Watson and colleagues[669] have optimized reaction conditions to improve the yield of cleavage products.

Douady and coworkers[670] have used *N*-chlorosuccinimide in acetic acid to cleave peptide bonds in the major polypeptide component of the light-harvesting complex from a brown alga (*Laminaria saccharina*). Droste and colleagues[671] used *N*-chlorosuccinimide for the fragmentation of adenyl cyclase type I in a study on the identification of an ATP-binding site. Pliszka and coworkers[672] used *N*-chlorosuccinimide to identify EDC cross-links in subfragment 1 or skeletal muscle myosin. The gel slices were first washed with water for 20 min with one change of solvent. The gel slices were then washed with urea/H_2O/acetic acid (1 g/1 mL/1 mL) for 20 min with one change of solvent. Cleavage is accomplished with *N*-chlorosuccinimide (15 mM) in urea/H_2O/acetic acid for 30 min.

This has been a brief overview of the use of chemical modification for proteins. The examples selected have been chosen for utility in the applications discussed in the following chapters. It must be emphasized that there is no substitute for a solid background in organic chemistry and physical chemistry in the application of solution chemistry technology to various biotechnology applications.

REFERENCES

REFERENCES FOR TABLE 1.1

1. Giedroc, D.P., Puett, D., Sinha, S.K., and Brew, K., Calcium effects on calmodulin lysine reactivation, *Arch. Biochem. Biophys.* 252, 136–144, 1987.
2. Illy, C., Thielens, N.M., and Arlaud, G.J., Chemical characterization and location of ionic interactions involved in the assembly of the C1 complex of human complement, *J. Protein Chem.* 12, 771–781, 1993.
3. Che, F.Y. and Fricker, L.D., Quantitation of neuropeptides in Cpe(fat)/Cpe(fat) mice using differential isotopic tags and mass spectrometry, *Anal. Chem.* 74, 3190–3198, 2002.
4. Turner, B.T., Jr., Sabo, T.M., Wilding, D., and Maurer, M.C., Mapping of factor XIII solvent accessibility as a function of activation state using chemical modification methods, *Biochemistry* 43, 9755–9765, 2004.
5. Nam, H.W., Lee, G.Y., and Kim, Y.S., Mass spectrometric identification of K210 essential for rat malonyl-CoA decarboxylase catalysis, *J. Proteome Res.* 5, 1398–1406, 2006.
6. Scherer, H.J., Karthein, R., Strieder, S., and Ruf, H.H., Chemical modification of prostaglandin endoperoxide synthase by *N*-acetylimidazole. Effect on enzyme activities and EPR spectroscopic properties, *Eur. J. Biochem.* 205, 751–757, 1992.
7. Cymes, G.D., Igelesias, M.M., and Wolfenstein-Todel, C., Chemical modification of ovine prolactin with *N*-acetylimidazole, *Int. J. Pept. Protein Res.* 42, 33–28, 1993.
8. Vazeux, G., Iturrioz, X., Corvol, P., and Llorens-Cortes, C., A tyrosine residue essential for catalytic activity in aminopeptidase A, *Biochem. J.* 327, 883–889, 1997.
9. Pal, J.K., Bera, S.K., and Ghosh, S.K., Acetylation of α-crystallin with *N*-acetylimidazole and its influence upon the native aggregate and subunit reassembly, *Curr. Eye Res.* 19, 358–367, 1999.
10. Zhang, F., Gao, J., Weng, J. et al., Structural and functional differentiation of three groups of tyrosine residues by acetylation of *N*-acetylimidazole in manganese stabilizing protein, *Biochemistry* 44, 719–725, 2005.
11. Zeitler, H.J. and Kulitz, M., Improved preparation and structural elucidation of the tryptophanyl cleavage reagent 2-(2′-nitro-phenylsulfenyl)-3-methyl-3-bromoindolenine (BNPS-skatole), *J. Clin. Chem. Clin. Biochem.* 16, 669–674, 1978.
12. Russell, J., Katzhendler, J., Kowalski, K. et al., The single tryptophan residue of human placental lactogen. Effects of modification and cleavage on biologic activity and protein conformation, *J. Biol. Chem.* 256, 304–307, 1981.

13. Xue, H., Xue, Y., Doublie, S., and Carter, C.W., Jr., Chemical modification of *Bacillus subtilis* tryptophanyl-tRNA synthetase, *Biochem. Cell Biol.* 75, 709–715, 1997.
14. Rahali, V. and Gueguen, J., Chemical cleavage of bovine β-lactoglobulin by BNPS-skatole for preparative purposes: Comparative study of hydrolytic procedures and peptide characterization, *J. Protein Chem.* 18, 1–12, 1999.
15. Kibbey, M.M., Jameson, M.J., Eaton, E.M., and Rosenzweig, S.A., Insulin-like growth factor binding protein-2: Contributions of the C-terminal domain to insulin-like growth factor-1 binding, *Mol. Pharmcol.* 69, 833–845, 2006.
16. Horn, A., Vandenberg, C.A., and Lange, K., Statistical analysis of single sodium channels. Effects of N-bromoacetamide, *Biophys. J.* 45, 323–335, 1984.
17. Pallotta, B.S., N-Bromosuccinimide removes a calcium-dependent component of channel opening from calcium-activated potassium channels in rat skeletal muscle, *J. Gen. Physiol.* 86, 601–611, 1985.
18. Huang, R.C., Novel pharmacological properties of transient potassium currents in central neurons revealed by *N*-bromosuccinimide and other chemical modifiers, *Mol. Pharmacol.* 48, 451–458, 1995.
19. Qi, X., Lee, S.H., and Kwon, J.Y., Aminobromination of unsaturated phosphonates, *J. Org. Chem.* 68, 9140–9143, 2003.
20. Wang, H., Vath, G.M., Gleason, K.J. et al., Probing the mechanism of hamster arylamine *N*-acetyltransferase 2 acetylation by active site modification, site-directed mutagenesis, and pre-steady state and steady state kinetic studies, *Biochemistry* 43, 8234–8246, 2004.
21. Glick, D.M., Goren, H.J., and Barnard, E.A., Concurrent bromoacetate reaction at histidine and methionine residues in ribonuclease, *Biochem. J.* 102, 7c–10c, 1967.
22. Lennette, E.P. and Plapp, B.V., Kinetics of carboxymethylation of histidine hydantoin, *Biochemistry* 18, 3933–3988, 1979.
23. Shapiro, R., Strydom, D.J., Weremowicz, S., and Vallee, B.L., Sites of modification of human angiogenin by bromoacetate at pH 5.5, *Biochem. Biophys. Res. Commun.* 156, 530–536, 1988.
24. Schelte, P., Boeckler, C., Frisch, B., and Schuber, F., Differential reactivity of maleimide and bromoacetyl functions with thiols: Application to the preparation of liposomal diepitope constructs, *Bioconjug. Chem.* 11, 118–123, 2000.
25. Chatani, E., Tanimizu, N., Ueno, H., and Hayashi, R., Structural and functional changes in bovine pancreatic ribonuclease A by the replacement of Phe120 with other hydrophobic residues, *J. Biochem.* 129, 917–922, 2001.
26. Okazaki, K., Yamada, H., and Imoto, T., A convenient *S*-2-aminoethylation of cysteinyl residues in reduced proteins, *Anal. Biochem.* 149, 516–520, 1985.
27. Planas, A. and Kirsch, J.F., Sequential protection-modification method for selective sulfhydryl group derivatization in proteins having more than one cysteine, *Protein Eng.* 3, 625–628, 1990.
28. Bochar, D.A., Tabernero, L., Stauffacher, C.V., and Rodwell, V.W., Aminoethylcysteine can replace the function of the essential active site lysine of *Pseudomonas mevalonii* 3-hydroxy-3-methylglutaryl coenzyme A reductase, *Biochemistry* 38, 8879–8883, 1999.
29. Thevis, M., Ogorzalek Loo, R.R., and Loo, J.A., In-gel derivatization of proteins for cysteine-specific cleavages and their analysis by mass spectrometry, *J. Proteome Res.* 2, 163–172, 2003.
30. Hopkins, C.E., Hernandez, G., Lee, J.P., and Tolan, D.R., Aminoethylation in model peptides reveals conditions for maximizing thiol specificity, *Arch. Biochem. Biophys.* 443, 1–10, 2005.
31. McAllister, K.A., Marrone, L., and Clarke, A.J., The role of tryptophan residues in substrate binding to catalytic domains A and B of xylanase C from *Fibrobacter succinogenes* S85, *Biochim. Biophys. Acta* 1400, 342–352, 2000.

32. Takita, T., Nakagoshi, M., Inouye, K., and Tonomura, B., Changes observed in the amino acid activation reaction, *J. Mol. Biol.* 325, 677–685, 2003.
33. Sargisova, Y., Pierfederici, F.M., Scire, A. et al., Computational, spectroscopic, and resonant mirror biosensor analysis of the interaction of adrenodoxin with native and tryptophan-modified NADPH-adrenodoxin reductase, *Proteins* 57, 302–310, 2004.
34. Faridmoayer, A. and Scaman, C.H., Binding residues and catalytic domain of soluble *Saccharomyces cerevisiae* processing α-glucosidase I, *Glycobiology* 15, 1341–1348, 2005.
35. Kumar, A., Tyagi, N.K., and Kinne, R.K., Ligand-mediated and conformational changes and positioning of tryptophans in reconstituted human sodium/*d*-glucose cotransporter (hSGLT1) probed by tryptophan fluorescence, *Biophys. Chem.* 127, 69077, 2007.
36. Leitner, A. and Lindner, W., Functional probing of arginine residues in proteins using mass spectrometry and an arginine-specific covalent tagging concept, *Anal. Chem.* 77, 4481–4488, 2005.
37. Foettinger, A., Leitner, A., and Lindner, W., Solid-phase capture and release of arginine peptides by selective tagging and boronate affinity chromatography, *J. Chromatog. A.* 1079, 187–196, 2005.
38. Saraiva, M.A., Borges, C.M., and Florencio, M.H., Reactions of a modified lysine with aldehydic and diketonic dicarbonyl compounds: An electrospray mass spectrometry structure/activity study, *J. Mass Spectrom.* 41, 216–228, 2006.
39. Holm, A., Rise, F., Sessler, N. et al., Specific modification of peptide-bound citrulline residues, *Anal. Biochem.* 352, 68–76, 2006.
40. Leitner, A., Amon, S., Rizzi, A., and Lindner, W., Use of the arginine-specific butanedione/phenylboronic acid tag for analysis of peptides and protein digests using matrix-assisted laser desorption/ionization mass spectrometry, *Rapid Commun. Mass Spectrom.* 21, 1321–1330, 2007.
41. de Cuyper, M., Hodenius, M., Lacava, E.G. et al., Attachment of water-soluble proteins to the surface of (magnetizable) phospholipid colloids via NeutraAvidin-derivatized phospholipids, *J. Colloid Interface Sci.* 245, 274–280, 2002.
42. Hosseinkhani, S., Ranjbar, B., Haderi-Manesh, H., and Nemat-Gorgani, M., Chemical modification of glucose oxidase: Possible formation of molten globule-like intermediate structure, *FEBS Lett.* 561, 213–216, 2004.
43. Habibib, A.E., Khajeh, K., and Nemat-Gorgani, M., Chemical modification of lysine residues in *Bacillus licheniformis* α-amylase: Conversion of an endo- to an exo-type enzyme, *J. Biochem. Mol. Biol.* 37, 642–647, 2004.
44. Dai, W., Sato, S., Ishizaki, M. et al., A new antigen retrieval method using citraconic anhydride for immunoelectron microscopy: Localization of surfactant pro-protein C (proSP-C) in the type II alveolar epithelial cells, *J. Submicrosc. Cytol. Pathol.* 36, 219–224, 2004.
45. Mossavarali, S., Hosseinkhani, S. Ranjbar, B., and Miroliaei, M., Stepwise modification of lysine residues of glucose oxidase with citraconic anhydride, *Int. J. Biol. Macromol.* 39, 192–196, 2006.
46. Griffey, R.H., Scavini, M., and Eaton, R.P., Characterization of the carbamino adducts of insulin, *Biophys. J.* 54, 295–300, 1988.
47. Kraus, L.M., Miyamura, S., Pecha, B.R., and Kraus, A.F., Jr., Carbamoylation of hemoglobin in uremic patients determined by antibody specific for homocitrulline (carbamoylated ε-*N*-lysine), *Mol. Immunol.* 28, 459–463, 1991.
48. Reyes, A.M., Bravo, M., Ludwig, H. et al., Modification of Cys-128 of pig kidney fructose 1,6-bisphosphatase with different thiol reagents: Size dependent effect on the substrate and fructose-2,6-bisphosphate interaction, *J. Protein Chem.* 12, 159–168, 1993.
49. Lapko, V.N., Smith, D.L., and Smith, J.B., Methylation and carbamylation of human gamma-crystallins, *Protein Sci.* 12, 1762–1774, 2003.

50. Jaisson, S., Lorimier, S., Ricard-Blum, S. et al., Impact of carbamylation of type I collagen conformational structure and its ability to activate human polymorphonuclear neutrophils, *Chem. Biol.* 13, 149–159, 2006.
51. Chang, L.S., Wu. P.F., Liou, J.C. et al., Chemical modification of arginine residues of *Notechis scutatus scutatus* notexin, *Toxicon* 44, 491–497, 2004.
52. Masuda, T., Ide, N., and Kitabatake, N., Structure-sweetness relationship in egg white lysozme: Role of lysine and arginine residues on the elication of lysozyme sweetness, *Chem. Senses* 30, 667–681, 2005.
53. Herrman, A., Svangard, E., Claeson, P. et al., Key role of glutamic acid for the cytotoxic activity of the cyclotide cycloviolacin O_2, *Chem. Mol. Life Sci.* 63, 235–245, 2006.
54. Schwartz, M.P., Barlow, D.E., Russell, J.N., Jr. et al., Semiconductor surface-induced 1,3-hydrogen shift: The role of covalent vs zwitterionic character, *J. Am. Chem. Soc.* 128, 11054–11061, 2006.
55. Daniel, J., Oh, T.J., Lee, C.M., and Kolattukudy, P.E., AccD6, a member of the Fas II locus, is a functional carboxyltranferase subunit of the acyl-coenzyme A carboxylase in *Mycobacterium tuberculosis*, *J. Bacteriol.* 189, 911–917, 2007.
56. Azzi, A., Casey, R.P., and Nalecz, M.J., The effect of *N,N′*-dicyclohexylcarbodiimide on enzymes of bioenergetic relevance, *Biochim. Biophys. Acta* 768, 209–226, 1984.
57. Dimroth, P., Matthey, U., and Kaim, G., Critical evaluation of the one- versus the two-channel model for the operation of the ATP synthase's F(o) motor, *Biochim. Biophys. Acta* 14589, 506–513, 2000.
58. Aresta, M., Dibenedetto, A., Fracchiolla, E. et al., Mechanism of formation of organic carbonates from aliphatic alcohols and carbon dioxide under mild conditions promoted by carbodiimides. DFT calculation an experimental study, *J. Org. Chem.* 70, 6177–6186, 2005.
59. Vgenopoulou, L., Gemperli, A.C., and Steuber, J., Specific modification of a Na^+ binding site in NADH: Quinine oxidoreductase from *Klebsiella pneumonia* with dicyclohexylcarbodiimide, *J. Bacteriol.* 188, 3264–3272, 2006.
60. Ogino, S., Sato, Y., Yamamoto, G. et al., Relation of the number of cross-links and mechanical properties of multi-walled carbon nanotube films formed by a dehydration condensation reaction, *J. Phys. Chem. B Condens. Matter Mater. Surf. Interfaces Biophys.* 110, 23159–23163, 2006.
61. Follmer, C. and Carlini, C.R., Effect of chemical modification of histidine on the copper-induced oligomerization of jack bean urease, *Arch. Biochem. Biophys.* 435, 15–20, 2005.
62. Colleluori, D.M., Reczkowski, R.S., Emig, F.A. et al., Probing the role of hyper-reactive histidine residue of arginase, *Arch. Biochem. Biophys.* 444, 15–26, 2005.
63. Runquist, J.A., and Miziorko, H., Functional contribution of a conserved mobile loop histidine of phosphoribulokinase, *Protein Sci.* 15, 837–842, 2006.
64. Wang, X.Y., Sun, M.L., Zhao, D.M., and Wang, M., Kinetics of inactivation of phytase (phy A) during modification of histidine residue by IAA and DEP, *Protein Pept. Lett.* 13, 565–570, 2006.
65. Nakanishi, N., Takeuchi, F., Okamoto, H. et al., Characterization of heme-coordinating histidyl residues of cytochrome b5 based on the reactivity with diethylpyrocarbonate: A mechanism for the opening of axial imidazole rings, *J. Biochem.* 140, 561–571, 2006.
66. Ghosh, M.K., Kildsig, D.O., and Mitra, A.K., Preparation and characterization of methotrexate-immunoglobulin conjugates, *Drug. Des. Deliv.* 4, 13–25, 1989.
67. Shen, X., Lagergard, T., Yang, Y. et al., Preparation and preclinical evaluation of experimental group B streptococcus type III polysaccharide-cholera toxin B subunit conjugate vaccine for intranasal immunization, *Vaccine* 19, 850–861, 2000.

68. Hafemann, B., Ghofrani, K., Gattner, H.G. et al., Cross-linking by 1-ethyl-3-(3-dimethylaminopropyl)-carbodiimide (EDC) of a collagen/elastin membrane meant to be used as a dermal substitute: Effects on physical, biochemical and biological features *in vitro*, *J. Mater. Sci. Mater. Med.* 12, 437–446, 2001.
69. Zhang, R., Tang, M., Bowyer, A. et al., A novel pH- and ionic-strength-sensitive carboxy methyl dextran hydrogel, *Biomaterials* 26, 4677–483, 2005.
70. Li, D., He, Q., Cui, Y. et al., Immobilization of glucose oxidase onto gold nanoparticles with enhanced thermostability, *Biochem. Biophys. Res. Commun.* 355, 488–493, 2007.
71. Owusu-Apenten, R., Colorimetric analysis of protein sulfhydryl groups in milk: Applications and processing effects, *Crit. Rev. Food Sci. Nutr.* 45, 1–23, 2005.
72. Laragione, T., Gianazza, E., Tonelli, R. et al., Regulation of redox-sensitive exofacial protein thiols in CHO cells, *Biol. Chem.* 387, 1371–1376, 2006.
73. Landino, L.M., Koumas, M.T., Mason, C.E., and Alston, J.A., Ascorbic acid reduction of microtubule protein disulfides and its relevance to protein *S*-nitrosylation assays, *Biochem. Biophys. Res. Commun.* 340, 347–352, 2006.
74. de Araujo, A.D., Palomo, J.M., Cramer, J. et al., Diels-Alder ligation of peptides and proteins, *Chemistry* 12, 6095–6109, 2006.
75. Cliff, M.J., Alizadeh, T., Jelinska, C. et al., A thiol labelling competition experiment as a probe for sidechain packing in the kinetic folding intermediate of N-PGK, *J. Mol. Biol.* 364, 810–823, 2006.
76. Mollinedo, F., Calafat, J., Janssen, H. et al., Combinatorial SNARE complexes modulate the secretion of cytoplasmic granules in human neutrophils, *J. Immunol.* 177, 2831–2841, 2006.
77. Rogers, L.K., Leinweber, B.L., and Smith, C.V., Detection of reversible protein thiol modifications in tissue, *Anal. Biochem.* 258, 171–184, 2006.
78. Kurono, S., Kurono, T., Komori, N. et al., Quantitative proteome analysis using D-labeled *N*-ethylmaleimide and ^{13}C-labeled iodoacetanilide by matrix-assisted laser desorption/ionization time-of-flight mass spectrometry, *Bioorg. Med. Chem.* 14, 8197–8209, 2006.
79. Togneri, J., Cheng, Y.S., Munson, M. et al., Specific SNARE complex binding mode of the Sec1/Munc-18 protein, Seclp, *Proc. Natl. Acad. Sci. USA* 103, 17730–17735, 2006.
80. Yan, L.J., Yang, S.H., Shu, H. et al., Histochemical staining and quantification of dihydrolipoamide dehydrogenase diaphorase activity using blue native PAGE, *Electrophoresis*, 28, 1036–1045, 2007.
81. Diaper, C.M., Sutherland, A., Pillai, B. et al., The stereoselective synthesis of aziridine analogues of diaminopimelic acid (DAP) and their interactions with dap epimerase, *Org. Biomol. Chem.* 3, 4402–4411, 2005.
82. Ponte-Sucre, A., Vicik, R., Schultheis, M. et al., Aziridine-2,3-dicarboxylates, peptidomimetic cysteine protease inhibitors with antileishmanial activity, *Antimicrob. Agents Chemother.* 50, 2439–2447, 2006.
83. Vicik, R., Busemann, M., Gelaus, C. et al., Aziridine-based inhibitors of cathepsin L: Synthesis, inhibition activity, and docking studies, *ChemMedChem* 1, 1126–1141, 2006.
84. Vicik, R., Buseman, M., Bauman, K., and Schirmeister, T., Inhibitors of cysteine proteases, *Curr. Top. Med. Chem.* 6, 331–353, 2006.
85. Mladenovic, M., Schirmeister, T., Thiel, S. et al., The importance of the active site histidine for the activity of epoxide- or aziridine-based inhibitors of cysteine proteases, *ChemMed. Chem.* 2, 120–128, 2007.
86. Brubaker, G., Peng, D.Q., Somerlot, B. et al., Apolipoprotein A-1 lysine modification: Effects on helical content, lipid binding and cholesterol acceptor activity, *Biochim. Biophys. Acta* 1761, 64–72, 2006.
87. Fu, Q. and Li, L., Fragmentation of peptides with *N*-terminal dimethylation and imine/methylol adduction at the tryptophan side-chain, *J. Am. Soc. Mass Spectrom.* 17, 859–866, 2006.

88. Xu, J. and Bowden, E.F., Determination of the orientation of adsorbed cytochrome c on carboxyalkanethiol self-assembled monolayers by *in situ* differential modification, *J. Am. Chem. Soc.* 128, 6813–6822, 2006.
89. Hsu, J.L., Huang, S.Y., and Chen, S.H., Dimethyl multiplexed labeling combined with microcolumn separation and MS analysis for time course study in proteomics, *Electrophoresis* 27, 3652–3660, 2006.
90. Walter, T.S., Meier, C., Assenberg, R. et al., Lysine methylation as a routine rescue strategy for protein crystallization, *Structure* 14, 1617–1622, 2006.
91. Torrance, L., Ziegler, A., Pittman, H. et al., Oriented immobilization of engineered single-chain antibodies to develop biosensors for virus detection, *J. Virol. Methods* 134, 164–170, 2006.
92. Li, Z.P., Duan, X.R., Liu, C.H., and Du, B.A., Selective determination of cysteine by resonance light scattering technique based on self-assembly of gold nanoparticles, *Anal. Biochem.* 351, 18–25, 2006.
93. Talib, J., Beck, J.L., and Ralph, S.F., A mass spectrometric investigation of the binding of gold antiarthritic agents and the metabolite $[Au(CN)_2]^-$ to human serum albumin, *J. Biol. Inorg. Chem.* 11, 559–570, 2006.
94. Gautier, C. and Burgi, T., Chiral *N*-isobutyryl-cysteine protected gold nanoparticles: Preparation, size selection, and optimal activity in the US-Vis and infrared, *J. Am. Chem. Soc.* 128, 11079–11087, 2006.
95. Urbina, R.D., Debaene, F., Jost, B. et al., Self-assembled small-molecule microarrays for protease screening and profiling, *ChemBioChem* 7, 1790–1797, 2006.
96. Strohalm, M., Kodicek, M., and Pechar, M., Tryptophan modification by 2-hydroxy-5-nitrobenzyl bromide studied by MALDI-TOF mass spectrometry, *Biochem. Biophys. Res. Commun.* 312, 811–816, 2003.
97. Strohalm, M., Santrucek, J., Hynek, R., and Kodicek, M., Analysis of tryptophan surface accessibility in proteins by MALDI-TOF mass spectrometry, *Biochem. Biophys. Res. Commun.* 323, 1134–1138, 2004.
98. Jung, J.W., Kuk, J.H. Kim, K.Y. et al., Purification and characterization of exo-β-D-glucosaminidase from *Aspergillus fumigatus* S-26, *Protein Expr. Purif.* 45, 125–131, 2006.
99. Tashima, I., Yoshida, T., Asada, Y., and Ohmachi, T., Purification and characterization of a novel L-2-amino-Δ_2-thiazoline-4-carboxylic acid hydroase from *Pseudomonas* sp. strain ON-4a expressed in *E. coli*, *Appl. Microbiol. Biotechnol.* 72, 499–507, 2006.
100. Ma, S.F., Nishikawa, M., Yabe, Y. et al., Role of tyrosine and tryptophan in chemically modified serum albumin on its tissue distribution, *Biol. Pharm. Bull.* 29, 1926–1930, 2006.
101. Smith, G.P., Kinetics of amine modification of proteins, *Bioconjug. Chem.* 17, 501–506, 2006.
102. Adden, K., Gamble, L.J., Castner, D.G. et al., Phosphonic acid monolayers for binding of bioactive molecules to titanium surfaces, *Langmuir* 22, 8197–8204, 2006.
103. Noti, C., de Paz, J.L., Polito, L., and Seeberger, P.H., Preparation and use of microarrays containing synthetic heparin oligosaccharides for the rapid analysis of heparin-protein interactions, *Chemistry* 12, 8664–8686, 2006.
104. Kenawy, el-R., el-Newehy, M., Abdel-Hay, F., and Ottenbrite, R.M., A new degradable hydroxamate linkage for pH-controlled drug delivery, *Biomacromolecules* 8, 196–201, 2007.
105. Pandey, P., Singh, S.P., Arya, S.K. et al., Application of thiolated gold nanoparticles for the enhancement of glucose oxidase activity, *Langmuir* 23, 3333–3337, 2007.
106. Gauvreau, V., Chevalier, P., Vallieres, K. et al., Engineering surfaces for bioconjugation: Developing strategies and quantifying the extent of the reactions, *Bioconjug. Chem.* 15, 1146–1156, 2004.

107. Balthasar, S., Michaelis, K., Dinauer, N. et al., Preparation and characterization of antibody modified gelatin nanoparticles as drug carrier system for uptake in lymphocytes, *Biomaterials* 26, 2723–2732, 2005.
108. Kommareddy, S. and Amiji, M., Preparation and evaluation of thiol-modified gelatin nanoparticles for intracellular DNA delivery in response to glutathione, *Bioconjug. Chem.* 16, 1423–1432, 2005.
109. Langoth, N., Kalbacher, H., Schoffmann, G. et al., Thiolated chitosans: Design and *in vivo* evaluation of a mucoadhesive buccal peptide drug delivery system, *Pharm. Res.* 23, 573–579, 2006.
110. Fowler, J.M., Stuart, M.C., and Wong, D.K., Self-assembled layer of thiolated protein G as an immunosensor scaffold, *Anal. Chem.* 79, 350–354, 2007.
111. Jao, S.C., English Ospina, S.M., Berdis, A.J. et al., Computational and mutational analysis of human glutaredoxin (thioltransferase): Probing the molecular basis of the low pKa of cysteine 22 and its role in catalysis, *Biochemistry* 45, 4785–4796, 2006.
112. Talib, J., Beck, J.L., and Ralph, S.F., A mass spectrometric investigation of the binding of gold antiarthritic agents and the metabolite $[Au(CN)_2]^-$ to human serum albumin, *J. Biol. Inorg. Chem.* 11, 559–570, 2006.
113. Rogers, L.K., Leinweber, B.L., and Smith, C.V., Detection of reversible protein thiol modification in tissues, *Anal. Biochem.* 358, 171–184, 2006.
114. Kurono, S., Kurono, T., Komori, N. et al., Quantitative proteome analysis using D-labeled *N*-ethylmaleimide and ^{13}C-labeled iodoacetamide by matrix-assisted laser desorption/ionization time-of-flight mass spectrometer, *Bioorg. Med. Chem.* 14, 8197–8209, 2006.
115. Yang, E. and Attygalle, A.B., LC/MS characterization of undesired products formed during iodoacetamide derivatization of sulfhydryl groups of peptides, *J. Mass Spectrom.* 42, 233–243, 2007.
116. Meng, T.C., Hsu, S.F., and Tonka, N.K., Development of a modified in-gel assay to identify protein tyrosine phosphatases that are oxidized and inactivated *in vivo*, *Methods* 35, 28–36, 2005.
117. Morty, R.E., Shih, A.Y., Fulop, V., and Andrews, N.W. Identification of the reactive cysteine residues in oligopeptidase B from *Trypanosoma brucci, FEBS Lett.* 579, 2191–2196, 2005.
118. Hasegawa, G., Kikuchi, M., Kobayashi, Y., and Saito, Y., Synthesis and characterization of a novel reagent containing dansyl group, which specifically alkylates sulfhydryl group: An example of application for protein chemistry, *J. Biochem. Biophys. Methods* 63, 33–42, 2005.
119. Atsriku, C, Scott, G.K., Benz, C.C., and Baldwin, M.A., Reactivity of zinc finger cysteines: Chemical modification with labile zinc fingers in estrogen receptors, *J. Am. Soc. Mass Spectrom.* 16, 2017–2026, 2005.
120. Chao, C.C., Chelius, D., Zhang, T. et al., Insight into the virulence of *Rickettsia prowazekii* by proteomic analysis and comparison with an avirulent strain, *Biochim. Biophys. Acta* 1774, 373–381, 2007.
121. Grigorian, A.L., Bustamante, J.J., Hernandez, P. et al., Extraordinary stable disulfide-linked homodimer of human growth hormone, *Protein Sci.* 14, 902–913, 2005.
122. Hedberg, J.J., Bjerneld, E.J., Cetinkaya, S. et al., A simplified 2-D electrophoresis protocol with the aid of an organic disulfide, *Proteomics* 5, 3088–3096, 2005.
123. Wojcik, A., Naumov, S., Marciniak, B., and Brede, O., Repair reactions of pyrimidine-derived radicals by aliphatic thiols, *J. Phys. Chem. B. Matter. Mater. Surf. Interact. Biophys.* 110, 12738–12748, 2006.
124. Okun, I., Malarchuk, S., Dubrovskaya, E. et al., Screening for caspace-3 inhibitors: Effect of a reducing agent on identified hit chemotypes, *J. Biomol. Screen.* 11, 694–703, 2006.

125. Okado-Matsumoto, A., Guan, E., and Fridovich, I., Modification of cysteine 111 in human Cu,Zn-superoxide dismutase, *Free. Radic. Biol. Med.* 41, 1837–1846, 2006.
126. Hisatome, I., Kurata, Y., Sasaki, N. et al., Block of sodium channels by divalent mercury: Role of specific cysteinyl residues in the P-loop region, *Biophys. J.* 79, 1336–1345, 2000.
127. Kinne-Saffran, E. and Kinne, R.K., Inhibition by mercuric chloride of Na-K-2Cl cotransport activity in rectal gland plasma membrane vesicles isolated from *Squalus scanthias*, *Biochim. Biophys. Acta* 1510, 442–451, 2001.
128. Taoka, S., Green, S.L., Loehr, T.M., and Banerjee, R., Mercuric chloride-induced spin or ligation state changes in ferric or ferrous human cystathionine β-synthase inhibit enzyme activity, *J. Inorg. Biochem.* 87, 253–259, 2001.
129. Alencar, J.L., Lobysheva, I., Geffard, M. et al., Role of *S*-nitrosylation of cysteine residues in long-lasting inhibitory effect of nitric oxide on arterial tone, *Mol. Pharmacol.* 63, 1148–1158, 2003.
130. Durand, A., Giardina, T., Villard, C. et al., Rat kidney acylase I: Further characterization and mutation studies on the involvement of Glu147 in the catalytic process, *Biochimie* 85, 953–962, 2003.
131. Liu, X., Alexander, C., Serrano, J. et al., Variable reactivity of an engineered cysteine at position 338 in cystic fibrosis transmembrane conductance regulator reflects different chemical states of the thiol, *J. Biol. Chem.* 281, 8275–8285, 2006.
132. Audia, J.P., Roberts, R.A., and Winkler, H.H., Cysteine-scanning mutagenesis and thiol modification of the *Rickettsia prowazerkii* ATP/ADP translocase: Characterization of the TMs IV-VII and IX-XII and their accessibility to the aqueous translocation pathway, *Biochemistry* 45, 2648–2656, 2006.
133. Tombolato, F., Ferrarini, A., and Freed, J.H., Modeling the effects of structure and dynamics of the nitroxide side chain on the ESR spectra of spin-labeled proteins, *J. Phys. Chem. B. Condens. Matter Mater. Surf. Interfaces Biophys.* 110, 26260–26271, 2006.
134. Karala, A.R., and Ruddock, L.W., Does *S*-methyl methanethiosulfonate trap the thiol-disulfide state of proteins?, *Antioxid. Redox. Signal* 9, 527–531, 2007.
135. Thonon, D., Jacques, v., and Desreux, J.F., A gadolinium triacetic monoamide DOTA derivative with a methanesulfonate anchor group. Relaxivity properties and conjugation with albumin and thiolated particles, *Contrast Media Mol. Imaging* 2, 24–34, 2007.
136. Ishikawa, Y., Yamamoto, Y., Otsubo, M. et al., Chemical modification of amine groups on PS II protein(s) retards photoassembly of the photosynthetic water-oxidizing complex, *Biochemistry* 41, 1972–1980, 2002.
137. Shortreed, M.R., Lamos, S.M., Frey, B.L., Ionizable isotopic labeling reagent for relative quantification of amine metabolites by mass spectrometry, *Anal. Chem.* 78, 6398–6403, 2006.
138. Poon, S.F., Stock, N., Payne, N.K. et al., Novel approach to pro-drugs of lactones: Water soluble imidate and ortho-ester derivatives of a furanone-based COX-2 selective inhibitor, *Bioorg. Med. Chem. Lett.* 15, 2259–2263, 2005.
139. Xu, J., Degraw, A.J., Duckworth, B.P. et al., Synthesis and reactivity of 6,7-dihydrogeranylazides: Reagents for primary amine incorporation into peptides and subsequent Staudinger ligation, *Chem. Biol. Drug Des.* 68, 85–96, 2006.
140. Takaku, H., Sato, J., Ishida, H.K. et al., A chemical synthesis of UDP-LacNAc and its regioisomer for finding "oligonucleotide transferases," *Glycoconj. J.* 23, 565–573, 2006.
141. Leane, M.M., Nankervis, R., Smith, A., and Illum, L., Use of the ninhydrin assay to measure the release of chitosan from oral solid dosage forms, *Int. J. Pharm.* 271, 241–249, 2004.
142. Drochioiu, G., Mangalagiu, I., Avram, E. et al., Cyanide reaction with ninhydrin: Elucidation of reaction and interference mechanisms, *Anal. Sci.* 20, 1443–1447, 2004.

143. Hansen, D.B. and Joullie, M.M., The development of novel ninhydrin analogues, *Chem. Soc. Rev.* 34, 408–417, 2005.
144. Wu, Y., Hussain, M., and Fassihi, R., Development of a simple analytical methodology for determination of glucosamine release from modified release matrix tablets, *J. Pharm. Biomed. Anal.* 38, 263–269, 2005.
145. Lipscomb, I.P., Pinchin, H.E., and Collin, R., The sensitivity of approved ninhydrin and biuret tests in the assessment of protein contamination on surgical steel as an aid to prevent iatrogenic prion transmission, *J. Hosp. Infect.* 64, 288–292, 2006.
146. Marche, P., Girma, J.P., Morgat, J.L., and Fromageot, P., Specific tritiation of indole derivatives by catalytic desulfenylation. Application to the labelling of tryptophan-containing peptides, *Eur. J. Biochem.* 50, 375–382, 1975.
147. Sasagawa, T. ,Titani, K., and Walsh, K.A., Selective isolation of tryptophan-containing peptides by hydrophobicity modulation, *Anal. Biochem.* 134, 224–229, 1983.
148. Hassani, O., Mansuella, P., Cestele, S. et al., Role of lysine and tryptophan residues in the biological activity of toxin VII (Ts γ) from the scorpion *Tityus serrulatus*, *Eur. J. Biochem.* 260, 76–86, 1999.
149. Ou, K., Kesuma, D., Ganesan, K. et al., Quantitative profiling of drug-associated proteomic alterations by combined 2-nitrophenylsulfenyl chloride (NBS) isotope labeling and 2DE/MS identification, *J. Proteome Res.* 5, 2194–2206, 2006.
150. Matsunaga, H. and Haginaka, J., Investigation of chiral recognition mechanism on chicken α(1)-acid glycoprotein using separation system, *J. Chromatog. A* 1106, 124–130, 2006.
151. Matthiesen, R., Bauw, G., and Welinder, K.G., Use of performic acid oxidation to expand the mass distribution of tryptic peptides, *Anal. Chem.* 76, 6848–6852, 2004.
152. Dai, J., Wang, J., Zhang, Y. et al., Enrichment and identification of cysteine-containing peptides from tryptic digests of performic oxidized proteins by strong cation exchange LC an MALDI-TOF/TOF MS, *Anal. Chem.* 77, 7594–7604, 2005.
153. Cao, J., Wijaya, R., and Leroy, F., Unzipping the cuticle of the human hair shaft to obtain micron/nano keratin filaments, *Biopolymers* 83, 614–618, 2006.
154. Kulkarni, A.D, Rai, D., Bartolotti, L.F., and Pathak, R.K., Interaction of performic acid with water molecules: A first-principles study, *J. Phys. Chem. A Mol. Spectros. Kinet Environ. Gen. Theory* 110, 11855–11861, 2006.
155. Bosch, L., Algeria, A., and Farre, R., Amino acid contents of infant foods, *Int. J. Food Sci. Nutr.* 57, 212–218, 2006.
156. Johans, M., Milanesi, E., Franck, M. et al., Modification of permeability transition pore arginine(s) by phenylglyoxal derivatives in isolated mitochondria and mammalian cells. Structure-function relationship of arginine ligands, *J. Biol. Chem.* 280, 12130–12136, 2005.
157. Saraiva, M.A., Borges, C.M., and Florencio, M.H., Reactions of a modified lysine with aldehydic and diketonic dicarbonyl compounds: An electrospray mass spectrometry structure/activity study, *J. Mass Spectrom.* 41, 216–228, 2006.
158. Takazaki, S., Abe, Y., Kang, D. et al., The functional role of arginine 901 at the *C*-terminus of the human anion transporter band 3 protein, *J. Biochem.* 139, 903–912, 2006.
159. Greig, N., Wyllie, S., Vickers, T.J., and Fairlamb, A.N., Trypanothione-dependent glyoxylase I in *Trypanosoma cruzi*, *Biochem. J.* 400, 217–223, 2006.
160. Ye, M. and English, A.M., Binding of polyaminocarboxylate chelators to the active-site copper inhibits the GSNO-reductase activity but not the superoxide dismutase activity of Cu, Zn-superoxide dismutase, *Biochemistry* 45, 12723–12732, 2006.
161. Peelen, D. and Smith, L.M., Immobilization of the amine-modified oligonucleotides on aldehyde-terminated alkanethiol monolayers on gold, *Langmuir* 21, 266–271, 2005.

162. Kim, H.S. and Wainer, I.W., The covalent immobilization of microsomal uridine diphospho-glucuronosyltransferase (UDPGT): Initial synthesis and characterization of an UDPGT immobilized enzyme reactor for the on-line study of glucuronidation, *J. Chromatog. B Anal. Technol. Biomed. Life Sci.* 823, 158–166, 2005.
163. Wildsmith, K.R., Albert, C.J., Hsu, F.F. et al., Myeloperoxidase-derived 2-chlorohexadecanal forms Schiff bases with primary amines of ethanolamine glycerophospholipids and lysine, *Chem. Phys. Lipids* 139, 157–170, 2006.
164. Mirzaei, H. and Regnier, F., Enrichment of carbonylated peptides using Girard P reagent and strong cation exchange chromatography, *Anal. Chem.* 78, 770–778, 2006.
165. Hsu, J.L., Huang, S.Y., and Chen, S.H., Dimethyl multiplexed labeling combined with microcolumn separation and MS analysis for time course study in proteomics, *Electrophoresis* 27, 3652–3660, 2006.
166. Damodaran, S., Estimation of disulfide bonds using 2-nitro-5-thiosulfobenzoic acid: Limitations, *Anal. Biochem.* 145, 200–204, 1985.
167. Martin de Llano, J.J. and Gaviulanes, J.G., Increased electrophoretic mobility of sodium sulfite-treated jack bean urease, *Electrophoresis* 13, 300–304, 1992.
168. Emerson, D. and Ghiorse, W.C., Role of disulfide bonds in maintaining the structural integrity of the sheath of the *Leptothrix discophora* SP-6, *J. Bacteriol.* 175, 7819–7827, 1993.
169. Mukhopadhyay, A., Reversible protection of disulfide bonds followed by oxidative folding render recombinant hCGβ highly immunogenic, *Vaccine* 18, 1802–1810, 2000.
170. Raftery, M.J., Selective detection of thiosulfate-containing peptides using tandem mass spectrometry, *Rapid Commun. Mass Spectrom.* 19, 674–682, 2005.
171. Sakoh, M., Okazaki, T., Nagaoka, Y., and Asami, K., *N*-terminal insertion of alamethicin in channel formation studied using its covalent dimer *N*-terminally linked by disulfide bond, *Biochim. Biophys. Acta.* 1612, 117–121, 2003.
172. Tie, J.K., Jin, D.Y., Loiselle, D.R. et al., Chemical modification of cysteine residues is a misleading indicator of their status as active site residues in the vitamin K-dependent γ-glutamyl carboxylation reaction, *J. Biol. Chem.* 279, 54079–54087, 2004.
173. Voslar, M., Matejka, P., and Schreiber, I., Oscillatory reactions involving hydrogen peroxide and thiosulfate-kinetics of the oxidation of tetrathionate by hydrogen peroxide, *Inorg. Chem.* 45, 2824–2834, 2006.
174. Hahn, S.K., Kim, J.S., and Shimoobouji, T., Injectable hyaluronic acid microhydrogels for controlled release formulation of erythropoietin, *J. Biomed. Mater. Res. A.* 80, 916–924, 2007.
175. Hahn, S.K., Park, J.R., Tomimatsu, T., and Shimoboji, T., Synthesis and degradation test of hyaluronic acid hydrogels, *Int. J. Biol. Macromol.* 40, 374–380, 2007.
176. Cline, D.J., Redding, S.E., Brohawn, S.G. et al., New water-soluble phosphines as reductants of peptide and protein disulfide bonds: Reactivity and membrane permeability, *Biochemistry* 43, 15195–15203, 2004.
177. Xu, G., Kiselar, J., He, Q., and Chance, M.R., Secondary reactions and strategies to improve quantitative protein footprinting, *Anal. Chem.* 77, 3029–3037, 2005.
178. Willis, M.S., Hogan, J.K., Prabhakar, P. et al., Investigation of protein refolding using a fractional factorial screen: A study of reagent effects and interactions, *Protein Sci.* 14, 1818–1826, 2005.
179. Valcu, C.M. and Schlink, K., Reduction of proteins during sample preparation and two-dimensional gel electrophoresis of woody plant samples, *Proteomics* 6, 1599–1605, 2006.
180. Rogers, L.K., Lienweber, B.L., and Smith, C.V., Detection of reversible thiol modifications in tissues, *Anal. Biochem.* 358, 171–184, 2006.
181. Santrucek, J., Strohalm, M., Kadlcik, V. et al., Tyrosine residue modification studied by MALDI-TOF mass spectrometry, *Biochem. Biophys. Res. Commun.* 323, 1151–1156, 2004.

182. Negrerie, M., Martin, J.L., and Njgiem, H.O., Functionality of nitrated acetylcholine receptor: The two-step formation of nitrotyrosines reveals their differential role in effector binding, *FEBS Lett.* 579, 2643–2647, 2005.
183. Carven, G.J. and Stern, L.J., Probing the ligand-induced conformational change in HLA-DR1 by selective chemical modification and mass spectrometric mapping, *Biochemistry* 44, 13625–13637, 2005.
184. Gruijthuijsen, Y.K., Grieshuber, I., Stocklinger, A. et al., Nitration enhances the allergic potential of proteins, *Int. Arch. Allergy Immunol.* 141, 265–275, 2006.
185. Ghesquiere, B., Goethals, M., van Damme, J. et al., Improved tandem mass spectrometric characterization of 3-nitrotyrosine sites in peptides, *Rapid Commun. Mass Spectrom.* 20, 2885–2893, 2006.
186. Korn, C., Scholz, S.R., Gimadutdinow, O. et al., Involvement of conserved histidine, lysine, and tyrosine residues in the mechanism of DNA cleavage by the caspase-3 activated DNase CAD, *Nucleic Acids Res.* 30, 1325–1332, 2002.
187. Metz, B., Jiskoot, W., Hennink, W.E. et al., Physicochemical and immunochemical techniques predict the quality of diphtheria toxoid vaccines, *Vaccine* 22, 156–167, 2003.
188. Lin, J.C., Chen, Q.X., Shi, Y. et al., The chemical modification of the essential groups of β-*N*-acetyl-D-glucosaminidase from *Turbo corutus Solander*, *IUBMB Life* 55, 547–552, 2003.
189. Jagtap, S. and Rao, M., Conformation and microenvironment of the active site of a low-molecular weight 1,4-β-D-glucan glucanohydrolase from an alkalothermophilic *Thermomonospora* sp.: Involvement of lysine and cysteine residues, *Biochem. Biophys. Res. Commun.* 347, 428–432, 2006.
190. Chang, L.S., Cheng, Y.C., and Chen, C.P., Modification of Lys-6 and Lys-65 affects the structural stability of Taiwan cobra phospholipase A_2, *Protein J.* 25, 127–134, 2006.
191. Bingham, J.P., Broxton, N.M., Livett, B.G. et al., Optimizing the connectivity in disulfide-rich peptides: α-conotoxin SII as a case study, *Anal. Biochem.* 338, 48–61, 2005.
192. Maeda, K., Finnie, C., and Svensson, B., Identification of thioredoxin h-reducible disulphides in proteomes by differential labelling of cysteines: Insight into recognition and regulation of proteins in barley seeds by thioredoxin h, *Proteomics* 5, 1634–1644, 2005.
193. Winnik, W.M., Continuous pH/salt gradient and peptide score for strong cation exchange chromatography in 2D-nano-LC/MS/MS peptide identification for proteomics, *Anal. Chem.* 77, 4991–4998, 2005.
194. Okado-Matsumoto, A., Guan, E., and Fridovich, I., Modification of cysteine 111 in human Cu, Zn-superoxide dismutase, *Free Radic. Biol. Med.* 41, 1837–1846, 2006.
195. Chowdhury, S.M., Munske, G.R., Ronald, R.C., and Bruce, J.E., Evaluation of low energy CID and ECD fragmentation behavior of mono-oxidized thio-ether bonds in peptides, *J. Am. Soc. Mass Spectrom.* 18, 493–501, 2007.
196. Salhany, J.M., Sloan, R.L., and Cordes, K.S., The carboxyl side chain of glutamate 681 interacts with a chloride binding modifier site that allosterically modulates the dimeric conformational state of band 3 (AE1). Implications for the mechanism of anion/proton cotransport, *Biochemistry* 42, 1589–1602, 2003.
197. Kosters, H.A. and de Jongh, H.H., Spectrophotometric tool for the determination of the total carboxylate content in proteins: Molar extinction coefficient of the enol ester from Woodward's reagent K reacted with protein carboxylates, *Anal. Chem.* 75, 2512–2516, 2003.
198. Carvajal, N., Uribe, E., Lopez, V., and Salas, M., Inactivation of human liver arginase by Woodward's reagent K: Evidence for reaction with His141, *Protein J.* 23, 179–183, 2004.
199. Jennings, M.L., Evidence for a second binding/transport site for chloride in erythrocyte anion transporter AE1 modified at glutamate 681, *Biophys. J.* 88, 2681–2691, 2005.

200. SinhaRoy, S., Banerjee, S., Ray, M., and Ray, S., Possible involvement of glutamic and/or aspartic residue(s) and requirement of mitochrondrial integrity for the protective effect of creatine against inhibition of cardiac mitochondrial respiration by methylglyoxal, *Mol. Cell. Biochem.* 271, 167–176, 2005.

Further Reading for Table 1.1

Antos, J.M. and Francis, M.B., Transition metal catalyzed methods for site-selective protein modification, *Curr. Opin. Chem. Biol.* 10, 253–262, 2006.

Kellam, B., de Bank, P.A., and Shakesheff, K.M., Chemical modification of mammalian cell surfaces, *Chem. Soc. Rev.* 32, 327–337, 2003.

Leitner, A. and Lindner, W., Chemistry meets proteomics: The use of chemical tagging reactions for MS-based proteomics, *Proteomics* 6, 5418–5434, 2006.

Lundblad, R.L., *Chemical Reagents for Protein Modification*, CRC Press, Boca Raton, FL, 2004.

Chapter References

1. Cheung, J.K. Raverkar, P.S., and Truskett, T.M., Analytical model for studying how environmental factors influence protein conformational stability in solution, *J. Chem. Phys.* 125, 224903, 2006.
2. Kostareva, I., Hung, F., and Campbell, C., Purification of antibody heteropolymers using hydrophobic interaction chromatography, *J. Chromatog.* 1177, 254–264, 2008.
3. Kallias, A., Bachmann, M., and Janke, W., Thermodynamics and kinetics of a Gō proteinlike heteropolymer model with two-state folding characteristics, *J. Chem. Phys.* 128, 055102, 2008.
4. Patel, B.A., Debenedetti, P.G., Stillinger, F.H., and Rossky, P.J., The effect of sequence on the conformational stability of a model heteropolymer in explicit water, *J. Chem. Phys.* 128, 175102, 2008.
5. Boström, M., Tavares, F.W., Finet, S. et al., Why forces between proteins follow different Hofmeister series for pK above and below pI, *Biophys. Chem.* 117, 217–224, 2005.
6. Xu, L.C., Vadillo-Rodriguez, V., and Logan, B.E., Residence time, loading force, pH, and ionic strength affect adhesion forces between colloids and biopolymer-coated surfaces, *Langmuir* 21, 7491–7500, 2005.
7. Sherrat, M.J., Baldock, C., Morgan, A., and Kielty, C.M., The morphology of adsorbed extracellular matrix assemblies is critically dependent on solution calcium concentration, *Matrix Biol.* 26, 156–166, 2007.
8. Hirano, A., Hamada, H., Okubo, T. et al., Correlation between thermal aggregation and stability of lysozyme with salts described by molar surface tension increment: An exceptional propensity of ammonium salts as aggregation suppressor, *Protein J.* 26, 423–433, 2007.
9. Ananthapadmanabhan, K.P., Lips, A., Vincent, C. et al., pH-induced alterations in stratum corneum properties, *Int. J. Cosmet. Sci.* 25, 103–112, 2003.
10. García-Moreno, B., Dwyer, J.J., GIttis, A.G. et al., Experimental measurement of the effective dielectric in the hydrophobic core of a protein, *Biophys. Chem.* 64, 211–224, 1997.
11. Hnízda, A., Šantrůček, J., Šanda, M. et al., Reactivity of histidine and lysine side-chains with diethylpyrocarbonate—A method to identify surface exposed residues in proteins, *J. Biochem. Biophys. Methods* 70, 1091–1097, 2008.
12. Richardson, G.M., The principle of formaldehyde, alcohol, and acetone titrations. With a discussion of the proof and implication of the zwitterionic conception, *Proc. Roy. Soc. B.* (London) 115, 121–141, 1934.

13. Duggan, E.L. and Schmidt, C.L.A., The dissociation of certain amino acids in dioxane-water mixtures, *Arch. Biochem.* 1, 453–471, 1943.
14. Canova-Davis, E. and Carpenter, F.H., Semisynthesis of insulin: Specific activation of arginine carboxyl group of the B chain of desoctapeptide—(B23–36)-insulin (Bovine), *Biochemistry* 20, 7053–7058, 1981.
15. Canova-Davis, E., Kessler, T.J., and Ling, V.T., Transpeptidation during the analytical proteolysis of proteins, *Anal. Biochem.* 196, 39–45, 1991.
16. Plapp, B.V., Application of affinity labeling for studying structure and function in enzymes, *Methods Enzymol.* 87, 469–499, 1982.
17. Myers, B. 2nd and Glazer, A.N., Spectroscopic studies of the exposure of tyrosine residues in proteins with special references to the subtilisins, *J. Biol. Chem.* 26, 412–419, 1971.
18. Skov, K., Hofmann, T., and Williams, G.R., The nitration of cytochrome c, *Can. J. Biochem.* 47, 750–752, 1969.
19. Battghyány, C., Souza, J.M., Durán, R. et al., Time course and site (s) of cytochrome c tyrosine nitration by peroxynitrite, *Biochemistry* 44, 8038–8046, 2005.
20. Huang, L., Jacob, R.J., Pegg, S.C. et al., Functional assignment of the 20 S proteasome from *Trypanosoma soma* using mass spectrometry and new bioinformatics approaches, *J. Biol. Chem.* 276, 28327–28339, 2001.
21. Payne, A.H. and Glish, G.L., Tandem mass spectrometry in quadrupole ion trap and ion cyclotron resonance mass spectrometers, *Methods Enzymol.* 402, 109–148, 2005.
22. Lin, D., Tabb, D.L., and Yates, J.R., III, Large-scale protein identification using mass spectrometry, *Biochim. Biophys. Acta* 1646, 1–10, 2003.
23. Bennett, K.L. et al., Rapid characterization of chemically-modified proteins by electrospray mass spectrometry, *Bioconjug. Chem.* 7, 16–22, 1996.
24. Kelleher, N.L., Zubarev, R.A., Bush, K. et al., Localization of labile posttranslational modifications by electron capture dissociation: The case of γ-carboxyglutamic acid, *Anal. Chem.* 71, 4250–4253, 1999.
25. Fligge, T.A., Kast, J., Bruns, S., and Przybylski, M., Direct monitoring of protein-chemical reactions by utilizing nanoelectrospray mass spectrometry, *J. Am. Soc. Mass Spectrom.* 10, 112–118, 1999.
26. Jahn, O., Hofmann, B., Brauns, O. et al. The use of multiple ion chromatograms in on-line HPLC-MS for the characterization of post-translational and chemical modifications of proteins, *Int. J. Mass Spectrom.* 214, 37–51, 2002.
27. Leite, J.F. and Cascio, M., Probing the topology of the glycine receptor by chemical modification coupled to mass spectrometry, *Biochemistry* 41, 6140–6148, 2002.
28. Fenaille, F., Guy, P.A., and Tabet, J.C., Study of protein modification by 4-hydroxy-2-nonenol and other short chain aldehydes analyzed by electrospray ionization tandem mass spectrometry, *J. Am. Soc. Mass Spectrom.* 14, 215–226, 2003.
29. Hakansson, S., Viljanen, J., and Broo, K.S., Programmed delivery of novel functional groups to the alpha class glutathione transferases, *Biochemistry* 42, 10260–10268, 2003.
30. Standing, K.G., Peptide and protein de novo sequencing by mass spectrometry, *Curr. Opin. Struct. Biol.* 13, 595–601, 2003.
31. Goshe, M.B., Characterizing phosphoproteins and phosphoproteomes by mass spectrometry, *Brief. Funct. Genomic. Proteomic.* 4, 363–376, 2006.
32. Butt, Y.K. and Lo, S.C., Detecting nitrated proteins by proteomic technologies, *Methods Enzymol.* 440, 17–31, 2008.
33. Crestfield, A.M., Stein, W.H., and Moore, S., Alkylation and identification of the histidine residues at the active site of ribonucleae, *J. Biol. Chem.* 238, 2413–2419, 1963.
34. Yamada, H., Imoto, T., Fujita, K. M. et al., Selective modification of aspartic acid-101 in lysozyme by carbodiimide reaction, *Biochemistry* 20, 4836–4832, 1981.
35. Ray, W.J., Jr. and Koshland, D.E., Jr., A method for characterizing the type and numbers of groups involved in enzyme action, *J. Biol. Chem.* 236, 1973–1979, 1961.

36. Tsou, C.-L., Relation between modification of functional groups of proteins and their biological activity. I. A graphical method for the determination of the number and type of essential groups, *Sci. Sinica* 11, 1535–1558, 1962.
37. Tsou, C.-L., Kinetics of substrate reaction during irreversible modification of enzyme activity, *Adv. Enzymol.* 61, 381–436 1988.
38. Zhou, J.-M., Liu, C., and Tsou, C.-L., Kinetics of trypsin inhibition by its specific inhibitors, *Biochemistry* 28, 1070–1076, 1989.
39. Horiike, K. and McCormick, D.B., Correlations between biological activity and the number of functional groups chemically modified, *J. Theoret. Biol.* 79, 403–414, 1979.
40. Holbrook, J.J. and Ingram, V.A., Ionic properties of an essential histidine residue in pig heart lactate dehydrogenase, *Biochem. J.* 131, 729–738, 1973.
41. Bloxham, D.P., The chemical reactivity of the histidine-195 residue in lactate dehydrogenase thiomethylated at the cysteine-165 residue, *Biochem. J.* 193, 93–97, 1981.
42. Horiike, K., Tsuge, H., and McCormick, D.B., Evidence for an essential histidyl residue at the active site of pyridoxamine (pyridoxine)-5′-phosphate oxidase from rabbit liver, *J. Biol. Chem.* 254, 6638–6643, 1979.
43. Rakitzis, E.T., Kinetics of protein modification reactions, *Biochem. J.* 217, 341–351, 1984.
44. Rakitzis, E.T., Kinetic analysis of regeneration by dilution of a covalently modified protein, *Biochem. J.* 268, 669–670, 1990.
45. Page, M.G., The reaction of cephalosporins with penicillin-binding protein 1bγ from *Escherichia coli*, *Biochim. Biophys. Acta* 1205, 199–206 1994.
46. Dubus, A., Normark, S., Kania, M., and Page, M.G.P., Role of Asparagine 152 in Catalysis of β-lactam hydrolysis by *Escherichia coli* ampC β-lactamase studied by site-directed mutagenesis, *Biochemistry* 34, 7757–7764, 1995.
47. Yang, S.J., Jiang, S.S., Tzeng, C.M. et al., Involvement of tyrosine residue in the inhibition of plant vacuolar H^+-pyrophosphatase by tetranitromethane, *Biochim. Biophys. Acta* 1294, 89–97, 1996.
48. Chu, C.L., Hsiao, Y.Y., Chen, C.H. et al., Inhibition of plant vacuolar H^+-ATPase by diethylpyrocarbonate, *Biochim. Biophys. Acta* 1506, 12–22, 2001.
49. Hsiao, Y.Y., Van, R.C., Hung, H.H., and Pan, R.L., Diethylpyrocarbonate inhibition of vacuolar H^+-pyrophosphatase possibly involves a histidine residue, *J. Protein Chem.* 21, 51–58, 2002.
50. Yang, S.-H., Wu, C.-H., and Lin, W.-Y., Chemical modification of aminopeptidase isolated from Pronase, *Biochem. J.* 302, 595–600, 1994.
51. Kelleher, N.L., Lin, H.Y., Valaskovic, G.A. et al., Top down versus bottom up protein characterization by tandem high-resolution mass spectrometry, *J. Am. Chem. Soc.* 121, 806–812, 1999.
52. Samyn, B., Sergeant, K., and Van Beeumen, J., A method for C-terminal sequence analysis in the proteomic era (proteins cleaved by cyanogen bromide), *Nat. Prot.* 1, 318–323, 2006.
53. Kadlik, V., Strohalm, M., and Kodicek, M., Citraconylation—a simple method for high protein sequence coverage in MALDI-TOF mass spectrometry, *Biochem. Biophys. Res. Commun.* 305, 1091–1093, 2003.
54. Pál, G., Santamaria, F., Kossiakoff, A.A., and Lu, W., The first semi-synthetic serine protease made by native chemical ligation, *Prot. Express. Purif.* 29, 185–192, 2003.
55. Dawson, P.E. et al., Synthesis of proteins by native chemical ligation, *Science* 266, 776, 1994.
56. Kuliopulos, A. and Walsh, C.T., Production, purification, and cleavage of tandem repeats of recombinant peptides, *J. Am. Chem. Soc.* 116, 4599–4607, 1994.
57. Karamloo, F., Scheurer, S., Wangosch, A. et al., Pyr c 1, the major allergen from pear (*Pyrus communis*), is a new member of the Bet v 1 allergen family, *J. Chromatog.* 756, 281–293, 2001.

58. Fairlie, W., Uboldi, A.D., De Souza, D.P. et al., A fusion protein system for the recombinant production of short disulfide-containing peptides, *Prot. Express. Purif.* 26, 171–178, 2002.
59. Rodríguez, J.C., Wong, L., Jennings, P.A. et al., The solvent in CNBr cleavage reactions determines the fragmentation efficiency of ketosteroid isomerase fusion proteins used in the production of recombinant peptides, *Prot. Express. Purif.* 28, 224–231, 2003.
60. Jenny, R.J., Mann, K.G., and Lundblad, R.L., A critical review of the methods for cleavage of fusion proteins with thrombin and factor Xa, *Prot. Express. Purif.* 31, 1–11, 2003.
61. Shlyapnikov, Y.M., Andreev, Y.A., Kozlov, S.A. et al., Bacterial production of latarin 2a, a potent antimicrobial peptides from spider venom, *Protein Express Purif.* 60, 89–95, 2008.
62. Bergman, L.W. and Kuehl, W.M., Co-translational modification of nascent immunoglobulin heavy and light chains, *J. Supramol. Struct.* 11, 9–24, 1979.
63. Hovland, R., Døskeland, A.P., Eikhom, T.S. et al., cAMP induces co-translational modification of proteins in IPC-81 cells, *Biochem. J.* 342, 369–377, 1999.
63a. Bradshaw, R.A. and Yi, E., Methionine aminopeptidases and angiogenesis, *Essays Biochem.* 38, 65–78, 2002.
63b. de Ruijter, A.J., van Gennip, A.H., Caron, H.N. et al., Histone deactylases (HCACs): Characterization of the classical HDAC family, *Biochem. J.* 370, 737–749, 2003.
64. Walsh, C.T., Garneau-Tsodikova, S., and Gatto, G.J., Jr. Protein posttranslational modifications: The chemistry of proteome diversifications, *Angew. Chem. Int. Ed. Engl.* 44, 7342–7372, 2005.
65. Johnson, D. and Travis, J., The oxidative inactivation of human alpha-1-proteinase inhibitor. Further evidence for methionine at the reactive center, *J. Biol. Chem.* 254, 4022–4026, 1979.
66. Matheson, N.R. and Travis, J., Differential effects of oxidizing agents on human plasma alpha 1 proteinase inhibitor and human neutrophils myeloperoxidase, *Biochemistry* 24, 1941–1945, 1985.
67. Ueda, M., Mashiba, S., and Uchida, K., Evaluation of oxidized alpha-1-antitrypsin in blood as an oxidative stress marker using anti-oxidative alpha1-AT monoclonal antibody, *Clin. Chim. Acta* 317, 125–131, 2002.
68. Vogt, W., Oxidation of methionyl residues in proteins: Tools, targets, and reversal, *Free Radic. Biol. Med.* 18, 93–105, 1995.
69. Maleknia, S.D., Brenowitz, M., and Chance, M.R., Millisecond radiolytic modification of peptides by synchrotron X-rays identified by mass spectrometry, *Anal. Chem.* 71, 3965–3673, 1999.
70. Heyduk, E. and Heyduk, T., Mapping protein domains involved in macromolecular interactions: A novel protein footprinting approach, *Biochemistry* 33, 9643–9650, 1994.
71. Rashidzadeh, H. et al., Solution structure and interdomain interactions of the *Saccharomyces cerevisiae* "TATA binding protein" (TBP) probed by radiolytic protein footprinting, *Biochemistry* 42, 3655–3665, 2003.
72. Kiselar, J.G., Janmey, P.A., Almo, S.C., and Chance, M.R., Structural analysis of gelsolin using synchrotron protein footprinting, *Molec. Cell. Proteomics* 2, 1120–1132, 2003.
73. Gay, C.A. and Gebicki, J.M., Measurement of protein and lipid hydroperoxides in biological systems by the ferric-xylenol orange method, *Anal. Biochem.* 315, 29–35, 2003.
74. Hawkins, C.L. and Davies, M.J., Hypochlorite-induced oxidation of proteins in plasma: Formation of chloramines and nitrogen-centered radicals and their role in protein fragmentation, *Biochem. J.* 340, 539–548, 1999.
75. Refsgaard, H.H., Tsai, L., and Stadtman, E.R., Modifications of proteins by polyunsaturated fatty acid peroxidation products, *Proc. Natl. Acad. Sci. USA* 97, 611–616, 2000.
76. Hidalgo, F.J., Alaiz, M., and Zamora, R., A spectrophotometric method for the determination of proteins damaged by oxidized lipids, *Anal. Biochem.* 262, 129–136, 1998.

77. Woods, A.A., Linton, S.M., and Davies, M.J., Detection of HOCl-mediated protein oxidation products in the extracellular matrix of human atherosclerotic plaques, *Biochem. J.* 370, 729–735, 2003.
78. Dong, J., Atwood, C.S., Anderson, V.E. et al., Metal binding and oxidation of amyloid-β within isolated senile plaque cores: Raman microscopic evidence, *Biochemistry* 42, 2768–2773, 2003.
79. Stadtman, E.R., Covalent modification reactions are marking steps in protein turnover, *Biochemistry* 29, 6323–6331, 1990.
80. Maillard, L.C., Action des acides amines sur les sucres: Formation des melanoidines par voie methodique, *C. R. Hebd. Seanes Acad. Sci.* 154, 66, 1912.
81. Ulrich, P. and Cerami, A., Protein glycation, diabetes, and aging, *Recent Prog. Horm. Res.* 56, 1–21, 2001.
82. Biemel, K.M. and Lederer, M.O., Site-specific quantitative evaluation of the protein glycation product N(6) 0 (2,3-Dihydroxy-5,6-dioxohexyl-1-lysinate by LC-(ESI) MS peptide mapping: Evidence for its key role in AGE formation, *Bioconjug. Chem.* 14, 619–628, 2003.
83. Tessier, F.J., Monnier, V.M., Sayre, L.M., Kornfield, J.A. et al., Triosidines: Novel Maillard reaction products and cross-links from the reaction of triose sugars with lysine and arginine residues, *Biochem. J.* 369, 705–719, 2003.
84. Miller, A.G., Meade, S.J., and Gerrard, J.A., New insights into protein crosslinking via the Maillard reaction: Structural requirements, the effects on enzyme function, and predicted efficacy of crosslinking inhibitors as anti-aging therapeutics, *Bioorg. Med. Chem.* 11, 843–852, 2003.
85. Degenhardt, T.P., Thorpe, S.R., and Baynes, J.W., Chemical modification of proteins by methylglyoxal, *Cell. Mol. Biol.* 44, 1139–1145, 1998.
86. Oya, T., Hattori, N., Mizuno, Y. et al., Methylglyoxal modification of protein. Chemical and immunochemical characterization of methylglyoxal-arginine adducts, *J. Biol. Chem.* 274, 18492–18502, 1999.
87. Seidler, N.W. and Kowalewski, C., Methylglyoxal-induced glycation affects protein topography, *Arch. Biochem. Biophys.* 410, 149–154, 2003.
88. Rao, A.G. and Neet, K.E., Tryptophan residues of the gamma subunit of 7S nerve growth factor: Intrinsic fluorescence, solute quenching, and N-bromosuccinimide oxidation, *Biochemistry* 21, 6843–6850, 1982.
89. Davies, K.J. and Delsignore, M.E., Protein damage and degradation by oxygen radicals III. Modification of secondary and tertiary structure, *J. Biol. Chem.* 262, 9908–9913, 1987.
90. Okajima, T., Kawata, Y., and Hamaguchi, K., Chemical modification of tryptophan residues and stability changes in proteins, *Biochemistry* 29, 9168–9175, 1990.
91. Suckau, D., Mak, M., and Przybylski, M., Protein surface topology-probing by selective chemical modification and mass spectrometric peptide mapping, *Proc. Natl. Acad. Sci. USA* 89, 5630–5634, 1992.
92. Gettins, P.G.W., Fan, B., Crews, B.C., Turko, I.V., Olson, S.T., and Streusand, V.J., Transmission of conformational change from the heparin binding site to the reactive center of antithrombin, *Biochemistry* 32, 8385–8389, 1993.
93. Buechler, J.A., Vedvick, T.A., and Taylor, S.S., Differential labeling of the catalytic subunit of cAMP-dependent protein kinase with acetic anhydride: Substrate-induced conformational changes, *Biochemistry* 28, 3018–3024, 1989.
94. Mykkanen, H.M. and Wasserman, R.H., Reactivity of sulfhydryl groups in the brush-border membranes of chick duodena is increased by 1,25-dihydroxycholecalciferol, *Biochim. Biophys. Acta* 1033, 282–286, 1990.
95. Landfear, S.M., Evans, D.R., and Lipscomb, W.N., Elimination of cooperativity in aspartate transcarbamylase by nitration of a single tyrosine residue, *Proc. Natl. Acad. Sci. USA* 75, 2654–2658, 1978.

96. Kumar, G.K., Beegen, N., and Wood, H.G., Involvement of tryptophans at the catalytic site and subunit-binding domains of transcarboxylase, *Biochemistry* 27, 5972–5978, 1988.
97. Winkler, M.A., Fried, V.A., Merat, D.L., and Cheung, W.Y., Differential reactivities of lysines in calmodulin complexed to phosphatase, *J. Biol. Chem.* 262, 15466–15471, 1987.
98. Salhany, J.M., Sloan, R.L., and Cordes, K.S., The carboxyl side chain of glutamate 681 interacts with a chloride binding modifier site that allosterically modulates the dimeric conformational state of Band 3 (AE1). Implications for the mechanism of anion/proton cotransport, *Biochemistry* 42, 1589–1602, 2003.
99. D'Ambrosio, C., Talamo, C., Vitale, R.M. et al., Probing the dimeric structure of porcine aminoacylase 1 by mass spectrometric and modeling procedures, *Biochemistry* 42, 4430–4443, 2003.
100. Li, J. and Bigelow, D.J., Phosphorylation by cAMP-dependent protein kinase modulates the structural coupling between the transmembrane and cytosolic domains of phospholamban, *Biochemistry* 42, 10674–10682, 2003.
101. Kirtley, M.E. and Koshland, D.E., Jr., The introduction of a "reporter" group at the active site of glyceraldehyde-3-phosphate dehydrogenase, *Biochem. Biophys. Res. Commun.* 23, 810–815, 1966.
102. Loudon, G.M. and Koshland, D.E., Jr., The chemistry of a reporter group: 2-hydroxy-5-nitrobenzyl bromide, *J. Biol. Chem.* 245, 2247–2254, 1970.
103. Riordan, J.F., Sokolovsky, M., and Vallee, B.L., Environmentally sensitive tyrosyl residues. Nitration with tetranitromethane, *Biochemistry* 6, 358–361, 1967.
104. Quaroni, L. and Smith, W.E., The nitro stretch as a probe of the environment of nitrophenols and nitrotyrosines, *J. Raman Spectros.* 30, 537–542, 1999.
105. Waggoner, A., Covalent labeling of proteins and nucleic acids with fluorophores, *Methods Enzymol.* 246, 362–373, 1995.
106. Bech, L.M., Branner, S., Hastrup, S., and Breddam, K., Introduction of a free cysteine residue at position 68 in the subtilisin Savinase, based on homology with proteinase K, *FEBS Lett.* 297, 164–166, 1992.
107. Kunkel, T.A., Rapid and efficient site-specific mutagenesis without phenotypic selection, *Proc. Natl. Acad. Sci. USA* 82, 488–492, 1985.
108. Hogrefe, H.H., Cline, J., Youngblood, G.L., and Allen, R.M., Creating randomized amino acid libraries with the QuikChange® multi site-directed mutagenesis kit, *BioTechniques* 33, 1158–1165, 2002.
109. Karlin, A. and Akabas, M.H., Substituted-cysteine accessibility method, *Methods Enzymol.* 293, 123–145, 1998.
110. Ratner, V., Kahana, E., Eichler, M., and Haas, E., A general strategy for site-specific double labeling of globular proteins for kinetic FRET studies, *Bioconjug. Chem.* 13, 1163–1170, 2002.
111. Heyduk, T., Measuring protein conformational changes by FRET/LRET, *Curr. Opin. Biotechnol.* 13, 292–296, 2002.
112. Watrob, H.M., Pan, C.P., and Barkley, M.D., Two-step FRET as a structural tool, *J. Am. Chem. Soc.* 125, 7336–7343, 2003.
113. Rhoades, E., Gussakovsky, E., and Haran, G., Watching proteins fold one molecule at a time, *Proc. Natl. Acad. Sci. USA* 100, 3197–3202, 2003.
114. Buschmann, V., Weston, K.D., and Sauer, M., Spectroscopic study and evaluation of red-absorbing fluorescent dyes, *Bioconjug. Chem.* 14, 195–204, 2003.
115. Kondo, T., Seike, M., Mori, Y. et al., Application of sensitive fluorescent dyes in linkage of laser microdissection and two-dimensional gel electrophoresis as a cancer proteomic study tool, *Proteomics* 3, 1758–1766, 2003.
116. Franklin, J.G. and Leslie, J., Some enzymatic properties of trypsin after reaction with 1-dimethylaminonaphthalene-5-sulfonyl chloride, *Can. J. Biochem.* 49, 516–521, 1971.

117. Wagner, R., Podestá, F.E., González, D.H., and Andreo, C.S., Proximity between fluorescent probes attached to four essential lysyl residues in phosphoenolpyruvate carboxylase—a resonance energy transfer study, *Eur. J. Biochem.* 173, 561–568, 1988.
118. Park, S.J., Song, J.S., and Kim, H.J., Dansylation of tryptic peptides for increased sequence coverage in protein identification by matrix-assisted laser desorption/ionization time-of-flight mass spectrometric peptide mass fingerprinting, *Rapid Commun. Mass Spectrom.* 19, 3089–3096, 2005.
119. Amoresano, A., Chipappetta, G., Pucci, P. et al., Bidimensional tandem mass spectrometry for selective identification of nitration sites in proteins, *Anal. Chem.* 79, 2109–2017, 2007.
120. Cirulli, C., Marino, G., and Amoresano, A., Membrane proteins in *Escherichia coli* probed by MS3 mass spectrometry: A preliminary report, *Rapid Commun. Mass Spectrom.* 21, 2389–2397, 2007.
121. Haugland, R.P., Molecular probes. *Handbook of Fluorescent Probes and Research* Chemicals, Molecular Probes, Eugene, OR, 1989, 37.
122. Tuls, J., Geren, L., and Millett, F., Fluorescein isothiocyanate specifically modifies lysine 338 of cytochrome P-450scc and inhibits adrenodoxin binding, *J. Biol. Chem.* 264, 16421–16425, 1989.
123. Miki, M., Interaction of Lys-61 labeled actin with myosin subfragment-1 and the regulatory proteins, *J. Biochem.* (Tokyo) 106, 651–655, 1989.
124. Bellelli, A., Ippoliti, R., Brunori, M. et al., Binding and internalization of ricin labelled with fluorescein isothiocyanate, *Biochem. Biophys. Res. Commun.* 169, 602–609, 1990.
125. Turner, D.C. and Brand, L., Quantitative estimation of protein binding site polarity. Fluorescence of N-arylaminonaphthalenesulfonates, *Biochemistry,* 7, 3381–3390, 1968.
126. Berliner, L.J., Ed., *Spin Labeling: Theory and Applications*, Academic Press, New York, 1975.
127. Likhtenshtein, G.I., *Biophysical Labeling Methods in Molecular Biology*, Cambridge University Press, Cambridge, 1993.
128. Hemminga, M.A. and Berlinger, L.J., *ESR Spectroscopy in Membrane Biophysics*, Springer, New York, 2007.
129. Spooner, P.J., Friesen, R.H., Knol, J. et al., Rotational mobility and orientational stability of a transport protein in lipid membranes, *Biophys. J.* 79, 756–766, 2000.
130. Sammalkorpi, M. and Lazaridis, T., Modeling a spin-labeled fusion peptide in a membrane: Implications for interpretation of EPR experiments, *Biophys. J.* 92, 10–22, 2007.
131. Stimson, L., Dong, L., Karttunen, M. et al., Stearic acid spin labels in lipid bilayers: Insight through aromatic simulations, *J. Phys. Chem. B* 111, 12447–12253, 2007.
132. Campbell, I.D. and Dwek, R.A., *Biological Spectroscopy*, Benjamin Cummings Publishing, Menlo Park, CA, 1984.
133. Gerson, F. and Huber, W., *Electron Spin Resonance Spectroscopy of Organic Radicals,* Wiley-VCH, Weinhemi, Germany, 2003.
134. Marsh, D., Reaction fields and solvent dependence of the EPR parameters of nitroxides: The microenviroment of spin labels, *J. Magn. Reson.* 190, 60–67, 2008.
135. Clatxton, D.P., Zou, P., and Mchaourab, H.S., Structure and orientation of T4 lysozyme bound to the small heat shock protein α-crystallin, *J. Mol. Biol.* 375, 1026–1039, 2008.
136. Alexander, N., Bortolus, M., Al-Mestasihi, A. et al., *De Novo* high-resolution protein structure determination from sparse spin-labeling, *Structure* 16, 191–195, 2008.
137. Timofeev, V.P., Novikov, V.V., Tkachev, Y.V. et al., Spin label method reveals barnase-barstar interaction: A temperature and viscosity dependence approach, *J. Biomol. Struct. Dyn* 25, 525–534, 2008.
138. Xu, Q., Kim, M., Ho, D. et al., Membrane hydrocarbon thickness modulates the dynamics of a membrane transport protein, *Biophys. J.*, 95, 2849–2858, 2008.

139. Chen, C.A. and Manning, D.R., Regulation of G proteins by covalent modification, *Oncogene* 20, 1643–1652, 2001.
140. Gibney, B.R., Johansson, J.S., Rabanal, F. et al., Global topology and stability and local structure and dynamics in a synthetic spin-labeled four-helix bundle protein, *Biochemistry* 36, 2798–2806, 1997.
141. Frazier, A.A., Roller, C.R., Havelka, J.J. et al., Membrane-bound orientation and position of the synaptotagmin I C2A domain by site-directed spin labeling, *Biochemistry* 42, 96–105, 2003.
142. Shafer, A.M., Kálai, T., Bin Liu, S.Q. et al., Site-specific insertion of spin-labeled L-amino acids in *Xenopus oocytes*, *Biochemistry* 43, 8470–8482, 2004.
143. Pantusa, M., Sportelli, L., and Bartucci, R., Spectroscopic and calorimetric studies on the interaction of human serum albumin with DPPC/PEG: 2000–DPPE membranes, *Eur. Biophys. J.* 37, 961–973, 2008.
144. D'Errico, G., D'Ursi, A.M., and Marsh, D., Interaction of a peptide derived from glycoprotein gp36 of feline immunodeficiency virus and its lipoylated analogue with phospholipid membranes, *Biochemistry* 47, 5317–5327, 2008.
145. Dmitriev, O.Y., Freedman, K.H., Hemolin, J., and Fillingame, R.H., Interaction of transmembrane helices in ATP synthase subunit a in solution as revealed by spin label difference NMR, *Biochim. Biophys. Acta* 1777, 227–237, 2008.
146. Kim, M., Xu, Q., Murray, D., and Cafiso, D.S., Solutes alter the conformation of the ligand binding loops in outer membrane transporters, *Biochemistry* 47, 670–679, 2008.
147. Berliner, L.J. and Wong, S.S., Evidence against two "pH locked" conformations of phosphorylated trypsin, *J. Biol. Chem.* 248, 1118–1120, 1973.
148. Berliner, L.J. and Wong, S.S., Spin-labeled sulfonyl fluorides as active site probes of protease structure. I. Comparison of the active site environments in α-chymotrypsin and trypsin, *J. Biol. Chem.* 249, 1668–1677, 1974.
149. Wong, S.S. et al., Spin-labeled sulfonyl fluorides as active site probes of protease structure. II. Spin label synthese and enzyme inhibition, *J. Biol. Chem.* 249, 1678–1682, 1974.
150. Bauer, R.S., Chang, T.L., and Berliner, L.J., Stability differences between high coagulant (alpha) and noncoagulant (gamma) human thrombins. Denaturation, *J. Biol. Chem.* 255, 5900–5903, 1980.
151. Musci, G., Berliner, L.J., and Esmon, C.T., Evidence for multiple conformational changes in the active center for thrombin induced by complex formation with thrombomodulin: An analysis employing nitroxide spin-labels, *Biochemistry* 27, 769–773, 1988.
152. Nienaber, V.L. and Berliner, L.J., Atomic structures of two nitroxide spin labels complexed with human thrombin: Comparison with solution studies, *J. Protein Chem.* 19, 129–137, 2000.
153. Twining, S.S., Sealy, R.C., and Glick, D.M., Preparation and activation of spin-labelled pepsinogen, *Biochemistry* 20, 1267–1272, 1981.
154. Taylor, J.C. and Markham, G.D., Conformational dynamics of the active site loop of *S*-adenosylmethionine synthetase illuminated by site-directed spin labeling, *Arch. Biochem. Biophys.* 415, 164–171, 2003.
155. Morozzo della Rocca, B., Lauria, G., Venerini, F. et al., The mitochondrial oxoglutarate carrier: Structural and dynamic properties of transmembrane segment IV studied by site-directed mutagenesis, *Biochemistry* 42, 5493–5499, 2003.
156. Smrnov, A.I., Ruuge, A., Reznikov, V.A. et al., Site-directed electrostatic measurements with a thiol-specific pH-sensitive nitroxide: Differentiating local p*K* and polarity effects by high-field EPR, *J. Am. Chem. Soc.* 126, 8872–8873, 2004.
157. Voinov, M.A., Ruuge, A., Reznikov, V.A. et al., Mapping local protein electrostatics by EPR of pH-sensitive thiol-specific nitroxide, *Biochemistry* 47, 5626–5637, 2008.

158. Altenbach, C., Kusnetzow, A.K, Ernst, O.P. et al., High-resolution distance mapping in rhodopsin reveals the pattern of helix movement due to activation, *Proc. Nat. Acad. Sci. USA* 105, 7439–7444, 2008.
159. Froncisz, W., Camenish, T.G., Ratke, J.J. et al., Saturation recovery EPR and ELDOR at W-band for spin labels, *J. Magn. Res.* 193, 297–304, 2008.
160. Schmidt, D.F., Jr. and Westheimer, F.H., pka of the lysine amino group of acetoacetate decarboxylase, *Biochemistry* 10, 1249–1253, 1971.
161. Highbarger, L.A., Gerlt, J.A., and Kenyou, G.L., Mechanism of the reaction catalyzed by acetoacetate decarboxylase. Importance of lysine 116 in determining the p*K*a of active-site lysine 115, *Biochemistry* 35, 41–46, 1996.
162. Mattingly, J.R., Jr., Farach, H.A., Jr., and Martinez-Carrion, M., Properties of the active site lysyl residues of mitochrondrial aspartate aminotransferase in solution, *J. Biol. Chem.* 258, 6243–6249, 1983.
163. Ludwig, H.C., Herrera, R., Reyes, A.M. et al., Suppression of kinetic AMP cooperativity of fructose-1,6-bisphosphatase by carbamylation of lysine 50, *J. Protein Chem.* 18, 533–545, 1999.
164. Zhang, G., Marzurkie, A.S., Dunaway-Mariano, D., and Allen, K.N., Kinetic evidence for a substrate-induced fit in phoshonoacetaldehyde hydrolase, *Biochemistry* 41, 13370–13377, 2002.
165. Mukouyama, E.B., Oguchi, M., Kodera, Y. et al., Low pKa lysine residues at the active site of sarcosine oxidase from *Corynebacterium* sp. U-96, *Biochem. Biophys. Res. Commun.* 320, 846–851, 2004.
166. Li, C. and Gershon, P.D., p*K*a of the mRNA cap-specific 2′-*O*-methyltransferase catalytic lysine by HSQC NMR detection of a two-carbon probe, *Biochemistry* 45, 907–917, 2006.
167. Zhang, W., Shi, Q., Meroueh, O.H. et al., Catalytic mechanism of penicillin-binding protein 5 of *Escherichia coli*, *Biochemistry* 46, 10113–10121, 2007.
168. Crnogorac, M.M., Ullmann, G.M., and Kostić, N.M., Effects of pH on protein association: Modification of the proton-linkage model and experimental verification of the modified model in the case of cytochrome c and plastocyanin, *J. Am. Chem. Soc.* 123, 10789–10798, 2001.
169. Dao-pin, S., Anderson, D.E., Baase, W.A. et al., Structural and thermodynamic consequences of burying a charged residue within the hydrophobic core of T4 lysozyme, *Biochemistry* 30, 11521–11529, 1991.
170. Harms, M.J., Schesssman, J.L., Chimenti, M.S. et al., A buried lysine that titrates with a normal p*K*a: Role of conformational flexibility at the protein-water interface as a determinant of p*K*a values, *Protein Sci.* 17, 833–845, 2008.
171. Kellam, B., De Bank, P.A., and Shakesheff, K.M., Chemical modification of mammalian cell surfaces, *Chem. Soc. Rev.* 32, 327–337, 2003.
172. Félix, L., Hernández, J., Argüelles-Monal, W.M., and Goycoolea, F.M., Kinetics of gelation and thermal sensitivity of *N*-isobutyryl chitosan hydrogels, *Biomacromolecules* 6, 2408–2415, 2005.
173. Izumi, S., Kaneko, H., Yamazaki, T. et al. Membrane topology of guinea pig cytochrome P450 17α revealed by a combination of chemical modification and mass spectrometry, *Biochemistry* 42, 14663–14669, 2003.
174. Nikfarjam, L., Izumi, S., Yamazaki, T., and Kominami, S., The interaction of cytochrome P450 17α with NADPH-cytochrome P450 reductase, investigated using chemical modification and MALDI-TOF mass spectrometry, *Biochim. Biophys. Acta* 1764, 1126–1131, 2006.
175. Gudiiksen, K.L., Gitlin, I., Yang, J. et al., Eliminating positively charged lysine ε-NH_3^+ groups on the surface of carbonic anhydrase has no significant influence on its folding from sodium dodecyl sulfate, *J. Am. Chem. Soc.* 127, 4707–4714, 2005.
176. Miyazaki, K. and Tsugita, A., *C*-Terminal sequencing method for peptides and proteins by the reaction with a vapor of perfloric acid in acetic anhydride, *Proteomics* 4, 11–19, 2004.

177. Miyazaki, K. and Tsugita, A., *C*-Terminal sequencing method for proteins in polyacrylamide gel by the reaction of acetic anhydride, *Proteomics* 6, 2026–2033, 2006.
178. Pan, Y., Wan, J., Roginski, H. et al., Comparison of the effects of acylation and amidation on the antimicrobial and antiviral properties of lactoferrin, *Lett. Appl. Microbiol.* 44, 229–234, 2007.
179. Sanchez, A., Ramos, Y., Solano, Y. et al., Double acylation for identification of amino-terminal peptides of proteins isolated by polyacrylamide gel electrophoresis, *Rapid Commun. Mass Spectrom.* 21, 2237–2244, 2007.
180. Higashimoto, Y., Sugishima, M., Sato, H., Mass spectrometric identification of lysine residues of heme oxygenase-1 that are involved in its interaction with NADPH-cytochrome P450 reductase, *Biochem. Biophys. Res. Commun.* 367, 852–858, 2008.
181. Calvete, J.J. et al., Characterisation of the conformation and quaternary structure-dependent heparin-binding region of bovine seminal plasma protein PDC-109, *FEBS Lett.* 444, 260–264, 1999.
182. Taralp, A. and Kaplan, H., Chemical modification of lyophilized proteins in nonaqueous environments, *J. Protein Chem.* 16, 183–193, 1997.
183. Vakos, H.T., Kaplan, H., Black, B. et al., Use of the pH memory effect in lyophilized proteins to achieve preferential methylation of α-amino groups, *J. Protein Chem.* 19, 231–237, 2000.
184. Govindarajan, R., Chatterjee, K., and Gatlin, L., Impact of freeze-drying on ionization of sulfonephthalein probe molecules in trehalose-citrate system, *J. Pharm. Sci.* 95, 498–510, 2006.
185. Gould, A.R. and Norton, R.S., Chemical modification of cationic groups in the polypeptide cardiac stimula anthopleurin-A, *Toxicon* 33, 187–199, 1995.
186. Becker, L. et al., Identification of a critical lysine residue in apolipoprotein B-100 that mediates noncovalent interaction with apolipoprotein (a), *J. Biol. Chem.* 276, 36155–36162, 2001.
187. Liu, J.Z, Wang, T.L, Huang, M.T. et al., Increased thermal and organic solvent tolerance of modified horseradish peroxidase, *Protein Eng. Des. Sel* 19, 169–173, 2006.
188. Lee, Y., Fukushima, S., Bae, Y. et al., A protein nanocarrier from charge-conversion polymer in response to endosomal pH, *J. Am. Chem. Soc.* 129, 5362–5363, 2007.
189. Mossavarali, S., Hosseinkhani, S., Ranjbar, B., and Miroliaei, M., Stepwise modification of lysine residues of glucose oxidase with citraconic anhydride, *Int. J. Biol. Macromol.* 39, 192–196, 2006.
190. Liu, J.Z. and Wang, M., Improvement of activity and stability of chloroperoxidase by chemical modification, *BMC Biotechnol.* 7, 23, 2007.
191. Shi, S.R., Liu, C., Young, L, and Taylor, C., Development of an optimal antigen retrieval protocol for immunohistochemistry of retinoblastoma protein (pRB) in formal fixed, paraffin sections based on comparison of different methods, *Biotech. Histochem.* 82, 301–309, 2007.
192. Wink, M.R. et al., Effect of protein-modifying reagents on ecto-apyrase from rat brain, *Int. J. Biochem. Cell. Biol.* 32, 105–113, 2000.
193. Ehrhard, B. et al., Chemical modification of recombinant HIV-1 capsid protein p24 leads to the release of a hidden epitope prior to changes of the overall folding of the protein, *Biochemistry* 35, 9097–9105, 1996.
194. Paetzel, M. et al., Use of site-directed chemical modification to study an essential lysine in *Escherichia coli* leader peptidase, *J. Biol. Chem.* 272, 994–10003, 1997.
195. Liu, S., Variero, M.M., Fraser, S., and Jenkins, A.T., Control of attachment of bovine serum albumin to pulse plasma-polymerized maleic anhydride by variation of pulse conditions, *Langmuir* 21, 8572–8575, 2005.
196. Yoshifuji, A., Noishiki, Y., Wada, M. et al., Esterification of β-chitin via intercalation by carboxylic anhydrides, *Biomacromolecules* 7, 2878–2881, 2006.

197. Chen, X., Zheng, Y., and Shen, Y., Natural products with maleic anhydride structure: Nonadrides, tautomycin, chaetomellic anhydride, and other compounds, *Chem. Rev.* 107, 1777–1830, 2007.
198. Alcalde, M. et al., Succinylation of cyclodextrin glucosyltransferase from *Thermoanaerobacter* s501 enhances its transferase activity using starch as a donor, *J. Biotechnol.* 86, 71–80, 2001.
199. Swart, P.J. et al., Lactoferin. Antiviral activity of lactoferrin, *Adv. Exp. Med. Biol.* 443, 205–213, 1998.
200. Zhao, Y., Ma, C.Y., Yuen, S.N., and Phillips, D.L., Study of succinylated food proteins by Raman spectroscopy, *J. Agric. Food Chem.* 52, 1815–1823, 2004.
201. Ebersold, M.F. and Zydney, A.L. Separation of protein charge variants by ultrafiltration, *Biotechnol. Prog.* 20, 543–549, 2004.
202. Habibi, A.E. Khajeh, K., and Nemat-Gorgani, M., Chemical modification of lysine residues in *Bacillus lichenformis* α-amylase: Conversion of an endo- to an exo-type enzyme, *J. Biochem. Mol. Biol.* 37, 642–647, 2004.
203. Ali, J. and Younus, H., Effect of succinylation of antibodies on their conformation and interaction with antigen, *Biochemistry* (Moscow) 71, 1336–1340, 2006.
204. An, Y., Chen, M., Xue, Q., and Liu, W., Preparation and self-assembly of carboxylic acid-functionalized silica, *J. Colloid Interface Sci.* 311, 507–513, 2007.
205. Fundueanu, G., Constantin, M., and Ascenzi, P., Preparation and characterization of pH- and temperature-sensitive pullan microspheres for controlled release of drugs, *Biomaterials* 29, 2767–2775, 2008.
206. Sheng, H. and Ye, B.C., Different strategies of covalent attachment of oligonucleotide probe onto glass beads and the hybridization properties, *Appl. Biochem. Biotechnol.* 152, 54–65, 2008.
207. Castelli, F., Sarpietro, G.M., Micieli, D. et al., Differential scanning calorimetry study on drug release from an inulin-based hydrogel and its interaction with a biomembrane model: pH and loading effect, *Eur. J. Pharm. Sci.* 35, 76–85, 2008.
208. Neurath, A.R. et al., Blocking of CD4 cell receptors for the human immunodeficiency virus type 1 (HIV-1) by chemically modified milk proteins: Potential for AIDS prophylaxis, *J. Mol. Recognit.* 8, 204–216, 1995.
209. Hopkins, J.E, Naisbitt, D.J., Kitteringham, N.R. et al., Selective haptenation of cellular or extracellular protein by chemical allergens: Association with cytokine polarization, *Chem. Res. Toxicol.* 18, 375–381, 2005.
210. Valstar, D.L., Shijf, M.A., Stelekati, E. et al., Trimellitic anhydride-conjugated serum albumin activated rat alveolar macrophages *in vitro*, *J. Occup. Med. Toxicol.* 1, 13, 2006.
211. Trimukhe, K.D., Bachate, S., Gokhale, D.V. et al., Metal complexes of cross-linked chitosans Part II. An investigation of their hydrolysis to chitooligosacchardies using chitosanase, *Int. J. Biol. Macromol.* 41, 491–496, 2007.
212. Kurata, S. and Uemoto, K., Synthesis of new silane coupling agents with a trimellitic anhydride group and applications as primers for ceramics and alloys, *Dent Mater. J.* 26, 800–804, 2007.
213. Swart, P.J. et al., Antiviral effects of milk proteins: Acylation results in polyanionic compounds with potent activity against human immunodeficiency virus types 1 and 2 *in vitro*, AIDS *Res. Human Retroviruses* 12, 769–775, 1996.
214. Reményi, J., Balázs, B., Tóth, S. et al., Isomer-dependent daunomycin release and in vitor antitumour effect of *cis*-aconityl-daunomycin, *Biochem. Biophys. Res. Commun.* 303, 556–601, 2003.
215. Gauvreau, V., Chevallier, P., Vallières, K. et al., Engineering surfaces for bioconjugation: Developing strategies and quantifying the extent of the reactions, *Bioconjug. Chem.* 15, 1146–1156, 2004.

216. Fletcher, S., Jorgensen, M.R. and Miller, A.D., Facile preparation of an orthogonally protected, pH-sensitive, bioconjugate linker for therapeutic applications, *Org. Lett.* 6, 4245–4248, 2004.
217. Martynov, A.V. and Smelyanskaya, M.V., Antiproliferative properties of chemical modified recombinant IFN-α2b, *J. Interferon Cytokine Res.* 25, 414–417, 2005.
218. Jonsson, B.A. et al., Lysine adducts between methyltetrahydrophthalic anhydride and collagen in guinea pig lung, *Toxicol. Appl. Pharmacol.* 135, 156–167, 1995.
219. Lindh, C.H. and Jonsson, B.A., Human hemoglobin adducts following exposure to hexhydrophthalic anhydride and methylhexahydrophthalic anhydride, *Toxicol. Appl. Pharmacol.* 153, 152–160, 1998.
220. Kristiansson, M.H., Jonsson, B.A., and Lindh, C.H., Mass spectrometric characterization of human hemoglobin adducts formed *in vivo* by hexahydrophthalic anhydride, *Chem. Res. Toxicol.* 15, 562–569, 2002.
221. O'Brien, A.M., Smith, H.T., and O'Fagain, C., Effects of phthalic anhydride modification on horse radish peroxidase.stability and activity, *Biotechnol. Bioeng.* 81, 233–240, 2003.
222. O'Connell, L.I., Bell, E.T., and Bell, J.F., 3,4,5,6-Tetrahydrophthalic anhydride modification of glutamate dehydrogenase: The construction and activity of heterohexamers, *Arch. Biochem. Biophys.* 263, 315–322, 1988.
223. Mok, H., Park, J.W., and Park, T.G., Enhanced intracellular delivery of quantum dot and adenovirus nanoparticles triggered by acidic pH via surface charge reversal, *Bioconjug. Chem.* 19, 797–801, 2008.
224. Kaplan, H., Stevenson, K.J., and Hartley, B.S., Competitive labeling, a method for determining the reactivity of individual groups in proteins. The amino groups of porcine elastin, *Biochem. J.* 124, 289–299, 1971.
225. Bosshard, H.R., Koch, G.L.E., and Hartley, B.S., The aminoacyl tRNA synthetase-tRNA complex: Detection by differential labeling of lysine residues involved in complex formation, *J. Mol. Biol.* 119, 377–389, 1978.
226. Richardson, R.H. and Brew, K., Lactose synthase. An investigation of the interaction site of alpha-lactalbumin for galactosyltransferase by differential kinetic labeling, *J. Biol. Chem.* 255, 3377–3385, 1980.
227. Rieder, R. and Bosshard, H.R., The cytochrome c oxidase binding site on cytochrome c. Differential chemical modification of lysine residues in free and oxidase-bound cytochrome c, *J. Biol. Chem.* 253, 6045–6053, 1978.
228. Hitchcock, S.E., Zimmerman, C.J., and Smalley, C., Study of the structure of troponin-T by measuring the relative reactivities of lysines with acetic anhydride, *J. Mol. Biol.* 147, 125–151, 1981.
229. Hitchcock, S.E., Study of the structure of troponin-C by measuring the relative reactivities of lysines with acetic anhydride, *J. Mol. Biol.* 147, 153–173, 1981.
230. Hitchcock-De Gregori, S.E., Study of the structure of troponin-I by measuring the relative reactivities of lysine with acetic anhydride, *J. Biol. Chem.* 257, 7372, 1982.
231. Giedroc, D.P., Sinha, S.K., Brew, K., and Puett, D., Differential trace labeling of calmodulin: Investigation of binding sites and conformational states by individual lysine reactivities. Effects of beta-endorphin, trifluoroperazine, and ethylene glycol bis(beta-aminoethyl ether)-N,N,N′N-tetraacetic acid, *J. Biol. Chem.* 260, 13406–13413, 1985.
232. Wei, Q., Jackson, A.E., Pervaiz, S. et al., Effects of interactions of with calcineurin of the reactivities of calmodulin lysines, *J. Biol. Chem.* 263, 19541–19444, 1988.
233. Winkler, M.A., Fried, V.A., Merat, D.L., and Cheung, W.Y., Differential reactivities of lysines in calmodulin complexed to phosphatase, *J. Biol. Chem.* 262, 15466–15471, 1987.
234. Hitchcock-De Gregori, S.E., Lewis, S.F., and Mistrik, M., Lysine reactivities of tropomyosin complexed with troponin, *Arch. Biochem. Biophys.* 264, 410–416, 1988.
235. Gurd, F.R.N., Carboxymethylation, *Meth. Enzymol.* 11, 532–541, 1967.

236. Boja, E.S. and Fales, H.M., Overalkylation of a protein digest with iodoacetamide, *Anal. Chem.* 73, 3575–3582, 2001.
237. Nielsen, M.I., Vermeulen, M., Bonaldi, T. et al., Iodoacetamide-induced artifact mimics ubiquitinylation in mass spectrometry, *Nat. Meth.* 5, 459–460, 2008.
238. Sanger, F. and Tuppy, H., The amino acid sequence in the phenylalanyl chain of insulin. I. The identification of lower peptides from partial hydrolysates, *Biochem. J.* 49, 463–481, 1951.
239. Carty, R.P. and Hirs, C.H.W., Modification of bovine pancreatic ribonuclease A with 4-sulfonyloxy-2-nitrofluorobenzene, *J. Biol. Chem.* 243, 5254–5365, 1968.
240. Watanabe, J., Sasajima, N., Aramaki, A., and Sonoyama, K., Consumption of fructo-oligosaccharide reduces 2,4-dinitrofluorobenzene-induced contact hypersensitivity in mice, *Br. J. Nutr.* 100, 339–346, 2008.
241. Watanabe, H., Gehrke, S., Contassot, E. et al., Danger signaling through the inflammasome acts as a master switch between tolerance and sensitization, *J. Immunol.* 180, 5826–5832, 2008.
242. Pae, H.O., Ae Ha, Y., Chai, K.Y., and Chung, H.T., Heme oxygenase-1 attenuates contact hypersensitivity induced by 2,4-dintrofluorobenzene in mice, *Immunopharmacol. Immunotoxicol.* 30, 207–216, 2008.
243. Holmdahl, M., Ahlfors, S.R., Holmdahl, R., and Hansson, C., Structure-immune response relationships of hapten-modified collagen II peptides in a T-cell model of allergic contact dermatitis, *Chem. Res. Toxicol.* 21, 1514–1523, 2008.
244. Stark, G.R., Stein, W.H., and Moore, S., Reaction of the cyanate present in aqueous urea with amino acids and proteins, *J. Biol. Chem.* 235, 3177–3181, 1960.
245. Metwalli, A.A., Lammers, W.L., and Van Boekel, M.A., Formation of homocitrulline during heating of milk, *J. Dairy Res.* 65, 579–589, 1988.
246. Lin, M.F., Williams, C., Murray, M.V. et al., Ion chromatographic quantification of cyanate in urea solutions: Estmation of the efficiency of cyanate scavengers for use in recombinant protein manufacturing, *J. Chromatog. B. Anal. Technol. Biomed. Life Sci.* 803, 353–362, 2004.
247. Park, K.D., Mun, K.C., Chang, E.J. et al., Inhibition of erythropoietin activity by cyanates, *Scand. J. Urol. Nephrol.* 38, 69–82, 2004.
248. Jaisson, S., Lorimier, S., and Ricard-Blum, S. et al., Impact of carbamylation on type I collagen conformational structure and its ability to activate human polymorphonuclear neutrophils, *Chem. Biol.* 13, 149–159, 2006.
249. Apostolov, E.O., Shah, S.V., Ok, E., and Basnakian, A.G., Quantification of carbamylated LDL in human sera by a new sandwich ELISA, *Clin. Chem.* 51, 719–728, 2005.
250. Jaisson, S., Larrreta-Garde, V., Bellon, G. et al., Carbamylation differentially alters type I collage sensitivity to various collagenases, *Matrix Biol.* 26, 190–196, 2007.
251. Wang, Z., Nicholls, S.J., Rodriguez, E.R. et al., Protein carbamylation links inflammation, smoking, uremia and atherogenesis, *Nat. Med.* 13, 1176–1184, 2007.
252. McCarthy, J., Hopwood, F., Oxley, D. et al., Carbamylation of proteins in 2-D electrophoresis—myth or reality? *J. Proteome Res.* 2, 239–242, 2003.
253. Righetti, P.G., Real and imaginary artifacts in proteome analysis via two-dimensional maps, *J. Chromatog. B. Anal. Technol. Biomed. Life Sci.* 841, 14–22, 2006.
254. Angel, P.M. and Orlando, R., Quantitative carbamylation as a stable isotopic labeling method for comparative proteomics, *Rapid Commun. Mass Spectrom.* 21, 1623–1634, 2007.
255. Stark, G.R., Modification of proteins with cyanate, *Methods Enzymol.* 25, 579–584, 1972.
256. Shaw, D.C., Stein, W.H., and Moore, S., Inactivation of chymotrypsin by cyanate, *J. Biol. Chem.* 239, 671–673, 1964.
257. Shen, W.-C. and Colman, R.F., Cyanate modification of essential lysine residues of the diphosphopyridine nucleotide-specific isocitrate dehydrogenase of pig heart, *J. Biol. Chem.* 250, 2873–2978, 1975.

258. Plapp, B.V., Moore, S., and Stein, W.H., Activity of bovine pancreatic deoxyribonuclease A with modified amino groups, *J. Biol. Chem.* 246, 939–945, 1971.
259. Plapp, B.V., Enhancement of the activity of horse liver alcohol dehydrogenase by modification of amino groups at the active sites, *J. Biol. Chem.* 245, 1727–1735, 1970.
260. Shapiro, S., Enser, M., Pugh, E., and Horecker, B.L., The effect of pyridoxal phosphate on rabbit muscle aldolase, *Arch. Biochem. Biophys.* 128, 554–562, 1968.
261. Schnackerz, K.D. and Noltmann, E.A., Pyridoxal-5′-phosphate as a site-specific protein reagent for a catalytically critical lysine residue in rabbit muscle phosphoglucose isomerase, *Biochemistry* 10, 4837–4843, 1971.
262. Havran, R.T. and du Vigneaud, V., The structure of acetone-lysine vasopressin as established through its synthesis from the acetone derivative of S-benzyl-l-cysteinyl-l-tyrosine, *J. Am. Chem. Soc.* 91, 2696–2998, 1969.
263. Ohsawa, H. and Gualerzi, C., Structure-function relationship in *Escherichia coli* inhibition factors. Identification of a lysine residue in the ribosomal binding site of initiation factor by site-specific chemical modification with pyridoxal phosphate, *J. Biol. Chem.* 256, 4905–4912, 1981.
264. Bürger, E. and Görisch, H., Evidence for an essential lysine at the active site of l-histidinol: NAD oxidoreductase, a bifunctional dehydrogenase, *Eur. J. Biochem.* 118, 125–130, 1981.
265. Kluger, R. and Tsue, W.-C., Methyl acetyl phosphate. A small anionic acetylating agent, *J. Org. Chem.* 45, 2723–2724, 1980.
266. Ueno, H., Pospischil, M.A., Manning, J.M., and Kluger, R., Site-specific modification of hemoglobin by methyl acetyl phosphate, *Arch. Biochem. Biophys.* 244, 795–800, 1986.
267. Ueno, H., Pospischil, M.A., and Manning, J.M., Methyl acetyl phosphate as a covalent probe for anion-binding sites in human and bovine hemoglobins, *J. Biol. Chem.* 264, 12344–12351, 1989.
268. Raibekas, A.A., Bures, E.J., Siska, C.C. et al., Anion binding and controlled aggregation of human interleukin-1 receptor antagonist, *Biochemistry* 44, 9871–9879, 2005.
269. Means, G.E., Reductive alkylation of amino groups, *Methods Enzymol.* 47, 469–478, 1977.
270. Jentoft, N. and Dearborn, D.G., Labeling of proteins by reductive methylation using sodium cyanoborohydride, *J. Biol. Chem.* 254, 4359–4365, 1979.
271. Dottavio Martin, D. and Ravel, J.M., Radiolabeling of proteins by reductive alkylation with [14C] formaldehyde and sodium cyanoborohydride, *Anal. Biochem.* 87, 562–565, 1978.
272. Dick, L.R., Geraldes, C.F.G.C., Sherry, A.D., Gray, C.W., and Gray, D.M., ^{13}C NMR of methylated lysines of fd gene 5 protein: Evidence for a conformational change involving lysine 24 upon binding of a negatively charged lanthanide chelate, *Biochemistry* 28, 7896–7904, 1989.
273. Brown, E.M., Pfeffer, P.E., Kumosinski, T.F., and Greenberg, R., Accessibility and mobility of lysine residues in β-lactoglobulin, *Biochemistry* 27, 5601–5610, 1988.
274. Fretheim, K., Iwai, S., and Feeney, R.F., Extensive modification of protein amino groups by reductive addition of different sized substituents, *Int. J. Pept. Protein Res.* 14, 451, 1979.
275. Fretheim, K., Edelandsdal, B., and Harbitz, O., Effect of alkylation with different size substituents on the conformation of ovomucoid, lysozyme and ovotransferrin, *Int. J. Pept. Protein Res.* 25, 601–607, 1985.
276. Goldfarb, A.R., A kinetic study of the reactions of amino acids and peptides with trinitrobenzenesulfonic acid, *Biochemistry* 5, 2570–2574, 1966.
277. Goldfarb, A.R., Heterogeneity of amino groups in proteins. I. Human serum albumin, *Biochemistry* 5, 2574–2578, 1966.
278. Habeeb, A.F.S.A., Determination of free amino groups in proteins by trinitrobenzenesulfonic acid, *Anal. Biochem.* 14, 328–336, 1966.
279. Fields, R., The rapid determination of amino groups with TNBS, *Methods Enzymol.* 25, 464–468, 1972.

280. Kotaki, A. and Satake, K., Acid and alkaline degradation of the TNP-amino acids and peptides, *J. Biochem.* 56, 299–307, 1964.
281. Cayot, P. and Tainturier, G., The quantification of protein amino groups by the trinitrobenzenesulfonic acid method: A reexamination, *Anal. Biochem.* 249, 184–200, 1997.
282. Coffee, C.J., Bradshaw, R.A., Goldin, B.R., and Frieden, C., Identification of the sites of modification of bovine liver glutamate dehydrogenase reacted with trinitrobenzenesulfonate, *Biochemistry* 10, 3516–3526, 1971.
283. Goldin, B.R. and Frieden, C., Effects of trinitrophenylation of specific lysyl residues on the catalytic, regulatory and molecular properties of bovine liver glutamate dehydrogenase, *Biochemistry* 10, 3527–3534, 1971.
284. Bates, D.J., Goldin, B.R., and Frieden, C., A new reaction of glutamate dehydrogenase: The enzyme-catalyzed formation of trinitrobenzene from TNBS in the presence of reduced coenzyme, *Biochem. Biophys. Res. Commun.* 39, 502–507, 1970.
285. Means, G.E., Congdon, W.I., and Bender, M.L., Reactions of 2,4,6-trinitrobenzenesulfonate ion with amines and hydroxide ion, *Biochemistry* 11, 3564–3571, 1972.
286. Salem, N., Jr., Lauter, C.J., and Trams, E.G., Selective chemical modification of plasma membrane ectoenzymes, *Biochim. Biophys. Acta* 641, 366–376, 1981.
287. Haniu, M., Yuan, H., Chen, S. et al., Structure-function relationship of NAD(P)H:quinone reductase: Characterization of NH2-terminal blocking group and essential tyrosine and lysine residues, *Biochemistry* 27, 6877–6883, 1988.
288. Anderson, G.W., Callahan, F.M., and Zimmerman, J.E., Synthesis of *N*-hydroxysuccinimide esters of acyl peptides by the mixed anhydride method, *J. Am. Chem. Soc.* 89, 178–179, 1967.
289. Smith, G.P., Kinetics of amine modification of proteins, *Bioconjug. Chem.* 17, 501–506, 2006.
290. Abello, N., Kerstjens, H.A., Postma, D.S., and Bischoff, R., Selective acylation of primary amines in peptides and proteins, *J. Proteome Res.* 6, 4770–4776, 2007.
291. Yem, A.W. et al., Biotinylation of reactive amino groups in native recombinant human interleukin-1β, *J. Biol. Chem.* 264, 17691–17697, 1989.
292. Lombardi, V.C. and Schooley, D.A., A method for selective conjugation of an analyte to enzymes without unwanted enzyme-enzyme cross-linking, *Anal. Biochem.* 331, 40–45, 2005.
293. Morpurgo, M. Bayer, E.A., and Wilchek, M., *N*-Hydroxysuccinimide carbonates and carbamates are useful reactive coupling ligands to lysines on proteins, *J. Biochem. Biophys. Methods* 38, 17–28, 1999.
294. Cooper, M., Ebner, A., Briggs, M. et al., Cy3B™: Improving the performance of cyanine dyes, *J. Fluorescence* 14, 145–150, 2004.
295. Deng, Y., Hou, Z., Wang, L. et al., Role of lysine 411 in substrate carboxyl group binding to the human reduced folate carrier, as determined by site-directed mutagenesis and affinity inhibition, *Mol. Pharmacol.* 73, 1274–1281, 2008.
296. Cuatrecasas, P. and Parikh, I., Adsorbents for affinity chromatography. Use of *N*-hydroxysuccinimide esters of growth, *Biochemistry* 11, 2291–2299, 1972.
297. Khan, W., Kapoor, M., and Kumar, N. Covalent attachment of proteins to functionalized polypyrrole-coated metallic surfaces for improved biocompatibility, *Acta Biomater.* 3, 541–549, 2007.
298. Lockett, M.R., Phillips, M.F., Jarecki, J.L. et al., A tetrafluorophenyl activated ester self-assembled monolayer for the immobilization of amine-modified oligonucleotides, *Langmuir* 24, 69–75, 2008.
299. Yang, M., Teeuwen, R.L., Giesbera, M. et al., One-step photochemical attachment of NHS-terminated monolayers onto silicon surfaces and subseqnent functionalization, *Langmuir*, 24, 7931–7938, 2008.

300. Takeda, S., Tsukiji, S., Ueda, N. et al., Covalent split protein fragment-DNA hybrids generated through *N*-terminus specific modification of proteins by oligonucleotides, *Org. Biomol. Chem.* 6, 2187–2194, 2006.
301. Bolton, A.E. and Hunter, W.M., The labelling of proteins to high specific radioactivities by conjugation to a 125I-containing acylating agent, *Biochem. J.* 133, 529–538, 1973.
302. Greenstein, J.P., Studies of multivalent amino acids and peptides. II. Synthesis of certain derivatives of lysyl-glutamic acid, *J. Biol. Chem.* 169, 541–544, 1935.
303. Cupo, P., El-Deiry, W., Whitney, P.L., and Awad, W.M., Jr., Stabilization of proteins by guanidination, *J. Biol. Chem.* 255, 10828–10833, 1980.
304. Siddiqui, K.S., Poljak, A., Guilhaus, M. et al., Role of lysine versus arginine in enzyme cold-adaptation: Modifying lysine to homo-arginine stabilizes the cold adapted α-amylase from *Pseudoalteramonas haloplanktis*, *Prot. Struct. Funct. Bioinform.* 64, 486–501, 2006.
305. Hundle, B.S. and Richards, W.R., Use of a new membrane-impermeant guanidinating reagent, 2-*S*-[^{14}C]thiuroniumethylsulfonate, for the labeling of intracytoplasmic membrane proteins in *Rhodbacter sphaeroides*, *Biochemistry* 25, 4505–4511, 1987.
306. Donavan, J.W., The spectrophotometric titration of the sulfhydryl and phenolic groups of aldolase, *Biochemistry* 3, 67–74, 1964.
307. Weber, G. and Teale, F.W.J., Interaction of proteins with radiation, in *The Proteins*, Ed. H. Neurath, Academic Press, New York, 1965.
308. Donovan, J.W., Changes in ultraviolet absorption produced by alteration of protein conformation, *J. Biol. Chem.* 244, 1961–1967, 1969.
309. Markland, F.S., Phenolic hydroxyl ionization in two subtilisins, *J. Biol. Chem.* 244, 694–700, 1969.
310. Mayberry, W.E. and Hockert, T.J., Kinetics of iodination. VI. Effect of solvent on hydroxyl ionization and iodination of L-tyrosine and 3-iodo-L-tyrosine, *J. Biol. Chem.* 245, 697–700, 1970.
311. Laws, W.R. and Shore, J.D., Spectral evidence for tyrosine ionization linked to a conformational change in liver alcohol dehydrogenase ternary complex, *J. Biol. Chem.* 254, 2582–2584, 1979.
312. Kuramitso, S. et al., Ionization of the catalytic groups and tyrosyl residues in human lysozyme, *J. Biochem.* 87, 771–778, 1980.
313. Demchenko, A., *Ultraviolet Spectroscopy of Proteins*, Springer-Verlag, Berlin, Germany, 1981.
314. Kobayashi, J., Hagashijima, T., and Miyazawa, T., Nuclear magnetic resonance analyses of side chain conformations of histidine and aromatic acid derivatives, *Int. J. Pept. Protein Res.* 24, 40–47, 1984.
315. Poklar, N., Vesnaver, G., and Laponje, S., Studies by UV spectroscopy of thermal denaturation of beta-lactoglobulin in urea and alkylurea solutions, *Biophys. Chem.* 47, 143–151, 1993.
316. Roholt, O.A. and Pressman, D., Iodination-isolation of peptides from the active site, *Methods Enzymol.* 25, 438, 1972.
317. Tsomides, T.J. and Eisen, H.N., Stoichiometric labeling of peptides by iodination on tyrosyl or histidyl residues, *Anal. Biochem.* 210, 129, 1993.
318. Rosenfeld, R., Philo, J.S., and Haniu, M. et al, Sites of iodination in recombinant human brain-derived neurotrophic factor and its effect on neurotrophic activity, *Protein Sci.* 2, 1664–1674, 1993.
319. Durr, J.A., Blankenship, M., Chauhan, S.S., and Pennington, M.W., Targeted tyrosine iodination in a multi-tyrosine vasopressin analog, *J. Pept. Sci.* 13, 756–761, 2007.
320. Mukai, T., Arana, Y., Nishida, Y. et al., Species differences in radioactivity elimination from liver parenchymal cells after injection of radiolabled protein, *Nucl. Med. Biol.* 26, 281–289, 1999.

321. Li, J., Xia, Y., and Kuter, D.J., Interaction of thrombopoietin with platelet c-mpl receptor in plasma: Binding, internalization, stability, and pharmacokinetics, *Brit. J. Haematol.* 106, 345–356, 1999.
322. Zhang, B., Shimoji, E., Tanaka, H., and Saku, K., Evaluation of apolipoprotein A-I kinetics in rabbits *in vivo* using *in situ* and exogenous radioidonation methods, *Lipids* 38, 209–218, 2003.
323. Keen, H.G., Dekker, B.A., Dislev, L. et al., Imaging apoptosis *in vivo* using 124I-annexin V and PET, *Nucl. Med. Biol.* 32, 395–402, 2005.
324. Braschi, S. et al., Role of the kidney in regulating the metabolism of HDL in rabbits: Evidence that iodination alters the catabolism of apolipoprotein A-1 by the kidney, *Biochemistry* 39, 5441–5449, 2000.
325. Li, H.S., Jiang, H.Y., and Carayanniotis, G., Modifying effects of iodine on the immunogenicity of thyroglobulin peptides, *J. Autoimmun.* 28, 171–174, 2007.
326. Sohoel, A. Plum, A., Frokjaer, S., and Thygesen, P., 125I used for labelling of proteins in an absorption model changes the absorption rate of insulin aspart, *Int. J. Pharm.* 330, 114–120, 2007.
327. Riordan, J.F. and Vallee, B.L., Acetylation, *Methods Enzymol.* 11, 565–576, 1967.
328. Karibian, D., Jones, C., Gertler, A., Dorrington, K.J., and Hofmann, T., On the reaction of acetic and maleic anhydrides with elastase. Evidence for a role of the NH2-terminal valine, *Biochemistry* 13, 2891–2897, 1974.
329. Ohnishi, M., Suganuma, T., and Hiromi, K., The role of a tyrosine residue of bacterial liquefying α-amylase in the enzymatic hydrolysis of linear substrates as studied by chemical modification with acetic anhydride, *J. Biochem.* (Tokyo) 76, 7–13, 1974.
330. Bernad, A., Nieto, M.A., Vioque, A., and Palacian, E., Modification of the amino groups and hydroxyl groups of lysozyme with carboxylic acid anhydrides: A comparative study, *Biochim. Biophys. Acta* 873, 350–355, 1986.
331. Simpson, R.T., Riordan, J.F., and Vallee, B.L., Functional tyrosyl residues in the active center of bovine pancreatic carboxypeptidase A, *Biochemistry* 2, 616–622, 1963.
332. Riordan, J.F., Wacker, W.E.C., and Vallee, B.L., N-Acetylimidazole: A reagent for determination of "free" tyrosyl residues of proteins, *Biochemistry* 4, 1758–1765, 1965.
333. Myers, B., II and Glazer, A.N., Spectroscopic studies of the exposure of tyrosine residues in proteins with special reference to the subtilisins, *J. Biol. Chem.* 246, 412–419, 1971.
334. Fife, T.H., Steric effects in the hydrolysis of N-acylimidazoles and ester of p-nitrophenol, *J. Am. Chem. Soc.* 87, 4597–4600, 1965.
335. Lee, J.P., Bembi, R., and Fife, T.H., Steric effects in the hydrolysis reactions of N-acylimidazoles. Effect of aryl substitution in the leaving group, *J. Org. Chem.* 62, 2872–2876, 1997.
336. Cronan, J.E., Jr. and Klages, A.L., Chemical synthesis of acyl thioesters of acyl carrier protein with native structure, *Proc. Natl. Acad. Sci. USA* 78, 5440–5444, 1981.
337. El Kebbaj, M.S. and Latruffe, N., Chemical reagents of polypeptide side chains. Relationship between solubility properties and ability to cross the inner mitochondrial membranes, *Cell. Mol. Biol.* 40, 781–786, 1994.
338. Martin, Wu, D., and Graves, D.J., Chemical influences on the specificity of tyrosine phosphorylation, *J. Biol. Chem.* 265, 7108–7111, 1990.
339. Riordan, J.F. and Vallee, B.L., O-Acetyltyrosine. *Methods Enzymol.* 25, 500–506, 1972.
340. Herriott, R.M., Reactions of native proteins with chemical reagents, *Adv. Protein Chem.* 3, 169, 1947.
341. Riordan, J.F., Sokolovsky, M., and Vallee, B.L., Tetranitromethane. A reagent for the nitration of tyrosine and tyrosyl residues in proteins, *J. Am. Chem. Soc.* 88, 4104–4105, 1966.
342. Sokolovsky, M., Harell, D., and Riordan, J.F., Reaction of tetranitromethane with sulfhydryl groups in proteins, *Biochemistry* 8, 4740–4745, 1969.

343. Riordan, J.F. and Vallee, B.L., Nitration with tetranitromethane, *Methods Enzymol.* 25, 515–521, 1972.
344. Cuatrecasas, P., Fuchs, S., and Anfinsen, C.B., The tyrosyl residues at the active site of staphylococcal nuclease. Modifications by tetranitromethane, *J. Biol. Chem.* 243, 4787–4798, 1968.
345. Sokolovsky, M., Fuchs, M., and Riordan, J.F., Reaction of tetranitromethane with tryptophan and related compounds, *FEBS Lett.* 7, 167–170, 1970.
346. Boesel, R.W. and Carpenter, F.H., Crosslinking during the nitration of bovine insulin with tetranitromethane, *Biochem. Biophys. Res. Commun.* 38, 678–682, 1970.
347. Nadeau, O.W., Traxler, K.W., and Carlson, G.M., Zero-length crosslinking of the beta subunit of phosphorylase kinase to the N-terminal half of its regulatory alpha subunit, *Biochem. Biophys. Res. Commun.* 251, 637–641, 1998.
348. Hugli, T.E. and Stein, W.H., Involvement of a tyrosine residue in the activity of bovine pancreatic deoxyribonuclease A, *J. Biol. Chem.* 246, 7191–7200, 1971.
349. Crow, J.P. and Ishiropoulos, H., Detection and quantitation of nitrotyrosine residues in proteins: *In vivo* marker of peroxynitrite, *Methods Enzymol.* 269, 185–194, 1996.
350. Greenacre, S.A.B. and Ischeriopoulos, H., Tyrosine nitration: Localization, quantification, consequences for protein function and signal transduction, *Free Radic. Res.* 34, 541–481, 2001.
351. Schmidt, P., Youhnovski, N., Daiber, A. et al., Specific nitration at tyrosine 430 revealed by high resolution mass spectrometry as basis for redox regulation of bovine prostacyclin synthase, *J. Biol. Chem.* 278, 12813–12819, 2003.
352. Petersson, A.-S., Steen, H., Kalume, D.E. et al., Investigation of tyrosine nitration in proteins by mass spectrometry, *J. Mass Spectrom.* 36, 616–625, 2001.
353. Willard, B.B., Ruse, C.I., Keightley, J.A. et al., Site-specific quantitation of protein nitration using liquid chromatography/tandem mass spectrometry, *Anal. Chem.* 75, 2370–2376, 2003.
354. Daiber, A., Bachschmid, M., Kavakli, C. et al., A new pitfall in detecting biological end products of nitric oxide—nitration, nitros(yl)ation and nitrite/nitrate artifacts during freezing, *Nitric Oxide* 9, 44–52, 2003.
355. Irie, Y., Saeki, M., Kamisaki, Y. et al., Histone H1.2 is a substrate for dinitrase, an activity that reduces nitrotyrosine immunoreactivity in proteins, *Proc. Nat. Acad. Sci.* USA 100, 5634–5639, 2003.
356. Nikov, G., Bhat, V., Wishnok, J.S. et al., Analysis of nitrated proteins by nitrotyrosine-specific affinity probes and mass spectrometry, *Anal. Biochem.* 320, 214–222, 2003.
357. Ogino, K., Nakajima, M., Kodama, N. et al., Immunohistochemical artifact for nitrotyrosine in eosinophils or eosinophil containing tissue, *Free Radic. Res.* 36, 1163–1170. 2002.
358. Miyagi, M., Sakaguchi, H., Darrow, R.M. et al., Evidence that light modulates protein nitration in rat retina, *Mol. Cell. Proteomics* 1, 293–303, 2003.
359. Zu, Y, Strong, M., Huang, Z., and Beckman, J.S., Antibodies that recognize nitrotyrosine, *Methods Enzymol.* 269, 201–209, 1996.
360. Sokolovsky, M., Riordan, J.F., and Vallee, B.L., Conversion of 3-nitrotyrosine to 3-aminotyrosine in peptides and proteins, *Biochem. Biophys. Res. Commun.* 27, 20–25, 1967.
361. Shovov, V.S., Dremina, E.S., Pennington, J.S. et al., Selective fluorogenic derivatization of 3-nitrotyrosine and 3,4-dihydroxyphenylalanine in peptides: A method designed for quantitative proteomics analysis, *Methods Enyzmol.* 441, 19–32, 2008.
362. Haas, J.A., Frederick, M.A., and Fox, B.G., Chemical and post-translational modifications of *Escherichia coli* acyl carrier protein for preparation of dansyl carrier protein, *Prot. Exp. Purif.* 20, 274–284, 2000.
363. Haas, J.A. and Fox, B.G., Fluorescence anisotropy studies of enzyme-substrate complex formation in stearoyl-ACP-desaturase, *Biochemistry* 41, 14472–14481, 2002.

364. Beckman, J.S. et al., Oxidative chemistry of peroxynitrite, *Methods Enzymol.* 233, 229–240, 1994.
365. Pryor, W.A., Cueto, R., Jin, X. et al., A practical method for preparing peroxynitrite solutions of low ionic strength and free of hydrogen peroxide, *Free Rad. Biol. Med.* 18, 75–83, 1995.
366. Uppu, R.M., Squadrito, G.L., Cueto, R., and Pryor, W.L., Selecting the most appropriate synthesis of peroxynitrite, *Methods Enzymol.* 269, 285–295, 1996.
367. Ischiropoulos, H., Biological tyrosine nitration: A pathophysiological function of nitric oxide and reactive oxygen species, *Arch. Biochem. Biophys.* 356, 1–11, 1998.
368. Ischiropoulos, H., Biological selectivity and functional aspects of protein tyrosine nitration, *Biochem. Biophys. Res. Commun.* 305, 776–783, 2003.
369. Szabo, C., Ischiropoulos, H., and Radi, R., Peroxynitrite: Biochemistry, pathophysiology, and development of therapeutics, *Nat. Rev. Drug. Discov.* 6, 662–680, 2007.
370. Zhang, H., Joseph, J., Feix, J. et al., Nitration and oxidation of a hydrophobic tyrosine probe by peroxynitrite in membranes: Comparison with nitration and oxidation by peroxynitrite in aqueous solution, *Biochemistry* 40, 7675–7686, 2001.
371. Bartesaghi, S., Petuffa, G., Zhang, H. et al., Tyrosine nitration, dimerization, and hydroxylation by peroxynitrite in membranes as studied by the hydrophobic probe N-t-BOC-L-tyrosine-t-butyl ester, *Methods Enzymol.* 411, 217–236, 2008.
372. Zhang, H., Bhargava, K., Keszler, A. et al., Transmigration nitration of hydrophobic tyrosyl peptides. Localization, characterization, mechanisms of nitration, and biological implications, *J. Biol. Chem.* 278, 8969–8978, 2003.
373. Shao, B., Bergt, C., and Fio, X., Tyrosine 192 in apolipoprotein A-I is the major site of nitration and chlorination by myeloperoxidase, but only chlorination markedly impairs ABCA1-dependent cholesterol transport, *J. Biol. Chem.* 280, 5983–5993, 2005.
374. Bartesaghi, S., Ferrer-Sueta, G., Peluffo, G. et al., Protein tyrosine nitration in hydrophilic and hydrophobic environments, *Amino Acids* 32, 501–515, 2007.
375. Lennon, C.W., Cox, H.D., Hennelly, S.P. et al., Probing structural differences in prion protein isoforms by tyrosine nitration, *Biochemistry* 46, 4850–4860, 2007.
376. Kurihara, K., Horinishi, H., and Shibata, K., Reaction of cyanuric halides with proteins. I. Bound tyrosine residues of insulin and lysozyme as identified with cyanuric fluoride, *Biochim. Biophys. Acta* 74, 678–687, 1963.
377. Gorbunoff, M.J., Cyanuration, *Methods Enzymol.* 25, 506–514, 1972.
378. Gorbunoff, M.J. and Timasheff, S.N., The role of tyrosines in elastase, *Arch. Biochem. Biophys.* 152, 413–422, 1972.
379. Coffe, G. and Pudles, J., Chemical reactivity of the tyrosyl residues in yeast hexokinase. Properties of the nitroenzyme, *Biochim. Biophys. Acta.* 484, 322–335, 1977.
380. Ferguson, S.J., Lloyd, W.J., Lyons, M.H., and Radda, G.K., The mitochondrial ATPase. Evidence for a single essential tyrosine residue, *Eur. J. Biochem.* 54, 117, 1975.
381. Ferguson, S.J., Lloyd, W.J., and Radda, G.K., The mitochondrial ATPase. Selective modification of a nitrogen residue in the β-subunit, *Eur. J. Biochem.* 54, 127–135, 1975.
382. Andrews, W.W. and Allison, W.S., 1-Fluoro-2,4-dinitrobenzene modifies a tyrosine residue when it inactivates the bovine mitochondrial F1-ATPase, *Biochem. Biophys. Res. Commun.* 99, 813–819, 1981.
383. De Vendittis, E.M, Ursby, T., Rullo, R. et al., Phenylmethylsulfonyl fluoride inactivates an archael superoxide dismutase by chemical modification of a specific tyrosine residue. Cloning, sequencing and expression of the gene coding for Sulfolobus solfataricus superoxide dismutase, *Eur. J. Biochem.* 268, 1794–1801, 2001.
384. Means, G.E. and Wu, H.L., The reactive tyrosine residue of human serum albumin: Characterization of its reaction with diisopropylfluorophosphate, *Arch. Biochem. Biophys.* 194, 526–530, 1979.

385. Riordan, J.F. and Vallee, B.L., Diazonium salts as specific reagents and probes of protein conformation, *Methods Enzymol.* 25, 521–531, 1972.
386. Tabachnick, M. and Sobotka, H., Azoproteins. I. Spectrophotometric studies of amino acid azo derivatives, *J. Biol. Chem.* 234, 1726–1730, 1959.
387. Tabachnick, M. and Sobotka, H., Azoproteins. II. A spectrophotometric study of the coupling of diazotized arsanilic acid with proteins, *J. Biol. Chem.* 235, 1051–1754, 1960.
388. Fairclough, G.F., Jr. and Vallee, B.L., Arsanilazochymotrypsinogen. The extrinsic Cotton effects of an arsanilazotyrosyl chromophore as a conformation probe of zymogen activation, *Biochemistry* 10, 2470–2477, 1971.
389. Johansen, J.T., Livingston, D.M., and Vallee, B.L., Chemical modification of carboxypeptidase A crystals. Azo coupling with tyrosine-248, *Biochemistry* 11, 2584–2588, 1972.
390. Harrison, L.W. and Vallee, B.L., Kinetics of substrate and product interactions with arsanilazotyrosine-248 carboxypeptidase A, *Biochemistry* 17, 4359–4363, 1978.
391. Cueni, L. and Riordan, J.F., Functional tyrosyl residues of carboxypeptidase A. The effect of protein structure on the reactivity of tyrosine-198, *Biochemistry*, 17, 1834–1842, 1978.
392. Liao, T.-H., Ting, R.S., and Young, J.E., Reactivity of tyrosine in bovine pancreatic deoxyribonuclease with p-nitrobenzenesulfonyl fluoride, *J. Biol. Chem.* 257, 5637–5644, 1982.
393. Friedman, M., *The Chemistry and Biochemistry of the Sulfhydryl Group in Amino Acids, Peptides, and Proteins*, Pergammon Press, Oxford, UK, 1973.
394. Ohno A. and Oae, S., Thiols, in *Organic Chemistry of Sulfur,* Ed. S. Oae, Plenum Press, New York, 1977, Chapter 4.
395. Sluyterman, L.A.A., The rate-limiting reaction in papain action as derived from the reaction of the enzyme with chloroacetic acid, *Biochem. Biophys. Acta.* 151, 178–187, 1968.
396. Cecil, R. and McPhee, J.R., The sulfur chemistry of proteins, *Adv. Prot. Chem.* 14, 255–389, 1959.
397. Liu, T.-Y., The role of sulfur in proteins, in *The Proteins*, Vol. 3, 3rd ed., Neurath, H.H. and Hill, R.L., Eds., Academic Press, New York, 1977.
398. Torchinsky, Y.M., *Sulfur in Proteins*, Pergammon Press, Oxford, 1981.
399. Creighton, T.E., Chemical nature of polypeptides, in *Proteins. Structure and Molecular Principles*, W.H. Freeman and Company, New York, Chapter 1, 1983.
400. Modena, G., Paradisi, C., and Scorrano, G., Solvation effects on basicity and nucleophilicity, in *Organic Sulfur Chemistry. Theoretic and Experimental Advances*, Ed. F. Bernardi, I.G. Csizmadia, and A. Mongini, Elsevier, Amersterdam, Netherlands, 1985.
401. Britto, P.J., Knipling, L., and Wolff, J., The local electrostatic environment determines cysteine reactivity of tubulin, *J. Biol. Chem.* 277, 29018–29027, 2002.
402. Gerwin, B.I., Properties of the single sulfhydryl group of streptococcal proteinase. A comparison of the rates of alkylation by chloroacetic acid and chloroacetamide, *J. Biol. Chem.* 242, 451, 1967.
403. Chiappe, C., Pieraccini, D., and Saullo, P., Nucleophilic displacement reactions in ionic liquids: Substrates and solvent effect in the reaction of of NaN_3 with alkyl halides and tosylates, *J. Org. Chem.* 68, 6710–6715, 2003.
404. Evans, B.L.B., Knopp, J.A., and Horton, H.R., Effect of hydroxynitrobenzylation of tryptophan-177 on reactivity of active site cysteine-25 in papain, *Arch. Biochem. Biophys.* 206, 362–371, 1981.
405. Dyson, H.J., Jeng, M.-F., and Ennant, L.L., Effects of buried charged groups of cysteine thiol ionization and reactivity in *Escherichia coli* thioredoxin: Structural and functional characterization of mutants of Asp 26 and Lys 57, *Biochemistry* 36, 2622–2636, 1997.
406. Nelson, K.J., Day, A.E., and Zeng, B.-B., Isotope-coded iodoacetamide-based reagent to determine individual cysteine p*K*a values by matrix-assisted laser desorption/ionization time-of-flight mass spectrometry, *Anal. Biochem.* 375, 187–195, 2008.

407. Kanaya, S., Kimura, S., Katsuda, C., and Ikehara, M., Role of cysteine residues in ribonuclease H from *Escherichia coli*. Site-directed mutagenesis and chemical modification, *Biochem. J.* 271, 59, 1990.
408. Kato, H., Tanaka, T., Nishioka, T., Kimura, A., and Oda, J., Role of cysteine residues in glutathione synthetase from *Escherichia coli B*. Chemical modification and oligonucleotide site-directed mutagenesis, *J. Biol. Chem.* 263, 11646, 1988.
409. Bech, L.M. and Breddam, K., Chemical modification of a cysteinyl residue introduced in the binding site of carboxypeptidase Y by site-directed mutagenesis, *Carlsberg Res. Commun.* 53(Abstr.), 381, 1988.
410. Minard, P., Desmadril, M., Ballery, N., Perahia, D., Mouawad, L., Hall, L., and Yon, J.M., Study of the fast-reacting cysteines in phosphoglycerate kinase using chemical modification and site-directed mutagenesis, *Eur. J. Biochem.* 185, 419–423, 1989.
411. Turina, P., Structural changes during ATP hydrolysis activity of the ATP synthase from *Escherichia coli* as revealed by fluorescent probes, *J. Bioenerg. Biomembr.* 32, 373–381, 2000.
412. Ahmed, S.A., Kawasaki, H., and Bauerle, R., Site-directed mutagenesis of the alpha subunit of tryptophan synthase from *Salmonella* typhimurium, *Biochem. Biophys. Res Commun.* 151, 672–678, 1988.
413. Thrower, A.R., Byrd, J., and Tarbet, E.B., Effect of mutation of cysteinyl residues in yeast Cu-metallothionein, *J. Biol. Chem.* 263, 7037–7242, 1988.
414. Winther, J.R. and Breddam, K., The free sulfhydryl group (CYS341) of carboxypeptidase Y: Functional effects of mutational substitutions, *Carlsberg Res. Commun.* 52, 263–273, 1987.
415. Seal, R.P., Leighten, B.H., and Amora, S.G., A model for the topology of excitatory amino acid transporters determined by the extracellular accesssiblity of substituted cysteines, *Neuron* 25, 695–706, 2000.
416. Toedt, G.H., Krishnan, R., and Friedhoff, P., Site-specific protein modification to identify the MutL interface of MutH, *Nucl. Acids Res.* 31, 819–825, 2003.
417. Venkatesan, P., Hu, Y., and Kaback, H.R., Site-directed sulfhydryl labeling of the lactose permease of *Escherichia coli*: N-ethylmaleimide-sensitive face of Helix II, *Biochemistry* 29, 10649–10655, 2000.
418. Chan, B.S., Santriano, J.A., and Schuster, V.L., Mapping the substrate binding site of the prostaglandin transporter PGT by cysteine scanning mutagenesis, *J. Biol. Chem.* 274, 25564–25570, 1999.
419. Frillingos, S., Sahin-Toth, M., Wu, J., and Kaback, H.R., Cys-scanning mutagenesis: A novel approach to structure-function in polytopic membrane proteins, *FASEB J.* 12, 1281–1299, 1998.
420. Yagur-Kroll, S. and Amster-Choder, O., Dynamic membrane topology of he *Escherichia coli* β-glucoside transporter BglF, *J. Biol. Chem.* 280, 19306–19318, 2005.
421. Want, Y., Toel, M., and Forgac, M., Analysis of the membrane topology of transmembrane segments in the *C*-terminal hydrophobic domain of the yeast vacuolar ATPase subunits a (vphlp) by chemical modification, *J. Biol. Chem.* 283, 20696–20702, 2008.
422. Smith, D.J., Maggio, E.T., and Kenyon, G.L., Simple alkane thiol groups for temporary blocking of sulfhydryl groups of enzymes, *Biochemistry* 14, 766–771, 1975.
423. Kluger, R. and Tsue, W.-C., Amino group reactions of the sulfhydryl reagent methyl methanethiosulfonate. Inactivation of D—3-hydroxybutyrate and reaction with amines in water, *Can. J. Biochem.* 58, 629–632, 1980.
424. Pathak, R., Hendrickson, T.L., and Imperiali, B., Sulfhydryl modification of the yeast Wbp1p inhibits oligosaccharide transferase activity, *Biochemistry* 34, 4179–4181, 1995.
425. Roberts, D.D., Lewis, S.D., and Ballou, D.P., Reactivity of small thiolate anions and cysteine-25 in papain toward methyl methanethiosulfonate, *Biochemistry* 25, 5595–5601, 1986.

426. Stauffers, D.A. and Karlin, A., Electrostatic potential of the acetylcholine binding site in the nicotinic receptor probed by reaction of binding-site cysteines with charged methanethiosulfonates, *Biochemistry* 33, 6840–6849, 1994.
427. Kenyon, G.L. and Bruice, T.W., Novel sulfhydryl reagents, *Methods Enzymol.* 47, 407–430, 1977.
428. Bruice, T.W. and Kenyon, G.L., Novel alkyl alkanethiolsulfonate sulfhydryl reagents. Modification of derivatives of l-cysteine, *J. Protein Chem.* 1, 47–58, 1982.
429. Toronto Research Chemicals, Inc.; http://www.trc-canada.com/white_papers.lasso: Methanethiosulfonate reagents: Application to the study of protein topology and ion channels.
430. Lobo, I.A., Maseia, M.P., Trudell, J.P., and Harris, R.A., Channel gating of the glycine receptor changes accessibility to residues implicated in receptor potentiation by alcohols and anesthetics, *J. Biol. Chem.* 279, 33919–33928, 2004.
431. Enkvetchakul, D., Jeliazkova, I., Bhattacharyya, J., and Nichols, C.G., Control of inward rectifier K channel activity by lipid tethering of cytoplasmic domains, *J. Gen. Physiol.* 130, 329–334, 2007.
432. Dahl, K.S. and McKinley-McKee, J.S., The reactivity of affinity labels: A kinetic study of the reaction of alkyl halides with thiolate anions—a model reaction for protein alkylation, *Bioorg. Chem.* 10, 329–341, 1981.
433. Crestfield, A.M., Moore, S., and Stein, W.H., The preparation and enzymatic hydrolysis of reduced and S-carboxymethylated proteins, *J. Biol. Chem.* 238, 622–627, 1963.
434. Friedman, M., Krull, L.H., and Cavins, J.F., The chromatographic determination of cystine and cysteine residues in proteins as *S*-β-(4-pyridyl-ethyl) cysteine, *J. Biol. Chem.* 245, 3868–3871, 1970.
435. Mak, A.S. and Jones, B.L., Application of S-pyridylethylation of cysteine to the sequence analysis of proteins, *Anal. Biochem.* 84, 432–440, 1978.
436. Plouq, M., Stoffer, B., and Jensen, A.L., *In situ* alkylation of cysteine residues in a hydrophobic membrane protein immobilized on polyvinylidene difluoride membranes by electroblotting prior to microsequence and amino acid analysis, *Electrophoresis* 13, 148–153, 1992.
437. Lundell, N. and Schreitmüller, T., Sample preparation of peptide-mapping—A pharmaceutical quality-control perspective, *Anal. Biochem.* 266, 31–47, 1999.
438. Sechi, S. and Chait, B.T., Modification of cysteine residues by alkylation. A tool in peptide mapping and protein identification, *Anal. Chem.* 70, 5150–5158, 1998.
439. Görg, A. Obermaier, C., Boguth, G. et al, The current state of two-dimensional electrophoresis with immobilized pH gradients, *Electrophoresis* 21, 1037–1053, 2000.
440. Santos, H.M., Rial-Otero, R., Fernandes, L. et al., Improving sample treatment for in-solution identification by peptide mass fingerprinting using matrix-assisted laser desorption/ionization time-of-flight mass spectrometry, *J. Proteome Res.* 6, 3393–3399, 2007.
441. Herbert, B. et al., Reduction and alkylation of proteins in preparation of two-dimensional map analysis. Why, when, and how?, *Electrophoresis* 22, 2046–2057, 2001.
442. Hoving, S. et al., Preparative two-dimensional gel electrophoresis at alkaline pH using narrow range immobilized pH gradients, *Proteomics* 2, 127–134, 2002.
443. Shaw, M.M. and Riederer, B.M., Sample preparation for two-dimensional gel electrophoresism, *Proteomics* 3, 1408–1417, 2003.
444. Herbert, B. et al., β-elimination: An unexpected artifact in proteome analysis, *Proteomics* 3, 826–831, 2003.
445. Rabilloud, T., Giraudel, A., and Lunardi, T., Improvement of the solubilization of proteins in two-dimensional electrophoresis with immobilized pH gradients, *Electrophoresis* 18, 307–316, 1997.
446. Galvani, M., Rovatti, L., Hamdan, M. et al., Protein alkylation in the presence/absence of thiourea in proteome analysis: A matrix assisted laser desorption/ionization—time of flight—mass spectrometry investigation, *Electrophoresis* 22, 2066–2074, 2001.

447. Tyagarajon, K., Pretzer, E., and Wiktorowicz, J.E., Thiol-reactive dyes for fluorescence labeling of proteomic samples, *Electrophoresis* 24, 2348–2358, 2003.
448. Plowman, J.E. Flanagan, L.M., and Paton, L.N., The effect of oxidation or alkylation on the separation of wool keratin proteins by two-dimensional gel electrophoresis, *Proteomics* 3, 942–950, 2003.
449. Galvani, M., Hamdan, M., Herbert, B., and Righetti, P.G., Alkylation kinetics of proteins in preparation for two-dimensional maps. A matrix, *Electrophoresis* 22, 2058–2065, 2001.
450. Friedman, M., Application of the S-pyridylethylation reaction to the elucidation of the structure and function of proteins, *J. Prot. Chem.* 20, 431–453, 2001.
451. Sebastiano, R., Citterio, A., Lapadula, M., and Righetti, P.G., A new deuterated alkylating agent for quantitative proteomics, *Rapid Commun. Mass Spectrom.* 17, 2380–2386, 2003.
452. Gregory, J.D., The stability of N-ethylmaleimide and its reaction with sulfhydryl groups, *J. Am. Chem. Soc.* 77, 3922–3923, 1955.
453. Leslie, J., Spectral shifts in the reaction of N-ethylmaleimide with proteins, *Anal. Biochem.* 10, 162–167, 1965.
454. Gorin, G., Martic, P.A., and Doughty, G., Kinetics of the reaction of N-ethylmaleimide with cysteine and some congeners, *Arch. Biochem. Biophys.* 115, 593–597, 1966.
455. Bednar, R.A., Reactivity and pH dependence of thiol conjugation to *N*-ethylmaleimide: Detection of a conformational change in chalcone isomerase, *Biochemistry* 29, 3684–3690, 1990.
456. Smyth, D.G., Blumenfeld, O.O., and Konigsberg, W., Reaction of N-ethylmaleimide with peptides and amino acids, *Biochem. J.* 91, 589–595, 1964.
457. Gehring, H. and Christen, P., A diagonal procedure for isolating sulfhydryl peptides alkylated with N-ethylmaleimide, *Anal. Biochem.* 107, 358–361, 1980.
458. Bordini, E., Hamdan, M., and Righetti, P.G., Probing acrylamide alkylation sites in cysteine-free proteins by matrix-assisted laser desorption/ionization time-of-flight, *Rapid Commun. Mass Spectrom.* 14, 840–848, 2000.
459. Mineki, R., Taka, H., Fujimura, T. et al., *In situ* alkylation with acrylamide for identification of cysteinyl residues in proteins during one-and two-dimensional sodium dodecyl sulphate-polyacrylamide gel electrophoresis, *Proteomics* 2, 1672–1681, 2002.
460. Cahill, M.A., Wozny, W., Schwall, G. et al., Analysis of relative isopologue abundances for quantitative profiling of complex protein mixtures labelled with acrylamide/D-3-acrylamide alkylation tag system, *Rapid Commun. Mass Spectrom.* 17, 1283–1290, 2003.
461. Ellman, G.L., A colorimetric method for determining low concentrations of mercaptans, *Arch. Biochem. Biophys.* 74, 443–450, 1958.
462. Riddles, P.W., Blakeley, R.L., and Zerner, B., Ellman's reagent: 5,5′-dithiobis(2-nitrobenzoic acid)—a reexamination, *Anal. Biochem.* 94, 75–81, 1979.
463. Riddles, P.W., Blakeley, R.L., and Zerner, B., Reassessment of Ellman's reagent, *Methods Enzymol.* 91, 49–60, 1983.
464. Fernandez-Diaz, M.D., Barsotti, L., Dumay, E., and Cheftel, J.C, Effects of electric fields on ovalbumin solutions and dialyzed egg white, *J. Agric. Food Chem.* 48, 2332–2339, 2000.
465. Helten, A. and Koch, K.W., Calcium-dependent conformational changes in guanylate cyclase-activting protein 2 monitored by cysteine accessibility, *Biochem. Biophys. Res. Commun.* 356, 687–692, 2007.
466. Okonjo, K.O., Bello, O.S., and Babalola, J.O., Transition of hemoglobin between two tertiary conformations: The transition constant differs significantly for the major and minor hemoglobins of the Japanese quail (*Corunix cortunix japonica*), *Biochim. Biophys. Acta* 1784, 464–471, 2008.
467. Jönsson, T.J., Ellils, H.R., and Poole, L.B., Cysteine reactivity and thiol-disulfide interchange pathways in AhpF and AhpC of the bacterial alkyl hydroperoxide reductase system, *Biochemistry* 46, 5709–5721, 2007.

468. Grassetti, D.R. and Murray, J.F., Jr., Determination of sulfhydryl groups with 2,2′ or 4,4′-dithiodipyridine, *Arch. Biochem. Biophys.* 199, 41–49, 1967.
469. Talgoy, M.M., Bell, A.W., and Duckworth, H.W., The reactions of *Escherichia coli* citrate synthase with the sulfhydryl reagents 5,5′-dithiobis-(2-nitrobenzoic acid) and 4,4′-dithiodipyridine, *Can. J. Biochem.* 57, 822–833, 1979.
470. Hansen, R.E., Østergaard, H., Nørgaard, P., and Winther, J.R., Quantification of protein thiols and dithiols in the picomolar range using sodium borohydride and 4,4-dithiodipyridine, *Anal. Biochem.* 363, 77–82, 2007.
471. Kimura, T., Matsueda, R., Nakagawa, Y., and Kaiser, E.T., New reagents of the introduction of the thiomethyl group at sulfhydryl residues of proteins with concomitant spectrophotometric titration of the sulfhydryl: methyl 3-nitro-2-pyridyl disulfide and methyl 2-pyridyl disulfide, *Anal. Biochem.* 122, 274–282, 1982.
472. Hellerman, L., Chinard, F.P., and Deitz, V.R., Protein sulfhydryl groups and the reversible inactivation of the enzyme urease. The reducing groups of egg albumin and the enzyme urease, *J. Biol. Chem.* 147, 443–462, 1943.
473. Boyer, P.D., Spectrophotometric study of the reaction of protein sulfhydryl groups with organic mercurials, *J. Am. Chem. Soc.* 76, 4331–4337, 1954.
474. Rothstein, A., Sulfhydryl groups in membrane structure and function, *Curr. Top. Memb. Transport* 1, 135–176, 1970.
475. Vanstevenick, J., Weed, R.J., and Rothstein, A., Localization of erythrocyte membrane sulfhydryl groups essential in glucose transport, *J. Gen. Physiol.* 48, 617–632, 1964–1965.
476. Fahn, S., Hurley, M.R., Koval, G.J. et al., Sodium-potassium-activated adenosine triphosphatase of *Electrophorus* electric organ. II. Effects of *N*-ethylmaleimide and other sulfhydryl reagents, *J. Biol. Chem.* 241, 1890–1895, 1966.
477. Ding, Z., Kim, S., Dorsam, R.T. et al., Inactivation of the human $P2Y_{12}$ receptor by thiol reagents requires interaction with both extracellular cysteine residues, Cys17 and Cys 270, *Blood* 101, 3908–3914, 2003.
478. Chiang, W.-C. and Knowles, A.F., Inhibition of human NTPDase 2 by modification of an intramembrane cysteine by *p*-chloromercuriphenylsulfonate and oxidative cross-linking of the transmembrane domains, *Biochemistry* 47, 8447–8785, 2008.
479. Neet, K.E. and Koshland, D.E., Jr., The conversion of serine at the active site of subtilisin to cysteine: A "chemical mutation," *Proc. Natl. Acad. Sci. USA* 56, 1606–1611, 1966.
480. Polgar, L. and Bender, M.L., The reactivity of thiol-subtilisin, an enzyme containing a synthetic functional group, *Biochemistry* 6, 610–620, 1967.
481. Slade, A., Horrocks, A.J., and Lindsay, C.D., Site-directed chemical conversion of serine to cysteine in penicillin acylase from *Escherichia coli* ATCC 11105. Effect of conformation and catalytic activity, *Eur. J. Biochem.* 197, 75–80, 1991.
482. Finley, J.W. and Friedman, M., New amino acid derivatives formed by alkaline treatment of proteins, *Adv. Exp. Med. Biol.* 86B, 123–130, 1977.
483. Bernardes, G.J.L., Chalker, J.M., Errey, J.C., and Davis, B.G., Facile conversion of cysteine and alkyl cysteines to dehydroalanine on protein surfaces: Versatile and switchable access to functionalized proteins, *J. Am. Chem. Soc.* 130, 5052–5053, 2008.
484. Levengood, M.R. and van der Donk, W.A., Dehydroalanine-containing peptides: Preparation from phenylselenocysteine and utility in convergent ligation strategies, *Nat. Protoc.* 3, 3001–3010, 2006.
485. Bar-Or, R., Rael, L.T., and Bar-Or, D., Dehydroalanine derived from cysteine is a common post-translational modification in human serum albumin, *Rapid Commun. Mass Spectrom.* 22, 711–716, 2008.
486. Stark, G.R., On the reversible reaction of cyanate with sulfhydryl groups and the determination of NH_2-terminal cysteine and cystine in proteins, *J. Biol. Chem.* 239, 1411–1414, 1964.

487. Jacobson, G.R., Schaffer, M.H., Stark, G.R., and Vanaman, T.C., Specific chemical cleavage in high yield at the amino peptide bonds of cysteine and cystine residues, *J. Biol. Chem.* 248, 6583–6591, 1973.
488. Wu, J., Gage, D.A., and Watons, J.T., A strategy to locate cysteine residues in proteins by specific chemical cleavage followed by matrix-assisted laser desorption/ionization time-of-flight mass spectrometry, *Anal. Biochem.* 235, 161–174, 1996.
489. Tang, H.Y. and Speicher, D.W., Identification of alternative products and optimization of the 2-nitro-5-thiocyanatobenzoic acid cyanylation and cleavage at cysteine residues, *Anal. Biochem.* 334, 48–61, 2004.
490. Fafarman, A.T. Webb, L.J., Chuang, J.I., and Boxere, S.G., Site-specific conversion of cysteine thiols into thiocyanate creates an IR probe for electric fields in proteins, *J. Am. Chem. Soc.* 128, 13356–13357, 2006.
491. Luo, D., Smith, S.W., and Anderson, B.D., Kinetics and mechanism of the reaction of cysteine and hydrogen peroxide in aqueous solution, *J. Pharm. Sci.* 94, 304–316, 2005.
492. Ikegaya, K. et al., Kinetic analysis of enhanced thermal stability of an alkaline protease with engineered twin disulfide bridges and calcium-dependent stability, *Biotechnol. Bioeng.* 81, 187–192, 2003.
493. Reiter, Y., Brinkman, U., Lee, B., and Pastan, I., Engineering antibody Fv fragments for cancer detection and therapy: Disulfide-stabilized Fv fragments, *Nat. Biotechnol.* 14, 1239–1245, 1996.
494. Craik, D.J., Daly, N.L., and Waine, C., The cysteine know motif in toxins and implications for drug design, *Toxicon* 39, 43–60, 2001.
495. Li, W.F., Zhou, X.X., and Lu, P., Structural features of thermozymes, *Biotechnol. Adv.* 23, 271–281, 2005.
496. Craik, D.J. and Adams, D.J., Chemical modification of conotoxins to improve stability and activity, *ACS Chem. Biol.* 2, 457–468, 2007.
497. Cemazar, M., Gruber, C.W., and Craik, D.J., Oxidative folding of cyclic cystine knot proteins, *Antioxid. Redox Signal.* 10, 103–111, 2008.
498. Gorin, G. and Godwin, W.E., The reaction of iodate with cystine and with insulin, *Biochem. Biophys. Res Commun.* 25, 227–232, 1966.
499. Moore, S., On the determination of cystine as cysteic acid, *J. Biol. Chem.* 239, 235–237, 1963.
500. Cole, R.D., Sulfitolysis, *Meth. Enzymol.* 11, 206–208, 1967.
501. Pihl, A. and Lange, R., The interaction of oxidized glutathione, cystamine monosulfoxide, and tetrathionate with the –SH groups of rabbit muscle D-glyceraldehyde 3-phosphate dehydrogenase, *J. Biol. Chem.* 237, 1356–1362.
502. Thannhauser, T.W., Konishi, Y., and Scheraga, H.A., Sensitive quantitative analysis of disulfide bonds in polypeptides and proteins, *Anal. Biochem.* 138, 181–188, 1984.
503. Kella, N.K.D. and Kinsella, J.E., A method for the controlled cleavage of disulfide bonds in proteins in the absence of denaturants, *J. Biochem. Biophys. Meth.* 11, 251–263, 1985.
504. Nilsson, J. et al., Integrated production of human insulin and its C-peptide, *J. Biotechnol.* 48, 241–259, 1996.
505. Mukhopadhyay, A., Reversible protection of disulfide bonds followed by oxidative folding render recombinant hCGβ highly immunogenic, *Vaccine* 18, 1802–1810, 2000.
506. Tikhonov, R.V. et al., Recombinant human insulin. VII. Isolation of fusion protein-*S*-sulfonate, biotechnological precursor of human insulin from the biomass of transformed *Escherichia coli* cells, *Prot. Exp. Purif.* 21, 176–182, 2001.
507. Gorin, G., Fulford, R., and Deonier, R.C., Reaction of lysozyme with dithiothreitol and with other mercaptans, *Experientia* 24, 26–27, 1968.
508. Cleland, W.W., Dithiothreitol, a new protective reagent for SH groups, *Biochemistry* 3, 480–482, 1964.

509. Iyer, K.S. and Klee, W.A., Direct spectrophotometric measurement of the rate of reduction of disulfide bonds. The reactivity of the disulfide bonds of bovine α-lactalbumin, *J. Biol. Chem.* 248, 707, 1973.
510. Neumann, H. and Smith, R.L., Cleavage of the disulfide bonds of cystine and oxidized glutathione by phosphorothioate, *Arch. Biochem. Biophys.* 122, 354–361, 1967.
511. Borman, C.D., Wright, C., Twitchett, M.D. et al. Pulse radiolysis studies on galactose oxidase, *Inorganic Chem.* 41, 2158–2163, 2002.
512. Vanhooren, A., Devreese, B., Vanhee, K. et al., Photoexcitation of tryptophan groups induces reduction of two disulfide bonds in goat α-lactalbumin, *Biochemistry* 41, 11035–11043, 2002.
513. Permyakov, E.A., Permyakov, S.E., Deikus, G.Y. et al., Ultraviolet illumination-induced reduction of α-lactalbumin disulfide bridges, *Prot. Struct. Funct. Genet.* 51, 498–503, 2003.
514. Miller, B.L., Hageman, M.J., Thamann,T.J. et al., Solid-state photodegradation of bovine somatotropin (bovine growth hormone): Evidence for tryptophan-mediated photooxidation of disulfide bonds, *J. Pharm. Sci.* 92, 1698–1709, 2003.
515. Alphey, M.S., Fairlamb, A.H., Bond, C.S., and Hunter, W.N., Tryporedoxins from *Crithidia fasciculata* and *Trypanosoma brucei*: Photoreduction of the redox disulfide using synchrotron radiation and evidence for a conformational switch implicated in function, *J. Biol. Chem.* 278, 25919–25925, 2003.
516. Ravelli, R.B.G. and McSweeney, S.M., The "fingerprint" that x-rays can leave on structures, *Structure* 8, 315–328, 2000.
517. Overman, L.E., Matzinger, D., O'Connor, D.M., and Overman, J.D., Nucleophilic cleavage of the sulfur-sulfur bond by phosphorus nucleophiles. Kinetic study of the reduction of aryl disulfides with triphenylphosphine and water, *J. Am. Chem. Soc.* 96, 6081–6089, 1975.
518. Rüegg, U.T., Reductive cleavage of *S*-sulfo groups with tributylphosphine, *Methods Ezymol.* 47, 123–126, 1977.
519. Rüegg, U.T. and Rudinger, J., Reductive cleavage of cystine disulfides with tributylphosphine, *Methods. Enzymol* 47, 111–116, 1977.
520. Grayson, M. and Farley, C.E., in *Chimie Organique du Phosphoros, Colloq. Int. C. N. R. S.*, No. 182, p. 275, 1969.
521. Burns, J.A., Butler, J.C., Moran, J., and Whitesides, G.M., Selective reduction of disulfides by tris-(2-carboethoxyethyl)-phosphinc, *J. Org. Chem.* 56, 2648–2650, 1991.
522. Gray, W.R., Disulfide structures of highly bridged peptides: A new strategy for analysis, *Protein Sci.* 2, 1732–1748, 1993.
523. Donovan, J.W., Spectrophotometric observation of the alkaline hydrolysis of protein disulfide bonds, *Biochem. Biophys. Res. Commun.* 29, 734, 1967.
524. Florence, T.M., Degradation of protein disulphide bonds in dilute alkali, *Biochem. J.* 189, 507–520, 1980.
525. Jensen, J.L., Kolvenbach, C., Roy, S., and Schöneich, C., Metal-catalyzed oxidation of bran-derived neurotrophic factor (BDNF): Analytical challenge for the identification of modified sites, *Pharm. Res.* 17, 190–196, 2000.
526. Duenas, E.T., Keck, R., De Vox, A. et al., Comparison between light induced and chemically induced oxidation of rhVEGF, *Pharm Res.* 18, 1455–1460, 2001.
527. Shapiro, R.I., Wen, D., Levesque, M. et al., Expression of sonic hedgehog-Fc fusion protein in *Pichia pastoris*. Identification and control of post-translational, chemical, and proteolytic modifications, *Prot. Express. Purif.* 29, 272–283, 2003.
528. Wood, M.J., Prieto, J.H., and Komives, E.A., Structural and functional consequences of methionine oxidation in thrombomodulin, *Biochim. Biophys. Acta.* 1703, 141–147, 2005.
529. Bakhtiar, R. and Guan, Z., Electron dissociation mass spectrometry in characterization of peptides and proteins, *Biotechnol. Lett.* 28, 1047–1059, 2006.

530. Jenkins, N., Murphy, L., and Tyther, R., Post-translational modifications of recombinant proteins: Significance for biopharmaceuticals, *Mol. Biotechnol.* 39, 113–118, 2008.
531. Tien, M., Berlett, B.S., Levine, R.L. et al., Peroxynitrite-mediated modification of protein at physiological carbon dioxide concentration: pH dependence of carbonyl formation, tyrosine nitration, and methionine oxidation, *Proc. Nat. Acad. Sci. USA* 96, 7809–7814, 1999.
532. Imlay, J.A., Pathways of oxidative damage, *Annu. Rev. Microbiol.* 57, 395–418, 2003.
533. Vogt, W., Oxidation of methionyl residues in proteins: Tools, targets, and reversal, *Free Radic. Biol. Med.* 18, 93–105, 1995.
534. Houghten, R.A. and Li, C.H., Reduction of sulfoxides in peptides and proteins, *Anal. Biochem.* 98, 36–46, 1979.
535. Weissbach, H., Etienne, F., Hoshi, T. et al., Peptide methionine sulfoxide reductase: Structure, mechanism of action, and biological function, *Arch. Biochem. Biophys.* 397, 172–178, 2002.
536. Stadtman, E.R., Moskovitz, J., Berlett, B.S., and Levine, R.L., Cyclic oxidation and reduction of protein methionine residues is an important antioxidant mechanism, *Mol. Cell. Biochem.* 234–235, 3–9, 2002.
537. Boschi-Muller, S., Gand, A., and Branlant, G., The methionine sulfoxide reductases: Catalysis and substrates specificities, *Arch. Biochem. Biophys.* 474, 266–273, 2008.
538. Hirs, C.H.W., Performic acid oxidation, *Methods Enzymol.* 11, 197–199, 1967.
539. Sharp, J.S., Becker, J.M., and Hettich, R.L., Protein surface mapping by chemical oxidation: Structural analysis by mass spectrometry, *Anal. Biochem.* 313, 216–225, 2003.
540. Trout, G.E., The estimation of microgram amounts of methionine by reaction with chloroamine-T, *Anal. Biochem.* 93, 419, 1979.
541. Li, C., Takazaki, S., Jin, X. et al., Identification of oxidized methionine sites in erythrocyte membrane protein by liquid chromatography/electrospray mass spectrometry peptide mapping, *Biochemistry* 45, 12117–12124, 2006.
542. Corless, S. and Cramer, R., On-target oxidation of methionine residues using hydrogen peroxide for composition-restricted matrix-assisted laser desorption/ionization peptide mass-mapping, *Rapid Commun. Mass Spectrom.* 17, 1212–1215, 2003.
543. Caldwell, P., Luk, D.C., Weissbach, H., and Brot, N., Oxidation of the methionine residues of *Escherichia coli* ribosomal protein L12 decreases the protein's biological activity, *Proc. Natl. Acad. Sci. USA.* 75, 5349, 1978.
544. Spande, T.F. and Witkop, B., Determination of the tryptophan content of proteins with N-bromosuccinimide, *Methods Enzymol.* 11, 498–506, 1967.
545. Keck, R.G., The use of *t*-butyl hydroperoxide as a probe for methionine oxidation in proteins, *Anal. Biochem.* 236, 56–62, 1996.
546. Liu, J.L, Lu, K.V., Eris, T. et al., *In vitro* methionine oxidation of recombinant human leptin, *Pharm. Res.* 15, 632–640, 1998.
547. Lu, H.S., Fausset, P.R., Narhi, L.O. et al., Chemical modification and site-directed mutagenesis of methionine residues in recombinant human granulocyte colony-stimulating factor: Effect on stability and biological activity, *Arch. Biochem. Biophys.* 362, 1–11, 1999.
548. Gundlach, H.G., Moore, S., and Stein, W.H., The reaction of iodoacetate with methionine, *J. Biol. Chem.* 234, 1761–1764, 1959.
549. Gurd, F.R.N., Carboxymethylation, *Methods Enzymol.* 11, 532–541, 1967.
550. Schroeder, W.A., Shelton, J.R., and Robberson, B., Modification of methionyl residues during aminoethylation, *Biochim. Biophys. Acta*, 147, 590–592, 1967.
551. Naider, F. and Bohak, Z., Regeneration of methionyl residues from their sulfonium salts in peptides and proteins, *Biochemistry*, 11, 3208–3211, 1972.

552. Kleanthous, C., Campbell, D.G., and Coggins, J.R., Active site labeling of the shikimate pathway enzyme, dehydroquinase. Evidence for a common substrate binding site within dehydroquinase and dehydroquinate synthase, *J. Biol. Chem.*, 265, 10929–10934, 1990.
553. Kleanthous, C. and Coggins, J.R., Reversible alkylation of an active site methionine residue in dehydroquinase, *J. Biol. Chem.*, 265, 10935–10939, 1990.
554. Weinberger, S.R., Viner, R.J., and Ho, P., Tagless extraction-retentate chromatography: A new global protein digestion strategy for monitoring differential protein expression, *Electrophoresis* 23, 3182–3192, 2002.
555. Grunert, T., Pock, K., Buchacher, A., and Allmaier, G., Selective solid-phase isolation of methionine-containing peptides and subsequent matrix-assisted laser desorption mass spectrometric detection of methionine- and methionine-sulfoxide-containing tryptic peptides, *Rapid Commun. Mass Spectrom.* 17, 1815–1824, 2003.
556. Shen, M., Guo, L., Wallace, A. et al., Isolation and isotope labeling of cysteine—and methionine-containing tryptic peptides, *Molec. Cell. Proteomics* 2, 315–324, 2003.
557. Hachimori, Y., Horinishi, H., Kurihara, K., and Shibata, K., States of amino residues in proteins. V. Different reactivities with H_2O_2 of tryptophan residues in lysozyme, proteinases and zymogens, *Biochim. Biophys. Acta* 93, 346, 1964.
558. Musatov, A., Herbert, E., Carroll, C.A. et al., Specific modification of two tryptophans within the nuclear-encoded subunits of bovine cytochrome c oxidase by hydrogen peroxide, *Biochemistry* 43, 1003–1009, 2004.
559. Gibbons, N.C., Wood, J.M., Rokos, H., and Schallreuter, K.U., Computer simulation of native epidermal enzyme structures in the presence and absence of hydrogen peroxide (H_2O_2): Potential and pitfalls, *J. Invest. Dermatol.* 126, 2576–2582, 2006.
560. Froelich, J.M. and Reid, G.E., The origin and control of *ex vivo* oxidative peptide modifications prior to mass spectrometry analysis, *Proteomics* 8, 1334–1345, 2008.
561. Reubsaet, J.L.E. et al., Analytical techniques used to study the degradation of proteins and peptides: Chemical instability, *J. Pharm. Biomed. Anal.* 17, 955–978, 1998.
562. Simat, T., Meyer, K., and Steinhart, H., Syntheses and analysis of oxidation and carbonyl condensation compounds of tryptophan, *J. Chromatog. A* 661, 93–99, 1994.
563. Mach, H., Middaugh, C.R., and Lewis, R.V., Statistical determination of the average values of the extinction coefficients of tryptophan and tyrosine in native proteins, *Anal. Biochem.* 200, 74–80, 1992.
564. Weng, J., Tan, C., Shen, J.R. et al., pH-Induced conformation changes in the soluble manganese-stabilizing protein of photosystem II, *Biochemistry* 43, 4855–4861, 2004.
565. Mahesha, H.G., Singh, S.A., Srinivasan, N., and Rao, A.G., A spectroscopic study of the interaction of isoflavones with human serum albumin, *FEBS J.* 273, 451–467, 2006.
566. Kumar, A., Tyagi, N.K., and Kinne, R.K., Ligand-mediated conformational changes and positioning of tryptophans in reconstituted human sodium/D-glucose cotransporter1 (hSGLT1) probed by tryptophan fluorescence, *Biophys. Chem.* 127, 69–77, 2007.
567. Daniel, V.W., III and Trowbridge, C.G., The effect of N-bromosuccinimide upon trypsinogen activation and trypsin catalysis, *Arch. Biochem. Biophys.*, 134, 506, 1969.
568. Ramachandran, L.K. and Witkop, B., N-Bromosuccinimide cleavage of peptides, *Methods Enzymol.* 11, 283–299, 1967.
569. Feldhoff, R.C. and Peters, T., Jr., Determination of the number and relative position of tryptophan residues in various albumins, *Biochem. J.* 159, 529, 1976.
570. Koshland, D.E., Jr., Karkhanis, Y.D., and Latham, H.G., An environmentally-sensitive reagent with selectivity for the tryptophan residue in proteins, *J. Am. Chem. Soc.* 86, 1448–1450, 1964.
571. Horton, H.R. and Koshland, D.E., Jr., A highly reactive colored reagent with selectivity for the tryptophanyl residue in proteins, 2-hydroxy-5-nitrobenzyl bromide, *J. Am. Chem. Soc.* 87, 1126–1132, 1965.

572. Lundblad, R.L. and Noyes, C.M., Observations on the reaction of 2-hydroxy-5-nitrobenzyl bromide with a peptide-bound tryptophanyl residue, *Anal. Biochem.* 136, 93-100, 1984.
573. Strohalm, M., Kodíĉek, M., and Pechar, M., Tryptophan modification by 2-hydroxy-5-nitrobenzyl bromide studied by MALDI-TOF mass spectrometry, *Biochem. Biophys. Res. Commun.* 312, 811–816, 2003.
574. Horton, H.R. and Koshland, D.E., Jr., Reactions with reactive alkyl halides, *Meth. Enzymol.* 11, 556–565, 1967.
575. Horton, H.R. and Young, G., 2-Acetoxy-5-nitrobenzyl chloride. A reagent designed to introduce a reporter group near the active site of chymotrypsin, *Biochim. Biophys. Acta* 194 272–278, 1969.
576. Fontana, A. and Scoffone, E., Sulfenyl halides as modifying reagents for polypeptides and proteins, *Methods Enzymol.* 25, 482–494, 1972.
577. Shechter, Y., Burstein, Y., and Patchornik, A., Sulfenylation of tryptophan-62 in hen egg-white lysozyme, *Biochemistry* 11, 653–660, 1972.
578. Wilchek, M. and Miron, T., The conversion of tryptophan to 2-thioltryptophan in peptides and proteins, *Biochem. Biophys. Res. Commun.* 47, 1015–1020, 1972.
579. Chersi, A. and Zito, R., Isolation of tryptophan-containing peptides by adsorption chromatography, *Anal. Biochem.* 73, 471–476, 1976.
580. Rubinstein, M., Schechter, Y., and Patchornik, A., Covalent chromatography—the isolation of tryptophanyl containing peptides by novel polymeric reagents, *Biochem. Biophys. Res. Commun.* 70, 1257–1263, 1976.
581. Kuyama, H., Watanabe, M., Toda, C. et al., An approach to quantitative proteome analysis by labeling tryptophan residues, *Rapid Commun. Mass Spectrom.* 17, 1642–1650, 2003.
582. Hansen, K.C., Schmitt-Ulms, G., Chalkley, R. J. et al., Mass spectrometric analysis of protein mixtures at low levels using cleavable ^{13}C-isotope-coded affinity tag and multi-dimensional chromatography, *Molec. Cell. Proteomics* 2, 299–314, 2003.
583. Li, C., Gawandi, V., Protos, A. et al., A matrix-assisted laser desorption/ionization compatible for tagging tryptophan residues, *Eur. J. Mass Spectrom.* 12, 213–221, 2006.
584. Iida, T., Kuyama, H., Watanabe, M. et al., Rapid and efficient MALDI-TOF MS peak detection of 2-nitrobenzylsulfenyl-labeled peptides using the combination of of HPLC and an automatic spotting apparatus, *J. Biomol. Tech.* 17, 333–341, 2006.
585. Ueda, K., Katagiri, T., Shimada, T. et al., Comparative profiling of serum glycoproteome by sequential purification of glycoproteins and 2-nitrobenzylsulfenyl (NBS) stable isotope labeling: A new approach for the novel biomarker discovery for cancer, *J. Proteome Res.* 6, 3475–3483, 2007.
586. Takahashi, K., The reaction of phenylglyoxal with arginine residues in proteins, *J. Biol. Chem.* 243, 6171–6179, 1968.
587. Yankeelov, J.A., Jr., Mitchell, C.D., and Crawford, T.H., A simple trimerization of 2,3-butanedione yielding a selective reagent for the modification of arginine in proteins, *J. Am. Chem. Soc.* 90, 1664–1666, 1968.
588. Patthy, L. and Smith, E.L., Reversible modification of arginine residues. Application to sequence studies by restriction of tryptic hydrolysis to lysine residues, *J. Biol. Chem.* 250, 557–564, 1975.
589. Xu, G., Takamoto, K., and Chance, M.R., Radiolytic modification of basic amino acid residues in peptides: Probes for examining protein-protein interactions, *Anal. Chem.* 75, 6995–7007, 2003.
590. Leitner, A. and Linder, W., Probing of arginine residues in peptides and proteins using selective tagging and electrospray ionization mass spectrometry, *J. Mass. Spect.* 38, 891–899, 2003.
591. Cotham, W.E., Metz, T.O., Ferguson, P.L. et al., Proteomic analysis of arginine adducts on glyoxal-modified ribonuclease, *Mol. Cell. Proteomics* 3, 1145–1153, 2004.

592. Brock, J.W., Cotham, W.E., Thorpe, S.R. et al., Detection and identification of arginine modifications on methylglyoxal-modified ribonuclease by mass spectrometric analysis, *J. Mass Spectrom.* 42, 89–100, 2007.
593. Deval, J., D'Abramo, C.M., Zhao, Z. et al., High resolution footprinting of the hepatitis C virus polypeptide NS5B in complex, *J. Biol. Chem.* 282, 16907–16916, 2007.
594. Borders, C.L., Jr., Pearson, L.J., McLaughlin, A.E. et al., 4-Hydroxy-3-nitrophenylglyoxal. A chromophoric reagent for arginyl residues in proteins, *Biochim. Biophys. Acta,* 568, 491–495, 1979.
595. Borders, C.L., Jr. and Johansen, J.T., Identification of Arg-143 as the essential arginine residue in yeast Cu, Zn superoxide dismutase by the use of a chromophoric arginine reagent, *Biochem. Biophys. Res. Commun.* 96, 1071–1078, 1980.
596. Borders, C.L., Jr. and Riordan, J.F., An essential arginyl residue at the nucleotide binding site of creatinine kinase, *Biochemistry* 14, 4699–4704, 1975.
597. Yamasaki, R.B., Vega, A., and Feeney, R.E., Modification of available arginine residues in proteins by p-hydroxyphenylglyoxal, *Anal. Biochem.* 109, 32–40, 1980.
598. Linder, M.D. et al., Ligand-modulation of the permeability transition pore by arginine modification. Opposing effects of *p*-hydroxyphenylglyoxal and phenylglyoxal, *J. Biol. Chem.* 277, 937–942, 2002.
599. Cheung, S.-T. and Fonda, M.L., Reaction of phenylglyoxal with arginine. The effect of buffers and pH, *Biochem. Biophys. Res. Commun.* 90, 940–947, 1979.
600. Branlant, G., Tritsch, D., and Biellmann, J.-F., Evidence for the presence of anion-recognition sites in pig-liver aldehyde reductase. Modification by phenylglyoxal and p-carboxyphenyl glyoxal of an arginyl residue located close to the substrate-binding site, *Eur. J. Biochem.*, 116, 505–512, 1981.
601. Eun, H.-M., Arginine modification by phenylglyoxal and (p-hydroxyphenyl)glyoxal: Reaction rates and intermediates, *Biochem. Int.* 17, 719–727, 1988.
602. Yankeelov, J.A., Jr., Modification of arginine in proteins by oligomers of 2,3-butanedione, *Biochemistry* 9, 2433–2439, 1970.
603. Riordan, J.F., Functional arginyl residues in carboxypeptidase A. Modification with butanedione, *Biochemistry* 12, 3915–3923, 1973.
604. Toi, K., Bynum, E., Norris, E., and Itano, H.A., Studies on the chemical modification of arginine. I. The reaction of 1,2-cyclohexanedione with arginine and arginyl residues of proteins, *J. Biol. Chem.*, 242, 1036–1037, 1967.
605. Calvete, J.J. et al., Characterization of the conformation and quaternary structure-dependent heparin-binding region of bovine seminal plasma protein PDC-109, *FEBS Lett.* 444, 260–264, 1999.
606. Bond, J.S., Francis, S.H., and Park, J.H., An essential histidine in the catalytic activities of 3-phosphoglyceraldehyde dehydrogenase, *J. Biol. Chem.* 245, 1041–1053, 1970.
607. Chang, S.H., Teshima, G.M., Milby, T. et al., Metal-catalyzed photooxidation of histidine in human growth hormone, *Anal. Biochem.* 244, 221–227, 1997.
608. Agon, V.V., Bubb, W.A., Wright, A. et al., Sensitizer-mediated photooxidation of histidine residues: Evidence for the formation of reactive side-chain peroxidase, *Free Radic. Biol. Med.* 40, 698–710, 2006.
609. Rashidzaden, H. et al., Solution structure and interdomain interactions of the *Saccharomyces cerevesiae* "TATA binding protein" (TBP) proved by radiolytic protein footprinting, *Biochemistry* 42, 3655–3665, 2003.
610. Xu, G., Takamoto, K., and Chance, M.R., Radiolytic modification of basic amino acid residues in peptides: Probes for examining protein-protein interactions, *Anal. Chem.* 75, 6995–7007, 2003.
611. Inagami, T. and Hatano, H., Effect of alkylguanidines on the inactivation of trypsin by alkylation and phosphorylation, *J. Biol. Chem.* , 244, 1176–1182, 1969.

612. Stark, G.R., Stein, W.H., and Moore, S., Relationships between the conformation of ribonuclease and its reactivity toward iodoacetate, *J. Biol. Chem.* 236–442, 436, 1961.
613. Heinrikson, R.L., Stein, W.H., Crestfield, A.M., and Moore, S., The reactivities of the histidine residues at the active site of ribonuclease toward halo acids of different structures, *J. Biol. Chem.*, 240, 2921–2934, 1965.
614. Fruchter, R.G. and Crestfield, A.M., The specific alkylation by iodoacetamide of histidine 12 in the active site of ribonuclease, *J. Biol. Chem.*, 242, 5807–5812, 1967.
615. Kettner, C. and Shaw, E., Inactivation of trypsin-like enzymes with peptides of arginine chloromethyl ketone, *Methods Enzymol.* 80, 826–842, 1981.
616. Williams, E.B., Krishnaswamy, S., and Mann, K.G., Zymogen/enzyme discrimination using peptide chloromethyl ketones, *J. Biol. Chem.*, 264, 7536–7545, 1989.
617. Glick, B.R., The chemical modification of *Escherichia coli* ribosomes with methyl *p*-nitrobenzenesulfonate. Evidence for the involvement of a histidine residue in the functioning of the ribosomal peptidyl transferase, *Can. J. Biochem.*, 58, 1345–1347, 1980.
618. Melchior, W.B., Jr. and Fahrney, D., Ethoxyformylation of proteins. Reaction of ethoxyformic anhydride with α-chymotrypsin, pepsin and pancreatic ribonuclease at pH 4, *Biochemistry* 9, 251–258, 1970.
619. Wolf, B., Lesnaw, J.A., and Reichmann, M.E., A mechanism of the irreversible inactivation of bovine pancreatic ribonuclease by diethylpyrocarbonate. A general reaction of diethylpyrocarbonate with proteins, *Eur. J. Biochem.*, 13, 519–525, 1970.
620. Miles, E.W., Modification of histidyl residues in proteins by diethylpyrocarbonate, *Methods Enzymol.*, 47, 431–442, 1977.
621. Krell, T., Chackrewarthy, S., Pitt, A.R. et al., Chemical modification monitored by electrospray mass spectrometry: A rapid and simple method for identifying and studying functional residues in enzymes, *J. Pept. Res.* 51, 201–209, 1998.
622. Qin, K., Yang, Y., Mastrangelo, P., and Westaway, D., Mapping Cu(II) binding sites in prion protein by diethyl pyrocarbonate modification of matrix-assisted laser desorption time-of-flight (MALDI-TOF) mass spectrometric footprinting, *J. Biol. Chem.* 277, 1981–1990, 2002.
623. Willard, B.B. and Kintes, M., Effects of internal histidine residues on the collision-induced fragmentation of triply protonated tryptic peptides, *J. Am. Soc. Mass. Spectrom.* 12, 1262–1271, 2001.
624. Dage, J.L., Sun, H., and Halsall, H.B., Determination of diethyl pyrocarbonate-modified amino acid residues in alpha-1-acid glycoprotein by high-performance liquid chromatography electrospray ionization mass spectrometry and matrix-assisted laser desorption/ionization time-of-flight mass spectrometry, *Anal. Biochem.* 257, 176–185, 1998.
625. Altman, J., Lipka, J.J., Kuntz, I., and Waskell, L., Identification by proton nuclear magnetic resonance of the histidines in cytochrome b5 modified by diethyl pyrocarbonate, *Biochemistry* 28, 7516–7523, 1989.
626. Hoare, D.G. and Koshland, D.E., Jr., A method for the quantitative modification and estimation of carboxyl groups in proteins, *J. Biol. Chem.* 242, 2447–2453, 1967.
627. George, A.L., Jr. and Border, C.L., Jr., Essential carboxyl groups in yeast enolase, *Biochem. Biophys. Res. Commun.* 87, 59–65, 1979.
628. Khorana, H.G., The chemistry of carbodiimides, *Chem. Rev.* 53, 145–166, 1953.
629. Kunkel, G.R., Mehrabian, M., and Martinson, H.G., Contact-site cross-linking agents, *Mol. Cell. Biochem.* 34, 2–13, 1981.
630. Iwamoto, H. et al., States of thin filament regulatory proteins as revealed by combining cross-linking/x-ray diffraction techniques, *J. Mol. Biol.* 317, 707–720, 2002.
631. Cook, G.M. et al., Purification and biochemical characterization of the F_1F_0 ATP synthase from thermophilic *Bacillus* sp. strain TA2.A1, *J. Bacteriol.* 185, 4442–4449, 2003.
632. Das, A. and Ljungdahl, L.G., *Clostridium pasteurianum* F_1F_0 ATP synthase: Operon, composition and some properties, *J. Bacteriol.* 185, 5527–5535, 2003.

633. Sheehan, J.C. and Hlavka, J.J., The use of water-soluble and basic carbodiimides in peptide synthesis, *J. Org. Chem.* 21, 439–440, 1956.
634. Sheehan, J.C. and Hlavka, J.J., The cross-linking of gelatin using a water-soluble carbodiimide, *J. Am. Chem. Soc.* 79, 4528–4529, 1957.
635. Riehm, J.P. and Scheraga, H.A., Structural studies on ribonuclease. XXI. The reaction between ribonuclease and a water-soluble carbodiimide, *Biochemistry* 5, 99–115, 1966.
636. Hoare, D.G. and Koshland, D.E., Jr., A procedure for the selective modification of carboxyl groups in proteins, *J. Am. Chem. Soc.* 88, 2057, 1966.
637. Gilles, M.A., Hudson, A.Q., and Borders, C.L., Jr., Stablity of water-soluble carbodiimides in aqueous solution, *Anal. Biochem.* 184, 244–248, 1990.
638. Lei, P.Q. et al., Kinetic studies on the rate of hydrolysis of *N*-ethyl-*N'*-(dimethylaminopropyl)carbodiimide I aqueous solution using mass spectrometry and capillary electrophoresis, *Anal. Biochem*, 310, 122–124, 2002.
639. Wrobel, N., Schinkinger, M., and Mirsky, V.M., A novel ultraviolet assay for testing side reactions of carbodiimide, *Anal. Biochem.* 305, 135–138, 2003.
640. Sehgal, D. and Vijay, I.K., A method for the high efficiency of water-soluble carbodiimide-mediated amidation, *Anal. Biochem.* 218, 87–91, 1994.
641. Wang, T.-T. and Young, N.M., Modification of aspartic acid residues to induce trypsin cleavage, *Anal. Biochem.*, 91, 696, 1978.
642. Lin, C., Mihal, K.A., and Krueger, R.J., Introduction of sulfhydryl groups into proteins at carboxyl sites, *Biochim. Biophys. Acta*, 1038, 382, 1990.
643. Pho, D.B. et al., Evidence for an essential glutamyl residue in yeast hexokinase, *Biochemistry* 16, 4533–4537, 1977.
644. Desvages, G. et al., Structural studies on yeast 3-phosphoglycerate kinase identification by immuno-affinity chromatography of one glutamyl residue essential for yeast 3-phosphoglycerate kinase activity. Its location in the primary structure, *Eur. J. Biochem.* 105, 259–266, 1980.
645. Takahashi, K., Stein, W.H., and Moore, S., The identification of a glutamic acid residue as part of the active site of ribonuclease T1, *J. Biol. Chem.*, 242, 4682–4690, 1967.
646. Erlanger, B.F., Vratsanos, S.M., Wassermann, M., and Cooper, A.G., Specific and reversible inactivation of pepsin, *J. Biol. Chem* 240, 3447–3448, 1965.
647. Gross, E. and Morell, J.L., Evidence for an active carboxyl group in pepsin, *J. Biol. Chem.* 241, 3638–3639, 1966.
648. Woodward, R.B., Olofson, R.A., and Mayer, H., A new synthesis of peptides, *J. Am. Chem. Soc.*, 83, 1010–1012, 1961.
649. Dunn, B.M., Anfinsen, C.B., and Shrager, R.I., Kinetics of Woodward's Reagent K hydrolysis and reaction with staphylococcal nuclease, *J. Biol. Chem.* 249, 3717–3723, 1974.
650. Bodlaender, P., Feinstein, G., and Shaw, E., The use of isoxazolium salts for carboxyl group modification in proteins. Trypsin, *Biochemistry*, 8, 4941–4949, 1969.
651. Gross, E., The cyanogen bromide reaction, *Methods Enzymol.*, 11, 238–255, 1967.
652. Spande, T.F., Selective cleavage and modification of peptides and proteins, *Adv. Protein Chem.* 24, 97–260, 1970.
653. Smith, B.J., Chemical cleavage of proteins, *Meth. Mol. Biol.* 32, 197–309, 1994.
654. Schmoldt, H.U., Wentzel, A., Becker, S., and Kolmar, H., A fusion protein system for the recombinant production of short disulfide bond rich cysteine knot peptides using barnase as a purification handle, *Protein Express Purif.* 39, 82–89, 2005.
655. Patwa, T.H., Wang, Y., Simeone, D.M., and Lubman, D.M., Enhanced detection of autoantibodies on protein microarray using a modified protein digestion technique, *J. Proteome Res.* 7, 2553–2561, 2008.
656. Tristram, G.R. and Smith, R.H., Amino acid composition of certain proteins, in *The Proteins*, 2nd ed., Neurath, H., Ed., Academic Press, New York, 1963.

657. Morrison, J.R., Fidge, N.N., and Grego, H., Studies on the formation, separation, and characterization of CNBr fragments of human A1 apolipoprotein, *Anal. Biochem.* 186, 145–152, 1990.
658. Shively, J.E., Reverse-phase HPLC isolation and microsequence analysis, in *Methods of Protein Microcharacterization*, Shively, J.E., Ed., Humana Press, Clifton, NJ, 1986.
659. Liao, T.-H., Salnikow, J., Moore, S., and Stein, W.H., Bovine pancreatic deoxyribonuclease A. Isolation of CNBr peptides; complete covalent structure of the polypeptide chain, *J. Biol. Chem.* 248, 1489–1495, 1973.
660. van Montfort, B.A., Canas, B., Duurkens, R. et al., Improved in-gel approaches to generate peptide maps of integral membrane proteins with matrix-assisted laser desorption/ionization time-of-flight mass spectrometry. *J. Mass. Spectrom.* 37, 322–330, 2002.
661. van Montfort, B.A., Doeven, M.K., Canas, B. et al., Combined in-gel tryptic digestion and CNBr cleavage for the generation of peptide maps of an integral membrane protein with MALDI-TOF mass spectrometry, *Biochim. Biophys. Acta* 1555, 111–115, 2002.
662. Quach, T.T., Li, N., Richards, D.P. et al., Development and applications of in-gel CNBr/tryptic digestion combined with mass spectrometry for analysis of membrane proteins, *J. Proteome Res.* 2, 543–552, 2003.
663. Weerasekera, R., She, Y.M., and Markham, K.A., Interaction and interface protocol (2IP): A novel strategy for high sensitivity topology mapping of protein complexes, *Proteomics* 7, 3835–3852, 2007.
664. Hulmes, J.D. and Pan, Y.-C.E., Selective cleavage of polypeptides with trifluoroacetic acid: Applications for microsequencing, *Anal. Biochem.* 197, 368–376, 1991.
665. Ye, J.M. et al., Degradation of antiflammin 2 under acidic conditions, *J. Pharm. Sci.* 85, 695–699, 1996.
666. Murakami, T., Natsuka, S., Nakakita, S., and Hase, S., Structure determination of a sulfated *N*-glycans, candidate for a precursor of the selectin ligand in bovine lung, *Glycoconj. J* 24, 195–206, 2007.
667. Lu, H.S. and Gracy, R.W., Specific cleavage of glucosephosphate isomerase at cysteinyl residues using 2-nitro-5-thiocyanobenzoic acid: Analyses of peptides eluted from polyacrylamide gels and localization of active site histidyl and lysyl residues, *Arch. Biochem. Biophys.*, 212, 347, 1981.
668. Qi, J., Wu.,J., Somkuti, G.A., and Watson, J.T., Determination of the disulfide structure of sillucin, a highly knotted, cysteine-rich peptide, by cyanylation/cleavage mass mapping, *Biochemistry* 40, 4531–4538, 2001.
669. Gallegos-Pérez, J.-L., Rangel-Ordóñez, L., Bowman, S.R. et al., Study of primary amines for nucleophilic cleavage of cyanylated cystinyl proteins in mass mapping methodology, *Anal. Biochem.* 346, 311–319, 2005.
670. Douady, D., Rousseau, B., and Caron, L., Fucoxanthin-chlorophyll a/c light-harvesting complexes of *Laminaria saccharina*—partial amino acid sequences and arrangement in the thylakoid membranes, *Biochemistry* 33, 3165–3170, 1994.
671. Droste, M., Mollner, S., and Pfeuffer, T., Localisation of an ATP-binding site on adenylyl cyclase type I after chemical and enzymatic fragmentation, *FEBS Lett.* 391, 208–214, 1996.
672. Pliszka, B., Karczewska, E., and Wawro, B., Nucleotide-induced movements in the myosin head near the converter region, *Biochim. Biophys. Acta* 1481, 55–62, 2000.

2 Application of Solution Protein Chemistry to the Study of Biopharmaceutical Conformation

Most biopharmaceuticals are proteins or protein conjugates and are considered to be biopolymers. Proteins have a unique conformation in solution, which is the product of diverse covalent and noncovalent interactions. It is generally accepted that the primary structure of proteins dictates their secondary and tertiary structures, and the final conformation is stabilized by the aforementioned covalent and noncovalent interactions. These interactions can be intramolecular or intermolecular; intramolecular interactions dominate at low protein concentrations, whereas intermolecular interactions are more significant at higher protein concentrations, where such forces are involved in processes such as aggregation. This is not to say that intermolecular interactions are not important at low protein concentrations; however, such interactions are usually driven by specific multivalent interactions.[1]

The study of protein conformation has been of great interest for the study of the relationship between protein structure and function[3,4] for some time and for the study of protein folding.[5,6] The emergence of biosimilars in commercial biotechnology[7–14] has increased interest in the use of protein conformation study in comparability studies.[15–18] Comparability is also of importance when there are process changes, formulation changes, and changes in source material.[19–25]

Protein conformation is the combination of secondary structure (helix, pleated sheet)[26–30] and tertiary structure.[31–37] It is generally accepted that primary structure drives the secondary structure, which in turn drives the formation of tertiary structure.[38–40]

The characterization of a protein therapeutic is a critical part of the drug development and approval process. Classical methods such as sequence analysis, compositional analysis, solution behavior with particular emphasis on the formation of aggregates, and, more recently, analysis by mass spectrometry are used in the evaluation of protein structure for use of the protein as a biopharmaceutical. The question then is, what quality attributes are critical for product performance and what physical or chemical techniques would effectively measure these attributes? It is generally accepted that immunogenicity is the most significant problem. Issues with glycosylation, which influence product half-life and may influence immunogenicity, are also of importance. The problem of immunogenicity is discussed in the following text and elsewhere in detail,[41–46] as are techniques for the evaluation of glycosylation.[47–53] Glycosylation is a bit of a challenge: although glycosylation is important for circulatory half-life (specifically, the covering of galactose/galactosamine by sialic acid), there is precious little evidence to suggest a true functional role for glycosylation.

Most solution protein chemistry characterization assays for biologics focus on chemical structure and biological activity. There is somewhat less interest in the use of conformational analysis. There are several reasons for this. First, to a certain extent, conformational analyses for purposes of identity or comparability are useful only if there is no change; if there is change, it is usually, but not always, difficult to quantitate as compared, for example, to a chemical modification in the peptide chain. However, there are a variety of techniques that can be used to study protein conformation.[54]

Analytical techniques such as amino acid analysis and mass spectrometry provide information regarding the chemical structure of the product. Techniques such as electrophoresis, chromatography, and size-exclusion chromatography provide information about purity and can, in selected situations, provide insight into conformation and chemical structure. Hydrophobic interaction chromatography[55–59] can also be useful in the study of conformational changes in proteins.[60–67]

The past 40 years have witnessed an increase in the sophistication of the technologies available to measure conformational change in proteins; however, there has not been an increase in the number of parameters measured. Kauzmann[68] proposed a classification system for the levels of conformation similar to the general classification of primary, secondary, tertiary, and quaternary structure, which separated conformation issues into shape properties and short-range properties. Shape (long-range) properties are parameters dependent on the overall shape (globular, rod, etc.), which might be relatively insensitive to changes in the immediate vicinity of amino acids and peptide bonds. Short-range properties include parameters defined by the immediate environment around individual amino acid residues. Although this is an imperfect separation, it does prove useful. Schellman and Schellman reviewed the problem of conformation change in proteins in 1964[69] and, as observed by Cantor and Timasheff,[70] there had been no change in the some 20 years between the two reviews. There has been a marked increase in the sophistication of the instrumentation used. Schellman and Schellman extended Kauzmann's earlier suggestions. Shape properties include hydrodynamic parameters such as frictional coefficient and viscosity changes, and solution properties such as fluorescence depolarization and flow birefringence. Also included in shape properties are electron microscopy, dipole moments, and diffusion through controlled pore membranes (thin-film dialysis).[71–73] Short-range properties are, to some extent, "micro" properties as compared to the "macro" properties of shape. Schellman and Schellman include optical properties such as absorbance (IR, UV) and circular dichroism and chemical properties such as side chain reactivity (trace labeling, chemical footprinting), individual p*K*a's, hydrogen isotope exchange, biological activity, and immunogenicity as short-range properties. Also included in short-range properties are nuclear magnetic resonance (NMR) and binding of small molecules such as dyes. This division admittedly is imperfect; for example, immunogenicity could be more accurately defined as a shape property, but reactivity is dependent on epitopic change.

Most of the techniques used for the conformational analysis of protein were developed either for the study of protein denaturation or, more recently, for the study of protein folding. The focus of this chapter is the application of solution protein chemistry to the study of conformational change associated with the processing of biotechnology products. These changes can be considered as more closely related to denaturation than to protein folding. Denaturation can be considered to be associated with the change in the spatial arrangement of the polypeptide chains in a protein (tertiary structure) from the native, ordered structure to a more disordered structure in an irreversible process. Denaturation is usually, but not always, associated with loss in solubility. Denaturation is not usually associated with the cleavage of the peptide chain. There are, however, situations that seem to be exceptions to this, such as the conversion of fibrinogen to fibrin and the cleavage of peptide chains. Protein denaturation is frequently, but not always, associated with the loss of biological activity as it has long been accepted that configuration is important for biological activity.[75,76] Protein denaturation is not necessarily irreversible,[77–82] but there can be a divergence in the quality of structure recovery dependent on measurement.[83–85] The key to renaturation is, in part, dependent on the quality of protein; for example, some zymogens (e.g., pepsinogen) can be reversibly inactivated under conditions in which active enzymes (e.g., pepsin) are irreversibly inactivated.[86–88] On the other hand, trypsin can be reversibly denatured[89,90] and may be more stable than trypsinogen to denaturation.[91–93] Denaturation is also discussed in Chapter 6 as applied to tissue soldering and tissue welding.

Techniques such as light scattering[94–99] and analytical ultracentrifugation[100–105] provide information about the shape and solution behavior of the material (tertiary and quaternary structures). These two techniques, together with size-exclusion chromatography, are critical for the evaluation of aggregation in pharmaceutical preparations. There is also reason to consider measurement of the second virial coefficient. The second virial coefficient is a factor used to correct for the nonideal behavior of a particle. Virial coefficients were originally developed as a series of coefficients of inverse powers of V in a polynomial series to approximate the quantity of pV/RT in an equation of state of an ideal gas or a similar collection of particles.[106,107] From a practical perspective, the second virial coefficient is related to the excluded volume of a particle[108,109] and is important in accounting for protein–protein interactions and molecular crowding.[110–113] The excluded volume of any particle depends on shape and can be defined as the volume surrounding and including a given object that is excluded to another object.[108] The second virial coefficient is mentioned most often in the study of the osmotic pressure of proteins but is generally used for the study of protein–protein interaction.[114–128] The reader is recommended to read articles on protein shape,[129–133] as this attribute is frequently overlooked in favor of the more sophisticated approaches discussed in the following text.

Techniques such as circular dichroism,[134–141] optical rotatory dispersion,[70,142–148] Fourier transform infrared spectroscopy (near NIR),[149–158] nuclear magnetic resonance,[159–171] intrinsic fluorescence,[172–183] binding of fluorescent probes,[184–193] hydrogen–deuterium exchange,[194–208] differential scanning calorimetry,[209–225] Ramen spectroscopy,[226–242] protein footprinting,[243–248] limited proteolysis,[249–265] and trace labeling[266–269] can provide information about secondary and tertiary structures. Near infrared (NIR) spectroscopy is also useful for noninvasive determination of moisture.[270–277]

One of the major problems with the use of most of these techniques is the requirement for substantial amounts of protein. This can be an issue with therapeutic proteins, which are biologically active at the microgram level, and the use of a destructive analytical technique is difficult to justify. However, the use of mass spectrometry for analysis enhances the sensitivity and therefore the value of hydrogen isotope exchange and trace labeling. An analysis of the literature indicates that optical rotatory dispersion is of limited value today as compared to other analytical technologies.

The key question is—what is the question that you wish to answer? Each of the aforementioned techniques has the potential to show changes in conformation secondary to changes in solvent environment. However, what is the relationship of these changes to biological activity, *in vivo* clearance, or immunogenicity? In the case of a biopharmaceutical, if you lose activity, you lose product. Creation of new epitopes (increase or change in immunogenicity; neoantigenicity) results either in an unfortunate immunological response or increased product clearance.[278–280] Changes in the immunological properties of biotherapeutic proteins can be identified by established immunoassays.[281–285] Changes in glycosylation such as the loss of sialic acid (exposure of galactose/galactosamine) can also increase the rate of product clearance. The demonstration of conformation change in a protein does not necessarily predict a loss of activity or neoantigenicity but can provide insight into the chemistry responsible for such changes. The following text describes the relationship between the chemical modification of a protein and changes in secondary or tertiary structures.

First, although the following discussion emphasizes changes in protein conformation (secondary and tertiary structure), primary structure and quaternary structure should also be briefly considered. Changes in primary structure can be divided into two categories: (1) the modification of individual amino acid residues, which is covered in great detail in Chapter 1 and (2) the cleavage of peptide bonds mostly by proteolytic enzymes or chemical means (i.e., cleavage at asparagine[286–290]). Changes in primary structure will be discussed because such changes can influence secondary and tertiary structure; changes in quaternary structure are in turn driven by changes in secondary and tertiary structure. This author is not aware of any biotherapeutic proteins where quaternary structure is a practical issue; however, general issues of protein–protein interaction, which are important for quaternary structure, are important in the action of most protein biotherapeutics.

Protein conformation can be influenced by both physical and chemical agents. Table 2.1 lists some examples of the effect of chemical modification on protein conformation. As an aside, one of the more lively discussions of 40 years ago concerned the effect of site-specific modification of protein on protein conformation with respect to the elucidation of the relationship between chemical structure and biological function. Without belaboring the detail, it was generally accepted that it was possible to accomplish the site-specific chemical modification of a protein without gross conformational change, but it was always useful to evaluate such potential changes.[291–293]

TABLE 2.1
Some Studies on the Effect of Chemical Modification on Protein Conformation

Protein	Reagent	Results	Reference
IgG	Citraconic anhydride	CD[a] showed loss of β-structure (positive CD band at 202 nm became negative); sedimentation coefficient decreased; immunoresponse to goat antiserum was lost. All changes were reversed by removal of citraconyl groups (dialysis at pH 4.0).	1[b]
IgG	KCNO/pH 7.7 carbamylation	No change in CD spectra; no change in sedimentation coefficient; no loss of immunoreactivity.	1[b]
Wheat-germ agglutinin	NBS[c]/pH 3.9–8.0 M urea or NBS/pH 6	Two tryptophans modified at pH 6.0 with loss of 85% intrinsic fluorescence (λ_{excit} 280 nm); third tryptophan modified only with pH 3.9/8.0 M urea. No major change in CD spectra (210–260 nm; 260–340 nm); oxidation of one tryptophan resulted in total loss of hemagglutinating activity.	2
HEWL[d]	I_2	^{13}C NMR[e] is used to demonstrate modification of tyrosine residues; modification is pH dependent.	3
Prolactin	2-Nps-Cl[f] in 70% formic acid	Loss of receptor binding activity; CD shows some loss of α-helical content and some increase in β-sheet content.	4
Basic pancreatic trypsin inhibitor	Reduced/carboxy-methylated	Sequence-specific assignments for ^1H-NMR shifts show changes near the modification site and internal hydrogen bonds are preserved.	5
HEWL	Cross-linkage; β-aspartyl formation	Difference in H/D exchange rates of *N*-1 hydrogens.	6
Trypsin	Asparagine deamidation	Tertiary structure as a major determinant of asparagine deamidation from neutron crystallographic analysis.	7
HEWL	Ozone	Oxidation of Trp 62 to kynurenine; little change in CD; decrease in thermal stability.	8
RNase T_1	Ozone	Oxidation of Trp 59 to kynurenine; little change in CD; decrease in thermal stability: this study also used the constant fragment from a λ-immunoglobulin light chain.	8
Plasma fibronectin	Chloramine-T	Decreased binding of collagen to oxidized (methionine) fibronectin; no change in intrinsic fluorescence or CD spectra.	9
Amyloid protein	Formic acid	NMR (1D and 2D) demonstrates formation of the formate ester of serine.	10

TABLE 2.1 (CONTINUED)
Some Studies on the Effect of Chemical Modification on Protein Conformation

Protein	Reagent	Results	Reference
Alanine peptides	N/A	Use of CD and amide hydrogen exchange to measure helix–coil transition.	11
Human IgG_1 Fc (recombinant in *Escherichia coli*)	H_2O_2 in phosphate (pH 7.0) or acetate (pH 5.0)	H_2O_2 oxidized methionine residues in recombinant Fc domain resulting in changes in secondary and tertiary structure measured by CD and NMR. There were also changes in DSC analysis.	12
BSA, lysozyme, IgG	Succinic anhydride	Results differed with the protein studied. In general, there was an increase in Stoke's radius (determined from analytical ultracentrifugation) and a variable difference in the reactivity of disulfide bonds. The sedimentation coefficient for BSA was dependent on sample concentration for the succinylated protein but not for the native protein.	13
BPTI	Haloacetamide modification of cysteine residue in disulfide bond	NMR measurements detected changes only in the vicinity of the modified residue.	14
Myoglobin, cytochrome C	Hydroxyl radical	Measurement of rate of γ-ray-mediated oxidation by "on-line" H-D exchange. *o*-Phosphoric acid increased rate of oxidation, whereas other denaturants such as urea decreased the rate by scavenging the radical.	15
Lysozyme[g]	Oxidation	Measurement of the rate of oxidation of native protein and oxidized protein using CD spectroscopy (198 nm). Provides a sensitive index of conformational change.	16
Myoglobin	Hydroxyl radical via γ-radiation; inclusion of O_2	Level of oxidative labeling (oxygen incorporation via mass spectrometry) depends on protein concentration; it is suggested that oxidative modification does not cause major changes in protein conformation.	17
SpoOF	Hydroxyl radical via γ-radiation; bicarbonate buffer	Measurement of rate of oxidation and observation of deviation from expected first-order rate constant permit evaluation of conformational change; CD supports loss of some helical structure but bulk of the structure is unchanged.	18

TABLE 2.1 (CONTINUED)
Some Studies on the Effect of Chemical Modification on Protein Conformation

Protein	Reagent	Results	Reference
3-Phospho-glycerate kinase	Tetranitromethane	Nitration of a single tyrosine inactivates enzyme; CD shows no change (200–260 nm); some small change in environment around aromatic residues at 260–400 nm.	19
Antithrombin	H_2O_s	Oxidation of two methionine residues (314 and 317) did not result in change in biological activity; oxidation of two more methionine residues (17 and 20) under more rigorous conditions results in some loss of heparin-binding affinity; oxidation did result in change in the behavior of RP-HPLC (decreased affinity); some change in 180–260 (far UV), and more pronounced changes in near UV (260–400 nm).	20
α-Syculein	4-HNE[h]	Modification of protein with 4-HNE caused major conformation changes (increase β-sheet) as judged by CD and FTIR spectroscopy; decreased aggregation of modified protein.	21
Catalase	H_2O_2	Oxidation yielded a form of enzyme with different catalytic properties and electrophoretic mobility but no gross conformational change (CD, intrinsic fluorescence).	22
Calmodulin	H_2O_2	Oxidation of methionine residues is a function of solvent exposure; lack of gross conformational changes by CD (small change at 208 nm); oxidized material has somewhat decreased thermal stability.	23
PAI-1[i]	NClSuc[j]	Oxidation of methionine resulted in CD change, which correlated with the loss of activity.	24
BSA	Iodoacetate, iodoacetamide, DTNB[k]	Modification of the single sulfhydryl group in BSA with either iodoacetate or DTNB yielded a derivative that demonstrated β-structure in guanidine (>3.0 M), whereas the iodoacetamide derivative demonstrated a random coil under such conditions (measurement by CD). The modification with any of the three reagents did demonstrate a conformational change; a change in conformation was observed with complete reduction of the disulfide bonds of the protein.	25

TABLE 2.1 (CONTINUED)
Some Studies on the Effect of Chemical Modification on Protein Conformation

Protein	Reagent	Results	Reference
Rhodanese	H_2O_2	Oxidized enzyme shows increased susceptibility to proteolysis by either trypsin or chymotrypsin.	26
N/A	N.A	Review of the use of IR spectroscopy to study conformational change in proteins.	27

[a] CD = circular dichroism.
[b] There is variability in the response of various proteins to different acylating and alkylating agents. (See Qasim, M.A. and Salahuddin, A., Changes in conformation and immunological activity of ovalbumin during its modification with different acid anhydrides, *Biochim. Biophys. Acta* 536, 50–63, 1978; Lakkis, J. and Villota, R., Effect of acylation on substructural properties of proteins: A study using fluorescence and circular dichroism, *J. Agric. Food Chem.* 40, 553–560, 1992; Mir, M.M., Fazili, K.M., and Abul Qasim, M., Chemical modification of buried lysine residues of bovine serum albumin and its influence on protein conformation and bilirubin binding, *Biochim. Biophys. Acta* 1119, 261–267, 1992).
[c] NBS = *N*-bromosuccinimide.
[d] HEWL = Hen-egg white lysozyme.
[e] NMR = nuclear magnetic resonance.
[f] Nps-Cl = 2-nitrophenylsulfenyl chloride.
[g] Turkey lysozyme, also β-lactoglobulin, ubiquitin, catalase.
[h] 4-HNE = 4-Hydroxy-2-nonenal.
[I] PAI = Plasminogen activator inhibitor-1.
[j] NClsuc = *N*-chlorosuccinimide.
[k] DTNB = 5,5-Dithio(bis)nitrobenzoic acid (Ellman's reagent).

Limited proteolysis has been used for the study of protein conformation for at least 60 years.[249,294] Linderstrom–Lange[249] observed that native proteins were slightly susceptible to proteolysis, and reversible denaturation increased this susceptibility. Somewhat later, Mihalyi[294] presented a comprehensive review of the proteolysis of proteins with an extensive discussion of the role of conformation. The susceptibility or rate of hydrolysis of a peptide bond is dependent on (1) the amino acids in the scissile peptide bond and the sequence of amino acids surrounding the scissile peptide bonds (primary structure effects) and (2) the environment around the peptide bond (long-range effects), which is a function of the secondary and tertiary structures that provide the environment around the scissile peptide bond. The latter consideration also involves solvent effects. It should be noted that a regulatory protease is more sensitive to primary structure effects than a digestive enzyme[295,296]; a digestive enzyme such as trypsin is more useful as a conformational probe[297–305] because the purpose is to identify peptide bonds that become exposed as a result of conformational change. Artificial metal–complex proteases have proved useful in limited proteolysis.[306–308] This approach is similar to protein footprinting.[309–312] Protein footprinting is an extension of competitive labeling (trace labeling), a chemical modification technique developed by Brian Hartley and colleagues in 1971,[313] which has proved to be a useful technique for the study of protein conformation and protein–protein interactions.[314–324]

Bovine pancreatic ribonuclease A (RNAase A) is resistant to tryptic hydrolysis at 23°C. Rupley and Scheraga[325] demonstrated that RNAase A was susceptible to chymotryptic hydrolysis at 50°C. The rate of hydrolysis was inhibited by phosphate and citrate; such polyanions had previously been demonstrated to stabilize RNAase A toward urea denaturation. The temperature (50°C) is at the bottom end of a conformational change as measured by optical rotation[326]; the transition temperature for RNAase A was 61.9°C in water and 66.1°C in deuterium oxide. Winchester and coworkers[327] determined a T_m of 60.5°C by DSC; alcohols lowered the transition temperature as determined by difference spectroscopy[328] and increased susceptibility to proteolysis.[329] Other studies using limited proteolysis extended these observations to the effect of temperature on RNAase conformation as assessed with limited proteolysis.[330–333] These studies provided early support for the use of limited proteolysis to study conformational change in proteins. There are also early studies on the effect of temperature on immunoglobulin structure and limited proteolysis.[334–338] The reader is directed to several recent reports on the development of limited proteolysis for the study of protein conformation.[339–345]

Consideration of the limited number of examples in Table 2.1 suggests that there may or may not be conformation secondary to chemical modification of protein primary structure. In general, reactions that result in charge reversal, such as the modification of lysine residues with organic acid anhydrides such as acetic anhydride or maleic anhydride, result in conformational changes, whereas changes in single amino acid residues that do not involve charge reversal do not necessarily result in any observed conformational change. An example is provided by the modification of tryptophan in wheat-germ agglutinin (Table 2.1), in which tryptophan residues are modified with no change in circular dichroism (CD) but with total loss of biological activity.

Selected studies on the effect of pressure on protein conformation are listed in Table 2.2. Table 2.3 lists some selected studies on the effect of temperature on protein conformation.

This discussion has focused on conformational change in proteins as a result of environmental effects. It has not been concerned with conformational changes of proteins in response to physiological influences such as those represented by allosteric interactions.[346–351]

TABLE 2.2
Some Studies on the Effect of Pressure on Protein Conformation

Protein	Reaction Conditions	Results	Reference
Lysozyme	30 to 2000 bar[a] change[b]. 50–100 mM deuterated glycine buffer, pD 2.0.	2D NMR; α-helical domain is compressed as the interdomain region; little effect on β-sheet.	1,2
RNase and BPTI[c]	1 GPa in 10 mM deuterated Tris-HCl, pD 7.6; 2-mercaptoethanol as reducing agent.	Reduction of all disulfide bonds could be accomplished with high pressure. FTIR[d] spectroscopy did not show differences between reduced and unreduced forms under pressure. Amorphous aggregates formed on decompression. It is suggested that high pressure results in the formation of aggregation-prone conformers.	3
N/A	Variable pressure NMR.	Review.	4
N/A	Pressure perturbation calorimetry.	Review.	5
Fibrinogen	Solid-state measurement on KBr pellets; pressure increase to 400 kg/cm².	FTIR spectroscopy demonstrates changes in secondary structure (transition from α-helix to β-sheet) and unfolding/denaturation of fibrinogen.[e]	6
Canine milk lysozyme	100 MPa; effect on thermal denaturation.	UV spectroscopy at pH 4.5 and 2.0.	7
N/A	Molecular simulation studies.	Structural, thermodynamic, and hydration changes as a function of temperature and pressure.	8
N/A	Review.	Pressure changes similar to changes induced by temperature.	9
Lysozyme	Molecular dynamics simulation at 1 and 3 kbar.	Simulation suggests an inversion of hydrophobic and hydrophilic solvent-accessible surface areas; also suggests that hydrophobic interactions are weaker at higher pressures.	10
Bacterial inclusion bodies	200 mPa in the presence of reducing agents, pH 8.0.[f]	Refolding of inclusion bodies assisted by high hydrostatic pressure. Additives such as arginine prevented aggregation.	11
BPTI	6000 bar (6 kbar), copper-beryllium high-pressure cell in 50 mM deuterated acetate buffer. Changes in protein structure evaluation by neutron scattering.[g]	Increasing pressure reduces radius of gyration; this reduction is not reversible. Slowing down of protein dynamics with increased pressure.	12
N/A	Simulation studies.	Random coil becomes more destabilized with increasing pressure.	13
Azurin	0–6 kbar in 2 mM Tris-HCl, pH 7.5.	Use of intrinsic fluorescence and phosphorescence to study protein denaturation; decrease in intrinsic fluorescence intensity on denaturation.	14

TABLE 2.2 (CONTINUED)
Some Studies on the Effect of Pressure on Protein Conformation

Protein	Reaction Conditions	Results	Reference
Equine serum albumin	0–110 mPa in 0.1 M NaCl at pD 4.4; 60°C	Study of the differential effect of pressure and heat on the denaturation of equine serum albumin using FTIR spectroscopy. Pressure prevents heat aggregation. Heat-induced aggregates with inter-molecular β-sheet and pressure-induced aggregates without intermolecular β-sheet.	15
Thioredoxin	0.1–400 mPa in 50 mM Tris, pH 7.5.	Measurement of intrinsic fluorescence as a function of pressure. There was an initial decrease in fluorescence, reflecting a more compact protein with quenching by a nearby disulfide; at higher pressure, the protein unfolds with an increase in fluorescence.	16
Ubiquitin	30–3000 atm.	NMR studies of water penetration into proteins with pressure.	17

[a] 1 atm = 1.01 bar = 101.3 kPa (kilopascal).

[b] Yamada, H., Pressure-resisting glass cell for high pressure, high-resolution NMR measurement, *Rev. Sci. Instrum.* 45, 640–642, 1974.

[c] RNase = Bovine pancreatic ribonuclease; BPTI = Bovine pancreatic trypsin inhibitor.

[d] FTIR = Fourier transform infrared.

[e] Fibrinogen is also subject to surface denaturation (Prokopowicz, M., Lukasiak, J., Banecki, B., and Przyjazny, A., In vitro measurement of conformational stability of fibrinogen adsorbed on siloxane, *Biomacromolecules* 6, 39–45, 2005; Santore, M.M. and Wertz, C.F., Protein spreading kinetics at liquid-solid interfaces via an adsorption probe, *Langmuir* 21, 10172–10178, 2005; Xu, L.C. and Siedlecki, C.A., Effects of surface wettability and contact time on protein adhesion to biomaterial surfaces, *Biomaterials* 28, 3273–3283, 2007).

[f] Alkaline pH is required for thiolate anion, which is required for disulfide reshuffling.

[g] Neutron scattering is a technique for the study of protein conformation (Harroun, T.A., Wignall, G.D., and Kataras, J., Neutron scattering in biology, in *Neutron Scattering in Biology*, ed. J. Fitter, T. Gutberlet, and J. Katsaras, Springer, Berlin, Germany, 2006). This technique requires access to a nuclear reactor or linear accelerator. Elastic scattering provides compositional information, whereas inelastic scattering measures molecular motion (Bucknall, Neutron scattering in analysis of polymers and rubbers, in *Encyclopedia of Analytical Chemistry*, ed. R.A. Meyers, John Wiley & Sons, Chichester, U.K., 2000). There is a large difference in scattering by hydrogen and deuterium, which is useful for the analysis of protein conformation; the ability to use neutron scattering for dry and hydrated samples is of great interest (Marconi, M., Conicchi, E., Onori, G., and Paciaroni, A., Comparative study of protein dynamics in hydrated powders and in solutions: A neutron scattering investigation, *Chemical Physics* 345, 224–229, 2008; Wood, K., Caronna, C., Fouquet, P. et al., A benchmark for protein dynamics: Ribonuclease A measured by neutron scattering in a large wave vector-energy transfer range, *Chemical Physics* 345, 305–314, 2008). There have been major technical advances in this field, which should facilitate greater use of this technology (Teixeira, S.C.M., Zaccai, G., Ankner, J. et al., New sources and instrumentation for neutrons in biology, *Chemical Physics* 345, 133–151, 2008).

TABLE 2.3
Some Studies on the Effect of Temperature on Protein Conformation

Protein	Reaction Conditions	Results	Reference
β-Lactoglobulin	Thermal denaturation	Use of capillary zone electrophoresis (CZE) to measure protein denaturation; CZE performed between 4–95°C.	1
β-Lactoglobulin	pH 7.0/heating	Synchrotron small-angle x-ray diffraction, Fourier transform IR spectroscopy; above 50°C, IR showed a loss of intramolecular β-sheet and α-sheet.	2
Horseradish peroxidase	Effect of temperature and pH	CD and intrinsic fluorescence; removal of calcium ions decreases stability.	3
Lysozyme	^{15}N-labeled human lysozyme (2 mM) in 50 mM KCl/D_2O, pH 3.8	Effect of temperature on heteronuclear multidimensional NMR spectroscopy; decrease in volume and surface area with decreasing temperature.	4
Cutinase (*Fusarium*)	Heating (protein melting) in 0.01 M acetate (pH 4.0), 0.01 M phosphate (pH 6.0, pH 8.0)	A single tryptophan residue; fluorescence quenched in native protein by cystine (disulfide bond); quenching decreased (fluorescence increased) on heating.	5
Lysozyme and ribonuclease A	Neutral pH (deuterated Tris-HCl, pD 7.6); approximately 3.5 mM protein	Use of FTIR to monitor thermal denaturation. Aggregate formation; dissociation of amorphous aggregates at higher temperatures preceded by conformation change.	6
N/A	Computer simulation using lattice model	Comparison between chemical (denaturant)- and temperature-induced protein denaturation. Results show a wider distribution of conformational states than temperature-induced denaturation.	7
Canine plasminogen	10 mM potassium phosphate–100 mM NaCl, pH 6.5	Large structural changes over the range of 4–20°C; Stokes' radius decreases for both the open and closed forms as measured by dynamic light scattering or analytical ultracentrifugation.	8
Bovine serum albumin	Glass transitions of aqueous solutions. Measurement of heat capacity and enthalpy relaxation with adiabatic calorimetry.	Relaxation with glass transition is from nonequilibrium to equilibrium state. Enthalpy relaxation rate depends on thermal history. There are three distinguishable glass transitions in the subzero range.	9
Bovine serum albumin	Dry and hydrated samples.	Nuclear magnetic transverse decay; proton second moment. There is a change in water at 170 K; but no change with D_2O. It is suggested that chains extend from proteins in hydrated state but not in dry state.	10

TABLE 2.3 (CONTINUED)
Some Studies on the Effect of Temperature on Protein Conformation

Protein	Reaction Conditions	Results	Reference
β-Lactoglobulin	Phenyl-Sepharose® vary salt concentration and temperature.	Two-conformation adsorption model; thermodynamic models to predict distribution between various conformers. They predict trends in retention strength and stability.	11
Poly-L-lysine	Perchlorate concentration and temperature.	Use of Raman spectroscopy to determine conformational energy landscape; at 1°C, 0.83 M perchlorate, 86% α-helix melting into extended conformation at 60°C.	12

REFERENCES

References for Table 2.1

1. Nakagawa, Y., Capetillo, S., and Jirgensons, B., Effect of chemical modification of lysine residues on the conformation of human immunoglobulin, G, *J. Biol. Chem.* 242, 5703–5708, 1972.
2. Privat, J.P., Lotan, R., Bouchard, P. et al., Chemical modification of the tryptophan residues of wheat-germ agglutinin. Effect on fluorescence and saccharide-binding properties, *Eur. J. Biochem.* 68, 563–572, 1976.
3. Norton, R.S. and Allerhand, A., Studies of chemical modifications of proteins by carbon 13 nuclear magnetic resonance spectroscopy. Reaction of hen egg white lysozyme with iodine, *J. Biol. Chem.* 251, 6522–6528, 1976.
4. Kochman, H., Garnier, J., and Kochman, K., Receptor binding and conformational properties of bovine and ovine prolactins after chemical modification of the two tryptophan residues, *Biochim. Biophys. Acta* 578, 125–134, 1979.
5. Stassinopoulou, C.I., Wagner, G., and Wüthrich, K., Two-dimensional ^{1}H NMR of two chemically modified analogs of the basic pancreatic trypsin inhibitor. Sequence-specific resonance assignments and sequence location of conformation changes relative to the native protein, *Eur. J. Biochem.* 145, 423–430, 1984.
6. Endo, T., Ueda, T., Yamada, H., and Imoto, T., pH dependence of individual tryptophan *N*-1 hydrogen exchange rates in lysozyme and its chemically modified derivatives, *Biochemistry* 26, 1838–1845, 1987.
7. Kossiakoff, A.A., Tertiary structure is a principal determinant to protein deamidation, *Science* 240, 191–194, 1988.
8. Okajima, T., Kawata, Y., and Hamaguchi, K., Chemical modification of tryptophan residues and stability changes in proteins, *Biochemistry* 29, 9168–9175, 1990.
9. Miles, A.M. and Smith, R.L., Functional methionines in the collagen/gelatin binding domain of plasma fibronectin: Effects of chemical modification by chloramine T, *Biochemistry* 32, 8168–8178, 1993.
10. Klunk, W.E., Xu, C.J., and Pettigrew, J.W., NMR identification of the formic acid-modified residue in Alzheimer's amyloid protein, *J. Neurochem.* 62, 349–354, 1994.
11. Rohl, C.A. and Baldwin, R.L., Comparison of NH exchange and circular dichroism as techniques for measuring the parameters of the helix-coil transition in peptides, *Biochemistry* 36, 8435–8442, 1997.

12. Liu, D., Ren, D., Huang, H. et al., Structure and stability changes on Human IgG1 Fc as a consequence of methionine oxidation, *Biochemistry*, 5088–5100, 2008.
13. Habeeb, A.F.S.A., Quantitation of conformation changes on chemical modification of proteins: Use of succinylated proteins as a model, *Arch. Biochem. Biophys.* 121, 652–664, 1967.
14. Yoshioka, S., Abe, H., Noguti, T. et al., Conformational change of a globular protein elucidated at atomic resolution. Theoretical and magnetic resonance study, *J. Mol. Biol.* 170, 1031–1036, 1983.
15. Tong, T., Wren, J.C., and Konermann, L., γ-Ray-mediated oxidative labeling for detecting protein conformational changes by electrospray mass spectrometry, *Anal. Chem.* 80, 2222–2231, 2008.
16. Venkatesh, S., Tomer, K.S., and Sharp, J.S., Rapid identification of oxidation-induced conformational changes by kinetic analysis, *Rapid Commun. Mass Spectrom.* 21, 3927–3936, 2007.
17. Tong, X., Wren, J.C., and Konermann, L., Effects of protein concentration on the extent of γ-ray-mediated oxidative labeling studies by electrospray mass spectrometry, *Anal. Chem.* 79, 6376–6382, 2007.
18. Sharpe, J.S., Sullivan, D.M., and Tomer, K.B., Measurement of multisite oxidation kinetics reveals as active site conformational change in SpoOF as a result of protein oxidation, *Biochemistry* 45, 6260–6266, 2006.
19. Markland, F.S., Bacharach, D.E., Weber, B.H. et al., Chemical modification of yeast 3-phosphoglycerate kinase, *J. Biol. Chem.* 259, 1301–1310, 1975.
20. Van Patten, S.M., Hanson, E., Bernasconi, R. et al., Oxidation of methionine residues in antithrombin. Effects on biological activity and heparin binding, *J. Biol. Chem.* 274, 10268–10276, 1999.
21. Qin, Z., Hu, D., Han, S. et al., Effect of 4-hydroxy-2-nonenal modification on α-synuclein aggregation, *J. Biol. Chem.* 282, 5862–5870, 2007.
22. Diaz, A., Muñoz-Clares, R.A., Rangel, P. et al., Functional and structural analysis of catalase oxidized by singlet oxygen, *Biochimie* 87, 205–214, 2005.
23. Gao, J., Yin, D.H., Yao, Y. et al., Loss of conformational stability in calmodulin upon methionine modification, *Biophys. J.* 74, 1115–1134, 1998.
24. Strandberg, L., Lawrence, D.A., Johansson, L.B., and Ny, T., The oxidative inactivation of plasminogen activator inhibitor type-1 results from a conformational change in the molecule and does not require the involvement of the P1' methionine, *J. Biol. Chem.* 266, 13852–13858, 1991.
25. Batra, P.P., Sasa, K., Ueki, T., and Takeda, K., Circular dichroic study of the conformational stability of sulfhydryl-blocked bovine serum albumin, *Int. J. Biochem.* 21, 857–862, 1989.
26. Horowitz, P.M. and Bowman, S., Oxidation increases the proteolytic susceptibility of a localized region in rhodanese, *J. Biol. Chem.* 262, 14544–14548, 1987.
27. Barth, A., Infrared spectroscopy of proteins, *Biochim. Biophys. Acta* 1767, 1073–1101, 2007.

References for Table 2.2

1. Akasaka, K., Tezuka, T., and Yamada, H., Pressure-induced changes in folded structure of lysozyme, *J. Mol. Biol.* 271, 671–678, 1997.
2. Reface, M., Tezuka, T., Akasaka, K., and Williamson, M.P., Pressure-dependent changes in the solution structure of hen egg-white lysozyme, *J. Mol. Biol.* 327, 857–865, 2003.
3. Meersman, F., and Heremans, K., High pressure induced the formation of aggregation-prone states of proteins under reducing conditions, *Biophys. Chem.* 104, 297–304, 2003.

4. Akasaka, K., Highly fluctuating protein structures revealed by variable-pressure nuclear magnetic resonance, *Biochemistry* 42, 10875–10885, 2003.
5. Ravindra, R. and Winter, R., Pressure perturbation calorimetry: A new technique provides surprising results on the effects of co-solvents on protein solvation and unfolding behavior, *Chemphyschem* 5, 566–571, 2004.
6. Lin, S.Y., Wei, Y.S., Hsieh, T.F., and Li, M.J., Pressure dependence of human fibrinogen correlated to the conformational alpha-helix to beta-sheet transition: A Fourier transform infrared study microspectroscopic study, *Biopolymers* 75, 393–402, 2004.
7. Watanabe, M., Aizawa, T., Demura, M., and Nitta, K., Effect of hydrostatic pressure on conformational changes of canine milk lysozyme between the native, molten globate, and unfolded states, *Biochim. Biophys. Acta* 1702, 129–136, 2004.
8. Paschek, D., Gnanakaran, S., and Garcia, A.E., Simulations of the pressure and temperature unfolding of an alpha-helical peptide, *Proc. Nat. Acad. Sci. USA* 102, 6765–6770, 2005.
9. Marchal, S., Torrent, J., Masson, P. et al., The powerful high pressure tool for protein conformational studies, *Braz. J. Med. Biol. Res.* 38, 1175–1183, 2005.
10. McCarthy, A.N. and Grigera, J.R., Effect of pressure on the conformation of proteins. A molecular dynamics simulation of lysozyme, *J. Mol. Graph. Model.* 24, 254–263, 2006.
11. Chang, B.S., Randolph, T.W., and Kim, Y.S., Effects of solutes on solubilization and refolding of proteins from inclusion bodies with high hydrostatic pressure, *Protein Sci.* 15, 304–313, 2006.
12. Appavou, M.S., Gibrat, G., and Billissent-Funel, M.C., Influence of pressure on structure and dynamics of bovine pancreatic trypsin inhibitor (BPTI) small angle and quasi-elastic neutron scattering studies, *Biochim. Biophys. Acta* 1764, 414–423, 2006.
13. Harano, Y. and Kinoshita, M., Crucial importance of translational entropy of water in pressure denaturation of proteins, *J. Chem. Phys.* 125, 24910, 2006.
14. Cioni, P., Role of protein cavities on unfolding volume change and on internal dynamics under pressure, *Biophys. J.* 91, 3390–3396, 2006.
15. Okuno, A., Kato. M., and Taniguchi, Y., Pressure effects on the heat-induced aggregation of equine serum albumin by FT-IR spectroscopic study: Secondary structure, kinetic and thermodynamic properties, *Biochim. Biophys. Acta* 1774, 652–660, 2007.
16. Ado, K. and Taniguchi, Y., Pressure effects on the structure and function of human thioredoxin, *Biochim. Biophys. Acta* 1774, 813–821, 2007.
17. Day, R. and Garcia, A.E., Water penetration in the low and high pressure native states of ubiquitin, *Proteins* 70, 1175–1184, 2008.

References for Table 2.3

1. Rochu, D., Ducret, G., and Masson, P., Measuring conformational stability of proteins using an optimized temperature-controlled capillary electrophoresis approach, *J. Chromatog. A* 838, 157–165, 1999.
2. Panick, G., Malessa, R., and Winter, R., Differences between the pressure- and temperature-induced denaturation and aggregation of beta-lactoglobulin A, B, and AB monitored by FT-IR spectroscopy and small-angle X-ray scattering, *Biochemistry* 38, 6512–6519, 1999.
3. Chattopadhyay, K. and Mazumdar, S., Structural and conformational stability of horseradish peroxidase: Effect of temperature and pH, *Biochemistry* 39, 263–270, 2000.
4. Kumeta, H., Miura, A., Kobashigawa, Y. et al., Low-temperature-induced structural changes in human lysozyme elucidated by three-dimensional NMR spectroscopy, *Biochemistry* 42, 1209–1216, 2003.
5. Martinho, J.M., Santos, A.M., Fedorov, A. et al., Fluorescence of the single tryptophan of cutinase: Temperature and pH effect on protein conformation and dynamics, *Photochem. Photobiol.* 78, 15–22, 2003.

6. Meersman, F. and Heremans, K., Temperature-induced dissociation of protein aggregates: Accessing the denatured state, *Biochemistry* 42, 14234–14241, 2003.
7. Choi, H.S., Huh, J., and Jo, W.H., Comparison between denaturant- and temperature-induced unfolding pathways of protein: A lattice Monte Carlo simulation, *Biomacromolecules* 5, 2289–2296, 2004.
8. Kornblatt, J.A., and Schuck, P., Influence of temperature on the conformation of canine plasminogen: An analytical ultracentrifugation and dynamic light scattering study, *Biochemistry* 44, 13122–13131, 2005.
9. Kawai, K., Suzuki, T., and Oguni, M., Low-temperature glass transitions of quenched and annealed bovine serum albumin aqueous solutions, *Biophys. J.* 90, 3732–3738, 2006.
10. Goddard, Y.A., Korb, J.P., and Bryant, R.G., Structural and dynamical examination of the low-temperature glass transition in serum albumin, *Biophys. J.* 91, 3841–4847, 2006.
11. Xiao, Y., Rathore, A., O'Connell, J.P., and Fernandez, E.J., Generalizing a two-conformation model for describing salt and temperature effects on protein retention and stability in hydrophobic interactions chromatography, *J. Chromatog. A* 1157, 197–206, 2007.
12. Ma, L., Ahmed, Z., Mikhonin, A.V., and Asher, S.A., UV resonance Raman measurements of poly-L-lysine's conformational energy landscapes: Dependence on perchlorate concentration and temperature, *J. Phys. Chem. B111*, 7675–7680, 2007.

Chapter References

1. Stevens, F.J., Analysis of protein–protein interaction by simulation of small-zone size-exclusion chromatography: Application to an antibody–antigen association, *Biochemistry* 25, 981–993, 1986.
2. Sung, M., Poon, G.M.K., and Gariépy, J., The importance of valency in enhancing the import and cell routing potential of protein transduction domain-containing molecules, *Biochim. Biophys. Acta* 1758, 355–363, 2006.
3. Schellman, J.A. and Schellman, C., The conformation of polypeptide chains in proteins, in *The Proteins*, 2nd ed., Ed. H. Neurath, Volume 2, Chapter 7, pp. 1–137, Academic Press, New York, 1964.
4. Fasman, G.D., *Prediction of Protein Structure and the Principles of Protein Conformation,* Plenum Press, New York, 1989.
5. Nall, B.T. and Dill, K.A., *Conformation and Forces in Protein Folding,* AAAS, Washington DC, 1991.
6. Merz, K.M. and Le Grand, S.M., *The Protein Folding Problem and Tertiary Structure Prediction*, Birkhäuser, Boston, MA, 1994.
7. Schellekens, H., Biosimilar therapeutic agents: Issues with bioequivalence and immunogenicity, *Eur. J. Clin. Invest.* 341, 797–799, 2004.
8. Kessler, M., Goldsmith, D., and Schellekens, H., Immunogenicity of biopharmaceuticals, *Nephrol. Dial. Transplant.* 21(Suppl. 5), v9–v12, 2006.
9. Gerrazani, A.A., Biggio, G,. Caputi, A.P. et al., Biosimilar drugs: Concerns and opportunities, *BioDrugs* 21, 351–356, 2007.
10. Roger, S.D. and Mikhail, A., Biosimilars: Opportunity or cause for concern?, *J. Pharm. Pharm. Sci.* 10, 405–410, 2007.
11. Pavlovic, M., Girardiin, E., Kapetaneovic, L. et al., Similar biological medicinal products containing recombinant human growth hormone: European regulation, *Horm. Res.* 69, 14–21, 2008.
12. Ma, L., Ahmed, X., Mikhonin, A.V., and Asher, S.A., UV resonance Raman measurements of poly-L-lysine; S conformational energy landscapes. Dependence on perchlorate concentration and temperature, *J. Phys. Chem. B111,* 7675–7680, 2007.

13. Kawanishi, T., Regulatory perspective from Japan—comparability of biopharmaceuticals, *Biologicals* 34, 65–68, 2006.
14. Lubiniecki, A.S. and Federici, M.M., Comparability is not just analytical equivalence, *Biologicals* 34, 45–47, 2006.
15. Kuhlmann, M. and Covic, A., The protein science of biosimilars, *Nephrol. Dial. Transplant.* 21(Suppl. 5), v4–v8, 2006.
16. Deechongkit, S., Aoki, K.H., Park, S.S., and Kerwin, B.A., Biophysical comparability of the same protein from different manufacturers: A case study using Epoetin alfa from Epogen and Eprex, *J. Pharm. Sci.* 95, 1931–1943, 2006.
17. Heavner, G.A., Arakawa, T., Philo, J.S. et al., Protein isolated from biopharmaceutical formulations cannot be used for comparative studies: Follow-up to a "case study" using Epoetin Alpha Form Epogen and EPREX, *J. Pharm. Sci.* 96, 3214–3225, 2007.
18. ICH 5QE: Comparability of biotechnological/biological products subject to changes in their manufacturing process, International Committee on Harmonisation, http://www.ich.org; http://www.ich.org/cache/compo/276-254-1.html; 2005.
19. DeFelippis, M.R. and Larimore, F.S., The role of formulation in insulin comparability studies, *Biologicals* 34, 49–54, 2006.
20. Petriccciani, J., A global view of comparability concepts, *Dev. Biol.* (Basel) 106, 9–13, 2002.
21. Moos, M., Jr., Regulatory philosophy for comparability protocols, *Dev. Biol.* (Basel) 109, 53–56, 2002.
22. Chirino, A.J., and Mire-Sluis, A., Characterizing biological products and assessing comparability following manufacturing changes, *Nat. Biotechnol.* 22, 1383–1391, 2004.
23. Simek, S.L., Characterization of gene therapy products and the impact of manufacturing changes on product comparability, *Dev. Biol. (Basel)* 122, 139–144, 2005.
24. Robertson, J.S., Changes in biological source material, *Biologicals* 34, 61–63, 2006.
25. Sewerin, K., Shacter, E., Robertson, J., and Wallerius, C., Changes in biological source materials, *Biologicals* 34, 71–72, 2006.
26 Johnson, W.C., Jr., Protein secondary structure and circular dichroism: A practical guide, *Proteins* 7, 205–214, 1990.
27. Yada, R.Y., Jackman, R.L., and Nakai, S., Secondary structure prediction and determination of proteins—A review, *Int. J. Pept. Protein Res.* 31, 98–108, 1988.
28. Andersen, C.A. and Rost, B., Secondary structure assignment, *Methods Biochem. Anal.* 44, 341–363, 2003.
29. Pelton, J.T. and McLean, L.R., Spectroscopic methods for analysis of protein secondary structure, *Anal. Biochem.* 277, 167–176, 2000.
30. Xie, M. and Schowen, R.L., Secondary structure and protein deamidation, *J. Pharm. Sci.* 88, 8–13, 1999.
31. Blow, D.M., Chymotrypsin: Tertiary structure and enzymic activity, *Biochem. J.* 110, 2P, 1968.
32. Crippen, G.M., Correlation of sequence and tertiary structure in globular proteins, *Biopolymers* 16, 2189–2201, 1977.
33. Richardson, J.S., Describing patterns of protein tertiary structure, *Methods Enzymol.* 115, 341–358, 1985.
34. Barton, G.J. and Sternberg, M.J., A strategy for the rapid multiple alignment of protein sequence. Confidence levels from tertiary structure comparisons, *J. Mol. Biol.* 198, 327–337, 1987.
35. Wang, C.X., Shi, Y.Y., and Huang, F.H., Fractal study of tertiary structure of proteins, *Phys. Rev. A.* 41, 7043–4078, 1998.
36. Meiler, J. and Baker, D., Coupled prediction of protein secondary and tertiary structure, *Proc. Natl. Acad. Sci. USA* 100, 12105–12110, 2003.

37. Shen, B.W., Spiegel, P.C., Chang, C.H. et al., The tertiary structure and domain organization of coagulation factor VIII, *Blood* 111, 1240–1247, 2008.
38. Sela, M., Anfinsen, C.B., and Harrington, W.F., The correlation of ribonuclease activity with specific aspects of tertiary structure, *Biochim. Biophys. Acta* 26, 502–512, 1957.
39. Anfinsen, C.B., The tertiary structure of ribonuclease, *Brookhaven Symp. Biol.* 15, 184–198, 1962.
40. Anfinesen, C.B., The formation of the tertiary structure of proteins, *Harvey Lect.* 61, 95–116, 1967.
41. Mire-Sluis, A.R., Challenges with current technology for the detection, measurement, and characterization of antibodies against biological therapeutics, *Dev. Biol. (Basel)* 109, 59–69, 2002.
42. Hermeling, S., Crommelin, D.J.A., Schellekens, H., and Jiskout, W., Structure-immunogenicity relationships of therapeutic proteins, *Pharm. Res.* 21, 897–803, 2004.
43. Sampaio, C., Costa, J., and Ferreira, J.J., Clinical comparability of marketed formulations of botulinum toxin, *Mov. Disord.* 19 (Suppl. 8), S129–S136, 2004.
44. Frost, H., Antibody-mediated side effects of recombinant proteins, *Toxicology* 209, 155–160, 2005.
45. Thorpe, R. and Swanson, S.J., Current methods for detecting antibodies against erythropoietin and other recombinant proteins, *Clin. Diag. Lab. Immunol.* 12, 28–39, 2005.
46. Romer, T., Peter, F., Saenger, P. et al., Efficacy and safety of a new ready-to-use recombinant human growth hormone solution, *J. Endocrinol. Invest.* 30, 578–589, 2007.
47. Henderson, C.J., Holme, M.J., and Aitken, R.J., Analysis of the biological properties of antibodies raised against native and deglycosylated porcine zonae pellucidae, *Gamete Res.* 16, 323–341, 1987.
48. Aouffen, M., Paquin, J., De Grandpre, E. et al., Deglycosylated ceruloplasmin maintains its enzymatic, antioxidant, cardioprotective, and neuroprotective properties, *Biochem. Cell Biol.* 79, 489–497, 2001.
49. Raju, T.S., Briggs, J.B., Chamow, S.M. et al., Glycoengineering of therapeutic glycoproteins: *In vitro* galactosylation and sialylation of glycoproteins with terminal *N*-acetylglucosamine and galactose residues, *Biochemistry* 31, 8868–8876, 2001.
50. Jefferis, R., Glycosylation of recombinant antibody therapeutics, *Biotechnol. Prog.* 21, 11–16, 2005.
51. Walsh, G. and Jefferis, R., Post-translational modifications in the context of therapeutic proteins, *Nat. Biotechnol.* 24, 1241–1252, 2006.
52. Jefferis, R., Antibody therapeutics: Isotype and glycoform selection, *Expert Opin. Biol. Ther.* 7, 1401–1413, 2007.
53. Temporini, C., Calleri, E., Massolini, G., and Caaccialanza, G., Integrated analytical strategies for the study of phosphorylation and glycosylation in proteins, *Mass Spectrom. Rev.* 27, 207–236 2008.
54. Crommelin, D.J.A., Storm, G., Verrijk, R. et al., Shifting paradigms: Pharmaceutical versus low molecular weight drugs, *Int. J. Pharmaceut.* 266, 3–16, 2003.
55. Kato, Y., High-performance hydrophobic interaction chromatography of proteins, *Adv. Chromatog.* 26, 97–115, 1987.
56. Arakawa, T. and Narhi, L.O., Solvent modulation in hydrophobic interaction chromatography, *Biotechnol. Appl. Biochem.* 13, 151–172, 1991.
57. Wu, S.-L. and Karger, B.L., Hydrophobic interaction chromatography of proteins, *Methods Enzymol.* 270, 27–47, 1996.
58. Hemström, P. and Irgum, K., Hydrophilic interaction chromatography, *J. Sep. Sci.* 29, 1784–1821, 2006.
59. Lienqueo, M.E., Mahn, A., Salgado, J.C., and Asenjo, J.A., Current insights on protein behavior in hydrophobic interaction chromatography, *J. Chromatog. B. Anal. Technol. Biomed. Life Sci* 849, 53–68, 2007.

60. Tarvers, R.C., Calcium-dependent changes in properties of human prothrombin: A study using high-performance size-exclusion chromatography and gel-permeation chromatography, *Arch. Biochem. Biophys.* 241, 639–648, 1985.
61. Wu, S.L., Figueroa, A., and Karger, B.L., Protein conformational effects in hydrophobic interaction chromatography. Retention characterization and the role of mobile phase additives and stationary phage hydrophobicity, *J. Chromatog.* 371, 3–27, 1986.
62. Withka, J., Moncuse, P., Baziotis, A., and Maskiewicz, R., Use of high-performance size-exclusion, ion-exchange, and hydrophobic interaction chromatography for the measurement of protein conformational change and stability, *J. Chromatog.* 398, 175–202, 1987.
63. Krull, I.S., Stuting, H.H., and Krzysko, S.C., Conformational studies of bovine alkaline phosphatase in hydrophobic interaction and size-exclusion chromatography with linear diode array and low-angle laser light scattering detection, *J. Chromatog.* 442, 29–52, 1988.
64. Lundblad, R.L., A hydrophobic site in human prothrombin present in a calcium-stabilized conformer, *Biochem. Biophys. Res. Commun.* 157, 295–300, 1988.
65. Haezebrouck, P., Noppe, W., van Dacl, H., and Hanssens, I., Hydrophobic interaction of lysozyme and α-lactalbumin from equine milk whey, *Biochim. Biophys. Acta* 1122, 305–310, 1992.
66. Bjerrun, O.J. Bjerrum, M.J., and Heegaard, N.H., Electrophoretic and chromatographic differentiation of two forms of albumin in equilibrium at neutral pH: New screening techniques for determination of liquid binding to albumin, *Electrophoresis* 16, 1401–1407, 1995.
67. To, B.C.S. and Lenhoff, A.M. Hydrophobic interaction chromatography of proteins. II. Solution thermodynamic properties as a determinant of retention, *J. Chromatog. A* 1141, 235–243. 2007.
68. Kauzmann, W., Some factors in the interpretation of protein denaturation, *Adv. Prot. Chem.* 14, 1– 63, 1959.
69. Schellman, J.A. and Schellman, C., The conformation of polypeptide chains in proteins, in *The Proteins*, 2nd ed., Ed. H. Neurath, Academic Press, New York, Chapter 7, pp. 1–137, 1964.
70. Cantor, C.R. and Timasheff, S.N., Optical spectroscopy of proteins, in *The Proteins*, 3rd ed., Ed. H. Neurath and R.L. Hill, Academic Press, New York, Vol. 5, pp. 145–306, 1982.
71. Craig, L.C. and Chen, H.C., On a theory for the passive transport of solute through semipermeable membranes, *Proc. Natl. Acad. Sci. USA* 69, 702–705, 1972.
72. Chen, H.C., Craig, L.C. and Stoner, E., On the removal of residual carboxylic acid groups from cellulose membranes and Sephadex, *Biochemistry* 11, 3559–3564, 1972.
73. Harris, M.J. and Craig, L.C., A study of the parameters which determine the conformation of linear polypeptides in solution by synthesis of models and determination of thin film dialysis rates, *Biochemistry* 13, 1510–1515, 1974.
74. Putman, F.W., Protein denaturation, in *The Proteins*, Eds. H. Neurath and K. Bailey, Academic Press, New York, Chapter 9, pp. 808–892, 1953.
75. Porter, R.R., The relation of chemical structure to the biological activity of proteins, in *The Proteins*, Eds. H. Neurath and K. Bailey, Academic Press, New York, Chapter 11, pp. 973–1015, 1973.
76. Putnam, F.W., The chemical modification of proteins, in *The Proteins*, Eds. H. Neurath and K. Bailey, Academic Press, New York, Chapter 10, pp. 893–972, 1953.
77. Kodicek, M., Infanzón, A., and Karpenko, V., Heat denaturation of human orosomucoid in water/methanol mixtures, *Biochim. Biophys. Acta* 1246, 10–16, 1995.
78. Pico, G., Thermodynamic aspects of the thermal stability of human serum albumin, *Biochem. Mol. Biol. Int.* 36, 1017–1023, 1995.
79. Herberhold, H. and Winter, R., Temperature- and pressure-induced unfolding and refolding of ubiquitin: A static and kinetic Fourier transform infrared spectroscopy study, *Biochemistry* 41, 2396–2401, 2002.

80. Roychaudhuri, R., Sarath, G., Zeece, M., and Markwell, J., Reversible denaturation of the soybean Kunitz trypsin inhibitor, *Arch. Biochem. Biophys.* 412, 20–26, 2003.
81. Mehta, R., Kundu, A., and Kishore, N., 4-Chlorobutanol induces unusual reversible and irreversible thermal unfolding of ribonuclease A: Thermodynamic, kinetic, and conformational characterization, *Int. J. Biol. Macromol.* 34, 13–20, 2004.
82. Kragh-Hansen, U., Saito, S., Nishi, K. et al., *Biochim. Biophys. Acta* 1747, 81–88, 2005.
83. Ryhähen, L., Zaragoza, E.J., and Uitto, J., Conformational stability of type I collagen triple helix: Evidence for temporary and local relaxation of the protein conformation using a proteolytic probe, *Arch. Biochem. Biophys.* 223, 562–571, 1983.
84. Isaacs, B.S., Brew, S.A., and Ingham, K.C., Reversible unfolding of the gelatin-binding domain of fibronectin: Structural stability in relation to function, *Biochemistry* 28, 842–850, 1989.
85. Mizuno, K., and Hayashi, T., Peculiar effect of urea on the interaction of type I collagen with heparin on chromatography, *J. Biochem.* 116, 1257–1263, 1994.
86. Ahmad, F. and McPhie, P., Thermodynamics of the denaturation of pepsinogen by úrea, *Biochemistry* 17, 241–246, 1979.
87. Ahmad, F. and McPhie, P., Characterization of a stable intermediate in the unfolding of diazoacetylglycine ethyl ester—pepsin by urea, *Biochemistry* 17, 241–246, 1979.
88. Konno, T., Kamatari, Y.O., Tanaka, N. et al., A partially unfolded structure of the alkaline-denatured state of pepsin and its implication for stability of the zymogen-derived protein, *Biochemistry* 39, 4182–4290, 2000.
89. Ruan, K., Lange, R., Meersman, F. et al., Fluorescence and FTIR study of the pressure-induced denaturation of bovine pancreas trypsin, *Eur. J. Biochem.* 265, 79–85, 1999.
90. Brumano, M.H., Rogana, E., and Swaisgood, H.E., Thermodynamics of unfolding of beta-trypsin at pH 2.8, *Arch. Biochem. Biophys.* 382, 57–62, 2000.
91. Hopkins, T.R. and Spikes, J.D., Denaturation of proteins in 8M urea as monitored by tryptophan fluorescence: Trypsin, trypsinogen and some derivatives, *Biochem. Biophys. Res. Commun.* 30, 540–545, 1968.
92. Delaage, M. and Lazdunski, M., Trypsinogen, trypsin, trypsin-substrate and trypsin-inhibitor complexes in urea solutions, *Eur. J. Biochem.* 4, 378–384, 1968.
93. Otlewski, J., Sywula, A., Kaolasinski, M., and Krowarsch, D., Unfolding kinetics of bovine trypsinogen, *Eur. J. Biochem.* 242, 601–607, 1996.
94. Carlson, F.D. The application of intensity fluctuation spectroscopy to molecular biology, *Annu. Rev. Biophys. Bioeng.* 4, 243–264, 1975.
95. Tinoco, I. Jr., Michols, W., Maestrae, M.F., and Bustamante, C., Absorption, scattering, and imaging of biomolecular structures with polarized light, *Annu. Rev. Biophys. Biophys. Chem.* 16, 319–349, 1987.
96. Eisenberg, H., Thermodynamics and the structure of biological macromolecules. Rozhinkes mit mandlen, *Eur. J. Biochem.* 187, 7–22, 1990.
97. *Laser Light Scattering in Biochemistry*, Eds. S.E. Herding, D.B. Sattelle, and V.A. Bloomfield, Royal Society of Chemistry, Cambridge, U.K., 1992.
98. Li-Chain, E.C., Methods to monitor process-induced changes in food proteins. An overview, *Adv. Exp. Med. Biol.* 434, 5–23, 1998.
99. Georgilis, Y. and Saenger, W., Light scattering studies on supersaturated protein solutions, *Sci. Prog.* 82, 271–294, 1998.
100. Aune, K.C., Molecular weight measurements by sedimentation equilibrium: Some common pitfalls and how to avoid them, *Method Enzymol.* 48, 163–185, 1978.
101. Schuster, T.M. and Toedt, J.M., New revolutions in the evolution of analytical ultracentrifugation, *Curr. Opin. Struct. Biol.* 6, 650–658, 1996.
102. Hensley, P., Defining the structure and stability of macromolecular assemblies in solution: The re-emergence of analytical ultracentrifugation as a practical tool, *Structure* 4, 367–373, 1996.

103. Eisenberg, H., Analytical ultracentrifugation in a Gibbsian perspective, *Biophys. Chem.* 88, 1–9, 2000.
104. Lebowiz, J., Lewis, M.S., and Schuck, P., Modern analytical ultracentrifugation in protein science: A tutorial review, *Protein Sci.* 11, 2067–2079, 2002.
105. Howlett, G.J., Minton, A.P., and Rivas, G., Analytical ultracentrifugation for the study of protein association and assembly, *Curr. Opin. Chem. Biol.* 10, 430–436, 2006.
106. *Oxford English Dictionary*, Oxford University Press, Oxford, U.K., 2008.
107. H. K. Onnes 1901, in *Arch. neérlandaises des Sci. exactes & nat.* VI. 874, as cited in *Oxford English Dictionary*, 2008.
108. Flory, P.J., *Principles of Polymer Chemistry*, Cornell University Press, Ithaca, NY, 1953.
109. McCabe, W.C. and Fisher, H.F., Measurement of the excluded volume of protein molecules by differential spectroscopy in the near infra red, *Nature* 207, 1274–1276, 1965.
110. Minton, A.P., Influence of excluded volume upon macromolecular structure and associations in "crowded" media, *Curr. Opin. Biotechnol.* 8, 65–69, 1997.
111. Minton, A.P., Implications of macromolecular crowding for protein assembly, *Curr. Opin. Struct. Biol.* 10, 34–39, 2000.
112. Despa, F., Orgill, D.P., and Lee, R.C., Molecular crowding effects on protein stability, *Ann. N. Y. Acad. Sci.* 1066, 54–66, 2005.
113. Konopka, M.C., Weisshaar, J.C,. and Record, M.T., Jr., Methods of changing biopolymer volume fraction and cytoplasmic solute concentrations for *in vivo* biophysical studies, *Methods Enzymol.* 428, 487–504, 2007.
114. Nichol, J.W., Janado, M., and Winzor, D.J., The origin and consequences of concentration dependence in gel chromatography, *Biochem. J.* 133, 15–22, 1973.
115. Wan, P.J. and Adams, E.T., Jr., Molecular weights and molecular weight distribution from ultracentrifugation of nonideal solutions, *Biophys. Chem.* 5, 207–241, 1976.
116. Tang, L.H., Powell, D.R., Escott, B.M., and Adams, E.T., Jr., Analysis of various indefinite self-associations, *Biophys. Chem.* 7, 121–139, 1977.
117. Neal, B.L., Asthagiri, D., and Lenhoff, A.M., Molecular origins of osmotic second virial coefficients of proteins, *Biophys. J.* 75, 2469–2477, 1998.
118. Weatherly, G.T. and Pielak, G.J., Second virial coefficients as a measure of protein-osmolyte interactions, *Protein Sci.* 10, 12–16, 2001.
119. Ruppert, S., Sandler, S.L, and Lenhoff, A.M. Correlation between the second virial coefficient and the solubility of proteins, *Biotechnol. Prog.* 17, 182–187, 2001.
120. Tessier, P.M. and Lenhoff, A.M., Measurements of protein self-association as a guide to crystallization, *Curr. Opin. Biotechnol.* 14, 512–516, 2003.
121. Sear, R.P., Solution stability and variability in a simple model of globular proteins, *J. Chem. Phys.* 120, 998–1005, 2004.
122. Valente, J.J., Payne, R.W., Manning, M.C. et al., Colloidal behavior of proteins: Effects of the second virial coefficient on solubility, crystallization and aggregation of proteins in aqueous solution, *Curr. Pharm. Biotechnol.* 6, 427–436, 2005.
123. Paliwal, A., Asthagiri, D. Abras, D. et al., Light-scattering studies of protein solutions: Role of hydration in weak protein-protein interactions, *Biophys. J.* 89, 1564–1573, 2005.
124. Ruckenstein, E. and Shulgin, I.L., Effect of salts and organic additives on the solubility of proteins in aqueous solutions, *Adv. Colloid Interface Sci.* 123–126, 97–103, 2006.
125. Payne, R.W., Nayar, R., Tarantino, R. et al., Second virial coefficient determination of a therapeutic peptide by self-interaction chromatography, *Biopolymers* 84, 527–533, 2006.
126. Bajaj, H., Sharma, V.K., Badkar, A. et al., Protein structural conformation and not second virial coefficient relate to long-term irreversible aggregation of a monoclonal antibody and ovalbumin in solution, *Pharm. Res.* 23, 1382–1394, 2006.
127. Winzor, D.J., Dezczynski, M., Harding, S.E., and Wills, P.R., Nonequivalence of second virial coefficients from sedimentation equilibrium and static light scattering studies of protein solutions, *Biophys,. Chem.* 128, 46–55, 2007.

128. Dumetz, A.C., Chockla, A.M., Kaler, E.W., and Lenhoff, A.M., Effects of pH on protein-protein interactions and implications for protein phase behavior, *Biochim. Biophys. Acta* 1784, 600–610, 2008.
129. Blattler, D.P. and Reisthel, F.J., Molecular weight determinations and the influence of gel density and protein shape in polyacrylamide gel electrophoresis, *J. Chromatog.* 46, 286–292, 1970.
130. Chae, K.S., and Lenhoff, A.M., Computation of the electrophoretic mobility of proteins, *Biophys. J.* 68, 1120–1127, 1995.
131. Røgen, P. and Bohr., H., A new family of global protein shape descriptors, *Math. Biosci.* 182, 167–181, 2003.
132. He, L. and Niemeyer, B., A novel correlation for protein diffusion coefficients based on molecular weight and radius of gyration, *Biotechnol. Prog.* 19, 544–548, 2003.
133. Chang, B.H. and Bae, Y.C., Salting-out in the aqueous single-protein solution: The effect of shape factor, *Biophys. Chem.* 104, 523–533, 2003.
134. Yang, J.T., Wu, C.S.C., and Martinez, H.M., Calculation of protein conformation from circular dichroism, *Methods Enzymol.* 130, 208–270, 1986.
135. Woody, R.W., Circular dichroism, *Methods Enzymol.* 246, 34–71, 1995.
136. Johnson, W.C., Jr., Protein secondary structure and circular dichroism: A practical guide, *Proteins, Structure, Function, and Genetics* 7, 205–214, 1990.
137. Bayer, T.S., Booth, L.N., Knudsen, S.M., and Ellington, A.D., Arginine-rich motifs present multiple interfaces for specific binding by RNA, *RNA* 11, 1848–1857, 2005.
138. Harrington, A., Darboe, N., Kenjale, R. et al., Characterization of the interaction of single tryptophan containing mutants of IpaC from *Shigella flexneri* with phospholipid membranes, *Biochemistry* 45, 626–636, 2006.
139. Paramonov, S.E., Jun, H.W., and Hartgerink, J.D., Modulation of peptide-amphiphile nanofibers via phospholipid inclusions, *Biomacromolecules* 7, 24–26, 2006.
140. Miles, A.J. and Wallace, R.A., Synchrotron radiation circular dichroism spectroscopy of proteins and applications in structural and functional genomics, *Chem. Soc. Rev.* 35, 39–51, 2006.
141. Steinberg, I.Z., Circularly polarized luminescence, *Methods Enzymol.* 49, 179–199, 1971.
142. Raghavendra, K. and Ananthanarayanan, V.S., Beta-structure of polypeptides in non-aqueous solutions, I. Spectral characteristics of the polypeptide backbone, *Int. J. Pept. Protein Res.* 17, 412–419, 1981.
143. Hvidt, S., Rodgers, M.E., and Harrington, W.F., Temperature-dependent optical rotatory dispersion properties of helical muscle proteins and homopolymers, *Biopolymers* 24, 1647–1662, 1985.
144. Walji, F., Rosen, A., and Hider, R.C., The existence of conformationally labile (preformed) drug binding sites in human serum albumin as evidenced by optical rotatory measurements, *J. Pharm. Pharmacol.* 45, 551–558, 1993.
145. Parkhusrt, L.J., A nanosecond ORD study of hemoglobin, *Biophys. J.* 68, 399–400, 1995.
146. Galvani, M., Hamdan, M., and Righetti, P.G., Probing protein unfolding through monitoring cysteine alkylation by matrix-assisted laser desorption/ionization mass spectrometry, *Rapid Commun. Mass Spectrom.* 14, 1925–1931, 2000.
147. Majewski, A.J., Sanzari, M., Cui, H.L., and Torzilli, P., Effects of ultraviolet radiation on the type-I collagen protein triple helical structure: A method for measuring structural changes through optical activity, *Phys. Rev. E. Stat. Nonlin. Soft Matter Phys.* 65:031920, 2002.
148. Sakurai, K. and Goto, Y., Dynamics and mechanism of the Tanford transition of bovine beta-lactaglobulin studied using heteronuclear NMR spectroscopy, *J. Mol. Biol.* 356, 483–496, 2006.
149. Susi, H., Infrared spectroscopy—conformation, *Methods Enzymol.* 26, 455–472, 1972.
150. Susi, H. and Byler, D.M., Resolution-enhanced Fourier-transform infrared spectroscopy of enzymes, *Methods Enzymol.* 130, 290–311, 1986.

151. Siebert, F., Infrared spectroscopy applied to biochemical and biological problems, *Methods Enzymol.* 246, 501–526, 1995.
152. Shaw, R.A. and Mantsch, H.H., Near-IR spectrophotometers, in *Encyclopedia of Spectroscopy and Spectrometry*, ed. J.C. Lindon, G.E. Tauter and J.C. and Holmes, Academic Press, New York, 2000.
153. Yuan, B., Murayama, K., Tsenkova, R. et al., Temperature-dependent near-infrared spectra of bovine serum albumin in aqueous solutions: Spectral analysis by principal component analysis and evolving factor analysis, *Appl. Spectros.* 57, 1223–1229, 2003.
154. Bai, S., Nayar, R., Carpenter, J.F., and Manning, M.C., Noninvasive determination of protein conformation in the solid state using near infrared (NIR) spectroscopy, *J. Pharm. Sci.* 94, 2030–2038, 2005.
155. Izutsu, K., Fujimaki, Y., Kuwabara, A. et al., Near-infrared analysis of protein structure in aqueous solutions and freeze-dried solids, *J. Pharm. Sci.* 95, 781–789, 2006.
156. Bruun, S.W., Holm, J., Hansen, S.I., and Jacobsen, S., Application of near-infrared and Fourier transform infrared spectroscopy in the characterization of ligand-induced conformation changes in folate binding protein purified from bovine milk: Influence of buffer type and pH, *Appl. Spectrosc.* 60, 737–746, 2006.
157. Bruun, S.W., Søndergaard, I., and Jacobsen, S., Analysis of protein structures and interactions in complex food by near-infrared spectroscopy. 2. Hydrated gluten, *J. Agric. Food Chem.* 55, 7244–7251, 2007.
158. Schiro, G. and Cupane, A., Quaternary relaxations in sol-gel encapsulated hemoglobin studied via NIR and UV spectroscopy, *Biochemistry* 46, 11568–11576, 2007.
159. Redfield, A.G., Proton nuclear magnetic resonance in aqueous solutions, *Methods Enzymol.* 49, 253–270, 1978.
160. Crespi, H.L. and Katz, J.J, Preparation of deuterated proteins and enzymes, *Methods Enzymol.* 26, 627–637, 1972.
161. Wagner, G. and Wuthrich, K., Observation of internal mobility of proteins by nuclear magnetic resonance in solution, *Methods Enzymol.* 131, 307–328, 1986.
162. Song, J., Laskowski, M., Jr., Qasim, M.A., and Markley, J.L., NMR determination of pK_a values for asp, glu, his, and lys mutants at each variable contiguous enzyme-inhibitor contact position of the turkey ovomucoid third domain, *Biochemistry* 42, 2847–2856, 2003.
163. Geyer, M., Wilde, C., Selzer, J. et al., Glucosylation of Ras by *Clostridium sordellii* lethal toxin: Consequences for effector loop conformations observed by NMR spectroscopy, *Biochemistry* 42, 11951–11959, 2003.
164. Samson, A.O., Chill, J.H., and Anglister, J., Two-dimensional measurement of protein $T_{1\rho}$ relaxation in unlabeled proteins: Mobility changes in α-bungarotoxin upon binding of an acetylcholine receptor peptide, *Biochemistry* 44, 10926–10934, 2005.
165. Mittelmaier, A. and Kay, L.E., New tools provide new insights in NMR studies of protein dynamics, *Science* 312, 224–228, 2006.
166. Jarymowycz, V.A. and Stone, M.J., Fast time scale dynamics of protein backbones: NMR relaxation studies, *Chem. Rev.* 106, 1624–1671, 2006.
167. Zartler, E.R. and Shapiro, M.J., Protein NMR-based screening in drug discovery, *Curr. Pharm. Des.* 12, 3963–3972, 2006.
168. Foster, M.P., McElroy, C.A., and Amero. C.D., Solution NMR of large molecules and assemblies, *Biochemistry* 46, 331–340, 2007.
169. Frieden, C., Protein aggregation processes: In search of the mechanism, *Protein Sci.* 16, 2334–2344, 2007.
170. Pazgier, M., Li, X., Lu, W., and Lubkowski, J., Human defensins: Synthesis and structural properties, *Curr. Pharm. Des.* 13, 3096–3118, 2007.
171. Spiess, H.W., NMR spectroscopy: Pushing the limits of sensitivity, *Angew. Chem. Int. Ed. Engl.* 47, 639–642, 2008.

172. Righetti, P.G. and Verzola, B., Folding/unfolding/refolding of proteins: Present methodologies in comparison with capillary zone electrophoresis, *Electrophoresis* 22, 2359–2374, 2001.
173. Cioni, P. and Strambini, G.B., Tryptophan phosphorescence and pressure effects on protein structure, *Biochim. Biophys. Acta* 1595, 116–130, 2002.
174. Cioni P., Role of protein cavities on unfolding volume change and on internal dynamics under pressure, *Biophys. J.* 91, 3390–3396, 2006.
175. Kumar, A., Tyagi, N.K., and Kinne, R.K., Ligand-mediated conformation changes and positioning of tryptophans in reconstituted human sodium/D-glucose cotransporter1 (hsGLT1) probed by tryptophan fluorescence, *Biophys. Chem.* 127, 69–77, 2007.
176. Harvey, B.J., Bell, E., and Brancaleon, L., A tryptophan rotamer located in a polar environment probes pH-dependent conformational changes in bovine β-lactoglobulin A, *J. Phys. Chem. B.* 111, 2610–2620, 2007.
177. Daly. S.M., Pryzybycien, T.M., and Tilton, R.D., Aggregation of lysozyme and of poly(ethylene glycol)-modified lysozyme after adsorption to silica, *Colloid Surf. B. Biointerfaces* 57, 81–88, 2007.
178. Ghasemi, A., Khajeh, K., and Ranjbar, B., Stabilization of *Bacillus licheniformis* α-amylase by specific antibody which recognizes the N-terminal fragment of the enzyme, *Int. J. Biol. Macromol.* 41, 162–167, 2007.
179. Fan, H., Vitharana, S.N., Chen, T. et al., Effects of pH and polyanions on the thermal stability of fibroblast growth factor 20, *Mol. Pharm.* 4, 232–240, 2007.
180. Benesch, J., Hungerford, G., Shuling, K. et al., Fluorescence probe techniques to monitor protein adsorption-induced conformation changes on biodegradable polymers, *J. Colloid Interface Sci.* 312, 193–200, 2007.
181. Boudier, C., Bonsquet, J.A., Schauinger, S. et al., Reversible inactivation of serpins at acidic pH, *Arch. Biochem. Biophys.* 466, 155–163, 2007.
182. Lee, J. and Tripathi, A., Measurements of label free protein concentration and conformational change using a microfluidic UV-LED method, *Biotechnol. Prog.* 23, 1506–1512, 2007.
183. Ramachander, R., Jiang, Y., Li, C. et al., Solid state fluorescence of lyophilized proteins, *Anal. Biochem.*, 376, 173–182, 2008.
184. Kanaoka, Y., Organic fluorescent reagents in the study of enzymes and proteins, *Angew. Chem. Int. Ed. Engl.* 16, 137–147, 1977.
185. Selvin, P.R., The renaissance of fluorescence resonance energy transfer, *Nat. Struct. Biol.* 7, 730–734, 2000.
186. Heyduk, T., Measuring protein conformational changes by FRET/LRET, *Curr. Opin. Biotechnol.* 13, 292–296, 2002.
187. Giepmans, B.N., Adams, S.R., Ellisman, M.H., and Tsien, R.Y., The fluorescent toolbox for assessing protein location and function, *Science* 312, 217–224, 2006.
188. Royer, C.A., Probing protein folding and conformational transitions with fluorescence, *Chem. Rev.* 106, 1769–1784, 2006.
189. Gull, N., Sen, P., and Kabir-Ud-Din, K.R.H., Effect of physiological concentrations of urea on the conformation of human serum albumin, *J. Biochem.* 141, 261–268, 2007.
190. Mazzini, A., Polverini, E., Parisi, M. et al., Dissociation and unfolding of bovine odorant binding protein at acidic pH, *J. Struct. Biol.* 159, 82–91, 2007.
191. Muriznieks, L.D. and Weiss, A.S., Flexibility in solution structure of human tropoelastin, *Biochemistry* 46, 8196–8205, 2007.
192. Vetri, V., Librizzi, F., Leone, M., and Militello, V., Thermal aggregation of bovine serum albumin at different pH: Comparison with human serum albumin, *Eur. Biophys. J.* 36, 717–725, 2007.
193. Rezaei-Ghaleh, N., Ramshini, H., Ebrahim-Habibi, A. et al., Thermal aggregation of α-chymotrypsin: Role of hydrophobic and electrostatic interactions, *Biophys. Chem.* 132, 23–32, 2008.

194. Hvidt, A. and Linderstrom-Lange, K., *C. R. Trav. Lab. Carlsburg.* 29, 395. 1954.
195. Englander, S.W. and Englander, J.J., Hydrogen-tritium exchange, *Methods Enzymol.* 49, 24–39, 1978.
196. Englander, J.J., Rogero, J.R., and Englander, S.W., Identification of an allosterically sensitive unfolding unit in hemoglobin, *J. Mol. Biol.* 169, 325–344, 1983.
197. Englander, J.J., Rogero, J.R., and Englander, S.W., Protein hydrogen exchange studied by the fragment separation method, *Anal. Biochem.* 147, 234–244, 1985.
198. Gregory, R.B., and Rosenberg, A., Protein conformational dynamics measured by hydrogen isotope exchange techniques, *Methods Enzymol.* 131, 448–508, 1986.
199. Rogero, J.R., Englander, J.J., and Englander, S.W., Individual breathing reactions measured by functional labeling and hydrogen exchange methods, *Methods Enzymol.* 131, 508–517, 1986.
200. Bierzyński, A., Methods of peptide conformation studies, *Acta Biochim. Pol.* 48, 1091–1099, 2001.
201. Hoofnagle, A.N., Resing, K.A., and Ahn, N.G., Protein analysis by hydrogen exchange mass spectrometry, *Annu. Rev. Biophys. Biomol. Struct.* 32, 1–23, 2003.
202. Busenlehner, L.S. and Armstrong, R.N., Insights into enzyme structure and dynamics elucidated by amide H/D exchange mass spectrometry, *Arch. Biochem. Biophys.* 433, 34–46, 2005.
203. Wales, T.E. and Enge, J.R., Hydrogen exchange mass spectrometry for the analysis of protein dynamics, *Mass Spectrom. Rev.* 25, 158–170, 2006.
204. Maier, C.S. and Deinzer, M.L., Protein conformations, interactions, and H/D exchange, *Methods Enzymol.* 402, 312–360, 2005.
205. Kaveti, S. and Engen, J.R., Protein interactions probed with mass spectrometry, *Methods Mol. Biol.* 316, 179–197, 2006.
206. Englander, S.W., Hydrogen exchange and mass spectrometry: A historical perspective, *J. Am. Soc. Mass Spectrom.* 17, 1481–1489, 2006.
207. Tsutsui, Y. and Wintrode, P.L., Hydrogen/Deuterium exchange-mass spectrometry: A powerful tool for probing protein structure, dynamics and interactions, *Curr. Med. Chem.* 14, 2344–2358, 2007.
208. Li, Y., Williams, T.D., and Topp, E.M., Effects of excipients on protein conformation in lyophilized solids by hydrogen/deuterium exchange mass spectrometry, *Pharm. Res.* 25, 259–267, 2008.
209. Krishnan, K.S. and Brandts, J.F., Scanning calorimetry, *Methods Enzymol.* 49, 3–14, 1978.
210. Biltonen, R.L. and Freire, E., Thermodynamic characterization of conformational states of biological macromolecules using differential scanning calorimetry, *CRC Crit. Rev. Biochem.* 5, 85–124, 1978.
211. Brandts, J.F. and Lin, L.N., Study of strong to ultratight protein interactions using differential scanning calorimetry, *Biochemistry* 29, 6927–6940, 1990.
212. Plum, G.E. and Breslauer, K.J., Calorimetry of proteins and nucleic acids, *Curr. Opin. Struct. Biol.* 5, 682–690, 1995.
213. Weber, P.C. and Salemme, F.R., Applications of calorimetric methods to drug discovery and the study of protein interactions, *Curr. Opin. Struct. Biol.* 13, 115–121, 2003.
214. Matheus, S., Friess, W., and Mahler, H.C., FTIR and nDSC as analytical tools for high-concentration protein formulations, *Pharm. Res.* 23, 1350–1363, 2006.
215. Tavirani, M.R., Moghaddamnia, S.H., Ranjbar, B. et al., Conformational study of human serum albumin in pre-denaturation temperatures by differential scanning calorimetry, circular dichroism and UV spectroscopy, *J. Biochem. Mol. Biol.* 39, 530–536, 2006.
216. Arakawa, T., Kita, Y., Ejima, D. et al., Aggregation suppression of proteins by arginine during thermal unfolding, *Protein Pept. Lett.* 13, 921–927, 2006.

217. de Groot, J., Kosters, H.A, and de Jongh, H.H., Deglycosylation of ovalbumin prohibits formation of a heat-stable conformer, *Biotechnol. Bioeng.* 97, 735–741, 2007.
218. Ejima, D., Tsumoto, K., Fukada, H. et al., Effects of acid exposure on the conformation, stability, and aggregation of monoclonal antibodies, *Proteins* 66, 954–962, 2007.
219. Wakankar, A.A., Lin, J., Vandervelde, D. et al., The effect of cosolutes on the isomerization of aspartic acid residue and conformational stability in a monoclonal antibody, *J. Pharm. Sci.* 96, 1708–1718, 2007.
220. Garber, E. and Demarest, S.J., A broad range of Fab stabilities within a host of therapeutic IgGs, *Biochem. Biophys. Res. Commun.* 355, 751–757, 2007.
221. Kar, K. and Kishore, N., Enhancement of thermal stability and inhibition of protein aggregation by osmolytic effect of hydroxyproline, *Biopolymers* 87, 339–351, 2007.
222. Han, Y., Jin, B.S., Lee, S.B., Effects of sugar additives on protein stability of recombinant human serum albumin during lyophilization and storage, *Arch. Pharm. Res.* 30, 1124–1131, 2007.
223. Tsybovsky, Y., Shubenok, D.V., Kravchuk, Z.I., and Martsev, S.P., Folding of an antibody variable domain in two functional conformations *in vitro* calorimetric and spectroscopic study of the anti-ferritin antibody VL domain, *Protein. Eng. Des. Sel.* 20, 481–490, 2007.
224. Sedlák, E., Zoldák, G., and Wittung-Stafshede, P., Role of copper in thermal stability of human ceruloplasmin, *Biophys. J.* 94, 1384–1391, 2008.
225. Bellezza, F., Cipiciani, A., Quotadamo, M.A. et al., Structure, stability, and activity of myoglobin adsorbed onto phosphate-grafted zirconia nanoparticles, *Langmuir* 23, 13007–13012, 2007.
226. Tobin, M.C., Ramen spectroscopy, *Methods Enzymol.* 26, 473–497, 1972.
227. Van Wart, H.E. and Scheraga, H.A., Raman and resonance Ramen spectroscopy, *Methods Enzymol.* 49, 67–149, 1978.
228. Williams, R.W., Protein secondary structure analyzed with Ramen amide I and amide II spectra, *Methods Enzymol.* 130, 311–331, 1986.
229. Hudson, B. and Mayne, L., Ultraviolet resonance Raman spectroscopy of biopolymers, *Methods Enzymol.* 130, 331–350, 1986.
230. Vass, E., Hollósi, M., Besson, F. and Buchet, R., Vibrational spectroscopic detection of β- and γ-turns in synthetic and natural peptides and proteins, *Chem. Rev.* 103, 1917–1954, 2003.
231. Pimenov, K.V., Bykov, S.V., Mikhonin, A.V., and Asher, S.A., UV Raman examination of α-helical peptide water hydrogen bonding, *J. Am. Chem. Soc.* 127, 2840–2841, 2005.
232. Zhu, F., Issacs, N.W., Hecht, L., and Barron, L.D., Raman optical activity: A tool for protein structure analysis, *Structure* 13, 1409–1419, 2005.
233. Zhu, F., Issacs, N.W., Hecht, L. et al., Raman optical activity of proteins, carbohydrates and glycoproteins, *Chirality* 18, 103–115, 2006.
234. Hédoux, A., Ionov, R., Willart, J.F. et al., Evidence of a two-stage thermal denaturation process in lysozyme: A Raman scattering and differential scanning calorimetry investigation, *J. Chem. Phys.* 124:14703, 2006.
235. Jaisson, S., Lorimier, S., Ricard-Blum, S. et al., Impact of carbamylation on type I collagen conformational structure and its ability to activate polymorphonuclear neutrophils, *Chem. Biol.* 13, 149–159, 2006.
236. Thawornchinsombut, S., Park, J.W., Meng. G., and Li-Chan, E.C., Raman spectroscopy determines structural changes associated with gelation properties of fish proteins recovered at alkaline pH, *J. Agric. Food Chem.* 54(6), 2178–2187, 2006.
237. Podstawka, E., Mak, P.J. Kincaid, J.R., and Proniewicz, L.M., Low frequency resonance Raman spectra of isolated alpha and beta subunits of hemoglobin and their deuterated analogues, *Biopolymers* 83, 455–466, 2006.

238. Dàvila, E., Parés, D., and Howell, N.K., Fourier transform Raman spectroscopy study of heat-induced gelation of plasma proteins as influenced by pH, *J. Agric. Food Chem.* 54, 7890–7897, 2006.
239. Xu, M., Shashilov, V.A., Ermolenkov, V.V. et al., The first step of hen egg white lysozyme fibrillation, irreversible partial unfolding, is a two-state transition, *Protein Sci.* 16, 815–832, 2007.
240. Ma, L., Ahmed, Z., Mikonin, A.V., and Asher, S.A., UV resonance Raman measurements of poly-L-lysine's conformational energy landscapes: Dependence on perchlorate concentration and temperature, *J. Phys. Chem. B* 111, 7675–7680, 2007.
241. Strachan, C.J., Rades, T., Gordon, K.C., and Rantanen, J., Ramen spectroscopy for quantitative analysis of pharmaceutical solids, *J. Pharm. Pharmacol.* 59, 179–192, 2007.
242. Wen, Z.Q., Raman spectroscopy of protein pharmaceuticals, *J. Pharm. Sci.* 96, 2861–2878, 2007.
243. Heyduk, T., Baichoo, N., and Heyduk, E., Hydroxyl radical footprinting using metal ion complexes, *Met. Ions Biol. Syst.* 38, 255–287, 2001.
244. Kiselar, J.G. Janmey, P.A., Almo, S.C., and Chance, M.R., Structural analysis of gelsolin using synchrotron protein footprinting, *Mol. Cell. Proteomics* 2, 1120–1132, 2003.
245. Guan, J.Q. and Chance, M.R., Structural proteomics of macromolecular assemblies using oxidative footprinting and mass spectrometry, *Trends Biochem. Sci.* 30, 583–592, 2005.
246. Takamoto, K. and Chance, M.R., Radiolytic protein footprinting with mass spectrometry to probe the structure of macromolecular complexes, *Annu. Rev. Biophys. Biomol. Struct.* 35, 251–276, 2006.
247. Gómez, G.E., Cauerhff, A., Craig, P.O. et al., Exploring protein interfaces with a general photochemical reagent, *Protein Sci.* 15, 744–752, 2006.
248. Shchebakova, I., Mitra, S., Beer, R.H., and Brenowitz, M., Fast Fenton footprinting: A laboratory-based method for the time-resolved analysis of DNA, RNA, and proteins, *Nucleic Acids Res.* 34:e48, 2006.
249. Linderstrom-Lange, K., Globular proteins and proteolytic enzymes, *Proc. Royal Soc. B* 127, 17–19, 1939.
249a. Wilson, W.D. and Foster, J.F., Conformation dependent limited proteolysis of bovine plasma albumin preparations, *Biochemistry* 10, 1772–1780, 1971.
250. Peters, T., Jr. and Feldhoff, R.C., Fragments of bovine serum albumin produced by limited proteolysis. Isolation and characterization of tryptic fragments, *Biochemistry* 14, 3384–3391, 1975.
251. Reed, R.B., Feldhoff, R.C., Clute, O.L., and Peters, T., Jr., Fragments of bovine serum albumin produced by limited proteolysis. Conformation and ligand binding, *Biochemistry* 14, 4578–4583, 1975.
252. Dombrádi, V., Gergely, P., and Bot, G., Limited proteolysis by subtilisin reveals structural differences between phosphorylase a and b, *Int. J. Biochem.* 15, 1088–1092, 1983.
253. Sheshberadaran, H., and Payne, L.G., Protein antigen-monoclonal antibody contact sites investigated by limited proteolysis of monoclonal antibody-bound antigen: Protein "footprinting," *Proc. Nat. Acad. Sci. USA* 85, 1–5, 1988.
254. Miedel, M.C., Hulmes, J.D., and Pan, Y.C., Limited proteolysis of recombinant human soluble interleukin-2 receptor. Identification of an interleukin-2 binding core, *J. Biol. Chem.* 264, 21097–21105, 1989.
255. Olson, M.O., Kirstein, M.N., and Wallace, M.O., Limited proteolysis as a probe of the conformation and nucleic binding regions of nucleolin, *Biochemistry* 29, 5682–5686, 1990.
256. Wilson, J.E., The use of monoclonal antibodies and limited proteolysis in elucidation of structure-function relationships in proteins, *Methods Biochem. Anal.* 35, 207–250, 1991.
257. Gentile, F. and Salvatore, G., Preferential sites of proteolytic cleavage of bovine, human, and rat thyroglobulin. The use of limited proteolysis to detect solvent-exposed regions of the primary structure, *Eur. J. Biochem.* 218, 603–621, 1993.

258. Fontana, A., Zambonin, M., Polverino de Laureto, P. et al., Probing the conformational state of apomyoglobin by limited proteolysis, *J. Mol. Biol.* 266, 223–230, 1997.
259. Hubbard, S.J., The structural aspects of limited proteolysis of native proteins, *Biochim. Biophys. Acta* 1382, 191–206, 1998.
260. Polverino de Laureto, P., Scaramella, E., Frigo, M. et al., Limited proteolysis of bovine α-lactalbumin: Isolation and characterization of protein domains, *Protein Sci.* 8, 2290–2303, 1999.
261. Yang, S.A. and Klee, C., Study of calcineurin structure by limited proteolysis, *Methods Mol. Biol.* 172, 317–334, 2002.
262. Polverino de Laureto, P., Frare, E. et al., Partly folded states of members of the lysozyme/lactalbumin superfamily: A comparative study by circular dichroism spectroscopy and limited proteolysis, *Protein Sci.* 11, 2932–2946, 2002.
263. Fontano, A. de Laureto, P.P., Spolaore, B. et al., Probing protein structure by limited proteolysis, *Acta Biochim. Pol.* 51, 299–321, 2004.
264. Stroh, J.G., Loulakis, P., Lanzetti, A.J., and Xie, J., LC-mass spectrometry analysis of *N*- and C-terminal boundary sequences of polypeptide fragments by limited proteolysis, *J. Am. Soc. Mass Spectrom.* 16, 38–45, 2005.
265. Williams, J.G., Tomer, K.B., Hice, C.E. et al., The antigenic determinants on HIV p24 for CD4+ T cell inhibiting antibodies as determined by limited proteolysis, chemical modification and mass spectrometry, *J. Am. Soc. Mass Spectrom.* 17, 1560–1590, 2006.
266. Sonoda, S. and Schlamowitz, M., Studies of I^{125} trace labeling of immunoglobulin G by chloramine-T, *Immunochemistry* 7, 885–898, 1970.
267. Giedroc, D.P., Sinha, S.K., Brew, K., and Puett, D., Differential trace labeling of calmodulin: Investigation of binding sites and conformational states by individual lysine reactivities. Effects of beta-endorphin, trifluoperazine, and ethylene glycol bis(beta-aminoethyl ether)-N,N,N′,N′-tetraacetic acid, *J. Biol. Chem.* 260, 13406–13413, 1985.
268. Jackson, A.E., Carraway, K.L., 3rd, Puett, D., and Brew, K., Effects of the binding of myosin light chain kinase on the reactivities of calmodulin lysines, *J. Biol. Chem.* 261, 12226–12232, 1986.
269. Giedroc, D.P., Puett, D., Sinha, S.K., and Brew, K., Calcium effects on calmodulin lysine reactivities, *Arch. Biochem. Biophys.* 252, 136–144, 1987.
270. Lin, T.P. and Hsu, C.C., Determination of residual moisture in lyophilized protein pharmaceuticals using a rapid and non-invasive method: Near infrared spectroscopy, *PDA J. Pharm. Sci. Technol.* 56, 196–205, 2002.
271. Bruun, S.W., Søndergaard, I., and Jacobsen, S., Analysis of protein structures and interaction in complex foods by near-infrared spectroscopy. I. Gluten powder, *J. Agric. Food Chem.* 55, 7234–7243, 2007.
272. Hermida, M., Rodriguez, N., and Rodriguez-Otero, J.L., Determination of moisture, starch, protein, and fat in common beans (*Phaseolus vulgaris* L.) by near infrared spectroscopy, *J. AOAC Int.* 89, 1039–1041, 2006.
273. Berntsson, O., Zackrisson, G., and Ostling, G., Determination of moisture in hard gelatin capsules using near-infrared spectroscopy applications to at-line process control of pharmaceutics, *J. Pharm. Biomed. Anal.* 15, 895–900, 1997.
274. Savage, M., Torres, J., Franks, L. et al., Determination of adequate moisture content for efficient dry-heat viral inactivation in lyophilized factor VIII by loss on drying and by near infrared spectroscopy, *Biologicals* 26, 119–124, 1998.
275. Zheng, Y., Lai, X., Bruun, S.W. et al., . Determination of moisture content of lyophilized allergen vaccines by NIR spectroscopy, *J. Pharm. Biomed. Anal.* 46, 592–596, 2008.
276. Brulls, M., Folestad, S., Sparén, A. et al., Applying spectral peak area analysis in near-infrared spectroscopy moisture assays, *J. Pharm. Biomed. Anal.* 44, 127–136, 2007.
277. Nagarajan, R., Singh, P., and Mehrotra, R., Direct determination of moisture in powder milk using near infrared spectroscopy, *J. Autom. Methods Manag. Chem.* 2006: 51342, 2006.

278. Konrad, M., Immunogenicity of proteins administered to humans for therapeutic purposes, *Trends Biotechnol.* 7, 175–179, 1989.
279. Hermeling, S., Crommelin, D.J.A., Schellekens, H., and Jiskott, W., Structure-immunogenicity relationships of therapeutic proteins, *Pharm. Res.* 21, 897–903, 2004.
280. Frost, H., Antibody-mediated side effects of recombinant proteins, *Toxicology* 209, 155–160, 2005.
281. Wadhwa, M., Bird, C., Dilger, P. et al., Strategies for detection, measurement and characterization of unwanted antibodies induced by therapeutic biologicals, *J. Immunol. Methods* 278, 1–17, 2003.
282. Thorpe, R. and Swanson, S.J., Current methods for detecting antibodies against erythropoietin and other recombinant proteins, *Clin. Diag. Lab. Immunol.* 12. 28–39. 2005.
283. Geng, D., Shankar, G., Schantz, A. et al., Validation of immunoassays used to assess immunogenicity to therapeutic monoclonal antibodies, *J. Pharm. Biomed. Anal.* 39, 364–375, 2005.
284. Doyle, J.W., Johnson, G.L., Eshhar, N., and Hammond, D., The use of rabbit polyclonal antibodies to assess neoantigenicity following viral reduction of an alpha-1-proteinase inhibitor preparation, *Biologicals* 34, 199–297, 2006.
285. Gupta, S., Indelicato, S.R., Jethwa, V. et al., Recommendations for the design, optimization, and qualification of cell-based assays used for the detection of neutralizing antibody responses elicited to biological therapeutics, *J. Immunol. Methods* 321, 1–18, 2007.
286. Brennen, T.V. and Clarke, S., Mechanism of cleavage at Asn 148 during the maturation of jack bean concanavlin A, *Biochem. Biophys. Res. Commun.* 193, 1031–1037, 1993.
287. Capasso, S., Mazzarella, L., Sorrentino, G. et al., Kinetics and mechanism of the cleavage of the peptide bond next to asparagine, *Peptides* 17, 1075–1077, 1996.
288. Mathys, S., Evans, T.C., Chute, I.C. et al., Characterization of a self-splicing mini-intein and its conversion into autocatalytic *N*- and C-terminal cleavage elements: Facile production of protein building blocks for protein ligation, *Gene* 231, 1–13,1988.
289. Paulus, H., Protein splicing and related forms of protein autoprocessing, *Annu. Rev. Biochem.* 69, 447–496, 2000.
290. Li, B., Gorman, E.M., Moore, K.D. et al., Effects of acidic N +1 residues on asparagine deamidation rates in solution and in the solid state, *J. Pharm. Sci.* 94, 666–675, 2005.
291. Kawauchi, H., Bewley, T.A., and Li, C.H., Studies on prolactin. 40. Chemical modification of tyrosine residues in the ovine hormone, *Biochim. Biophys. Acta* 493, 380–392, 1977.
292. Yokosawa, H. and Ishii, S., Anydrotrypsin and trypsin: Subtle differences in the active-site conformations detected by chemical modification and CD spectroscopy, *J. Biochem.* 81, 657–663, 1977.
293. Atassi, M.Z. and Zablocki, W., Conformation, enzymic activity, and immunochemistry of a lysozyme derivative modified at tryptophan 123 by reaction with 2,3-dioxo-5-indolinesulfonic acid, *J. Biol. Chem.* 251, 1653–1658, 1976.
294. Mihalyi, E., *Application of Proteolytic Enzymes to Protein Structure Studies*, 2nd ed., CRC Press, West Palm Beach, FL, 1978.
295. Neurath, H., Proteolytic enzymes, past and present, *Fed. Proc.* 44, 2907–2913, 1085.
296. Friedrich, P. and Bozóky, Z., Digestives versus regulatory proteases: On calpain action *in vivo, Biol. Chem.* 386, 609–612, 2005.
297. Egelund, R., Petersen, T.E., and Andreasen, P.A., A serpin-induced extensive proteolytic susceptibility of urokinase-type plasminogen activator implicates distortion of the proteinase pocket and oxyanion hole in the serpin inhibitory mechanism, *Eur. J. Biochem.* 268, 673–685, 2001.
298. Reid, J., Kelly, S.M., Watt, K. et al., Conformational analysis of the androgen receptor amino-terminal domain involved in transactivation. Influence of structure-stabilizing solutes and protein-protein interactions, *J. Biol. Chem.* 277, 22079–20086, 2002.

299. Varne, A., Muthukumaraswamy, K., Jatiani, S.S., and Mittal, R., Conformational analysis of the GTP-binding protein MxA using limited proteolysis, *FEBS Lett.* 516, 129–132, 2002.
300. Bito, R., Shikano, T., and Kawabata, H., Isolation and characterization of denatured serum albumin from rats with endotoxicosis, *Biochim. Biophys. Acta* 1646, 100–111, 2003.
301. Stiuso, P., Marabotti, A., Facchiano, A. et al., Assessment of the conformational features of vasoactive intestinal peptide in solution by limited proteolysis experiments, *Biopolymers* 81, 110–119, 2006.
302. Wehbi, Z., Pérez, M.D., Salgalarrondo, M. et al., Study of ethanol-induced conformational changes of holo and apo alpha-lactalbumin by spectroscopy and limited proteolysis, *Mol. Nutr. Food Res.* 50, 34–43, 2006.
303. Manea, M., Mezo, G., Hudecz, F., and Przybylski, M., Mass spectrometric identification of the trypsin cleavage pathway in lysyl-proline containing oligotuftsin peptides, *J. Pept. Sci.* 13, 227–236, 2007.
304. Liu, H., Gaza-Bulseco, G., Xiang, T., and Chumsae, C., Structural effect of deglycosylation and methionine oxidation of a recombinant monoclonal antibody, *Mol. Immunol.* 45, 701–708, 2008.
305. Kathir, K.M., Ibrahim, K., Rajalingam, D. et al., S100A13-lipid interactions-role in the non-classical release of the acidic fibroblast growth factor, *Biochim. Biophys. Acta* 1768, 3080–3089., 2007.
306. Datwyler, S.A. and Meares, C.F., Artificial iron-dependent proteases, *Met. Ions Biol. Syst.* 38, 213–254, 2001.
307. Milovic, N.M., Dutca, L.M., and Kostic, N.M., Combined use of platinum(II) complexes for selective cleavage of peptides and proteins, *Inorg. Chem.* 42, 4036–4045, 2003.
308. Milovic, N.M., Dutca, L.M., and Kostic, N.M., Transition metal complexes as enzyme-like reagents for protein cleavage. Complex *cis*$[Pt(en)(H_2O)_2]^{2+}$ as a new methionine-specific protease, *Chemistry—A European J.* 9, 5097–5106, 2003.
309. Dutca, L.M., Ko, K.S., Pohl, N.L., and Kostic, N.M., Platinum(II) complex as an artificial peptidase: Selective cleavage of peptides and a protein by *cis*$[Pt(en)(H_2O)_2]^{2+}$ ion under ultraviolet and microwave irradiation, *Inorg. Chem.* 44, 5141–4146, 2005.
310. Heyduk, T., Baichoo, N., and Heyduk, E., Hydroxyl radical footprinting of proteins using metal ion complexes, *Met. Ions Biol. Syst.* 38, 255–287. 2001.
311. Gian, J.Q. and Chance, M.R., Structural proteomics of macromolecular assemblies using oxidative footprinting and mass spectrometry, *Trends Biochem. Sci.* 30, 583–592, 2005.
312. Takamoto, K. and Chance, M.R., Radiolytic protein footprinting with mass spectrometry to probe the structure of macromolecular complexes, *Annu. Rev. Biophys. Biomol. Struct.* 35, 215–276, 2006.
313. Kaplan, B., Stevenson, K.J., and Hartley, B.S., Competitive labelling, a method for determining the reactivity of individual groups in proteins, *Biochem. J.* 124, 289–299, 1971.
314. Cruikshank, W.H. and Kaplan, H., Competitive labeling method for determining ionization constants and reactivity of individual histidine residues in proteins—histidines of alpha-chymotrypsin, *Biochem. J.* 130, 1125–1131, 1972.
315. Malchy, B. and Kaplan, H., Reactive properties of amino-groups of histones in calf thymus chromatin, *J. Mol. Biol.* 82, 537–540, 1974.
316. Bosshard, H.R., Koch, G.L.E., and Hartley, B.S., Aminoacyl-tRNA synthetase-tRNA complex—detection by differential labeling of lysine residues involved in complex formation, *J. Mol. Biol.* 119, 377–389, 1978.
317. Bosshard, H.R. and Zurrier, M., The conformation of cytochrome-C in solution—localization of a conformational difference between ferricytochrome-C and ferrocytochrome-C on the surface of the molecule, *J. Biol. Chem.* 255, 6694–6699, 1980.
318. Degregori-Hitchcock, S.E., Study of the structure of troponin-I by measuring the relative reactivities of lysines with acetic-anhydride, *J. Biol. Chem.* 257, 7372–7380, 1982.

319. Kaplan, H., Hefford, M.A., Chan, A.M.L., and Oda, G., Chemical-reactivity of the functional-groups of insulin—concentration-dependence studies, *Biochem. J.* 217, 135–143, 1985.
320. Giedroc, D.P., Sinha, S.K., Brew, K., and Puett, D., Differential trace labeling of calmodulin—investigation of binding sites and conformational states by individual lysine reactivities—effects of beta-endorphin, trifluoperazine, and ethylene glycol bis (β-aminoethyl ether)-*N,N,N′,N′*-tetraacetic acid, *J. Biol. Chem.* 260, 13406–13413, 1985.
321. Winkler, M.A., Fried, V.A., Merat, D.L., and Cheung, W.Y., Differential reactivities of lysines in calmodulin complexed to phosphatase, *J. Biol. Chem.* 262. 15466–15471, 1987.
322. Giedroc, D.P., Puett, D., Sinha, S.K., and Brew, K., Calcium effects on calmodulin lysine reactivities, *Arch. Biochem. Biophys.* 252, 136–144, 1987.
323. Wei, Q., Jackson, A.E., Pervaiz, S. et al., Effects of interaction with calcineurin on the reactivities of calmodulin lysines, *J. Biol. Chem.* 263, 19541–19544, 1988.
324. Takata, Y. and Fujioka, M., Recombinant rat guanidinoacetate methyltransferase—study of the structure by trace labeling lysine residues with acetic anhydride, *Int. J. Biochem.* 22, 1333–1339, 1990.
325. Rupley, J.A. and Scheraga, H.A., Digestion of ribonuclease A with chymotrypsin and trypsin at high temperatures, *Biochim. Biophys. Acta* 44, 191–193, 1960.
326. Hermans, J., Jr. and Scheraga, H.A., The thermally induced configurational change of ribonuclease I water and deuterium, *Biochim. Biophys. Acta* 36, 534–535, 1959.
327. Winchester, B.G., Mathias, A.P., and Robin, B.R., Study of the thermal denaturation of ribonuclease A by differential thermal analysis an susceptibility to proteolysis, *Biochem. J.* 117, 299–307, 1970.
328. Schrier, E.E. and Scheraga, H.A., The effect of aqueous alcohol solutions on the thermal transition of ribonuclease, *Biochim. Biophys. Acta* 64, 406–408, 1962.
329. Ooi, T. and Scheraga, H.A., Structural studies of ribonuclease XIV. Tryptic hydrolysis of ribonuclease in propyl alcohol solution, *Biochemistry* 3, 1209–1213, 1964.
330. Ooi, T. and Scheraga, H.A., Structural studies of ribonuclease XII. Enzymic hydrolysis of active tryptic modifications of ribonuclease, *Biochemistry* 3, 641–647, 1964.
331. Ooi, T. and Scheraga, H.A., Structural studies of ribonuclease 13. Physicochemical properties of tryptic modification of ribonuclease, *Biochemistry* 3, 648–652, 1964.
332. Ooi, T., Rupley, J.A., and Scheraga, H.A., Structural studies of ribonuclease VIII. Tryptic hydrolysis of ribonuclease A at elevated temperatures, *Biochemistry* 3, 432–437, 1963.
333. Rupley, J.A. and Scheraga, H.A., Structural studies of ribonuclease VII. Chymotryptic hydrolysis of ribonuclease at elevated temperatures, *Biochemistry* 2, 421–431, 1963.
334. Seon, B.-K., Roholt, O.A., and Pressman, D., Differences in the enzymatic digestability of the variable and constant halves of Bence-Jones protein with the temperature, *J. Biol. Chem.* 247, 2151–2155, 1972.
335. Seon, B.-K. and Pressman, D., Fragment from the constant portion of IgG obtained by peptic digestion at high-temperatures, *J. Immunol.* 113, 1190–1198, 1974.
336. Edmundson, A.B., Ely, K.R., Abola, E.E. et al., Rotational allosterism and divergent evolution of domains in immunoglobulin light-chains, *Biochemistry* 14, 3953–3961, 1974.
337. Pascual, D.W. and Clem, L.W. Low temperature pepsin proteolysis. An effective procedure for mouse IgM F(ab′)$_2$ fragment production, *J. Immunol. Methods* 146, 249–255, 1992.
338. Krói, M., Roterman, I., Piekarska, B. et al., Local and long-range structural effects caused by the removal of the N-terminal polypeptide fragment from immunoglobulin L chain lamba, *Biopolymers* 69, 189–200, 2003.
339. Hubbard, S.J., Eisensenger, F., and Thornton, J.M., Modeling studies of the change in conformation required for cleavage of limited proteolytic sites, *Protein Sci.* 3, 757–768, 1994.
340. Hubbard, S. and Beynon, R.J., Proteolysis of native proteins as a structural probe, in *Proteolytic Enzyme, A Practical Approach*, 2nd ed., Eds. R.J. Benyon and J. Bond, Oxford University Press, Oxford, U.K., Chapter 10, pp. 233–264, 2001.

341. Tsai, C.J., Polverino de Laureto, P., Fontana, A., and Nussinov, R., Comparison of protein fragments identified by limited proteolysis and by computational cutting of proteins, *Protein Sci.* 11, 1753–1770, 2002.
342. Fontana, A., de Laureto, P.P., Spolaore, B. et al., Probing protein structure by limited proteolysis, *Acta Biochim. Pol.* 51, 299–321, 2004.
343. Park, C. and Marqusee, S., Pulse proteolysis: A simple method for quantitative determination of protein stability and ligand binding, *Nat. Methods* 2, 207–212, 2005.
344. Breitender, H. and Mills, E.N., Molecular properties of food allergens, *J. Allergy Clin. Immunol.* 155, 14–23, 2005.
345. Reyda, M.R., Dippold, R., Dotson, M.E., and Jarrett, J.T., Loss of iron-sulfur clusters from biotin synthase as a result of catalysis promotes unfolding and degradation, *Arch. Biochem. Biophys.* 471, 32–41, 2008.
346. Otsuka, J. and Kunisawa, T., Conformational change and cooperative ligand binding in hemoglobin, *Adv. Biophys.* 11, 53–92, 1978.
347. Saibil, H.R., Horwich, A.L, and Fenton, W.A., Allostery and protein substrate conformational change during GroEl/GroES-mediated protein folding, *Adv. Protein Chem.* 59, 45–72, 2001.
348. Goh, C.S., Milburn, D., and Gerstein, M., Conformational changes associated with protein-protein interactions, *Curr. Opin. Struct. Biol.* 14, 104–109, 2004.
349. Kern, D. and Zuiderweg, E.R., The role of dynamics in allosteric regulation, *Curr. Opin. Struct. Biol.* 13, 748–757, 2003.
350. Lewis, M., The lac repressor, *C. R. Biol.* 328, 521–548, 2005.
351. Wiechert, W., Schweissgut, O., Takanaga, H., and Frommer, W.B., Fluxomics: Mass spectrometry versus quantitative imaging, *Curr. Opin. Plant. Biol.* 10, 323–330, 2007.

3 Chemistry of the Attachment of Proteins and Peptides to Solid Surfaces

DNA AND PROTEIN MICROARRAY

Advances in analytical technologies, including microarray analysis (including surface-plasmon resonance technology), are of importance to proteomics and other "omics" technology and for diagnostics. DNA microarray technology has become a well-accepted analytical technology that has proved its utility in the study of gene expression.

Unlike DNA microarray technology, where the construction of multiple oligonucleotide arrays with specificity is easy to obtain because of knowledge obtained from the genomic sequence, the construction of protein microarrays is more difficult.[1–4] There are 4 monomer units instead of the 20 monomer units for proteins. DNA microarrays are physically stronger than protein microarrays as it is difficult to irreversibly denature DNA, permitting the facile regeneration of DNA microarrays.[5–9] The interaction of DNA with RNA or DNA uses base-pairing and is less sensitive to conformation issues than is protein–protein interaction. As a result of these factors, there is considerably more experience with DNA microarrays.

DNA microarray technology employs a technique referred to as *Southern blotting,* which is based on the hybridization between DNA and RNA.[10–14] The terminology dates to the development of the Southern blot, where the labeled RNA was the probe to label specific DNA sequences on the electrophoretogram.[11] DNA microarrays are composed of synthetic oligonucleotides or intact cDNA, which are used to analyze RNA probes prepared from the cDNA of samples.[15–17] Early hybridization experiments used radiolabeled probes; current technology uses fluorescent dyes. The probes are either synthesized *in situ* on the matrix with oligonucleotides or "spotted" in the case of cDNA.[18–35]

In addition to the now classic DNA microarray, nucleic acids can be bound to beads by several mechanisms,[36,37] including the binding of biotin-labeled oligonucleotides to streptavidin-coated magnetic beads.[38–40] Another approach used the covalent attachment of 5′-aminohexyl oligonucleotides to polyethyleneimine-coated nylon beads.[41] Glass slides are cleaned with piranha solution (33% 30% H_2O_2–67% concentrated H_2SO_4).[42–45] The glass slides are derivatized with silane derivatives to yield amino, carboxyl, or aldehyde derivatives[46] (Figure 3.1).

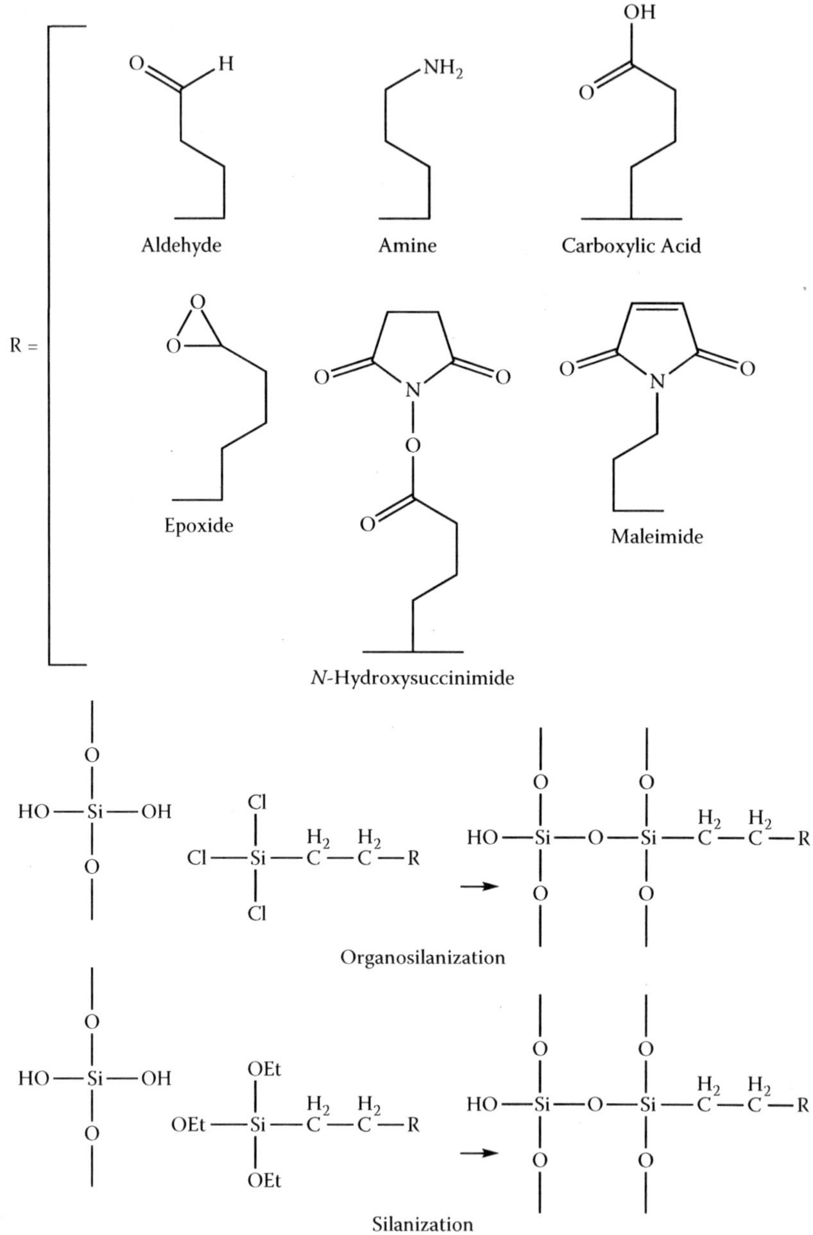

FIGURE 3.1 Silane derivatives for microarrays. The derivatives are prepared by coupling a trifunctional organic silane to a glass matrix. (See Weetall, H.H., Preparation of immobilized proteins covalently coupled through silane coupling agents to inorganic supports, *Appl. Biochem. Biotechnol.* 41, 157–188, 1993; Matyska, M.T. and Pesek, J.J., Comparison of silanization/hydrosilation and organosilanization modification procedures on etched capillaries for electrokinetic chromatography, *J. Chromatog. A* 1079, 366–371, 2005.)

Specific base-pairing is well understood as the basis for oligonucleotide/polynucleotide interaction; similar specific phenomena are not available for protein–protein or protein–peptide interactions. Protein–protein interactions are more complex than, for example, DNA–DNA interactions, partially because of the number of different amino acids involved compared with the number of purine and pyrimidine bases but also as a result of the number of interaction modalities, including hydrogen-bonding, van der Waals forces, and hydrophobic interactions. Although it is possible to use combinatorial chemistry or phage display to obtain "protein" microarrays similar to the DNA microarrays in binding site numbers, the rationale for site design is missing because absolute predictive rules for protein–protein interaction are still being developed. Furthermore, the primary structure complexity must be combined with co- and posttranslational modifications in the use of such rules. Nevertheless, protein microarray technology markedly increased from one publication in 1999 to more than 600 in 2006 (either SciFinder Scholar or PubMed).

SOLID-PHASE MATRICES (INCLUDING BEADS) FOR ATTACHMENT OF PROTEIN PROBES

Protein and other biological macromolecules bind to plastic (polystyrene, polypropylene, polyvinyl chloride) and other materials (glass, steel) in a "nonspecific" manner.[47–87] The adsorption of protein to plastic surfaces is frequently associated with conformational change/denaturation.[57,72,77–79,83] The binding of protein to plastic can be influenced by solvent conditions.[88–92] The binding of protein to microplate wells can markedly influence cell-based assays.[93–96] The nonspecific binding of protein to various biomaterials can have positive and negative consequences.[96–99]*

PROTEIN INTERACTION WITH STEEL[100–108] AND TITANIUM[109–115]

The studies on titanium are concerned with biocompatibility studies as are many of the studies with steel. The studies with stainless steel are important for biotechnology, given the importance of 316 stainless steel in the use of reaction (fermentation, cell culture) vessels, piping, and chromatographic columns in biopharmaceutical manufacture,[116–120] thus requiring an understanding of cleaning validation.[121–123] Proteins can bind directly to gold via cysteine where a covalent bond is formed[124–136] (Figure 3.2). Although direct binding of protein to gold surfaces is useful, the use of self-assembled monolayers (SAMs) is also popular.[137–142] In this technique, a functionalized alkyl thiol or disulfide (i.e., long-chain ω-hydroxylalkanethiols, dithio-bis(succinimidylundecanoate), ω-mercaptohexadecanoic acid, ω-mercaptohexadecylamine, long-chain mercaptoalkylphosphonic acid derivatives, diothio-bis-butyrylamino-*m*-phenylboronic acid) is

* The literature in this area is voluminous and only several recent citations from the literature are presented. There are several large reference works in this area (*Hydrogels and Biodegradable Polymers for Bioapplications*, Eds. R.M. Ottebrite and S.J. Huang, American Chemical Society, Washington, DC, 1996; Barbucci, R., *Integrated Biomaterials Science*, Kluwer Academic/Plenum, New York, 2002; Dumitriu, S., *Polymeric Biomaterials*, Marcel Dekker, New York, 2002; *An Introduction to Biomaterials*, Ed. S.A. Guelcher and J.O. Hollinger, CRC Press, Boca Raton, FL, 2006; *Biodegradable Systems in Tissue Engineering an Regenerative Medicine*, CRC Press, Boca Raton, FL, 2006).

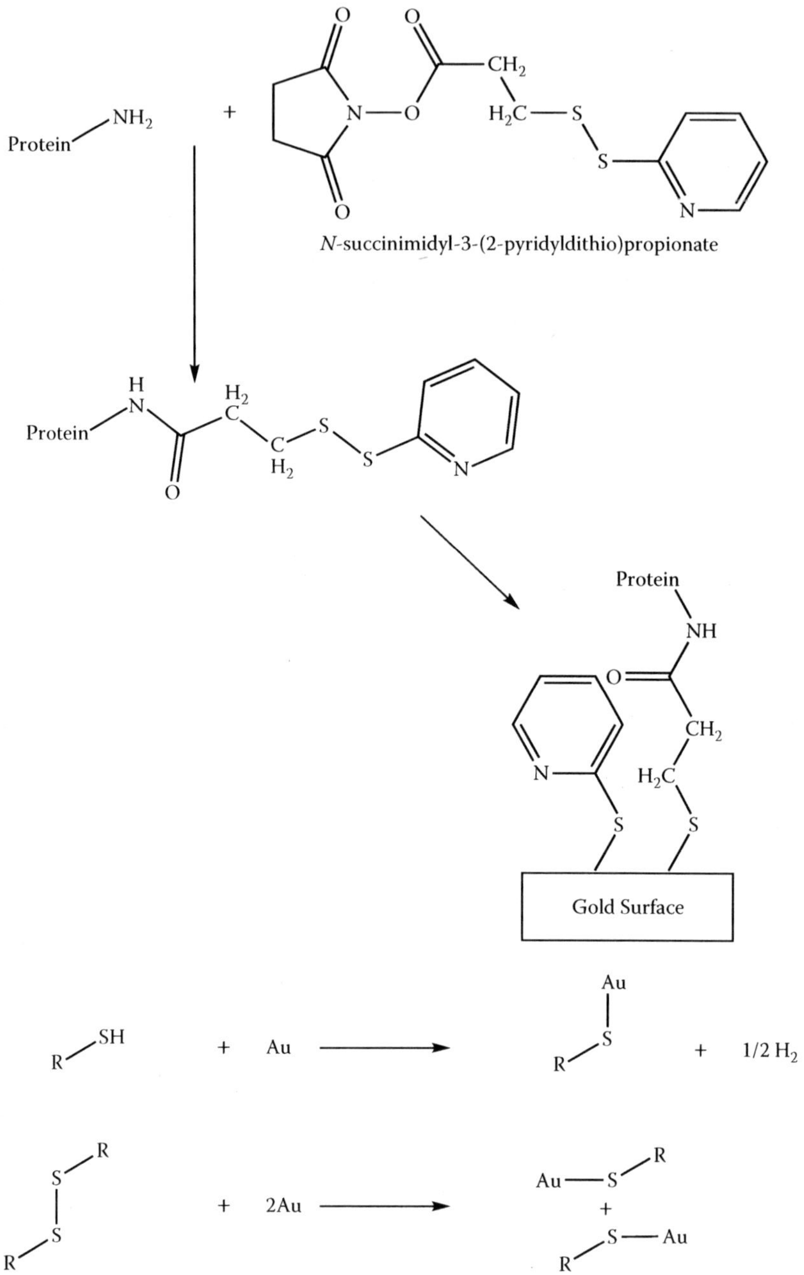

FIGURE 3.2 Possible mechanisms for the binding of protein to gold surfaces. (See Franzman, M.A. and Barrios, A.M., Spectroscopic evidence for the formation of goldfingers, *Inorg. Chem.* 47, 3928–3930, 2008). For the application of the pyridylthio propionate, see Kohli, N., Hassler, B.L., Parthasarathy, L. et al., Tethered lipid bilayers on electrolessly deposited gold for bioelectronic applications, *Biomacromolecules* 7, 3327–3335, 2006.

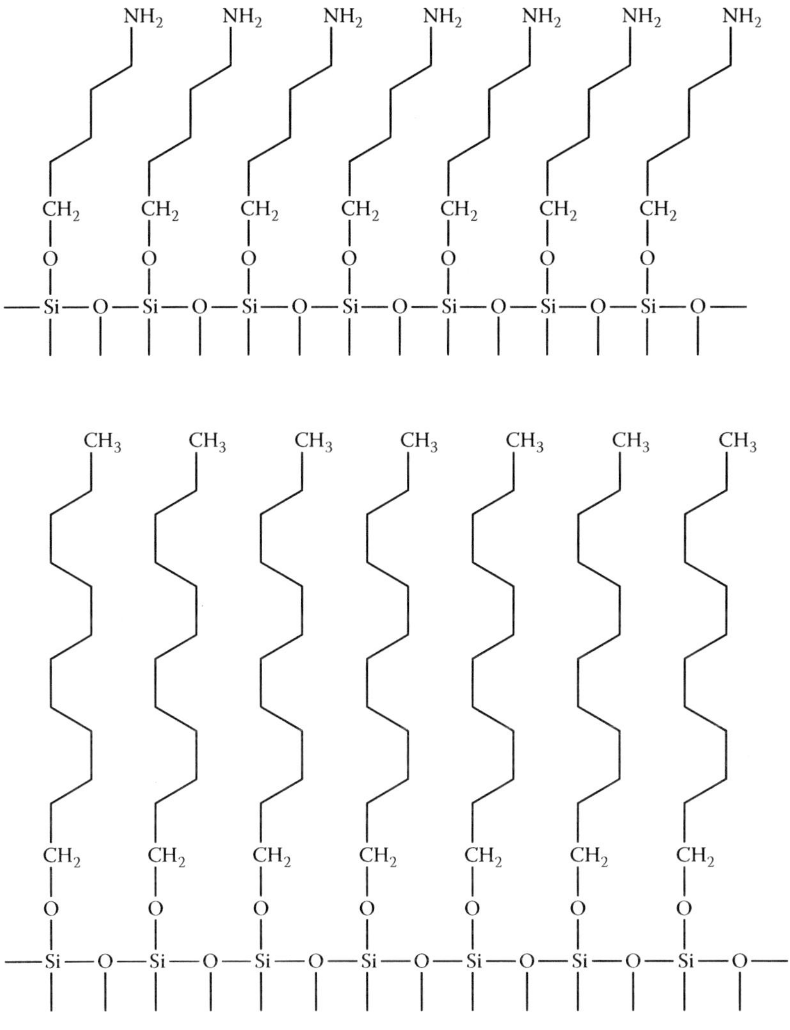

FIGURE 3.3 Self-assembled monolayers. (See Acarturk, T.O., Peel, M.M., Petrosko, P. et al., Control of attachment, morphology, and proliferation of skeletal myoblasts on silanized glass, *J. Biomed. Mater. Res.* 44, 355–370, 1999; Tsai, P.-S., Yang, Y.-M., and Lee, Y.-L., Fabrication of hydrophobic surfaces by coupling of Langmuir-Blodgett deposition and a self-assembled monolayer, *Langmuir* 22, 5660–5665, 2006.)

bound to the gold surface. There has been considerable interest in the use of self-assembling monolayers on gold surfaces as used in surface plasmon resonance.[143–150] Self-assembled monolayers can also be formed on glass surfaces with the use of functionalized alkylsilane derivatives[151–159] (Figure 3.3).

CHEMISTRY FOR ATTACHMENT OF PROTEINS AND PEPTIDES TO SOLID-PHASE MATRICES

The preceding discussion illustrates that reproducible performance with DNA microarrays requires covalent linkage to the surface. The discussion on the interaction of proteins with plastics shows that proteins can be tightly bound to surfaces without covalent interaction and used for analytical purposes such as ELISA assays.[160–163] The most common approach is to noncovalently bind an antibody to the surface of a well in a microplate, where a variety of technical approaches such as competitive immunoassay or sandwich immunoassay can be used to measure an analyte. Microplate or microarray surfaces and other surfaces such as beads may be modified to modulate binding.[164–177] Another approach used the precoating of surfaces with nonspecific materials such as polylysine.[178–185] Proteins with a special affinity for immunoglobulins such as protein A[186–192] and protein G[193–203] have been used to bind antibody matrices. Because protein A and protein G bind to the Fc domain of immunoglobulin, the bound antibody has the CDR region available for interaction with the analyte or target. Other approaches used protein engineering to attach affinity labels such as hexahistidine.[204–208]

Covalent linkage of proteins to matrices has been of use for the past 50 years, dating to the work of Campbell and associates[209] on the coupling of bovine serum albumin to diazotized cellulose (prepared by reaction of *p*-aminobenzyl-cellulose with $NaNO_2/HCl$). The resulting albumin cellulose derivative was used for the isolation of antibodies to albumin. This chemistry continues to be useful as diazo coupling (Figure 3.4) has recently been used for the coupling of protein to a microarray surface.[210] The early work on the development of covalent-insolubilized proteins has been reviewed by Silman and Katchalski.[211] Sipehia and coworkers used

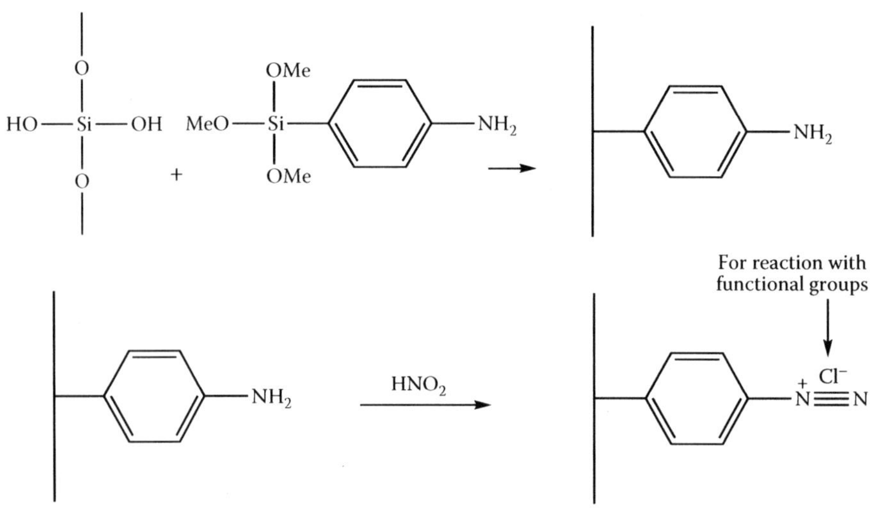

FIGURE 3.4 The formation of a diazo function on a matrix which can be coupled to protein (See Wu, Y., Buranda, T., Metzenberg, R.I. et al., Diazo coupling method for covalent attachment of proteins to solid surfaces, *Bioconjug.Chem.* 17, 359–365, 2006)

anhydrous ammonia to introduce amino groups onto polypropylene bead surfaces.[212] Glutaraldehyde was used to covalently link enzymes (glucose oxidase or peroxidase) to the modified polypropylene beads. There has been continued use of this technical approach.[213,214] Another approach introduced the 1-fluoro-2-nitrobenzyl function onto polypropylene or polyethylene surfaces by the UV-photolysis of 1-fluoro-2-nitro-4-azidobenzene[215] (Figure 3.9). The reactive nitrene inserts into the polymer surface, leaving the 1-fluoro-2-nitrobenzyl function to couple with an appropriate nucleophile.[216,217] Hughes and coworkers[218–220] used matrix-bound salicylhydroxamic acid to bind with immunoglobulins modified with phenylboronic acid (Figure 3.5). This technology has also proved useful for the coupling of nucleic acids. Phenylboronic acid has proved useful for the binding of a variety of biological materials, largely based on affinity for carbohydrates.[221–223] The Staudinger ligation has been demonstrated to be useful for protein immobilization.[224] In these studies, bovine pancreatic ribonuclease was modified via an intein to provide a C-terminal azido derivative that was then coupled to a matrix-bound phosphinothioate (Figure 3.6).

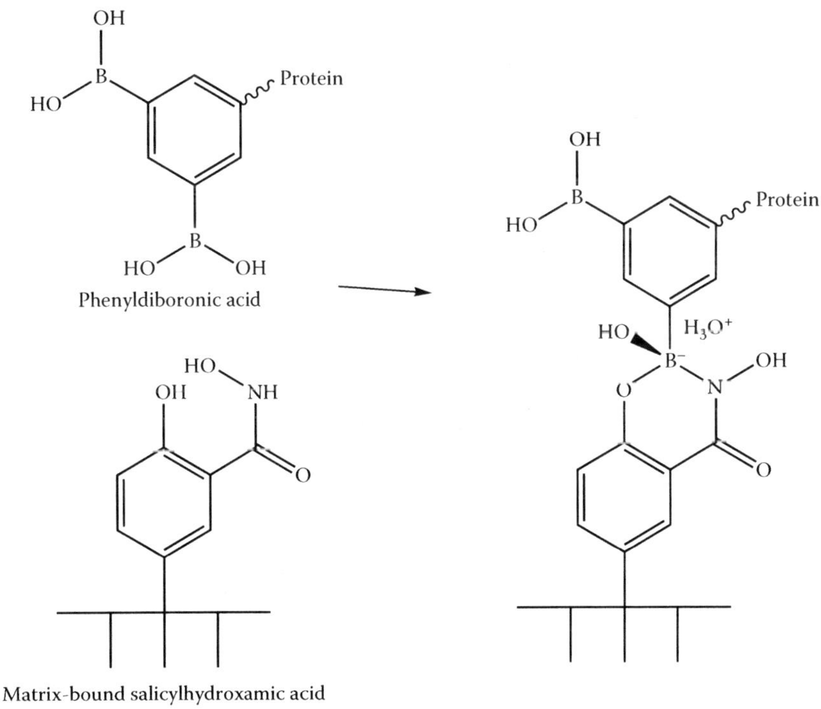

FIGURE 3.5 Phenylboronic acid coupling. (See Wiley, J.P., Hughes, K.A., Kaiser, R.J. et al., Phenylboronic acid-salicylhydroxamic acid bioconjugates 2. Polyvalent immobilization of protein ligands for affinity chromatography, *Bioconjug. Chem.* 12, 240–250, 2001; Spinger, A.L., Gall, A.S., Hughes, K.A. et al., Salicylhydroxamic acid functionalized affinity membranes for specific immobilization of proteins and oligonucleotides, *J. Mol. Recognit.* 14, 183–190, 2003.)

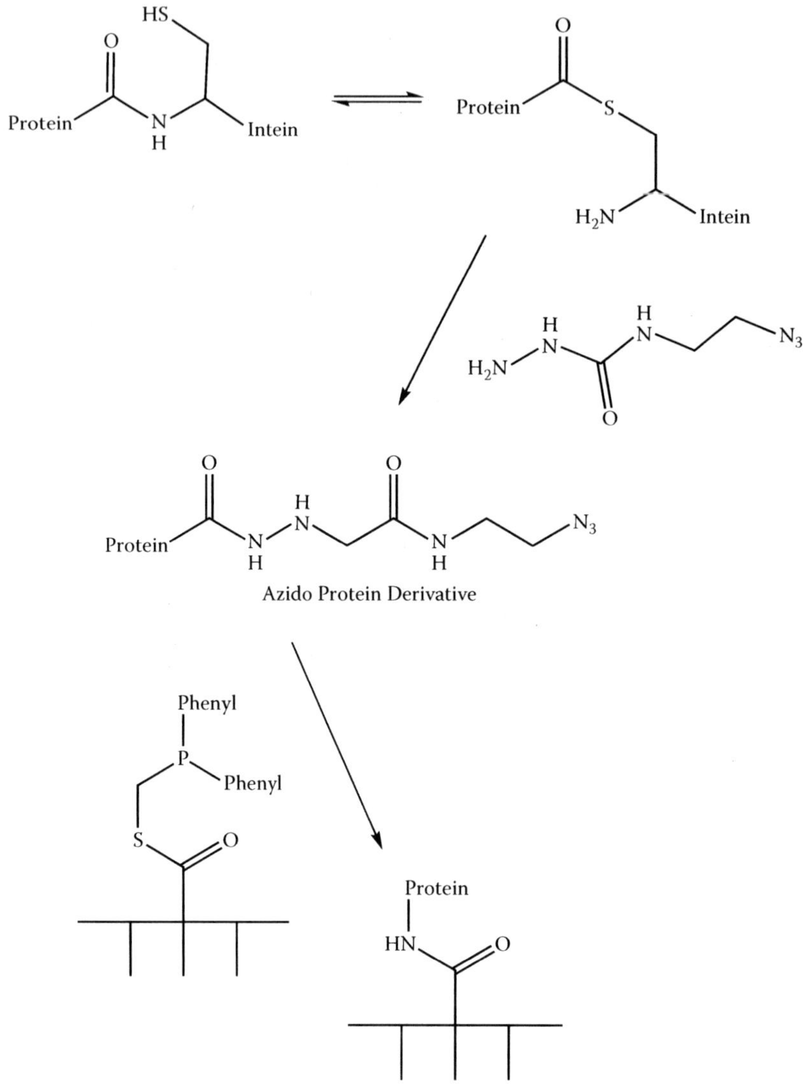

FIGURE 3.6 Staudinger ligation for coupling. (See Kalis, J., Abbott, N.L., and Raines, R.T., General method for the site-specific protein immobilization by Staudinger ligation, *Bioconjug. Chem.* 18, 1064–1069, 2007.)

Early work on the immobilization of proteins and nucleic acids focused on bonding to celluosic and agarose matrices. Cyanuric chloride (2,4,6-trichloro-1,3,5-triazine; Figure 3.7) has been used for the coupling of proteins to hydroxy polymers[225–227] and for the cross-linking of proteins.[228] The use of cyanogen bromide for coupling proteins and nucleic acids to agarose matrices has been the dominant technology. Cyanogen bromide was introduced by Porath and coworkers 40 years ago,[229,230] and although the chemistry is not completely understood, the technology remains

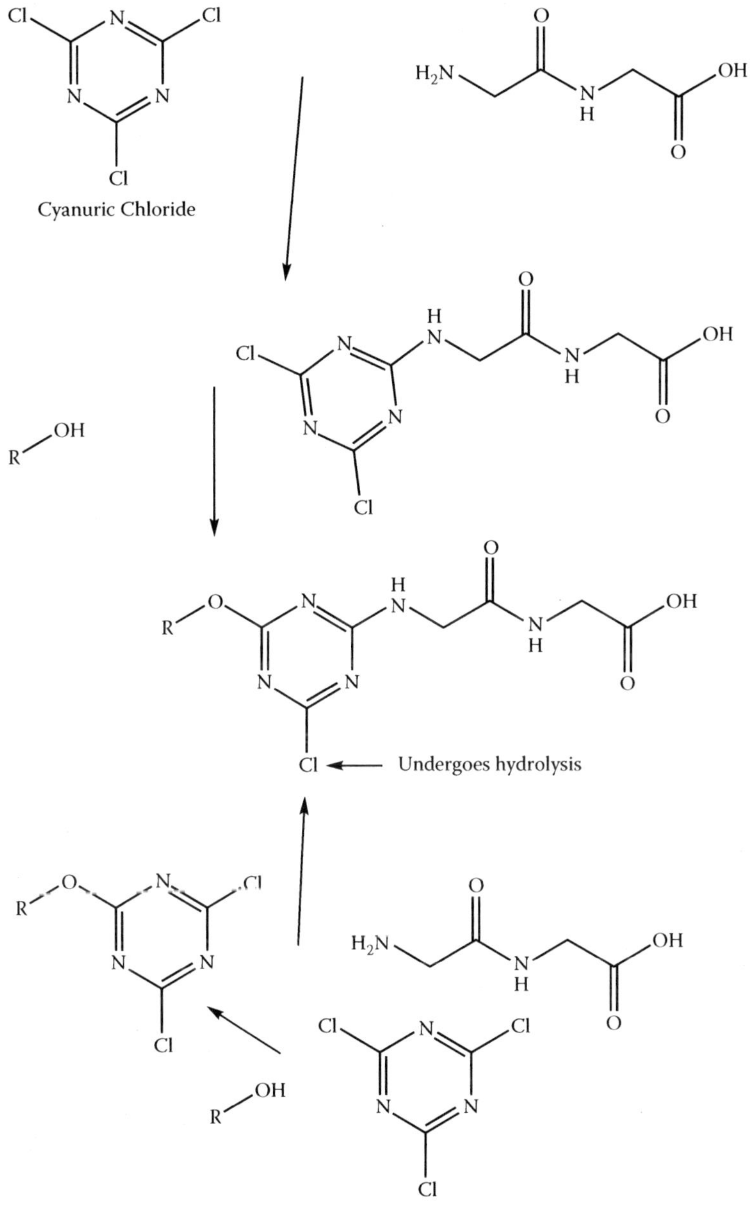

FIGURE 3.7 Cyanuric chloride coupling. Coupling can also occur between two amino groups or two hydroxyl functions. (See Stanková, M. and Lebl, M., Library generation through successive substitution of trichlorotriazine, *Mol. Divers.* 2, 75–80, 1996; Bendas, G., Krause, A., Bakowsky, U. et al., Targetability of novel immunoliposomes prepared by a new antibody conjugation technique, *Int. J. Pharm.* 181, 79–93, 1999; Abuknesha, R.A., Luk, C.Y., Griffiths, H.H. et al., Efficient labeling of antibodies with horseradish peroxidase using cyanuric chloride, *J. Immunol. Methods* 306, 211–217, 2005.)

popular for the coupling of proteins and nucleic acids to agarose (Figure 3.8) and other hydroxyl-containing matrices.[231–240] A recent example is the use of cyanogen bromide coupling to link a targeting antibody to a polylactic acid microparticle containing paclitaxel.[240] Although agarose is the more frequently used matrix for cyanogen bromide activation, cyanogen bromide has also been used to couple proteins and nucleic acids to silica diol matrices (Figure 3.9)[236,237,241,242] permitting application

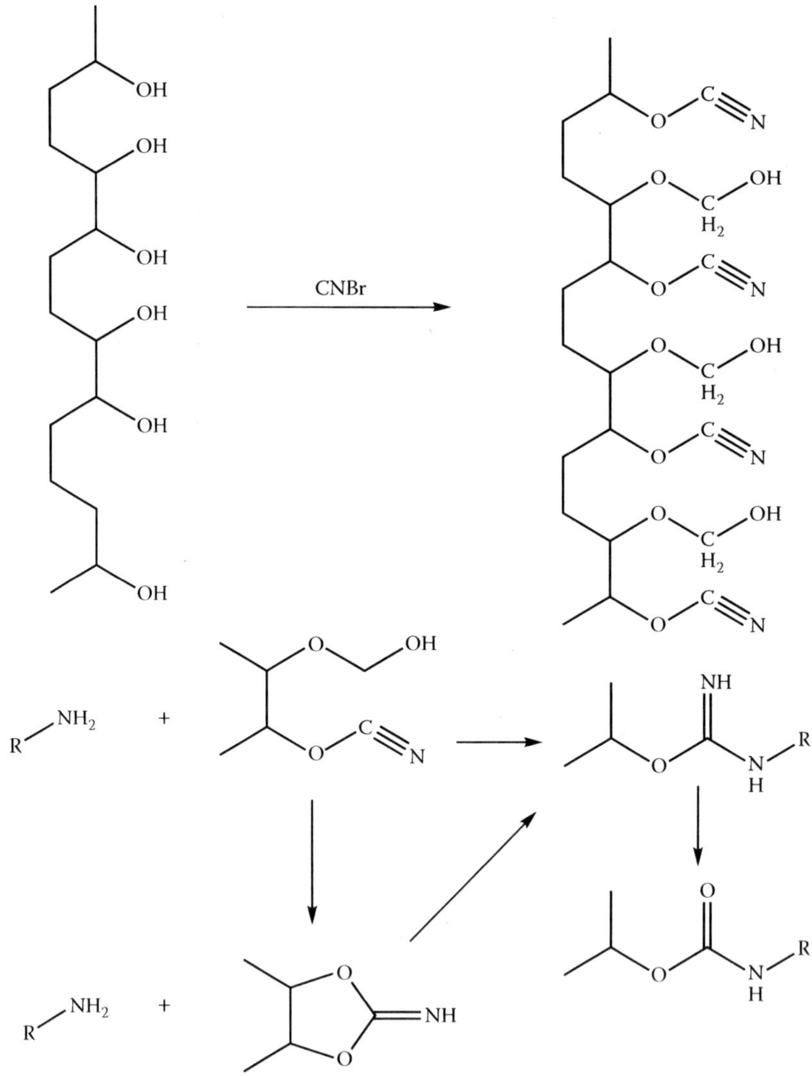

FIGURE 3.8 CNBr coupling. (See Porath, J. and Axen, R., Immobilization of enzymes to sugars, agarose, and Sephadex supports, *Methods Enzymol.* 44, 19–45, 1976; Robberson, D.L. and Davidson, N., Covalent coupling of ribonucleic acid to agarose, *Biochemistry* 11, 533–537, 1972.)

Cyanogen Bromide Activation of Silica Diol Group

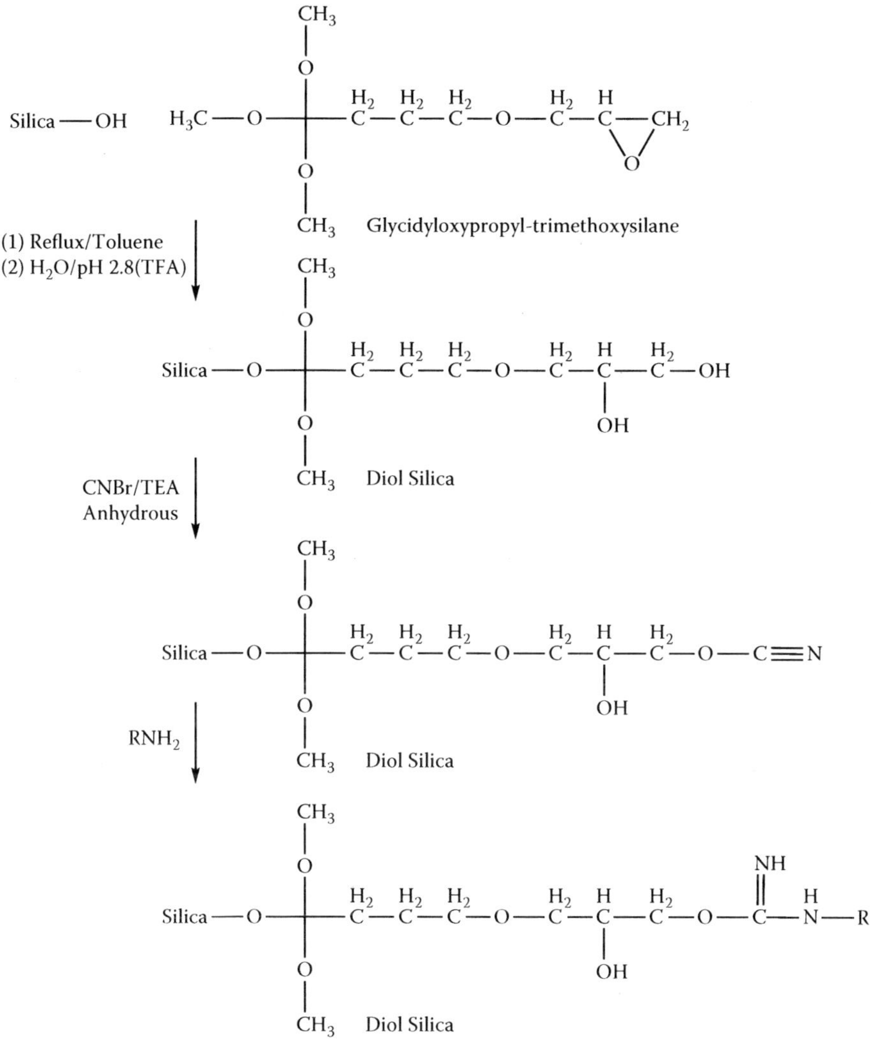

FIGURE 3.9 CNBr coupling to silica diol. (See Jurado, L.A., Mosley, J., and Jarrett, H.W., Cyanogen bromide activation and coupling of ligands to diol-containing silica for high-performance affinity chromatography. Optimization of conditions, *J. Chromatog.* 971, 95–104, 2002.)

of this technology to the manufacture of a high-pressure/high-performance matrix. Robberson and Davidson[243] coupled ε-aminocaproic methyl ester (6-aminohexanonic acid, methyl ester) to agarose with alkaline CNBr and converted the ester group to a hydrazine function. The hydrazine derivative was coupled to periodate-oxidized RNA (Figure 3.10) to yield the hydrazone derivative. DNA has been coupled to cyanogen bromide-activated agarose using 5′aminoethyl derivatives;[244–246] the same derivative

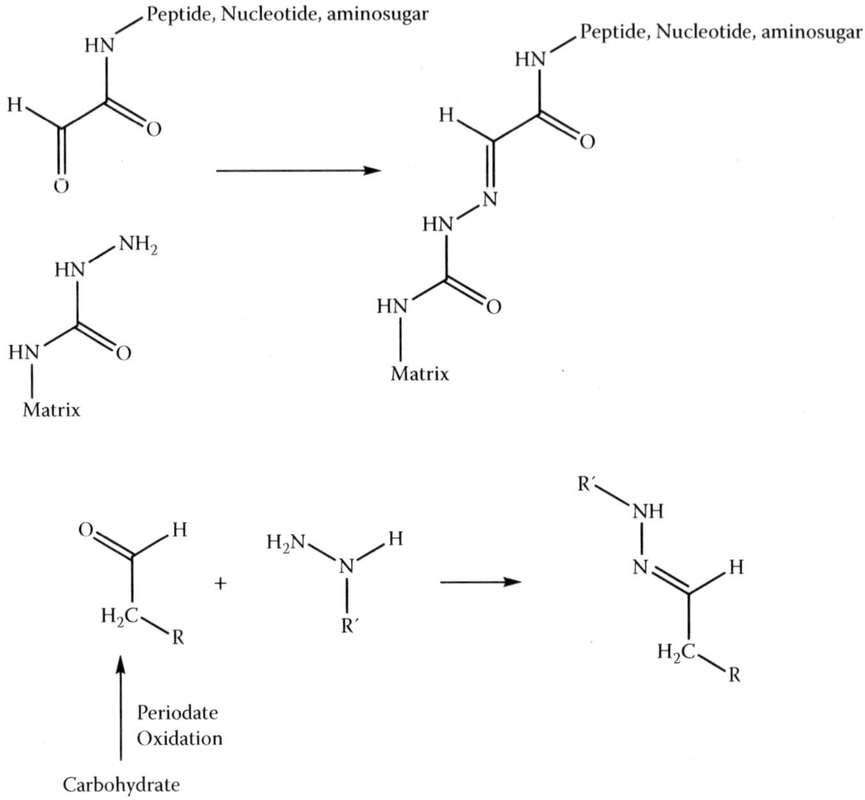

FIGURE 3.10 Coupling of the glyoxal derivative of an amino compound to a semicarbazide matrix. (See Dubureq, X., Olivier, C., Malingue, F. et al., Peptide-protein microarrays for the simultaneous detection of pathogen infections, *Bioconjug. Chem.* 15, 307–316, 2004.) Also, hydrazine derivative coupling to aldehyde. (See Andreson, H., Zarse, K., Grötzinger, C. et al., Development of peptide microarrays for epitope mapping of antibodies against the human TSH receptor, *J. Immunol. Methods* 315, 11–18, 2006; Schatterer, J.C., Stuhlmann, F., and Jäschke, A., Stereoselective synthesis using immobilized Diels-Alderase ribozymes, *ChemBioChem* 4, 1089–1092, 2003.)

has been coupled to silica using a carbodiimide and *N*-hydroxysuccinimide.[246] CNBr has also been used for the chemical ligation of oligodeoxyribonucleotides.[247–251] The nature of the reactive intermediate formed with cyanogen bromide required the immediate proximity of a hydroxyl or other nucleophile for a successful synthetic reaction. Other early work coupled nucleic acids directly to agarose in the presence of 2-(*N*-morpholino) ethyl sulfonic acid.[252,253] The choice of solvent is important for the stability of reactive intermediates, as demonstrated for carbodiimides.[254] Table 3.1 contains selected examples of the use of cyanogen bromide in coupling proteins and nucleic acids to matrices. These examples have been selected to demonstrate coupling conditions and novel approaches; it is by no means inclusive (a PUBMED

TABLE 3.1
Selected Studies on the Use of Cyanogen Bromide for Coupling Proteins to Agarose Matrices

Protein Coupled	Conditions and Comment	Reference
9-*O*-acetylneuraminic acid specific lectin (AchatininH)	Purification of sheep submaxillary mucin.	1
Mannose-6-phosphate receptor protein	Phosphomannan purification.	2
Protein disulfide isomerase	Coupled protein retains activity for bioprocessing application.	3
N/A	Method evaluation for agarose bead activation.	4
Ovalbumin	Use of matrix for "affinity protection" of antiovalbumin antibodies during dye labeling (chemical modification).	5
Monoclonal antibody	Evaluation of experimental conditions (pH, CNBr concentration) for efficacy of coupling; CNBr is more effective than *N*-hydroxysuccinimide ester-mediated coupling. High-density coupling is not productive.	6
Monoclonal antibodies	Evaluation of coupled antibody density on performance of affinity column.	7
N/A	Analysis of CNBr-agarose activation products.	8
N/A	Analysis of activated agarose.	9
Antibody fragments (Fv, VH, paralog peptides)	Evaluation of the binding specificity of the coupled fragments.	10
IgM (Murine)	Mannan-binding protein.	11
N/A	Importance of agarose as matrix for cyanogen bromide activation.	12
UDP-glucuronyl transferase	CNBr coupling to hemocompatible agarose beads; potential use in extracorporeal liver assist device.	13
N/A	Effect of reaction conditions on the quality of coupled product.	14
FITC conjugates	CNBr-coupled antigen; used to evaluate immunoreactivity of the conjugate.	15
N/A	Spectrophotometric method for estimating immobilized ligand concentration.	16
Albumin	Evaluation of nonspecific adsorption to matrix.	17
Human IgG`	Use of beaded composites of agarose or kieselguhr-agarose for the manufacture of coupled products for use in fluidized bed chromatography.	18
N/A	Comparison of CNBr coupling with divinyl sulfone and tresyl chloride; extent of coupling is similar with the three approaches but product with divinyl sulfone and tresyl chloride is more stable.	19

TABLE 3.1 (CONTINUED)
Selected Studies on the Use of Cyanogen Bromide for Coupling Proteins to Agarose Matrices

Protein Coupled	Conditions and Comment	Reference
IgG	Comparison of CNBr and hydrazide-mediated (following oxidation of IgG carbohydrate to aldehyde with periodic acid); Antibody coupled via carbohydrate more effective.	20
IgG	Evaluation of coupling technologies (CNBr, periodate activation of matrix, activation of matrix with carbonyl-diimidazole) for coupling to magnetizable cellulose particles.	21-23
		22
		23
N/A	Technical review of CNBr coupling technology.	24

search obtained 286 citations for a search that combined cyanogen bromide and protein and agarose). Although CNBr-coupled antibodies have been extremely useful in biopharmaceutical manufacturing, there have been stability issues[255–260] that must be considered in the use of these products.

General approaches to the coupling of proteins and nucleic acids to matrices have been discussed earlier. The chemistry is adapted from material covered in Chapter 1 and Chapter 2 in detail. Although there are many possibilities, the majority of microarrays are prepared by coupling of molecules to matrices via aldehyde, amino, succinimide, maleimide, epoxy, and carboxyl groups.[261–270] Other approaches use the modification of protein as, for example, by the oxidation of carbohydrate with periodate to form aldehyde functions that are in turn coupled to matrices with hydrazide or amino functions or by the insertion of a functional group either by chemical or genetic means. Table 3.2 presents selected studies on the modification of proteins to provide linker functions for attachment to matrices.

Table 3.3 presents selected studies on the coupling of antibodies (or related macromolecules) to planar or bead matrices.

TABLE 3.2
Modification of Proteins for Attachment to Matrices

Functional Group	Comment	Reference
Thiol	3-Mercaptopropionic acid coupled to protein via lysine residues using carbodiimide.	1
Aldehyde formed by periodate oxidation	Coupling of proteins to hydrazide or amino matrices.	2
Biotin	Biotinylation using maleimide derivative at sulfhydryl group in antibody hinge region (after reduction with 2-mercaptoethanol) or at lysine residues with *N*-hydroxysuccinimide biotin derivatives.	3
Metal-binding domain	A peptide of five glutamic and six histidine residues was either coupled or engineered into an antibody fragment.[a]	4
Cross-linking reagent for coupling to various carbon-containing surfaces	Method for immobilization of proteins and oligonucleotides to a variety of supports, including polypropylene, nylon, and agarose. Linkage to the protein or nucleic acid occurs via amino or sulfhydryl function; coupling to polymer platform is via photochemical linkage.	5
Protein Thiol	Soluble scFv C-terminal free thiol.	6
C-terminal or N-terminal hexalysine sequence	Coupling to maleic anhydride matrix.	7
Aldehyde	Oxidation of carbohydrate with periodate to provide oriented antibody coupling to matrix for immunochromatography.	8
Aldehyde	Kinetic model for oxidation of antibody carbohydrate by periodate.	9

[a] Goel, A., Colcher, D., Koo, J.S. et al., Relative position of the hexahistidine tags effects binding properties of a tumor-associated single-chain Fv construct, *Biochim. Biophys. Acta* 1523, 13–20, 2000.

TABLE 3.3
Selected Studies on the Coupling of Antibodies and Related Proteins to Matrices

Antibody or Related Protein	Comment	Reference
Fab′, F(ab′), IgG	SAM[a](thiols), random coupling via protein amino groups or specific attachment via protein sulfhydryl.	1
Biotinylated antibody	Productive-oriented binding to streptavidin-coated plates.	2
IgG, Fab′	Used random biotinylated (lysine) or specific (carbohydrate oxidation/coupling to resultant aldehyde to biotin hydrazide or free sulfhydryl in Fab′) Specific orientation provided greater system efficacy.	3
N/A	Use of "histag"—Protein A to immobilized IgG.	4
N/A	Evaluation of commercial membranes for the manufacture of antibody microarrays.	5
N/A	Purification of monospecific polyclonal antibodies for use in antibody microarrays.	6
N/A	Autoantibody profiling microarray; comparison with multiplex beads.	7
N/A	Phage versus phagemid libraries for generation of human monoclonal antibodies.	8
Antibodies to CD antigen	Measure leukocyte binding to microplate as index of CD expression.	9
N/A	Internal control for antibody microarray.	10
Antibodies to IL-1β, IL-1ra, IL-6, IL-8, MCP-1, TNFα	Piezoelectric application of antibody to conventional 96-well polystyrene microplate, ELISA-based assay, and validation procedure.	11
"Normal" proteins	Reverse-capture for detection of autoantibodies.	12
N/A	Comparison of bead and planar array technologies; rapid evaluation of antibody specificity.	13
Candidate and control antigens	Profiling of autoantibodies in rheumatoid arthritis.	14
N/A	Recombinant antibody-binding protein (hydrophobic domain fused with antibody-binding domain) for attachment of antibody to matrix.	15
Lipopolysaccharide (LPS)	Use for analysis of anti-LPS antibodies.	16

[a] SAM, self-assembled monolayer.

REFERENCES

References for Table 3.1

1. Indra, D., Ganesh, J., Ramaligaim, K. et al., Immunological significance of metal induced conformational changes in the mitogenic AchatininH binding to carbohydrate ligands, *Comp. Biochem. Physiol. Pt. C. Toxicol. Pharmacol* 127C, 177–183, 2000.

2. Yerramalla, U. L. and Nadimpalli, S.K., Affinity purification of mannose-6-phosphate receptor proteins. Purification and partial characterization of goat liver receptors, *Biochem. Mol. Biol. Int.* 40, 815–821, 1996.
3. Morjana, N.A. and Gilbert, H.F., Catalysis of protein folding by agarose-immobilized protein disulfide isomerase, *Protein Expr. Purific.* 5, 144–148, 1994.
4. Kümel, G., Daus, H., and March, H., Improved method for the cyanogen bromide activation of agarose beads, *J. Chromatog.* 172, 221–226, 1979.
5. Attiya, S., Dickinson-Laing, T., Cesarz, J. et al., Affinity protection chromatography for efficient labeling of antibodies for use in affinity capillary electrophoresis, *Electrophoresis* 23, 750–758, 2002.
6. Pfeiffer, N.E., Wylie, D.E., and Schoster, S.M., Immunoaffinity chromatography utilizing monoclonal antibodies. Factors which influence antigen-binding capacity, *J. Immunol. Methods* 97, 1–9, 1997.
7. Gómez, L., Hernández, R., Ibarra, N. et al., Comparison of different ligand densities for the manufacture of CB .Hep-1 immunosorbents, *J. Biochem. Biophys. Methods* 52, 151–159, 2002.
8. Kohn, J. and Wilchek, M., Procedures for the analysis of cyanogen bromide-activated Sepharose or Sephadex by quantitative determination of cyanate esters and imidocarbonates, *Anal. Biochem.* 115, 375–382, 1981.
9. Weber, M.M., Nucleophilicity of isourea linkages in substituted agaroses. A direct method for the determination of alkylamino groups coupled to CNBr-activated agarose, *Anal. Biochem.* 76, 177–183, 1976.
10. Berry, M.J. and Davies, J., Use of antibody fragments in immunoaffinity chromatography. Comparison of Fv fragments, VH fragments and paralog peptides, *J. Chromatog.* 597, 239–245, 1992.
11. Nevens, J.P., Mallia, A.K., Wendt, M.W., and Smith, P.K., Affinity chromatographic purification of immunoglobulin M antibodies utilizing mannan-binding protein, *J. Chromatog.* 597, 247–256, 1992.
12. Jennissen, H.P., Cyanogen bromide and tresyl chloride chemistry revisited: The special reactivity of agarose as a chromatographic and biomaterial support for immobilizing novel chemical groups, *J. Mol. Recognit.* 8, 116–124, 1995.
13. Brummer, G., Holloway, C.J., and Lössen, H., Large agarose beads for extracorporeal detoxification systems. Preparation and enzymatic properties of agarose-bound UDP-glucuronyltransferase, *Int. J. Artif. Organs* 2, 163–169, 1979.
14. Schnnar, R.L., Sparks, T.F., and Roseman, S., Cyanogen bromide activation of polysaccharides. Effects of reaction conditions on cationic charge and ligand content, *Anal. Biochem.* 79, 513–525, 1977.
15. Scales, R.W., Jacobs, N.F., Jr., and Skaggs, R., Use of immunoglobulin coupled to agarose beads for examining the specificity of conjugates, *J. Clin. Microbiol.* 2, 292–295, 1977.
16. Miyagawa, A. and Okuyama, T., An improved estimation method for amino alkyl ligand immobilized on matrix beads, *J. Biochem.* 81, 1715–1720, 1977.
17. Heinzel, W., Rahimi-Laridjani, I., and Grimminger, H., Immunoadsorbents: Non-specific binding of proteins to albumin-Sepharose, *J. Immunol. Methods* 9, 337–344, 1976.
18. Desai, M.A. and Lyddiatt, A., Comparative studies of agarose and kieselguhr-agarose composites for the preparation and operation of immunoadsorbents, *Bioseparation* 1, 43–58, 1990.
19. Ubrich, N., Hubert, P., Regnault, V. et al., Compared stability of Sepharose-based immunoadsorbent prepared by various activation methods, *J. Chromatog.* 584, 17–22, 1992.
20. Orthner, C.L., Highsmith, F.A., Therakan, J. et al., Comparison of the performance of immunosorbents prepared by site-directed or random coupling of monoclonal antibodies, *J. Chromatog.* 558, 55–70, 1991.

21. al-Abdulla, I.H., Melor, G.W., Chiderstone, M.S. et al., Comparison of three different activation methods for coupling antibodies to magnetizable cellulose particles, *J. Immunol. Methods* 122, 253–258, 1989.
22. Stage, D.E. and Mannik, M., Covalent binding of molecules to CNBr-activated agarose: Parameters relevant to the activation and coupling reactions, *Biochim. Biophys. Acta* 343, 382–391, 1974.
23. King, M., Alaye, N., and Augustin, R., Demonstration of reaginic antibodies on human basophils by immune adherence to allergen-coated Sepharose beads, *Clin. Allergy* 6, 339–348, 1976.
24. Pepper, D.S., Improved cyanogen bromide activation, in *Laboratory Methods in Immunology*, Volume II, Ed. H. Zola, CRC Press, Boca Raton, FL, Chapter 9, pp. 161–167, 1990.

References for Table 3.2

1. Pyun, J.C., Kim, S.D., and Chung, J.W., New immobilization method for immunoaffinity biosensors by using thiolated proteins, *Anal. Biochem.* 347, 227–233, 2005.
2. O"Shannessy, D.J. and Quarles, R.H., Labeling of the oligosaccharide moieties of immunoglobulins, *J. Immunol. Methods* 99, 153–161, 1987.
3. Cho, H.-H., Pack, E.-H., Lee, H. et al., Site-directed biotinylation of antibodies for controlled immobilization on solid surfaces, *Anal. Biochem.* 365, 14–23, 2007.
4. Malecki, M., Hsu, A., Truong, L., and Sanchez, S., Molecular immunolabeling with recombinant single-chain variable fragment (scFv) antibodies designed with metal-binding domains, *Proc. Natl. Acad. Sci. USA* 99, 213–218, 2002.
5. Kumar, P., Agarwal, S.K., and Gupta, K.C., *N*-(3-trifluoroethanesulfonyloxypropyl) anthraquinone-2-carboxamide: A new heterobifunctional reagent for immobilization of biomolecules on a variety of polymer surfaces, *Bioconjug. Chem.* 15, 7–11, 2004.
6. Albrecht, H., Burke, P.A., Natarajan, A. et al., Production of soluble scFvs with C-terminal-free thiol for site-specific conjugation or stable dimeric scFvs on demand, *Bioconjug. Chem.* 15, 16–26, 2004.
7. Allard, L., Cheyne, V., Oriol, G. et al., Antigenicity of recombinant proteins after regioselective immobilization onto polyanhydride-based copolymers, *Bioconjug. Chem.* 15, 458–466, 2004.
8. Vankova, R., Gaudinova, A., Sussenekova, H. et al., Comparison of oriented and random antibody immobilization in immunoaffinity chromatography of cytokines, *J. Chromatog. A* 811, 77–84, 1998.
9. Hage, D.S., Wolfe, C.A., and Oates, M.R., Development of a kinetic model to describe the effective rate of antibody oxidation by periodate, *Bioconjug. Chem.* 8, 914–920, 1997.

References for Table 3.3

1. Bonray, K., Frederix, F., Reekmans, G. et al., Comparison of random and oriented immobilisation of antibody fragments on mixed self-assembled monolayers, *J. Immunol. Mehods.* 312, 167–181, 2006.
2. Davies, J., Roberts, C.J., Dawkes, A.C. et al., Use of scanning probe microscopy and surface plasmon resonance as analytical tools in the study of antibody-coated microtiter wells, *Langmuir* 10, 2654–2661, 1994.
3. Peluso, P., Wilson, D.S., Do. D. et al., Optimizing antibody immobilization strategies for the construction of protein microarrays, *Anal. Biochem.* 312, 113–124, 2003.
4. Johnson, C.P., Jensen, I.E., Prakasam, A. et al., Engineered protein A for the orientational control of immobilized proteins, *Bioconjug. Chem.* 14, 974–978, 2003.

5. Huong, R.-P., Detection of multiple proteins in an antibody-based protein microarray system, *J. Immunol. Methods* 255, 1–13, 2001.
6. Agaton, C., Falk, R., Guthenberg, I.H. et al., Selective antibody-based proteomics efforts, *J. Chromatog. A.* 1043, 33–40, 2004.
7. Hueber, W., Utz, P.J., Steinman, L., and Robinson, W.H., Autoantibody profiling for the study and treatment of autoimmune disease, *Arthritis Res.* 4, 290–295, 2002.
8. O'Connell, D.O., Becerrill, B., Ray-Burman, A. et al., Phage versus phagemid libraries for the generation of human monoclonal antibodies, *J. Mol. Biol.* 321, 49–56, 2002.
9. Lai, S., Lui, R., Nguyen, L. et al., Increases in leukocyte cluster of differentiation antigen expression during cardiopulmonary bypass in patients undergoing heart transplantation, *Proteomics* 4, 1916–1926, 2004.
10. Olie, E.W., Sreekumar, A., Warner, R.L. et al., Development of an internally controlled antibody microarray, *Mol. Cell. Proteomics* 4, 1664–1672, 2005.
11. Urbanowska, T., Managialaio, S., Zickler, C. et al., Protein microarray platform for multiple analysis of biomarkers in human sera, *J. Immunol. Methods* 316, 1–7, 2006.
12. Qin, S., Qin, W., Ehrlich, J.R. et al., Development of a "reverse capture" autoantibody microarray for studies of antigen profiling, *Proteomics* 6, 3199–3209, 2006.
13. Schwenk, J.M., Lindberg, J., Sundberg, M. et al., Determination of binding specificities in highly multiplexed bead-based assays for antibody proteomics, *Mol. Cell. Proteomics* 6, 125–132, 2007.
14. Hueber, W., Kidd, B.A., Tomooka, B.H. et al., Antigen microarray profiling of autoantibodies in rheumatoid arthritis, *Arthritis Rheumat.* 53, 2645–2655, 2005.
15. Sugihara, T., Seong, G.H., Kobatake, E., and Alzawa, M., Genetically synthesized antibody binding proteins self-assembled on hydrophobic matrix, *Bioconjug. Chem.* 11, 789–794, 2000
16. Thirunmalapura, N.R., Morton, R.J., Ramachandran, A., and Malayer, J.R., Lipopolysaccharide microarrays for the analysis of antibodies, *J. Immunol. Methods* 298, 73–81, 2005.

Chapter References

1. Lee, P.S. and Lee, K.H., Genomic analysis, *Curr. Opin. Biotechnol.* 11, 171, 2000.
2. Blohm, D.H. and Guiseppi-Elie, A., New developments in microarray technology, *Curr. Opin. Biotechnol.* 12, 41, 2001.
3. Olson, J.A., Application of microarray profiling to clinical trials in cancer, *Surgery* 136, 519, 2004.
4. Geschwind, D.H., DNA Microarrays: Translation of the genome from laboratory to clinic, *Lancet Neurol.* 2, 275, 2003.
5. Hu, Z., Troester, M., and Perou, C.M., High reproducibility using sodium hydroxide-stripped long oligonucleotide DNA microarrays, *Biotechniques* 38, 121–124, 2005.
6. Jung, A., Stemmler, I., Brecht, A., and Gauglitz, G., Covalent strategy for immobilization of DNA-microspots suitable for microarrays with label-free and time-resolved optical detection of hybridization, *Fresenius J. Anal. Chem.* 371, 128–136, 2001.
7. Dolan, P.L., Wu. Y., Ista, L.K. et al., Robust and efficient synthetic method for forming DNA microarrays, *Nucleic Acids Res.* 29, E107, 2001.
8. Consolandi, C., Castiglioni, B., Bordoni, R. et al., Two efficient polymeric chemical platforms for oligonucleotide microarray preparation, *Nucleos. Nucleot. Nucleic Acids* 21, 561–580, 2002.
9. Hahnke, K., Jacobsen, M., Gruetzkau, A. et al., Striptease on glass: Validation of an improved stripping procedure for *in situ* microarrays, *J. Biotechnol.* 128, 1–13, 2007.

10. Gillespie, D. and Speigelman, S., A quantitative assay for DNA-RNA hybrids with DNA immobilized on a membrane, *J. Mol. Biol.* 12, 829–842, 1965.
11. Southern, E.M., Detection of specific sequences among DNA fragments separated by gel electrophoresis, *J. Mol. Biol.* 98, 503–517, 1975.
12. Thompson, J. and Gillespie, D., Molecular hybridization with RNA probes in concentrated solutions of guanidine thiocyanate, *Anal. Biochem.* 163, 281–291, 1987.
13. Southern, E.M., Case-Green, S.C., Elder, J.K. et al., Array of complementary oligonucleotides for analyzing the hybridization behavior of nucleic acids, *Nucleic Acids Res.* 22, 1368–1373, 1994.
14. Southern, E.M., DNA microarrays. History and Overview, *Methods Mol. Biol.* 170, 1–15, 2001.
15. *Microarray Analysis*, Ed. M. Schena, Wiley-Liss, Hoboken, NJ, 2001.
16. Simon, R.M., *Design and Analysis of DNA Microarray Investigations*, Springer, New York, 2003.
17. *DNA Microarrays: A Molecular Cloning Manual*, Eds. D. Bowtell and J. Sambrook, Cold Spring Harbor Laboratory Press, Cold Spring Harbor, NY, 2003.
18. Stec, J., Wang, J., Coombes, K. et al., Comparison of the predictive accuracy of DNA array-based multigene classifers across cDNA arrays and Affymetrix GeneChips, *J. Mol. Diagn.* 7, 357–367, 2005.
19. Stahl, Y.C.M., van Nerwijnen, M.H.M., van Schacten, F.J., and van Delft, J.H.M., Application of four dyes in gene expression analysis by microarrays, *BMC Genomics* 6, 101, 2005.
20. Mahajan, S., Swarmi, A., Sethi, D., et al., Oligonucleotide microarrays with stem-loop probes: Enhancing the hybridization of nucleic acids for sensitive analysis, *Bioorg. Med. Chem. Lett.* 18, 3585–3588, 2008.
21. Anderson, K., Hess, K.R., Kapoor, M. et al., Reproducibility of gene expression signature-based predictions in replicate experiments, *Clin. Cancer Res.* 12, 1721–1727, 2006.
22. Nilsson, B., Andersson, A., Johansson, M., and Fioretos, T., Cross-platform classification in microarray-based leukemia diagnostics, *Haematologica* 91, 821–824, 2006.
23. Zhang, L., Yoder, S.J., and Enkermann, S.A., Identical probes on different high-density oligonucleotide microarrays can produce different measurements of gene-expression, *BMC Genomics* 7, 153, 2006.
24. Dalma-Weiszhausz, D.D., Warrington, J., Tanimoto, E.Y., and Miyada, C.G., The Affymetrix GeneChip® platform: An overview, *Methods Enzymol.* 410, 3–28, 2006.
25. Woblber, P.K., Collins, P.J., Lucas, A.B. et al., The Aligent *in-situ*-synthesized microarray platform, *Methods Enzymol.* 410, 28–57, 2006.
26. Hager, J., Making and using spotted DNA microarrays in an academic core laboratory, *Methods Enzymol.* 410, 125–168, 2006.
27. Canales, R.D., Luo, Y., Willey, J.C. et al., Evaluation of DNA microarray results with quantitative gene expression platforms, *Nat. Biotechnol.* 24, 1115–1122, 2006.
28. Cheadle, C., Becker, K.G., Cho-Chung, Y.S. et al., A rapid method for microarray cross platform comparisons using gene expression signatures, *Mol. Cell. Probes* 21, 35–46, 2007.
29. Guo, L., Lobenhofer, F.M., Wang, C. et al., Rat toxicogenomic study reveals analytical consistency across microarray platforms, *Nat. Biotechnol.* 24, 1162–1169, 2006.
30. Tan, D.S., Lambros, M.B., Natrajan, R., and Reis-Filho, J.S., Getting it right: Designing microarray (and not 'microawry') comparative genomic hybridization studies for cancer research, *Lab. Invest.* 87, 737–754, 2007.
31. Zakhrabekova, S., Gough, S.P., Lundqvist, U., and Hansson, M., Comparing two microarray platforms for identifying mutated genes in barley (*Hordeum vulgare* L.), *Plant Physiol. Biochem.* 45, 617–622, 2007.

32. Jiang, A., Pan., W., Milbauer, L.C et al., A practical question based on cross-platform microarray data normalization: Are BOEC more like large vessel or microvascular endothelial cells or neither of them?, *J. Bioinform. Comput. Biol.* 5, 875–893, 2007.
33. Yauk, C.L. and Berndt, M.L., Review of the literature examining the correlation among DNA microarray technologies, *Environ. Mol. Mutagenesis* 48, 380–394, 2007.
34. Bruland, T., Anderssen, E., Doseth, B. et al., Optimization of cDNA microarrays procedures using criteria that do not rely on external standards, *BMC Genomics* 8, 377, 2007.
35. Koltai, H. and Weingarten-Baror, C., Specificity of DNA microarray hybridization: Characterization, effectors and approaches for data correction, *Nucleic Acids Res.* 36, 2395–2405, 2008.
36. Kuhn, K., Baker, S.C., Chudin, E. et al., A novel, high-performance random array platform for quantitative gene expression profiling, *Genome Res.* 14, 2347–2356, 2004.
37. Degenkolbe, T., Hannah, M.A., Freund, S. et al., A quality-controlled microarray method for gene expression profiling, *Anal. Biochem.* 346, 217–224, 2005.
38. Schibler, U., Rifat, D., and Lavery, D.J., The isolation of differentially expressed mRNA sequences by selective amplification via biotin and restriction-mediated enrichment, *Methods* 24, 3–14, 2001.
39. Xu, Z., Jablonx, D.M., and Gruenert, D.C., Expression sequence tag-specific full-length cDNA cloning: Actin cDNAs, *Gene* 263, 265–272, 2001.
40. Liu, G. and Lin, Y., Electrochemical quantification of single-nucleotide polymorphisms using nanoparticle probes, *J. Am. Chem. Soc.* 129, 10394–10401, 2007.
41. Van Ness, J., Kalbfleisch, S., Petrie, C.R. et al., A versatile solid support system for oligonucleotide probe-based hybridization assays, *Nucleic Acids Res.* 19, 3345–3350, 1991.
42. Guo, W. and Ruckenstein, E., Crosslinked glass fiber affinity membrane chromatography and its application to fibronectin separation, *J. Chromatog. B.* 795, 61–72, 2003.
43. Sirghi, L., Kylian, G., Gilliland, B. et al., Cleaning and hydrophilization of atomic force microscopy silicon probes, *J. Phys. Chem. B.* 110, 25975–25981, 2006.
44. Petrovykh, D.Y., Kimura-Suda, H., Opdahl, A. et al., Alkanethiols on platinum: Multicomponent self-assembled monolayers, *Langmuir* 22, 2578–2787, 2006.
45. Kang, J. and Rowntree, P.A., Gold film preparation for self-assembled monolayer studies, *Langmuir* 23, 509–516, 2007.
46. Zammatteo, N., Jeanmart, L., Hamels, S. et al., Comparison between different strategies of covalent attachment of DNA to glass surfaces to build DNA microarrays, *Anal. Biochem.* 280, 143–150, 2000.
47. Hollemann, O. and Czeslik, C., Characterization of a planar poly(acrylic acid) brush as a materials coating for controlled protein immobilization, *Langmuir* 22, 3300–3305, 2006.
48. Gan, B.K., Kondyurin, A., and Bilek, M.M., Comparison of protein surface attachment on untreated and plasma immersion ion implantation treated polystyrene: Protein islands and carpet, *Langmuir* 23, 2741–2746, 2006.
49. Urban, K., Redford, D., and Larson, D.F., Insulin binding to the cardiopulmonary bypass biomaterials, *Perfusion* 22, 207–210, 2007.
50. Gray, J.J., The interaction of proteins with solid surfaces, *Curr. Opin. Struct. Biol.* 14, 110–115, 2004.
51. Rosado, E., Caroll, H., Sánchez, O., and Peniche, C., Passive adsorption of human antirabic immunoglobulin onto a polystyrene surface, *J. Biomater. Sci. Polym. Ed.* 16, 435–448, 2005.
52. Raffaini, G. and Ganazzoli, F., Sequential adsorption of proteins and the surface modification of biomaterials: A molecular dynamics study, *J. Mater. Sci. Mater. Med.* 18, 309–316, 2007.
53. Fang, F., Satulovsky, J., and Szleifer, I., Kinetics of protein adsorption and desorption on surfaces with grafted polymers, *J. Biophys.* 89, 1516–1533, 2005.

54. Inouye, S., Nonspecific adsorption of proteins to microplates, *Appl. Microbiol.* 25, 279–283, 1973.
55. Lu, D.R., Lee, S.J., and Park, K., Calculation of solvation interaction energies for protein adsorption on polymer surfaces, *J. Biomater. Sci. Polym. Ed.* 3, 127–147, 1991.
56. Suzawa, T. and Shirahama, H., Adsorption of plasma proteins onto polymer lattices, *Adv. Colloid Interface Sci.* 35, 139–172, 1991.
57. Butler, J.E., Ni, L., Nessler, R. et al., The physical and functional behavior of capture antibodies adsorbed on polystyrene, *J. Immunol. Methods* 150, 77–90, 1992.
58. Buteler, J.E., Solid supports in enzyme-linked immunosorbent assay and other solid-phase immunoassays, *Methods* 22, 4–23, 2000.
59. Inouye, S., Nonspecific adsorption of proteins in microplates, *Appl. Microbiol.* 25, 279–283, 1973.
60. Piskin, E., Tuncel, A., Denizli, A., and Ayhan, H., Monosize microbeads based on polystyrene and their modified forms for some selected medical and biological applications, *J. Biomater. Sci. Polym. Ed.* 5, 451–471, 1994.
61. Nieto, A., Gays, A., Moreno, C. et al., Adsorption-desorption of antigen to polystyrene plates, *Ann. Inst. Pasteur Immunol.* 137C, 161–172, 1986.
62. Cantarero, L.A., Butler, J.E., and Osborne, J.W., The absorptive characteristics of proteins for polystyrene and their significance in solid-phase immunoassays, *Anal. Biochem.* 105, 375–382, 1980.
63. Pesce, A.J., Ford, D.J., Gaizutis, M., and Pollak, V.E., Binding of protein to polystyrene in solid-phase immuoassays, *Biochim. Biophys. Acta* 492, 399–407, 1977.
64. Kochanowaka, I.E., Rapak, A., and Szewczuk, A., Effect of pretreatment of wells in polystyrene plates on adsorption of some human serum proteins, *Arch. Immunol. Ther. Exp.* 42, 135–139, 1994.
65. Rosado, E., Caroll, H., Sanchez, O., and Peniche, C., Passive adsorption of human antirabic immunoglobulin onto a polystyrene surface, *J. Biomater. Sci. Polym. Ed.* 16, 435–448, 2005.
66. Gosslau, B. and Barrach, H.J., Enzyme-linked immunosorbent microassay for quantification of specific antibodies to collagen type I, II, III, *J. Immunol. Methods* 29, 71–77, 1979.
67. Kenny, G.E. and Dunsmoor, C.L., Effectiveness of detergents in blocking nonspecific binding of IgG in the enzyme-linked immunosorbent assay (ELISA) depends upon the type of polystyrene used, *Isr. J. Med. Sci.* 23, 732–734, 1987.
68. Stevens, P.W., Hansberry, M.R., and Kelso, D.M., Assessment of adsorption and adhesion of proteins to polystyrene microwells by sequential enzyme-linked immunosorbent-assay analysis, *Anal. Biochem.* 225, 197–205, 1995.
69. Nygren, H., Werthén, M., and Stenberg, M., Kinetics of antibody binding to solid-phase-immobilized antigen—effect of diffusion rate limitation and steric interaction, *J. Immunol. Methods* 101, 63–71, 1987.
70. Werthén, M. and Nygren, H., Effect of antibody dilution on the isotherm of antibody binding to surface-immobilized antigen, *J. Immunol. Methods* 115, 71–78, 1988.
71. Hollmann, O. and Czeslik, C., Characterization of a planar poly(acrylic acid) brush as a materials coating for controlled protein immobilization, *Langmuir* 22, 3300–3305, 2006.
72. McGinlay, P.B. and Bardsley, W.G., The kinetics of adsorption of human immunoglobulin G to poly(vinyl chloride) enzyme-linked-immunoadsorbent-assay vessel walls, *Biochem. J.* 261, 715–720, 1989.
73. Underwood, P.A. and Steele, J.G., Practical limitations of estimation of protein adsorption to polymer surfaces, *J. Immunol. Methods* 142, 83–94, 1991.
74. Zalazar, F.E., Chaibrando, G.A., Aldao, M.A., and Vides, M.A, Parameters affecting the adsorption of ligands to polyvinyl chloride plates in enzyme immunoassays, *J. Immunol. Methods* 152, 1–7, 1992.

75. Brash, J.L. and Ten Hove, P., Protein adsorption studies on "standard" polymeric materials, *J. Biomater. Sci. Polym. Ed.* 4, 591–599, 1993.
76. Römer, W. and Ruaterberg, K., Enzyme-linked immunosorbent assay (ELISA) with covalently bound protein on glass tubes: 1. Stable antigenicity and binding of IgG as a model antigen after repeated use, *Immunobiology* 166, 24–34, 1984.
77. Chang, H.Y. and Andrade, J.D., Immunochemical detection by specific antibody to thrombin of prothrombin; conformational changes upon adsorption to artificial surfaces, *J. Biomed. Mater. Res.* 19, 913–925, 1985.
78. Kilshaw, P.J., McEwan, E.J., Baker, K.C., and Cant, A.J., Studies on the specificity of antibodies to ovalalbumin in normal human serum: Technical considerations in the use of ELISA methods, *Clin. Exp. Immunol.* 66, 481–489, 1986.
79. Hylton, D.M., Shalaby, S.W., and Latour, R.A., Jr., Direct correlation between adsorption-induced changes in protein structure and platelet adhesion, *J. Biomed. Mater. Res. A.* 73, 349–358, 2005.
80. Stevenson, B., DiSpirito, A.A., and Schmidt, T.M., Reduction of enzyme adsorption to polypropylene surfaces in the presence of a nonionic detergent, *Biotechniques* 17, 1048–1050, 1994.
81. Johnston, T.P., Adsorption of recombinant human granulocyte colony stimulating factor (rHG-CSF) to polyvinyl chloride, polypropylene, and glass: Effect of solvent additives, *PDA J. Pharm. Sci. Technol.* 50, 238–245, 1996.
82. Kanna, K., Mulkerrin, M.G., Zhang, M. et al., Rapid analytical tryptic mapping of a recombinant chimeric monoclonal antibody and method validation challenges, *J. Pharm. Biomed. Anal.* 16, 631–640, 1997.
83. Kawamoto, N., Mori, H., Yui, N., and Terano, M., Mechanistic aspects of blood—contacting properties of polypropylene surfaces—from the viewpoint of macromolecular entanglement and hydrophobic interaction via water molecules, *J. Biomater. Sci. Polym. Ed.* 9, 543–559, 1998.
84. Lu, D.R., Glucagon adsorption on polymer surfaces with alpha-helical and extended beta-strand conformations: A computational approach, *J. Biomater. Sci. Polym. Ed.* 4, 323–335, 1993.
85. Waugh, D.F., Lippe, J.A., and Freund, Y.R., Interactions of bovine thrombin and plasma albumin with low-energy surfaces, *J. Biomed. Mater. Res.* 12, 599–625, 1978.
86. Livesey, J.H. and Donald, R.A., Prevention of adsorption losses during radioimmuoassay of polypeptide hormones: Effectiveness of albumins, gelatin, caseins, Tween 20, and plasma, *Clin. Chim. Acta* 123, 193–198, 1982.
87. Horne, M.K., 3rd, The adsorption of thrombin to polypropylene tubes: The effect of polyethylene glycol and bovine serum albumin, *Thromb. Res.* 37, 201–212, 1985.
88. Jørgensen, P.E., Eskildsen, L., and Nexø, E., Adsorption of EGF receptor ligands to test tubes—a factor with implications of studies on the potency of these peptides, *Scand. J. Clin. Invest.* 59, 191–197, 1999.
89. Sutjita, M, Hohmannm, A, Boey, M.L., and Bradley, J., Microplate ELISA for detection of antibodies to DNA in patients with systemic lupus erythematosus: Specificity and correlation with Farr radioimmunoassay, *J. Clin. Lab. Anal.* 3, 34–40, 1989.
90. Paczuski, R., Determination of von Willebrand factor activity with collagen-binding assay and diagnosis of von Willebrand disease: Effect of collagen source and coating conditions, *J. Lab. Clin. Med.* 140, 250–254, 2002.
91. Shrivastav, T.G., Bass, A., and Karlya, K.P., Substitution of carbonate buffer by water for IgG immobilization in enzyme linked immunosorbent assay, *J. Immunoassay Immunochem.* 24, 191–203, 2003.
92. Cavazzana, A., Ruffatti, A., Tonello, M. et al., An analysis of experimental conditions influencing the anti-β_2-glycoprotein I Elisa assay results, *Ann. N. Y. Acad. Sci.* 1109, 484–492, 2007.

93. Clinchy, B., Youssefi, M.R., and Håkansson, L., Differences in adsorption of serum proteins and production of IL-1ra by human monocytes incubated in different tissue culture microtiter plates, *J. Immunol. Methods* 282, 53–61, 2003.
94. Clinchy, B., Gunnerås, M., Håkansson, A., and Håkansson, L., Production of IL-1Ra by human mononuclear blood cells *in vitro*: Influence of serum factors, *Cytokine* 34, 329–330, 2006.
95. Allen, L.T., Tosetto, M., Miller, I.S. et al., Surface-induced changes in proteins adsorption and implications for cellular phenotypic responses to surface interaction, *Biomaterials* 27, 3096–3108, 2006.
96. Thevenot, P., Hu, W., and Tang, L., Surface chemistry influences implant biocompatility, *Curr. Top. Med. Chem.* 8, 270–280, 2008.
97. Schönmeyr, B.H., Wong, A.K., Li. S. et al., Treatment of hydroxyapatite scaffolds with fibronectin and fetal calf serum increases osteoblast adhesion and proliferation *in vitro*, *Plast. Reconstr. Surg.* 121, 751–762, 2008.
98. Schroeder, A.C., Scmidbauer, J.M., Sobke, A. et al., Influence of fibronectin on the adherence of *Staphylococcus epidermis* to coated and uncoated intraocular lenses, *J. Cataract Refract. Surg.* 34, 497–504, 2008.
99. Diniz Oliveira, H.F., Weiner, A.A., Majunder, A., and Shastri, V.P., Non-covalent surface engineering of an alloplastic polymeric bone graft material for controlled protein release, *J. Control. Release* 126, 237–245, 2008.
100. Williams, R.L., Brown, S.A., and Merritt, K., Electrochemical studies on the influence of proteins on the corrosion of implant alloys, *Biomaterials* 9, 181–186, 1988.
101. Merritt, K. and Chang, C.C., Factors influencing bacterial adherence to biomaterials, *J. Biomater. Appl.* 5, 185–203, 1991.
102. Shannon, C., Thull, R. and von Recum, A., Types I and III collagen in the tissue capsules of titanium and stainless-steel implants, *J. Biomed. Mater. Res.* 34, 401–408, 1997.
103. Kilpadi, K.L., Chang, P.L., and Bellis, S.L., Hydroxylapatite binds more serum proteins, purified integrins, and osteoblast precursor cells than titanium or steel, *J. Biomed. Mater. Res.* 57, 258–267, 2001.
104. Weber, N., Pesnell, A., Bolikal, D. et al., Viscoelastic properties of fibrinogen adsorbed to the surface of biomaterials used in blood-contacting medical devices, *Langmuir* 23, 3298–3304, 2007.
105. Khan, W., Kapoor, M., and Kumar, N., Covalent attachment of proteins to functionalized polypyrrole-coated metallic surfaces for improved biocompatibility, *Acta Biomater.* 3, 541–549, 2007.
106. Gettens, R.T. and Gilbert, J.L., Quantification of fibrinogen adsorption onto 316L stainless steel, *J. Biomed. Mater. Res. A* 81, 46–473, 2007.
107. Kang, C.K. and Lee, Y.S., The surface modification of stainless steel and the correlation between surface properties and protein adsorption, *J. Mater. Sci. Mater. Med.* 18 1389–1398, 2007.
108. Desroches, M.J., Chaudhary, N., and Omanovic, S., PM-IRRAS investigation of the interaction of serum albumin and fibrinogen with a biomedical-grade stainless steel 316LVM surface, *Biomacromolecules* 8, 2836–2844, 2007.
109. Arvidsson, S., Askendal, A., and Tengvall, P., Blood plasma contact activation on silicon, titanium and aluminum, *Biomaterials* 28, 136–1354, 2007.
110. Imamura, K., Kawasaki, Y., Nagayasu, T. et al., Adsorption characteristics of oligopeptides composed of acidic and basic amino acids on titanium surfaces, *J. Biosci. Bioeng.* 103, 7–12, 2007.
111. Sela, M.N., Badihi, L, Rosen, G. et al., Adsorption of human plasma proteins to modified titanium surfaces, *Clin. Oral Implants Res.* 18, 630–638, 2007.
112. Sousa, S.R., Brás, M.M., Moradas-Ferreira, P., and Barbosa, M.A., Dynamics of fibronectin adsorption on TiO_2 surfaces, *Langmuir* 23, 7046–7054, 2007.

113. van den Dolder, J. and Jansen, J.A., The response of osteoblast-like cells towards collagen type I coating immobilized by *p*-nitrophenylchloroformate to titanium, *J. Biomed. Mater. Res. A*. 83, 712–719, 2007.
114. Ito, Y., Hasuda, H., Sakuragi, M., and Tsuzuki, S., Surface modification of plastic, glass and titanium by photoimmobilization of polyethylene glycol for antibiofouling, *Acta Biomater.* 3, 1024–1032, 2007.
115. Protivinský, J., Appleford, M., Strnad, J. et al., Effect of chemically modified titanium surfaces on protein adsorption and osteoblast precursor cell behavior, *Int. J. Oral Maxillofac. Implants* 22, 542–550, 2007.
116. Sadak, P.C., Carr, P.W, Bowers, L.D., and Haddad, L.C., A radiochemical study of irreversible protein loss on high-performance liquid chromatography column frits, *Anal. Biochem.* 144, 128–131, 1985.
117. Friesen, A.D., Chromatographic methods of fractionation, *Dev. Biol. Stand.* 67, 3–13, 1987.
118. Denner, L.A., Weigel, N.L., Schrader, W.T., and O'Malley, B.W., High-yield high-performance liquid chromatographic analysis of steroid hormone receptors on glass columns, *Anal. Biochem.* 161, 291–299, 1987.
119. Lam, X.M., Yang, J.Y., and Cleland, J.L., Antioxidants for prevention of methionine oxidation in recombinant monoclonal antibody HER2, *J. Pharm. Sci.* 86, 1250–1255, 1997.
120. Chen, B., Bautista, R., Yu, K. et al., Influence of histidine on the stability and physical properties of a fully human antibody in aqueous and solid forms, *Pharm. Res.* 20, 1952–1960, 2003.
121. Baffi, R., Dolch, G., Garnick, R. et al., A total organic carbon analysis method for validating cleaning between products in biopharmaceutical manufacturing, *J. Parenter. Sci. Technol.* 45, 13–19, 1991.
122. Burgoyne, R.F., Priest, M.C., Roche, K.L., and Vella, G., Systematic development and validation of sanitization protocols for a chromatographic system designed for biotherapeutics purification, *J. Pharm. Biomed. Anal.* 11, 1317–1325, 1993.
123. Rathore, N., Qi, W., and Ji, W., Cleaning characterization of protein drug products using UV-VIS spectroscopy, *Biotechnol. Prog.*, 24, 684–690, 2008.
124. Safer, D., Bolinger, L., and Leigh, J.S., Jr., Undecagold clusters for site-specific labeling of biological macromolecules: Simplified preparation and model applications, *J. Inorg. Biochem.* 26, 77–91, 1986.
125. Sasaki, Y.C., Yasuda, K., Suzuki, Y. et al., Two-dimensional arrangement of a functional protein by cysteine-gold interaction: Enzyme activity and characterization of a protein monolayer on a gold substrate, *Biophys. J.* 72, 1842–1848, 1997.
126. Dawson, S.L. and Tirrell, D.A., Peptide-derived self-assembled monolayers: Adsorption of *N*-stearoyl-L-cysteine methyl ester on gold, *J. Mol. Recognit.* 10,18–25, 1997.
127. Possari, R., Caravalhal, R.F., Mendes, R.K., and Kubota, L.T., Electrochemical determination of cysteine in a flow system based on reductive desorption of thiols from gold, *Anal. Chim. Acta* 575, 172–179, 2006.
128. Prisco, J., Leung, C., Xirouchaki, C. et al., Residue-specific immobilization of protein molecules by size-selected clusters, *J. R. Soc. Interface* 2, 169–175, 2005.
129. Torrance, L., Ziegler, A., Pittman, H. et al., Oriented immobilization of engineered single-chain antibodies to develop biosensors for virus detection, *J. Virol. Methods* 134, 164–170, 2006.
130. Palmer, R.E. and Leung, C., Immobilization of proteins by atomic clusters on surfaces, *Trends Biotechnol.* 25, 48–55, 2007.
131. Andolft, L., Caroppi, P., Bizzarri, A.R. et al., Nanoscopic and redox characteristics of engineered horse cytochrome C chemisorbed on a bare gold electrode, *Protein J.* 26, 271–279, 2007.

132. Baltus, R.E., Carmon, K.S., and Luck, L.A., Quartz crystal microbalance (QCM) with immobilized protein receptors: Comparison of response to ligand bindng for direct protein immobilization and protein attachment via disulfide linker, *Langmuir* 23, 3880–3885, 2007.
133. Lee, J.M., Park, H.K., and Jung, Y., Direct immobilization of protein G variants with various numbers of cysteine residues on a gold surface, *Anal. Chem.* 79, 2680–2687, 2007.
134. Schranz, M., Noll, F., and Hampp, N., Oriented purple membrane monolayers covalently attached to gold by multiple thiole linkages analyzed by single molecule force spectroscopy, *Langmuir* 23, 11134–11138, 2007.
135. Staii, C., Wood, D.W., and Scoles, G., Verification of biochemical activity for proteins nanografted on gold surfaces, *J. Am. Chem. Soc.* 130, 640–646, 2008.
136. Andreescu, S. and Luck, L.A., Studies of the binding and signaling of surface-immobilized periplasmic glucose receptors on gold nanoparticles: A glucose biosensor application, *Anal. Biochem.* 375, 282–290, 2008.
137. Vericat, C., Vela, M.E., and Salvarezza, R.C., Self-assembled monolayers of alkanethiols on AU(111): Surface structures, defects and dynamics, *Phys. Chem. Chem. Phys.* 7, 3258–3268, 2005.
138. Mrksich, M. and Whitesides, G.M., Using self-assembled monolayers to understand the interactions or made-made surfaces with proteins and cells, *Annu. Rev. Biophys. Biomol. Struct.* 25, 55–78, 1996.
139. Mrksich, M., Tailored substrates for studies of attached cell culture, *Cell. Mol. Life. Sci.* 54, 653–662, 1998.
140. Shah, D.S., Thomas, M.B., Phillips, S. et al., Self-assembling layers created by membrane proteins on gold, *Biochem. Soc. Trans.* 35, 522–526, 2007.
141. Ariga, K., Nakanishi, T., and Michinobu, T., Immobilization of biomaterials to nano-assembled films (self-assembled monolayers, Langmuir-Blodgett films, and layer-by-layer assemblies) and their related functions, *J. Nanosci. Nanotechnol.* 6, 2278–2301, 2006.
142. Chaki, N.K. and Vijayamohanan, K., Self-assembled monolayers as a tunable platform for biosensor applications, *Biosens. Bioelectron.* 17, 1–12, 2002.
143. Sigal, G.B., Bamdad, C., Barberis, A. et al., A self-assembled monolayer for the binding and study of histidine-tagged proteins by surface plasmon resonance, *Anal. Chem.* 68, 490–497, 1996.
144. Duschi, C., Sevin-Landais, A.F., and Vogel, H., Surface engineering: Optimization of antigen presentation in self-assembled monolayers, *Biophys. J.* 70, 1985–1995, 1196.
145. Disley, D.M., Morrill, P.R., Sproule, K., and Lowe, C.R., An optical biosensor for monitoring recombinant proteins in process, *Biosens. Bioelectron.* 14, 481–493, 1999.
146. Mark, S.S., Sandhyarani, N., Zhu, C. et al., Dendrimer-functionalized self-assembled monolayers as a surface plasmon resonance sensor surface, *Langmuir* 20, 6808–6817, 2004.
147. Moon, J., Kang, T., Oh, S. et al., *In situ* sensing of metal ion adsorption to a thiolated surface using surface plasmon resonance spectroscopy, *J. Colloid Interface Sci.* 298, 543–549, 2006.
148. Ayela, C., Roquet, F., Valera, L. et al., Antibody-antigenic peptide interactions monitored by SPR and QCM-D. A model for SPR detection of IA-1 autoantibodies in human serum, *Biosens. Bioelectron.* 22, 3113–3119, 2007.
149. Lee, K.H., Su, Y.D., Chen, S.J. et al., Microfluidic systems integrated with two-dimensional surface plasmon resonance phase imaging systems for microarray immunoassay, *Biosens. Bioelectron.* 23, 466–472, 2007.
150. Jans, K., Bonroy, K., De Palma, R. et al., Stability of mixed PEO-thiol SAMs for biosensing applications, *Langmuir* 24, 3949–3954, 2008.
151. Sukenik, C.N., Balachander, N., Culp, L.A. et al., Modulation of cell adhesion by modification of titanium surfaces with covalently attached self-assembled monolayers, *J. Biomed. Mater. Res.* 24, 1307–1323, 1990.

152. Margel, S., Vogler, E.A., Firment, L. et al., Peptide, protein, and cellular interactions with self-assembled monolayer model surfaces, *J. Biomed. Mater. Res.* 27, 1463–1476, 1993.
153. Acarturk, T.O., Peel, M.M., Petrosko, P. et al., Control of attachment, morphology, and proliferation of skeletal myoblasts on silanized glass, *J. Biomed. Mater. Res.* 44, 355–370, 1999.
154. Faucheux, N., Schweiss, R., Lützow, K. et al., Self-assembled monolayers with different terminating groups as model substrates for cell adhesion studies, *Biomaterials* 25, 2721–2730, 2004.
155. Wu, Y., Buranda, T., Metzenberg, R.L. et al., Diazo coupling method for covalent attachment of proteins to solid substrates, *Bioconjug. Chem.* 17, 359–365. 2006.
156. Tsai, P.S., Yang, Y.M., and Lee, Y.L., Fabrication of hydrophobic surfaces by coupling of Langmuir-Blodgett deposition and a self-assembled monolayer, *Langmuir* 22, 5660–5665, 2006.
157. Wayment, J.R. and Harris, J.M., Controlling binding site densities on glass surfaces, *Anal. Chem.* 78, 7841–7849, 2006.
158. Weiss, E.A., Kaufman, G.K., Kriebel, J.K. et al., Si/SiO_2-templated formation of ultraflat metal surfaces on glass, polymer, and solder supports: Their use as substrates for self-assembled monolayers, *Langmuir* 23, 9686–9694, 2007.
159. Mehne, J., Markovic, G., Pröll, F. et al., Characterization of morphology of self-assembled PEG monolayers: A comparison of mixed and pure coatings optimized for biosensor applications, *Anal. Bioanal. Chem.*, 391, 1783–1791, 2008.
160. Kemeny, D.M. and Challacombe, S.J., *ELISA and Other Solid Phase Immunoassays: Theoretical and Practical Aspects*, Wiley, New York, 1988.
161. Law, B., *Immunoassay: A Practical Guide*, Taylor & Francis, London, 1996.
162. *Immunochemical Protocols*, Ed. M.M. Manson, Humana Press, Totowa, NJ, 1992.
163. *ELISA: Theory and Practice*, Ed. J.R. Crowther, Humana Press, Totowa, NJ, 1995.
164. Litauszki, L., Howard, L., Salvati, L., and Tarcha, P.J., Surfaces modified with PEO by the Williamson reaction and their affinity for proteins, *J. Biomed. Mater. Res.* 35, 1–8, 1997.
165. Sugihara, T., Seong, G.H., Kobatake, E., and Aizawa, M., Genetically synthesized antibody-binding protein self-assembled on hydrophobic matrix, *Bioconjug. Chem.* 11, 789–794, 2000.
166. Bonen, M.R., Hoffman, S.A., and García, A.A., Silver ion microplates for immunoassays, *Biotechniques* 30, 1340–1344, 2001.
167. Avseenko, N.V., Morozova, T.Ya., Ataullakhanov, F.I., and Morozov, V.N., Immobilization of proteins in immnochemical microarrays fabricated by electrospray deposition, *Anal. Chem.* 73, 6047–6052, 2001.
168. Bai, Y., Huang, W.C., and Yang, S.T., Enzyme-linked immunosorbent assay of *Escherichia coli* 0157: H7 in surface enhanced poly(methyl methacrylate) microchannels, *Biotechnol. Bioeng.* 98, 328–339, 2007.
169. Muir, E.W., Barden, M.C., Collett, S.P. et al., High-throughput optimization of surfaces for antibody immobilization using metal complexes, *Anal. Biochem.* 363, 97–107, 2007.
170. Goto, Y., Matsuno, R., Konno, T. et al., Polymer nanoparticles covered with phosphorylcholine groups and immobilized with antibody for high-affinity separation of proteins, *Biomacromolecules* 9, 828–833, 2008.
171. Suzuki, N., Quesenberry, M.S., Wang, J.K. et al., Efficient immobilization of proteins by modification of plate surface with polystyrene derivatives, *Anal. Biochem* 247, 412–416, 1997.
172. Zouali, M. and Stollar, B.D., A rapid ELISA for measurement of antibodies to nucleic acid antigens using UV-treated polystyrene microplates, *J. Immunol. Methods* 90, 105–110, 1986.
173. Larsson, H., Johansson, S.G., Hult, A., and Göthe, S., Covalent binding of proteins to grafted plastic surfaces suitable for immunoassays, *J. Immunol. Methods* 98, 129–135, 1987.

174. Boudet, F., Thèze, J., and Zouali, M., UV-treated polystyrene microtitre plates for use in an ELISA to measure antibodies against synthetic peptides, *J. Immmunol. Methods* 142, 73–82, 1991.
175. Yuan, S., Szakalas-Gratzl, G., Ziats, N.P. et al., Immobilization of high-affinity heparin oligosaccharides to radiofrequency plasma-modified polyethylene, *J. Biomed. Mater. Res.* 27, 811–819, 1993.
176. Dagenais, P., Desprez, B., Albert, J., and Escher, E., Direct covalent attachment of small peptide antigens to enzyme-linked immunosorbent assay plates using radiation and carbodiimide activation, *Anal. Biochem.* 222, 149–155, 1994.
177. Goldberg, J.S., Wagenknecht, D.R., and McIntrye, J.A., Alteration of the aPA ELISA by UV exposure of polystyrene microtiter plates, *J. Clin. Lab. Anal.* 10, 243–249, 1996.
178. Varani, J., Inman, D.R., Fligiel, S.E., and Hillegas, W.J., Use of recombinant and synthetic peptides as attachment factors for cells on microcarriers, *Cytotechnology* 13, 89–98, 1993.
179. Towler, M.J. and Weathers, P.J., Adhesion of plant roots to poly-L-lysine coated polypropylene substrates, *J. Biotechnol.* 101, 147–155, 2003.
180. Csucs, G., Michel, R., Lussi, J.W. et al., Microcontact printing of co-polymers in combination with proteins for cell-biological applications, *Biomaterials.* 24, 1713–1720, 2003.
181. Ivanova, E.P., Pham, D.K., Brack, N. et al., Poly(L-lysine)-mediated immobilization of oligonucleotides or carboxy-rich polymer surfaces, *Biosens. Bioelectron.* 19, 1363–1370, 2004.
182. Minigo, G., Scholzen, A., Tang, C.K. et al., Poly-L-lysine-coated nanoparticles: A potent delivery system to enhance DNA vaccine efficacy, *Vaccine* 25, 1316–1327, 2007.
183. Harnett, E.M., Alderman, J., and Wood, T., The surface energy of various biomaterials coated with adhesion molecules used in cell culture, *Colloids Surf. B. Biointerfaces* 55, 90–97, 2007.
184. Greene, G., Radhakrishna, H., and Tannenbaum, R., Protein binding properties of surface-modified porous polyethylene membranes, *Biomaterials* 26, 5972–5982, 2005.
185. Bai, Y., Koh, C.G., Boreman, M. et al., Surface modification for enhancing antibody binding on polymer-based microfluidic device for enzyme-linked immunosorbent assay, *Langmuir* 22, 9458–9467, 2006.
186. O'Neill, H.C. and Parish, C.R., A rapid, automated colorimetric assay for measuring antibody binding to cell surface antigens, *J. Immunol. Methods.* 64, 257–268, 1983.
187. Choi, C.J. and Cunningham, B.T., A 96-well microplate incorporating a replica molded microfluidic network integrated with photonic crystal biosensors for high throughput kinetic biomolecular interaction analysis, *Lab Chip* 7, 550–556, 2007.
188. Renberg, B., Shiroyama, I., Engfeldt, T. et al., Affibody capture microarrays: Synthesis and evaluation of random and directed immobilization of affibody molecules, *Anal. Biochem.* 341, 334–343. 2005.
189. Matson, R.S., Milton, R.C., Rampal, J.B. et al, Overprint immunoassay using protein A microarrays, *Methods Mol. Biol.* 382, 273–286, 2007.
190. Ogi, H., Motohisa, K., Hatanaka, K. et al., Concentration dependence of IgG-protein A affinity studied by wireless-electrodeless QCM, *Biosens. Bioelectron.* 22, 3238–3242, 2007.
191. Ngo, T.T. and Narinesingh, D., Kosmotropes enhance the yield of antibody purified by affinity chromatography, *J. Immunoassay Immunochem.* 29, 105–115, 2008.
192. Lu, L. and Wang, Y., Immunoprecipitation alert: DNA binding proteins directly bind to protein A/G without any antibody as the bridge, *Cell Cycle* 7, 417–418, 2008.
193. Akerström, B., Brodin, T., Reis, K., and Björk, L., Protein G: A powerful tool for binding and detection of monoclonal and polyclonal antibodies, *J. Immunol.* 135, 2589–2592, 1985.
194. Backer, E.T. and Harff, G.A., Autoantibodies to lactate dehydrogenase in serum identified by use of immobilized protein G and immobilized jacalin, a jackfruit lectin, *Clin. Chem.* 35, 2190–2195, 1989.

195. Faulmann, E.L., Otten, R.A., Barrett, D.J., and Boyle, M.D., Immunological applications of type III Fc binding proteins. Comparison of different sources of protein G., *J. Immunol. Methods* 123, 269–281, 1989.
196. Pilcher, J.B., Tsang, V.C., Zhou, W., Black, C.M., and Sidman, C., Optimization of binding capacity and specificity for protein G on various solid matrices for immunoglobulins, *J. Immunol. Methods* 136, 279–286, 1991.
197. Desai, S. and Dworecki, B.R., Coated microplate-based affinity purification of antigens, *Anal. Biochem.* 328, 162–165, 2004.
198. Bae, Y.M., Oh, B.K., Lee, W. et al., Study on orientation of immunoglobulin G or protein G layer, *Biosens. Bioelectron.* 21, 103–110, 2005.
199. Tanaka, G., Funabashi, H., Mie, M., and Kobatake, E., Fabrication of an antibody microwell array with self-adhering antibody binding protein, *Anal. Biochem.* 350, 298–303, 2006.
200. Fowler, J.M., Stuart, M.C., and Wong, D.K., Self-assembled layer of thiolated protein G as an immunosensor scaffold, *Anal. Chem.* 79, 350–354, 2007.
201. Ha, T.H., Jung, S.O., Lee, J.M. et al., Oriented immobilization of antibodies with GST-fused multiple Fc-specific B-domains on a gold surface, *Anal. Chem.* 79, 546–556, 2007.
202. Lee, J.M., Park, H.K., Jung, Y. et al., Direct immobilization of protein g variants with various numbers of cysteine residues on a gold surface, *Anal. Chem.* 79, 2680–2687, 2007.
203. Jung, Y., Lee, J.M., Jung, H., and Chung, B.H., Self-directed and self-oriented immunobilization of antibody by protein G-DNA conjugates, *Anal. Chem.* 79, 6534–6541, 2007.
204. Kato, K., Sato, H., and Iwata, H., Immobilization of histidine-tagged recombinant proteins onto micropatterned surfaces for cell-based functional assays, *Langmuir* 21, 7071–7075, 2005.
205. Wu, Y., Buranda, T., Metzenberg, R.L. et al., Diazo coupling method for covalent attachment of proteins to solid substrates, *Bioconjug. Chem.* 17, 359–365, 2006.
206. Wilton, R., Yousef, M.A., Saxena, P. et al., Expression and purification of recombinant human receptor for advanced glycation endproducts in *Escherichia coli*, *Protein Expr. Purif.* 47, 25–35, 2006.
207. Simons, P.C., Young, S.M., Gibaja, V. et al., Duplexed, bead-based competitive assay for inhibitors of protein kinases, *Cytometry A* 71, 451–459, 2007.
208. Cressey, R., Pimpa, S., Chewaskulyong, B. et al., Simplified approach for the development of an ELISA to detect circulating autoantibodies to p53 in cancer patients, *BMC Biotechnol.* 8, 16, 2008.
209. Campbell, D.H., Luescher, E., and Lerman, L.S., Immuologic absorbents I. Isolation of antibody by means of a cellulose protein antigen, *Proc. Natl. Acad. Sci. USA* 37, 575–578, 1951.
210. Wu, Y., Buranda, T., and Metzenberg, R.L., Diazo coupling method for covalent attachment of proteins to solid substrates, *Bioconjug. Chem.* 17, 359–365. 2006.
211. Silman, I.H. and Katchalski, E., Water-insoluble derivatives of enzymes, antigens, and antibodies, *Annu. Rev. Biochem.* 35, 873–908, 1966.
212. Sipehia, R., Chawla, A.S., Daka, J., and Chang, T.M.S., Immobilization of enzymes on polypropylene bead surfaces by anhydrous ammonia gaseous plasma technique, *J. Biomed. Mater. Res.* 22, 417–422, 1988.
213. Hayat, U., Tinsley, A.M., Calder, M.R., and Clarke, D.J., ESCA investigation of low-temperature ammonia plasma-treated polyethylene substrate for immobilization of protein, *Biomaterials* 13, 801–806, 1992.
214. Meyer-Plath, A. A., Schroder, K., Finke, B., and Ohl, A., Current trends in biomaterial surface functionalization—nitrogen-containing plasma assisted processes with enhanced selectivity, *Vacuum* 71, 391–406, 2003.

215. Naqvi, A., Nahar, P., and Gandhi, R.P., Introduction of functional groups onto polypropylene and polyethylene surfaces for immobilization of enzymes, *Anal. Biochem.* 306, 74–78, 2002.
216. Erdtmann, M., Keller, R., and Bauman, H., Photochemical immobilization of heparin, dermatan sulphate, dextran sulphate and endothelial cell surface heparan sulphate onto cellulose membranes for the preparation of athrombogenic and antithrombogenic polymers, *Biomaterials* 15, 1043–1048, 1994.
217. Naqvi, A. and Nahar, P., Photochemical immobilization of proteins on microwave-synthesized photoreactive polymers, *Anal. Biochem.* 327, 68–73, 2004.
218. Hughes, K.A., Booth, L.R., Kaiser, R.J. et al., Optimization of a protein microarray platform based on a small molecule chemical affinity system, in *Protein Microarray Technology*, Ed. D. Kambhampahi, Wiley-VCH, Weinheim, Germany, Chapter 3, pp. 39–55, 2004.
219. Stolowitz, M.L., Ahelm, C., Hughes, K.A. et al., Phenylboronic acid-salicylhydroxamic acid bioconjugates. 1. A novel boronic acid complex for protein immobilization, *Bioconjug. Chem.* 12, 229–239, 2001.
220. Wiley, J.P., Hughes, K.A., Kaiser, R.J. et al., Phenylboronic acid-salicylhydroxamic acid bioconjugates. 2. Polyvalent immobilization of protein ligands for affinity chromatography, *Bioconjug. Chem.* 12, 240–250, 2001.
221. Li, F., Zhao, X., Wang,W., and Xu, G., Synthesis of silica-based benzeneboronic acid affinity materials and application as pre-column in coupled-column high-performance liquid chromatography, *Anal. Chim. Acta* 580, 181–187, 2006.
222. Gontarev, S., Shmanai, V., Frey, S.K. et al., Application of phenylboronic acid modified hydrogel affinity chips for high-throughput mass spectrometric analysis of glycated proteins, *Rapid Commun. Mass Spectrom.* 21, 1–6, 2007.
223. Polsky, R., Harper, D.R., Arango, D.C., and Brozik, S.M., Electrically addressable cell immobilization using phenylboronic acid diazonium salts, *Angew. Chem. Int. Ed. Engl.* 47, 2631–2634, 2008.
224. Kalia, J., Abbott, N.L., and Raines, R.T., General method for site-specific protein immobilization by Staudinger ligation, *Bioconjug. Chem.* 18, 1064–1069, 2007.
225. Das, K., Dunnill, P., and Lilly, M.D, Affinity chromatography of enzyme cofactors: The separation of NAD on immobilized dehydrogenase columns, *Biochim. Biophys. Acta* 397, 277–287, 1975.
226. Kennedy, J.H., Kricka, L.J., and Wilding, P., Protein-protein coupling reactions and the application of protein conjugates, *Clin. Chim. Acta* 70, 1–31, 1976.
227. Zhang, Q., Zou, K., Chen, X. et al., Synthesis and characterization of the human serum albumin-triazine chiral stationary phase, *Chirality* 12, 714–719, 2000.
228. Zerlotti, E., Cross-linking of rat tail tendons with chloro-*s*-triazine, *Nature* 214, 1304–1306, 1967.
229. Axén, R., Porath, J., and Ernback, S., Chemical coupling of peptides and proteins to polysaccharides by means of cyanogen halides, *Nature* 214, 1302–1304, 1967.
230. Porath, J., Axén, R., and Ernback, S., Chemical coupling of proteins to agarose, *Nature* 215, 1491–1492, 1967.
231. Chockalingam, P.S., Gadgil, H., and Jarrett, H.W., DNA-support coupling for transcription factor purification factor purification. Comparison of aldehyde, cyanogen bromide and *N*-hydroxysuccinimide chemistries, *J. Chromatog. A* 942, 167–175, 2002.
232. Attiya, S., Dickinson-Laing, T., Cearz, J. et al., Affinity protection chromatography for efficient labeling of antibodies for use in affinity capillary electrophoresis, *Electrophoresis* 23, 750–758, 2002.
233. Jurado, L.A., Mosley, J., and Jarrett, H.W., Cyanogen bromide activation and coupling of ligands to diol-containing silica for high-performance affinity chromatography optimization of conditions, *J. Chromatog. A* 971, 95–104, 2002.

234. Costa, S.A. and Reis, R.L., Immobilization of catalase on the surface of biodegradable starch-based polymers as a way to change its surface characteristics, *J. Mater. Sci. Mater. Med.* 15, 335–342, 2004.
235. Hernández, R., Plana, L., Gómez, L. et al., Optimization of the coupled monoclonal antibody density for recombinant hepatitis B virus surface antigen immunopurification, *J. Chromatog. B. Anal. Technol. Biomed. Life Sci.* 816, 1–6, 2005.
236. Mitra, S., Jarrett, H.W., and Jurado, L.A., High-performance catalytic chromatography. The adaptor approach, *J. Chromatog. A.* 1076, 71–82, 2005.
237. Nagore, L.I., Mitra, S., and Jiang, D., Cyanogen bromide-activated coupling: DNA catalytic chromatography purification of EcoRI endonuclease, *Nat. Protoc.* 1, 2909–2915, 2006.
238. Gontarev, S., Shanai, V., Frey, S.K. et al., Application of phenylboronic acid modified hydrogel affinity chips for high-throughput mass spectrometric analysis of glycated proteins, *Rapid Commun. Mass Spectrom.* 21, 1–6, 2007.
239. Yang, T.H. and Feng, C.L., New ligand coupling procedure for formation of an immunoadsorption wall, *ASAIO J.* 53, 201–206, 2007.
240. Lu, J., Jackson, J.K., Gleave, M.E., and Burt, H.M., The preparation and characterization of anti-VEGFR2 conjugated, paclitaxel-loaded PLLA or PLGA microspheres for the systemic targeting of human prostate tumors, *Cancer Chemother. Pharmacol.* 61, 997–1005, 2008.
241. Jurado, L.A., Mosley, J., and Jarrett, H.W., Cyanogen bromide activation and coupling of ligands to diol-containing silica for high-performance affinity chromatography optimization of conditions, *J. Chromatog. A* 971, 95–104, 2002.
242. Jurado, L.A. and Jarrett, H.W., In flow activation of diol-silica with cyanogen bromide and triethylamine for preparing high-performance affinity chromatographic columns, *J. Chromatog. A* 984, 9–17, 2003.
243. Robberson, D.L. and Davidson, N., Covalent coupling of ribonucleic acid to agarose, *Biochemistry* 11, 533–537, 1972.
244. Gadgil, H. and Jarrett, H.W., Heparin elution of transcription factors from DNA-Sepharose columns, *J. Chromatog. A* 848, 131–138, 1999.
245. Jarrett, H.W., Temperature dependence of DNA affinity chromatography of transcription factors, *Anal. Biochem.* 279, 209–217, 2000.
246. Chockalkingam, P.S., Jurado, L.A., and Jarrett, H.W., DNA affinity chromatography, *Mol. Biotechnol.* 19, 189–199, 2001.
247. Sokolova, N.I., Ashırbekova, D.T., Dolinnaya, N.G., and Shabarova, Z.A., Chemical reactions within DNA duplexes. Cyanogen bromide as an effective oligodeoxyribonucleotide coupling agent, *FEBS Letters* 1, 153–155, 1999.
248. Carroero, S. and Damha, M.J., Template-mediated synthesis of lariat RNA and DNA, *J. Org. Chem.* 68, 8328–8338, 2003.
249. Silverman, A.P. and Kool, E.T., Detecting RNA and DNA with templated chemical reactions, *Chem. Rev.* 106, 3775–3789, 2006.
250. Zhang, L.C., Long, H., Schatz, G.C. et al., Synthesis and properties of nicked dumbbell DNA conjugates having stilbenedicarboximide linkers, *Org. Biomol. Chem.* 5, 450–456, 2007.
251. Dolinnaya, N.G., Sokolova, N.I., Ashirbekova, D.T. et al., The use of BrCN for assembling modified DNA duplexes and DNA-RNA hybrids: Comparison with water-soluble carbodiimide, *Nucleic. Acids Res.* 19, 3067–3072, 1991.
252. Wagner, A.F., Bugianesi, R.L., and Shen, T.Y., Preparation of Sepharose-bound Poly(rI:rC), *Biochem. Biophys. Res. Commun.* 45, 184–189, 1971.
253. Gilboa, E., Prives, C.L., and Aviv, H., Purification of SV-40 messenger RNA by hybridization to SV-40 DNA covalently bound to Sepharose, *Biochemistry* 14, 4215–4220, 1975.
254. Gilles, M.A., Hudson, A.Q., and Borders, C.L., Jr., Stability of water-soluble carbodiimides in aqueous solution, *Anal. Biochem.* 184, 244–248, 1990.

255. Lasch, J., Ligand-leakage in affinity chromatography: A second note on the mathematical approach, *Experentia* 31, 1125–1126, 1975.
256. Sato, H., Kidaka, T., and Hori, M., Leakage of immobilized IgG from therapeutic immunoadsorbents, *Appl. Biochem. Biotechnol.* 15, 145–158, 1987.
257. Hagen, M. and Strejan, G.H., Antigen leakage from immunosorbents. Implications for the detection of site-directed auto-anti-idiotypic antibodies, *J. Immunol. Methods* 100, 47–57, 1987.
258. Goldberg, M., Knudsen, K.L., Platt, D. et al., Specific interchain cross-linking of antibodies using bimaleimides. Repression of ligand leakage in immunoaffinity chromatography, *Bioconjug. Chem.* 2, 275–280, 1991.
259. Hernández, R., Chong, K., Morales, R. et al., Stirrer tank: An appropriate technology to immobilize the CB.Hep-1 monoclonal antibody for immunoaffinity purification, *J. Chromatog. B Biomed. Sci. Appl.* 754, 77–83, 2001.
260. Wang, T., Yang, Z., Emregul, E. et al., Strategies for improving the functionality of an affinity bioreactor, *Int. J. Pharm.* 306, 132–141, 2005.
261. Schaeferling, M. and Kambhampahi, D., Protein microarray surface chemistry and coupling scheme, in *Protein Microarray Technology*, Ed. D. Kambhampahi, Wiley-VCH, Weinheim, Germany, Chapter 2, pp. 11–38, 2004.
262. Mann, C.J., Stephens, S.K., and Burke, J.E., Production of protein microarrays, in *Protein Microarray Technology*, Ed. D. Kambhampahi, Wiley-VCH, Weinheim, Germany, Chapter 8, pp. 165–194, 2004.
263. Seuryck-Servoss, White, A.M., Baird, C.L. et al., Evaluation of surface chemistries for antibody microarrays, *Anal. Biochem.* 371, 105–115, 2007.
264. Mahajan, S., Vaijayanthi, B., Rembhotkar, G. et al., Choice of polymer matrix, its functionalization and estimation of functional group density for preparation of biochips, *Methods Mol. Biol.* 381, 165–187, 2007.
265. Cutler, P., Protein arrays: The current state-of-the-art, *Proteomics* 3, 3–18, 2003.
266. Schweitzer, B. and Kingsmore, S.F., Measuring proteins on microarrays, *Curr. Opin. Biotechnol.* 13, 14–19, 2002.
267. Zhu, H. and Snyder, M., Protein chip technology, *Curr. Opin. Chem. Biol.* 7, 55–63, 2003.
268. Gauvreau, V., Chevallier, P., Vallières, K. et al., Engineering surfaces for bioconjugation: Developing strategies and quantifying the extent of reactions, *Bioconjug. Chem.* 15, 1146–1156, 2004.
269. Seurynck-Servoss, S.L., White, A.M., Baird, C.L. et al., Evaluation of surface chemistries for antibody microarrays, *Anal. Biochem.* 371, 105–115, 2007.
270. Lovrinovic, M. and Niemeyer, C.M., Rapid synthesis of DNA-cysteine conjugates for expressed protein ligation, *Biochem. Biophys. Res. Commun.* 335, 943–948, 2005.

4 Protein Conjugates

INTRODUCTION

The term *bioconjugate* is well accepted in the scientific literature but undefined in, for example, *The Oxford Dictionary of the English Language*; yet, it is beyond definition as a neologism. Meares[1] defined a bioconjugate as the joining of two molecular functions by either chemical or biological means. Bioconjugates are most frequently the combination of a well-known protein such as albumin, immunoglobulin, or immunoglobulin fragment with a nonprotein moiety such as cytotoxic agents.[2–4] The term *bioconjugate* has also been defined as a coupled protein reagent.[5] Having said this, a compound is a bioconjugate in the eye of the viewer.

The term bioconjugate has been used to describe a broad variety of compounds obtained by chemical ligation of biological compounds and would appear to represent an effort to add value by combination of differing chemical and functional "biological" molecules. A brief literature search revealed a somewhat eclectic group of studies. Martins and coworkers[6] modified L-asparaginase with palmitoyl chloride. Approximately 30% of the lysine residues were modified with retention of catalytic activity. The modified enzyme was more hydrophobic (octanol–water partition coefficient increased from 0.13 to 1.78 on modification with palmitoyl chloride). Pretargeting is a strategy in which a bifunctional protein is prepared. The protein may be a bispecific antibody (e.g., a diabody) or an antibody "fused" to streptavidin. In the latter case, a biotinylated radiolabeled compound is then targeted to the streptavidin, which has been targeted by the antibody–streptavidin.[7] Biotin derivatives are subject to hydrolysis in serum by biotinidase, and a variety of derivatives have been prepared for conjugation to proteins to yield more stable bioconjugates for use in pretargeting.[8] A conjugate of paclitaxel and oxytocin has been prepared.[9] A complex of heparin and a novel thermoresponsive cationic polymer acting as a surfactant has been formed, yielding a bioconjugate functioning as a thromboresistant coating.[10] Here, the term bioconjugate is used to describe the combination of two large molecules enhancing the functional qualities of one of the two components. Chemically cleavable bioconjugates have been developed. One example is the reversible PEGylation of proteins using coupling (dithiobenzyl urethane, Figure 4.1), which can be cleaved by mild reduction.[11] This chemistry has been used for the construction of biodegradable drug delivery systems.[12–14] Disulfide linkages have also been used to impart mineral-binding ability through the coupling of thiol derivatives of bisphosponates (Figure 4.2) to hydroxylapatite matrices.[15] Coupling of thiazole orange, a DNA-sensitive fluorophore, to a peptide from the Tc3 transposase DNA-binding domain yielded a bioconjugate (Figure 4.3) that served as a probe for DNA.[16] Poole and coworkers[17] have described a novel bioconjugate composed

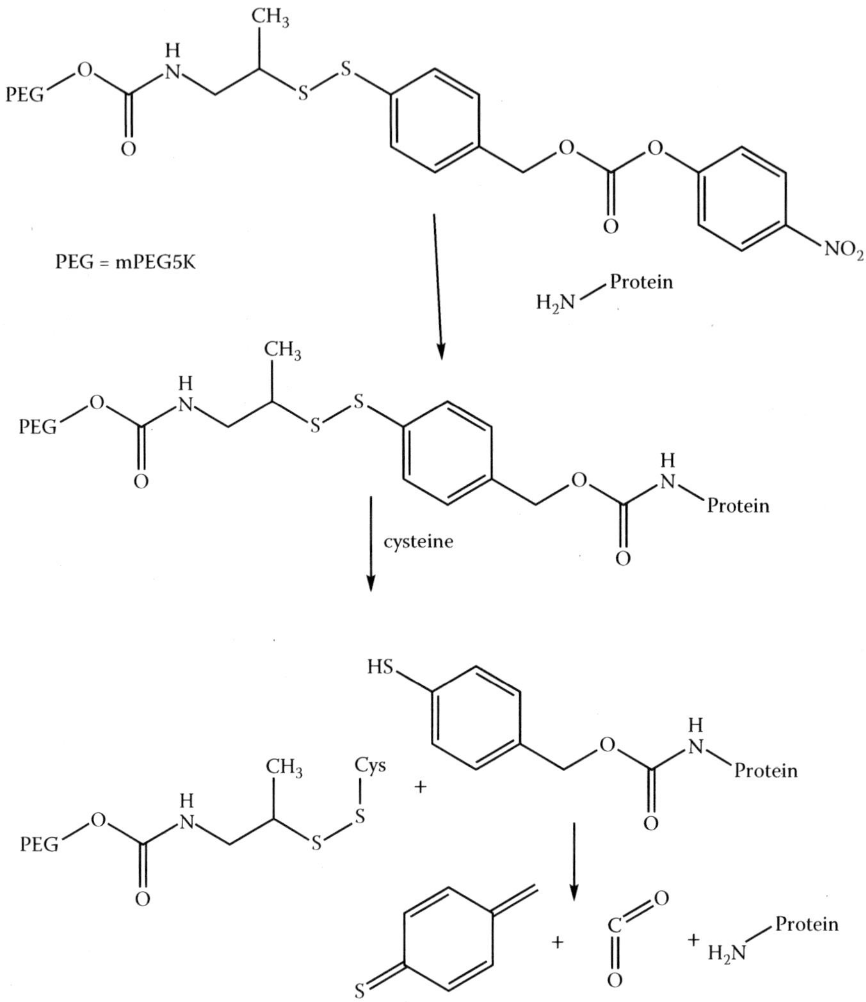

FIGURE 4.1 The reversible PEGylation of proteins using dithiobenzyl urethane. (See Zalipsky, S., Mullah, N., Engbers, C. et al., Thiolytically cleavable dithiobenzyl urethane-linked polymer–protein conjugates as macromolecular prodrugs: Reversible PEGylation of proteins, *Bioconjug. Chem.* 18, 1869–1878, 2007.)

of a dimedone analog, 1,3,-cyclohexanedione, and fluorophores (Figure 4.4) for the detection of cysteine sulfenic acid. More complex bioconjugates have been derived from coacervate core micelles with lysozyme-modified corona[18] or dendritic displays (cross-linked hemoglobin).[19] Chen and coworkers[20] describe the synthesis of a porphyrinmaleimide for the formation of bioconjugates with cysteine and related sulfhydryl compounds. These few selected studies are intended to illustrate the breadth of chemical derivatives considered to be bioconjugates. The reader is directed to

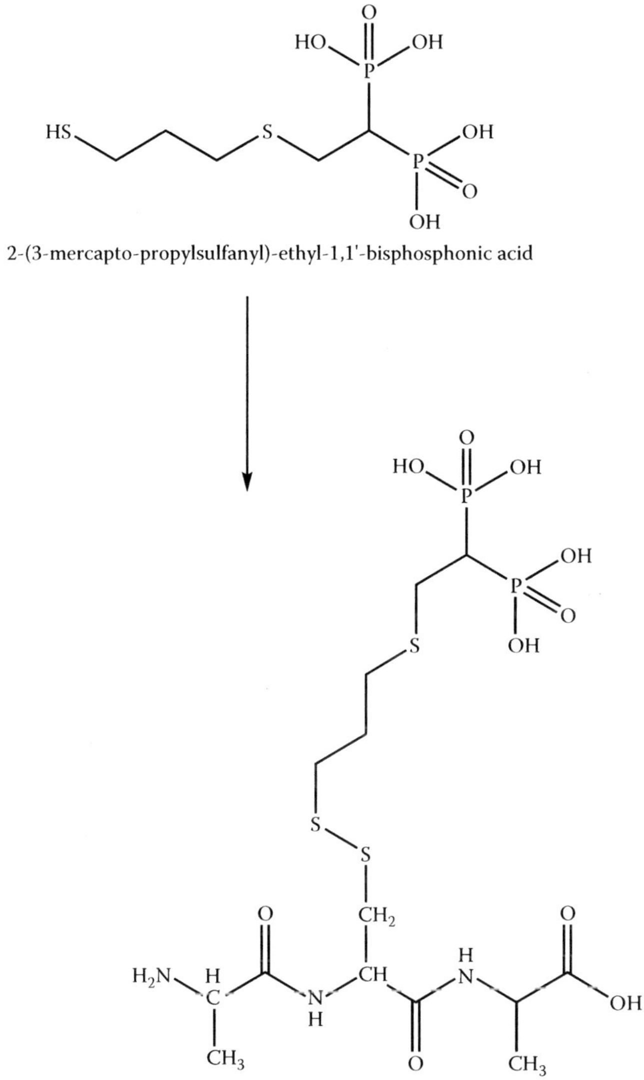

FIGURE 4.2 Thiol derivatives of bisphosphonates. (See Bansal, G., Wright, J.E.I., Zhang, S. et al., Imparting mineral affinity to proteins with thiol-labile disulfide linkages, *J. Biomed. Mater. Res.* 74A, 618–628, 2005.)

Bioconjugate Chemistry (American Chemical Society) for more illustrative examples of bioconjugates.

This chapter will be concerned with the use of solution protein chemistry to form bioconjugates. The basic chemistry underlying bioconjugation is discussed in Chapters 1 and 2 and various review articles.[21–26] The reader is also referred to Chapter 5 (Protein Hydrogels) and Chapter 9 on the use of chemical modification for the manufacture of biotherapeutics.

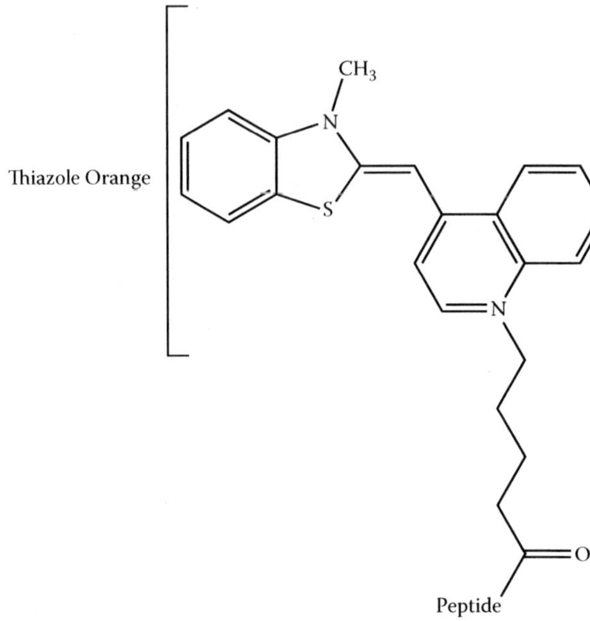

FIGURE 4.3 Thiazole orange peptide probe for DNA. (See Thompson, M., Spectral properties and DNA targeting features of a thiazole orange–peptide bioconjugate, *Biomacromolecules* 8, 3628–3633, 2007.)

PROTEIN CONJUGATES

This section will focus on protein conjugates as distinguished from fusion proteins,[27–34] which are created by genetic engineering. Fusion proteins may be engineered in such a fashion to include a site for facile chemical modification such as the tetracysteine sequence.[35] Current fusion proteins are usually engineered for a function such as the inclusion of segment to aid purification or binding, such as the hexahistdine sequences (hexaHis), whereas earlier fusion proteins were engineered to "ask" basic questions of cellular function.[34–38] It is noted that the term *hybrid protein* was used earlier to describe such molecular constructs[39–41] and is still in vogue today.[43,43] The term *fusion protein* was also used earlier to describe proteins involved in viral function.[44] There appears to be greater use of fusion proteins and some suggestion that genetic fusion may be better than protein conjugation.[45] However, there is the issue of obtaining good expression of fusion proteins.

Large peptides and proteins have proved difficult to join by chemical means. There are two primary methods. The first involved reaction with a sulfhydryl by, for example, the use of a maleimide function, whereas the second is expressed protein ligation (chemical ligation). A single cysteine residue may be inserted into a recombinant protein or a sulfhydryl function added to the protein as, for example, by reaction with 2-iminothiolane.

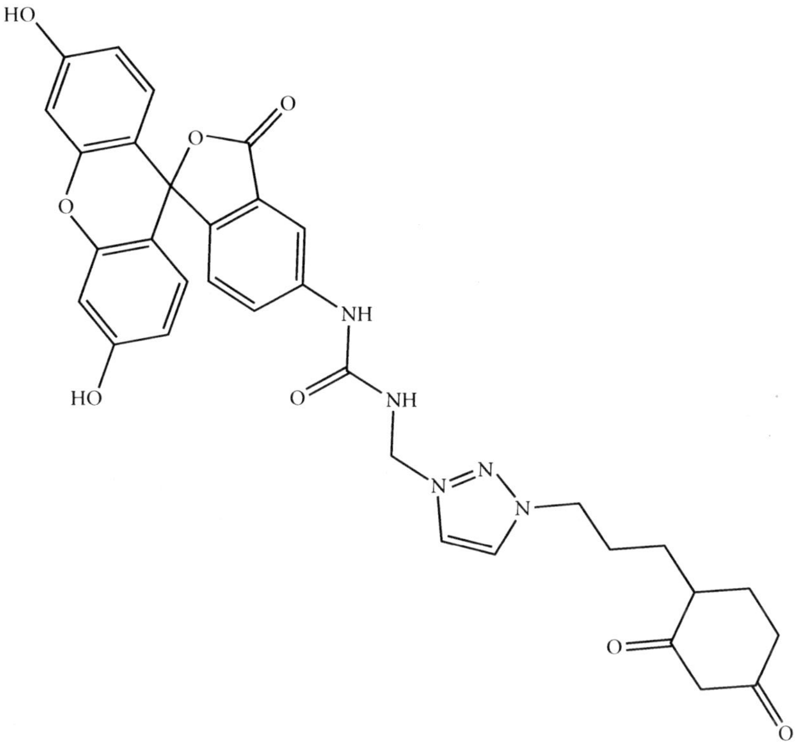

FIGURE 4.4 Cyclohexanedione fluorophore for detection of cysteine sulfenic acid. (See Poole, L.B., Klomsiri, C., Knaggs, S.A. et al., Fluorescent and affinity-based tools to detect cysteine sulfenic acid formation in proteins, *Bioconjug. Chem.* 18, 2004–2017, 2007.)

There are some other approaches that are novel, including the use of Diels–Alder condensation of an N-terminal maleimide with a C-terminal 2,4-hexadienyl ester to yield a cycloadduct[46] linking the two blocks (Figure 4.5). A maleimide function is also one of the several functional groups in a trifunctional "generic" building block that could "like" a C-terminal and an N-terminal and a fluorescent probe.[47] The use of a heterobifunctional reagent, ε-(maleimidocaproyloxy)sulfosuccinimide ester (Figure 4.6), was used to prepare a conjugate between an antigen (parasite protein Pfs25) and *Neisseria meningitis* outer membrane protein for an immunogen[48]; this approach results in a high-titer response; covalent linkage of the two proteins was required for the high response. Other studies have shown that the linker molecule can result in an immune response independent from the conjugate partners or modulate the response to the conjugate pair.[49–49c]

Chemical cross-linking (see Chapter 1) can be used, but generally lacks the specificity required for providing a defined bioconjugate. There have been exceptions, but with unique circumstances. The catalytic B-chain of human urokinase (the catalytic domain) was obtained by limited reduction of the parent two-chain protein and coupled to a hydrophobic surfactant protein.[50] The cross-linking reaction with sulfosuccinimidyl 4-(*p*-maleimidophenylbutyrate) was performed in 1-propanol.

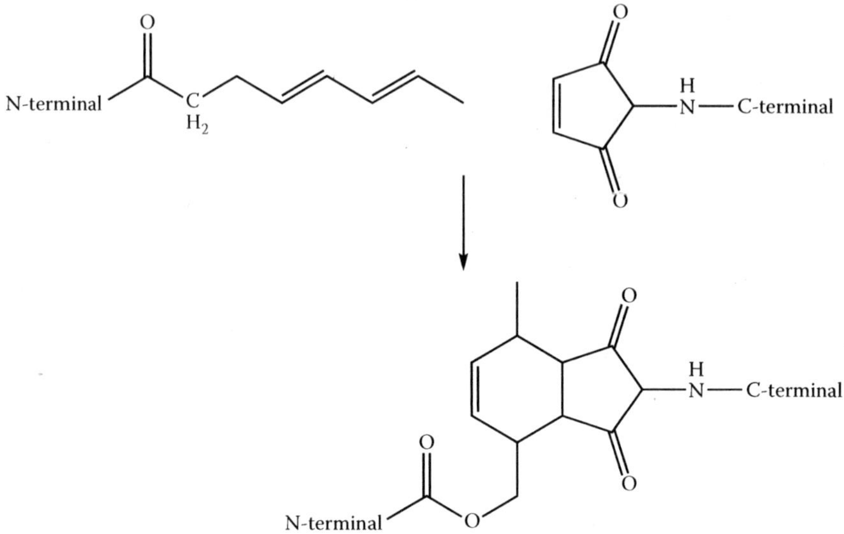

FIGURE 4.5 Maleimide for the modification of C-terminal diene. (See Dantes de Araujo, A., Palomo, J.M., Cramer, J. et al., Diels–Alder ligation of peptides and proteins, *Chem. Eur. J.* 12, 6095–6109, 2006.)

Immobilization of one protein component by coupling to a matrix has been proposed[51,52] to improve specificity. Divinyl sulfone (vinyl sulfone) and iodoacetyl succinimide (Figure 4.7) have been suggested for use in the preparation of protein conjugates.[53–55]

The sulfhydryl group of cysteine is (usually) the most nucleophilic in a protein, and it is relatively easy to modify. It is somewhat more difficult to modify one of several sulfhydryl groups (see Chapter 1). Disulfide exchange has proved useful in the preparation of bioconjugates.[56–60] It is extremely difficult to form specific heteromolecular disulfide bonds when one or both partners have multiple sulfhydryl groups. One of the more well-known examples of this difficulty is the attempt to synthesize insulin[61] before Steiner's discovery of proinsulin.[62–64]

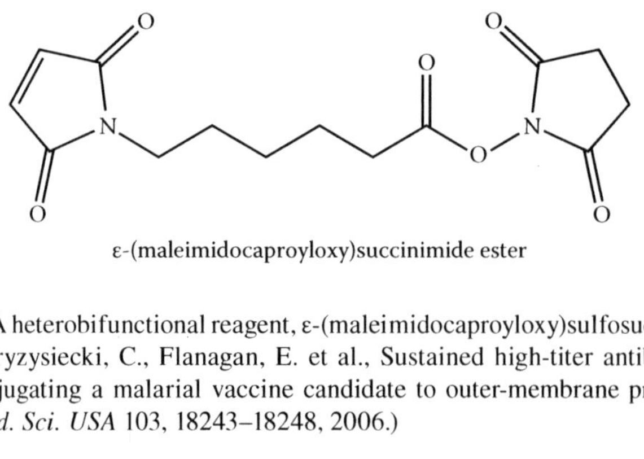

FIGURE 4.6 A heterobifunctional reagent, ε-(maleimidocaproyloxy)sulfosuccinimide ester. (See Wu, Y., Pryzysiecki, C., Flanagan, E. et al., Sustained high-titer antibody responses induced by conjugating a malarial vaccine candidate to outer-membrane protein complex, *Proc. Natl. Acad. Sci. USA* 103, 18243–18248, 2006.)

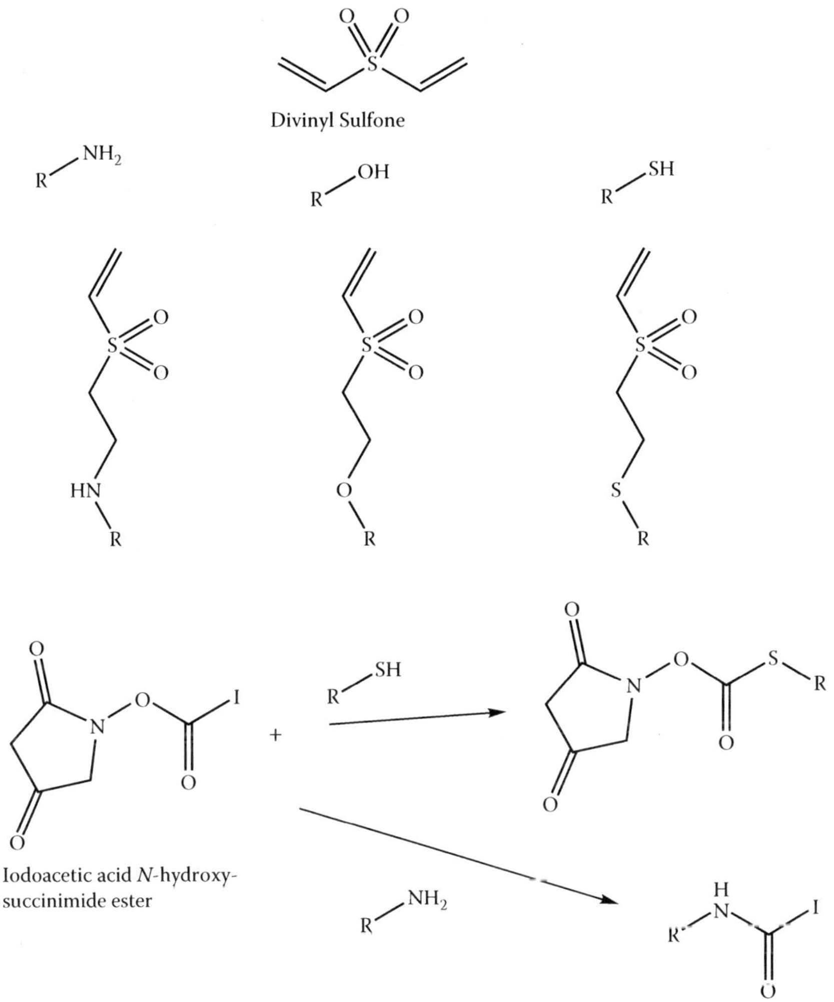

FIGURE 4.7 Divinyl sulfone and iodoacetylsuccinimide. (See Houen, G. and Jensen, O.M., Conjugation to preactivated proteins using divinylsulfone and iodoacetic acid, *J. Immunol. Methods* 181, 187–200, 1995.)

Coupling at a sulfhydryl group with a maleimide derivative has been cited earlier for the preparation of a porphyrinmaleimide,[20] and this chemistry (Michael addition) has been used to prepare other bioconjugates.[65–69] The maleimide function is highly specific for sulfhydryl groups in proteins.

The discovery of inteins,[70–74] self-splicing sequences in proteins (Figure 4.8), resulted in the development of two closely related approaches for joining two peptide/protein chains in a selective manner. Dawson and coworkers[75] described native chemical ligation (Figure 4.9) as a technique that joined the C-terminal thioester of a peptide or protein with the N-terminal cysteine residues; the initial product

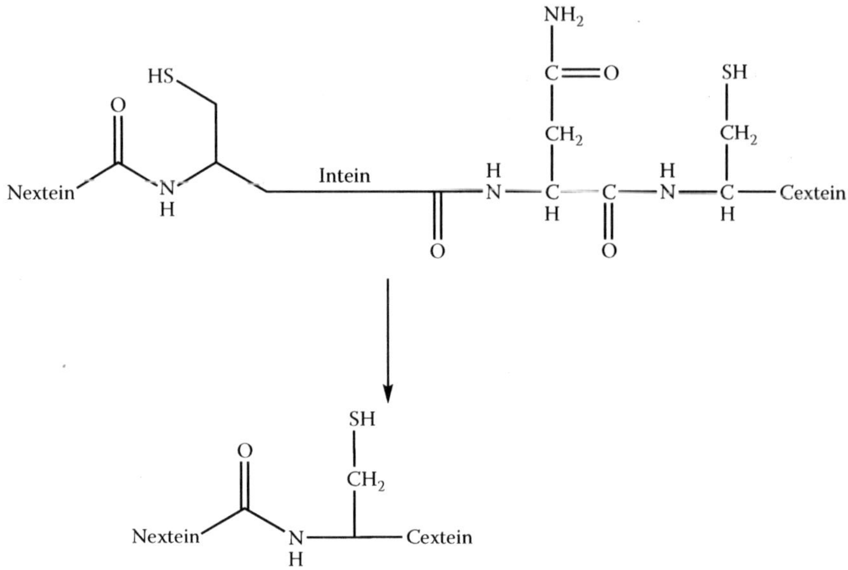

FIGURE 4.8 Intein, a self-splicing mechanism for proteins. (Adapted from Xu, M.-Q. and Evans, T.C., Jr., Recent advances in protein splicing: Manipulating proteins *in vitro* and *in vivo*, *Curr. Opin. Biotechnol.*16, 440–446, 2005.)

is a thioester that rearranges to a peptide bond. This technique permits the facile coupling of two peptide or protein fragments without the necessity of protecting functional groups. Variations on this approach have included the use of Staudinger ligation, in which an azide replaces the amino-terminal cysteine with the reaction performed in the presence of phosphinobenzene thiol.[76–78] Clippingdale and coworkers[79] have described an improved method for generation of peptide thioesters that uses Fmoc (fluorenylmethoxy-carbonyl) in place of other protecting groups (Figure 4.10). Native chemical ligation has been used for the synthesis of proteins, including bovine pancreatic trypsin inhibitor,[80] human matrix Gla protein,[81] a serine protease,[82] a glycosylated IL-2,[83] and a human tau-construct.[84] The synthesis of cyclic peptides containing a disulfide knot (cyclotides) was achieved by microwave-assisted synthesis, starting with coupling of a thioester to the first amino acid and terminating with cysteine.[85] Native chemical ligation has also been used to attach proteins to a solid surface[86,87] (Figure 4.11).

Expressed protein ligation is a method of producing the reactive C-terminal thioester by recombinant DNA technology. In its current form, the amino-terminal protein/peptide fragment is obtained from an intein protein by cleavage with a thiol to yield the C-terminal thioester (Figure 4.12).[88] The term *expressed protein ligation* was first used by Severinov and Muir[89] to describe formation of the C-terminal thiol ester via an intein–chitin binding domain protein using a pCVB expression vector[90] and the condensation of the thiol ester with an N-terminal cysteine residue. At the time, expressed protein ligation was described as a tool for introducing sequences containing unnatural amino acids, protein probes, and sequences with

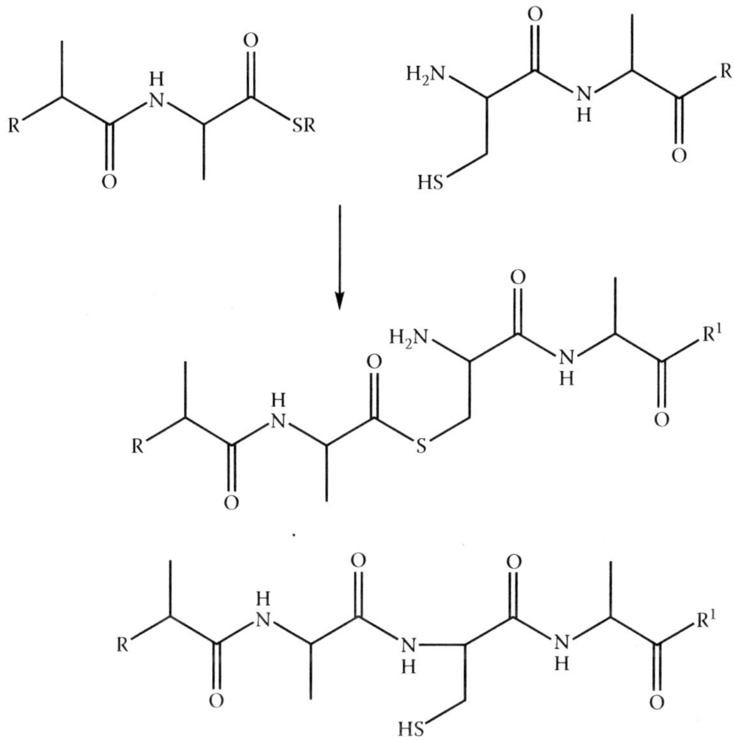

FIGURE 4.9 A scheme for native chemical ligation. (Adapted from Dawson, P.E., Muir, T.W., Clark-Lewis, I., and Kent, S.B.H., Synthesis of proteins by native chemical ligation, *Science* 266, 776–779, 1994.)

posttranslational modifications into proteins.[91] The use of expressed protein ligation has increased and been shown to be extremely useful in preparing hybrid or chimeric proteins.[92–96] This chemistry has been used to join cysteine-derivatized oligonucleotides to proteins via a recombinant intein protein.[97] This is a simple approach to preparing a protein–nucleic acid bioconjugate.[98–100]

Click chemistry (Figure 4.13) has been a major advance in the ligation of macromolecules.[101–105] Click chemistry is an extension of the Huisgen reaction[106] and has been used for the preparation of bioconjugates.[107–112] Parrish and colleagues[107] used click chemistry to prepare conjugates of polyesters with PEG or peptides (Figure 4.14). Aucagne and Leigh[108] used a trimethylsilyl group to block an alkyne, permitting the formation of successive triazole linkages (Figure 4.15). Danishefsky and coworkers[109] used click chemistry in work on the synthesis of carbohydrate-based anticancer vaccines (Figure 4.16). Haddleton and coworkers[111] used click chemistry for the synthesis of glycoprotein mimics. DeNardo and coworkers[112] used click chemistry to form a covalent dimer of single-chain Fv regions of IgG. Lutz and Zarafshani[113] have a recent comprehensive review on the use of click chemistry in biotechnology applications.

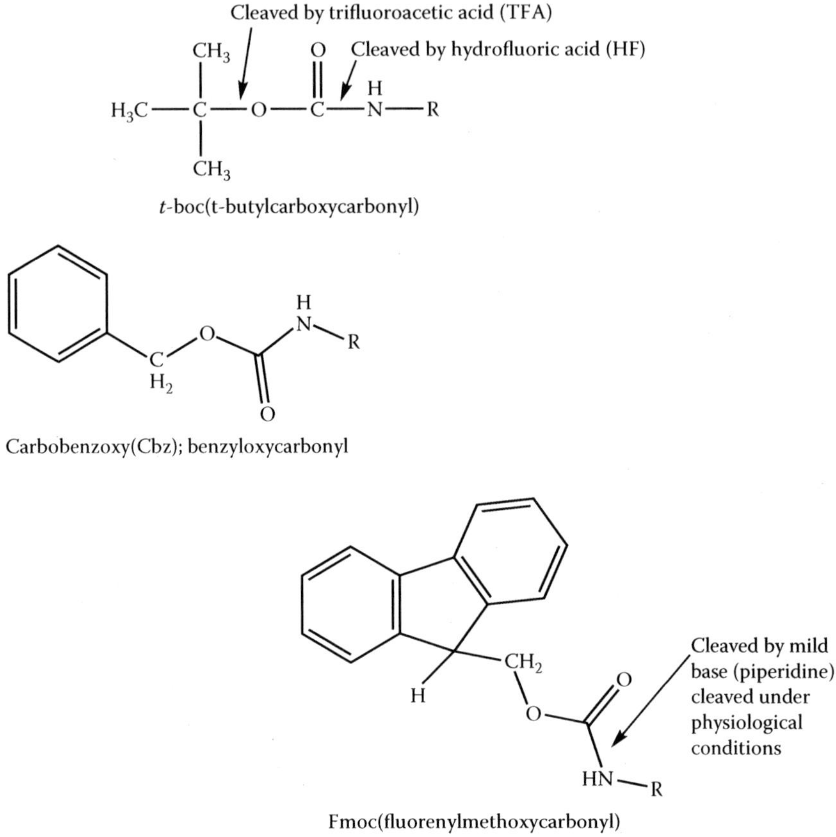

FIGURE 4.10 Protecting groups for amine function. Shown are some of the various blocking groups used to protect amine functions in peptide and protein chemistry. (Adapted from *Synthetic Peptides: A User's Guide*, 2nd ed., Ed. G.A. Grant, Oxford University Press, Oxford, 2002.)

ALBUMIN BIOCONJUGATES

Albumin has been used as a protein bioconjugate partner. There is extensive and clinical understanding of this protein (see Chapter 6), and it is available in large quantities in pure form. Albumin has also been used as a fusion partner for recombinant DNA-derived chimeric proteins.[114–121] McCurdy and coworkers[122] showed that an albumin–albumin fusion protein (two engineered [C34A] albumin molecules joined by a hexaglycine spacer and having an amino-terminal hexahistidine sequence) had a somewhat shorter half-life than the parent monomer protein with or without the hexahistidine sequence. However, Matsushita and coworkers[123] showed that an albumin–albumin fusion protein with a different spacer sequence (GGGS),[2] having cysteine 34, and without a hexahistidine sequence had a prolonged half-life compared to the monomer. The differences in pharmacokinetics might reflect differences in the animal models used for the half-life determination, as noted by Sheffield and

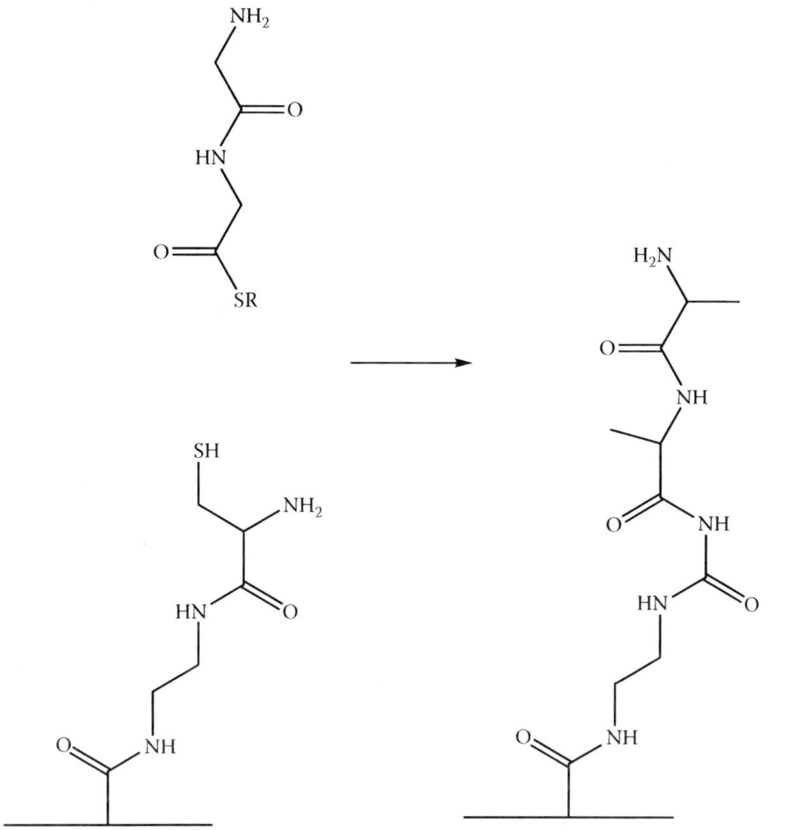

FIGURE 4.11 Native chemical ligation to a matrix. (See Helms, B., van Baal, I., Marks, M., and Meijer, E.W., Site-specific protein and peptide immobilization on a biosensor surface by pulse native chemical ligation, *ChemBioChem* 8, 1790–1794, 2007.)

coworkers.[124] It is of interest that an albumin dimer formed by the cross-linking of a recombinant human serum albumin via the cysteine residues with 1,2-bis(maleimido) hexane had an extended circulatory half-life.[125] These investigators also reported that the extinction coefficient (280 nm) for the dimer was precisely twice that of the monomer and that the antigenic epitopes were preserved. Possible contributions of the linker region (either recombinant or chemical) are frequently ignored. Although there is much literature in this area, neoantigen expression has been noted with the use of a synthetic adjuvant, *N*-acetylmuramyl-L-alanyl-D-isoglutamine, covalently linked to a polypeptide antigen.[126] There has been some discussion of the contribution of less complex linkers to immunogenicity of conjugates earlier.[49–49c]

Human serum albumin contains a single sulfhydryl group at cysteine 34. This sulfhydryl group is considered to have a low p*K*a of approximately 5–7[127,128] based on reaction with nucleophiles such as dithiopyridine. However, the modification of the sulfhydryl group is complex as the pH dependence curve for the reaction of the albumin sulfhydryl group is bimodal, decreasing to a minimum at pH 4 and then rapidly

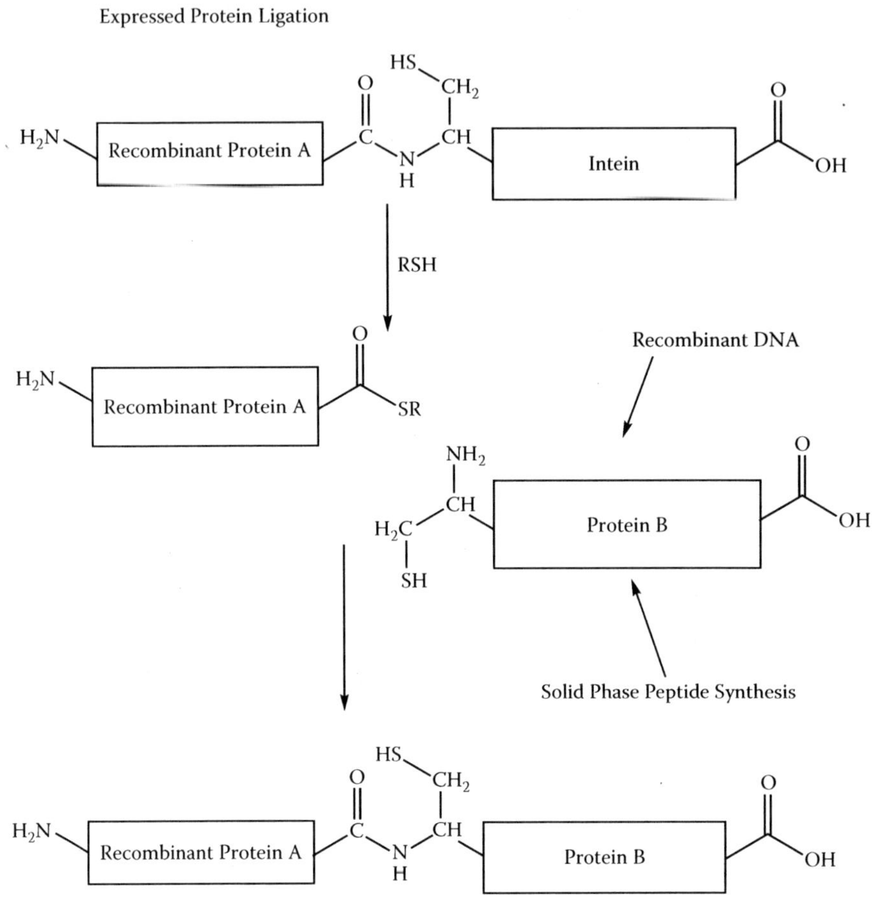

FIGURE 4.12 A scheme for expressed protein ligation. (See Muralidharan, V. and Muir, T.W., Protein ligation: An enabling technology for the biophysical analysis of proteins, *Nat. Methods* 3, 429–438, 2006.)

increasing with increasing pH. It is thought that a conformational transition (N→F) is important for the observed pH dependence.[129,130] The rate of reaction of dithiopyridine is as rapid at pH 2.6 as it is at pH 6.6.[129] Other studies[128,130,131] on the modification of the sulfhydryl groups suggest that the p*K*a of the cysteine sulfhydryl is 5. Pedersen and Jacobsen[127] obtained a value of 7, whereas the earlier study of Svenson and Carlsson[129] suggests an approximate value of 8.5 (bovine serum albumin).

The p*K*a for the sulfhydryl group of free cysteine is 8.14.[132] The value for protein-bound sulfhydryl groups outside of enzyme active sites is 8–10,[133] whereas lower values are seen for sulfhydryl residues at active sites.[134–137] The values for the sulfhydryl p*K*a at enzyme active sites is usually 7–8, although much lower values (pH 4)[136] have been observed, reflecting the influence of local environment. The lower p*K*a value for cysteine 34 would imply greater reactivity, but this is not thought to be the situation[130] although large amounts of data are not available. Kharitonov and colleagues[138]

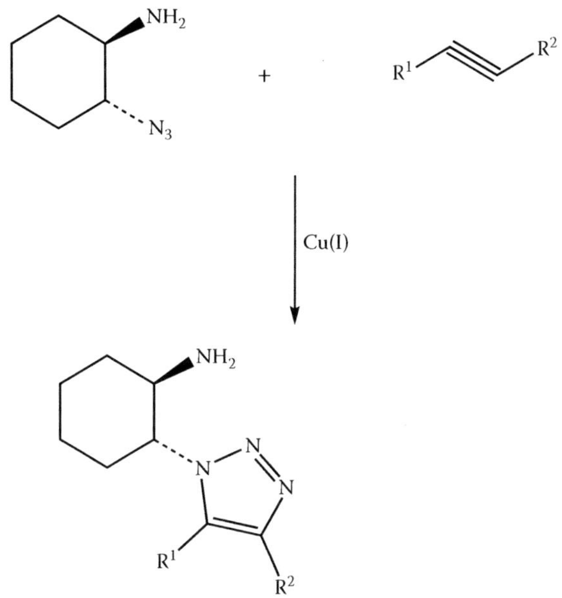

FIGURE 4.13 Click chemistry. (See Kolb, H.C., Fiun, M.G., and Sharpless, K.B., Click chemistry: Diverse chemical function from a few good reactions, *Angew. Chem. Int. Ed.* 44, 2004–2023, 2001.)

observed that the sulfhydryl of human albumin reacts with nitric oxide (N_2O_3) more slowly ($0.3 \times 10^5\ M^{-1}s^{-1}$) than with glutathione ($2.9 \times 10^5\ M^{-1}\ s^{-1}$) or *N*-acetylcysteine ($1.5 \times 10^5\ M^{-1}\ s^{-1}$). In earlier studies, Knight and Green[139] observed that rates of reaction of *N*-(*N*-dinitrophenyl-aminoalkyl) maleimides with bovine serum albumin were faster than those observed for reaction of maleimides with simple thiols.[140] It can be concluded that the reactivity of the sulfhydryl group in albumin is complex, depending on the conformation of the albumin and the modifying reagent.

Maleimide chemistry has been used for preparation of insulin–albumin conjugates via modification of the sulfhydryl group.[141,142] One group[141] used an Fmoc (9-fluorenymethoxycarbonyl) derivative (Figure 4.17) to couple insulin (via lysine) to the sulfhydryl group of albumin. This derivative has the advantage of slow hydrolysis of the Fmoc linkage to the amino group of the conjugate partner permitting slow drug release.[143] The Fmoc had been modified to a sulfo derivative (FMS; 9-hy droxymethyl-7-sulfofluorene) to improve solubility.[143] Use of this reagent provides a mechanism for the reversible PEGylation of proteins (Figure 4.18).[144] Coupling of the succinimidyl derivative (2-sulfonyl-9-fluorenylmethoxycarbonyl-*N*-hydroxysuccinimide) to the amino groups of a protein has the potential of forming a prodrug, which is activated upon hydrolysis.[143] Thibaudeau and coworkers[142] coupled insulin to the sulfhydryl group of albumin using a maleimide–succinimide or nitrophenyl cross-linkers (Figure 4.19). Using the differences in the p*Ka* values of the several amino groups in insulin, it was possible to obtain single modifications at GlyA1, LysB29, or PheB1. The best results were obtained with linkage at the B1 phenylalanyl residue.

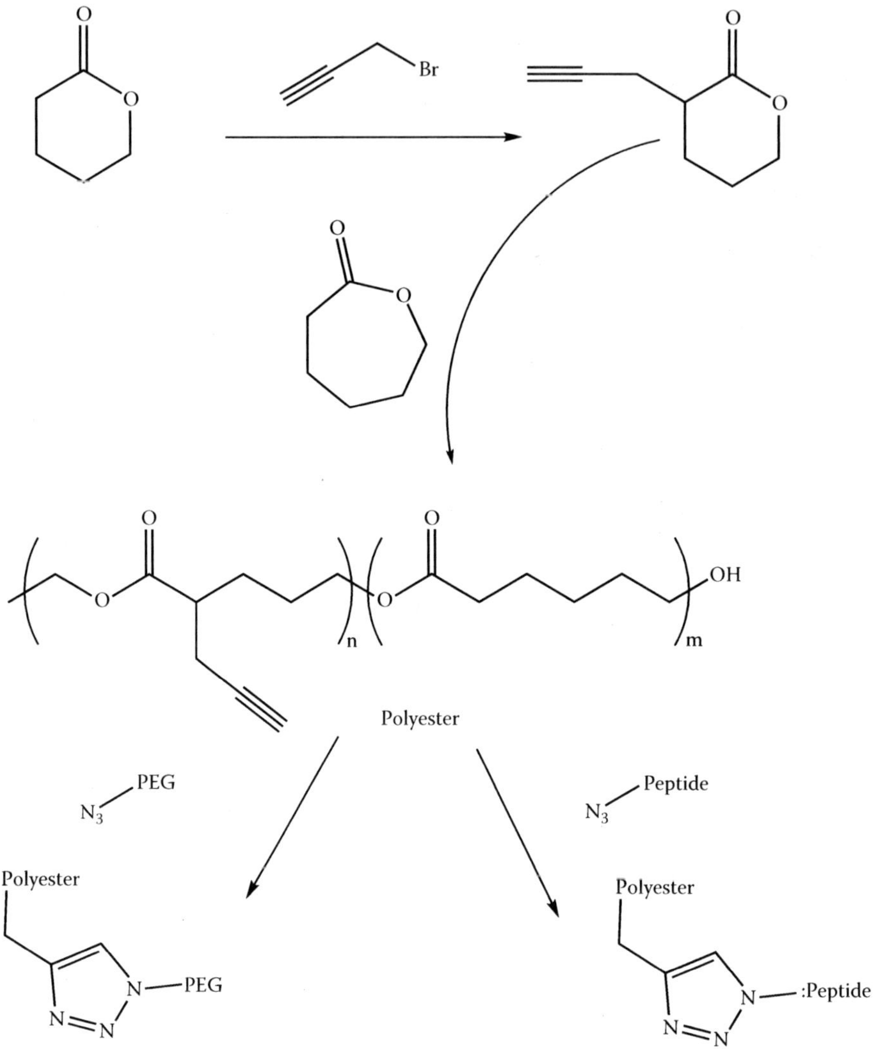

FIGURE 4.14 Click chemistry and polyester linkage. (See Parrish, B., Breitenkamp, R.B., and Emrick, T., PEG- and peptide-grafted polyesters by click chemistry, *J. Am. Chem. Soc.* 127, 7404–7410, 2005.)

Maleimide chemistry was used to prepare an albumin–carboplatinin conjugate (Figure 4.20) that functions as a prodrug.[145] Léger and colleagues[146] prepared a conjugate of albumin with atrial natriuretic peptide (ANP) using maleimide chemistry. The maleimide derivative of the ANP (Figure 4.21) was prepared during solid-phase peptide synthesis and subsequently coupled to human serum albumin. Lysine residues were inserted at three positions (C-terminal extension, Met12Lys, and Ala17Lys) to yield four possible maleimide derivatives (including the amino-terminal serine). The amino-terminal (serine) and the carboxyl-terminal extensions were the most

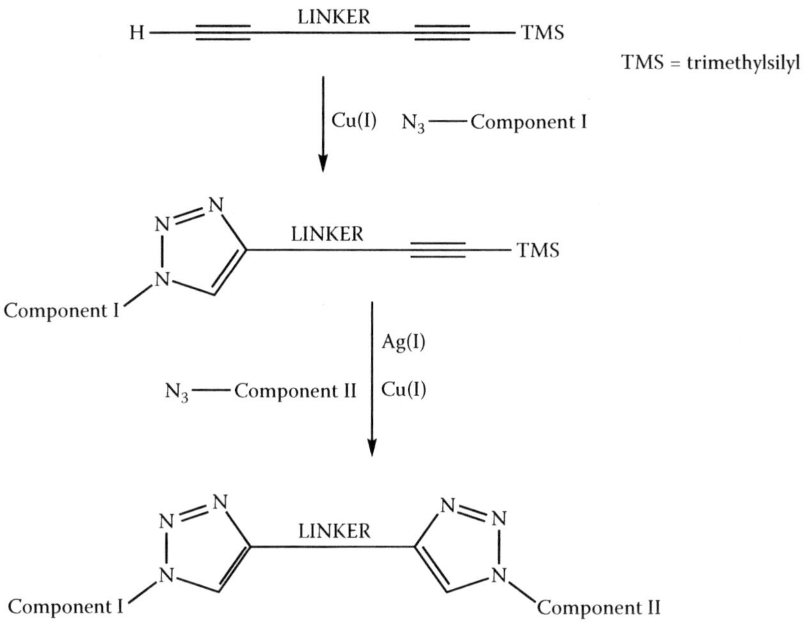

FIGURE 4.15 Formation of successive triazole linkages with click chemistry. (See Aucagne, V. and Leigh, D.A., Chemoselective formation of successive triazole linkages in one pot: "Click-Click" chemistry, *Org. Lett.* 8, 4505–4507, 2006.)

active, whereas the internal substitutions (Met12, Ala17) were much less effective; the amino-terminal conjugate had threefold less activity than the unmodified ANP and showed a marked improvement in stability compared to the unmodified ANP in plasma. Pozsgay and coworkers[147] used the maleimide function in a Diels–Alder reaction to conjugate carbohydrate to albumin (Figure 4.22). In these studies, 38 of the 54 amino groups of albumin were modified. This approach is not unique to albumin and can be applied to other proteins[148] for a variety of purposes, including surface immobilization (Figure 4.23)[149] and the preparation of protein–nucleic acid conjugates (Figure 4.24).[150]

ANTIBODY–PROTEIN CONJUGATES

Bioconjugates based on antibodies or antibody fragments represent the most extensive use of this technology.[121–126] Antibodies are used to target "partners" to sites; antibodies provide the selectivity of binding, whereas the partner, such as radioisotope, therapeutic, or cytotoxic agent, provide the "signal." The bioconjugates are referred to as *immunoconjugates*,[151–156] and when combined with a cytotoxic agent, the term *immunotoxin* may be used.[160–162] There are limited studies on bioconjugates composed of an antibody or antibody fragment and another protein. Gelatin nanoparticles, which had been functionalized by avidin by coupling via 2-iminothiolane[163] and sulfo-maleimidobenzoyl sulfosuccinimide (sulfo-MBS; Figure 4.25),[164] were coupled with biotinylated anti-CD3 antibodies for use in drug uptake in

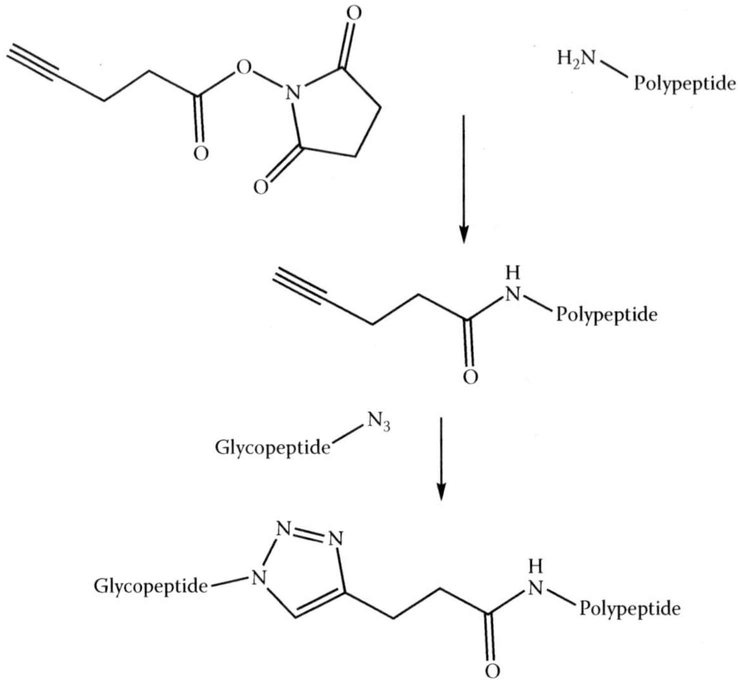

FIGURE 4.16 Click chemistry and carbohydrate linkage. (See Wan, Q., Chan, J., Chen, G., and Danishefsky, S.J., A potentially valuable advance in the synthesis of carbohydrate-based anticancer vaccines through extended cycloaddition chemistry, *J. Org. Chem.* 71, 8244–8249, 2006.)

lymphocytes.[165] This technical approach was used to prepare a dendrimer–monoclonal antibody immunoconjugate for use as a delivery system for neutron capture therapy.[166] The use of dendrimers has an amplifying effect.[167–170] Sulfomaleimidobenzoyl sulfosuccinimide has also been used to couple therapeutic monoclonal antibodies to thiolated poly(lactic acid) nanoparticles.[171] *N*-succinimidyl 3-(2-pyridyldithio) propionate has been used for the preparation of a variety of antibody conjugates[172–176] and for the preparation of a cholera toxin–insulin conjugate.[177]

Antibodies and receptors have similar binding characteristics, and bioconjugates have been prepared using the ligand (e.g., peptide growth factor)[178] as the specificity component.[179–184] Basic fibroblast growth factor (basic FGF) was conjugated with saporin, a ribosome-inactivating protein.[179,182,183] Prior work on the development of conjugates of other ribosome-inactivating protein with monoclonal antibodies demonstrated that the function of protoxin was dependent on a cleavage disulfide bond Lambert et al.[186] In these experiments, thiol groups were introduced into the ribosomal inactivating protein with 2-iminothiolane. Linkage of the modified ribosomal inactivating protein with the antibody via a maleimide linkage resulted in a 70% loss of activity, whereas linkage via a dithiopyridyl linkage provided a derivative where activity could be restored by reduction (thiolytic cleavage). Lappi and colleagues[179] used *N*-succinimidyl-3(2-pyridyldithio)propionate (Figure 4.26) to modify

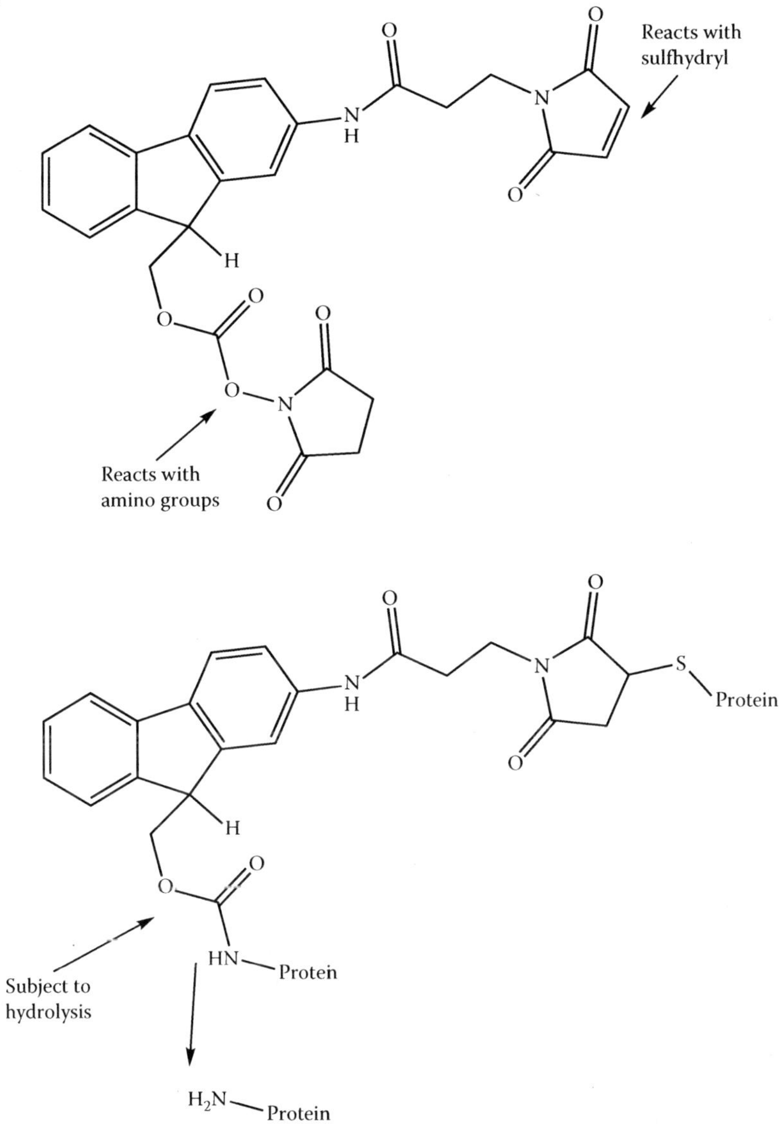

FIGURE 4.17 Coupling of insulin and albumin using Fmoc chemistry. (See Schechter, Y., Mirochik, M., Rubinraut, S. et al., Albumin-insulin conjugate releasing insulin slowly under physiological conditions: A new concept for long-acting insulin, *Bioconjug. Chem.* 16, 913–920, 2005; Schechter, Y., Mironchek, M., and Saul, A., New technologies to prolong life-time of peptide and protein drugs *in vivo*, *Int. J. Peptide Res. Therapeut.* 13, 105–117, 2007.)

the saporin and then used disulfide exchange to react with free sulfhydryl groups on the FGF protein.[187,188] The presence of several free sulfhydryl groups on the FGF surface resulted in a heterogeneous product; a homogeneous product was obtained by replacing one of the two reactive cysteine residues with a serine.[182,184] More recent

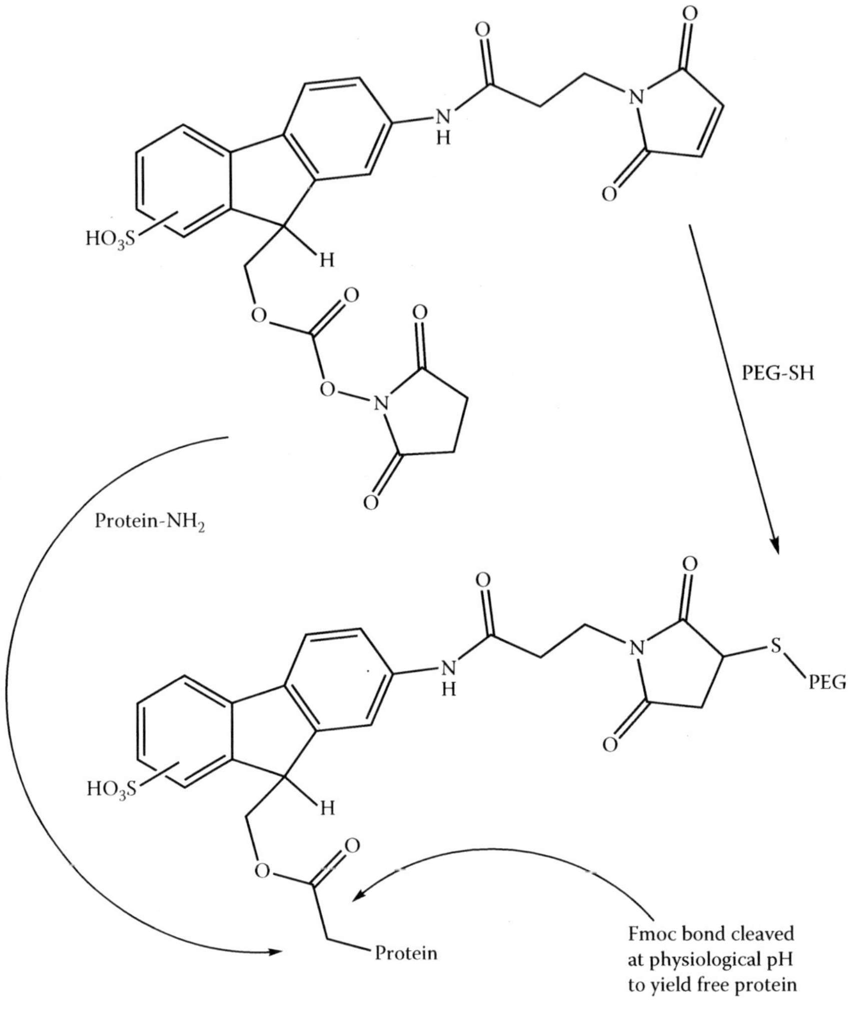

FIGURE 4.18 Sulfo derivatives of the Fmoc protecting group. (See Tsubery, H., Mironchik, M., Fridkin, M., and Schechter, Y., Prolonging the action of protein and peptide drugs by a novel approach of reversible polyethylene glycol modification, *J. Biol. Chem.* 279, 38118–38124, 2004.)

approaches have used genetic engineering to prepare fusion proteins.[189–192] Peptide growth factors can also be used to form bioconjugates that function as radiopharmaceuticals.[193–203] Some studies directly radiolabel the protein with an isotope, such as the reaction of chloramine-T and iodine isotopes,[194,199] whereas others link a chelating group to the protein (Figure 4.27). This example is *S*-acetylmercaptoacetyltriglycine-*N*-hydroxysuccinimide ester (NHS-MAG3), which was developed for the radiolabeling of nucleic acids[204] and later for peptides and proteins.[205] This compound has seen continued use for the labeling of both peptides[206–211] and nucleic acids.[212–215]

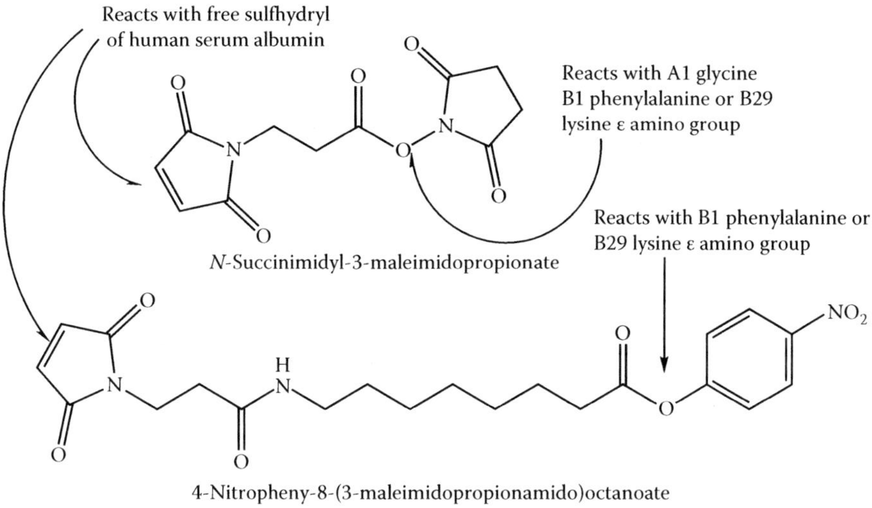

FIGURE 4.19 Coupling of insulin to the sulfhydryl group of albumin using a maleimide-succinimide or nitrophenyl cross-linkers. (See Thibaudeau, K., Leger, R., Huang, X. et al., Synthesis and evaluation of insulin-human serum albumin conjugates, *Bioconjug. Chem.* 16, 1000–1008, 2005.)

DIRECT LABELING OF ANTIBODIES WITH RADIOISOTOPES

As with peptides, antibodies can be directly labeled with radioisotopes[216–238] such as ^{131}I or ^{99}Tc(m), or a metal-binding function such as the chelates described earlier is added to carry the metal ion. The presence of sulfhydryl groups assists with technetium binding,[238–243] which is obtained by limited reduction of the IgG. Alternatively, a suitable sulfhydryl-containing function may be added by chemical modification (Figure 4.28)[244,245] for use in chelating metal ions such as technetium.

ANTIBODY–DRUG

The selectivity of antibodies has been used to target drug delivery.[246–250] It is not our intent to discuss this area in great detail, but we wish to focus on the coupling chemistry. Drugs bound to antibodies should have the potential to be released after reaching the target area.[251] In some cases, antibody–drug conjugates, which are stable at pH 7, are taken into the cell via endocytosis and the drug is released in the lysosome.[252–255] The disulfide bond can also be used to create a conjugate that, although stable in the circulation, should be labile in more reducing intracellular environment,[58,59] but there may be complications because of the oxidizing environment of lysosomes.[256]

Braslawsky and coworkers[257] reported that hydrazone conjugates (Figure 4.29) of adriamycin (doxorubicin) with antibody required internationalization and intracellular hydrolysis for antitumor activity. Later, Froesch and coworkers[258] reported on the synthesis of an acid-labile monoclonal antibody conjugate with

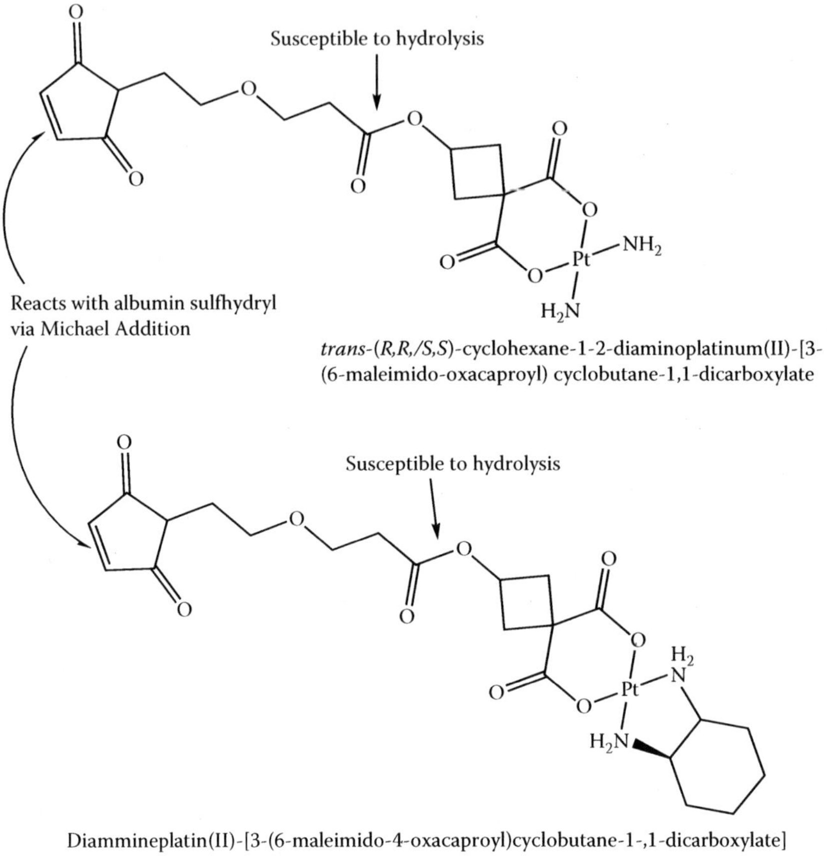

FIGURE 4.20 Maleimide chemistry was used to prepare an albumin-carboplatinin conjugate which functions as a prodrug. (See Warnecke, A., Fichtner, I., Garmann, D. et al., Synthesis and biological activity of water-soluble maleimide derivatives of the anticancer drug carboplatin designed as albumin-binding prodrugs, *Bioconjug. Chem.* 15, 1349–1359, 2004.)

doxorubicin (Figure 4.29). The hydrazone linker is stable at pH 7.0 but labile at pH 5.0. These are two examples of the use of an acid-labile bond for the intracellular release of a drug after targeted delivery. The Braslawsky chemistry[257] is in current use.[259] Disulfide linkages are also used in the preparation of antibody–drug conjugates[260–264] but are not as common as the other labile linkages. The issue of the oxidizing environment of the lysosome has been discussed as a complicating factor in the use of such reagents.[256] Other labile linkers are peptide and ester linkages, which are susceptible to enzymatic and nonenzymatic hydrolysis[265–270] (Figure 4.30). It is noted that one of the examples cited[269] concerns the conjugate of 3′-azido-3′-deoxythymidine (AZT) with histone for delivery to brain tissue; the stability data in this study is useful for antibody–drug conjugates as well. The

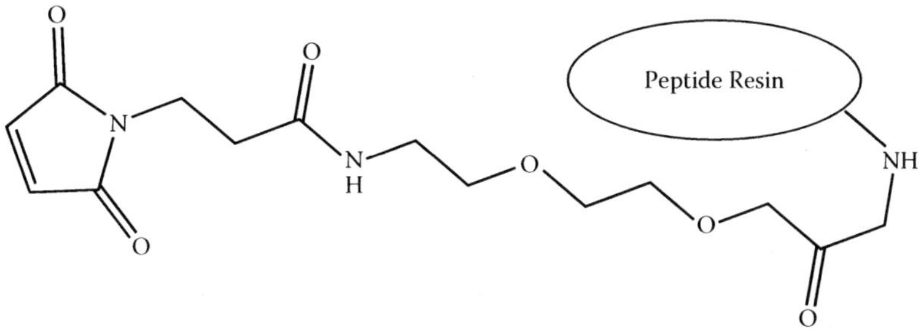

FIGURE 4.21 The maleimide derivative of the atrial natriuretic peptide prepared during solid phase peptide synthesis and subsequently coupled to albumin. (See Leger, R., Rotitaille, M., Quraishi, O. et al., Synthesis and *in vitro* analysis of atrial natriuretic peptide-albumin conjugates, *Bioorg. Med. Chem. Lett.* 13, 3571–3575, 2003.)

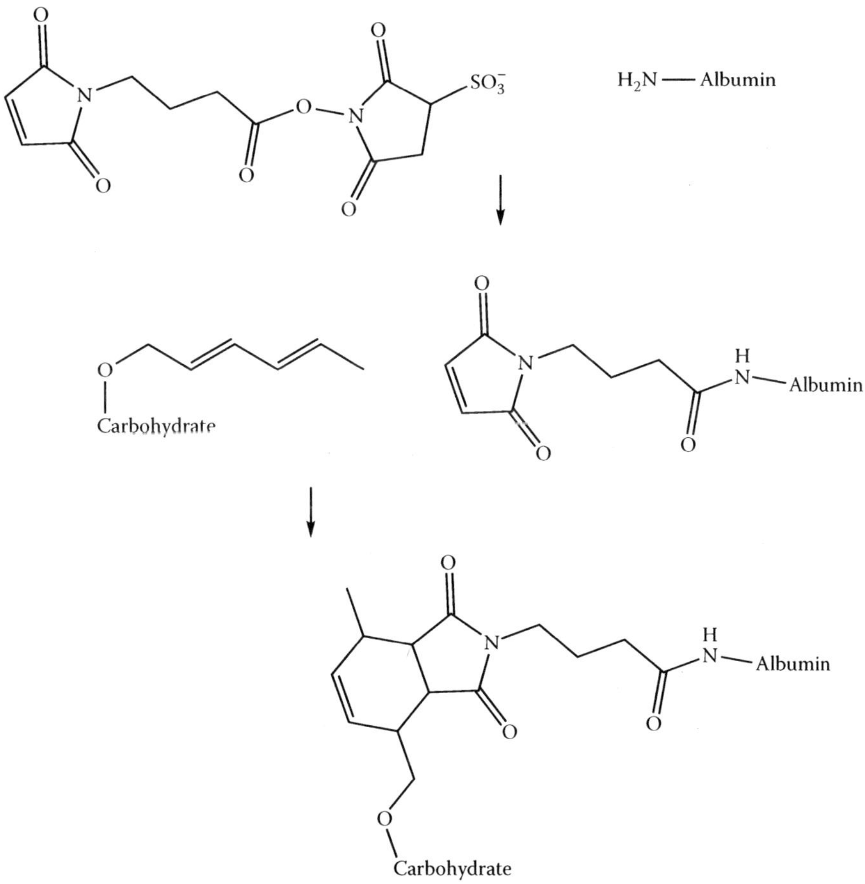

FIGURE 4.22 Diels–Alder reaction to conjugate carbohydrate to albumin. (See Pozsgay, V., Vieria, N.E., and Yergey, A., A method for bioconjugation of carbohydrates using Diels-Alder cycloaddition, *Org. Lett.* 4, 3191–3194, 2002.)

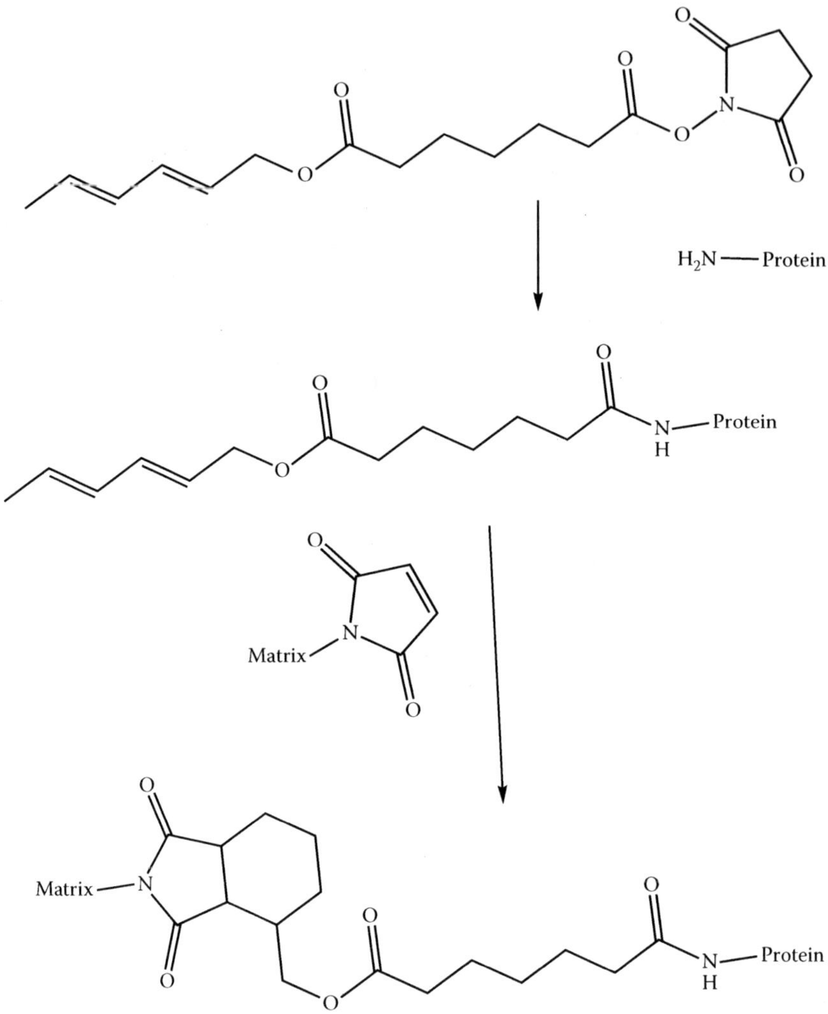

FIGURE 4.23 Use of the maleimide function in the Diels–Alder reaction for coupling protein to a matrix. (See de Araujo, A.D., Paloma, J.M., Cramer, J. et al., Diels–Alder ligation and surface immobilization of proteins, *Angew. Chem. Int. Ed.* 45, 296–301. 2006.)

reader is directed to an excellent recent review by Wolfenden[270a] on the rates of nonenzymatic hydrolysis of peptide bonds.

Quadri and Vriesendorf[271] study the *in vivo* distribution of radioimmunoconjugates as a function of linker structure (stable versus labile; Figure 4.31). The work on antibody–drug conjugates has been of great scientific interest, but there has been limited success in developing approved drug products based on this technology.[272]

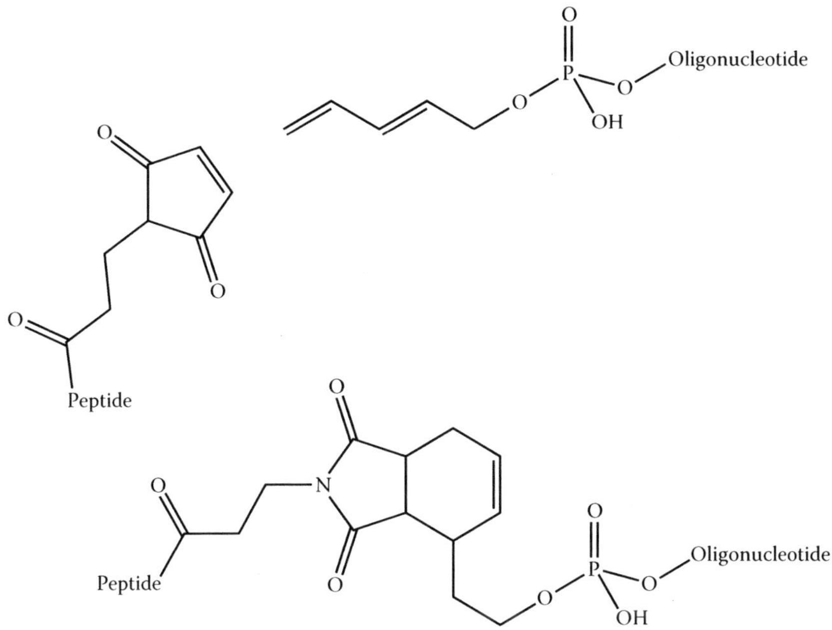

FIGURE 4.24 Use of the maleimide function in the Diels–Alder reaction to couple peptide/protein to nucleic acids. (See Marchan, V., Ortega, S., Pulido, D. et al., Diels-Alder cycloadditions in water for the straightforward preparation of peptide-oligonucleotide conjugates, *Nucl. Acids Res.* 34, e24, 2006.)

ANTIBODY–RADIOLABEL

Although antibodies can be directly labeled with radioisotopes as described earlier, a more useful approach is to attach a metal-chelating function to the antibody, such as ethylene dicysteine (Figure 4.32),[273,274] nitrilotriacetic acid (NTA),[275,276] diethylenetriaminepentaacetic acid (DTPA),[276–279] and 1,4,7,10-tetraazacyclododecane-*N,N',N'',N'''*-tetraacetic acid (DOTA; Figure 4.33).[280–284] A humoral response to DOTA has been observed,[285] but is not considered significant[286] and is possibly related to the protein rather than to the hapten.[287,288] Possible immune responses have not precluded continued development of interesting DOTA–antibody conjugates.[289]

PROTEIN–CARBOHYDRATE CONJUGATES

There have been a variety of protein–polysaccharide complexes (Table 4.1). Many of the conjugates have been formed via periodate oxidation of the carbohydrate[290–299] following reaction with an amino group on the protein to form a Schiff base, which is then stabilized by reduction with, for example, sodium cyanoborohydride (Figure 4.34). The carbohydrate located on the Fc domain of antibodies can be oxidized to the aldehyde[300–305] and coupled to an amine or hydrazide.[306] Periodate can also oxidize protein-bound *N*-hydroxylysine[307,308] and amino-terminal serine or threonine to yield aldehydes,[309,310] which can be used to link with a hydrazide to form a

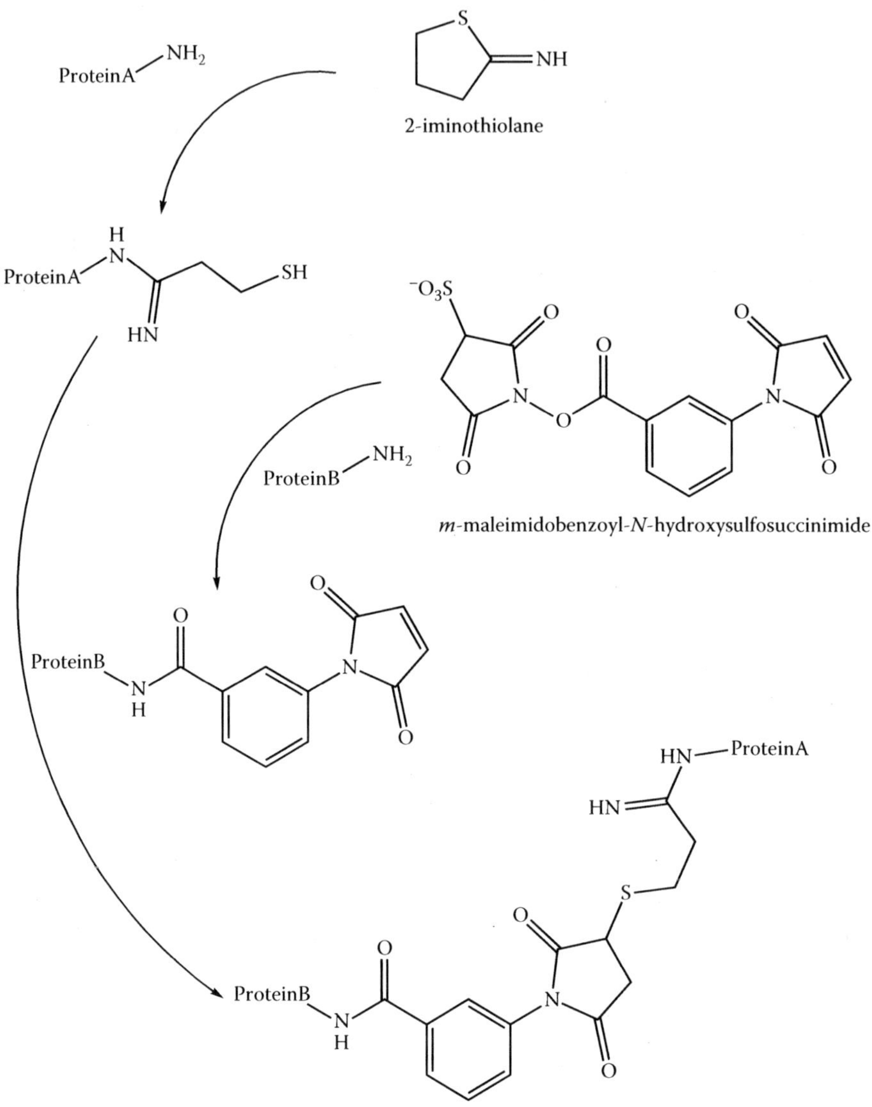

FIGURE 4.25 Use of 2-iminothiolane and maleimide for conjugate formation. (See Jue, R. Lambert, J.M., Pierce, L.R., and Traut, R.R., Addition of sulfhydryl groups to *Escherichia coli* ribosomes by protein modification with 2-iminothiolane (methyl-4-mercaptobutyrimidate, *Biochemistry* 17, 5399–5406, 1978; Barth, R.F., Adams, D.M., Soloway, A.H. et al., Boronated starburst dendrimer-monoclonal antibody immunoconjugates: Evaluation as a potential delivery system for neutron capture therapy, *Bioconjug. Chem.* 5, 58–66, 1994.)

hydrazone (Figure 4.35).[311] The modification of N-terminal serine or threonine by periodate oxidation has seen considerable application for selective modification of proteins and peptides.[312–318]

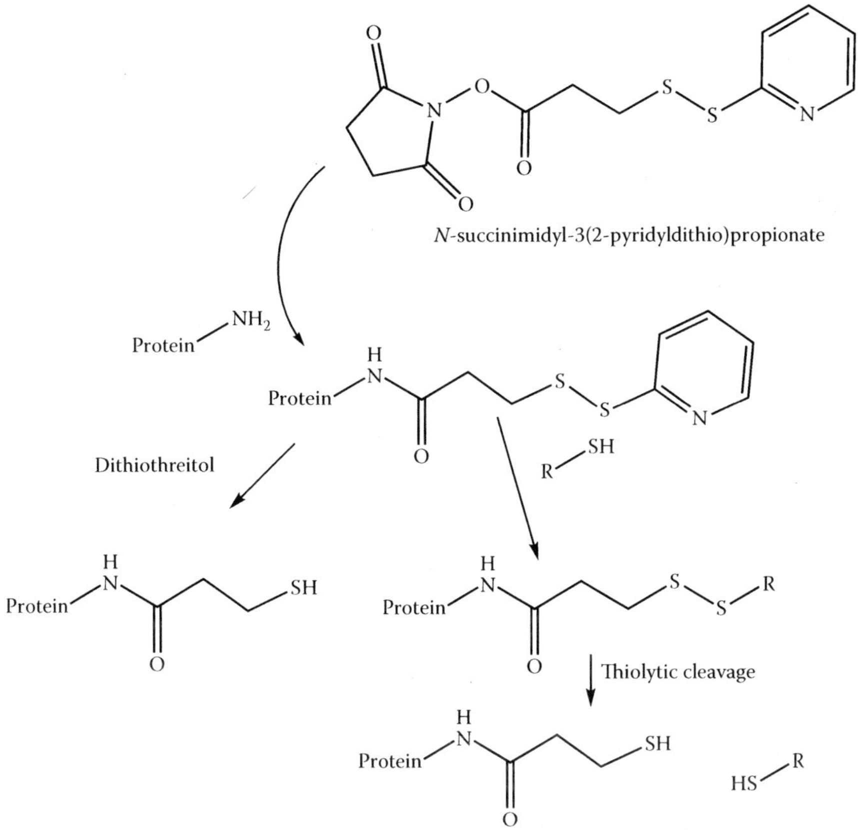

FIGURE 4.26 *N* succinimidyl-3(2-pyridyldithio)propionate. (See Carlsson, J., Drevin, H., and Axen, R., Protein thiolation and reversible protein-protein conjugation *N*-succinimidyl 3-(2-pyridyldithio)propionate, a new heterobifunctional reagent, *J. Biochem.* 173, 723–737, 1978.)

There has been recent interest in the chemical addition of colominic acid (polysialic acid) to proteins to improve pharmacokinetic properties.[319–322] Fernandes and Gregoriadis[319] modified asparaginase (*Erwinia carotovora*) with colominic acid (average molecular weight, 10 kDa) using periodic acid oxidation followed by sodium cyanoborohydride reduction. The modified enzyme retained approximately 85% activity with no significant change in K_m. The modified enzyme was more stable than the native enzyme in plasma, suggesting increased resistance to proteolytic degradation; the modified enzyme had a longer circulatory half-life in a mouse model. Subsequent studies[322] demonstrated reduced antigenicity of the modified protein. Polysialic acid is a cell surface glycan having an important role in nervous system function; it interacts with neural cell adhesion molecule (NCAM)[323,324] and is considered critical for plasticity, which in turn is critical for function.

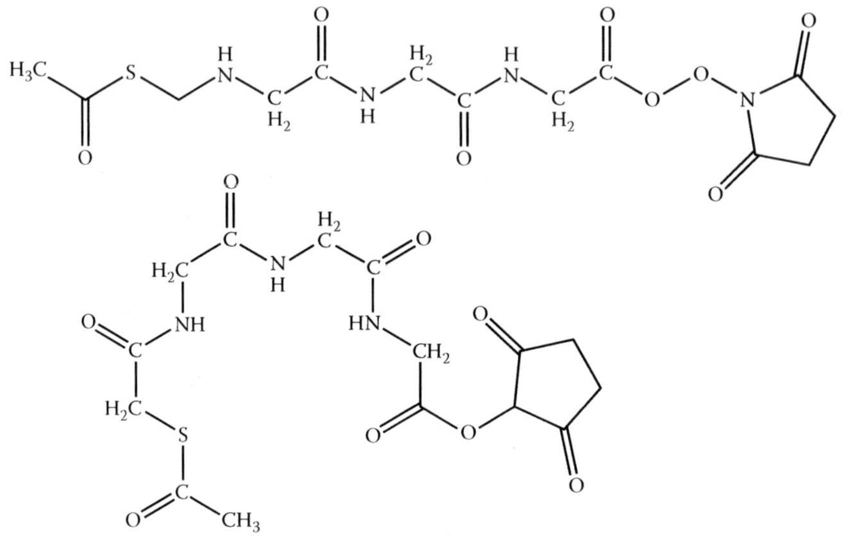

S-acetylmercaptoacetyltriglyine, *N*-hydroxysuccinmide ester (S-acetyl-NHS-MAG_3)

FIGURE 4.27 A chelating agent *S*-acetylmercaptoacetyltriglycine-*N*-hydroxysuccinimide ester (NHS-MAG3). (See Winnard, P., Jr., Chang, F., Ruschkowski, G. et al., Preparation and use of NHS-MAG_3 for technetium-99m of DNA, *Nuclear Med. Biol.* 24, 425–432, 1997; Wang, Y., Liu, X.R., and Hnatowich, D.J., An improved synthesis of NHS-MAG_3 for conjugation and radiolabeling of biomolecules with Tc-99m at room temperature, *Nat. Protoc.* 2, 972–978, 2007.)

POLYETHYLENE GLYCOL

The use of poly(ethylene glycol) [PEG] merits a separate section, reflecting its wide use in the preparation of a variety of bioconjugates. The chemistry is straightforward and has been reviewed elsewhere. Thus, consideration here is limited to some general comments and to some recent developments of interest.

Modification with PEG is the most popular approach for the chemical modification of biopharmaceuticals to improve efficacy. Abuchowski and colleagues introduced modification with PEG in 1977,[325] and there are a number of excellent reviews.[326,327] Successful modification of therapeutic proteins and peptides with PEG is associated with an extension of circulatory half-life and reduced or eliminated immunogenicity. It is thought that these properties arise from the physical blocking of the therapeutic from immunological surveillance and catabolic recognition.[328,329] The current concept is that the covalently attached chains of PEG are mobile and shield the surface of the biopharmaceutical. It can be argued that PEGylation is similar to glycosylation in "covering" antigenic sites.[330–334] On the other hand, carbohydrate moieties can be a critical determinant in the antigenicity of glycoproteins.[355–338]

PEGylation has been successful for the modification of enzymes that work on small substrates such as asparaginase or adenosine deaminase.[339–341] The effect is similar to the early observations on insoluble enzymes.[342] PEGylation has also been

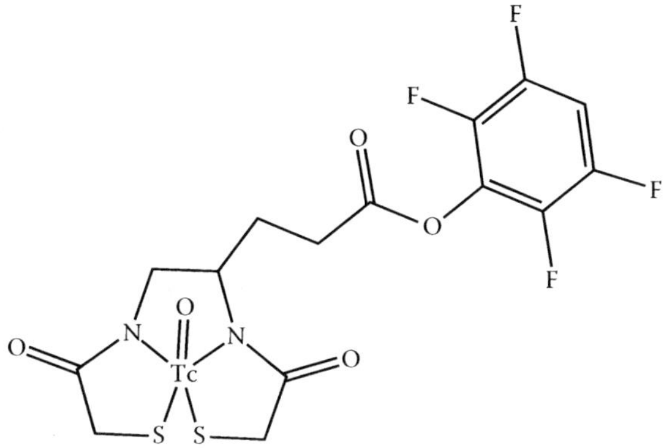

^{99m}Tc-4,5-bis(thioacetamido)pentanoate tetrafluorophenyl ester

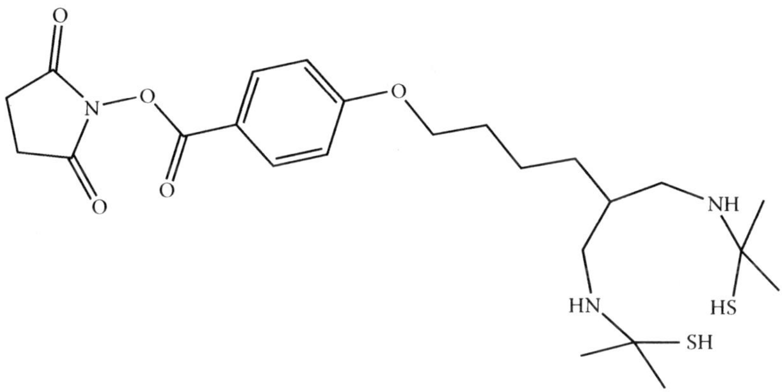

FIGURE 4.28 The chemical modification of a protein to add a chelating sulfhydryl function. (See Fritzberg, A.R., Abrams, P.G., Beaumier, P.L. et al., Specific and stable labeling antibodies with Technetium-99m with a diamide dithiolate chelating agent, *Proc. Natl. Acad. Sci. USA* 85, 4025–4029, 1988; Eisenhut, M., Miszfeldt, M., Lehmann, W.D., and Karas, M., Synthesis of a bis(aminoethanethiol) ligand with an activated ester for protein conjugation and ^{99m}Tc labeling, *J. Labelled Compounds Radiophamaceuticals* 29, 1281–1291, 1991.)

successful in the modification of relatively small protein therapeutics in which site-specific modification, usually monosubstitution, has been possible.[343–349] Where multiple potential sites of modification are possible, the site of modification can influence the extent of modification of biological properties.[346] Specific modification can be accomplished by engineering a potent nucleophile such as cysteine into the therapeutic protein,[343] by removing lysine residues permitting specific modification at the N-terminal residue.[348] Performing the modification at a lower pH will drive specificity toward modification of the primary amino group at the N-terminal.[349]

Concomitant change in the solution structure of proteins secondary to chemical modification is always a possibility.[349] There have been some studies on the effect of PEGylation on protein structure. In an elegant study, Dhalluin and coworkers[350] characterized several positional isomers of interferon-α modified at lysine residues.

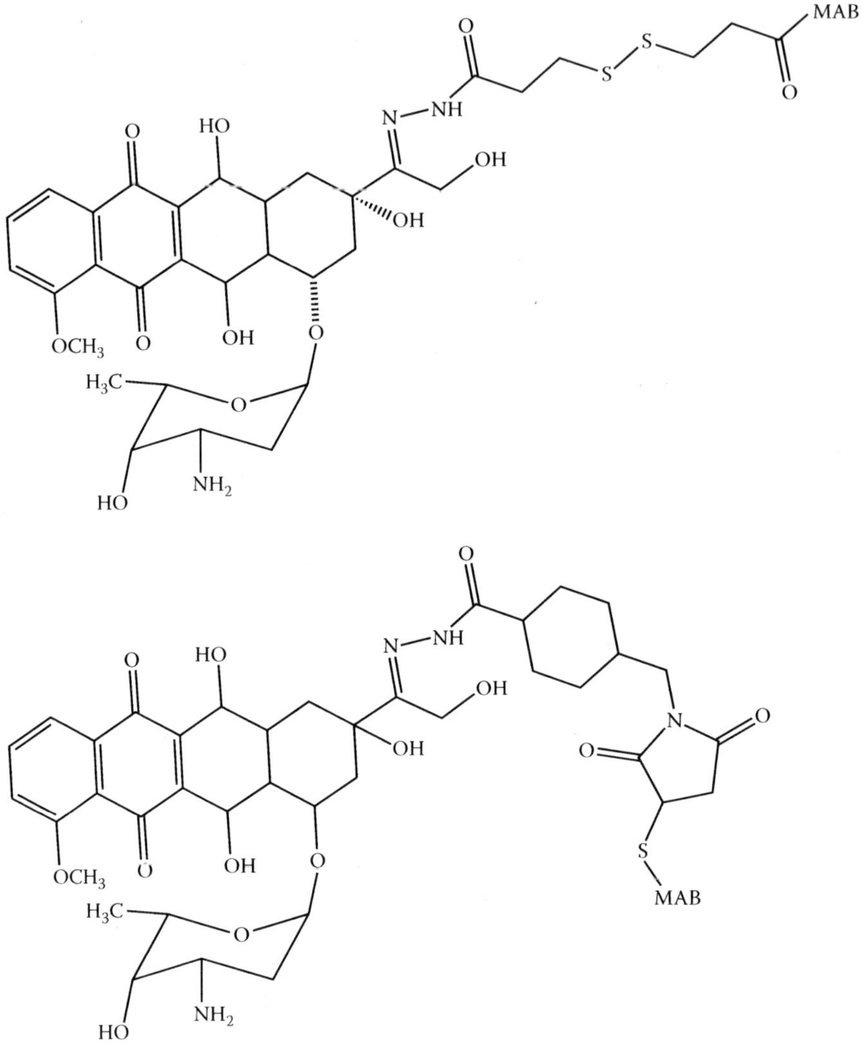

FIGURE 4.29 Acid labile linkers for doxorubicin and adriamycin (See Braslawksy, G.R., Kadow, K., Knipe, J. et al., Adriamycin (hydrazone)-antibody conjugates require internalization and intracellular acid hydrolysis for antitumor activity, *Cancer Immunol. Immunother.* 33, 367–374, 1991; Froesch, B.A., Stahel, R.A., and Zangemeister-Witte, U., Preparation and functional evaluation of new doxorubicin immunoconjugates containing an acid-sensitive linker on small-cell cancer cells, *Cancer Immunol. Immunother.* 42, 55–63 1996.)

Modification did not change structure as judged by a variety of physical measurements, including ultracentrifugation, CD, fluorescence, and differential scanning calorimetry. These investigators observed that the PEG moiety adopted a flexible mobile conformation, producing a shield without a permanent presence over any specific surface area.

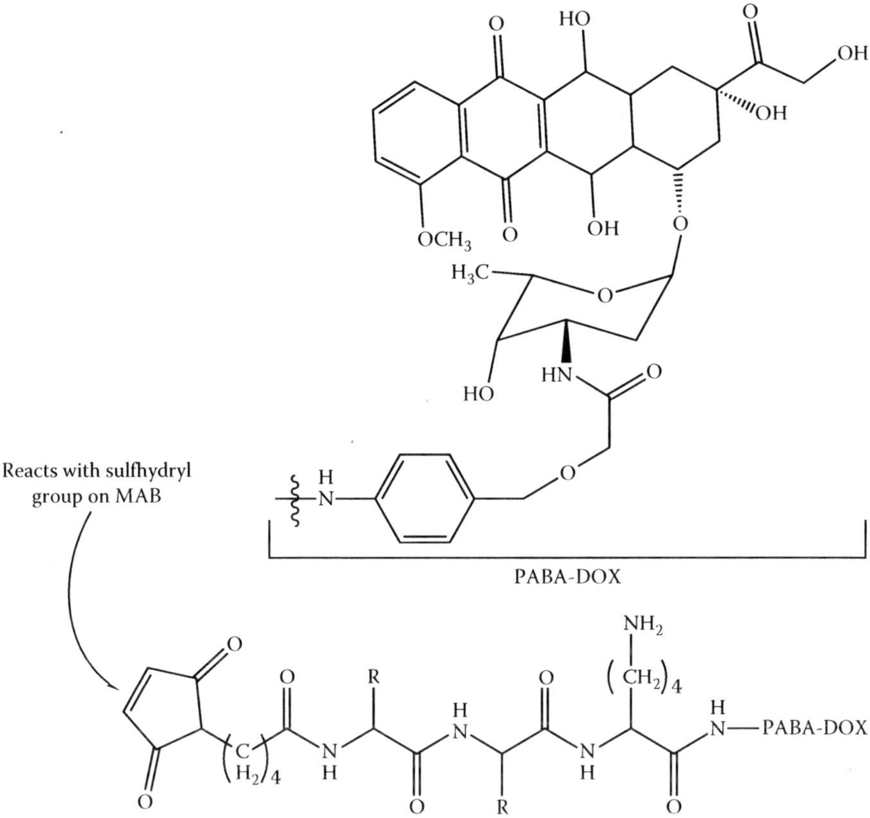

FIGURE 4.30 An example of a peptide linker for drug delivery. (See King, D.H., Dubowchik, G.M., Mastalerz, H. et al., Monoclonal antibody conjugates of doxorubicin prepared with branched peptide linkers: Inhibition of aggregation by methoxytriethyleneglycol chain, *J. Med. Chem.* 45, 4336–4343, 2002.)

Finn and coworkers[351] modified cowpea mosaic virus with PEG (modification accomplished via use of a *N*-succinimidyl derivative that reacts with protein amino groups). Modification of the virus surface proteins with PEG blocked subsequent reaction with an antibody directed against stilbene moieties conjugated to surface proteins. These experiments are part of a larger study examining the reaction of a blue fluorescent antibody[352] with the modified virus, which differentiated surface stilbene moieties from stilbene moieties in the interior. The blue color is derived from the formation of an excited state (an "exiplex") between stilbene and the antibody. These experiments demonstrated that PEGylation could block the physical association of an antibody with antigen.

PEGylation will physically block macromolecular interaction with less of an effect on interaction with smaller molecules. As noted, therapeutic success depends on the balance between basic activity loss and circulatory half-life increase; there are no examples in which PEGylation increases intrinsic biological activity. Although there

Stable and Labile Linkers

Labile Linkages

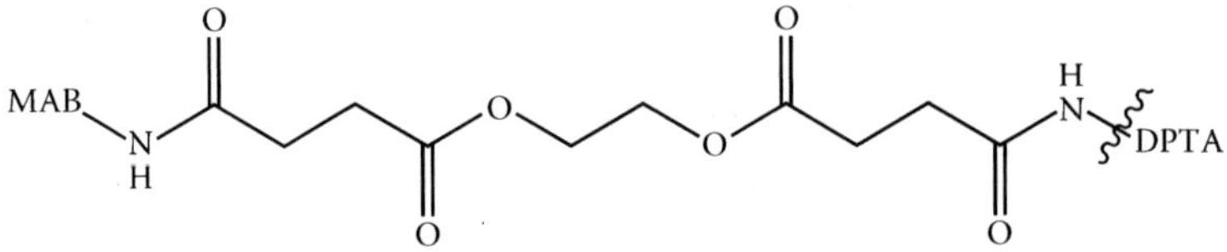

Diester linkage [ethyleneglycol-bis(succinimidylsuccinate)(EGS)

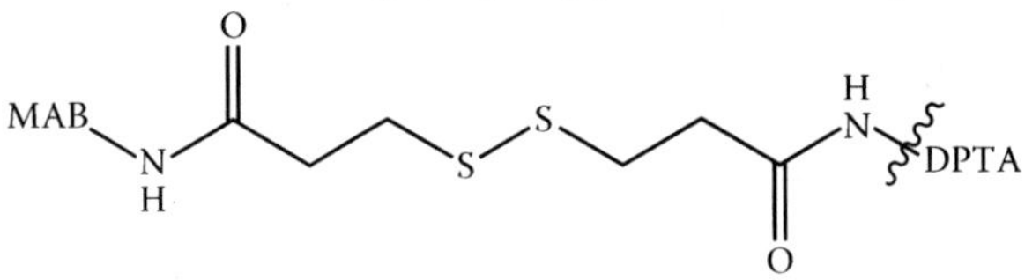

Disulfide linkage[dithiobis propionate (DSP)]

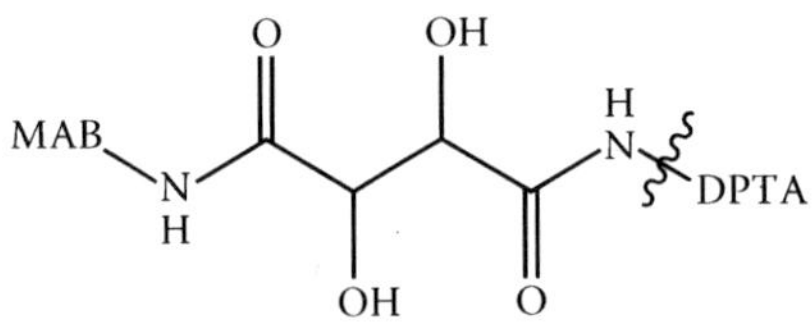

Tartaramid (DST)

Linker	Tumor/Blood 6 d
EGS	75.3
DSP	37.5
DST	14.8
DSS	3.9
ICTH	4.6
BSOCOES	6.8

Stable Linkages

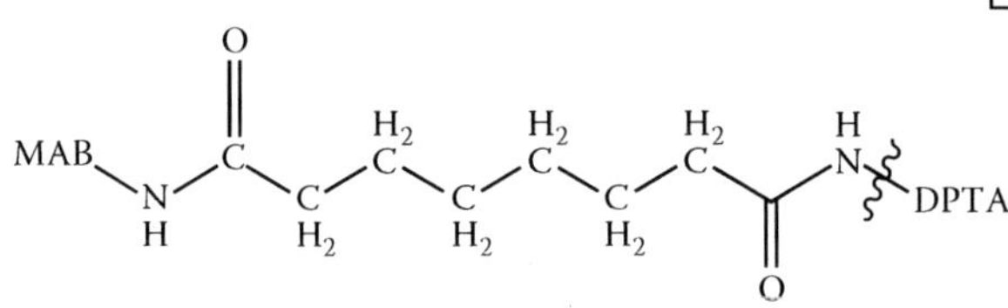

Hydrocarbon(suberate)(DSS)

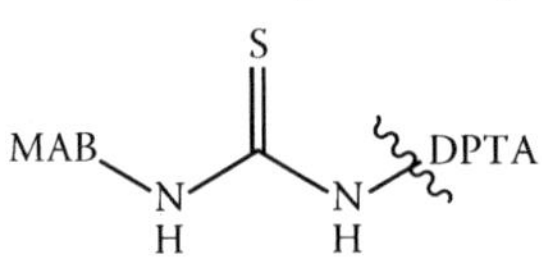

Thiourea (ICTH)

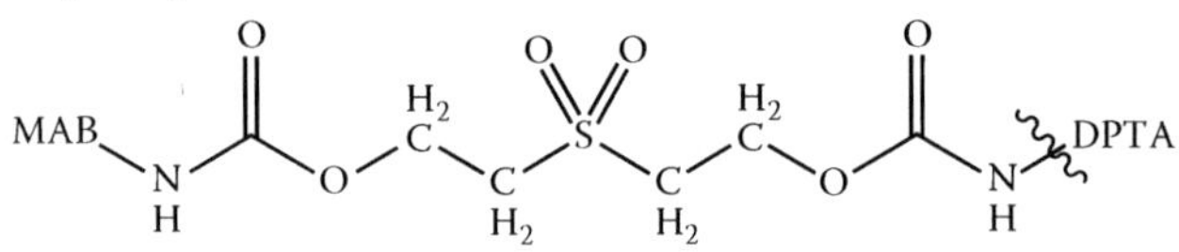

Sulfone (BSOCOES)

FIGURE 4.31 Stable and labile linker for antibody–drug conjugates. Shown are three labile linkers and three stable linkers. Also shown are the tumor-to-normal tissue distribution of radioactivity on day 6. The specific example is blood, but similar distribution was obtained for normal liver and kidney tissue. (Taken from Quadri, S.M. and Vriesendorf, H.M., Effects of linker chemistry on the pharmacokinetics of radioimmunoconjugates, *Q. J. Nucl. Med.* 42, 250–261, 1998.)

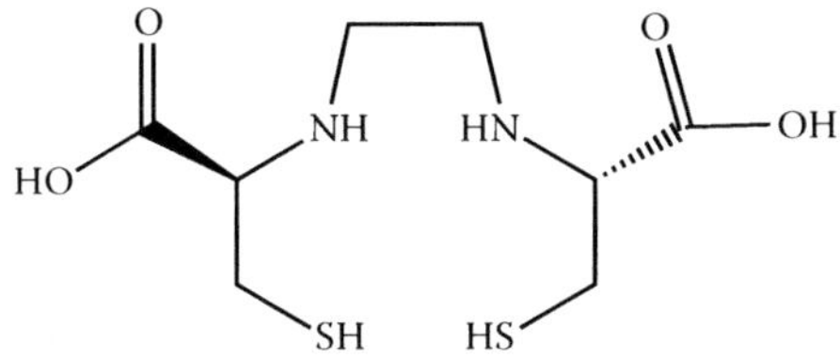

Ethylene Dicsyteine

FIGURE 4.32 Ethylene dicysteine for metal chelation.

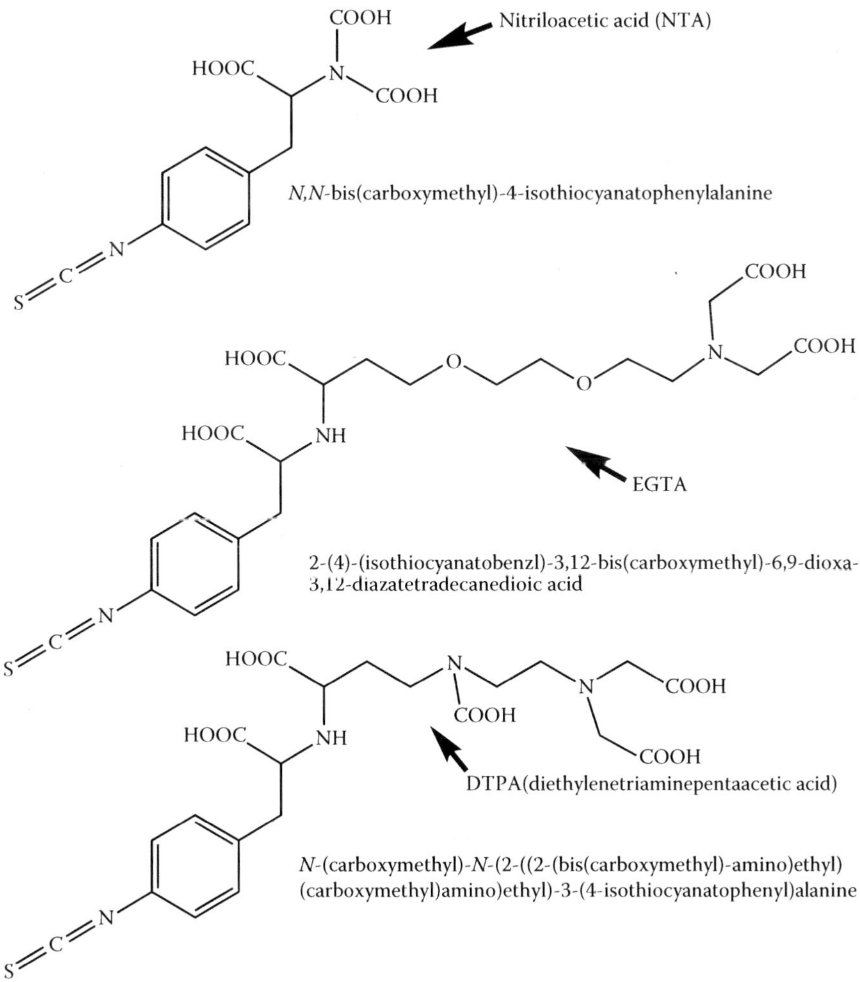

FIGURE 4.33 Some metal chelating agents.

TABLE 4.1
Some Polysaccharide Conjugates

Conjugate	Chemistry	References
Dextran–albumin	60°C/65% relative humidity/3 weeks; product more homogeneous than that obtained with CNBr coupling	1
Dextran–albumin and dextran–lysozyme	Low-angle light-scattering studies and high-power liquid chromatography (HPLC) gel filtration for molecular weight determination of conjugates formed by heating	2
Dextran–albumin	Use of 1-cyano-4-dimethylaminopryidinium tetrafluoroborate for activation of dextran	3
Dextran–protein (several proteins studied)	Use of dextran dialdehyde (periodate oxidation) or benzenetetracarboxylate-modified dextran (reaction with benzenetetracarboxylate anhydride)—Modification of factor IX with either derivative resulted in low activity owing to the modification of amino groups	4
Dextran–albumin	Heating in at 60°C/79% relative humidity/7 days (Maillard reaction)	5
Glycoconjugate vaccine (*Haemophilus influenzae*), type b	Periodate oxidation[a] /$NaCNBrH_3$	6
Glycoconjugate vaccine (*Haemophilus influenzae*), type b	Stability studies on conjugate vaccine	7
Carbohydrate vaccines	NMR spectroscopy; characterization of activated intermediates (periodate oxidation product) during carbohydrate–vaccine manufacturing	8
Oligosaccharide based bacterial vaccines	Review of various strategies for coupling of carbohydrate to protein for vaccine preparation	9
Heparin–superoxide Dismutase	Periodate oxidation	10
Hydrophilic polyols	Aminoxy cross-linkers	11
Trypsin–alginate	Noncovalent immobilization in MES buffer	12
Chitosan–peptide	Peptide mimetic based on mussel adhesive protein; noncovalent bonding to chitosan	13
Immuoconjugates	Coupling of oxidized carbohydrate (periodate) with protein using *S*-(2-thiopyridyl)-L-cysteine hydrazide	14
Laccase–chitosan	Coupling mediated by carbodiimide	15
Carbohydrate–protein	Staudinger ligation	16
Chitosan–gelatin	Linkage using genepin yielding fluorescent product	17
Xylose–albumin	Reductive amination to form derivative to be used as immunogen for generation of antibodies to xylitol	18
Synthetic oligosaccharide–ovalbumin	Ovalbumin modified with succinimidyl -4(maleimidomethyl)-cyclohexane-1-carboxylate, which was then used to couple to sulfhydryl function on synthetic oligosaccharide	19

TABLE 4.1 (CONTINUED)
Some Polysaccharide Conjugates

Conjugate	Chemistry	References
Polyvinylsaccharide–protein (growth factor, polylysine, RGDpeptide	Polymer based on 2-deoxy-2-methacrylamido-D-glucose was modified with protein using aldehyde function on polymer	20

[a] Spiro, R.G., Periodate oxidation of the glycoprotein fetuin, *J. Biol. Chem.* 239, 567–573, 1964; Williams, D.G., Comparison of three conjugation procedures for the formation of tracers for use in enzyme immunoassays, *J. Immunol. Methods* 72, 261–268, 1984; O'Shannessy, D.J. and Quarles, R.H., Labeling of the oligosaccharide moieties of immunoglobulins, *J. Immunol. Methods* 99, 153–161, 1987.

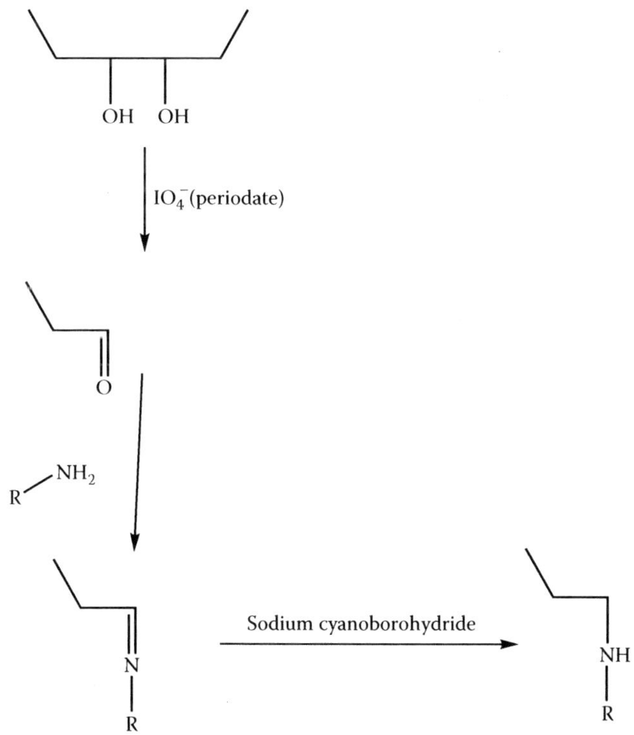

FIGURE 4.34 A mechanism for periodate oxidation. (See O'Shammessy, D.J. and Quarles, R.H., Labeling of the oligosaccharide moieties of immunoglobulins, *J. Immunol. Methods* 99, 153–161, 1987.)

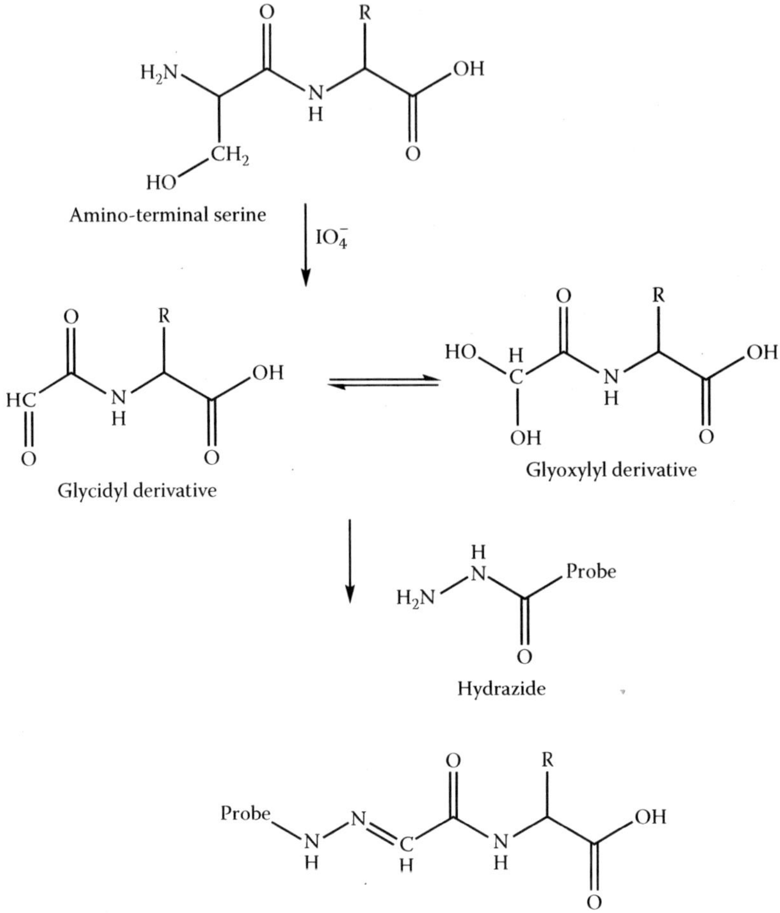

FIGURE 4.35 Hydrazide coupling to aldehyde function generated from N-terminal serine or threonine. (See Geoghegan, K.F. and Strob, J.G., Site-directed conjugation of nonpeptide groups to peptides and proteins via periodate oxidation of 2-amino alcohol. Application to modification at N-terminal serine, *Bioconjug. Chem.* 3, 138–146, 1992.)

are no specific examples, prolongation of circulatory half-life may be counterproductive. Although prolongation of the circulatory half-life of a single-chain Fv (scFv) antibody derivative[353] might be useful for therapy based on receptor occupancy,[354] it might not be useful for the delivery of radioisotope[355] when rapid clearance of unbound radioisotope is desirable. On the other hand, Shively and coworkers[356] showed the PEGylation (PEG 3400) of anti-CEA diabody resulted in an increase in Stokes radius and decreased renal clearance with longer circulatory life. Modification with a smaller poly(ethylene glycol)(PEG_{12}) yielded a smaller molecule with clearance intermediate between the native diabody, the PG34000-modified material.

Modification of proteins with PEG appears to reduce the native antigenicity of proteins and block the reaction with existing antibodies in serum. There is, however, some evidence to suggest that the PEG moiety and/or coupling chemistry utilized can produce an antigenic response. Nishimura and colleagues[357] modified uricase (uric acid oxidase from *Candida utilis*) employing PEG (PEG 5, Mr ca. 5000) and cyanuric chloride chemistry. Modification of 36 amino groups (98 total) eliminated binding to a rabbit antisera against the native enzyme with retention of 45% of enzyme activity. The *bis* PEG derivatives were more effective than the single-chain derivative. Decreased clearance was observed in a murine model. Tsugi and coworkers[358] demonstrated that although PEGylated (cyanuric chloride) uricase (*Candida utilis*) did not react with antisera to native uricase, it did elicit the formation of antibody to the PEGylated enzyme. The antisera developed against PEGylated uricase also reacted with PEGylated superoxide dismutase. This strongly suggests that an antibody has developed toward the PEG moiety. There was a smaller reaction to a PEGylated protein prepared with a succinimidyl PEG derivative, suggesting that there is a potential role of coupling chemistry in the development of neoantigen on PEGylation.[359] The PEGylated enzyme obtained with succinimidyl PEG has been subjected to further preclinical characterization,[360] which demonstrated a small antigenic response to chronic administration. The neoantigenic response was larger for the 5,000 MW PEG than with the 20,000 MW PEG. A PEGylated uricase is in Phase 3 clinical trials.[361] Uricase is used to treat gout, and its highly immunogenic nature is a problem.[362] There is one report on the successful use of PEG-uricase (uricase from *Arthrobacter protoformica*) to treat hyperuricemia in a single patient with non-Hodgkin's lymphoma.[362] There are no additional reports on the clinical development of uricase, which suggests that there have been problems in the development of this biopharmaceutical as of this date. Armstrong and colleagues[363] reported that a third of patients receiving PEGylated asparaginase demonstrated an increased rate of drug clearance. This increase in the rate of clearance was associated with an antibody against PEG; approximately 25% of the normal population have an antibody against PEG.

Croyle and coworkers have presented data suggesting the formation of antibodies following the administration of PEGylated E1-deleted adenovirus vectors.[364] Finally, Cheng and coworkers described the role of IgM in the accelerated clearance of PEGylated proteins.[365]

Although the preceding studies suggest that the PEG moiety has the potential to elicit antibody formation, there is no question that PEGylation will block-react with antibodies to the native protein, such as described for adeno-associated virus.[366] As noted by these investigators, there is a delicate balance between blocking antigen reactivity and abolishing viral function. Antigenicity of PEG may be more of an issue with liposome technology[367,368] perhaps indicating an adjuvant role for the liposomes.[367] Finally, it is possible to develop monoclonal antibodies to the PEG moiety of conjugates.[369,370]

There are several recent studies describing the preparation of PEG–protein conjugates that are reversible in that cleavage of the polymer can occur with regeneration of the native protein.[11,371] Fipula and colleagues[371] developed a novel PEG conjugate that underwent stepwise hydrolytic cleavage to yield the native protein. Zalipsky and colleagues[11] used a dithiobenzyl group with a urethane linkage to provide reversible PEGylation of proteins (Figure 4.1).

REFERENCES

REFERENCES FOR TABLE 4.1

1. Kato, A., Sasaki, Y., Furuta, R., and Kobayashi, K., Functional protein-polysaccharide conjugate prepared by controlled dry-heating of ovalbumin-dextran mixture, *Agric. Biol. Chem.* 54, 107–112, 1990.
2. Kato, A., Kamayama, K., and Takagi, T., Molecular weight determination and compositional analysis of dextran-protein conjugates using low-angle laser light scattering technique combined with high-performance gel chromatography, *Biochim. Biophys. Acta* 1159, 22–28, 1992.
3. Lees, A., Nelson, B.L., and Mond, J.J., Activation of soluble polysaccharides with 1-cyano-4-dimethylaminopyridinium tetrafluoroborate for use in protein-polysaccharide conjugate vaccines and immunological reagents, *Vaccine* 14, 190–198, 1996.
4. Zambaux, M.F., Bonneaux, F., and Dellacherie, E., Covalent fixation of soluble derivatized dextrans to model proteins in low-concentration medium: Application to factor IX and protein C, *J. Protein Chem.* 17, 273–284, 1998.
5. Jung, S.H., Choi, S.J., Kim, H.J., and Moon, T.W., Molecular characteristics of bovine serum albumin-dextran conjugates, *Biosci. Biotechnol. Biochem.* 70, 2064–2070, 2006.
6. Seid, R.C., Jr., Boykins, R.A., Liu, D.-F. et al,, Chemical evidence for covalent linkages of a semi-synthetic glycoconjugate vaccine for *Haemophilus influenzae* type B disease, *J. Glycoconjug.* 6, 489–498, 1989.
7. Bolgiano, B., Mawas, F., Yost, S.E. et al., Effect of physico-chemical modification on the immunogenicity of *Haemophilus influenzae* type b oligosaccharide—CRM_{197} conjugate vaccines, *Vaccine* 19, 3189–3200, 2001.
8. Jones, C., NMR assays for carbohydrate-based vaccines, *J. Pharmaceut. Biomed. Anal.* 38, 940–850, 2005.
9. Pozsgay, V., Recent developments in synthetic oligosaccharide-based bacterial vaccines, *Curr. Top. Med. Chem.* 8, 126–140, 2008.
10. Zhang, H.W., Wang, F.S., Shao, W. et al., Characterization and stability investigation of Cu,Zn-superoxide dismutase covalently modified by low molecular weight heparin, *Biochemistry(Moscow)* 71, S96–S100, 2006.
11. Yurovetskiv, A., Choi, S., Hiller, A. et al., Fully degradable hydrophilic polyals for protein modification, *Biomacromolecules* 6, 2648–2658, 2005.
12. Jain, S., Roy, I., and Gupta, M.N., A smart bioconjugate of trypsin with alginate, *Artif. Cells Blood Substit. Immobil. Biotechnol.* 32, 325–337, 2004.
13. Wang, J., Lin, C., Wei, J. et al., Synthesis and properties of chitosan/polypeptide bioconjugate composite, *Biomed. Mater.* 2, 32–38, 2007.
14. Zara, J.J., Wood, R.D., Boon, P. et al., A carbohydrate-directed heterobifunctional cross-linking reagent for the synthesis of immunoconjugates, *Anal. Biochem.* 194, 156–162, 1991.
15. Vazquez-Duhalt, R., Tinoco, R.D., Antonio, P.D. et al., Enzyme conjugation to the polysaccharide chitosan: Smart biocatalysts and biocatalytic hydrogels, *Bioconjug. Chem.* 12, 301–306, 2001.
16. Grandjean, C., Boutonnier, A., Guerreiro, C. et al., On the preparation of carbohydrate-protein conjugates using the traceless Staudinger ligation, *J. Org. Chem.* 70, 7123–7132, 2005.
17. Mi, F.-L., Synthesis and characterization of a novel chitosan-gelatin bioconjugate with fluorescence emission, *Biomacromolecules* 6, 975–987, 2005.
18. Sreenath, K. and Venkatesh, Y.P., Reductively aminated D-xylose-albumin conjugate as the immunogen for generation of IgG and IgM antibodies specific of D-xylitol, a haptenic allergen, *Bioconjug. Chem.* 18, 1995–2003, 2007.

19. Adams, E.W., Ratner, D.M., Seeberger, P.H. et al., Carbohydrate-mediated targeting of antigen to dendritic cells leads to enhanced presentation of antigen to T cells, *ChemBioChem* 9, 294–303, 2008.
20. Korzhikov, V., Rocker, S., Vlakh, E. et al., Synthesis of multifunctional polyvinylsaccharide containing controllable amounts of biospecific ligand, *Bioconjug. Chem.* 19, 617–625, 2008.

Chaper References

1. Meares, C.F., Introduction to bioconjugate chemistry, *Bioconjug. Chem.* 1, 1, 1990.
2. Rowlinson-Busza, G. and Epenetos, A.A., Targeted delivery of biologic and other antineoplastic agents, *Curr. Opin. Oncol.* 4, 1142–1148, 1992.
3. Nagy, A. and Schally, A.V., Targeting cytotoxic conjugates of somatostatin, luteinizing hormone-releasing hormone and bomesin to cancers expressing their receptors: A "smarter" chemotherapy, *Curr. Pharm. Des.* 11, 1167–1180, 2005.
4. Xie, H. and Blättler, W.A., *In vivo* behavior of antibody-drug conjugates for the targeted treatment of cancer, *Expert Opin. Biol. Ther.* 6, 281–291, 2006.
5. Aslam, M. and Dent, A., *Bioconjugation. Protein Coupling for the Biomedical Sciences*, Macmillan Reference, London, 1998.
6. Martins, M.B., Gonçalves, A,P., and Cruz, M.E., Biochemical characterization of an L-asparaginase bioconjugate, *Bioconjug. Chem.* 430–435, 1996.
7. Hamblett, K.J., Press, O.W., Meyer, D.L. et al., Role of biotin-binding affinity in streptavidin-based pretargeted radioimmunotherapy of lymphoma, *Bioconjug. Chem.* 16, 131–138, 2005.
8. Wilbur, D.S., Hamlin, D.K., and Chyan, M.-K., Biotin reagents for antibody pretargeting. 5. Additional studies of biotin conjugate design to provide biotinidase stability, *Bioconjug. Chem.* 12, 616–623, 2001.
9. Cavallaro, G., Maniscalco, L., Campisi, M. et al., Synthesis, characterization and *in vitro* cytotoxicity studies of macromolecular conjugate of paclitaxel bearing oxytocin as targeting moiety, *Eur. J. Pharm. Biopharm.* 66, 182–192, 2007.
10. Nakayama, Y., Okahashi, E., Iwai, R., and Uchida, K., Heparin bioconjugate with a thermoresponsive cationic branched polymer: A novel aqueous antithrombogenic coating material, *Langmuir* 23, 8206–8211, 2007.
11. Zalipsky, S., Mullah, N., Engbers, C. et al., Thiolytically cleavable dithiobenzyl urethane-linked polymer-protein conjugated as macromolecular prodrugs: Reversible pegylation of proteins, *Bioconjug. Chem.* 18, 1869–1878, 2007.
12. Zelikin, A.N., Quinn, J.F., and Caruso, F., Disulfide cross-linked polymer capsules: En route to biodeconstructable systems, *Biomacromolecules* 7, 27–30, 2006.
13. Romberg, B., Hennink, W.E., and Storm, G., Sheddable coatings for long-circulating nanoparticles, *Pharm. Res.* 25, 55–71, 2008.
14. Zelikin, A.N., Li, Q., and Caruso, F., Disulfide-stabilized poly(methacrylic acid) capsules: Formation, cross-linking, and degradation behavior, *Chem. Mater.* 20, 2655–2661, 2008.
15. Bansal, G., Wright, J.E.I., Zhang, S. et al., Imparting mineral affinity to proteins with thiol-labile disulfide linkages, *J. Biomed. Mater. Res. A* 74A, 618–628, 2005.
16. Thompson, M., Spectral properties and DNA targeting features of a thiazole orange-peptide bioconjugate, *Biomacromolecules*. 8, 3628–3633, 2007.
17. Poole, L.B., Klomsiri, C., Knaggs. S.A. et al., Fluorescent and affinity-based tools to detect cysteine sulfenic acid formation in proteins, *Bioconjug. Chem.* 18, 2004–2017, 2007.
18. Danial, M., Klok, H.-A., Norde, W., and Stuart, M.A.C., Complex coacervate core micelles with a lysozyme modified corona, *Langmuir* 23, 8003–8009, 2007.

19. Hu, D. and Kluger, R., Efficient generation of dendritic arrays of cross-linked hemoglobin: Symmetry and redundancy, *Org. Biomol. Chem.* 6, 151–156, 2008.
20. Chen, Y., Parr, T., Holmes, A.E., and Nakanishi, K., Porphyrinmaleimides: Towards thiol probes for cysteine residues in proteins, *Bioconjug. Chem.* 19, 5–9, 2008.
21. Profy, A.T. and Schimmel, P., Complementary use of chemical modification and site-directed mutgenesis to probe structure-activity relationships in enzymes, *Prog. Nucl. Acid Res. Mol. Biol.* 35, 1–26, 1988.
22. Means, G.E. and Feeney, R.E., Chemical modification of proteins: History and applications, *Bioconjug. Chem.* 1, 2–12, 1990.
23. Eyzaguirre, J., An overview on chemical modification of enzymes. The use of group-specific reagents, *Biol. Res.* 19, 1–11, 1996.
24. Kellam, B., De Bank, P.A., and Schesheff, K.M., Chemical modification of mammalian cell surfaces, *Chem. Soc. Rev.* 32, 327–337, 2003.
25. Tomasik, P. and Schilling, C.H., Chemical modification of starch, *Adv. Carbohydr. Chem. Biochem.* 59, 175–403, 2004.
26. Corey, D.R., Chemical modification: The key to clinical application of RNA interference?, *J. Clin. Invest.* 117, 3615–3622, 2007.
27. Helguera, G., Morrison, S.L., and Penichet, M.L., Antibody-cytokine fusion proteins: Harnessing the combined power of cytokines and antibodies for cancer therapy, *Clin. Immunol.* 105, 233–246, 2002.
28. Jenny, R.J., Mann, K.G., and Lundblad, R.L., A critical review of the methods for cleavage of fusion proteins with thrombin and factor Xa, *Protein Expr. Purif.* 31, 1–11, 2003.
29. Johnsson, N. and Johnsson, K., A fusion of disciplnes: Chemical approaches to exploit fusion proteins for functional genomics, *ChemBioChem* 4, 803–810, 2003.
30. Rohrbach, P., Broders, O., Toleikis, L., and Dübel, S., Therapeutic antibodies and antibody fusion proteins: Harnessing the combined power of cytokines and antibodies for cancer therapy, *Clin. Immunol.* 105, 233–246, 2002.
31. Teillaud, J.L., Engineering of monoclonal antibodies and antibody-fusion proteins: Successes and challenges, *Expert Opin. Biol. Ther.* 5(Suppl. 1), S15–S27, 2005.
32. Suga, U. and Haga, T., Ligand screening system using fusion proteins of G protein-coupled receptors with G protein α subunits, *Neurochem. Int.* 51, 140–164, 2007.
33. Zhang, K., Zhu, D., Kepley, C. et al., Chimeric human Fcγ allergen fusion proteins in the prevention of allergy, *Immunol. Allergy Clin. North Am.* 27, 93–103, 2007.
34. Khawli, L.A., Hu, P., and Epstein, A.L., Cytokine, chemokine, and co-stimulatory fusion proteins for the immunotherapy of solid tumors, *Handb. Exp. Pharmacol.* (181), 291–328, 2008.
35. Gronemeyer, T., Godin, G., and Johnsson, K., Adding value to fusion proteins through covalent labelling, *Curr. Opin. Biotechnol.* 16, 453–458, 2005.
36. Herrero, E., Jackson, M., Bassford, P.J. et al., Insertion of a MalE-β-galactosidase fusion protein into the envelope of *Escherichia coli* disrupts biogenesis of outer membrane proteins and processing of inner membrane proteins, *J. Bacteriol.* 152, 133–139, 1982.
37. Donoghue, D.J. and Hunter, T., Expression of a transforming region of Moloney murine sarcoma virus in *Escherichia coli* as a fusion protein with small tumor antigen of polyoma virus, *Proc. Natl. Acad. Sci. USA* 79, 800–804, 1982.
38. Keng, T. and Schimmel, P., Synthesis of two polypeptide subunits of an aminoacyl tRA synthetase as a single polypeptide chain, *J. Biomol. Struct. Dyn.* 1, 225–229, 1983.
39. Rajewsky, K., Rottländer, E., Peltre, G., and Müller, B., The immune response to a hybrid protein molecule: Specificity of secondary stimulation and of tolerance induction, *J. Exp. Med.* 126, 581–606, 1967.

40. Chang, T.M. and Neville, D.M., Jr., Artificial hybrid protein containing a toxic protein fragment and a cell membrane receptor-binding moiety in a disulfide conjugate. I. Synthesis of diphtheria toxin fragment A-S-S-human placental lactogen with methyl-5-bromovalerimmidate, *J. Biol. Chem.* 252, 1505–1514, 1977.
41. Chang, T.M., Dazord, A., and Neville, D.M., Jr., Artificial hybrid protein containing a toxic protein fragment and an a cell membrane receptor-binding moiety in a disulfide conjugate. II. Biochemical and biological properties of diphtheria toxin fragment A-S-S-human placental lactogen, *J. Biol. Chem.* 252, 1515–1522, 1977.
42. Vitte, A.L. and Jalinot, P., Intracellular delivery of peptides via association with ubiquitin or SUMO-1 coupled to protein transduction domains, *BMC Biotechnol.* 8: 24, 2008.
43. Andrews, J.A., Bligh, W.J., Chiodini, P.L. et al., The role of recombinant hybrid protein based ELISA for the serodiagnosis of *Onchocerca volvulus*, *J. Clin. Pathol.* 61, 347–351, 2008.
44. Gething, M.J., White, J.M., and Waterfield, M.D., Purification of the fusion protein of Sendai virus: Analysis of the NH_2-terminal sequence generated during precursor activation, *Proc. Natl. Acad. Sci. USA* 75, 2737–2740, 1978.
45. Pagel, J.M., Lin, Y., Hedin, N. et al., Comparison of a tetravalent single-chain antibody-streptavidin fusion protein and an antibody-streptavidin chemical conjugate for pretargeted anti-CD20 radioimmunotherapy of B-cell lymphomas, *Blood* 108, 328–336, 2006.
46. Dantes de Alraújo, A., Palomo, J.M., Cramer, J. et al., Diels-Alder ligation of peptides and proteins, *Chem. Eur. J.* 12, 6095–6109, 2006.
47. Watzke, A., Gutierrez-Rodriguez, M., Köhn, M. et al., A generic building block for C- and N-terminal protein-labeling and protein-immobilization, *Bioorg. Med. Chem.* 14, 6288–6306, 2006.
48. Wu, Y., Przysiecki, C., Flanagan, E. et al., Sustained high-titer antibody responses induced by conjugating a malarial vaccine candidate to outer-membrane protein complex, *Proc. Natl. Acad. Sci. USA* 103, 18243–18248, 2006.
49. Peeters, J.M., Hazendonk, T.G., Beuvery, E.C., and Tesser, G.L., Comparison of four bifunctional reagents for coupling peptides to proteins and the effect of the three moieties on the immunogenicity of the conjugates, *J. Immunol. Methods* 120, 133–143, 1989.
49a. Boeckler, C., Frisch, B., Muller, S., and Schuber, F., Immunogenicity of new heterofunctional cross-linking reagents used in the conjugation of synthetic peptides to liposomes, *J. Immunol. Methods* 191, 1-10, 1996.
49b. Kirkley, J.E., Goldstein, A.L., and Naylor, P.H., Effect of peptide-carrier coupling on peptide-specific immune responses, *Immunobiology* 203, 601-615, 2001.
49c. Buskas, T., Li, Y.H., and Boons, G.J., The immunogenicity of the tumor associated antigen Lewis(y) may be suppressed by a bifunctional cross-linker required for coupling to a carrier protein, *Chemistry* 10, 3517–3524, 2004.
50. Ruppert, C., Markart, P., Schmidt, R. et al., Chemical crosslinking of urokinase to pulmonary surfactant protein B for targeting alveolar fibrin, *Thromb. Haemost.* 89, 53–64, 2003.
51. Russell, J.C., Colpitts, T.L., Holets-McCormack, S.R. et al., Solid phase assembly of defined protein conjugates, *Bioconjug. Chem.* 13, 958–965, 2002.
52. Russell, J., Colpitts, T., Holets-McComrack, S.R. et al., Defined protein conjugates as signaling agents in immunoassays, *Clin. Chem.* 50, 1921–1929, 2004.
53. Houen, G. and Jensen, O.M., Conjugation to preactivated proteins using divinylsulfone and iodoacetic acid, *J. Immunol. Methods* 181, 187–200, 1995.
54. Li, L., Tsai, S.W., Anderson, A.L. et al., Vinyl sulfone bifunctional derivatives of DOTA allow sulfhydryl- or amino-directed coupling to antibodies: Conjugates retain immunoreactivity and have similar biodistributions, *Bioconjug. Chem.* 13, 110–115, 2002.
55. Houen, G., Olsen, D.T., Hansen, P.G. et al., Preparation of bioconjugates by solid-phase conjugation to ion exchange matrix-adsorbed carrier proteins, *Bioconjug. Chem.* 14, 75–79, 2003.

56. Tanaka, K., Einaga, K., Tsuchiyama, H. et al., Preparation and characterization of a disulfide-linked bioconjugate of annexin V with the B-chain of urokinase: An improved fibrinolytic agent targeted to phospholipid-containing thrombi, *Biochemistry* 35, 922–929, 1996.
57. Liang, K.W., Hoffman, E.P., and Huang, L., Targeted delivery of plasmid DNA to myogenic cells via a transferring-conjugated peptide nucleic acid, *Mol. Ther.* 1, 236–243, 2000.
58. Saito, G., Swanson, J.A., and Lee, K.D, Drug delivery strategy utilizing conjugation via reversible disulfide linkages: Role and site of cellular reducing activities, *Adv. Drug Deliv. Rev.* 55, 199–215, 2003.
59. Zalipsky, S., Mullah, N., Enbers, C. et al., Thiolytically cleavable dithiobenzyl urethane-linked polymer-protein conjugates as macromolecular prodrugs: Reversible PEGylation of proteins, *Bioconjug. Chem.* 18, 1969–1878, 2007.
60. Miao, Z., McCoy, M.R., Singh, D.D. et al., Cysteinylated protein as reactive disulfide: An alternative route to affinity labeling, *Bioconjug. Chem.* 19, 15–19, 2008.
61. Katsoyannis, P. and Ginos, J.Z., Chemical synthesis of peptides, *Annu. Rev. Biochem.* 38, 881–912, 1969.
62. Steiner, D.F., Cunningham, D., Spiegelman, L., and Aten, B., Insulin biosynthesis: Evidence for a precursor, *Science* 157, 697–700, 1967.
63. Steiner, D.F., Cho, S., Bayliss, C., and Hallund, O., On the isolation and some properties of bovine proinsulin, *Diabetes* 17, 309, 1968.
64. Steiner, D.F., Proinsulin and the biosynthesis of insulin, *N. Eng. J. Med.* 280, 1106–1113, 196945.
65. Chen, J. and Selvin, P.R., Thiol-reactive luminescent chelates of terbium and europium, *Bioconjug. Chem.* 10, 311–315, 1999.
66. Ni, J., Singh, S., and Wang, L.X., Synthesis of maleimide-activated carbohydrates as chemoselective tags for site-specific glycosylation of peptides and proteins, *Bioconjug. Chem.* 14, 232–238, 2003.
67. Léger, R., Thibaudeau, K., Robitaille, M. et al., Identification of CJC-I^{131}-albumin bioconjugate as a stable and bioactive GLP-1(7-36) analog, *Bioorg. Med. Chem. Lett.* 14, 4395–4398, 2004.
68. Slavica, A., Dib, I., and Nidetzky, B., Selective modification of surface-exposed thiol groups in *Trigonopsis variabilis* D-amino acid oxidase using poly(ethylene glycol) maleimide and its effect on activity and stability of the enzyme, *Biotechnol. Bioeng.* 96, 9–17, 2007.
69. Reetz, M.T., Rentzsch, M., Pletsch, A. et al., A robust protein host for anchoring chelating ligands and organocatalysts, *ChemBioChem* 9, 552–564, 2008.
70. Perler, F.B., Davis, E.O., Dean, G.E. et al., Protein splicing elements: Inteins and exteins—a definition of terms and recommended nomenclature, *Nucl. Acids Res.* 22, 1125–1127, 1994.
71. Pietrokovski, S., Conserved sequence features of inteins (protein introns) and their use in identifying new inteins and related proteins, *Protein Sci.* 3, 2340–2350, 1994.
72. Xy, M.Q. and Perler, F.B., The mechanism of protein splicing and its modulation by mutation, *Embo J.* 15, 5146–5153, 1996.
73. Xu, M-Q. and Evans, T.C., Jr., Recent advances in protein splicing: Manipulating proteins *in vitro* and *in vivo*, *Curr. Opin. Biotechnol.* 16, 440–446, 2005.
74. Saleh, L. and Perler, F.B., Protein splicing in *cis* and in *trans*, *Chem. Rec.* 6, 183–193, 2006.
75. Dawson, P.E., Muir, T.W., Clark-Lewis, I., and Kent, S.B.H., Synthesis of proteins by native chemical ligation, *Science* 266, 776–779, 1994.
76. Nilsson, B.L., Kiessling, L.L., and Raines, R.T., Staudinger ligation: A peptide from a thioester and azide, *Org. Lett.* 2, 1939–1941, 2000.
77. Nilsson, B.L., Hondal, R.J., Soelinger, M.B., and Raines, R.T., Protein assembly by orthogonal chemical ligation method, *J. Am. Chem. Soc.* 125, 5268–5269, 2003.

78. Tam, A., Soellner, M.B., and Raines, R.T., Water-soluble phosphinothiols for traceless Staudinger ligation and integration with expression protein ligation, *J. Am. Chem. Soc.* 129, 11421–11430, 2007.
79. Clippingdale, A.B., Barrow, C.J., and Wade, J.D., Peptide thioester preparation by Fmoc solid phase peptide synthesis for use in native chemical ligation, *J. Pept. Sci.* 6, 225–234, 2000.
80. Lu, W., Starovasnik, M.A., and Kent, S.B., Total chemical synthesis of bovine pancreatic trypsin inhibitor by native chemical ligation, *FEBS Lett.* 429, 31–35, 1998.
81. Hackeng, T.M., Rosing, J., Sprong, H.M., and Vermeer, C., Total chemical synthesis of human matrix Gla protein, *Protein Sci.* 10, 864–870, 2001.
82. Pál, G., Santamaria, F., Kossiakoff, A.A., and Lu, W., The first semi-synthetic serine protease made by native chemical ligation, *Protein Expr. Purif.* 29, 185–192, 2003.
83. Tolbert, T.J. and Wong, C.H., Conjugation of glycopeptide thioesters to expressed protein fragments: Semisynthesis of glycosylated interleukin-2, *Methods Mol. Biol.* 283, 255–266, 2004.
84. Singer, D., Herth, N., Kuhlman, J. et al., Mapping of phosphorylation-dependent anti-tau monoclonal antibodies in immunoblots using human tau-constructs synthesized by native chemical ligation, *Biochem. Biophys. Res. Commun.* 367, 318–322, 2008.
85. Čemažar, M. and Craik, D.J., Microwave-assisted Boc-solid phase peptide synthesis of cyclic cysteine-rich peptides, *J. Peptide Sci.* 14, 683–689, 2008.
86. Helms, B., van Baal, I., Mericx, M., and Meijer, E.W., Site-specific protein and peptide immobilization on a biosensor surface by pulsed native chemical ligation, *ChemBioChem* 8, 1790–1794, 2007.
87. Wieczerzak, E., Hamel, R., Jr., Chabot, V. et al., Monitoring of native chemical ligation on solid substrate by surface plasmon resonance, *Biopolymers* 90, 415–420, 2008.
88. Muralidhoran, V. and Muir, T.W., Protein ligation—an enabling technology for the biophysical analysis of proteins, *Nat. Methods* 3, 429–438, 2006.
89. Severinov, K. and Muir, T.W., Expression protein ligation: A novel method for studying protien-protein interactions in transcription, *J. Biol. Chem.* 273, 16205–16209, 1998.
90. Chong, S., Mersha, F.B., Comb, D.G. et al., Single-column purification of free recombinant proteins using a self-cleavable affinity tag derived from a protein splicing element, *Gene* 192, 271–281, 1997.
91. Muir, T.W., Sondhi, D., and Cole, P.A., Expressed protein ligation: A general method for protein engineering, *Proc. Natl. Acad. Sci. USA* 95, 6705–6710, 1998.
92. Richer, M.P. and Beck-Sickinger, A.G., Expressed protein ligation to obtain selectively modified aldo/keto reductases, *Protein Pept. Lett.* 12, 777–781, 2005.
93. Hondal, R.J., Incorporation of selenocysteine into proteins using peptide ligation, *Protein Pept. Lett.* 12, 757–764, 2005.
94. Schwarzer, D. and Cole, P.A., Protein semisynthesis and expressed protein ligation: Chasing a protein's tail, *Curr. Opin. Chem. Biol.* 9, 561–569, 2005.
95. Moroder, L., Isosteric replacement of sulfur with other chalcogens in peptides and proteins, *J. Pept. Sci.* 11, 187–214, 2005.
96. David, R., Richter, M.P., and Beck-Sickinger, A.G., Expressed protein ligation. Method and applications, *Eur. J. Biochem.* 271, 663–677, 2004.
97. Takeda, S., Tsukiji, S., and Nagume, T., Site-specific conjugation of oligonucleotides to the *C*-terminus of recombinant proteins by expressed protein ligation, *Bioorg. Med. Chem. Lett.* 14, 2407–2410, 2004.
98. Lovrinovic, M. and Nieymeýer, C.M., Microtiter plate-based screening for the optimization of DNA-protein conjugate synthesis by means of expressed protein ligation, *ChemBioChem* 8, 61–67, 2007.

99. Lovrinovic, M., Fruk, L., Schröder, H., and Niemeyer, C.M., Site-specific labeling of DNA-protein conjugates by means of expressed protein ligation, *Chem. Commun.* (4), 353–355, 2007.
100. Singh, Y., Spinelli, N., and Defranq, E., Chemical strategies for oligonucleotide-conjugate synthesis, *Curr. Org. Chem.* 12, 263–290, 2008.
101. Kolb, H.C., Finn, M.G., and Sharpless, K.B., Click chemistry: Diverse chemical function from a few good reactions, *Angew. Chem. Int. Ed. Engl.* 40, 2004–2021, 2001.
102. van Steenis, D.J., David, O.R., van Strijdonck, G.P. et al., Click-chemistry as an efficient synthetic tool for the preparation of novel conjugated polymers, *Chem. Commun. (Chem)* (34), 4333–4335, 2005.
103. Tron, G.C., Pirali, T., Billington, R.A. et al., Click chemistry reactions in medicinal chemistry: Applications of 1,3-dipolar cycloaddition between azides and alkynes, *Med. Res. Rev.* 28, 278–308, 2008.
104. Moses, J.E. and Moorhouse, A.D., The growing applications of click chemistry, *Chem. Soc. Rev.* 127, 7404–7410, 2007.
105. Service, R.F., Click chemistry clicks along, *Science* 320, 868–869, 2008.
106. Bräse, S., Gil, C., Knepper, K., and Zimmerman, U., Organic azides: An exploding diversity of a unique class of compounds, *Angew. Chem. Int. Ed. Engl.* 44, 5180–5240, 2006.
107. Parrish, B., Breitenkamp, R.B., and Emrick, T., PEG- and peptide-grafted aliphatic polyesters by click chemistry, *J. Am. Chem. Soc.* 127, 7404–7410, 2005.
108. Aucagne, V. and Leigh, D.A., Chemoselective formation of successive triazole linkages in one pot: "click-click" chemistry, *Org. Lett.* 8, 4505–4507, 2006.
109. Wan, Q., Chen, J., Chen, J., and Danishefsky, S.J., A potentially valuable advance in the synthesis of carbohydrate-based anticancer vaccines through extended cycloaddition chemistry, *J. Org. Chem.* 71, 8244–8249, 2006.
110. Micoine, K., Hasenknopf, B., Thorimbert, S. et al., A general strategy for ligation of organic and biological molecules to Dawson and Keggin polyoxotungstates, *Org. Lett.* 9, 3981–3984, 2007.
111. Geng, J., Mantovani, G., Tao, L. et al., Site-directed conjugation of "clicked" glycopolymers to form glycoprotein mimics: Binding to mammalian lectin and induction of immunological function, *J. Am. Chem. Soc.* 129, 15156–15163, 2007.
112. Natarajan, A., Du, W., Xiong, C.-Y. et al., Construction of di-scFv through a trivalent alkyne-azide 1,3-dipolar cycloaddition, *Chem. Commun.* 695–697, 2007.
113. Lutz, J.F. and Zarafshani, Z., Efficient construction of therapeutics, bioconjugates, biomaterials and bioactive surfaces using azide-alkyne "click" chemistry, *Adv. Drug. Deliv. Rev.* 60, 958–970, 2008.
114. Syed, S., Schuyler, P.D., Kulczycky, M., and Sheffield, W.P., Potent antithrombin activity and delayed clearance from the circulation characterize recombinant hirudin genetically fused to albumin, *Blood* 89, 3243–3252, 1997.
115. Halpern, W., Riccobene, T.A., Agostini, H. et al., Albugranin, a recombinant human granulocyte colony stimulating factor (G-CSF) genetically fused to recombinant human albumin induces prolonged myelopoietic effects in mice and monkeys, *Pharm. Res.* 19, 1720–1729, 2002.
116. Sheffield, W.P., Mamdani, A., Hortelano, G. et al., Effects of genetic fusion of factor IX to albumin on *in vivo* clearance in mice and monkeys, *Pharm. Res.* 19, 1720–1729, 2002.
117. Chuang, V.T., Kragh-Hansen, U., and Otagiri, M., Pharmaceutical strategies utilizing recombinant human serum albumin, *Pharm. Res.* 19, 569–573, 2002.
118. Yao, Z., Dai, W., Perry, J. et al., Effect of albumin-fusion on the biodistribution of interleukin-2, *Cancer Immunol. Immunother.* 53, 404–410, 2004.
119. Müller, D., Karle, A., Meissburger, B. et al., Improved pharmacokinetics of recombinant bispecific antibody molecules by fusion to human serum albumin, *J. Biol. Chem.* 282, 12650–12660, 2007.

120. Weimer, T., Wormsbächer, W., Kronthaler, U. et al., Prolonged in-vivo half-life of factor VIIa by fusion to albumin, *Thromb. Haemost.*, 99, 659–667, 2008.
121. Huang, Y.J., Lundy, P.M., Lazaris, A. et al., Substantially improved pharmacokinetics of recombinant human butyrylcholinesterase by fusion to human serum albumin, *BMC Biotechnol.* 8:50, 2008.
122. McCurdy, T.R., Gataiance, S., Eltringham-Smith, S., and Sheffield, W.P., A covalently linked recombinant albumin dimer is more rapidly cleared *in vivo* that are wild-type and mutant C34A albumin, *J. Lab. Clin. Med.* 143, 155–124, 2004.
123. Matsushita, S., Chuang, V.T.G., Kanazawa, M. et al., Recombinant human serum albumin dimer has high blood circulation activity and low vascular permeability in comparison with native human serum albumin, *Pharm. Res.* 23, 881–891, 2006.
124. Sheffield, W.P., Mandani, A., Hortelano, G. et al., Effects of genetic fusion of factor IX to albumin on *in vivo* clearance in mice and rabbits, *Brit. J. Haematol.* 126, 565–573, 2004.
125. Komatsu, T., Ogura, Y., Teramura, Y. et al., Physicochemical characterization of cross-linked human serum albumin dimer and its synthetic heme hybrid as an oxygen carrier, *Biochim. Biophys. Acta* 1675, 21–31, 2004.
126. Abehsira-Amar, O., Uzan, M., Audibert, F. et al., Covalent linkage of the synthetic adjuvant MDP to the synthetic polypeptide (T,G)-A-L changes the specificity of the immune response to the T and B cell level, *Mol. Immunol.* 24, 945–951, 1987.
127. Pedersen, A.O. and Jacobsen, J., Reactivity of the thiol group in human and bovine albumin at pH 3–9, as measured by exchange with 2,2′-dithiopyridine, *Eur. J. Biochem.* 106, 291–295, 1980.
128. Narazaki, R., Maruyama, T., and Otagiri, M., Probing the cysteine 34 residue in human serum albumin using fluorescence techniques, *Biochim. Biophys. Acta.* 1338, 275–281, 1997.
129. Svenson, A. and Carlsson, J., The thiol group of bovine serum albumin. High reactivity at acidic pH as measured by the reaction with 2,2'-dipyridyl disulfide, *Biochim. Biophys. Acta.* 400, 433–438, 1975.
130. Di Simplicio, P., Franconi, E., Frosali, S., and Di Giuseppe, D., Thiolation and nitrosylation of cysteines in biological fluids and cells, *Amino Acids* 25, 323–339, 2003.
131. Sengupta, S., Chen, H., Togawa, T. et al., Albumin thiolate anion is an intermediate in the formation of albumin-S-S-homocysteine, *J. Biol. Chem.* 276, 30111–30117, 2001.
132. *CRC Handbook of Chemistry and Physics*, Ed. D. Lide, CRC Press, Boca Raton, FL, 2008.
133. Bulaj, G., Kortemme, T., and Goldberg, D.P., Ionization-reactivity relationships for cysteine thiols in polypeptides, *Biochemistry* 37, 8965–8972, 1998.
134. Vinogradov, A.D., Gavrikova, E.V., and Zuevsky, V.V., Reactivity of the sulfhydryl groups of soluble succinate dehydrogenase, *Eur. J. Biochem.* 63, 365–371, 1976.
135. Kenney, W.C., The reaction of N-ethylmaleimide at the active site of succinate dehydrogenase, *J. Biol. Chem.* 250, 3089–3094, 1975.
136. Knap, A.K. and Pratt, R.F., Chemical modification of the RTEM-1 thiol β-lactamase by thiol-selective reagents: Evidence for activation of the primary nucleophile of the β-lactamase active site by adjacent functional groups, *Proteins* 6, 316–323, 1989.
137. Vohník, S. Hanson, C., Tuma, R. et al., Conformation, stability, and active-site cysteine titrations of *Escherichia coli* D25A thioredoxin probed by Ramen spectroscopy, *Protein Sci.* 7, 193–200, 1998.
138. Kharitonov, V.G., Sundquist, A.R., and Sharma, V.S., Kinetics of nitrosation of thiols by nitric oxide in the presence of oxygen, *J. Biol. Chem.* 270, 28158–28164, 1995.
139. Knight, C.G. and Green, N.M., The accessibility of protein-bound dinitrophenyl groups to univalent fragments of anti-dinitrophenyl antibody, *Biochem. J.* 159, 323–333, 1976.
140. Gorin, G., Martic, P.A., and Doughty, G., Kinetics of the reactions of N-ethylmaleimide with cysteine and some congeners, *Arch. Biochem. Biophys.* 115, 593–597, 196.

141. Shechter, Y., Mironchik, M., and Rubinraut, S. et al., Albumin-insulin conjugate releasing insulin slowly under physiological conditions: A new concept for long-acting insulin, *Bioconjug. Chem.* 16, 913–920, 2005.
142. Thibaudeau, K., Léger, R., Huang, X., Robitaille, M., Quarachi, O., Saucy, C., Bousquet-Gagnim, M., and van Wyck, P. et al., Synthesis and evaluation of insulin-human serum albumin conjugates, *Bioconjug. Chem.* 16, 1000–1008, 2005.
143. Schechter, Y., Mironchik, M., Saul, A. et al., New technologies to prolong life-time of peptide and protein drugs *in vivo*, *Int. J. Peptide Res. Therapeut.* 13, 105–117, 2007.
144. Tsubery, H., Mironchik, M., Fridkin, M., and Schechter, Y., Prolonging the action of protein and peptide drugs by a novel approach of reversible polyethylene glycol modification , *J. Biol. Chem.* 279, 38118–38124, 2004.
145. Warnecke, A., Fichtner, I., Garmann, D. et al., Synthesis and biological activity of water-soluble maleimide derivatives of the anticancer drug carboplatin designed as albumin-binding prodrugs, *Bioconjug. Chem.* 15, 1349–1359, 2004.
146. Léger, R., Robitaille, M., Quraishi, O. et al., Synthesis and *in vitro* analysis of atrial natriuretic peptide-albumin conjugates, *Bioorg. Med. Chem. Lett.* 13, 3571–3575, 2003.
147. Pozsgay, V., Vierira, N.E., and Yergey, A., A method for bioconjugation of carbohydrates using Diels-Alder cycloaddition, *Org. Lett.* 4, 3191–3194, 2002.
148. Langenhan, J.M. and Thorson, J.S., Recent carbohydrate-based chemoselective ligation applications, *Curr. Org. Synthesis* 2, 58–81, 2005.
149. de Araujo, A.D., Palomo, J.M., Cramer, J. et al., Diels-Alder ligation and surface immobilization of proteins, *Angew. Chem. Int. Ed.* 45, 296–301, 2006.
150. Marchan, V., Ortega, S., Pulido, D. et al., Diels-Alder cycloaddition in water for the straightforward preparation of peptide-oligonucleotide conjugates, *Nucl. Acids Res.* 34, e24, 2006.
151. Chen, J., Jaracz, S., Zhao, X. et al., Antibody-cytotoxic agent conjugates for cancer therapy, *Exp. Opin. Drug. Deliv.* 2, 873–890, 2005.
152. Wu, A.M. and Senter, P.D., Arming antibodies: Prospects and challenges for immunoconjugates, *Nat. Biotechnol.* 23, 1137–1146, 2005.
153. Jaracz, S., Chen, J, Kuznetsova, L.V., and Ojima, I., Recent advances in tumor-targeting anticancer drug conjugates, *Bioorg. Med. Chem.* 13, 5043–5054, 2005.
154. Filpula, D., Antibody engineering and modification technologies, *Biomol. Engineer.* 24, 201–215, 2007.
155. Boswell, C.A. and Brechbiel, M.W., Development of radioimmunotherapeutic and diagnostic antibodies: An inside-out view, *Nucl. Med. Biol.* 34, 757–778, 2007.
156. Juliano, R., Challenges to macromolecular drug delivery, *Biochem. Soc. Trans.* 35, 41–43, 2007.
157. Amadori, S. and Stasi, R., Monoclonal antibodies and immunoconjugates in acute myeloid leukemia, *Best Pract. Res. Clin. Haematol.* 19, 715–736, 2006.
158. Brumlik, M.J., Daniel, B.J., Waehler, R. et al., Trends in immunoconjugate and ligand-receptor based targeting development for cancer therapy, *Exp. Opin. Drug. Deliv.* 5, 87–103, 2008.
159. Amadori, S. and Stasi, R., Integration of monoclonal antibodies and immunoconjugates into the treatment of acute myeloid leukemia, *Curr. Opin. Hematol.* 15, 95–100, 2008.
160. Grossbard, M.L. and Nadler, L.M., Immunotoxin therapy of malignancy, *Important Adv. Oncol.* 111–135, 1992.
161. Knechtle, S.J., Treatment with immunotoxin, *Philos. Trans. R. Soc. B. Biol. Sci.* 356, 681–689, 2001.
162. Pastan, I., Hassan, R., Fitzgerald, D.J., and Kreitman, R.J., Immunotoxin treatment of cancer, *Annu. Rev. Med.* 58, 221–237, 2007.

163. Jue, R., Lambert, J.M., Pierce, L.R., and Traut, R.R., Addition of sulfhydryl groups to *Escherichia coli* ribosomes by protein modification with 2-iminothiolane (methyl-4-mercaptobutyrimidate), *Biochemistry* 17, 5399–5406, 1978.
164. Aithal, H.N., Knigge, K.M., Kartha, S. et al., An alternate method utilizing small quantities of ligand for affinity purification of monospecific antibodies, *J. Immunol. Methods* 112, 63–70, 1988.
165. Balthasar, S., Michaelis, K., Dinauer, N. et al., Preparation and characterization of antibody modified gelatin nanoparticles as drug carrier system for uptake in lymphocytes, *Biomaterials* 26, 2773–2732, 2005.
166. Barth, R.F., Adams, D.M., Soloway, A.H. et al., Boronated starburst dendrimer-monoclonal antibody immunoconjugate: Evaluation as a potential delivery system for neutron capture therapy, *Bioconjug. Chem.* 5, 58–66, 1994.
167. Ajikumar, P.K., Ng, J.K., Tang, Y.C. et al., Carboxyl-terminated dendrimer-coated bioactive interface for protein microarray: High-sensitivity detection of antigen in complex biological samples, *Langmuir* 23, 5670–5677, 2007.
168. Sheng, K.C., Kalkanidis, M., Pouniotis, D.S. et al., Delivery of antigen using a novel mannosylated dendrimer potentiates immunogenicity *in vitro* and *in vivo*, *Eur. J. Immunol.* 38, 424–436, 2008.
169. Mora, J., Zielinski, T., Nelson, B., and Getts, R., Protein detection enhanced by 3DNA dendrimer signal applification, *Biotechniques* 44, 815–818, 2008.
170. Wängler, C., Modenhäuser, G., Eisenhut, M. et al., Antibody-dendrimer conjugates: The number, not the size of the dendrimer, determines the immunoreactivity, *Bioconjug. Chem.* 19, 813–820, 2008.
171. Nobs, L., Buchegger, F., Gurny, R., and Allémann, E., Biodegradable nanoparticles for direct for two-step tumor immunotargeting, *Bioconjug. Chem.* 17, 139–145, 2006.
172. Cumber, A.J. and Wawrzynczak, E.J., Preparation of cytotoxic antibody-toxin conjugates, *Methods Mol. Biol.* 80, 135–144, 1998.
173. Foulon, C.F., Bigner, D.D., and Zalutsky, M.R. Preparation of anti-tenascin monoclonal antibody-streptavidin conjugates for pretargeting applications, *Bioconjug. Chem.* 10, 867–876, 1999.
174. Suwa, T., Ueda, M., Jinno, H. et al., Epidermal growth factor receptor-dependent cytotoxic effect of anti-EGRF antibody-ribonuclease conjugate on human cancer cells, *Anticancer Res.* 19, 4161–4165, 1999.
175. Thomas, A.C. and Campbell, J.H., Conjugation of an antibody to cross-linked fibrin for targeted delivery of anti-restenotic drugs, *J. Control Release* 100, 357–377, 2004.
176. Haugland, R.P. and Bhalgat, M.K., Preparation of avidin conjugates, *Methods Mol. Biol.* 418, 1–12, 2008.
177. Bergerot, I., Ploix, C., Petersen, J. et al., A cholera toxin-insulin conjugate as an oral vaccine against spontaneous autoimmune disease, *Proc. Natl. Acad. Sci. USA* 94, 4610–4614, 1997.
178. Englebienne, P., *Immune and Receptor Assays in Theory and Practice*, CRC Press, Boca Raton, FL, 2000.
179. Lappi, D.A., Martineau, D., and Baird, A., Biological and chemical characterization of basic FGF-saporin mitotoxin, *Biochem. Biophys. Res. Commun.* 160, 917–923. 1989.
180. Lappi, D.A. and Baird, A., Mitotoxins: Growth factor-targeted cytotoxic molecules, *Prog. Growth Factor Res.* 2, 223–236, 1990.
181. Lappi, D.A., Martineau, D., Sarmientos, P. et al., Characterization of a saporin mitotoxin specifically cytotoxic to cells bearing the granulocyte-macrophage colony-stimulating receptor, *Growth Factors* 9, 31–39, 1993.
182. Lappi, D.A., Matsunami, R., Martineau, D., and Baird, A., Reducing the heterogeneity of chemically conjugated targeted toxins—homogeneous basic FGF-saporin, *Anal. Biochem.* 212, 446–451, 1993.

183. Lappi, D.A., Tumor targeting through fibroblast growth factor receptors, *Semin. Cancer Biol.* 6, 279–288, 1995.
184. Buechler, Y.J., Sosnowski, B.A., and Victor, K.D., Synthesis and characterization of a homogeneous chemical conjugate between basic fibroblast growth factor and saporin, *Eur. J. Biochem.* 234, 706–713, 1995.
185. Chen, C., Li, J., Micko, C.J. et al., Cytotoxic effects of basic FGF and heparin binding EGF-conjugated with cytotoxin saporin on vascular cell cultures, *J. Surg. Res.* 75. 35–41, 1998.
186. Lambert, J.M., Senter, P.D., Yau-Young, A. et al., Purified immunotoxins that are reactive with human lymphoid cells. Monoclonal antibodies conjugated to the ribosome inactivating proteins gelonin and the pokeweed antiviral proteins, *J. Biol. Chem.* 260, 12035–12041, 1985.
187. Caccia, P., Nitti, G., Cletini, O. et al., Stabilization of recombinant human basic fibroblast growth factor by chemical modification of cysteine residues, *Eur. J. Biochem.* 204, 649–655, 1992.
188. Engleka, K.A. and Maciag, T., Inactivation of human fibroblast growth factor-1 (FGF-1) activity by interaction with copper ions involves FGF-1 dimer formation induced by copper-catalyzed oxidation, *J. Biol. Chem.* 267, 11307–11315, 1992.
189. Pastan, I.L. and Kreitman, R.J., Immunotoxins for targeted cancer therapy, *Adv. Drug Deliv. Res.* 6, 53–88, 1998.
190. Ippoliti, R., Lendaro, E., Benedetti, P.A. et al., Endocytosis of a chimera between human pro-urokinase and the plant toxin saporin: An unusual internalization mechanism, *FASEB J.* 14, 1335–1344, 2000.
191. Frankel, A.E., Bugge, T.H., Liu, S. et al., Peptide toxins directed at the matrix dissolution systems of cancer cells, *Protein Pept. Lett.* 9, 1–14, 2000.
192. Tcheniuk, S.O., Chroboczek, J., and Balakirev, M.Y., Construction of tumor-specific toxins using ubiquitin fusion technique, *Mol. Ther.* 11, 196–204, 2005.
193. Remy, S., Reilly, R.M., Sheldon, K., and Gariépy, J., A new radioligand for the epidermal growth factor receptor ^{111}In labeled human epidermal growth factor derivatized with a bifunctional metal-chelating peptide, *Bioconjug. Chem.* 6, 683–690, 1995.
194. van der Laken, C.J., Boerman, O.C., Oyen, W.J. et al., Different behavior of radioiodinated recombinant interleukin-1 and its receptor antagonist in an animal model of infection, *Eur. J. Nucl. Med.* 23, 1531–1535, 1996.
195. Reilly, R.M. and Gariépy, J., Factors influencing the sensitivity of tumor imaging with a receptor-binding radiopharmaceutical, *J. Nucl. Med.* 39, 1036–1043, 1998.
196. Reilly, R.M., Kiarash, R., Cameron, R.G. et al., ^{111}In-labeled EGF is selectively radiotoxic to human breast cancer cells overexpressing EGFR, *J. Nucl. Med.* 41, 429–438, 2000.
197. Boerman, O.C., Dams, E.T., Oyen, W.U. et al., Radiopharmaceuticals for scintigraphic imaging of infection and inflammation, *Inflamm. Res.* 50, 55–64, 2001.
198. Boerman, O.C., Rennen, H., Oyen, W.J., and Corstens, F.H., Radiopharmaceuticals to image infection and inflammation, *Semin. Nucl. Med.* 31, 285–295, 2001.
199. Li, S. ,Peck-Radosavljevic, M., Kienasst, O. et al., Iodine-123-vacular endothelial growth factor-165 (^{123}I-VEGF165). Biodistribution, safety and radiation dosimetry in patients with pancreatic carcinoma, *Q. J. Nucl. Med. Mol. Imaging* 48, 198–206, 2004.
200. Chan, C., Sandhu, J., Guha, A. et al., A human transferring-vascular endothelial growth factor (hnTf-VEGF) fusion protein containing an integrated binding site for ^{111}In for imaging tumor angiogenesis, *J. Nucl. Med.*46, 1745–1752, 2005.
201. Annovazzi, A., D'Alessandria, C., Bonanno, E. et al., Synthesis of ^{99}Tc-HYNIC-interleukin-12, a new specific radiopharmaceutical for imaging T lymphocytes, *Eur. J. Nucl. Med. Mol. Imaging* 33, 474–482, 2006.

202. Cai, W., Chen, K., Mohamedali, K.A. et al., PET of vascular endothelial growth factor expression, *J. Nucl. Med.* 47, 2048–2056, 2006.
203. Levashova, Z., Backer, M., Backer, J.M., and Blankenberg, F.G., Direct site-specific labeling of the cys-tag moiety in scVEGF with Technetium 99m, *Bioconjug. Chem.*, 19, 1049–1054, 2008.
204. Winnard, P., Jr., Chang, F., Rusckowski, M. et al., Preparation and use of NHS-MAG_3 for technetium-99m labeling of DNA, *Nucl. Med. Biol.* 24, 425–432, 1997.
205. Hnatowich, D.J., Qu, F., Chang, F. et al., Labeling peptides with technetium-99m using a bifunctional chelator of a *N*-hydroxysuccinimide ester of mercaptoacetylglycine, *J. Nucl. Med.* 39, 56–64, 1998.
206. Liu, S. and Edwards, D.S., Tc-99m-labeled small peptides as diagnostic radiopharmaceuticals, *Chem. Rev.* 99, 2235–2268, 1999.
207. Hunter, D.H. and Luyt, L.G., Lysine conjugates for the labelling of peptides with technetrium-99m and rhenium, *J. Labelled Compounds Radiopharamceuticals* 43, 403–412, 2000.
208. Signore, A., Annovazzi, A., Chianelli, M. et al., Peptide radiopharmaceuticals for diagnosis and therapy, *Eur. J. Nucl. Med.* 28, 1555–1565, 2001.
209. Qu, T., Wang, Y., Zhu, Z. et al., Different chelators and different peptides together influence the *in vitro* and mouse *in vivo* properties of Tc-99(m), *Nucl. Med. Commun.* 22, 203–215, 2001.
210. Lupetti, A., Pauwels, E.K.J., Nibbering, P.H., and Weling, M.M., Tc-99m-antimicrobial peptides: Promising candidates for infection imaging, *Q. J. Nucl. Med.* 47, 238–245, 2003.
211. Valderheyden, J.L., Liu, G.Z., He, J. et al., Evaluation of Tc-99m-MAG(3)-annexin V: Influence of the chelate on *in vitro* and *in vivo* properties in mice, *Nucl. Med. Biol.* 33, 135–144, 2006.
212. Zhang, Y.M., Wang, Y., Liu, N. et al., *In vitro* investigation of tumor targeting with 99)(m)Tc-labeled antisense DNA, *J. Nucl. Med.* 42, 1660–1669, 2001.
213. Liu, G.Z., Zhang, S.R., He., J. et al., Improving the labeling of *S*-acetyl-MAG(3)-conjugated morpholino oligomers, *Bioconjug. Chem.* 13, 893–897, 2002.
214. Rusckowski, M., Gupta, S., Liu, G.Z. et al., Investigations of a Tc-99m-labeled bacteriophage as a potential infection-specific imaging agent, *J. Nucl. Med.* 45, 1201–1208, 2004.
215. Li, Y.C., Tan. T.Z., Zheng, J.G. et al., Anti-sense oligonucleotide labeled with technetium-99m using hydrazinonictinamide derivative and *N*-hydroxysuccinimidyl-*S*-acetylmercaptoacetyltriglycine. A comparison of radiochemical behaviors and biological properties, *World J. Gastroenterol.* 14, 2235–2240, 2008.
216. Marchalonis, J.J., An enzymic method for the trace iodination of immunoglobulins and other proteins, *Biochem. J.* 113, 299–305, 1969.
217. Roholt, O.A. and Pressman, D., A differential method for determining the relative reactivity to iodination of different tyrosyl residues in a protein molecule, *Biochim. Biophys. Acta* 147, 1–14, 1967.
218. Miles, L.E. and Hales, C.N., The preparation and properties of 125-I-labelled antibodies to insulin, *Biochem. J.* 108, 611–618, 1968.
219. Shuma, K., Sawazaki, N., Tanaka, R. et al., Effect of an exposure to chloramine-T on the immunoreactivity of glucagon, *Endocrinology* 96, 1254–1260, 1975.
220. Cort, S. and McDougal, J.S., Isolation, lactoperoxidase catalyzed radioiodination, and recovery of proteins bound to insoluble immunoadsorbents, *J. Immunol. Methods* 18, 269–280, 1977.
221. Eckelman, W.C., Paik, C.H., and Reba, R.C., Radiolabeling of antibodies, *Cancer Res.* 40, 3036–3042, 1980.
222. Ferens, J.M., Krohn, K.A., Beauumier, P.L. et al., High-level iodination of monoclonal antibody fragements for radiotherapy, *J. Nucl. Med.* 25, 367–370, 1984.

223. Matzku, S., Kirchgessner, H., and Nissen, M., Iodination of monoclonal IgG antibodies at a sub-stoichiometric level: Immunoreactivity changes related to the site of iodine incorporation, *Int. J. Rad. Appl. Instrum. B.* 14, 451–457, 1984.
224. Ramjeesingh, M., Zywulko, M., Rothstein, A. et al., Antigen protection of monoclonal antibodies undergoing labelling, *J. Immunol. Methods* 133, 159–167, 1990.
225. Hussain, A.A., Jones, J.A., Yamada, A., and Dittert, LW., Chloramine-T in radiolabeling techniques. II. A nondestructive method for radiolabeling biomolecules by halogenation, *Anal. Biochem.* 224, 221–226, 1995.
226. Nikula, T.K., Bocchia, M., Curcio, M.J. et al., Impact of the high tyrosine fraction in complementarity determining regions measure and predicted effects of radioiodination on IgG immunoreactivity, *Mol. Immunol.* 32, 865–872, 1995.
227. Stein, R., Goldenberg, D.M., Thorpe, S.R., et al., Effects of radiolabeling monoclonal antibodies with a residualizing iodine radiolabel on the accretion of radioisotopes in tumors, *Cancer Res.* 5, 3132–3139, 1995.
228. Tashtoush, B.M., Traboulai, A.A., Dittert, L., and Hussain, A.A., Chloramine-T in radiolabeling techniques. IV. Penta-*O*-acetyl-*N*-chloro-*N*-methylgluamine as an oxidizing agent in radiolabeling techniques, *Anal. Biochem.* 288, 16–21, 2001.
229. Visser, G.W., Klok, R.P., Klein, J.W. et al., Optimal quality ^{131}I-monoclonal antibodies on high-dose labeling in a large reaction volume and temporary coating the antibody with IODO-GEN, *J. Nucl. Med.* 42, 509–519, 2001.
230. Ong, G.L., Elsamra, S.E., Goldenberg, D.M., and Marites, M.J., Single-cell cytotoxicity with radiolabeled antibodies, *Clin. Cancer Res.* 7, 192–201, 2001.
231. Behr, T.M., Gotthardt, M., Becker, W., and Béhé, M., Radioiodination of monoclonal antibodies. A review of standardized, reliable and safe procedures for clinical grade levels kBq to gBq in the Göttingen/Marburg experience, *Nuklearmedizin* 41, 71–79, 2002.
232. Li, H.S. and Carayanniotis, G., Iodination of tyrosyls in thyroglobulin generates neoantigenic determinants that cause thyroditis, *J. Immunol.* 176, 4479–4483, 2006.
233. Holmberg, M., Stibius, K.B., Ndoni, S. et al., Protein aggregation and degradation during iodine labeling and its consequences for protein adsorption to biomaterials, *Anal. Biochem.* 361, 120–125, 2007.
234. Peacock, R.D., *The Chemistry of Technetium and Rhenium*, Elsevier, Amsterdam, Netherlands, 1966.
235. Lebowitz, E. and Richards, P., Radionuclide generating systems, *Sem. Nucl. Med.* 4, 257–268, 1974.
236. Steigman, J. and Richards, P., Chemistry of technetium 99m, *Sem. Nucl. Med.* 4, 269–279, 1974.
237. Thakur, M.L. and DeFulvio, J.D., Technetium-99m-labeled monoclonal antibodies for immunoscintigraphy, *J. Immunol. Methods* 137, 217–224, 1991.
238. Griffiths, G.L., Goldenberg, D.M., Jones, A.L., and Hansen, H.J. Radiolabeling of monoclonal antibodies and fragments with technetium and rhenium, *Bioconjug. Chem.* 3, 91–99, 1992.
239. Thakur, M.L., DeFulvio, J., Richard, M.D., and Park, C.H., Technetium-99m labeled monoclonal antibodies: Evaluation of reducing agents, *Int. J. Rad. Appl. Instrum. B.* 18, 227–233, 1991.
240. Singh, A.K., Mishra, P., Kashyap, R., and Chauhan, U.P., A simplified kit for instant preparation of technetium-99m human immunoglobulin-G for imaging inflammatory foci, *Nucl. Med. Biol.* 21, 277–281, 1994.
241. Hnatowich, D.J., Virzi, F., Winnard, P., Jr. et al., Investigations of ascorbate for direct labeling of antibodies with technetium-99m, *J. Nucl. Med.* 35, 127–134, 1994.
242. John, E., Thakur, M.L., Wilder, S. et al., technetrium-99m-labeled monoclonal antibodies: Influence of Technetium-99m binding sites, *J. Nucl. Med.* 35, 876–881, 1994.

243. Qi, P., Muddukrishna, S.N., Torok-Both, R. et al., Direct 99mTc-labeling of antibodies by sodium dithionite reduction, and role of ascorbate as a stabilizer in cysteine challenge, *Nucl. Med. Biol.* 23, 827–835, 1996.
244. Fitzberg, A.R., Abrams, P.G., Beaumier, P.L. et al., Specific and stable labeling of antibodies with technetium-99m with a diamide dithiolate chelating agent, *Proc. Natl. Acad. Sci. USA* 85, 4025–4029, 1988.
245. Govidan, S.V., Goldenberg, D.M., Brebenau, R.C. et al., Thiolations, ^{99m}Tc labelings, and animal *in vivo* biodistributions of divalent monoclonal antibody fragments, *Bioconjug. Chem.* 7, 290–297. 1996.
246. Rodwell, J.D., *Antibody-Mediated Delivery Systems*, Marcel Dekker, New York, 1988.
247. Francis, G.E. and Delgado, C., *Drug Targeting: Strategies, Principles, and Applications,* Humana Press, Totowa, NJ, 2000.
248. Klussman, K., Mixan, B.J., Cerveny, C.G. et al., Secondary mAB-vcMMAE conjugates are highly sensitive reporters of antibody internalization via the lysosome pathway, *Bioconjug. Chem.* 15, 765–773, 2004.
249. Meibohm, B., *Pharmacokinetics and Pharmacodynamics of Biotech Drugs: Principles and Case Studies in Drug Development,* Wiley-VCH, Weinheim, Germany, 2006.
250. Safavy, A., Recent developments in taxane drug delivery, *Curr. Drug. Deliv.* 5, 42–54, 2008.
251. Ojima, I., Guided molecular missiles for tumor-targeting chemotherapy—case studies using second-generation taxoids as warheads, *Acc. Chem. Res.* 41, 108–119, 2008.
252. King, H.D., Yurgaitis, D., Willner, D. et al., Monoclonal antibody conjugates of doxorubicin prepared with branched linkers: A novel method for increasing the potency of doxorubicin immunoconjugates, *Bioconjug. Chem.* 10, 279–288, 1999.
253. Doronina, S.O., Toki, B.E., Torgov, M.Y. et al., Development of potent monoclonal antibody auristatin conjugates for cancer, *Nat. Biotechnol.* 21, 778–784, 2003.
254. Erickson, H.K., Park, P.U., Widdison, W.C. et al., Antibody-maytansinoid conjugates are activated in target cancer cells by lysosomal degradation and linker-dependent intracellular processing, *Cancer Res.* 66, 4426–4433, 2006.
255. Etrych, T., Mrkvan, T., Rihova, B., and Ulbrich, K., Star-shaped immunoglobulin-containing HPMA-based conjugates with doxorubicin for cancer therapy, *J. Control Release* 122, 31–38, 2007.
256. Austin, C.D., Wen, X., Gazzard, L. et al., Oxidizing potential of endosomes and lysosomes limits intracellular cleavage of disulfide-based antibody drug conjugates, *Proc. Natl. Acad. Sci. USA* 102, 17987–17992, 2005.
257. Braslawsky, G.R., Kadow, K., Knipe, J. et al., Adriamycin(hydrazone)-antibody conjugates require internalization and intracellular hydrolysis for antitumor activity, *Cancer Immunol. Immunother.* 33, 367–374, 1991.
258. Froesch, B.A., Stahel, R.A., and Zangemeister-Wittke, U., Preparation and functional evaluation of new doxorubicin immunoconjugates containing an acid-sensitive linker on small-cell lung cancer cells, *Cancer. Immunol. Immunother.* 42, 55–63, 1996.
259. Muldoon, L.L. and Neuwelt, E.A., BR96-D)X immunoconjugate targeting of chemotherapy in brain tumor models, *J. Neurooncol.* 65, 49–62, 2003.
260. Ramakrishnan, S. and Houston, L.L., Comparison of the selective cytotoxic effects of immunotoxins containing ricin A chain or pokeweed antiviral protein and anti-Thy1.1 monoclonal antibodies, *Cancer Res.* 44, 201–208, 1984.
261. Letvin, N.L., Goldmacher, V.S., Ritz, J. et al., *In vivo* administration of lymphocyte-specific monoclonal antibodies in nonhuman primates. *In vivo* stability of disulfide-linked immunotoxin conjugates, *J. Clin. Invest.* 77, 977–984, 1986.
262. Stein, S., Weiss, A., Adermann, K. et al., A disulfide conjugate between anti-tetanus antibodies and HIV (37–72)Tat neutralizes tetanus toxin inside chromaffin cells, *FEBS Lett.* 458, 383–386, 1999.

263. Yang, Y., Chen, H., and Vlahov, R., Evaluation of disulfide reduction during receptor-mediated endocytosis by using FRET imaging, *Proc. Nat. Acad. Sci. USA* 103, 13872–13877, 2006.
264. Manickam, D.S. and Oupicky, D., Polyplex gene delivery modulated by redox potential gradients, *J. Drug Target.* 14, 519–526, 2006.
265. Kukis, D.L., Novak-Hofer, I., and DeNardo, S.J., Cleavable linkers to enhance selectivity of antibody-targeted therapy of cancer, *Cancer Biother. Radiopharm.* 16, 457–467, 2001.
266. King, H.D., Dubowchik, G.M., Mastalerz, H. et al., Monoclonal antibody conjugates of doxorubicin prepared with branched peptide linkers: Inhibition of aggregation by methoxytriethyleneglycol chains, *J. Med. Chem.*, 45, 4336–4343, 2002.
267. Siantar, C.L.H., DeNardo, G.L., Lain, K. et al., Selecting an intervention time for intravascular enzymatic cleavage of peptide linkers to clear radioisotopes from normal issues, *Cancer Biother. Radiopharm.* 22, 556–563, 2007.
268. Peterson, J.J. and Meares, C.F., Enzymatic cleavage of peptide-linked radiolabels from immunoconjugates, *Bioconjug. Chem.* 10, 553–557, 1999.
269. Tadayoni, B.M., Friden, P.M., Walus, L.R., and Musso, G.F., Synthesis, *in vitro* kinetics, and *in vivo* studies on protein conjugates of AZT: Evaluation as a transport system to increase brain delivery, *Bioconjug. Chem.* 4, 139–145, 1993.
270. Doronina, S.O., Mendelsohn, B.A., Bovee, T.D. et al., Enhanced activity of monomethylauristatin F through monoclonal antibody delivery: Effects of linker technology on efficacy and toxicity, *Bioconjug. Chem.* 17, 114–124, 2005.
270a. Wolfenden, R., Degress of difficulty of water-consuming reactions in the absence of enzymes, *Chem. Rev.* 106, 3379-3396, 2006.
271. Quadri, S.M. and Vriesendorf, H.M., Effects of linker chemistry on the pharmacokinetics of radioimmunoconjugates, *Q. J. Nucl. Med.* 42, 250–261, 1998.
272. Kovtun, Y.V. and Goldmacher, V.S., Cell killing by antibody-drug conjugates, *Cancer Lett.* 255, 232–250, 2007.
273. Van Schepdael, A., Verbeke, K., Van Nerom et al., Capillary electrophoretic analysis of ethylene dicysteine, a precursor of the radiopharmaceutical ^{99m}Tc ethylene dicysteine, *J. Chromatog. B.* 697, 251–254, 1997.
274. Schechter, N.R., Yang, D.J, Azhdarinia, A. et al., Assessment of epidermal growth factor receptor with ^{99m}Tc-ethylenedicysteine-C225 monoclonal antibody, *Anti-Cancer Drugs* 14, 49–56, 2003.
275. Gokce, A., Nakamura, R.M., Tubis, M., and Wolf, W., Synthesis of indium-labeled antibody chelate conjugates for radioassays, *Int. J. Nucl. Med.* 9, 85–95. 1982.
276. Brandt, K.D., Schnobrich, K.E., and Johnson, D.K., Characterization of antibody-chelator conjugates: Determination of chelator content by terbium fluorescence titration, *Bioconjug. Chem.* 2, 67–70. 1991.
277. Krejacerek, G.E. and Tucker, K.L., Covalent attachment of chelating groups to macromolecules, *Biochem. Biophys. Res. Commmun.* 77, 581–585, 1977.
278. Westerberg, D.A., Carney, P.L., Rogers, P.E. et al., Synthesis of novel bifunctional chelators and their use in preparing monoclonal antibody conjugates for tumor targeting, *J. Med. Chem.*, 32, 236–243, 1989.
279. Brandt, K.D. and Johnson, D.K., Structure-function relationships in Indium-111 radioimmunoconjugates, *Bioconjug. Chem.*, 3, 119–125, 1991.
280. Kukis, D.L., DeNardo, S.J., DeNardo, G.L. et al., Optimized conditions for chelation of yttrium-90-DOA immunoconjugates, *J. Nucl. Med.* 39, 2105–2110, 1998.
281. Chappell, L.L. Ma.D., Milenic, D.E. et al., Synthesis and evaluation of novel bifunctional chelating agents based 1,4,7,10-tetraazacyclododecane-*N,N',N",N'"*-tetraacetic acid for radiolabeling proteins, *Nucl. Med. Biol.* 30, 581–595, 2003.

282. Lewis, M.R., Kao, J.Y., Anderson, A.-L. J., Shively, J.E., and Raubitschek, A., An improved method for conjugating monoclonal antibodies with *N*-hydroxysulfosuccinimidyl DOTA, *Bioconjug. Chem.* 12, 320–324, 2001.
283. Lu, S.X., Takach, E.J., Solomon, M. et al., Mass spectral analysis of labile DOTA-NHS and heterogeneity determination of DOTA or DMI conjugated anti-PSMA antibody for prostate antibody cancer therapy, *J. Pharm. Sci.* 94, 788–797, 2005.
284. Larsen, R.H., Borrebaek, J., Dahle, J. et al., Preparation of Th227-labeled radioimmunoconjugates, assessment of serum stability and antigen binding ability, *Cancer Biother. Radiopharm.* 22, 431–437, 2007.
285. Kosmas, C., Snook, D., Gooden, C.S. et al., Development of humoral immune response against a macrocylic chelating agent (DOTA) in cancer patients receiving radioimmunoconjugates for imaging and therapy, *Cancer Res.* 52, 904–911, 1992.
286. DeNardo, G.L., Mirak, G.R., Kroger, L.A. et al., Antibody responses to macrocycles in lymphoma, *J. Nucl. Med.* 37, 451–456, 1996.
287. Perico, M.E., Chinoi, M., Nacca, A. et al., The humoral response to macrocyclic chelating agent DOTA depends on the carrier molecule, *J. Nucl. Med.* 42, 1697–1703, 2001.
288. DeNardo, G.L., Bradt, B.M., Mirick, G.R., and DeNardo, S.I., Human antiglobulin response to foreign antibodies: Therapeutic benefit?, *Cancer Immunol. Immnother.* 52, 309–316, 2003.
289. Li, L, Yazaki, P.J., Anderson, A.-L. et al., Improved biodistribution and radioimmunoimaging with poly(ethylene glycol)-D0TA-conjugated anti-CEA diabody, *Bioconjug. Chem.* 17, 68–76, 2006.
290. Williams, D.G., Comparison of three conjugation procedures for the formation of tracers for use in enzyme immunoassays, *J. Immunol. Methods* 71, 261–268, 1984.
291. Takakura, Y., Kaneko, Y., Fujita, T. et al., Control of pharmaceutical properties of soybean trypsin inhibitor by conjugation with dextran. I. Synthesis and characterization, *J. Pharm. Sci.* 78, 117–121, 1989.
292. Banoub, J.H., Shaw, D.H., Nakhla, N.A., and Hodder, H.J., Synthesis of glycoconjugates derived from various lipopolysaccharides of the *Vitrionnaceae* family, *Eur. J. Biochem.* 179, 651–657, 1989.
293. Takakura, Y., Fujita, T., Hashida, M. et al., Control of pharmaceutical properties of soybean trypsin inhibitor by conjugation with dextran. II: Biopharmaceutical and pharmacological properties, *J. Pharm. Sci.* 78, 219–222, 1989.
294. Peeters, C., Tenbergen-Meekes, A.M., Poolmann, J. et al., Induction of anti-pneumococcal cell wall polysaccharide antibodies by type 4 pneumococcal polysaccharide-protein conjugates, *Med. Micriobiol. Immunol.* 181, 35–42, 1992.
295. Aron, L., Di Fabio, J., and Cabello, F.C., *Salmonella typhi O*:9,12 polysaccharide-protein conjugates: Characterization and immunoreactivity with pooled and individual normal sera, sera from patients with paratyphoid A and B and typhoid fever, and animal sera, *J. Clin. Microbiol.* 31, 975–978, 1993.
296. Jain, S., Hreczuk-Hirst, D.H., McCormack, B. et al., Polysialylated insulin: Synthesis, characterization and biological activity *in vivo*, *Biochim. Biophys. Acta* 1622, 42–49, 2005.
297. Eliyahu, H., Sianu, S., Azzam, T. et al., Relationships between chemical composition, physical properties and transfection efficiency of polysaccharide-spermine conjugates, *Biomaterials* 27, 1646–1655, 2006.
298. Devakumar, J. and Mookambesaran, V., A novel affinity-based controlled release system involving derivatives dextran with enhanced osmotic activity, *Bioconjug. Chem.* 18, 477–483, 2007.
299. Gudlavalleti, S.K., Lee, C.H., Norris, S.E. et al., Comparison of *Neisseria meningitides* W135 polysaccharide-tetanus toxoid conjugate vaccines made by periodate activation of *O*-acetylated, non-*O*-acetylated, and chemical de-*O*-acetylated.

300. O'Shannessy, D.J., Dobersen, M.J., and Quarles, R.H., A novel procedure of labeling immunoglobulins by conjugation to oligosaccharide moieties, *Immunol. Lett.* 8, 273–277, 1984.
301. Tijssen, P. and Kurstak, E., Highly efficient and simple methods for the preparation of peroxidase and active peroxidase-antibody conjugates for enzyme immunoassays, *Anal. Biochem.* 136, 451–457, 1984.
302. O'Shannessy, D.J. and Quarles, R.H., Specific conjugation reactions of the oligosaccharide moieties of immunoglobulins, *J. Appl. Biochem.* 7, 347–355, 1985.
303. Tang, V.C., Greene, R.M., and Pilcher, J.B., Optimization of the covalent conjugating procedure ($NAIO_4$) of horseradish peroxidase to antibodies for use in enzyme-linked immunosorbent assay, *J. Immunoassay* 16, 395–418, 1995.
304. Presentini, R. and Terrana, B., Influence of the antibody-peroxidase coupling methods on the conjugate stability and on the methodologies for the preservation of the activity in time, *J. Immunoassay* 16, 309–324, 1995.
305. Werlen, R.C., Lankinen, M., Rose, K. et al., Site-specific conjugation of an enzyme and an antibody fragment, *Bioconjug. Chem.* 5, 411–417, 1994.
306. O'Shannessy, D.J. and Quarles, R.H., Labeling of the oligosaccharides moieties of immunoglobulins, *J. Immunol. Methods* 99, 153–161, 1987.
307. Aronson, R.B., Sinex, F.M., Franzblau, C., and Van Slyke, D.D., The oxidation of protein-bound hydroxylysine by periodate, *J. Biol. Chem.* 242, 809–812, 1967.
308. Robbins, S.P. and Bailey, A.J., The chemistry of the collagen cross-links. The mechanism of reducible intermediate cross-links, *Biochem. J.* 149, 381–385, 1975.
309. Nicolet, B.H. and Shinn, L.A., The action of periodic acid on α-amino alcohols, *J. Am. Chem. Soc.* 61, 1615, 1939.
310. Dixon, H.B.F. and Fields, R., Specific modification of N-terminal residues by transamination, *Methods Enzymol.* 25, 409–419, 1972.
311. Geoghegan, K.F. and Stroh, J.G., Site-directed conjugation of nonpeptide groups to peptides and proteins via periodate oxidation of a 2-amino alcohol. Application to modification at N-terminal serine, *Bioconjug. Chem.* 3, 138–146, 1992.
312. Maasen, J.A., Thielen, T.P., and Möller, W., Synthesis and application of two reagents for the introduction of sulfhydryl groups into proteins, *Eur. J. Biochem.* 134, 327–330, 1983.
313. Gaertner, H.F., Rose, K., Cotton, R. et al., Construction of protein analogues by site-specific condensation of unprotected fragments, *Bioconjug. Chem.* 3, 262–268, 1992.
314. Geoghegan, K.F., Emery, M.J., Martin, W.H. et al., Site-directed double fluorescent tagging of human rennin and collagenase (MMP-1) substrate peptides using the periodate oxidation of N-terminal serine. An apparently general strategy for provision of energy-transfer substrates for proteases, *Bioconjug. Chem.* 4, 537–544, 1993.
315. Mikolajczyk, S.D, Meyer, D.L., Starling, J.J. et al., High-yield, site-specific coupling of N-terminally modified β-lactamase to a proteolytically cleaved single-sulfhydryl murine Fab', *Bioconjug. Chem.* 5, 636–646, 1994.
316. Gaertner, H.F. and Offord, R.E. Site-specific attachment of functionalized poly (ethylene glycol) to the amino terminus of proteins, *Bioconjug. Chem.* 7, 38–44, 1996.
317. Kawakami, T., Akaji, K., and Aimoto, S., Peptide bond formation mediated by 4,5-dimethoxy-2-mercaptobenzyamine after periodate oxidation of the N-terminal serine residue, *Org. Lett.* 3, 1403–1405, 2001.
318. Chelius, D. and Shaler, T.A., Capture of peptides with N-terminal serine and threonine: A sequence specific chemical method for peptide mixture simplification, *Bioconjug. Chem.* 14, 205–211, 2003.
319. Fernandes, A.I. and Gregoriadis, G., Polysialylated asparaginase: Preparation, activity and pharmacokinetics, *Biochim. Biophys. Acta* 1341, 26–34, 1997.
320. Fernandes, A.I. and Gregoriais, G., The effect of polysialylation on the immunogenicity of asparaginase: Implications in its pharmacokinetics, *Int. J. Pharm.* 217, 215–224, 2001.

321. Jain, S., Hreczuk-Hirst, D.H., and McCormack, B., Polysialylated insulin: Synthesis, characterization and biological activity *in vivo*, *Biochim. Biophys. Acta* 1622, 42–49, 2003.
322. Constantinou, A., Epenetos, A.A., Hreczuk-Hirst, D. et al., Modulation of antibody pharmacokinetics by chemical polysialylation, *Bioconjug. Chem.* 19, 643–650, 2008.
323. Hildebrandt, H., Mühlenhoff, M., Weinhold, B., and Gerardy-Schahn, R., Dissecting polysialic acid and NCMA function in brain development, *J. Neurochem.* 103(Suppl. 1), 56–64, 2007.
324. Rutishauser, U., Polysialic acid in the plasticity of the developing and adult vertebrate nervous system, *Nat. Rev. Neurosci.* 9, 26–35, 2008.
325. Abuchowski, A., McCoy, J.R., Palezuk, N.C., van Es, T., and Davis, F.F., Effect of covalent attachment of polyethylene glycol on immunogenicity and circulatory life of bovine liver catalase. *J. Biol. Chem.* 252, 3582–3586, 1977.
326. Veronese, F.M. and Harris, J.M., Introduction and overview of peptide and protein pegylation, *Adv. Drug Deliv. Rev.* 54, 453, 2002.
327. Harris, J.M., and Chess, R.B., Effect of pegylation on pharmaceuticals, *Nat .Rev. Drug Discov.* 2, 214, 2003.
328. Katre, N.V., Immunogenicity of recombinant IL-2 modified by covalent attachment of polyethylene glycol. *J. Immunol.* 144, 209–213, 1990.
329. Wang, Q., Krishnaswami, S.R., Janda, K.D., Lin, T., and Finn, M.G., Blue fluorescent antibodies as reporters for steric accessibility in virus conjugates. *Bioconjug. Chem.* 14 28–43, 2003.
330. Goffard, A. and Dubuisson, J., Gycosylation of hepatitis C virus envelope proteins, *Biochime* 85, 295–301, 2003.
331. Brooks, S.A., Appropriate glycosylation of recombinant proteins for human use: Implications for choice of an expression system, *Mol. Biotechnol.* 28, 241–255, 2004.
332. Abe, Y., Takashita, E., Sugawara, K. et al., Effect of the addition of oligosaccharides on the biological activities and antigenicity of influenza A/H3N2 virus hemagglutinin, *J. Virol.* 78, 9605–9611, 2004.
333. Jones, J. Krag, S.S., and Betenbaugh, M.J., Controlling *N*-linked glycan sites occupancy, *Biochim. Biophys. Acta* 1726, 121–137, 2005.
334. Rawling, J. and Melero, J.A. The use of monoclonal antibodies and lectins to identify changes in viral glycoproteins that are influenced by glycosylation: The case of human respiratory virus attachment (GO glycoprotein, *Methods Mol. Biol.* 379, 109–125, 2007.
335. Lorenzo, C., Last, J.A., and González-Sapienza, G.G., The immunogenicity of *Echinococcus granulosus* antigen 5 is determined by its post-translational modifications, *Paristology* 131, 669–677, 2005.
336. Dowling, W., Thompson, E., Badger, C. et al., Influences of glycosylation on antigenicity, immunogenicity, and protective efficacy of ebola virus GP DNA vaccines, *J. Virol.* 81, 1821–1837, 2007.
337. Siddiqui, C., Involvement of glycan chains in the antigenicity of *Rapana thomasiana* hemocyanin, *Biochem. Biophys. Res. Commun.* 361, 705–711, 2007.
338. Lam, J.S., Huang, H., and Levitz, S.M., Effect of differential *N*-linked and *O*-linked mannosylation on recognition of fungal antigens by dendritic cells, *PLoS ONE* 2:e1008, 2007.
339. Molineux, G., Pegylation: Engineering improved biopharmaceuticals for oncology. *Pharmacotherapy* 23, 3S-8S, 2003.
340. Kochendoerfer, G., Chemical and biological properties of polymer-modified proteins. *Exp. Opin. Biol. Ther.* 3, 1253–1261, 2003.
341. Schorr, R.G.L., Bentley, M., Zhao, S., Pouler, R., and Whittle, B., Polyethylene glycol conjugates of proteins and small molecule drugs. Past, present, and future, in *Drug Delivery Systems in Cancer Therapy*, Humana Press, Totowa, NJ, 2–4.
342. Goldstein, L., Kinetic behavior of immobilized systems, *Methods Enzymol.* 44, 397–443, 1976.

343. Doherty, D.H., Rosendahl, M., Smith, D.J., Hughes, J.M., Chlipala, E.A., and Cox, D.N., Site-specific PEGylation of engineered cysteine analogues of recombinant human granulocyte-macrophage colony-stimulating factor, *Bioconjug. Chem.* 16, 1291–1298, 2005.
344. Grace, M.J. and Cutler, D., Pegylating IPNs at His-34 improves the *in vitro* antiviral activity through the JAK/STAT pathway, *Antiviral Chem. Chemother.* 15, 287–297, 2004.
345. Campbell, R.M., Heimer, E.P., Ahmud, M., Esienbeis, H.G., Landros, T.J., Lee, Y., Miller, R.W., Shider, P.P., and Felix, A.M., Pegylated peptides V. Carboxyl-terminal PEGylated analogues of growth hormone-releasing factor (GRF) display enhanced duration of biological activity *in vivo*, *J. Peptide Res.* 49, 527–537, 1997.
346. Foser, S., Schaeler, A., Weyer, K.A., Brugger, D., Dietel, E., Marti, S., and Streitmuller, T., Isolation, structural characterization, and antiviral activity of position isomers of monopegylated interferon alpha-2a (PEGASYS), *Protein Exp. Purific.* 30, 78–87, 2003.
347. Ramm, J., Saez, V., Baez, R., Aldana, R., and Hardy, E., PEGylated interferon-alpha 2: A branched 40 K polyethylene glycol derivative, *Pharm. Res.* 22, 1374–1386, 2005.
348. Yamamoto, Y., Tsutoumi, Y., Yoshioka, Y., Nishibata, T., Kobayashi, K., Okaomoto, T., Mukai, Y., Shimizu, T., Nokayama, M.S., Nagata, S., and Mayumi, T., Site-specific pegylation of a lysine-deficient TNF-alpha with full bioactivity, *Nat. Biotechnol.* 21, 546–557, 2003.
349. Lundblad, R.L., *Chemical Reagent for the Modification of Proteins*, 3rd ed., CRC Press, Boca Raton, FL, 2004.
350. Dhalluin, C., Ross, A., Leuthold, C.A., Foser, S., Gsell, B., Muller, F., and Senn, H., Structural and biophysical characterization of the 40 kDA PEG-interferon-alpha-2a and its individual positional isomers, *Bioconjug. Chem.* 16, 504–517, 2005.
351. Wang, Q.,Rajo, K.S., Janda, K.D., Lin, T., and Finn, M.G., Blue fluorescent antibodies as reporters of steric accessibility in virus conjugates, *Bioconjug. Chem.* 14, 28–43, 2003.
352. Simeonov, A., Matsushita, M., Juban, E.A., Thompson, E.H.Z., Hoffman, T.Z., Beuscher, A.F., IV. et al., Blue-fluorescent antibodies, *Science* 290, 307–313, 2000.
353. Colcher, D., Paulinkova, G., Beresford, G., Booth, B.J.M., Choudhury, A., and Batra, S.K., Pharmacokinetics and biodistribution of genetically-engineered antibodies, *Q. J. Nucl. Med.* 42, 225–241, 1998.
354. Natarajan, A., Xiong, C.-Y., Albrecht, H., DeNardo, G.L., and DeNardo, S.J., Characterization of site-specific ScFv PEGylation for tumor-targeting pharmaceuticals, *Bioconjug. Chem.* 16, 113–121, 2005.
355. DeNardo, S.J., Radioimmunodetection and therapy of breast cancer, *Semin. Nucl. Med.* 35, 143–151, 2005.
356. Li, L., Yazaki, P.J., Anderson, A.L. et al., Improved biodistribution and radioimaging with poly(ethylene glycol)-DOTA-conjugated anti-CEA diabody, *Bioconjug. Chem.* 17, 68–76, 2006.
357. Nishimura, H., Matsushima, A., and Inada, Y., Improved modification of yeast uricase with polyethylene glycol, accompanied with nonimmunoreactivity toward anti-uricase serum and high enzyme activity, *Enzyme* 26, 49–53, 1981.
358. Tsugi, J., Hirose, K., Kasahana, E., Naitoh, M., and Yamamoto, I., Studies on antigenicity of the polyethylene glycol (PEG)–modified uricase, *Int. J. Immunopharm.* 7, 725–730, 1985.
359. Bomalaski, J.S., Holtsberg, F.W., Ensor, C.M., and Clark, M.A., Uricase formulated with polyethylene glycol (uricase PEG 20): Biochemical rationale and preclinical studies, *J. Rheumatol.* 29, 1942–1949, 2002.
360. Suresh, E., Diagnosis and management of gout: A rational approach, *Postgrad. Med. J.* 81, 572–579, 2005.
361. Bieber, J.D. and Terkettaub, R.A., Gout: On the brink of novel therapeutic options for an ancient disease, *Arthrit. Rheumat.* 50, 2400–2414, 2004.

362. Chua, C.C., Greenberg, M.L., Viau, A.T., Nucci, M., Brenkman, W.D., and Hershfield, M.S., Use of polyethylene glycol-modified uricase (PEG-uricase) to treat hyperuricemia in a patient with non-Hodgkins lymphoma, *Ann. Int. Med.* 109, 114–117, 1988.
363. Armstrong, J.K., Hempel, G., Kohling, S. et al., Antibody against poly (ethylene glycol) adversely affects PEG-asparaginase therapy in acute lymphoblastic leukemia patients, *Cancer* 110, 103–111, 2007.
364. Croyle, M.A., Chirmule, N., Zhang, Y., and Wilson, J.M., PEGylation of E1-deleted adenovirus vectors allows significant gene expression on readministration to liver, *Human Gene Therapy* 13, 1887–1900, 2002.
365. Cheng, T.-L., Wu, P.-Y., Wu, M-F., Chern, J.-W., and Raffler, S.R.(1999), Accelerated clearance of polyethylene glycol-mediated proteins by anti-polyethylene glycol IgM, *Bioconjug. Chem.* 10, 520–528, 1999.
366. Lee, C.K., Maheshiri, N., Kaspar, B., and Schafter, D.V., PEG conjugation moderately protects adeno-associated viral vectors against antibody neutralization, *Biotechnol. Bioengineer.* 92, 24–34, 2005.
367. Środa, K., Rydlewski, J., Langner, M., Kozubek, A., Grzybek, M., and Sikorski, A.F., Repeated injections of PEG-liposomes generate anti-PEG antibodies, *Cell. Mol. Biol. Letters* 10, 37–47, 2005.
368. Sample, S.C., Harasym, T.O., Clow, K.A., Ansell, S.M., Limuk, S.K., and Hope, M.J., Immunogenicity and rapid blood clearance of liposomes containing polyethylene glycol-lipid conjugates and nucleic acid, *J. Pharmacol. Exp. Ther.* 312, 1020–1026, 2005.
369. Cheng, T.-L., Cheng, C.-M., Chen, B.M., Taso, D.-A., Chuang, K.-H., Hsiao, S.-W., Lin, Y.-H., and Raffler, S.R., Monoclonal antibody-based quantitation of poly(ethylene glycol)-derivatized proteins, liposomes, and nanoparticles, *Bioconjug. Chem.* 16, 1225–1231, 2005.
370. Wundulich, D.A., MacDougall, M., Micrcz, D.V. et al., Generation and characterization of a monoclonal antibody to polyethylene glycol, *Hybridoma* 26, 168–172, 2007.
371. Zalipsky, S., Mullah, N., Engbers, C. et al., Thiolytically cleavage dithiobenzyl urethane-linked polymer-protein conjugates as macromolecular prodrugs: Reversible pegylation of proteins, *Bioconjug. Chem.* 18, 1969–1878, 2007.

5 Protein Hydrogels

Hydrogels are easily deformed pseudo-solid masses formed from largely hydrophilic colloids dispersed in an aqueous medium (dispersion medium or continuous phase).[1–10] Hydrogels are composed mostly of water within a three-dimensional (3D) polymer network. Aerogels are 3D polymer networks that contain air or gas in place of water.[11–20] Hydrogels are used for tissue engineering,[21–26] drug delivery,[27–31] and contact lenses.[32–35] The use of hydrogels for cell growth matrices[36–44] is related to the use of tissue engineering, but is of great value without that application. The use of hydrogels for cell growth allows 3D growth rather than two-dimensional (2D) growth, which is considered more representative of physiological conditions.[45–54] 3D cultures can be viewed as a method intermediate between 2D culture and organ or tissue culture.[54]

Hydrogels can be passive in response to environmental factors such as ionic strength, pH, and temperature, or "smart" where the network is responsive to such factors. Expansion of such networks would permit release of drug otherwise retained within the matrix.[55–60] Most hydrogels are based on hydrophilic organic polymers such as poly(ethylene glycol) (PEG), poly-L-lactic acid (PLA), poly lactide/glycolide, and polymer blends. The reader is directed to several recent IUPAC documents on polymer nomenclature.[61–65] Some examples of the applications of various polymers are presented in Tables 5.1–5.8.

Reactive monomer units may be included or the polymer backbone can be modified by chemical means to provide reactive groups, which can then bind covalently to biologically active materials. This permits the attachment of biologics to the matrix, which can enable biological activity of the hydrogel. Some examples are presented in Table 5.9. Biological hydrogels can also be formed from phospholipid-containing polymers,[66–69] chitosan,[70–77] and alginate.[78–87]

Proteins and peptides may be enclosed with hydrogels or covalently linked to the hydrogel polymers by chemistries described in Chapters 1 and 3 using reactive sites such as those listed in Table 5.9. Hydrogels may also be composed entirely of proteins or they may have a major protein constituent. The use of proteins for hydrogels is reasonable considering that proteins such as collagen and fibrin form *in vivo* hydrogel-like structures.[88] The reader is also directed to Chapter 6, which describes the use of collagen and fibrinogen/fibrin in tissue soldering. The key to the successful use of proteins such as collagen or fibrinogen is the homogeneity and reproducibility of the starting material and the use of a rigorous process for manufacture of the hydrogel product. Description of the use of proteins for hydrogel preparation is presented in Tables 5.10–5.13. Proteins such as collagen and fibrin have self-associated properties that can be stabilized by covalent cross-linking. It is possible to design proteins that contain integrin-binding domain.[89] Factor XIIIa is also used to cross-link a biomimetic peptide-PEG derivative to form a hydrogel.[90] The availability of albumin and its long successful clinical history would appear to make it a likely candidate for a

TABLE 5.1
Some Selected Examples of the Application of Poly-D,L-lactic Acid (poly-D,L-lactide) in Hydrogels

Polymer	Application	Reference
Poly(D,L-lactic acid)	Biodegradable granules for antigen release as adjuvant.	1
Poly(acroyl-hydroxyethyl starch)-poly(D,L-lactide-co-glycolide)[a]	Microsphere drug delivery system; insulin used as model protein.	2
Poly(D,L-lactic acid)	Coating implant surfaces for optimizing bone-implant contact ("osseointegration").	3
Poly(D,L-lactide-co-glycolide-b-ethylene-b-D,L-lactide-co-glycolide) (PLGA-PEG-PLGA) a triblock copolymer	Soluble at 23°C (room temperature) but forms hydrogel at 37°C (body temperature). Used for drug delivery.	4
Poly(D,L-lactide-co-glycolide)(PLGA) microspheres incorporated into PEG hydrogels (cross-linked with genipin)	Drug (TGF-β_1) delivery vehicle.	5
Poly(D,L-lactide-co-glycolide-b-ethylene-b-D,L-lactide-co-glycolide) (PLGA-PEG-PLGA) a triblock copolymer	Thermosensitive copolymer used for sustained release of bee venom peptide.	6
Poly(D,L-lactide-co-glycolide-b-ethylene-b-D,L-lactide-co-glycolide) (PLGA-PEG-PLGA) a triblock copolymer	Bee venom peptide delivery; interaction between bee venom peptide and hydrogel copolymer.	7
Poly(ethylene glycol-b-[D,L-lactic acid-co-glycolide]-b-ethylene glycol) (PEG-PGLA-PEG)	Thermosensitive hydrogel as wound dressing/ scaffold for engraftment of muscle stem cells.	8

[a] Poly(D,L-lactide-co glycolide) (PGLA); AcHES, acroyl-hydroxyethyl starch, acroyl-HES.

hydrogel; however, it has been put to very limited use for this purpose.[91–94] Cleavable peptides have been included as cross-linking agents providing for a sensitive biosensor.[95] In this study, Frisk and coworkers used a peptide cross-linker in an acrylamide hydrogel. The peptide cross-linker contained a unique sequence cleaved by botulinum neurotoxin type A. Cleavage of this cross-linker resulted in degradation of the gel, permitting its development as a unique biosensor. The hydrogel structure in this case is described as a *sacrificial* hydrogel or structure. There is another example of an individual structure of a hydrogel being sacrificed to provide a microstructure: a hydrogel containing interpenetrating gelatin fibers, which when removed by melting, leaves channels as small as 6 nm in diameter.[96]

The properties of engineered vascular tissues is modulated by the combination of extracellular matrix components such as fibrin and collagen together with mechanical stimulation,[97] suggesting the value of composite hydrogels modeled after *in vivo* situations. In this case, fibrin is replaced by collagen in the normal wound healing or remodeling process.[98–100]

TABLE 5.2
Some Selected Examples of the Use of Poly(L-lactide) or Poly(L-lactic acid) in Hydrogels

Polymer	Application	Reference
Poly(L-lactide)(PLLA)	Used as the underlayer in a bilayer matrix used as matrix for a cell seeded skin substitute.	1
Poly(L-lactide)-g-oligo(ethylene glycol)	Used in a blend with poly(D,L-lactic-co-glycolic acid) for fabrication of microsphere in development for protein drug delivery.	2
Poly(L-lactide) (PLLA)	Matrix for collagen hydrogel used for myocardial tissue engineering.	3
Polyoxyethylene[a]-poly(L-lactide)-polyoxyethylene	Used as a component of mixed suspension of enantiomeric block copolymers with polyoxyethylene-poly(D-lactide)[b]-polyoxyethylene in the development of temperature-sensitive copolymers.	4
Poly(L-lactide) (PLLA)	PLLA grafted with dextran (Dex-graft-PLA) used to prepare microsphere by water-in-oil-in-water emulsion solvent evaporation/extraction method.	5
L-lactide-PEG oligomer	Formation of biodegradable polymeric hydrogel tubes for neural guidance.	6
Poly(L-lactide) (PLLA)	Scaffold prepared by thermally induced phase separation used as analogs of extracellular matrix for chondrocytes.	7
Poly(lactide-co-ethylene oxide-co-fumarate)	Preparation of a biodegradable hydrogel for cell transplantation.	8
Poly(lactide-co-ethylene oxide-co-fumarate)	Preparation of an injectable hydrogel for cellbone marrow stromal cells; transplantation for treatment of osteochondral defects.	9
Poly(L-lactide-co-D,L-lactide)	Preparation of scaffold for bone growth. Scaffold contains BMP-2 and TGF-β_3 combined with RGD-alginate.	10

[a] More commonly known as poly(ethylene glycol).
[b] Poly(D-lactide) (PDLA).

Hydrogels may also be composed of biopolymers known for specific interaction with proteins. Specific examples are provided for hyaluronan (hyaluronic acid), a nonsulfated component of the extracellular matrix (Table 5.14) and heparin/heparan (Table 5.15).

The preparation of hydrogels from natural materials such as protein and polysaccharide usually means starting with heterogeneous starting materials. It is, therefore, necessary to exercise considerable control over the starting materials such that one is always starting with a consistently heterogeneous product. It is equally important to maintain process control over the hydrogel production process, whether at a small scale in the research laboratory or at large scale in development

TABLE 5.3
Some Selected Examples of the Use of Polyacrylamide in Hydrogels

Polymer	Application	Reference
Polyacrylamide	Polyacrylamide hydrogel-coated charcoal for treatment of hepatic coma.	1
Copolymer of acrylamide and vinylimidazole	Formation of a redox polymer for coupling redox enzymes to electrodes.	2
Composite polyacrylamide-agar hydrogel	Development of hydrogel dressing foils and gels.	3
Acrylamide-chitosan cross-linked with *N,N'*-bisacrylamide	Development of hydrogel for controlled release of antibiotics.	4
Polyacrylamide	Hydrogel for manufacture of antibody microarrays.	5
Polyacrylamide	Formation of a hydrogen with reversible DNA cross-links. Temperature-dependent viscosity and elastic modulus are functions of cross-link density.	6
Poly(methyl methacrylate)	Development of hydrophobic nanoparticles. Higher equilibrium swelling than polyacrylamide hydrogels.	7
Polyacrylamide	Preparation of highly cross-linked hydrogel films for antibody microarrays.	8
Polyacrylamide with grafted single-stranded DNA	Development of a hydrogel that "shrinks" on the addition of single-stranded DNA samples, with application as DNA-sensing devices or DNA-triggered devices.	9
Polyacrylamide	Review of antibody-containing hydrogels of microfluidic immunoassays.	10

and production. In the latter case, it is not unreasonable to consider a matrix such as that provided by FMEA (failure mode and effects analysis)[101] and the general principles of product development.[102]

TABLE 5.4
Some Selected Uses of Polycaprolactones in Hydrogels

Polymer	Application	Reference
Polycaprolactone (PCL)	PCL fibers were embedded in poly(2-hydroxyethyl methacrylate) (pHEMA) hydrogels—sonication in acetone dissolved PCL, leaving longitudinally oriented channels in the pHEMA hydrogel.	1
PCL polymer and PEG macromer[a]	PCL polymer and poly(ethylene glycol) macromonomer were cross-linked to form a biodegradable hydrogel. The product is intended for controlled release of drugs.	2
PCL	PCL and PLC-hydroxyapatite frameworks for the culture of bone mesenchymal progenitor cells.	3
PCL maleic acid	PCL maleic acid and poly(ethylene glycol)diacrylate were photo-cross-linked to form a three-dimensional network. This hydrogel was evaluated for protein drug delivery.	4
PEG/PCL	PEG and PCL were used to synthesize alternating block copolymers forming a hydrogel. Degradation of the hydrogel was accelerated by temperature or infrared radiation.	5
PCL-PEG-PCL	Poly(caprolactone)-co-poly(ethylene glycol)-co-(caprolactone)diacrylate and chitosan were irradiated in mild acid (1% HOAc) to form hydrogel for cell culture.	6
PCL-polyurethane	Bovine chondrocytes were added to a concentrated (100 mg/mL) solution of bovine fibrinogen and clotting with thrombin. The resulting gel was inserted into the scaffold.	7

[a] The term *macromer* refers to an oliogomer or polymer that has a functional group, usually at the end, allowing such a molecule to act as a monomer. The term macromer is discouraged; instead, the term macromonomer should be used. (Definitions of terms relating to reaction of polymers and to functional polymeric materials, IUPAC, Project1999-048-1-400, February, 2003.)

TABLE 5.5
Some Selected Applications of the Use of Poly(vinyl alcohol) for Hydrogels

Polymer	Application	Reference
Poly(vinyl alcohol)(PVA)	Hydrogel membranes prepared by radiation and chemical cross-linking of PVA were evaluated for protein permeability.	1
Poly(vinyl alcohol-vinyl acetate)	Hydrogels are prepared by blending aqueous solutions of poly(vinyl alcohol-vinyl acetate) with poly(acrylic acid) in various proportions. Covalent cross-linking was accomplished with glutaraldehyde or glyoxal.	2
Poly(vinyl alcohol)	Hydrogel nanoparticles prepared by freeze–thaw using a water-in-oil emulsion or cyclic freeze–thaw process. No cross-linking agent is required. Proposed to be used for controlled release of therapeutic proteins.	3
Poly(vinyl alcohol)	Evaluation of the effect of chitosan or dextran or PVA hydrogels prepared by freeze–thawing.	4
Poly(vinyl alcohol)	Preparation of hydrogel films containing protein by freeze–thaw process.	5
Poly(vinyl alcohol)	High-molecular-weight PVA was used to entrap pegylated-lipase using a freeze–thaw method.	6
Poly(vinyl alcohol)	Characterization of the physical properties of PVA hydrogels as a biomaterial for replacement of diseased or damaged articular cartilage.	7
Poly(vinyl alcohol)	PVA was cross-linked with ethylene glycol diglycidyl ether to form a hydrogel for use in controlled drug release.	8
Poly(vinyl alcohol)-poly(acrylic acid)	Evaluation of the effect of pH on complexation of poly(acrylic acid) with poly(vinyl alcohol) for subsequent cross-linkage with γ-irradiation.	9
Poly(vinyl alcohol)	Cell-adhesive domains were created on PVA-coated glass cover slips by sodium hypochlorite.	10
Poly(vinyl alcohol)	PEG is evaluated for maintaining pores in PVA hydrogel membranes.	11
Poly(vinyl alcohol)	PVA and DNA were subjected to high pressure (10,000 atm) to produce DNA-containing hydrogels for controlled release.	12

TABLE 5.6
Some Selected Applications of the Use of Poly (ethylene glycol) for Hydrogels

Polymer	Application	References
Chitosan-poly(ethylene glycol)	Chitosan was coupled to PEG using glutaraldehyde to form an interpenetrating network, which bound heparin.	1
Poly(ethylene glycol) with hexamethylene diisocyanate and 1,2,6-hexanetril	Preparation of a range of networks with different cross-linking densities. These networks showed differences in temperature in equilibrium swelling, which could be used for controlled drug release.	2
Poly(ethylene glycol)(PEG)-poly(L-glutamic acid)(PGA)	The diamino derivative of PEG was cross-linked to PGA using 2-isobutoxy-1-isobutoxy-carbonyl-1,2-dihydroquinoline[a] chemistry. The resulting hydrogel was hydrophilic, and swelling was pH-dependent (increased with increasing pH). The product was evaluated for protein (lysozyme) release.	3
Poly(polypropylene fumarate-co-ethylene glycol) [P(PF-co-EG)]	The block copolymer P(PF-co-EG) was prepared by the transesterification of mononethyoxy ethylene glycol and subsequently cross-linked with PEG diacrylate in the presence of ammonium persulfate and ascorbic acid. The resulting product is intended for use as an injectable hydrogel for tissue engineering.	4
Poly(ethylene glycol)di-[ethyl phosphatidyl (ethylene glycol) methacrylate (PhosPEG-dMA)	PhosPEG-dMA is a macromonomer that forms a hydrogel applicable for cartilage and bone tissue engineering.	5
Poly(ethylene glycol)	PEG was grafted onto a chitosan backbone to yield a temperature-sensitive hydrogel to use for protein drug delivery.	6
Poly(ethylene glycol)	PEG is used to adjust the size of pores in poly(*N*-isopropylacrylamide) hydrogels with adjustable size "cut-off" to be used for the immobilization of biological polymers such as proteins and nucleic acids.	7
Poly(ethylene glycol)-soy protein	PEG was cross-linked with soy protein using carbonate activation of the PEG. The product is a biomimetic hydrogel for wound dressing and controlled drug release.	8
Poly(ethylene glycol)	Amino PEG was used to prepare nanohydrogels by cross-linking with a focused electron beam. Proteins could be coupled to the gel.	9
Poly(ethylene glycol)	PEG was blended with poly(*N*-isopropylacrylamide) and chitosan to prepare a hydrogel film with both temperature and pH sensitivity.	10

[a] Belleau, B. and Malek, G., A new convenient reagent for peptide synthesis, *J. Am. Chem. Soc.* 90, 1651–1652, 1990.

TABLE 5.7
Some Selected Applications of the Use of Poly (*N*-isopropylacrylamide) for Hydrogels

Polymer	Applications	References
Poly(*N*-isopropylacrylamide) (NiPAAm)	Temperature-sensitive hydrogel beads prepared by inverse suspension copolymerization of *N*-isopropylacrylamide and acrylamide.	1
Copolymer of *N*-isopropyl-acrylamide and 4-(*N*-cinnamoylcarbamide)	Copolymer of *N*-isopropyl-acrylamide and 4-(*N*-cinnamoylcarbamide) was prepared and dissolved in toluene/1-butanol. This solution was allowed to evaporate on the surface of a polystyrene culture dish. This was cross-linked by UV-radiation. This is a temperature-sensitive hydrogel; cells attach and grow on this surface and are detached by a change in temperature.	2
Poly(*N*-isopropylacrylamide-co-acrylic acid)	Poly(*N*-isopropylacrylamide-co-acrylic acid) hydrogel microspheres were prepared by membrane emulsification. The microspheres showed temperature-dependent electrophoretic mobility.	3
Poly(*N*-isopropylacrylamide)	A PNIPAAm hydrogel was prepared by gamma radiation of *N*-isopropyl-acrylamide. The product was a pH-dependent hydrogel. Differential scanning calorimetry was used to determine the lower critical solution temperature (LCST).	4
Poly(*N*-isopropylacrylamide)	Nanoparticles (composite composition) were prepared from NiPAAm, potassium persulfate, *N,N'*-methylenebisacrylamide in the presence of SDS. A composite product was prepared from these nanoparticles and PEG-diacrylate. These particles are temperature sensitive and are being developed for protein drug delivery.	5
Poly(*N*-isopropylacrylamide)	Copolymer of poly(*N*-isopropyl-acrylamide) and hydroxymethyl-methacrylate developed for intravascular embolization.	6
Poly(*N*-isopropylacrylamide)	Hydrogel prepared by free radical polymerization of NiPAAm; a composite was formed with a polyurethane foam.	7

TABLE 5.8
Some Selected Applications of the Use of Poly (lactide-co-glycolide) for Hydrogels

Polymer	Application	References
Poly (lactide-co-glycolide) (PLGA)	Controlled drug delivery from nanoparticles.	1
Poly(lactide-co-glycolide) triblock copolymer with PEG	PEG-PGLA-PEG triblock copolymer (low-molecular-weight PEG) was synthesized. The copolymer is a solution at "room temperature" (23°C) and becomes a solid at "body temperature" (37°C). It is proposed to use this product for controlled drug delivery.	2
Poly (lactide-co-glycolide)	PLGA is used to encapsulate a poly(vinyl alcohol) hydrogel in a solvent evaporation technique. The inclusion of the PVA hydrogel increased the drug-loading capacity.	3
Poly (lactide-co-glycolide)-poly(ethylene glycol)	Copolymer of PLGA and PEG in aqueous solution exhibits temperature-dependent sol–gel transition moving to gel at higher temperatures.	4
Poly (lactide-co-glycolide)	PLGA microspheres containing insulin-like growth factor-1 were combined with alginate and tricalcium phosphate to form an injectable scaffold for bone implant.	5
Poly (lactide-co-glycolide)-monomethoxy poly (ethylene glycol)	PLGA-PEG diblock copolymer used to formulate a temperature-sensitive hydrogel with gelatin.	6
Poly (lactide-co-glycolide)	PLBA microspheres containing alginate lysase included in alginate hydrogels permit the timely dissolution of the alginate hydrogels.	7

TABLE 5.9
Chemical Modification of Hydrogel Polymers

Polymer	Modification	Reference
Poly(hydroxyethyl methacrylate)	Oxidation with sulfuric acid to form surface carboxyl groups (hydrolytic etching). Modified hydrogel supports cell growth.	1
2-Hydroxyethyl methacrylate-methylmethacrylate copolymer	Modification of polymer with ammonia in gaseous plasma.	2
Hyaluronic acid	Modification with carboxyl groups with hydrazide to provide derivatives for cross-linking.	3
Hyaluronic acid	Carbodiimide-modification of hyaluronic acid with bifunctional amine-containing compounds, which can be converted into aldehydes or amino groups for subsequent modification.	4
Guar gum	Cross-linking with sodium trimetaphosphate.	5,6
Alginate or hyaluronan	Formation of methacryl derivative by reaction with methacrylic acid anhydride. The methacryl derivatives form hydrogels upon photolysis.	7
Agarose gel	Modification of gel with benzophenone (oxidation of agarose hydroxyl to carboxylic acid with sodium hypochlorite and coupling with poly(allylamine) derivative of benzophenone) or modification of biomolecule (model ovalalbumin) with 4-benzoylbenzoic succinimidyl ester. Photoactivation for coupling reaction of activated protein to matrix or activated matrix to biomolecule.	8
Hyaluronan (hyaluronic acid)	Cross-linking with poly(ethylene glycol) diepoxide.	9
Graft copolymer of poly(ethylene glycol) and poly(vinyl alcohol)-heparin hydrogel	Oxidation of graft copolymer with dimethyl sulfoxide in acetic acid to form surface aldehydes groups that react with the hydroxyl groups on the PVA. Heparin could be released from the film by electrical stimulation (at 3.5 mA).	10
Hyaluronic acid	Methacrylate derivative cross-linked by Michael addition between vinyl function and sulfhydryl of dithiothreitol. A cysteine peptide linker (GCYKNRDCG) was also used as a cross-linker.	11
N-isopropylacrylamide copolymerized with *N*-acryloxysuccinimide	Modification of a carbonyl group on the polymer with cystamine to form a free sulfhydryl, which then reacts with poly(ethylene glycol)diacrylate via Michael addition.	12
Poly(ethylene glycol)	PEG modified to provide terminal sulfhydryl groups that would then react with maleimide derivatives of heparin.	13
Hyaluronan	The thioethyl derivative of hyaluronan was prepared by reaction with ethyl sulfide, which can be cross-linked to form some novel derivatives; the parent thiol derivative is stable to oxidation.	14
Gelatin (Type B)	Gelatin was modified with 2-iminothiolane (Traut's reagent; Jue, R., Lambert, J.M., Pierce, L.R., and Traut, R.R., Addition of sulfhydryl groups to *Escherichia coli* ribosomes by protein modification with 2-iminothiolane (methyl-4-mercaptobutyrimidate), *Biochemistry* 17, 5399–5406, 1978) to provide thiol-modified gelatin nanoparticles for intracellular DNA delivery.	15

TABLE 5.10
The Use of Collagen for the Preparation of Hydrogels

Collagen Type	Application	Reference
Collagen-hydroxyethyl-methacrylate (HEMA)	Culture of endothelial cells or fibroblasts.	1
Graft copolymerization of either hydroxylmethacrylate or methyl methacrylate or using different cross-linking agents	Implantation studies in rats; no untoward rejection of gels was observed.	2
Type I collagen hydrogels	Measurement of local shear moduli.	3
Collagen (injectable)	Injectable collagen is a dispersion of phase-separated collagen fibers in aqueous solutions. Gels formed from the polymers at a lower pH/higher temperature have a different structure than gels formed at higher pH/lower temperature.	4
Collagen	Collagen succinylated to provide matrix for endothelial cell growth. Rapid endothelial cell growth is possible with clean collagen.	5
Chitosan–collagen or chitosan–tropocollagen	Chitosan–collagen mixtures or chitosan–tropocollagen mixtures are spun into an aqueous ammonia solution with ammonium sulfate to provide fibers. The blended fiber was *N*-modified with various carboxylic acid anhydrides or aldehydes.	6
Type I collagen (rat tail)	Collagen was modified with methacrylic anhydride to yield the methacryl derivative. The derivatized collagen was cast into a gel containing smooth muscle cells and photo-cross-linked (visible light, 50 mW/cm^2).	6a
Collagen "vitrigel" prepared by incubation of type I collagen in a neutral salt solution. This process of vitrification forms a glassy material upon drying, which is hydrated into a gel membrane	Development of a three-dimensional scaffold for reconstruction of an epithelial-mesenchymal model or a hard connective tissue model.	7
Bovine type I collagen denatured with trifluoroacetic acid	Collagen denatured with trifluoroacetic acid (TFA) was compared to gelatin in the preparation of hydrogels (sponges) containing FGF-2 to support cell growth. TFA-denatured collagen has properties different from gelatin in that TFA-collagen appears to self-implement endothelial cell growth.	8
Collagen hydrogel (rat tail type I collagen)	Observed differences in neurite extension with 2D and 3D gels. Gels characterized by confocal microscopy and a potentiometric-sensing dye.	9

TABLE 5.10 (CONTINUED)
The Use of Collagen for the Preparation of Hydrogels

Collagen Type	Application	Reference
Acidic solution of type I collagen (porcine) and a neutral collagen gel	A gel was formed from the acidic solution by γ-irradiation and compared to a neutral gel also subjected to γ-irradiation. The irradiation produced cross-links. The neutral gel showed a 3D network; such a network was not seen in the acidic gel.	10
Type I collagen (porcine skin)	Type I collagen (porcine skin) was cross-linked to hyaluronic acid using a poly(ethylene glycol) diglycidyl cross-linking agent.	11
Collagen type I	Co-gels of collagen I and laminin proved useful for promoting peripheral nerve regeneration.	12
Porcine type I atelocollagen	Hydrogel prepared by cross-linking (pH 9.0) with oxidized collagen (periodate) for use in tissue engineering. Unreacted aldehyde groups responsible for some cytotoxicity (human fibroblast culture) were removed by reaction with sodium borohydride after the initial cross-linking reaction.	13
Bovine fibrillar collagen	The collagen was dispersed in water and lyophilized to yield a sponge that was then stabilized with formaldehyde cross-linking and sterilized with ethylene oxide. The stabilized sponges were soaked in recombinant bone morphogenic protein-2 (bhBMP-2) for evaluation as drug delivery vehicle.	14

TABLE 5.11
The Use of Gelatin in the Preparation of Hydrogels

Derivative	Application	References
Type B gelatin	Gelatin gels were subjected to drying at 105°C (oven) under reduced pressure (vacuum desiccator) for 5 days, resulting in the formation of cross-links between lysine residues (ε-amino) and carboxyl groups. This provided a hydrogel product.	1
Gelatin	A 3% gelatin solution was cross-linked with glutaraldehyde, resulting in the formation of a hydrogel. The hydrogel was lyophilized resulting in porous scaffolds that could be used for tissue engineering.	2
Type B gelatin from bovine skin	Methacryl derivatives of gelatin were prepared by reaction with methacrylic anhydride. Cross-linked hydrogels were obtained by photopolymerization in the presence of a water-soluble free radicant photoinitiator.	3
Type A porcine skin gelatin	Biocompatibility studies on glutaraldehyde-cross-linked gelatin or interpenetrating networks of poly(ethylene glycol)diacrylate photopolymerized around gelatin. Both products elicited an inflammatory response, which was delayed by dexamethasone.	4
Bovine skin gelatin (type B)	Covalent blends were prepared with hyaluronan. Both gelatin and hyaluron were modified with 3,3′-dithiobis(propionic hydrazide) in the presence of EDC to allow modification at carboxylic groups. Subsequent treatment with dithiothreitol yielded the free thiol derivatives of thc two polymers. The modified proteins were mixed and allowed to dry in air with the concomitant formation of disulfide bonds.	5
Gelatin (type B)	Interpenetrating chains of poly(acrylamide) and gelatin cross-linked with glutaraldehyde, forming hydrogels for tissue engineering.	6,7
Gelatin (unflavored)	A composite was obtained from the coprecipitation of calcium phosphate (prepared calcium hydroxide and *o*-phosphoric acid) and gelatin. Cross-linkage of the product with glutaraldehyde resulted in the assembly of individual fibers along the preferential growth direction, suggesting a large conformational change on reaction with glutaraldehyde.	8
Type B gelatin	Gelatin was modified with *trans*-4-nitrocinnamoyl chloride, providing a derivative with *p*-nitrocinnamate groups that can be reversibly cross-linked by low-intensity UV light (365 nm) to form a gel which can be cleaved by 254 nm light.	9
Type A gelatin (bovine skin), type B gelatin (porcine skin), and cold water fish skin gelatin	Hydrogels were formed from the gelatins by γ-irradiation and electron beam irradiation. The extent of cross-linkage increased with dose and gelatin concentration. Satisfactory hydrogels are formed from gelatins without the addition of chemical reagents.	10
From acid-treated porcine skin type I gelatin	"Cationized" gelatin is prepared by the carbodiimide-mediated coupling of ethylenediamine, putrescine, and spermidine. The hydrogel prepared from these modified gelatins is used for the controlled release of plasmid DNA. Prolonged gene expression was observed with this derivative.	11

TABLE 5.11 (CONTINUED)
The Use of Gelatin in the Preparation of Hydrogels

Derivative	Application	References
Recombinant gelatin	Recombinant gelatin containing a sequence from the α-chain of human type I collagen was used for the preparation of hydrogels after modification with methacrylate.	12
Type B bovine (skin) gelatin	Porous gelatin scaffolds (methacryl modified; methacrylic acid anhydride) with varying pore sizes were prepared by removal of water from frozen hydrogels by lyophilization. A pore size gradient in the hydrogel was established by variation in the cryogenic parameters (gelatin concentration, cooling rate).	13,14

TABLE 5.12
Application of Fibrinogen and Fibrin in Hydrogels

Protein Form	Application	References
Fibrin	Chondrocytes suspended in fibrin and injected into polycaprolactone-based polyurethane scaffold.	1
Fibrin	Review of the use of 3D fibrin matrices for stimulation of angiogenesis and in tissue engineering.	2,3
Fibrinogen	Cross-linked matrix formed by photoactivation (titanium:sapphire laser at 800 nm) in the presence of rose Bengal. Biological activity is retained by the cross-linked matrix.	3
Fibrin	An engineered form of vascular endothelial growth factor (VEGF) containing a factor XIIIa cross-linking sequence was coupled to fibrin by the action of factor XIIIa. The coupled form of VEGF retains biological activity.	4
Fibrinogen	Scaffolds were prepared from denatured fibrinogens cross-linked with PEG-acrylates via photolysis. The scaffolds were used to form a hydrogel containing cells. Smooth muscle cells were able to penetrate the hydrogel via proteolysis.	5–8
Fibrinogen	Fibrin gels containing chondrocytes were prepared by the action of thrombin on fibrinogen. It is intended to develop a product for cartilage engineering.	9
Fibrinogen	Rat aortic smooth muscle cells were added to a solution containing collagen and fibrinogen to form a composite matrix for tissue engineering. It is noted that collagen has an influence on fibrin polymerization.[a]	10
Fibrin	Fibrin scaffolds were formed around poly (methyl-methacrylate) beads followed by dissolution of the beads. The mechanical strengths of fibrin scaffolds are enhanced by cross-linking with genepin.	11

[a] Jones, M. and Gabriel, D.A., Influence of the subendothelial basement membrane components on fibrin assembly. Evidence for a fibrin binding site on type IV collagen, *J. Biol. Chem.* 263, 7043–7048, 1988.

TABLE 5.13
Application of Elastin in the Preparation of Hydrogels

Protein Form	Application	Reference
Recombinant elastin-like polypeptides	Engineered elastin-like polypeptides consisting of Val-Pro-Gly-X-Gly repeats, where X can be Lys. The resulting products are cross-linked with tris-succinimidyl aminotriacetate to form hydrogels.	1
Recombinant elastin-like protein	A recombinant protein consisting of repeating elastin-derived sequences (VPGIG) and CSr cell-binding domains was cross-linked with hexamethylene diisocyanate in dimethylsulfoxide to form a hydrogel.	2
Recombinant elastin-like protein	A recombinant protein consisting of repeating elastin-derived sequences (VPGIG) and CS5 cell-binding domains was cross-linked with glutaraldehyde to form a hydrogel. Some of the elastin-like sequences were engineered to provide the VPGKG sequence for cross-linking.	3
Recombinant elastin-like (elastin mimetic) protein (polypeptide)	The signature elastin sequence is VPGZG where Z can be any amino acid. A 30-amino-acid sequence containing 1 inserted lysine and 1 inserted glutamic acid residues—VPGKGVPGVGPVPGVGVPGEGVPGIG. A hexahistidine (hexahis) sequence was included to provide for facile purification of the recombinant protein (expressed in *Escherichia coli*). Cross-linkage to yield a hydrogel was provided by reaction with bis(sulfo-succinimidyl)suberate.	4
Elastin-like polypeptide	An elastin-like polypeptide (ELP) containing lysine residue was expressed in *Escherichia coli*. The engineered protein contained a terminal hexahis sequence that could be identified by reaction with anti-his antibodies. The elastin-like polypeptide could be cross-linked with β [tris(hydroxymethyl)phosphino] propionic acid.	5
Copolymer based on elastin-like polypeptide	Triblock copolymers based on elastin-like polypeptides were expressed in *Escherichia coli*. The sequences were hydrophilic (charged) and hydrophobic. Hexahistidine sequences were inserted in some of the hydrophilic blocks, yielding a product that formed a hydrogel in the presence of selected metal ions. The product has application for removal of toxic metals.	6

TABLE 5.14
Hyaluronan/Hyaluronic Acid in Hydrogels[a]

Polymer	Application	References
Hyaluronan	Derivatization of hyaluronic acid at the carboxyl group to provide amine derivatives, hydrazide derivatives, and acetal derivatives, which can be used for the manufacture of hydrogels.	1
Hyaluronan	3D hyaluronic acid strands (hyaluronic acid was esterified and drawn into strands, which were then treated with glutaraldehyde and then coated with polylysine) were prepared with or without keratinocytes and used to treat full-thickness skin incision wounds in rats. The hyaluronic acid gel with or without cells improved wound healing and reduced scar formation.	2,3
Hyaluronan	Hyaluronic acid is converted to adipic dihydrazide and cross-linked with poly(ethylene glycol)propionldehyde to give product polymer network. A solvent casting method was used to obtain a hyaluronan film, which could be rehydrated to a product hydrogel film.	4
Hyaluronan	Disulfide linkages were introduced into hyaluronan by reaction at glucuronic acid carboxyl groups with either dithiobis(propanoic dihydrazide) or dithiobis(butyric hydrazide) using carbodiimide chemistry. The disulfide bonds could be reduced with dithiothreitol and the resultant thiol derivatives isolated and reoxidized to form hydrogel films.	5–7
Hyaluronan	Preparation of hyaluronan hydrogel by cross-linking with poly(ethylene glycol) diepoxide. Collagen could be incorporated into the hydrogel and the hyaluronan could be "functionalized" by the linkage of biotin.	8
Hyaluronan	Hyaluronan is cross-linked with 1,2,7,8-diepoxyoctane and glutaraldehyde, resulting in cross-links between hydroxyl groups at alkaline pH and between carboxyl groups at acid pH. The highly cross-linked material demonstrated increased stability.[b]	9
Hyaluronan	Adipic acid hydrazide grafted hyaluronan was cross-linked with bis(sulfosuccinimidyl suberate) to prepare a hydrogel. The metharyl derivative of the hydrazide derivative hyaluronan was prepared by reaction with methacrylic acid anhydride and cross-linked with dithiothreitol (Michael addition) to prepare a hydrogel. A third derivative was prepared from the hydrazide derivative and reacted with 2-iminothiolane (Traut's reagent) to yield a thiol derivative, which was cross-linked by disulfide formation mediated by sodium tetrathionate. The stability of the hydrogels was evaluated; the methacrylic acid derivatives were found to be the most stable.	10
Hyaluronan	Hydrogel prepared by cross-linking hyaluronan with chitosan using carbodiimide [1-ethyl-3-(3-dimethylaminopropyl) carbodiimide] chemistry.	11
Hyaluronan	Hyaluronan is modified with reagents containing terminal alkyl azide or alkyne. A hydrogel between the two derivatives is formed in the presence of copper (click chemistry[c]).	12

TABLE 5.14 (CONTINUED)
Hyaluronan/Hyaluronic Acid in Hydrogels[a]

Polymer	Application	References
Hyaluronan	Hydrogels were formed by cross-linking with a water-soluble carbodiimide, most likely resulting in ester bond formation. The formation and dissolution of the hydrogel is pH dependent. Dynamic light scattering is used to characterize the hydrogel.	13
Hyaluronan	Hyaluronan is functionalized with the methacryloyl derivative of β-cyclodextrin to yield a derivative with enhanced drug-binding capability.	14

[a] There are other examples of the use of hyaluronan in hydrogels in other tables, most notably in Table 5.9 on chemical modification and hydrogels.

[b] This reference has a discussion of the stability of various cross-linked hyaluronan products.

[c] Moses, J.E. and Moorhouse, A.D., The growth applications of click chemistry, *Chem. Soc. Rev.* 36, 1249–1262, 2007.

TABLE 5.15
Derivatives of Heparin Useful in Hydrogel

Heparin Source	Derivative/Application	Reference
Heparin	Heparin was immobilized onto poly(vinyl alcohol) hydrogel.	1
Heparin from porcine intestinal mucosa	Heparin was modified with adipic acid dihydrazide to provide a hydrazide derivative. The derivatized heparin was cross-linked by reaction with bis succinimidyl derivatives.	2
Heparin (species not given, molecular weight (MW 18,000)	Heparin modified with methacrylamide was mixed with diacryl-Pluronic F127 to provide a solution of macromonomers, which formed a hydrogel on photolysis.	3
	A thiol function is added to heparin via cardodiimide-mediated coupling of cysteamine to the carboxyl groups of glucuronic acid or iduronic acid. This derivative can be linked to diacrylate-terminated polymers such as poly(ethylene glycol)diacrylate to form a hydrogel.	4
Porcine intestinal mucosal heparin (MW 12,000)	The maleimide derivative of heparin was obtained by modification of heparin with *N*-(2-ethylamino) maleimide and incorporated into hydrogels by reaction with thiol-PEG derivatives.	5
Heparan sulfate	Heparan sulfate, which is closely related to heparin, was incorporated into a fibrin matrix during the coagulation process. The fibrin matrix served as a controlled delivery vehicle for the heparan sulfate.	6

REFERENCES

REFERENCES FOR TABLE 5.1

1. Nakaoka, R., Tabata, Y., and Ikada, Y., Adjuvant effect of biodegradable poly(DL-lactic acid) granules capable for antigen release following intraperitoneal injection, *Vaccine* 14, 1671–1676, 1996.
2. Jiang, G., Qiu, W., and DeLuca, P.P., Preparation and *in vitro/in vivo* evaluation of insulin-loaded poly(acroyl-hydroxyethyl starch)-PLGA composite microspheres, *Pharm. Res.* 20, 452–459, 2003.
3. Bagno, A., Genovese, M., Luchini, A. et al., Contact porfilometry and correspondence analysis to correlate surface properties and cell adhesion *in vitro* of uncoated and coated Ti and Ti6A14V disks, *Biomaterials* 25, 2437–2445, 2004.
4. Qian, M., Chen, D., Ma, X., and Liu, Y., Injectable biodegradable temperature-responsive PLGA-PEG-PLGA copolymers: Synthesis and effect of copolymer composition on the drug release from the copolymer-based hydrogels, *Int. J. Pharm.* 294, 103–112, 2005.
5. DeFaul, A.J., Chu, C.R., Izzo, N., and Marra, K.G., Controlled release of bioactive TGF-β_1 from microspheres embedded within biodegradable hydrogels, *Biomaterials* 27, 1579–1585, 2006.
6. Qiao, M., Chen, D., Ma, X., and Hu, H., Sustained release of bee venom peptide from biodegradable thermosensitive PLGA-PEG-PLGA triblock copolymer-based hydrogels *in vitro*, *Pharmazie* 61, 199–202, 2006.
7. Qiao, M., Chen, D., Hao, T. et al., Effect of bee venom peptide-copolymer interactions on thermosensitive hydrogel delivery systems, *Int. J. Pharm.* 345, 116–124, 2007.
8. Lee, P.Y., Cobain, E., Huard, J., and Huang, L., Thermosensitive hydrogel PEG-PLGA-PEG enhanced engraftment of muscle-derived stem cells and promotes healing in diabetic wound, *Mol. Ther.* 15, 1189–1194, 2007.

REFERENCES FOR TABLE 5.2

1. Beurmer, G.J., van Bitterswijk, C.A., Bakker, D., and Ponec, M., A new biodegradable matrix as part of a cell seeded skin substitute for the treatment of deep skin defects: A physico-chemical characterization, *Clin. Mater.* 14, 21–27, 1993.
2. Cho, K.Y., Choi, S.H., Kim, C.H. et al., Protein release microparticles based on the blend of poly (D,L-lactic-co-glycolic acid) and oligo-ethylene glycol grafted poly(L-lactide), *J. Control. Release* 76, 275–284, 2001.
3. Krupnick, A.S., Kreisal, D., Engels, F.H. et al., A novel small animal model of left ventricular tissue engineering, *J. Heart Lung Transplant.* 21, 233–243, 2002.
4. Mukose, T., Fujiwara, T., Nakano, J. et al., Hydrogel formation between enantiomeric B–A-B-type block copolymers of polylactides (PLLA or PDLA:A) and polyoxyethylene (PEG:B); PEG-PLLA-PEG and PEG-PDLA-PEG, *Macromol. Biosci.* 4, 361–367, 2004.
5. Ouchi, T., Saito, T, Kontani, T., and Ohya, Y., Encapsulation and/or release behavior of bovine serum albumin within and from polylactide–grafted dextran microspheres, *Macromol. Biosci.* 4, 458–463, 2004.
6. Goraltchouk, A., Freier, T., and Shoicher, M.S., Synthesis of degradable poly(L–lactide–co-ethylene glycol) porous tubes by liquid-liquid centrifugal casting for use as nerve guidance channels, *Biomaterials* 26, 7555–7563, 2005.
7. Gong, Y., He, L., Li, J. et al., Hydrogel-filled polylactide porous scaffolds for cartilage tissue engineering, *J. Biomed. Mater. Res. B. Appl. Biomater.* 82, 192–204, 2007.
8. Sarvestani, A.S., He, X., and Jabbari, E., Viscoelastic characterization and modeling of gelation kinetics of injectable *in situ* cross-linkable poly(lactide-co-ethylene oxide-co-fumarate) hydrogels, *Biomacromolecules* 8, 406–415, 2007.

9. He, X. and Jabbari, E., Material properties and cytocompatibility of injectable MMP degradable poly(lactide ethylene oxide fumate) hydrogel as a carrier for marrow stromal cells, *Biomacromolecules* 8, 780–792, 2007.
10. Oest, M.E., Dupont, K.M., Kong, H.J. et al., Quantitative assessment of scaffold and growth factor-mediated repair of critically sized bone defects, *J. Orthop. Res.* 25, 941–950, 2007.

References for Table 5.3

1. Gelfand, W.C., Knepshield, J.H., Cohan, S. et al., Treatment of hepatic coma with hemoperfusion through polyacrylamide hydrogel-coated charcoal, *Kidney Int. Suppl.* (7), S239–243, 1976.
2. de Lumley-Woodyear, T., Rocca, P., Lindsay, J. et al., Polyacrylamide-based redox polymer for connecting redox centers of enzymes to electrodes, *Anal. Chem.* 67, 1332–1338, 1995.
3. Lukaszczyk, J., Investigations on preparation and properties of modified polyacrylamide hydrogels for application to wound dressing materials, *Polim. Med.* 25, 15–23, 1995.
4. Risbud, M.V. and Bhonde, R.R., Polyacrylamide-chitosan hydrogels: *in vitro* biocompatibility and sustained antibiotic release, *Drug Deliv.* 7, 69–75, 2000.
5. Angenendt, P., Glöcker, J., Murphey, D. et al., Toward optimized antibody microarrays: A comparison of current microarray support materials, *Anal. Biochem.* 309, 253–260, 2002.
6. Lin, D.C., Yurke, B., and Langrana, N.A., Mechanical properties of a reversible, DNA-crosslinked polyacrylamide hydrogel, *J. Biomech. Eng.* 126, 104–110, 2004.
7. Nuño-Donlucas, S.M., Sánchez-Diaz, J.C., Rabelero, M. et al., Microstructured polyacrylamide hydrogels made with hydrophobic nanoparticles, *J. Colloid Interface Sci.* 270, 94–98, 2004.
8. Charles, P.T., Godman, E.R., Rangasammy, J.G. et al., Fabrication and characterization of 3D hydrogel microarrays to measure antigenicity and antibody functionality for biosensor application, *Biosens. Bioelectron.* 20, 753–764, 2004.
9. Murakami, Y. and Maeda, M., DNA-responsive hydrogels that can shrink or swell, *Biomacromolecules* 6, 2927–2929, 2005.
10. Thomas, G., El-Giar, E.M., Locascio, L.E., and Tarlov, M.J., Hydrogel-immobilized antibodies for microfluidic immunoassays: Hydrogel immunoassay, *Methods Mol. Biol.* 321, 83–95, 2006.

References for table 5.4

1. Flynn, L., Dalton, P.D., and Shoichet, M.S., Fiber templating of poly(2-hydroxyethyl methacrylate) for neural tissue engineering, *Biomaterials* 24, 4265–4272, 2003.
2. Cho, C.S., Han, S.Y., Ha, J.H. et al., Clonazepam release from bioerodible hydrogels based on semi-interpenetrating, polymer networks composed of poly(ε-caprolactone) and poly(ethylene glycol), *Int. J. Pharm.* 181, 235–242, 1999.
3. Endres, M., Humacher, D.W., Salgado, A.J. et al., Osteogenic induction of human bone marrow-derived mesenchymal progenitor cells in novel synthetic polymer-hydrogen matrices, *Tissue Eng.* 9, 698–702, 2003.
4. Wu, D., Zhang, X., and Chu, C.C., Synthesis, characterization and drug release from three-arm poly(ε-caprolactone)maleic acid/poly(ethylene glycol)diacrylate hydrogels, *J. Biomater. Sci. Polym.Ed.* 14, 777–802, 2003.
5. Kim, J.H., Park, S.K., and Bae,Y.H., *In situ* accelerated degradation of polyoxyethylene/poly(ε-caprolactone) multiblock copolymer by moderate thermal treatment, *J. Biomater. Sci. Polym. Ed.* 14, 903–916, 2003.
6. Zhu, A.P. and Chan-Park, M.B., Cell viability of chitosan-containing semi-interpenetrated hydrogels based on PCL-PEG-PCL diacrylate macromer, *J. Biomater. Sci. Polym. Ed.* 16, 301–316, 2005.

7. Eyrich, D., Wiese, H., Maier, G. et al., *In vitro* and *in vivo* cartilage engineering using a combination of chondrocyte-seeded long term stable fibrin gels and polycaprolactone-based polyurethane scaffolds, *Tissue Eng.* 13, 2207–2218, 2007.

References for Table 5.5

1. Burczak, K., Fujisato, T., Hatada, M., and Ikada, Y., Protein permeation through poly(vinyl alcohol) hydrogel membranes, *Biomaterials* 15, 231–238, 1994.
2. Cauich-Rodriquez, J.V., Deb, S., and Smith, R., Effect of cross-linking agents on the dynamic mechanical properties of hydrogel blends of poly(acrylic acid)-poly(vinyl alcohol-vinyl acetate), *Biomaterials* 17, 2259–2264, 1996.
3. Li, J.K., Wang, N., and Wu, X.S., Poly(vinyl alcohol) nanoparticles prepared by freezing-thawing process for protein/peptide drug delivery, *J. Control Release* 56, 117–126, 1998.
4. Cascone, M.G., Maltinti, S., Barbani, N., and Laus, M., Effect of chitosan and dextran on the properties of poly(vinyl alcohol) hydrogels, *J. Mater. Sci. Mater. Med.* 10, 431–435, 1999.
5. Hassan, C.M., Stewart, J.E., and Pappas, N.A., Diffusional characteristics of freeze/thawed poly(vinyl alcohol)hydrogels: Application to protein controlled release from multilaminate devices, *Eur. J. Pharm. Biopharm.* 49, 161–165, 2000.
6. Veronese, F.M., Mammucari, C., Schiavon, F. et al., Pegylated enzyme entrapped in poly(vinyl alcohol) hydrogel for biocatalytic application, *Farmaco* 56, 541–547, 2001.
7. Stammen, J.A., Williams, S., Ku, D.N., and Guldberg, R.E., Mechanical properties of a novel PVA hydrogel in shear and unconfined compression, *Biomaterials* 22, 799–806, 2001.
8. Orienti, I., Treré, R., Luppi, B. et al., Hydrogels formed by crosslinked poly(vinyl alcohol) as sustained drug delivery, *Arch. Pharm. (Weuiheim)* 335, 89–93, 2002.
9. Nurkeeva, Z.S., Mun, G.A,. Dubolazov, A.V., and Khutoryanskiy, V.V., pH Effects on the complexation, miscibility and radiation-induced crosslinking in poly(acrylic acid)-poly(vinyl alcohol) blends, *Macromol. Biosci.* 5, 424–432, 2005.
10. Peterbauer, T., Heitz, J., Olbrich, M., and Hering, S., Simple and versatile methods for the fabrication of arrays of live mammalian cells, *Lab Chip.* 6, 857–863, 2006.
11. Bodugoz-Senturk, H., Choi, J., Oral, E. et al., The effect of polyethylene glycol on the stability of pores in polyvinyl alcohol hydrogels during annealing, *Biomaterials* 29, 141–149, 2008.
12. Fujisato, T. and Kishida, A., Preparation of poly(vinyl alcohol)/DNA hydrogels via hydrogen bonds formed on ultra-high pressurization and controlled release of DNA from the hydrogels for gene delivery, *J. Artif. Organs* 10, 104–108, 2007.

References for Table 5.6

1. Beena, M.S., Chandy, T., and Sharma, C.P., Heparin immobilized chitosan-poly ethylene glycol interpenetrating network: Antithrombogenicity, *Artif. Cells Blood Sustit. Immobil. Biotechnol.* 23, 175–192, 1995.
2. Iza, M., Stoianovici, G., Viora, L. et al., Hydrogels of poly(ethylene glycol) mechanical characterization and release of a model drug, *J. Control. Release* 52, 41–51, 1998.
3. Markland, P., Zhang, Y., Amidon, G.L., and Yang, V.C., A pH- and ionic strength-responsive polypeptide hydrogel: Synthesis, characterization, and preliminary protein release studies, *J. Biomed. Mater. Res.* 47, 595–602, 1999.
4. Shung, A.K., Behravesh, E., Jo, S., and Mikos, A.G., Crosslinking characteristics of and cell adhesion to an injectable poly(propylene fumarate-co-ethylene glycol) hydrogel using a water-soluble crosslinking system, *Tissue Eng.* 9, 243–254, 2003.
5. Wang, D.A., Williams, C.G., Li, Q. et al., Synthesis and characterization of a novel degradable phosphate-containing hydrogel, *Biomaterials* 24, 3969–3980, 2003.

6. Bhattarai, N., Ramay, H.R., Gunn, J. et al., PEG-grafted chitosan as an injectable thermosensitive hydrogel for sustained protein release, *J. Control. Release* 103, 609–624, 2005.
7. Fänger, C., Wack, H., and Ulbricht, M., Macroporous poly(*N*-isopropylacrylamide) hydrogels with adjustable size "cut-off" for the efficient and reversible immobilization of biomacromolecules, *Macromol. Biosci.* 6, 393–402, 2006.
8. Snyders, R., Shingel, K.I., Zabeida, O. et al., Mechanical and microstructural properties of hybrid poly(ethylene glycol)-soy protein hydrogels for wound dressing applications, *J. Biomed. Mater. Res. A.* 83, 88–97, 2007.
9. Saaem, I., Papasotiropoulos, V., Wang, T., Hydrogel-based protein nanoarrays, *J. Nanosci. Nanotechnol.* 7, 2623–2632, 2007.
10. Sun, G., Zhang, X.Z., and Chu, C.C., Formulation and characterization of chitosan based hydrogel films having both temperature and pH sensitivity, *J. Mater. Sci. Mater. Med.* 18, 1563–1577, 2007.

References for Table 5.7

1. Park, T.G. and Hoffman, A.S., Estimation of temperature-dependent pore size in poly(*N*-isopropylacrylamide) hydrogel beads, *Biotechnol. Prog.* 10, 82–86, 1994.
2. von Recum, H.A., Kim, S.W., Kikuchi, A. et al., Novel thermally reversible hydrogel as detachable cell culture substrate, *J. Biomed. Mater. Res.* 40, 631–639, 1998.
3. Makino, K., Agata, H., and Ohshima, H., Dependence of temperature-sensitivity of poly(*N*-isopropylacrylamide-co-acrylic acid) hydrogel microspheres upon their sizes, *J. Colloid Interface Sci.* 230, 128–134, 2000.
4. Pei, Y., Chen, J., Yang, L. et al., The effect of pH on the LCST of poly(*N*-isopropylacrylamide) and poly(*N*-isopropylacrylamide-co-acrylic acid), *J. Biomater. Sci. Polym. Ed.* 15, 585–594, 2004.
5. Ramanan, R.M., Chellamuthu, P., Tang, L., and Nguyen, K.T., Development of a temperature-sensitive composite hydrogel for drug delivery applications, *Biotechnol. Prog.* 22, 118–125, 2006.
6. Lee, B.H., West, B., McLemore, R. et al., In-situ injectable physically and chemically guided gelling NiPAAm-based copolymer system for embolization, *Biomacromolecules* 7, 2059 2064, 2006.
7. Liu, K., Ovaert, T.C., and Mason, J.J., Preparation and mechanical characterization of a PNiPA hydrogel composite, *J. Mater. Sci. Mater. Med.* 19(4): 1815–1821, April, 2008.

References for Table 5.8

1. McCarron, P.A., Woolfon, A.D., and Keating, S.M., Sustained release of 5-fluorouracil from polymeric nanoparticles, *J. Pharm. Pharmacol.* 52, 1451–1459, 2000.
2. Jeong, B., Bae, Y.H., and Kim, S.W., Drug release from biodegradable injectable thermosensitive hydrogel of PEG-PLGA-PEG triblock copolymers, *J. Control. Release* 63, 155–163, 2000.
3. Mandal, T.K., Bostanian, L.A., Graves, R., and Chapman, S.R., Poly(D,L-lactide-co-glycolide) encapsulated poly(vinyl alcohol) hydrogel as a drug delivery system, *Pharm. Res.* 19, 1713–1719, 2002.
4. Chung, Y.M., Simmons, K.L., Gutowska, A., and Jeong, B., Sol-gel transition temperature of PLGA-g-PEG aqueous solutions, *Biomacromolecules* 3, 511–516, 2002.
5. Luginbuehl, V., Wenk, E., Koch, A. et al., Insulin-like growth factor 1-releasing alginate-tricalciumphosphate composites for bone regeneration, *Pharm. Res.* 22, 940–950, 2005.
6. Yang, H. and Kao, W.J., Thermosensitive gelatin/monomethoxy poly(ethylene glycol)-poly(D,L-lactide) hydrogels: Formulation, characterization, and antibacterial drug delivery, *Pharm. Res.* 23, 205–214, 2006.

7. Ashton, R.S., Banerjee, A., Punyani, S. et al., Scaffolds based on degradable alginate hydrogels and poly(lactide-co-glycoide) microspheres for stem cell cultures, *Biomaterials* 28, 5518–5525, 2007.

References for Table 5.9

1. McAuslan, B.R. and Johnson, G., Cell responses to biomaterials. I: Adhesion and growth of vascular endothelial cells on poly(hydroxyethyl methacrylate) following surface modification by hydrolytic etching, *J. Biomed. Mater. Res.* 21, 921–935, 1987.
2. Sipehia, R., Garfinkle, A., Jackson, W.B., and Chang, T.M., Toward an artificial cornea: Surface modifications of optically clear, oxygen permeable soft contact lens materials by ammonia plasma modification technique for the enhanced attachment and growth of corneal epithelial cells, *Biomater. Artif. Cells Artif. Organs* 18, 643–655, 1990.
3. Prestwick, G.D., Marecak, D.M., Marecek, J.F. et al., Controlled chemical modification of hyaluronic acid: Synthesis, applications, and biodegradation of hydrazide derivatives, *J. Control. Release* 53, 93–103, 1998.
4. Bulpttt, P. and Aeschlimann, D., New strategy for chemical modification of hyaluronic acid: Preparation of functionalized derivatives and their use in the formation of novel biocompatible hydrogels, *J. Biomed. Mater. Res.* 47, 152–169, 1999.
5. Gliko-Kabir, I., Lack, S., Le Cerf, D. et al., Hyaluron-based hydrogel particles prepared by crosslinking with trisodium trimetaphosphate. Synthesis and characterization, *Carbohydrate Polymers* 47, 1–6, 2000.
6. Gliko-Kabir, I., Yagen, B., Baluom, M., and Rubenstein, A., Phosphate crosslinked guar for color-specific drug delivery. II. *In vitro* and *in vivo* evaluation in the rat, *J. Control. Release* 63, 129–134, 2000.
7. Smeds, K.A., Pfister-Serres, A., Miki, D. et al., Photocrosslinkable polysaccharides for *in situ* hydrogel formation, *J. Biomed. Mater. Res.* 54, 115–121, 2001.
8. Cao, X. and Shoichet, M.S., Photoimmobilization of biomolecules within a 3-dimensional matrix, *J. Biomater. Sci. Polym. Ed.* 13, 623–636, 2003.
9. Segura, T., Anderson, B.C., Chung, P.H. et al., Crosslinked hyaluronic acid hydrogels: A strategy to functionalize and pattern, *Biomaterials* 26, 359–371, 2005.
10. Li, Y., Neoh, K.G,. and Kang, E.T., Controlled release of heparin from polypyrrole-poly(vinyl alcohol) assembly by electrical stimulation, *J. Biomed. Mater. Res. A.* 73, 171–181, 2005.
11. Hahn, S.K., Oh, E.J., Miyamoto, H., and Shimobouji, T., Sustained release formulation of erythropoietin using hyaluronic acid hydrogels crosslinked by Michael addition, *Int. J. Pharm.* 322, 44–51, 2006.
12. Robb, S.A., Lee, B.H., McLemore, R., and Vernon, B.L., Simultaneously physically and chemically gelling polymer system using a poly(NIPAAm-co-cystamine)-based copolymer, *Biomacromolecules,* 8, 2294–2300, 2007.
13. Nie, T., Baldwin, A., Yamaguchi, N., and Kiick, K.L., Production of heparin-functionalized hydrogels for the development of responsive and controlled growth factor delivery systems, *J. Control. Release* 122, 287–296, 2007.
14. Serban, M.A., Yang, G., and Prestwich, G.D., Synthesis, characterization and chondroprotective properties of a hyaluronan thioethyl ether derivative, *Biomaterials* 29, 1388–1399, 2008.
15. Kommareddy, S. and Amiji, M., Preparation and evaluation of thiol-mediated gelatin nanoparticles for intracellular DNA delivery in response to glutathione, *Bioconjugate Chem.* 16, 1423–1432, 2005.

References for Table 5.10

1. Toselli, P., Mogayzel, P.J., Jr., Faris, B. et al., Mammalian cell growth on collagen-hydrogels, *Scan. Electron Microsc.* (pt 3), 1301–1303, 1984.
2. Amudeswari, S., Nagarajan, B., Reddy, C.R., and Joseph, K.T., Short-term biocompatibility studies of hydrogel-grafted collagen copolymers, *J. Biomed. Mater. Res.* 20, 1103–1109, 1986.

X. Nicolas, F.L. and Gagnieu, C.H., Denatured thiolated collagen: I. Synthesis and characterization, *Biomaterials* 18, 807–813, 1997.

X1. Nicolas, F.L. and Gagnieu, C.H., Denatured thiolated collagen: II. Cross-linked by oxidation, *Biomaterials* 18, 815–821, 1997.

3. Velegol, D. and Lanni, F., Cell traction forces on soft biomaterials: I. Microrheology of type I collagen gels, *Biophys. J.* 81, 1786–1792, 2001.
4. Rosenblatt, J., Devereux, B., and Wallace, D.G., Injectable collagen as a pH-sensitive hydrogel, *Biomaterials* 15, 985–995, 1994.
5. Noishiki, Y., Ma, X.H., Yamane, Y. et al., Succinylated collagen crosslinked by thermal treatment for coating vascular prostheses, *Artif. Organs* 22, 672–680, 1998.
6. Hinno, S., Zhang, M., Nakagawa, M., and Miyata, T., Wet spun chitosan-collagen fibers, their chemical *N*-modifications, and blood compatibility, *Biomaterials* 21, 997–1003, 2000.

6a. Brinkman, W.T., Nagapudi, K., Thomas, B.S., and Chaikof, E.L., Photo-cross-linking of type I collagen gels in the presence of smooth muscle cells: Mechanical properties, cell viability, and function, *Biomacromolecules* 4, 890–895, 2003.

7. Takezawa, T., Ozaki, K., Nitani, A. et al., Collagen vitrigel: A novel scaffold that can facilitate a three-dimensional culture for reconstructing organoids, *Cell Transplant.* 13, 463–473, 2004.
8. Cote, M.F., Laroche, G., Gagnon E. et al., Denatured collagen as support for a FGF-2 delivery system: Physicochemical characterization and *in vitro* release kinetics and bioactivity, *Biomaterials* 25, 3761–3772, 2004.
9. Mao, C. and Kisaalita, W.S., Characterization of 3-D collagen hydrogels for functional cell-based biosensing, *Biosens. Bioelectron.* 19, 1075–1088, 2004.
10. Inoue, N., Bessho, M., Furuta, M. et al., A novel collagen hydrogel cross-linked by gamma-ray irradiation in acidic pH conditions, *J. Biomater. Sci. Polym. Ed.* 17, 837–858, 2006.
11. Kim, J.K., Lee, J.S., Jung, H.J. et al., Preparation and properties of collagen/modified hyaluronic acid hydrogel for biomedical applications, *J. Nanosci. Nanotechnol.* 7, 3852–3856, 2007.
12. Deister, C., Aliabari, S., and Schmidt, C.E., Effects of collagen 1, fibronectin, laminin and hyaluronic acid concentration in multi-component gels on neurite extension, *J. Biomater. Sci. Polym. Ed.* 18, 983–997, 2007.
13. Rousseau, C.F. and Gagnieu, C.H., *In vitro* cytocompatibility of porcine type I atelocollagen crosslinked by oxidized protein, *Biomaterials* 23, 1503–1510, 2002.
14. Friess, W., Uludag, H., Foskett, S. et al., Characterization of absorbable collagen sponges as rhBMP-2 carriers, *Int. J. Pharmaceut.* 187, 91–99, 1999.

References for Table 5.11

1. Welz, M.M. and Ofner, C.M., 3rd, Examination of self-crosslinked gelatin as a hydrogel for controlled release, *J. Pharm. Sci.* 81, 85–90, 1992.
2. Kang, H.W., Tabata, Y., and Ikada, Y., Fabrication of porous gelatin scaffolds for tissue engineering, *Biomaterials* 20, 1339–1344, 1999.
3. Van Den Bulcke, A.L., Bodganov, B. et al., Structural and rheological properties of methacrylamide modified gelatin hydrogels, *Biomacromolecules* 1, 31–38, 2000.

4. Stevens, K.R., Einerson, N.J., Barmania, J.A, and Kao, W.J., *In vivo* biocompatibility of gelatin-based hydrogels and interpenetrating networks, *J. Biomater. Sci. Polym. Ed.* 13, 1353–1366, 2002.
5. Shu, X.Z., Liu, Y., Palumbo, F., and Prestwich, G.D., Disulfide-crosslinked hyaluronan-gelatin hydrogel films: A covalent mimic of the extracellular matrix for *in vitro* cell growth, *Biomaterials* 24, 3825–3834, 2003.
6. Burugapalli, K., Bhatia, D., Koul, V., and Choudhary, V., Interpenetrating polymer networks based on poly(acrylic acid) and gelatin. I: Swelling and thermal behavior, *J. Appl. Polym. Sci.* 82, 217–227, 2001.
7. Burugapalli, K., Koul, V., and Dinda, A., Effect of composition of interpenetrating polymer network hydrogels based on poly(acrylic acid) and gelatin on tissue response: A quantitative *in vivo* study, *J. Biomed. Mater. Res.* 68A, 210–218, 2004.
8. Chang, M.C., Ko, C.-C., and Douglas, W.H., Conformational change of hydroxyapatite/gelatin nanocomposite by glutaraldehyde, *Biomaterials* 24, 3087–3094, 2003.
9. Gattás-Asfura, K.M., Weisman, E., Andreopoulos, F.M. et al., Nitrocinnamate-functionalized gelatin synthesis and "smart" hydrogel formation, *Biomacromolecules* 6, 1503–1509, 2005.
10. Terao, K., Nagasawa, N., Nishida, H. et al., Reagent-free crosslinking of aqueous gelatin: Manufacture and characteristics of gelatin gels irradiated with gamma-ray and electron beam, *J. Biomater. Sci. Polym. Ed.* 14, 1197–1208, 2003.
11. Kushibiki, T., Tomoshige, R., Iwanaga, K. et al., Controlled release of plasmid DNA from hydrogels prepared from gelatin cationized by different amine compounds, *J. Control. Release* 112, 247–256, 2006.
12. Sutter, M., Siepmann, J., Hennink, W.E., and Jiskoot, W., Recombinant gelatin hydrogels for the sustained release of proteins, *J. Control. Release* 119, 301–312, 2007.
13. Vlierberghe, S.V., Cnudde, V., Dubruel, P. et al., Porous gelatin hydrogels: I. Cryogenic formation and structure analysis, *Biomacromolecules* 8, 331–337, 2007.
14. Dubruel, P., Unger, R., Vlierberghe, S.V. et al., Porous gelatin hydrogels: 2. *In vitro* cell interaction study, *Biomacromolecules* 6, 338–334, 2007.

References for Table 5.12

1. Eyrich, D., Wiese, H., Maier, G. et al., *In vitro* and *in vivo* cartilage engineering using a combination of chondrocyte-seeded long-term stable fibrin gels and polycaprolactone-based polyurethanes scaffolds, *Tissue Eng.* 13, 2207–2217, 2007.
2. Hall, H., Modified fibrin hydrogel matrices: Both, 3D-scaffolds and local and controlled release systems to stimulate angiogenesis, *Curr. Pharm. Design.* 13, 3597–1607, 2007.
3. Basu, S. and Campagnola, P.J., Properties of crosslinked protein matrices for tissue engineering applications synthesized by multiphoton excitation, *J. Biomed. Mater. Res.* 71A, 359–368, 2004.
4. Zisch, A.H., Schenk, U., Schense, J.C. et al., Covalently conjugated VEGF-fibrin matrices for endothelialization, *J. Control. Release* 72, 101–113, 2001.
5. Almany, L. and Seliktar, D., Biosynthetic hydrogel scaffolds made from fibrinogen and polyethylene glycol for 3D cell cultures, *Biomaterials* 26, 2467–2477, 2005.
6. Dikovsky, D., Bianco-Peled, H., and Seliktar, D., The effect of structural alterations of PEG-fibrinogen hydrogel scaffolds on 3-D cellular morphology and cellular migration, *Biomaterials* 27, 1496–1506, 2006.
7. Schmidt, O., Mizrahi, J., Elisseeff, J., and Seliktar, D., Immobilized fibrinogen in PEG hydrogels does not improve chondrocyte-mediated matrix deposition in response to mechanical stimulation, *Biotechnol. Bioeng.* 95, 1061–1069, 2006.
8. Peled, E., Boss, J., Bejar, J. et al., A novel poly (ethylene glycol) hydrogel for tibial segmental defect repair in a rat model, *J. Biomed. Mater. Res.* A 80, 874–884, 2007.

9. Eyrich, D., Brandl, F., Appel, B. et al., Long-term stable fibrin gels for cartilage engineering, *Biomaterials* 28, 55–65, 2007.
10. Rowe, S.L. and Stegemann, J.P., Interpenetrating collagen-fibrin composite matrices with varying protein contents and ratios, *Biomacromolecules* 7, 2742–2748, 2006.
11. Linnes, M.P., Ratner, B.D., and Giachelli, C.M., A fibrinogen-based precision microporous scaffold for tissue engineering, *Biomaterials* 28, 5298–5306, 2007.

References for Table 5.13

1. Trabbic-Carison, K., Setton, L.A., and Chilkoti, A., Swelling and mechanical behaviors of chemically cross-linked hydrogels of elastin-like polypeptides, *Biomacromolecules* 4, 572–580, 2003.
2. Nowatzki, P.J. and Tirrell, D.A., Physical properties of artificial extracellular matrix protein films prepared by isocyanate crosslinking, *Biomaterials* 25, 1261–1267, 2004.
3. Girotti, A., Reguera, A., Rodriguez-Cabello, J.C. et al., Design and bioproduction of a recombinant multi(bio)functional elastin-like protein polymer containing cell adhesion sequences for tissue engineering purposes, *J. Mater. Sci. Mater. Med.* 15, 479–484, 2004.
4. Junger, A., Kaufmann, D., Scheibel, T., and Weberskirch, R., Biosynthesis of an elastin-mimetic polypeptide with two different chemical functional groups within the repetitive elastin fragments, *Macromol. Biosci.* 5, 494–501, 2005.
5. Ong, S.R., Trabbie-Carlson, K.A., Nettles, D.L. et al., Epitope tagging for tracking elastin-like polypeptides, *Biomaterials* 27, 1930–1935, 2006.
6. Lao, U.L., Sun, M., Matsumoto, M. et al., Genetic engineering of self-assembled protein hydrogel based on elastin-like sequences with metal binding functionality, *Biomacromolecules* 8, 3736–3739, 2007.

References for Table 5.14

1. Bulpitt, P. and Aeschlimann, D., New strategy for chemical modification of hyaluronic acid: Preparation of functionalized derivatives and their use in the formation of novel biocompatible hydrogels, *J. Biomed. Mater. Res.* 47, 152–169, 1999.
2. Hu, M., Sabelman, E.E., Lai, S. et al., Polypeptide resurfacing method improves fibroblast's adhesion to hyaluronan strands, *J. Biomed. Mater. Res.* 47, 79–84, 1999.
3. Hu, M., Sabelman, E.E., Cao, Y. et al., Three-dimensional hyaluronic acid grafts promote healing and reduce scar formation in skin incisional wounds, *J. Biomed. Mater. Res.* 65B, 586–592, 2003.
4. Luo, Y., Kirker, K.R., and Prestwich, G.D., Cross-linked hyaluronic acid hydrogel films: New biomaterials for drug delivery, *J. Control. Release* 69, 169–184, 2000.
5. Shu, X.Z., Liu, Y., Roberts, M.C. et al., Disulfide cross-linked hyaluronan hydrogels, *Biomacromolecules* 3, 1304–1311, 2002.
6. Sue, X.Z., Palumbo, F., and Prestwich, G.D., Disulfide-crosslinked hyaluronan-gelatin hydrogel films: A covalent mixture of the extracellular matrix for *in vitro* cell growth, *Biomaterials* 24, 3825–3834, 2003.
7. Shu, X.Z., Liu, Y., Palumbo, F.S. et al., *In situ* crosslinked hyaluronan hydrogels for tissue engineering, *Biomaterials* 25, 1339–1348, 2004.
8. Segura, T., Anderson, B.C., Chung, P.H. et al., Crosslinked hyaluronic acid hydrogels: A strategy to functionalize and pattern, *Biomaterials* 26, 359–371, 2005.
9. Zhao, X., Synthesis and characterization of a novel hyaluronic acid hydrogel, *J. Biomater. Sci. Polym. Ed.* 17, 419–433, 2006.
10. Hahn, S.K., Park, J.K., Tomimatsu, T., and Shimoboji, T., Synthesis and degradation test of hyaluronic acid hydrogels, *Int. J. Biol. Macromolecules* 40, 374–380, 2007
11. Wang, W., A novel hydrogel crosslinked hyaluronan with glycol chitosan, *J. Mater. Sci. Mater. Med.* 17. 1259–1265, 2006.

12. Crescenzi, V., Cornelio, L., Di Meo, C. et al., Novel hydrogels via click chemistry: Synthesis and potential biomedical applications, *Biomacromolecules* 8, 1844–1850, 2007.
13. Maleki, A., Kjoniksen, A.L., and Nyström, B., Characterization of the chemical degradation of hyaluronic acid during chemical gelation in the presence of different cross-linker agents, *Carbohydr. Res.* 342, 2776–2792, 2007.
14. Zawko, S.A., Truong, Z., and Schmidt, C.D., Drug-binding hydrogels of hyaluronic acid functionalized with β-cyclodextrin, *J. Biomed. Mater. Res. A*, 87, 1044–1052, 2008.

References for Table 5.15

1. Llanos, G.R. and Sefton, M.V., Heparin-poly(ethylene glycol)-poly(vinyl alcohol)hydrogel: Preparation and assessment of thrombogenicity, *Biomaterials* 13, 421–424, 1992.
2. Tae, G., Scatena, M., Stayton, P.S., and Hoffman, A.S., PEG-cross-linked heparin is an affinity hydrogel for sustained release of vascular endothelial growth factor, *J. Biomater. Sci. Polym. Ed.* 17, 187–197, 2006.
3. Yoon, J.J., Chung, H.J., and Park. T.G., Photo-crosslinkable and biodegradable Pluronic/heparin hydrogels for local and sustained delivery of angiogenic growth factor, *J. Biomed. Mater. Res. A.* 83, 597–605, 2007.
4. Tae, G., Kim, Y.-J., Choi, W.-H. et al., Formation of a novel heparin-based hydrogel in the presence of heparin-binding biomolecules, *Biomacromolecules* 8, 1979–1986, 2007.
5. Nie, T., Baldwin, A., Yamaguchi, N., and Kiick, K.L., Production of heparin-functionalized hydrogels for the development of responsive and controlled growth factor delivery systems, *J. Control. Release* 122, 287–296, 2007.
6. Woodruff, M.A., Rath, S.N., Susanto, E. et al., Sustained release and osteogenic potential of heparan sulfate-doped fibrin glue scaffolds within a rat cranial model, *J. Mol. Histol.* 38, 425–433, 2007.

Chapter References

1. Young, S., Wong, M., Tabata, Y., and Mikos, A.G., Gelatin as a delivery vehicle for the controlled release of bioactive molecules, *J. Control. Release* 109, 256–274, 2005.
2. Dusek, K., *Responsive Gels; Volume Transitions*, Springer-Verlag, Berlin, 1993.
3. Zrinyl, N., *Gels,* Springer, Darmstadt, 1996; McCormick, C.L., *Stimuli-Responsive Water Soluble and Amphiphilic Polymers,* American Chemical Society, Washington, DC, 2001.
4. Dumitriu, S., *Polymeric Biomaterials,* Marcel Dekker, New York, 2002.
5. Jhon, M.S. and Andrade, J.D., Water and hydrogels, *J. Biomed. Mat. Res.* 7, 509–522, 1973.
6 Roorda, W., Do hydrogels contain different classes of water, *J. Biomater. Sci. Polym. Ed.* 5, 383–395, 1994.
7. Omidian, H., Rocca, J.G., and Park, K., Advances in superporous hydrogels, *J. Control. Release* 102, 3–12, 2005.
8. Frokjaer, S. and Otzen, D.E., Protein drug stability: A formulation challenge, *Nat. Rev. Drug Discov.* 4, 298–306, 2005.
9. Kashyap, N., Kumar, N., and Kumar, K.N., Hydrogels for pharmaceutical and biomedical applications, *Crit. Rev. Ther. Drug Carrier Syst.* 22, 107–149, 2005.
10. Fairman, R. and Akerfeldt, K.S., Peptides as novel smart materials, *Curr. Opin. Struct. Biol.* 15, 453–463, 2005.
11. Tamon, H. and Ishizaka, H., SAXS study on gelation process in preparation of resorcinol-formaldehyde aerogel, *J. Colloid Interface Sci.* 206, 577–582, 1998.
12. Ying, T.Y., Yang, K.L., Yiacoumi, S., and Tsouris, C., Electrosorption of ions from aqueous solutions by nanostructured carbon aerogel, *J. Colloid Interface Sci.* 250, 18–27, 2002.
13. Yamamoto, T., Mukai, S.R., Endo, A. et al., Interpretation of structure formation during the sol-gel transition of a resorcinol-formaldehyde solution by population balance, *J. Colloid Interface Sci.* 264, 532–537, 2003.

14. Smirnova, I., Suttiruengwong, S., Seiler, M., and Arlt, W., Dissolution rate enhancement by adsorption of poorly soluble drugs on hydrophilic silica aerogels, *Pharm. Dev. Technol.* 9, 443–452, 2004.
15. Venkateswara Rao, A., Kulkarni, M.M., and Bhagat, S.D., Transport of liquids using superhydrophobic aerogels, *J. Colloid Interface Sci.* 285, 413–418, 2005.
16. Valentin, R., Horga, R., Bonelli, B. et al., FTIR spectroscopy of NH^3 on acidic and ionotropic alginate aerogels, *Biomacromolecules* 7, 877–882, 2006.
17. Carsula, M.F., Loche, D., Marras, S. et al., Role of urea in the preparation of highly porous nanocomposite aerogels, *Langmuir* 23, 3509–3512, 2007.
18. Steiner, S.A, 3rd, Baumann, T.F., Kong, J. et al., Iron-doped carbon aerogels: Novel porous substrates for direct growth of carbon nanotubes, *Langmuir* 23, 5161–5166, 2007.
19. Koziol, K., Vilatela, J., Moisala, A. et al., High-performance carbon nanotube fiber, *Science* 318, 1892–1895, 2007.
20. Gavillon, R. and Budtova, T., Aerocellulose: New highly porous cellulose prepared from cellulose prepared from cellulose-NaOH aqueous solutions, *Biomacromolecules* 9, 269–277, 2008.
21. Nguyen, K.T. and West, J.L., Photopolymerizable hydrogels for tissue engineering applications, *Biomaterials* 23, 4307–4314, 2002.
22. Boland, T., Xu, T., Damon, B., and Cui, X., Applications of inkjet printing to tissue engineering, *Biotechnol. J.* 1, 910–912, 2006.
23. Tessmar, J.K. and Gopferich, A.M., Customized PEG-derived copolymers for tissue engineering applications, *Macromol. Biosci.* 7, 23–39, 2007.
24. Fedorovich, N.E., Alblas, J., de Wijn, J.H. et al., Hydrogels as extracellular matrices for skeletal tissue engineering: State-of-the-art and novel applications in organ printing. *Tissue Eng.* 13, 1905–1925, 2007.
25. Baroli, B., Hydrogels for tissue engineering and delivery of tissue-inducing substances, *J. Pharm. Sci.* 96, 2197–2223, 2007.
26. Hynd, M.R., Turner, J.N., and Shain, W., Applications of hydrogels for neural cell engineering, *J. Biomater. Sci. Polym. Ed.* 18, 1223–1244, 2007.
27. Tiller, J.C., Increasing the local concentration of drugs by hydrogel formation, *Angew. Chem. Int. Ed. Engl.* 42, 3072–3075, 2003.
28. Sande, S.A., Pectin-based oral drug delivery to the colon, *Expert Opin. Drug Deliv.* 2, 441 450, 2005.
29. Ludwig, A., The use of mucoadhesive polymers in ocular drug delivery, *Adv. Drug Deliv. Rev.* 57, 1595 1639, 2005.
30. Coviello, T., Matricardi, P., and Alhaique, F., Drug delivery strategies using polysaccharide gels, *Expert Opin. Drug Deliv.* 3, 395–404, 2006.
31. Nanjawade, B.K., Manvi, F.V., and Manjappa, A.S., *In situ*-forming hydrogels for sustained ophthalmic drug delivery, *J. Control Release* 122, 119–134, 2007.
32. Wheeler, J.C., Woods, J.A., Cox, M.J. et al., Evolution of hydrogel polymers as contact lenses, surface coatings, dressings, and drug delivery systems, *J. Long Term. Eff. Med. Implants* 6, 207–217, 1996.
33. Yasuda, H., Biocompatibility of nanofilm-encapsulated silicone and silicone-hydrogel contact lenses, *Macromol. Biosci.* 6, 121–138, 2006.
34. Stapleton, F., Stretton, S., Papas, E. et al., Silicone hydrogel contact lenses and the ocular surface, *Ocul.Surf.* 4, 24–43, 2006.
35. Keay, L., Edwards, K., and Stapleton, F., An early assessment of silicone hydrogel safety: Pearls and pitfalls, and current status, *Eye Contact Lens.* 33, 358–361, 2007.
36. O'Connor, S.M., Andreadis, J.D., Shaffer, K.M. et al., Immobilization of neural cells in three-dimensional matrices for biosensor applications, *Biosens. Bioelectron.* 14, 871–881, 2000.

37. Behravesh, E. and Mikos, A.G., Three-dimensional culture of differentiating marrow stromal osteoblasts in biomimetic poly(propylene fumerate-co-ethylene glycol)-based macroporous hydrogels, *J. Biomed. Mater. Res. A* 66, 698–706, 2003.
38. Hwang, N. S., Kim, M. S., Sampattavanich, S. et al., Effects of three-dimensional culture and growth factors of the chondrogenic differentiation of murine embryonic stem cells, *Stem Cells* 24, 284–291, 2006.
39. Smith, L.E., Rimmer, S., and MacNeil, S., Examination of the effects of poly(*N*-vinylpyrrolidinone) hydrogels in direct and indirect contact with cells, *Biomaterials* 27, 2806–2812, 2006.
40. Li, Y.J., Chung, E.H., Rodriquez, R.T. et al., Hydrogels as artificial matrices for human embryonic stem cell self-renewal, *J. Biomed. Mater. Res. A.* 79, 1–5, 2006.
41. Leany, P., Prádny, M., Jendelová, P. et al., Macroporous hydrogels based on 2-hydroxyethyl methacrylate. Part 4: Growth of rat bone marrow stromal cells in three-dimensional hydrogels with positive and negative surface charges and in polyelectrolyte complexes, *J. Mater. Sci. Mater. Med.* 17, 829–833, 2006.
42. Liu, Y. and Wang, S., 3D inverted opal hydrogel scaffolds with oxygen sensing capability, *Colloids Surf. B Biointerfaces* 58, 8–13, 2007.
43. Prestwich, G.D., Evaluating drug efficiency and toxicology in three-dimensions: Using synthetic extracellular matrices in drug discovery, *Acc. Chem. Res.* 41, 139–148, 2008.
44. David, L., Dulong, V., Le Cerf. D. et al., Hyaluronan hydrogel: An appropriate three-dimensional model for evaluation of anticancer drug sensitivity, *Acta Biomater.* 4, 256–263, 2008.
45. Boxberger, H.J. and Meyer, T.F., A new method for the 3-D *in vitro* growth of RT112 bladder carcinoma cells using the alginate culture technique, *Biol. Cell.* 82, 109–119, 1994.
46. Xie, Y., Yang, S.T., and Kniss, D.A., Three-dimensional cell-scaffold constructs promote efficient gene transfection: Implications for cell-based gene therapy, *Tissue Eng.* 7, 585–598, 2001.
47. Edelman, D.B. and Keefer, E.W., A cultural renaissance: *In vitro* cell biology embraces three-dimensional context, *Exp. Neurol.* 192, 1–6, 2005.
48. Frisk, T., Rydhom, S., Andersson, H. et al., A concept for miniaturized 3-D cell culture using an extracellular matrix gel, *Electrophoresis* 26. 4751–4758, 2005.
49. Pinto, M., Azzam, K.I., and Howell, R.W., Bystander responses in three-dimensional cultures containing radiolabelled and unlabelled human cells, *Radiat. Prot. Dosimetry* 122, 252–255, 2006.
50. Papenburg, B.J., Vogelaar, L., Bolhuis-Versteeg, L.A. et al., One-step fabrication of porous micropatterned scaffolds to control cell behavior, *Biomaterials* 28, 1998–2009, 2007.
51. Boxhari, M., Carnachan, B.J., Cameron, N.R,, and Przyborski, S.A., Novel cell culture device enabling three-dimensional cell growth and improved cell function, *Biochem. Biophys. Res. Commun.* 354, 1095–1100, 2007.
52. Prestwich, G.D., Simplifying the extracellular matrix for 3-D cell culture and tissue engineering: A pragmatic approach, *J. Cell. Biochem.* 101, 1370–1383, 2007.
53. Boxhari, M. Carnachan, R.J., Cameron, N.R., and Przyborski, S.A., Culture of HepG2 cells on three dimensional polystyrene scaffolds enhances cell structure and function during toxicological challenge, *J. Anat.* 211, 567–576, 2007.
54. Frisk, T, Rydholm, B., Liebman, T. et al., A microfluidic device for parallel 3-D cell cultures in asymmetric environments, *Electrophoresis* 28, 4705–4712, 2007.
55. Graham, N.B., Hydrogels: Their future, Part I, *Med. Device Technol.* 9, 18–22, 1998.
56. Galaev, I.Y. and Mattiasson, B., "Smart" polymers and what they could do in biotechnology and medicine, *Trends Biotechnol.* 17, 335–340, 1999.
57. Tirelli, N., Lutolf, M.P., Napoli, A., and Hubbell, J.A., Poly(ethylene glycol) block copolymers, *J. Biotechnol.* 90, 3–15, 2002.
58. Kopecek, J. Smart and genetically engineered biomaterials and drug delivery systems, *Eur. J. Pharm. Sci.* 20, 1–16, 2003.

59. Lin, C.C. and Metters, A.T., Hydrogels in controlled release formulations: Network design and mathematical modeling, *Adv. Drug. Deliv. Rev.* 58, 1379–1408, 2006.
60. Tessmar, J.K. and Goperich, A.M., Customized PEG-derived copolymers for tissue-engineering applications, *Macrmol. Biosci.* 7, 23–39, 2007.
61. IUPAC Nomenclature Home Site. http://www.chem.qmul.ac.uk/iupac/
62. Ring, W., Mita, I., Jenkins, A.D., and Bikales, N.M., Source-based nomenclature for copolymers, *Pure Appl. Chem.* 57, 1427–1440, 1985.
63. Bareiss, R,E., Fox, R.B., Hatada, K. et al., Generic source-based nomenclature for polymers,. *Pure Appl. Chem.* 73, 1511–1519, 2001.
64. Horie, K., Fox, R.B., He, J. et al., Definition of terms relating to reactions of polymers and to functional polymeric materials, *Pure Appl. Chem.* 76, 889–906, 2004.
65. International Union of Pure and Applied Chemistry. http://www.iupac.org/
66. Nam, K.W., Watanabe, J., and Ishihara, K., Characterization of the spontaneously forming hydrogels composed of water-soluble phospholipid polymers, *Biomacromolecules* 3, 100–105, 2002.
67. Nam, K.W., Watanabe, J., and Ishihara, K., Modelling of swelling and drug release behavior of spontaneously forming hydrogels composed of phospholipid polymers, *Int. J. Pharm.* 275, 259–269, 2004.
68. Kimura, M., Fukumoto, K., Watanabe, J., and Ishihara, K., Hydrogen-bonding-driven spontaneous gelation of water-soluble phospholipid polymers in aqueous medium, *J. Biomater. Sci. Polym. Ed.* 15, 631–644, 2004.
69. Kimura, M., Takai, M., and Ishihara, K., Tissue-compatible and adhesive polyion complex hydrogels composed of amphiphilic phospholipid polymers, *J. Biomater. Sci. Polym. Ed.* 18, 623–640, 2007.
70. Zhang, Y., Guan, Y., and Zhou, S., Single component chitosan hydrogel microcapsule from a layer-by-layer approach, *Biomacromolecules* 6, 2365–2369, 2005.
71. Cho, J., Heurzey, M.C., Bégin, A., and Carreau, P.J., Physical gelation of chitosan in the presence of β-glycerophosphate: The effect of temperature, *Biomacrmolecules* 6, 3267–3275, 2005.
72. Hong, Y., Mao, Z., Wang, H. et al., Covalently crosslinked chitosan hydrogel formed at neutral pH and body temperature, *J. Biomed. Mater. Res. A* 79, 913–922, 2006.
73. Spinks, G.M., Lee, C.K., Wallace, G.G. et al., Swelling behavior of chitosan hydrogels in ionic liquid-water binary systems, *Langmuir* 22, 9375–9379, 2006.
74. Moura, M.J., Figueiredo, M.M., and Gil, M.H., Rheological study of genipin cross-linked chitosan hydrogels, *Biomacromolecules* 8, 3823–3829, 2007.
75. Jain, S.K., Jain, A., Gupta, Y., and Ahirwar, M., Design and development of hydrogel beads for targeted drug delivery in the colon, *AAPS PharmSciTech.* 8, E56, 2007.
76. Marsich, E,. Borgogna, M., Donata, I. et al., Alginate/lactose-modified chitosan hydrogels: A bioactive biomaterial for chondrocyte encapsulation, *J. Biomed. Mater. Res. A* 84, 364–376, 2008.
77. Schuetz, Y.B., Gurny, R., and Jordan, O., A novel thermoresponsive hydrogel based on chitosan, *Eur. J. Pharm. Biopharm.* 68, 19–25, 2008.
78. Rowley, J.A,. Madlambayan, G., and Mooney, D.J., Alginate hydrogels as synthetic extracellular materials, *Biomaterials* 20, 5, 45–53, 1999.
79. Kuo, C.K. and Ma, P.X., Ionically crosslinked alginate hydrogels as scaffolds for tissue engineering: Part 1. Structure, gelation rate and mechanical properties, *Biomaterials,* 22, 511–521, 2001.
80. Drury, J.L., Dennis, R.G., and Mooney, D.J., The tensile properties of alginate hydrogels, *Biomaterials* 25, 3187–3199, 2004.
81. Kong, H.J., Kaigler, D., Kim, K., and Mooney, D.J., Controlling rigidity and degradation of alginate hydrogels via molecular weight distribution, *Biomacromolecules* 5, 1720–1727, 2004.

82. Augst, A.D., Kong, H.J,. and Mooney, D.J., Alginate hydrogels as biomaterials, *Macromol. Biosci.* 6, 623–633, 2006.
83. Kobaslija, M. and McQuade, D.T., Removable colored coatings based on calcium alginate hydrogels, *Biomacromolecules* 7, 2357–2361, 2006.
84. Nunamaker, E.A., Purcell, E.K., and Kipke, D.R., *In vivo* stability and biocompatibility of implanted calcium alginate disks, *J. Biomed. Mater. Res. A.* 83, 1129–1137, 2007.
85. West, E.R., Xu, M., Woodruff, T.K., and Shea, L.D., Physical properties of alginate hydrogels and their effects on *in vitro* follicle development, *Biomaterials* 28, 4439–4448, 2007.
86. Ashton, R.S., Banerjee, A., Punyani, S. et al., Scaffolds based on degradable alginate hydrogels and poly(lactide-co-glycolide) microspheres for stem cell culture, *Biomaterials* 28, 5518–5525, 2007.
87. Kuo, C.K. and Ma, P.X., Maintaining dimensions and mechanical properties of ionically crosslinked alginate hydrogel scaffolds *in vitro*, *J. Biomed. Mater. Res. A* 84, 899–907, 2008.
88. Khor, E., Methods for the treatment of collagenous tissues for bioprosthesis, *Biomaterials* 18, 95–105. 1997.
89. Fischer, L.M.S., Chung, B., Sundelzcruz, S., and Huber, J.L., Self-assembling protein hydrogels with modular integrin binding domains, *Biomacromolecules* 7, 38–47, 2006.
90. Sanborn, T.J., Messersmith, P.B., and Barron, A.E., *In situ* crosslinking of a biomimetic peptide-PEG hydrogel via thermally triggered activation of factor XIII, *Biomaterials* 23, 2703–2710, 2002.
91. D'Urso, E.M., Jean-Francois, J., Doillon, C.J., and Fortier, G., Poly(ethylene glycol)-serum albumin hydrogel as matrix for enzyme immobilization: Biomedical applications, *Artif. Cells Blood Substit. Immobil. Biotechnol.* 23, 587–595, 1995.
92. Demers, N., Agostinelli, E., Averill-Bates, D.A., and Fortier, G., Immobilization of native and poly(ethylene glycol)-treated ("Pegylated") bovine serum amine oxidase into a biocompatible hydrogel, *Biotechnol. Appld. Biochem.* 33, 201–207, 2001.
93. Tada, D., Tanabe, T, Tachibana, A., and Yamauchi, K., Drug release from hydrogel containing albumin as crosslinker, *J. Biosci. Bioeng.* 100, 551–555, 2005.
94. Tada, D., Tanabe, T., Tachibana, A., and Yamanuchi, K., Albumin-crosslinked alginate hydrogels as sustained drug release carrier, *Mater. Sci. Eng. C Biomim. Supramol. Struct.* 27, 870–874, 2007.
95. Frisk, M.L., Tepp, W.H., Guangyun, L. et al., Substrate-modified hydrogels for autonomous sensing of Botulinum neurotoxin type A, *Chem. Mater.* 19, 5842–5844, 2007.
96. Golden, A.P. and Tien, J., Fabrication of microfluidic hydrogels using molded gelatin as a sacrificial element, *Lab on a Chip* 7, 720–725, 2007.
97. Cummings, C.L., Gawlitta, D., Nerem, R.M., and Stegermann, J.P., Properties of engineered vascular constructs made from collagen, fibrin, and collagen-fibrin mixtures, *Biomaterials* 25, 3699–3706, 2004.
98. Tuan, T.L. Song, A., Chang, S. et al., *In vitro* fibroplasia: Matrix contraction, cell growth, and collagen production of fibroblasts cultured in fibrin gels, *Exp. Cell. Res.* 223, 127–134, 1996.
99. Dobaczewski,M., Bujak, M., Zymek, P. et al., Extracellular matrix remodeling in canine and mouse myocardial infarcts, *Cell Tissue Res.* 324, 475–488, 2006.
100. Hartmann, A., Boukamp, P., and Friedl, P., Confocal reflection imaging of 3D fibrin polymers, *Blood Cells Mol. Dis.* 36, 191–193, 2006.
101. Willis, G., Failure modes and effects analysis in clinical engineering, *J. Clin. Eng.* 17, 59–63, 1992.
102. ICH Harmonised Guidelines, Pharmaceutical Development, Q8, International Committee on Harmonisation; http://www.ich.org/LOB/media/MEDIA1707.pdf.

6 Adhesives, Glues, and Sealants

A variety of biological and synthetic organic chemical materials are used for wound closure; they act as a direct substitute for sutures, permitting sutureless closure for suture support. These materials are described as adhesives, glues, and/or sealants. Glues and adhesives have a considerable history[1] dating back to at least 4000 B.C. An early definition of glue is as follows: an organic material with adhesive properties that is obtained from the skin and other nonmeat parts of cattle and sheep,[2,3] which are by-products of the meat packing industry. Glues were produced by boiling these parts and reducing the liquid extract (liquor). The first patent for glue was granted in the United Kingdom in 1850 for a product made from fish. Glue materials are also derived from insects such as bees.[4] The early preparation of glues was not unlike that of the preparation of gelatin and gelatin-like materials from animal tissues, and there are materials described as gelatin glues.[5–8]

Until approximately 1920, glues and adhesives were derived from natural sources as described previously. Although most glues and adhesives today are derived from synthetic organic chemistry, there is still considerable interest in adhesives from natural or renewable sources.[9] Examples of adhesives derived from natural resources include lignins, tannins, carbohydrates,[10] and proteins.[11–13] Carbohydrates used as adhesives include cellulose, starches, and dextrin (product of dry-roasting starch in the presence of an acid catalyst). In addition, natural gums (hydrophilic and hydrophobic polysaccharides) are adhesives. Other examples include the use of chitosan derivatives[14,15] and mucilaginous compounds.[16] Casein from milk and blood protein has been used as an adhesive.[17] Cereal flours have been included in glues.[18] These various glue products are used primarily in the manufacture of wood products, including veneer panels. It is generally thought that albumin is the active component from blood as it is present in the highest concentration. It is noted that bovine albumin is used with glutaraldehyde as a surgical adhesive and in suture support,[19] which is marketed as BioGlue®.[20,21] Blood and casein are compounded with other materials for the final adhesive product. The resulting glue products have the quality of being gap filling and cold setting, which are major attributes in the furniture industry. Soy and peanut powders and bovine blood have been used with phenol–formaldehyde for adhesive resins for wood composites.[22] Wheat, gluten, and soluble starch are used as extenders in KLF (Kraft-lignin-formaldehyde)-isocyanate adhesive.[23]

An adhesive has the property of being sticky or possessing tack.[24] Tack is defined as the ability of a material to adhere to a solid surface when brought into contact with light pressure.[25] Tack is more generally defined as that which fastens or attaches. The strength of tack is measured by debonding surfaces that are bound together with an adhesive. An adhesive is then defined as a substance that has the potential of holding two or more surfaces together in a strong and permanent manner. Thus, glue can be considered to be an adhesive in terms of use, but an adhesive is not necessarily a glue.[24a] As noted by Reece and coworkers,[24a] the term *glue* for tissue adhesives suggests expectations that are not reasonable, considering the various surgical products.

An adhesive is generally associated with a supporting matrix, as for example with a postage stamp or a Band-Aid. Thus, a surface may have an adhesive property without being a glue. Adhesives range in strength from relatively weak (pressure-sensitive adhesives of 0.1–1.0 kg/m^2) to strong (epoxides of 200–400 kg/m^2). Contact adhesives (5–50 kg/m^2) and cyanoacrylates (100–200 kg/m^2) are intermediate. Pressure-sensitive adhesives have extensive medical use. Pressure-sensitive adhesives are different from other adhesives in that heat solvency is not required to induce tackiness. Tackiness is a property that is given to adhesives by the addition of materials such as tackifying resins.[11] A resin can be defined as a solid or semisolid organic amorphous material of high molecular weight that can soften or melt over a range of temperatures.[11] Resins can be derived from natural sources such as fossil resins (asphalite) and secreted material from insects (shellacs).[11] Synthetic resins can be hydrocarbon resins similar to natural resins or purely synthetic resins.

Pressure-sensitive adhesives are sometimes referred to as *self-adhesive* materials. Biological adhesives would seem to resemble pressure-sensitive adhesives. Pressure-sensitive adhesives used in medical practice include medical tapes and wound dressings.[26–29] Sealants are also involved in wound dressings. Most of this discussion involves the fibrin product,[30–33] but other products are also described as sealants for wound dressings.[34,35] A sealant can be defined as a substance that is capable of attaching to at least two surfaces, filling the gap between the two surfaces and providing a barrier.[36,37] A sealant can also provide a protective coating. Adhesives and sealants share some common properties in that both are generally liquids which form bonds with a surface through molecular interactions. This is one of the suggested mechanisms for function of an adhesive: adhesive forces hold separate materials together at an interface. Other suggested mechanisms for adhesion include mechanical interlocking; electronic, weak boundary layers; adsorption (or thermodynamic); and diffusion.[38] Molecular interactions between molecules of the sealant or adhesive are also important. Such interactions are described as *cohesive forces*.[37] It should be emphasized that both adhesives and sealants combine with other components in an assembly to form a useful product.

Adhesion, which is a property serving as the basis for the action of glues, can be described as the process of sticking or holding two pieces of material together. The process then involves the molecular interaction of a substance (the glue or adhesive) with one or more different materials (attractions between molecules at an interface) while also interacting with itself. Adhesion does represent a different process to chemists, physicists, biologists, mathematicians, and engineers.[39] Adhesion is an important concept in health care. Adhesive bandages are important in medicine as examples of pressure-sensitive adhesives. Adhesion, referred to as *bioadhesion*, is important in drug delivery.[40] *Mucoadhesive* is a term used to describe the property of a material that enables it to adhere to epithelial cell surfaces.[41–44]

Adhesion is also a well-recognized concept involved in cell–cell interactions and cell–matrix interactions[45–47] as well as the adherence of bacteria to surfaces.[48–50] The process of eukaryotic cell interactions involves adhesive proteins.[51–59] Although adhesion is a critical process in normal cell–cell interactions such as leukocyte function in inflammatory response,[60] platelet aggregation in hemostatic response,[61] and tissue development,[62] adhesion can also be a medical problem in surgery.[63–65] There are strategies to prevent surgical adhesions including the use of oxidized cellulose films,[66–70] hyaluronic acid films[71–75] and fibrin sealant.[76] The last several decades have seen a significant increase in our understanding of adhesive forces in protein interactions as a result of studies at the single molecule level.[77–85]

TISSUE SOLDERING

Tissue soldering or laser soldering is a wound closure technology that uses an exogenous macromolecule, usually a protein, as patch material or solder in sealing a wound. The use of a protein solder dates back to 1988 with studies by Poppas and coworkers[86] on urethral surgery. Subsequent work from another group established the value of including a dye to increase the sensitivity of the system.[87,88] The classic definition of a solder is a metallic alloy used for uniting metal surfaces or parts, whereas a patch is defined as a piece of material attached to something to repair a hole or a tear, so as to strengthen or protect a weak area.[89] Tissue soldering uses a protein that is "melted" to repair a lesion or strengthen a weak area. Tissue soldering has the advantage over laser welding[a] in providing greater bond strength, less collateral tissue damage, and a wider parameter window for providing a satisfactory bond or seal.[90]

The tissue solder must establish an adhesive bond with tissue and cohesive bonds within the solder. Establishment of an appropriate balance between adhesion and cohesion is essential to the formation of a strong bond between opposing tissue surfaces.

Solders are usually proteinaceous in nature, although there has been limited use of other materials such as chitosan.[91,92] Lauto and colleagues prepared stents from a chitosan film and welded the stents in place with laser irradiation.[91] The same group[92] used strips of chitosan film to repair intestinal tissue. Genipin is a cross-linking agent (Figure 6.1) that has increased the bond strength of albumin solder welds.[93] It is suggested that strong bonds are formed between chitosan and collagen in the tissue; it is not known as to whether there are covalent bonds formed between amino groups on chitosan and collagen. Covalent cross-linking of chitosan and collagen has been achieved with carbodiimide[94,95] and glutaraldehyde.[96,97] It is not unreasonable to suggest that covalent bonds are formed between chitosan and collagen, but additional research is required. The strong noncovalent interaction between chitosan and tissue facilitates the subsequent laser soldering.

The specific proteins that are used as solder materials will be discussed in greater detail in the following text. The primary requisite for a solder is the ability to form strong adhesive or cohesive bonds in response to laser irradiation or some other thermal challenge. The observed physical effect of laser irradiation or heat is to denature or melt the protein. Protein denaturation can be defined as a physical, intramolecular change in the native protein structure.[98] Denaturation is not a chemical

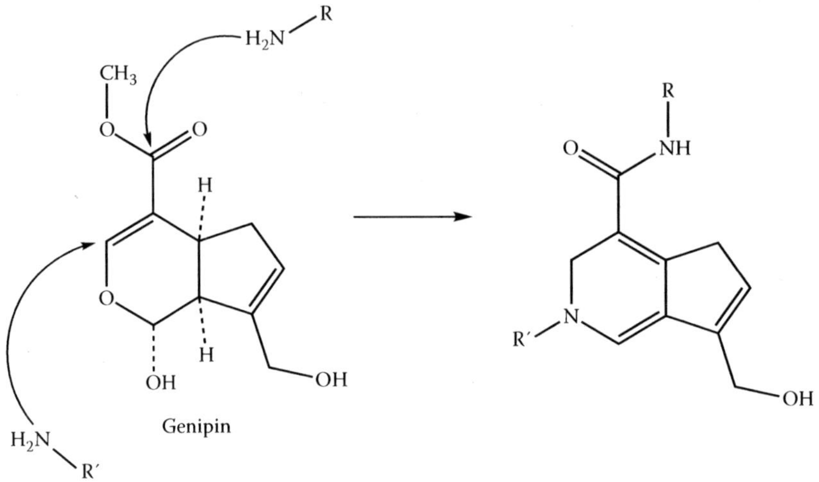

FIGURE 6.1 The structure of genipin cross-links: Cross-links may be formed between primary amines supplied by proteins and carbohydrates such as glucosamine. (See Butler, M.F., Ng, Y.-F., and Pudney, P.D.A., Mechanism and kinetics of the cross-linking reaction between biopolymers containing primary amine groups and genipin, *J. Polym. Sci. Part A*, 41, 3941–3953, 2003.)

change, but it is a conformational change in the protein and is not associated with the cleavage of protein bonds. However, the use of the term, even in 1954, had become so broad as to obfuscate its value.[99] Denaturation is not an irreversible process and, in fact, is used in the processing of recombinant proteins expressed as inclusion bodies in bacterial systems.[100–103] Aggregation and subsequent loss of solubility and loss of biological activity are the most commonly observed manifestations of protein denaturation. Aggregation results from the exposure of aromatic amino acids or hydrophobic regions,[104–107] which then interact with other proteins via hydrophobic interactions.[108–112] Aggregation is the property of protein denaturation that is important in laser soldering. The key in laser soldering is the precise application of energy to denature the protein to produce a useful aggregate with sufficient adhesive and cohesive properties.[113] An issue, therefore, in the development of this technology is measurement of "melting of the solder" with respect to the application of energy. Measurement of physical changes such as aggregation or loss of activity is likely to occur too late in the melting process to be of value.

The stability of proteins is frequently measured by differential scanning calorimetry.[114–121] Differential scanning calorimetry (DSC)[122–126] is a powerful tool for measuring structural change in a variety of materials. DSC is a technique for measuring both conformational change in proteins[127–139] and the integrity of formulated biopharmaceutical protein products.[140–164] DSC measures heat flow into (endothermic) and out of (exothermic) a sample as a function of temperature (change in heat capacity; C_p). An endothermic reaction occurs with melting, glass transition, and protein denaturation, whereas freezing is an exothermic reaction. In the most common format, changes in heat capacity (dH/dt) are measured as a function of temperature.

This is an enthalpic change and is thus a state function as opposed to a thermodynamic function. The midpoint of the change in heat flow is designated as T_m. As T_m increases, protein stability increases.[165–170] DSC can be used for the evaluation of excipient efficiency and protein engineering in the improvement of therapeutic protein stability. However, DSC does not necessarily measure the aggregation and precipitation that occurs secondary to the initial conformational change.

PROTEINS AS TISSUE SOLDER MATERIAL

Three proteins have been used as tissue solders: collagen, albumin, and fibrinogen.

COLLAGEN

Collagen undergoes a complex cross-linking process as it participates in the development of biological structures[171–174] such as tendons and skin. This involves extensive posttranslational modification of collagen with the formation of amino acid derivatives such as hydroxylysine and hydroxyproline.[175–178] These cross-linking processes are important in the development of supramolecular structures, including tendons and skin. These posttranslational modifications are mediated by enzymes such as lysyl oxidase, lysyl hydroxylase, and prolyl-4-hydroxylase.[179–181] These are time-dependent processes that occur before the formation of the various complex cross-links in mature collagen. The glycation of collagen also results in cross-links via Maillard and related reactions.[182–184] In addition to internal protein–protein interactions, which result in complex helix formation and subsequent formation of fibers,[185–190] collagen interacts with a variety of other proteins during development and subsequent functioning. For example, collagen interacts with various proteoglycans.[191–196] Fibromodulin interacts with collagen during fibril formation,[197] after formation of the extracellular matrix,[198] and in the structure of tumor stroma.[199] It is also well documented that mature collagen interacts with various cell types[200–202] and that such interactions are dependent on the mature collagen helical form.[203–205]

It is suggested that collagen is involved in photocoagulation or laser welding processes.[206–213] However, the mechanisms involved in the participation of collagen in the welding process are not clear. It is assumed that there is covalent bond formation between various cellular constituents, but this has not been documented. Given the covalent bonds that are formed with collagen during the development of ligament and hard tissue, it is not unreasonable that collagen would be a major participant. Several of the various studies report a decrease in collagen concentration at the site of laser welding.[207,208,210] This decrease in collagen concentration is followed by an increase in collagen concentration during the healing process, in which the scar tissue is remodeled.[214] Laser irradiation has been reported to have a favorable effect on the wound healing process.[215–217] There are some studies suggesting that there is dependence on the wavelength of the incident laser irradiation. Al-Watban and associates[218] suggest that irradiation at 633 nm (neon:helium) was optimal in enhancing wound healing in diabetic rats; irradiation at other wavelengths (532, 810, and 980 nm) did provide some enhancement but less than that observed at 633 nm. Hawkins and Abrahamse[219] studied fibroblast proliferation in response to laser irradiation in the dark, in broad-spectrum light, and in infrared light. They showed that wounded cells (human fibroblast monolayers with physical induction of a wound) responded optimally to 633 nm (He:Ne) irradiation or 1064 nm (Nd:YAG) in the dark or 830 nm (diode laser) in the light. Optimal results were obtained with 633 nm radiation in the light.

Hwang and coworkers[220] observed that laser welding (carbon dioxide laser) improved nerve regeneration in facial nerves of rats. The lasers used most frequently in tissue welding include the carbon dioxide laser (10600 nm), Nd:YAG (neodymium:yttrium-aluminum-garnet; 1064 nm), Ne:He (neon: helium; 633 nm), and argon (488 nm).[221–223] Laser irradiation from other sources is being evaluated.[224] Dyes are used in some laser welding,[225] but are more often used in laser soldering.[226,227]

The photocoagulation or laser welding process most likely involves the selective application of energy to tissues resulting in local heating as opposed to any specific spectral interaction of solutes (proteins, carbohydrates, lipids, etc.) with the incident radiation. Photocoagulation or laser welding is manifested grossly as the formation of a tissue coagulum that solidifies.[228] The protein chemistry of laser welding is clearly complex and most likely involves a number of proteins and other membrane and cytoplasmic constituents.[229,230] It has been shown that laser irradiation of pure collagen does not result in covalent bond formation between collagen chains.[231] Gayen and coworkers[232] showed that welding strength of aortic welds was a function of incident wavelength using a chromium:YAG(Cr^{4+}:YAG) laser. The binding in the tissue weld is suggested to be due to collagen resulting from thermal, photothermal, and photochemical reactions via the absorption of the incident laser illumination by water. It has been suggested that noncovalent bonds between denatured collagen molecules are important in tissue welding.[233] Denaturation of proteins exposes hydrophobic regions that were previously buried in the interior of a protein, which can result in aggregation.[234–236] A better knowledge of the mechanism involved would help solve the problems of reproducibility and cost-effectiveness.[237]

Albumin

Albumin is used more frequently in tissue soldering than either fibrinogen or collagen. Albumin is a protein that is most notably derived from plasma or serum and secondarily from egg (ovalbumin). It is the most abundant protein in blood/plasma, constituting approximately half of the total plasma protein. Its function is to establish plasma colloid strength, which preserves the fluid balance between the intravascular and extravascular space.[238,239] Albumin was the first protein biopharmaceutical[240,241] and is used for a variety of clinical indications,[242] including use in extracorporeal circulation as a "bridge-to-transplant."[243,244] The abundance of albumin as a material combined with its extensive clinical experience as a parenteral drug makes it a reasonable choice for a solder material to be used in tissue welding. As noted elsewhere, blood has been used as glue in the wood industry with some success, and it is likely that blood's albumin content was responsible for its adhesive property.[245] Albumin is also noted for its ability to interact with various dyes, and the binding of bromocresol green and bromocresol purple are used for the clinical assay of albumin.[246] This is likely to be an advantage when dyes are included to improve energy transfer from lasers.

Albumin is a general designation to describe a fraction of simple proteins, which is soluble in water and dilute salt solutions, as opposed to the globulin fraction, which is insoluble in water but soluble in dilute salt solutions. This is an old classification and has many exceptions.[247] Albumins also migrate faster than globulins on electrophoresis; this resulted in the development of the classification of plasma proteins as albumins and globulins.[248]

Blood was first used as a solder for vascular anastomoses[249] but was too weak to be useful. Blood was replaced by egg albumin, which was functionally useful but, due to its source, was not considered for therapeutic use. Albumin (also collagen and fibrinogen) has been considered as a candidate for tissue solder. The successful therapeutic use of serum albumin without significant adverse reaction made it a logical candidate for tissue solder. The ready availability of an approved liquid product was an additional advantage.

The ability of albumin (or any protein) to serve as a tissue solder is based on the ability of the protein to "melt" on denaturation and subsequently congeal (aggregate). Differing from, for example, metals, proteins can aggregate without cooling. As such, you need a protein that will melt at a temperature which will not cause collateral tissue damage during the process of coaptation.[250] The melting of albumin can be measured by DSC. The T_m for defatted (fatty-acid-free) human albumin is approximately 65°C, and there are species differences in albumin thermal stability, with bovine albumin considered less stable (lower T_m).[251,252] The binding of fatty acids to albumin increases thermal stability (higher T_m).[252,253] There is heterogeneity in ligand binding, which results in broad, multicomponent thermograms.[251,254] Although it is possible that the heterogeneous thermograms represent multiple domains, this is not considered likely. Protein concentration (and scan rate) also influences the observed T_m for human albumin, which decreases with increasing protein concentration. The effect is modest compared to the effect of ligand and likely reflects processes other than the accepted one-step transition from native to denatured proteins. Processes such as aggregation and dissociation of oligomeric proteins can influence T_m. An effect of protein concentration on the DSC behavior of albumin has been observed.[252] The effect of protein concentration on T_m has been observed by other investigators with other proteins.[255,256] Solvent composition can also have a profound effect on the thermograms of protein.[257]

The goal of tissue soldering or tissue welding is to establish a strong bond with satisfactory coaptation and acceptable collateral tissue damage. Reduction of thermal damage is a major part of this goal. Thus, the use of fatty-acid-free albumin is highly desirable. Second, the use of a homogeneous product will produce a "tighter" T_m, which will allow more uniform heating or denaturation of the solder. Finally, the inclusion of a suitable dye in the solder permits lower energy input.[258] Indocyanine green is the best example,[259] with an adsorption maximum of 805 nm; it is used with diode laser (808 nm).

Application of albumin solders to human clinical surgery has been limited to urology studies,[260] which seem to have been successful. A particularly intriguing application has been the fabrication and use of stents using congealed albumin.[261,262] In these studies, the 25% commercial human albumin is concentrated by ultrafiltration to 50%

and then further concentrated to yield a gel that can be fabricated into a structure. Another group has also developed an albumin stent referred to as Bioweld®.[263,264]

Fibrinogen

Fibrinogen is a large (340 kDa) plasma protein whose function can be simply stated as the ability to be converted from a soluble protein to an insoluble matrix.[265] The product of the denaturation of fibrinogen by heating is not dissimilar from the product obtained by the clotting of fibrinogen by thrombin; both are insoluble products that can be solubilized in chaotropic solvents such as urea or guanidine; the exception is that fibrin cross-linked by factor XIIa is insoluble.[266] A recent study on the use of heat-denaturated fibrinogen as a biological matrix provides support for use of thermally treated fibrinogen as a patch.[267] Scanning electron microscopy showed that the heat-denatured, precipitated fibrinogen was composed of aggregates of more than 3000 fibrinogen monomers.

There has been only limited use of fibrinogen as a tissue solder. Oz and coworkers[268] used a fibrinogen solder with indocyanine green (a tricarbocyanine dye). The dye (absorption maximum at 805 nm) was included to absorb the diode laser (808 nm) energy (4.8 W/cm^2) to minimize the collateral tissue damage. The weld created in the presence of fibrinogen had greater burst strength (330 ± 75 mm Hg) than a weld created in the absence of fibrinogen (262 ± 29 mm Hg). Treatment of the weld with urokinase did not significantly decrease burst strength (290 ± 74 mm Hg), suggesting resistance to fibrinolysis. A normal intimal surface was regenerated at the weld sites, and there was no foreign body reaction.

Wider and coworkers studied the use of fibrinogen–dye solders for the closure of skin incisicion.[269] These investigators used a diode laser (808 nm; 9.55 W/cm^2) with indocyanine dye and fluorescein isothiocyanate (absorption maximum, 495 nm) with an argon laser (488–515 nm; 4.78 W/cm^2). Initial analysis suggested stronger bonds with the argon laser system, but later analysis suggested that the diode laser system was stronger. Comparison with suture closure showed both laser systems were stronger than suture closure on initial analysis, whereas later analysis showed the diode laser was stronger than the suture and the argon laser system was the weakest. The argon system produced more cosmetically acceptable results.

Mueller and coworkers[270] studied the use of the argon laser for bovine heterograft anastomoses. Solder proteins included fibrin sealant (prepared *in situ* from bovine fibrinogen and bovine thrombin) and collagen. The grafts were welded (7.5 W/cm^2; 75 s/5 mm length of anastomosis) in the presence or absence of solder protein. The experiments were performed in the absence of a dye. The laser welded with the fibrin sealant, and the sutured control did not show statistical difference in burst pressure. The use of collagen for a solder resulted in decreased burst pressure. Shohet and colleagues[271] also reported poor results in ex vivo tensile measurements with collagen as solder (diode laser with indocyanine green), whereas significant tensile strength was obtained with a fibrinogen solder.

Khadem and coworkers[272] have studied the performance of a mixture of fibrinogen and riboflavin-5-phosphate in soldering central corneal incisions using an argon laser. It is suggested that there is cross-linkage between corneal collagen and the fibrinogen solder. Inhibition of the reaction with the inclusion of sodium azide provides support for a mechanism involving singlet oxygen with riboflavin-5-phosphate as a photosensitizer. Riboflavin is an effective photosensitizer[273] and has been shown to cross-link scleral collagen.[274] Cross-linkage is suggested to proceed via oxidative deamination of a lysine residue with the formation of the aldehyde (6-semialdehyde-2-amino-adipic acid; allysine), which then reacts with another proximal lysine residue to form a Schiff base. This reaction is similar to that catalyzed by lysine oxidase, which results in Schiff base cross-links in collagen and other proteins.[275]

FIBRIN SEALANT

Fibrin sealant or fibrin glue is the product obtained from the reaction of thrombin with a fibrinogen preparation in a suitable solvent that contains calcium ions. The presence of blood coagulation factor XIII is considered an advantage, contributing to the tensile strength of the final product. Early sealant preparations were composed of a fibrinogen-rich fraction from plasma and, usually, bovine topical thrombin. Early commercial products contained antiplasmin agents such as aprotinin or 6-aminocaproic acid (EACA) for clot stability.

- Fibrinogen is a plasma protein, which is converted by thrombin into an insoluble form in the final phase of the hemostatic response.
- Factor XIII (fibrin stabilizing factor) is critical for the formation of a stable clot. Factor XIII is converted to factor XIIIa by thrombin in the presence of calcium ions.
- Calcium ions are required for the activation of factor XIII by thrombin and for the polymerization of the fibrin monomer.
- Thrombin is an enzyme that catalyzes the cleavage of peptide bonds in fibrinogen, resulting in the formation of fibrin, the "activation of other blood coagulation factors including factor XIII, factor IX, factor VIII, factor V, and the aggregation of blood platelets. Contact of blood with thrombin causes the formation of a clot, and reaction with platelets results in the release of diverse growth factors. Thrombin is a free-standing biological product as well as a component of fibrin sealant.[276]

Thrombin and fibrin sealant are different products with different indications. Unlike thrombin, fibrin sealant can form a polymer network that functions as a glue or sealant independent of contact with blood or tissue. In this respect, fibrin sealant is similar to cyanoacrylate glues. Fibrin sealant is considered to be a biological glue, whereas cyanoacrylates are nonbiological glues. In principle, fibrin sealant will not polymerize until the two components are mixed and extruded from the delivery device. Cyanoacrylates will spontaneously polymerize upon contact with water. Fibrinogen has been used as a solder for laser welding.[277] Heat-denatured fibrinogen has also been shown to be a matrix for cell growth.[278] Fibrinogen bound to a surface has immunological characteristics similar to fibrin.[279] As noted earlier in the section on tissue soldering, the denaturation of fibrinogen has characteristics similar to the formation of fibrin from fibrinogen.[280,281]

Fibrin sealant or glue evolved from an effort to duplicate the *in vivo* effectiveness of blood to seal wounds. There were a number of early studies on the use of blood or plasma to function as a glue or gel when combined with thrombin.[282–286] Although plasma clots or gels do not have the tensile strength of clots or gels formed from concentrated fibrinogen solutions,[277,278] such products have useful hemostatic properties.[282] Blood plasma, mostly as platelet-rich plasma (PRP) or less often as platelet-poor plasma (PPP), is still used for a variety of therapeutic purposes.[287–293] The platelets present in PRP are a rich source of growth factors, which are released upon activation by thrombin.[294–296] PRP is prepared by low-speed centrifugation, which removes red cells and most leukocytes. PPP is usually prepared by high-speed centrifugation. PRP is used with thrombin to form platelet gel,[288,297–299] whereas PPP can be used as a source of fibrinogen for use in fibrin sealant. However, it is more common to use PRP for the preparation of fibrin glue[300–310] as the growth factors derived from blood platelets are considered beneficial. The advantage from a regulatory perspective is that the inclusion of exogenous components into an approved biological product does not require justification. This latter consideration may well be a moot point because regulatory approval is not required for the intrainstitutional (and hence within a state) use of this product. It is noted that there is individual variability in the fibrinogen content of cryoprecipitate preparation,[311] and fibrinogen concentration is a quality attribute for the use of cryoprecipitate as biological glue.[312] It is noted that it is possible to analyze fibrin clot structure with a microplate reader.[313] This would permit the determination of the most effective ratio of thrombin to fibrinogen prior to use of the cryoprecipitate in a glue.

Proteolysis of fibrinogen by thrombin yields fibrin monomer that polymerizes in a staggered overlapping manner to form fibrin protofibrils; the protofibrils subsequently coalesce into fibers.[314–317] The protofibrils are extended in the staggered chain formation and undergo lateral assembly into fibrin fibers, which then form a three-dimensional matrix. These interactions are noncovalent in nature, and the structure formed is not stable. The fibrin structure is stabilized by the formation of covalent bonds in a transaminase reaction catalyzed by factor XIIIa.[318]

The fibrin clot is subject to proteolytic dissolution by components of the fibrinolytic system.[319–325] The susceptibility of a fibrin clot to dissolution depends on the structure of the fibrin clot. Gabriel and coworkers[322] have shown that as fibrin fiber size (diameter) decreases, the rate of fibrinolysis decreases. As fibrin monomer assembly rate in protofibrils increases, fiber diameter increases. The fibrin fiber size (diameter) can be controlled by a variety of exogenous factors such as dextran.[326] From a practical point of view, and addressing the issue of variability in fibrinogen concentration in autologous cryoprecipitate preparations,[311] the ratio of thrombin to fibrinogen would be the easiest variable to control to ensure consistent fibrin product with variable cryoprecipitate preparations. As thrombin concentration relative to fibrinogen increases, fiber diameter increases, and as thrombin concentration relative to fibrinogen concentration decreases, fiber diameter decreases. Excluding all other factors, a more stable clot or patch would be obtained with a low thrombin concentration. However, the stability of the final product would have to be balanced with extended time of formation at low thrombin concentrations for effective product use in a given clinical situation. There are other factors that could be used to influence fibrin product quality.

Calcium ions appear to have a structural role in fibrinogen.[317,327–332] Calcium ions appear to have a modest effect on thrombin-catalyzed fibrinopeptide release from fibrinogen,[333–336] but a large effect on the polymerization process.[313,337–345] Carr and Powers report that the presence of calcium ions results in more rapid polymerization with the formation of thicker fibrin fibers.[346] For practical use, the presence of calcium ions leads to the more rapid formation of a fibrin gel, which would presumably be more sensitive to dissolution by fibrinolytic agents.

Fibrinogen concentration is also a quality attribute for a fibrin sealant product. The effect of the ratio of thrombin to fibrinogen has been discussed earlier. There is a correlation between fibrinogen concentration and tensile strength of the fibrin sealant product.[347–352] The tensile strength is proportional to the fibrinogen concentration.

Factor XIII is the last component of what would be considered the minimal fibrin sealant. Factor XIII, originally known as *fibrin stabilizing factor*, is converted to its activated form, factor XIIIa, by thrombin in the presence of fibrinogen/fibrin.[353–358] Factor XIIIa, a transglutaminase, catalyzes the formation of cross-linking between fibrin chains. The cross-link reaction occurs between lysine and glutamine residues. The original term, *fibrin stabilized factor*, was based in part on the observation that the action of factor XIIIa created a urea-insoluble clot. Factor XIII is a quality attribute and has been demonstrated to be important for hemostatic effectiveness[359–361] and clot strength.[362,363] It is less clear that factor XIII/XIIIa is required for adhesion to the wound interface as an argument can be made for the participation of tissue transglutaminase in this process.[362] The participation of tissue transglutaminase may depend on the presence of other proteins such as fibronectin in the fibrinogen component of sealant.[364] Although fibronectin has little effect on fibrin polymerization,[365,366] it is incorporated into the gel structure of the fibrin clot by covalent cross-linkage catalyzed by factor XIIIa.[367–373] This does result in increased fiber diameter[365] and is important for interaction with cells.[330,371,373–375]

In addition to the components that would be considered critical quality attributes of the fibrin sealant product, there are a number of other substances that influence the fibrin polymerization pathway and, hence, the quality of the final fibrin clots. These are going to be considered in the following text and in summary in Table 6.1.

Unfractionated heparin appears to accelerate fibrin polymerization by increasing lateral association of protofibrils to yield thicker fibrin fibers.[376] Such clots showed an increase in sensitivity to lysis by t-PA. Fibrin clots prepared in the presence of low-molecular-weight heparin (LMWH) showed a lesser effect on lysis.[377] Subsequent work from another laboratory[378,379] confirmed the effect of unfractionated heparin and showed that the presence of LMWH resulted in the formation of long, thin fibrin fibers, which would be expected to be more resistant to lysis. These investigators also demonstrated that whereas both heparin species inhibited b-FGF stimulation of microvascular endothelial cells, fibrin clots formed in the presence of LMWH were less permissive to invasion by capillary-forming endothelial cells. In other studies, serum amyloid p component and heparin appear to act synergistically to the lateral aggregation of fibrin protofibrils to form fibrin fibers.[380,381] Heparin has

TABLE 6.1
Effect of Various Materials on Fibrin Polymerization and Gel Structure

Component	Effect	References
Fibronectin	Some positive effect on polymerization; incorporated into gel structure by cross-linkage mediated by factor XIIIa. Considered important for wound retraction.	364–375
Calcium ions	Enhances polymerization process by accelerated lateral association of protofibrils into fibers; required for thrombin activation of factor XIII.	313, 318, 321–346
Factor XIII/XIIIa	Increases fiber size; stabilizes final fibrin gel structure as shown by increase in tensile strength. Bulk of evidence supports role in cohesive force of gel but likely not in adhesive forces.	353–363
Heparin	Unfractionated heparin enhances lateral polymerization of fibrin protofibrils to form fibrin fibers; low-molecular-weight heparin inhibits lateral polymerization and results in the formation of thin fibers. The effect of unfractionated heparin is potentiated by serum amyloid P component.	376–389
Collagen	The effect depends on the type of collagen. Type IV collagen enhances lateral polymerization or protofibrils. Collagen incorporated into a clot enhances resistance to fibrinolysis, which would imply reorganization into thin fibers or more likely a result of incorporation of collagen into the clot by cross-linkage to fibronectin. Collagen monomer appears to increase elasticity but decrease adhesive strength.	365, 390–412
Platelet factor 4	Enhances lateral association of fibrin protofibrils to form fibrin fibers.	417–424
Hyaluronan	Binds to fibrinogen and accelerates fibrin polymerization with the development of thicker fibrin fibers.	425–437
IgG	Inhibits lateral polymerization, forming thinner fibers. Resulting gel structure is formed more slowly and is more rigid.	412–418

been included in fibrin with fibroblast growth factor.[382–384] Fibrin sealant has been considered a drug delivery vehicle.[385–389]

Collagen is an important partner in the action of fibrin sealant. The previous discussion of fibronectin considers the role of covalent cross-linkage with collagen in a reaction catalyzed by tissue transglutaminase in the adhesion of sealant to tissue.[365] It should be noted that collagen is not a homogeneous species.[390–392] There are 27 different human collagens. Mature collagen is found as a triple helix (both homotrimers and heterotrimers) that has the ability to form larger structures, resulting in materials such as bone, skin, cartilage, and tendon, and has an important role in the extracellular matrix.[393] The triple helical structure is critical for the biological activities of collagen.[394] The interaction of collagen with other materials such as proteoglycans is frequently a characteristic of the surface of fibrillar collagen and not of individual collagen molecules. The majority of collagen found in the extracellular matrix is fibrillar collagen. Type I collagen is found in tendon, bone, and skin, whereas type II collagen, which is structurally quite similar to type I collagen, is found in cartilage. These two collagens are considered fibrillar collagens. Type IV collagen is a network-forming collagen, which is a component of basement membrane. Type I collagen and type IV collagen have different binding characteristics and biological activities.[390,395,396] Jones and Gabriel[397] demonstrated a binding site for fibrin on type IV collagen. The presence of type IV collagen resulted in thick fibrin fibers, suggesting enhancement of lateral polymerization. Laminin or dermatan sulfate, both basement membrane constituents, decreased the mass/length ratio, resulting in thinner fibers. Laminin decreased the ability of type IV collagen to bind fibrinogen; however, it was concluded that the net effect of the basement membrane was to increase fibrin fiber diameter. Incorporation of collagen into fibrin clots after formation is suggested to be associated with inhibition of fibrinolysis.[398,399] Atelocollagen is a derivative of skin collagen, which is prepared by pepsin digestion.[400] Atelocollagen is a proprietary product, and there is little in the literature describing any of its characteristics. Nomori and coworkers[401] observed that the inclusion of atelocollagen into fibrin sealant resulted in a product with increased elasticity but decreased adhesive strength. Pepsin digestion of skin collagen at acid pH (the pH optimum for pepsin is 2 with a small secondary peak at 5; at about pH 6, pepsin undergoes irreversible denaturation) results in the solubilization of collagen into monomeric alpha chains.[402–405] There are several fibrin sealant applications that include the combination with or inclusion of collagen or collagen products.[406–411] As noted earlier, fibrin sealant products are thought to bind to collagen on the basis of fibronectin content.[365]

Gabriel and coworkers demonstrated that IgG influenced the assembly of the fibrin clot.[412] These investigators observed that the mass/length ratio of fibrin was decreased by the inclusion of polyclonal IgG; monoclonal IgG (paraprotein isolated from a myeloma patient) was found to be more effective than polyclonal IgG. These results suggest that IgG inhibited the lateral association of fibrin protofibrils into fibrin fibers. McDonagh and coworkers subsequently demonstrated that fibrin fiber assembly process is delayed in the presence of paraprotein, and the resulting fibrin clot is more rigid.[413] Other investigators have reported inhibition of fibrin formation in the presence of myeloma protein.[414–418]

Platelet factor 4 (PF4), a small (subunit Mr 9600) chemokine characterized by an isoelectric point of 7.6 that is secreted by platelets, has the ability to neutralize heparin,[419] and is used as a marker for platelet activation.[420,421] Carr and coworkers[422] showed that PF4 enhanced fibrin polymerization. This enhancement is based on the ability of PF4 to increase the mass/length ratio, suggesting increased lateral polymerization of fibrin protofibrils[423] and emphasizing the importance of kinetic control of the polymerization process. Le Bonniec and coworkers[424] mention the importance of PF4 in the formation of a sealed fibrin network.

Hyaluronan (hyaluronic acid) is a large (500–3000 kDa) polyanionic glycosaminoglycan composed of alternating residues of *N*-acetylglucosamine and glucuronic acid,[425] which has an important role in cartilage, vitreous fluid, and extracellular matrices.[426–429] Hyaluronan binds to fibrinogen[430,431] and accelerates the rate of fibrin polymerization by enhancement of the rate of lateral aggregation of protofibrils to form thick fibrin fibers.[432] There is interest in the use of fibrin gels containing hyaluronan as matrices for cell growth.[433–437]

Albumin has been suggested to have an effect on fibrin polymerization,[438,439] but the effect is considered negligible.[440] However, there is no question that fibrin clots formed in plasma are markedly different from those formed by purified fibrinogen.[438,441–443]

The effect of some selected components on fibrin polymerization and subsequent gel structure are presented in Table 6.1.

GELATIN–RESORCINOL–FORMALDEHYDE AND GELATIN–RESORCINOL–FORMALDEHYDE–GLUTARALDEHYDE

The combination of resorcinol and formaldehyde is used as a glue in wood fabrication. The use of resorcinol is based on the early work on the production of polymers from phenol and formaldehyde. The inclusion of gelatin was intended to make the polymer more useful as a biological product. Glutaraldehyde is included to enhance the strength of the polymer. The reaction products are complex as the aldehydes react with both the phenolic compounds and the proteins, and the products of the reaction between the aldehydes and the phenolic compounds can react with the protein components.

Gelatin is a protein product derived from the treatment of collagen at elevated temperature at acid pH; thus, gelatin is a denatured form of collagen, which forms a gel on cooling.[444,445] The process of cooling and gel formation is renaturation with the recovery of some secondary structure,[446–449] which provides a basis for use of gelatin in hydrogels[450,451] (Chapter 5). The physical properties of the gel also permit use as capsules and films.[452–456] There has been concern about transmission of bovine spongiform encephalopathy by gelatin preparation, but this does not seem to be a problem.[457,458] Separate from its formulation with resorcinol and formaldehyde/glutaraldehyde, there has been independent interest in gelatin as an adhesive product.[444,459–461]

Phenolic resins, which are obtained by the condensation of phenol or phenol-like compounds with formaldehyde, were developed by Baekeland in 1909.[462] Baekeland referred to these polymeric materials as *Novolak resins*. These materials were developed as the Bakelite® resins. The early development of these materials has been reviewed by Ellis[463] and Megson.[464] Phenolic resins have a wide variety of applications. The degree of hardness is a product of the ratio of formaldehyde to phenol, reaction temperatures, and solution concentrations of reagents and catalysts. There is a somewhat confusing nomenclature for the phenolic resins. *Resol resins* (*A resins*) are phenolic resins that have been "hardened" by heat but will melt (retain fusibility), and they are soluble in acetone. Resol resins may be prepared from other sources such as tannins derived from pine.[465] *Resolite resins* (*B resins*) have been further heat-hardened to the extent that it will no longer melt but will soften; it is not soluble in acetone but will swell in this solvent. The swelling of polymers in solvents in an approximation of the extent of cross-linkage in the material: the greater the swelling, the smaller the extent of cross-linkage. *Resite resins* (*class C resins*) are hardened further beyond resolite resins; these materials do not melt or soften, and acetone has no effect on them. Novolak phenol resins are prepared with a low ratio of formaldehyde (supplied as paraformaldehyde); they remain fusible and cannot be hardened. The condensation reaction can be catalyzed by either acid or base, and the products from the two systems are different. The properties of resins could be modified by inclusion of materials such as gelatin; the inclusion of gelatin provided a product with enhanced elasticity,[4,66] that could be used for laminating glass.[467]

Phenolic compounds other than phenol are used with aldehydes for the preparation of resins. Resorcinol (*m*-dihydroxybenzene; 1,3-benzenediol; 3-hydroxyphenol) is extensively used for the preparation of resins with formaldehyde (RF resins).[468,469] Resorcinol formaldehyde resins (RF resins) are best known as adhesive material in the plywood industry. Formaldehyde reacts more rapidly with resorcinol than with phenol; the formation of a gel is more rapid below pH 2 and above pH 4.5. As with phenol, there are differences between the acid-catalyzed polymerization pathway and the base-catalyzed polymerization pathway.

The process of the adaptation of basic phenolic resin technology to a gelatin-resorcinol-formaldehyde (GRF) is somewhat unclear. Most of the work was done prior to the development of procedures such as Design Control,[470–472] where the development of devices is circumscribed. Many studies trace the GRF tissue glue to the work of Guilmet and colleagues reported in 1977[473] and extended over the next 4 years.[474,475] There was a report by Droegemueller and colleagues at the University of Colorado on the use of GRF glue in obstetrical surgery during this period.[476] As a result of the work of Guilmet and colleagues, the GRF glue is referred to as the "French" glue.[477–479] These studies were preceded by the earlier work of Braunwald and colleagues in the development of a GRF surgical glue in 1966.[480–482] The GRF glue was developed by Richard Falb and colleagues at the Battelle Memorial Institute[483,484] in collaboration with Nina Braunwald and her colleagues at the National Institutes of Health in Bethesda, Maryland. As with the later work in this area, the gelatin and resorcinol (three parts of gelatin and one part resorcinol) were combined and the formaldehyde (a few drops) added to initiate the polymerization/cross-linkage reaction.[483,484] A later study[482] used a ratio of 5 parts of gelatin to 1 part of resorcinol. Precise quantitative relationships cannot be determined from these studies but assuming weight/weight ratios, there would be a substantial molar excess of resorcinol. The source of formaldehyde is usually not given but it is assumed that it is the commercial 37% solution (formalin, formal). It is likely that the positive experience with formaldehyde-cross-linked gelatin as a sponge[485] or film[486] played a role in the development of the GRF glue when combined with the prior incorporation of gelatin into phenolic resins,[466]. Later studies have incorporated glutaraldehyde into the GRF glue.[487] GRF glues have greater strength than fibrin sealant products[478,487] This product has had clinical success,[488–490] but it is associated with complications[491] and skill in application is clearly required.[492]

Considerations of the chemistry of the reaction of aldehydes with phenolic compounds (Figure 6.2) and of the reaction of formaldehyde and other aldehydes with proteins (Chapter 1) suggest that the GRF glue product is clearly complex and uniformity of product will be driven by reagent purity and process consistency. The issue of formaldehyde toxicity is also a potential problem. The product quality attributes for this product, given the indications, are (1) biocompatibility, (2) rapid

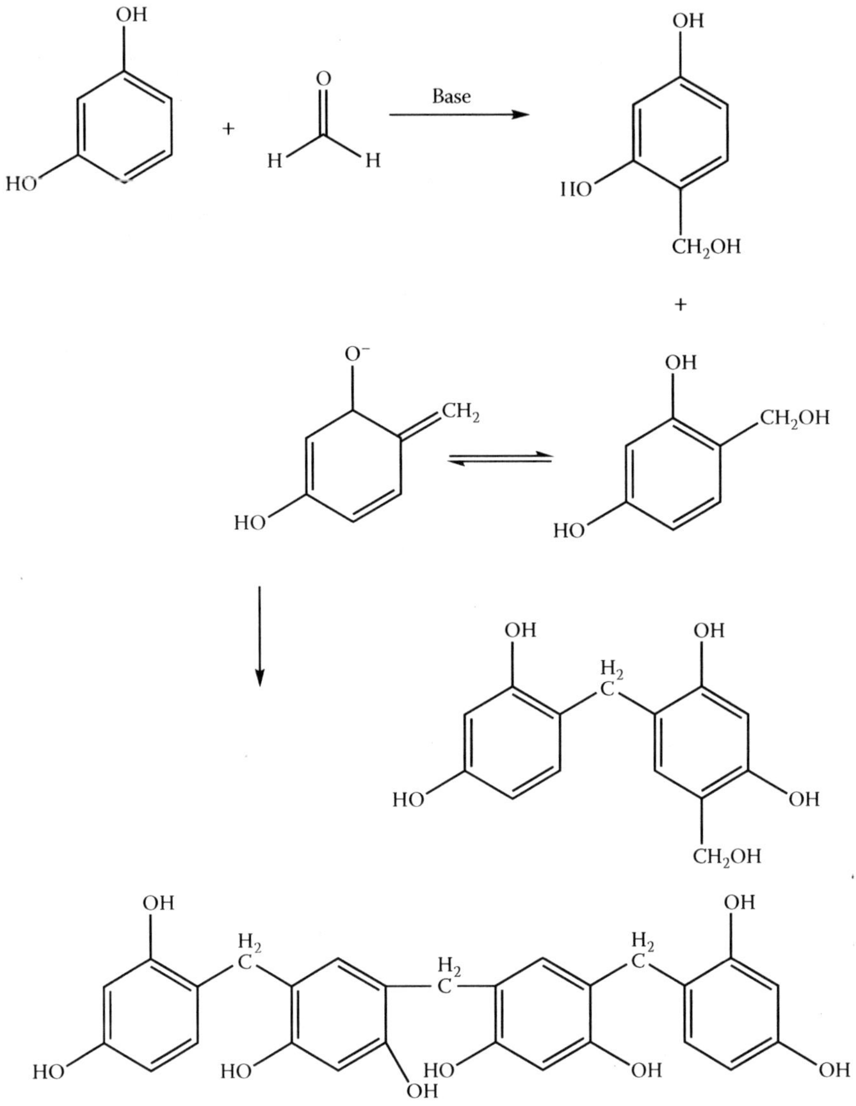

FIGURE 6.2 The reaction of aldehydes and phenolic compounds. (See Durairaj, R.B., *Resorcinol Chemistry, Technology and Applications*, Springer Verlag, Heidelberg, Germany, 2005.)

curing/hardening, and (3) good tensile strength and tissue adhesion. The original formulation meets these criteria but has the disadvantage of formaldehyde toxicity. There have been efforts to improve the product by replacing the formaldehyde; the inclusion of glutaraldehyde presumably allows a decrease in the amount of formaldehyde. The reaction of glutaraldehyde would be slower than formaldehyde and would act more to stabilize the glue rather than initiate the adhesion process. This, however, is an assumption and must account for the apparent success of glutaraldehyde with bovine serum albumin in BioGlue®. There was one report on the use of GRF adhesive as a matrix for bone growth,[493] but other materials have proved to be more useful.[494,495]

There have been efforts to improve the GRF resin product quality by using other cross-linking reagents. Ennker and colleagues[496] used glyoxal (ethanedial) and glutaraldehyde (1,5-pentanedial) with collagen and resorcinol to prepare a useful glue. Sung and colleagues[497] evaluated carbodiimides for the cross-linking of gelatin and alginate and genipin for the cross-linking of gelatin and polylysine.

BIOGLUE®

BioGlue® is a surgical adhesive prepared by the cross-linking of bovine serum albumin with glutaraldehyde.[498–501] An autologous glue has been prepared from plasma and glutaraldehyde[502] but, as with cryoprecipitate-based fibrin sealant products, it does have the tensile strength of the commercial product. Gelation of plasma with glutaraldehyde is used to estimate the combined immunoglobulin–fibrinogen level in cattle.[503] The cross-linking of proteins can occur via either intermolecular processes, intramolecular processes, or both. Intramolecular cross-linking is favored at low protein concentration and low reagent concentration, whereas intermolecular cross-linking is favored with higher reagent and higher protein concentrations.[504] With this and related products, intermolecular cross-linking is favored for the development of both cohesive and adhesive forces.[505] The reaction of a dialdehyde such as glutaraldehyde with a protein should be fairly straightforward with the formation of two Schiff bases (imine) between the aldehyde function and an amine (usually the ε-amino group of lysine) on the protein, and the properties of the modified proteins suggested a more complex reaction,[506] which has been confirmed by recent crystallographic analysis.[507] This complexity reflects the tendency of glutaraldehyde to polymerize in aqueous solutons.[508–511] Although important for the use of glutaraldehyde in protein structure studies,[512,513] such issues are not likely to be important for the use of glutaraldehyde as a component of a tissue glue or to preserve tissue for prosthetic use.[514–552]

MUSSEL ADHESIVE PROTEIN

The classic mussel adhesive protein is synthesized and secreted by *Mytilus edulis* L.[522–525] There are six proteins (Mefp-1 through Mefp-6), which vary in sequence and size.[526] The function of this protein is to provide anchorage of the organism to a solid support in an aqueous environment. Although adhesive proteins are well known in biological systems,[527–533] the interactions of these proteins with various substrates are generally weak, and in the absence of the establishment of covalent bonds (disulfide, Schiff base), such interactions are readily reversible. Despite considerable work on the mussel adhesive protein, the adhesive and cohesive mechanisms are still poorly understood.[534] It is suggested that the high content of 3,4-dihydroxyphenylalanine (DOPA) provides the basis for adhesion. The presence of the vicinal hydroxyl groups permits the formation of ion adducts, leading to protein cross-linking.[535] It is also proposed that cross-linkage occurs between the DOPA residues via the quinine form after oxidation. Covalent cross-linkage could occur at the *ortho*(2 position) to the *p*-hydroxyl on the phenol ring.[526] A study of DOPA-mediated cross-linkage in a different system suggested a role for lysine residues,[536] whereas nuclear magnetic resonance (NMR) studies with the mussel system indicate that lysine residues are not directly involved in the cross-linkage process.[537] The marine mussel protein can form a hydrogel,[535,538] which is a functional aspect of the adhesive plaque. The plaque or hydrogel can be "hardened" by oxidation (presumably involving the oxidation of the DOPA residues).[537,539] This oxidation process decreases the mass of the hydrogel through the removal of water.[536] In summary, the mussel and related marine organisms synthesize and secrete a protein that forms an adhesive hydrogel as part of the byssus threads, which can be stabilized by the formation of covalent cross-links. The molecular mechanism for the cross-link formation is not clear, but it does involve the DOPA residues. Given the structural relationship of the mussel protein and collagen,[522] the process of cross-link formation is likely quite complex.

The strength of the byssal bond formed by mussels in an aqueous environment has created interest in its application as a surgical sealant or adhesive.[540–543] As of early 2008, this interest has not extended beyond the research laboratory. Recently, Ninan and coworkers[543] did show that the bond strength obtained with mussel protein extract in a porcine intestinal model was less than that obtained with ethyl cyanoacrylate but greater than that obtained with octyl cyanoacrylate. The strength of the mussel protein bond could be increased with oxidation.

There has been some success in the use of the mussel adhesive protein to bind cells and proteins to surfaces.[544–548] Pretreatment of a microplate with this protein increased the binding of an antigen, human choroionic gonadotrophin, in an immunoassay screening for polyclonal and monoclonal antibodies.[546] The use of the mussel adhesive protein allows 50–100 times less antigen to be used (5–10 μg rather than 500 μg quantities). The use of the mussel adhesive protein matrix was essential for the single-cell studies comparing several cell lines at the same time and place.[547] Mussel adhesive protein has been used for the culture of *Plasmodium falciparum* gametocytes on erythrocytes immobilized on mussel adhesive protein.[548] An interesting hydrogel has been prepared from mussel adhesive protein by chelate cross-linking with iron.[535]

Biopolymers based on the repeated decapeptide sequence have been synthesized,[549,550] and recombinant engineered proteins are available.[551,552] This should increase the amount of protein available for research.

END NOTES

Tissue welding uses laser energy to bond tissue[1–3] and has been thought to involve the denaturation and cross-linkage of collagen.[4–7] Successful tissue welding involved the physical contact (coaptation) of the tissues such that a successful seal could be accomplished.[8] Laser welding for most tissues is still in the development stage[9–11] but there is a recent application for corneal welding in cataract surgery.[12] Although described by a different term, photocoagulation is also an example of tissue welding or, more accurately from a historical perspective, tissue welding is an extension of photocoagulation to other issues. Photocoagulation is one of several methods that have been developed for the formation of an adhesive bond between the retina and choroid and dates to the use of cauterization by Gonin for treatment of detached retina.[13] Cauterization was replaced by diathermy.[14] A review of methods for the treatment of retinal detachment has been presented by Hilton and colleagues.[15]

Photocoagulation is an example of tissue denaturation[16–19]; tissue denaturation is a loss of structure and function with associated tissue coagulation,[20] where coagulation is understood to mean the formation of a solid or coagulated mass.[21] Photocoagulation was developed as a technique in ophthalmic surgery for sealing retinal breaks.[22–24] The early application of photocoagulation involved the use of focused sunlight, but was later replaced with a high-intensity light source.[22,23] The advent of lasers for photocoagulation[25] greatly accelerated the maturation of this surgical treatment for retinal tears and detachment.[26–31] The mechanisms for tissue sealing with laser welding or photocoagulation are poorly understood, but most likely involve denaturation of protein with the exposure of hydrophobic regions causing association and the formation of complex cross-linked products[32,33] involving collagen.[34,35] Tissue soldering also involves the exposure of hydrophobic regions in proteins such as collagen, albumin, and fibrinogen.[36–38]

Tissue welding, predated by photocoagulation in retinal surgery, is a method for sutureless wound closure and dates to Jain and Gorisch, who used an Ne:YAG laser for repair of incisions in small vessels in the rat in 1976.[39] This application of laser technology is based on earlier work on the use of high-frequency electric current (electrocoagulation) for the closure of incisions in blood vessels (coaptive closure).[40] Tissue welding uses laser energy to bond tissue[41–43] and is thought to involve denaturation and renaturation with the formation of new noncovalent and covalent bonds.[44–47] There is little direct experimental evidence to support the involvement of collagen, and it is likely that other proteins and cell constituents such as carbohydrates are involved in the process of tissue welding. Indeed, work from tissue soldering would suggest that collagen bonding is ineffective in the bonding of tissues.[48,49]

Tissue welding is analogous to the process of fusion welding in metal manufacturing. Fusion welding is the process of joining two metals without the application of pressure. This involves the heating of a small region of the metal to a high temperature at which this small region is surrounded by a cold region with a volume much larger than the heated area.[50] Fusion welding of metals involves the heating of metal with the goal of joining them. Materials are considered to have good weldability only if they can be reliably welded on a production scale; critical attributes of a good weld include the welding process, environment, alloy composition, and joint design and size.[51] The melting point of a metal is of great importance in welding because together with the heat capacity of the metal, it determines the amount of energy required for a good weld.[52]

REFERENCES

References for End Notes

1. Bass, L.S. and Treat, M.R., Laser tissue welding: A comprehensive review of current and future clinical applications, *Lasers Surg. Med.* 17, 315–349, 1995.

2. Capan, A. and Mordon, S., Can thermal lasers promote skin wound healing?, *Am. J. Clin. Dermatol.* 4, 1–12, 2003.
3. Bleier, R.S., Grafton, M.A., Leibowitz, J.M., Laser-welded endoscopic endoluminal repair of iatrogenic esophageal perforation: an animal model, *Otolaryngol. Head Neck Surg.* 139, 713–717, 2008.
4. Bass, L.S., Moazami, N., Focsidio, J. et al., Changes in type I collagen following laser welding, *Lasers Surg. Med.* 12, 500–505, 1992.
5. Tang, J., Godlewski, G., Rouy, S., and Delacrétaz, G., Morphologic changes in collagen fibers after 830 nm diode laser welding, *Lasers Surg. Med.* 21, 438–443, 1997.
6. Tang, J., Zeng, F., Savage, H. et al., Fluorescence spectroscopic imaging to detect changes in collagen and elastin following laser tissue welding, *J. Clin. Laser. Med. Surg.* 18, 3–8, 2000.
7. Theodossiou, T., Rapti, G.S., Hovhannisyan, V. et al., Thermally induced irreversible conformational changes in collagen probed by optical second harmonic generation and laser-induced fluorescence, *Lasers Med. Sci.* 17, 34–51, 2002.
8. Bass, L.S. and Treat, M.R., Laser tissue welding: A comprehensive review of current and future clinical applications, in *Laser Surgery and Medicine*, ed. A. Puliafito, Wiley-Liss, New York, 1996.
9. Michekm, R.G., Weinstock, B.I., and Tsau, K., Safety and efficacy of pressure-assisted tissue-welding tonsillectomy: a preliminary evaluation, *Ear Nose Throat J.* 87, 100–105, 2008.
10. Gulsoy, M., Dereli, Z., Tabakoglu, H.O., and Bozkulak, O., Closure of skin incisions by 980-nm diode laser welding, *Lasers Med. Sci.* 21, 5–10, 2006.
11. Foyt, D., Slattery, W.H., 3rd and Carfrae, M.J., Underlay tympanoplasty with laser tissue welding, *Ear Nose Throat J.* 85, 247–248, 2006.
12. Menabuoni, L., Pini, R., Rossi, F. et al., Laser-assisted corneal welding in cataract surgery: Retrospective study, *J. Cataract Refract. Surg.* 33, 1608–1612, 2007.
13. Gonin, J., The treatment of detached retina by searing the retinal tears, *Archs. Ophthalmol.* 4, 621–625, 1930.
14. Parry, R., Some principles in the surgery of retinal separation, *Trans. Ophthalmol. Soc. United Kingdom,* 76, 443–452, 1956.
15. Hilton, G.F., McLean, E.B., and Chuang, E.L., *Retinal Detachment*, Am. Acad. Ophthalmol., San Francisco, CA, 1990.
16. Gabriel, E., Faion, H., and Dieckmann, G., Radio frequency pulsed coagulation: An improved method for controlled thermoelectrode tissue denaturation by determination of electrical and thermal conductivity changes, *Confin. Neurol.* 29, 213–219, 1967.
17. Beacco, C.M., Mordon, S.R., and Brunetaud, J.M., Development and experimental *in vivo* validation of mathematical modeling of laser coagulation, *Lasers Surg. Med.* 14, 362–373, 1994.
18. Duncan, A.C. and Boughner, D., Effect of dynamic glutaraldehyde fixation of the viscoelastic properties of bovine pericardial tissue, *Biomaterials* 19, 777–783, 1998.
19. Vakoc, B.J., Tearney, G.J., and Bouma, B.E., Real-time microscopic visualization of tissue response to laser thermal therapy, *J. Biomed. Opt.* 12:020501, 2007.
20. Hillenkamp, F., Laser radiation tissue interaction, *Health Phys.* 56, 613–616, 1989.
21. *Oxford English Dictionary Online*, Oxford University Press, Oxford, http://www.oup.com.
22. Meyer-Schwickerath, G., Lichtkoagulation: Eine Methode zur Behandlung und Verhütting der Netzhautblösung, *von Graefes Arch. Ophth.* 156, 2–34, 1954.
23. Meyer-Schwickerath, G., Prophylactic treatment of retinal detachment by light coagulation, *Trans. Opthal. Soc. United Kingdom* 76, 739–750, 1956.
24. McDonald, J.E., and Light, A., Photocoagulation of iris and retina, *Arch. Ophthalmol.* 63, 384–391, 1958.

25. Zaret, M.M., Ripps, H., Siegel, I.M., and Breinin, G.M., Laser photocoagulation of the eye, *Arch. Ophthalmol.* 69, 131–139, 1963.
26. Gitter, K.A. and Robinson, T.R., Photocoagulation by argon laser—advantages and limitations: An analysis, *Southern Med. J.* 65, 1208–1212, 1972.
27. Gratton, I., Gazocchi, M., Simonini, F. et al., Argon laser photocoagulation in the management of retinal detachment and predisposing lesions, *Laser Surg. Med.* 4, 337–344, 1984.
28. Folk, J.C., Sneed, S.R., Folberg, R. et al., Early retinal adhesion from laser photocoagulation, *Ophthalmology* 96, 1523–1525, 1989.
29. Greenberg, P.B. and Baumal, C.R., Laser therapy for rhegmatogenous retinal detachment, *Curr. Opin. Ophthalmol.* 12, 171–174, 2001.
30. Ip, M.O. and Puliafito, C.A., Laser photocoagulation, in *Ophthalmology*, Eds. M. Yanoff and J.S. Duker, Mosby, St. Louis, MO, Chapter 102, 2004.
31. Brucker, A.J. and Hopkins, T.B., Retinal detachment surgery: The latest in current management, *Retina* 26(Suppl.), S22–S33, 2006.
32. Murray, L.W., Su, L., Kopchok, G.E., and White, R.A., Crosslinking of extracellular matrix proteins: A preliminary report on a possible mechanism of argon laser welding, *Lasers Surg. Med.* 9, 490–496, 1989.
33. Guthrie, C.R., Murray, L.W., Kopchok, G.E. et al., Biochemical mechanisms of laser vascular tissue fusion, *J. Invest. Surg.* 4, 3–12, 1991.
34. Gayen, T.K., Katz, A., Savage, H.E. et al., Aorta and skin tissues welded by near-infrared Cr4+-YAG laser, *J. Clin. Laser Med. Surg.* 21, 259–269, 2003.
35. Matteini, P., Rossi, F., Menabuoni, L., and Pini, R., Microscopic characterization of collagen modifications induced by low-temperature diode-laser welding of corneal tissue, *Lasers Surg. Med.* 39, 597–604, 2007.
36. Poppas, D.P., Wright, E.J., Guthrie, P.D. et al., Human albumin solders for clinical application during laser tissue welding, *Lasers Surg. Med.* 19, 2–8, 1996.
37. Lanzafame, R.J., Soltz, B.A., Stadler, I. et al., Acute tensile strength analysis of collagen solder for mesh fixation to the peritoneal surface, *Surg. Endosc.* 19, 178–183, 2005.
38. Khadem, J., Truong, T., and Ernest, J.T., Photodynamic biologic tissue glue, *Cornea* 13, 406–410, 1994.
39. Jain, K.K. and Gorisch, W. Repair of small blood vessels with the neodymium-YAG laser: A preliminary report, *Surgery* 85, 684–688, 1979.
40. Sigel, B. and Dunn, M.R., The mechanism of blood vessel closure by high frequency electrocoagulation, *Surg. Obstetrics Gynecol.* 12, 823–831, 1965.
41. Bass, L.S. and Treat, M.R., Laser tissue welding: A comprehensive review of current and future clinical applications, *Lasers Surg. Med.* 17, 315–349, 1995.
42. Capan, A. and Mordon, S., Can thermal lasers promote skin wound healing? *Am. J. Clin. Dermatol.* 4, 1–12, 2003.
43. Flock, S.T. and Marchitto, K.S., Progress towards seamless tissue fusion for wound closure, *Otolaryngol. Clin. North Am.* 38, 295–305, 2005.
44. Bass, L.S., Moazami, N., Focsidio, J. et al., Changes in type I collagen following laser welding, *Lasers Surg. Med.* 12, 500–505, 1992.
45. Tang, J., Godlewski, G., Rouy, S., and Delacrétaz, G., Morphologic changes in collagen fibers after 830 nm diode laser welding, *Lasers Surg. Med.* 21, 438–443, 1997.
46. Tang, J., Zeng, F., Savage, H. et al., Fluorescence spectroscopic imaging to detect changes in collagen and elastin following laser tissue welding, *J. Clin. Laser. Med. Surg.* 18, 3–8, 2000.
47. Theodossiou, T., Rapti, G.S., Hovhannisyan, V. et al., Thermally induced irreversible conformational changes in collagen probed by optical second harmonic generation and laser-induced fluorescence, *Lasers Med. Sci.* 17, 34–51, 2002.
48. Mueller, M.P., Kopchok, G.E., Tabbara, M.R. et al., Argon laser-welded bovine heterograft anastomoses, *J. Clin. Laser Med. Surg.* 11, 1–5, 1993.

49. Shohet, J.A., Reinisch, L., and Ossoff, R.H., Prevention of pharyngocutaneous fistulas by means of laser-weld techniques, *Laryngoscope* 105, 717–722, 1995.
50. Deyev, G. and Detev, D., *Surface Phenomenon in Fusion Welding Process*, Taylor & Francis, Boca Raton, FL, 2006.
51. Easterling, K., *Introduction to the Physical Metallurgy of Welding,* Butterworth, London, 1983.
52. Davies, A.C., *The Sciences and Practice of Welding*, Cambridge University Press, Cambridge, 1989.

Chapter References

1. Keimel, F.A., Historical developments of adhesives and adhesive bonding, in *Handbook of Adhesive Technology*, 2nd ed., Eds. A. Pizzi and K.L. Mittal, Marcel Dekker, New York, 2003.
2. Taggard, J.A., *The Glue Book. How to Select, Prepare and Use Glue*, The Republican Publishing Company, Hamilton, Ohio, 1913.
3. Pearson, C.J., Animal glues and adhesives, in *Handbook of Adhesive Technology*, 2nd ed., Eds. A. Pizzi and K.L. Mittal, Marcel Dekker, New York, 2003.
4. Khalil, M.L. Biological activity of bee propalis in health and disease, *Asian Pac. J. Cancer Prev.* 7, 22–31, 1996.
5. Smith, P.E., *Glues and Gelatin*, Sir Issac Putman & Sons, Ltd., London, 1929.
6. Pursifull, N.E. and Morey, A.F., Tissue glues and nonsuturing techniques, *Curr. Opin. Urol.* 17, 396–401, 2006.
7. Sung, H.W., Huang, D.M., and Chang, W.H., Evaluation of gelatin hydrogel crosslinked with various crosslinking bioadhesives: *In vitro* study, *J. Biomed. Mater. Res.* 46, 520–530, 1999.
8. Bachet, J. and Guilmet, D., The use of biological glue in aortic surgery, *Cardiol. Clin.* 17, 779–796, 1999.
9. *Adhesives from Renewable Resources*, Eds. R.N. Hemingway, A.H. Conner, and S.J. Branham, American Chemical Society, Washington, DC, 1989.
10. Baumann, M.G.D. and Conner, A.H., Carbohydrate polymers as adhesives, in *Handbook of Adhesives*, 2nd ed., Eds. A. Pizzi and A.H. Conner, Marcel Dekker, New York, Chapter 21, pp. 479–494, 2003.
11. Mildenberg, R., Zander, M., and Collin, G., *Hydrocarbon Resins*, VCH Publishing, New York, 1997.
12. Gardziella, A., Pilato, L.A., and Knop, A., *Phenolic Resins*, 2nd ed., Springer-Verlag, Berlin, Germany, 2000.
13. Lambuth, A.L., Protein adhesives for wood, in *Handbook of Adhesive Technology*, 2nd ed., Eds. A. Pizzi and K.L. Mittal, Marcel Dekker, New York, Chapter 20, pp. 457–477, 2003.
14. Harding, S.E., Davis, S.S., Deacon, M.P., and Fiebrig, I., Biopolymer mucoadhesives, *Biotechnol. Genet. Eng. Rev.* 16, 41–86, 1999.
15. Chopra, S., Mahdi, S., Kaur, J. et al., Advances and potential applications of chitosan derivatives as mucoadhesive biomaterials in modern drug delivery, *J. Pharm. Pharmacol.* 59, 1021–1032, 2006.
16. Morten, J.E., Mucilagenous plants and their uses in medicine, *J. Ethnopharmacol.* 29, 245–266, 1990.
17. Detlefsen, W.D., Blood and casein adhesives for bonding woods, in *Adhesives from Renewable Resources*, Eds. R.N. Hemingway, A.H. Conner, and S.J. Branham, American Chemical Society, Washington, DC, Chapter 31, pp. 445–452, 1989.
18. López-Rice, R., Moneo, I., Rice, A. et al., Cereal α-amylase inhibitor causes occupational sensitization in the wood industry, *Clin. Exp. Allergy* 28, 1286–1291, 1998.

19. Hides, G., Kastin, A., Mullerad, M. et al., Sutureless nephron-sparing surgery: Use of albumin glutaraldehyde tissue adhesive (Bioglue), *Urology* 67, 697–700, 2006.
20. Passage, J., Jalali, H., Tam, R.K. et al., Bioglue surgical adhesive—an appraisal of its indications in cardiac surgery, *Ann. Thorac. Surg.* 74, 432–437, 2002.
21. Zehr, K.G., Use of bovine albumin-glutaraldehyde glue in cardiovascular surgery, *Ann. Thorac. Surg.* 84, 1048–1052, 2007.
22. Yang, I., Kuo, M., Myers, D.J., and Po, A., Comparison of protein-based adhesive resins for wood composites, *J. Wood Sci.* 52, 503–508, 2006.
23. Gama, M., Wood adhesives from natural raw materials, in *Wood Adhesives*, Ed. A. Pizzi (*J. Appl. Poly. Sci. Applied Polymer Symposium 40*), John Wiley & Sons, New York, 1983.
24. Satos, D., Pressure sensitive adhesives and adhesive products in the United States, in *Handbook of Pressure Sensitive Adhesive Technology*, 3rd ed., Ed. D. Satos, Satos & Associates, Warwick, RI, Chapter 1, pp. 1–21, 1999.
24a. Reece, T.B., Maxey, T.S., and Kron, I.L., A prospectus on tissue adhesives, *Am. J. Surg.* 182, 40S–44S, 2001
25. Satos, D., Tack, in *Handbook of Pressure Sensitive Adhesive Technology*, 3rd ed., Ed. D. Satos, Satos & Associates, Warwick, RI, Chapter 4, pp. 36–61, 1999.
26. Satos, D., Medical products, in *Handbook of Pressure Sensitive Adhesive Technology*, 3rd ed., Ed. D. Satos, Satos & Associates, Warwick, RI, Chapter 29, pp. 706–723, 1999.
27. Pfister, W.R. and Hsieh, D.S., Permeation enhancers compatible with transdermal drug delivery systems: Part II: System design considerations, *Med. Device Technol.* 1, 28–33, 1990.
28. Venkatraman, S. and Gale, R., Skin adhesives and skin adhesion. 1. Transdermal drug delivery systems, *Biomaterials* 19, 1119–1136, 1998.
29. van der Walle, G.A., de Koning, G.J., Weusthuius, R.A., and Eggink, G., Properties, modifications and applications of biopolyesters, *Adv. Biochem. Eng. Biotechnol.* 71, 263–291, 2001.
30. Bishara, S.E., Zeitler, D.L., and Kremenak, C.R., Effects of a fibrin-sealant dressing on the healing of full-thickness wounds of the hard palate: Preliminary report, *Cleft Palate J.* 23, 144–152, 1986.
31. Holcomb, J.B., Pusateri, A.E., Hess, J.E. et al., Implications of a new dry fibrin sealant technology for trauma surgery, *Surg. Clin. North Am.* 77, 943–952, 1997.
32. Drake, D.B. and Wong, L.G., Hemostatic effect of Vivostat patient-derived fibrin sealant on split-thickness skin graft donor sites, *Ann. Plast. Surg.* 50, 367–372, 2003.
33. Campbell, K., Woodbury, M.G., Whittle, H. et al., A clinical evaluation of 3M no sting barrier film, *Ostomy Wound Manage.* 46, 24–30, 2000.
34. Connolly, R.J., Application of the poly-*N*-acetyl glucosamine-derived rapid deployment hemostat trauma dressing in severe/lethal Swine hemorrhage trauma models, *J. Trauma* 57(Suppl. 1), S26–S28, 2004.
35. Brunkwall, J., Ruemenapf, G., FLorek, H.J. et al., A single arm, prospective study of an absorbable cyanoacrylate surgical sealant for use in vascular reconstructions as an adjunct to conventional techniques to achieve haemostasis, *Cardiovasc. Surg.* 48, 471–476, 2007.
36. Petrie, E.M., Introduction to adhesives and sealants, *Handbook of Adhesives and Sealants,* McGraw-Hill, New York, Chapter 1, 2000.
37. Petrie, E.M., Theories of adhesion, *Handbook of Adhesives and Sealants,* McGraw-Hill, New York, Chapter 2, 2000.
38. Schultz, J. and Nordin, M., Theories and mechanisms of adhesion, in *Handbook of Adhesion Technology*, 2nd ed., Eds. A. Pizzi and K.L. Mittal, Marcel Dekker, New York, Chapter 3, pp. 53–67, 2003.
39. *Adhesion*, Ed. D.D. Eley, Oxford University Press, Oxford, 1961.
40. Irons, B.K. and Robinson, J.R., Bioadhesion in drug delivery, in *Handbook of Adhesive Technology*, 2nd ed., Eds. A. Pizzi and K.L. Mittal, Marcel Dekker, New York, Chapter 48, pp. 957–969, 2003.

41. Bernkop-Schnürch, A., Hoffer, M.H., and Kafedjiiski, K., Thiomers for oral delivery of hydrophilic macromolecular drugs, *Expert. Opin. Drug Deliv.* 1, 87–98, 2004.
42. Smart, J.D., The basics and underlying mechanisms of mucoadhesion, *Adv. Drug Deliv. Rev.* 57, 1556–1568, 2005.
43. Salamat-Miller, N., Chittchang, M., and Johnston, T.P., The use of mucoadhesive polymers in buccal drug delivery, *Adv. Drug. Deliv. Rev.* 57, 1666–1691, 2005.
44. Bowman, K. and Leong, K.W., Chitosan nanoparticles for oral drug and gene delivery, *Int. J. Nanomedicine* 1, 117–128, 2006.
45. Glinsky, V.V., Intravascular cell-to-cell interactions and bone metastasis, *Cancer Metastasis Rev.* 25, 521–540, 2006.
46. Seymour, G.B., Tucker, G., and Leach, L., Cell adhesion molecules in plants and animals, *Biotechnol. Genet. Eng. Rev.* 21, 123–132, 2004.
47. Cereijido, M., Contreras, R.G., and Stoshami, L., Cell adhesions, polarity, and epithelia in the dawn of metazoans, *Physiol. Rev.* 84, 1229–1262, 2006.
48. Clark, S.R. and Foster, S.J., Surface adhesion of *Staphylococcus aureus*, *Adv. Microb. Physiol.* 51, 187–224, 2006.
49. Barnich, N. and Darfeuille-Michaud, A., Adherent-invasive *Escherichia coli* and Crohn's disease, *Curr. Opin. Gastroenterol.* 23, 16–20, 2007.
50. Lindsay, D. and von Holy, A., Bacterial biofilms within the clinical setting: What healthcare professionals should know, *J. Hosp. Infect.* 64, 313–325, 2006.
51. Delan, I. and Brown, N.H., Integrins and the actin cytoskeleton, *Curr. Opin. Cell Biol.* 19, 43–50, 2007.
52. Uchimuran, K. and Rosen, S.D., Sulfated L-selectin ligands as a therapeutic target in inflammation, *Trends Immunol.* 27, 559–565, 2006.
53. Xiao, K., Oas, R.G., Chiasson, C.M., and Kowalczyk, A.R., Role of p120-catenin in cadherin inflammation, *Biochim. Biophys. Acta* 1773, 8–16, 2007.
54. Lu, X., Lu. D., Scully, M.F., and Kakkar, V.V., Integrins in drug targeting-RGD templates in toxins, *Curr. Pharm. Des.* 12, 2749–2769, 2006.
55. *Adhesion Protein Protocols*, Ed. A.S. Coutts, Humana Press, Totowa, NJ, 2007.
56. Liu, W.F., Nelson, C.M., Tan, J.L., and Chen, C.S., Cadherins, RhoA, and Rac1 are differently required for stretch-mediated proliferation in endothelial versus smooth muscle cells, *Circ. Res.* 101, e44–e52, 2007.
57. Shima, Y., Kawaguchi, S.Y., Kosaka, K. et al., Opposing roles in neurite growth control by two seven-pass transmembrane cadherins, *Nat. Neurosci.* 10, 963–969, 2007.
58. Pokutta, S. and Weis, W.I., Structure and mechanism of cadherins and catenins in cell-cell contacts, *Annu. Rev. Cell Dev. Biol.* 23, 237–261, 2007.
59. Mbalaviele, G., Shin, C.S., and Civitelli, R., Cell-cell adhesion and signaling through cadherins: Connecting bone cells in their microenvironment, *J. Bone. Miner. Res.* 21, 1821–1827, 2006.
60. Etzioni, A. and Alon, R., Leukocyte adhesion deficiency III: A group of integrin activation defects in hematopoietic linkage cells, *Curr. Opin. Allergy Clin. Immunol.* 4, 485–490, 2005.
61. Ruggeri, Z.M., The role of von Willebrand factor and fibrinogen in the initiation of platelet adhesion to thrombogenic surfaces, *Thromb. Haemostas.* 74, 460–463, 1995.
62. Lecuit, T., Adhesion remodeling underlying tissue morphogenesis, *Trends Cell Biol.* 15, 34–42, 2005.
63. Ellis, H., The cause and prevention of postoperative intraperitoneal adhesions, *Surg. Gynecol. Obstet.* 133, 497–511, 1971.
64. Holtz, G., Prevention of postoperative adhesions, *J. Reprod. Med.* 24, 141–146, 1980.
65. Divilio, L.T., Surgical adhesion development and prevention, *Int. Surg.* 90(Suppl. 3), S6–S9, 2005.

66. Larsson, B., Efficacy of Interceed™ in adhesion prevention in gynecological surgery: A review in clinical studies, *J. Reprod. Med.* 41, 27–34, 1996.
67. Azziz, R., Murphy, A.A., Rosenberg, S.M., and Patton, G.W., Jr., Use of an oxidized, regenerated cellulose absorbable adhesion barrier at laparoscopy, *J. Reprod. Med.* 36, 479–482, 1991.
68. Ryan, C.K. and Sax, H.C., Evaluation of a carboxymethylcellulose sponge for prevention of postoperative adhesions, *Am. J. Surg.* 169, 154–159, 1995.
69. Ikeda, K., Yamauchi, D., and Tomita, K., Preliminary study for prevention of neural adhesion using an absorbable oxidized regenerated cellulose sheet, *Hand Surg.* 7, 11–14, 2002.
70. DeCherney, A.H. and di Zerega, G.S., Clinical problem of intraperitoneal postsurgical adhesion formation following general surgery and the use of adhesion prevention barriers, *Surg. Clin. North Am.* 77, 671–688, 1997.
71. Burns, J.W., Colt, M.J., Burgees, L.S., and Skinner, K.C., Preclinical evaluation of Separfilm bioresorbable membrane, *Eur. J. Surg.* (Suppl. 577), 40–48, 1997.
72. Mohri, Y., Uchida, K., Araki, T. et al., Hyaluronic acid-carboxycellulose membrane (Seprafilm) reduces early postoperative small bowel obstruction in gastrointestinal surgery, *Am. Surg.* 71, 861–863, 2005.
73. Takeuchi, H., Kitade, M., Kikuchi, I. et al., A novel instrument and technique for using seprafilm hyaluronic acid/carboxymethylcellulose membrane during laparoscopic myomectomy, *J. Laparoendosc. Adv. Surg. Tech. A* 16, 497–502, 2006.
74. Bristow, R.E., Santillan, A., Diaz-Montez, T.P. et al., Prevention of adhesion formation after radical hysterectomy using a sodium hyaluronate-carboxymethylcellulose (HA-CMC) barrier: A cost-effectiveness analysis, *Gynecol. Oncol.* 104, 739–746, 2006.
75. Johns, A., Evidence-based prevention of post-operative adhesions, *Hum. Reprod. Update* 7, 577–579, 2001.
76. Tsuji, S., Takahashi, K., Yomo, K. et al., Effectiveness of antiadhesion barriers in preventing adhesion after myomectomy in patients with uterine leiomyoma, *Eur. J. Obstet. Gynecol. Reprod. Biol.* 123, 244–248, 2005.
77. Armstrong, P.B., Cell sorting out: The self-assembly of tissue *in vitro*, *Crit. Rev. Biochem. Mol. Biol.* 24, 119–149, 1989.
78. Neu, T.R. and Mashall, K.C., Bacterial polymers: Physicochemical aspects of their interactions at interfaces, *J. Biomater. Appl.* 5, 107–133, 1990.
79. Hammer, D.A., Simulation of cell rolling and adhesion on surfaces in shear flow. Microvilli-coated hard spheres with adhesive springs, *Cell Biophys.* 18, 145–182, 1991.
80. Pierres, A., Benoliel, A.M., and Bongrand, P., Measuring bonds between surface-associated molecules, *J. Immunol. Methods* 196, 105–120, 1996.
81. Michalski, M.C., Desobry, S., and Hardy, J., Food materials adhesion: A review, *Crit. Rev. Food Sci. Nutr.* 37, 591–619, 1997.
82. Smith, B.L., The importance of molecular structure and conformation: Learning with scanning probe microscopy, *Prog. Biophys. Mol. Biol.* 74, 93–113, 2000.
83. Adams, J.C., *Methods in Cell-Matrix Interactions*, Academic Press, San Diego, CA, 2002.
84. Leckband, D., Nanomechanics of adhesion proteins, *Curr. Opin. Struct. Biol.* 14, 524–530, 2004.
85. Couts, A.S., *Adhesion Protein Protocols*, Humana Press, Totowa, NJ, 2007.
86. Poppas, D.P., Schlossberg, S.M., Richmond, I.L. et al., Laser welding in urethral surgery: Improved results with a protein solder, *J. Urol.* 139, 415–417.
87. Oz, M.C., Johnson, J.P., Paragi, S. et al., Tissue soldering by use of indocyanine green dye-enhanced fibrinogen with the near infrared diode laser, *J. Vasc. Surg.* 11, 718–725, 1990.
88. Mozami, N., Oz., M.C., Bass, L.S., and Treat, M.R., Reinforcement of colonic anastomoses with a laser and dye-enhanced fibrinogen, *Arch. Surg.* 125, 1452–1454, 1990.
89. *Oxford English Dictionary*, Oxford University Press, Oxford, 2006; http://www.oed.com.

90. Bass, L.S. and Treat, M.R., Laser tissue welding: A comprehensive review of current and future clinical applications, in *Laser Surgery and Medicine*, Ed. A. Puliafito, Wiley-Liss, New York, 1996.
91. Lauto, A., Ohebshalom, M., Esposito, M. et al., Self-expandable chitosan stent design and preparation, *Biomaterials* 22, 1869–1874, 2001.
92. Lauto, A., Stoodley, M., Marcel, H. et al., *In vitro* and *in vivo* tissue repair with laser-activated chitosan adhesive, *Lasers Surg. Med.* 39, 19–27, 2007.
93. Lauto, A., Foster, L.J.R., Ferris, L. et al., Albumin-genipin solder for laser tissue repair, *Lasers Surg. Med.* 35, 140–145, 2004.
94. Wang, X.H, Li, D.P., Wang, W.J. et al., Crosslinked collagen/chitosan matrix for artificial livers, *Biomaterials* 24, 3213–3220, 2003.
95. Lee, J.E., Kim, K.E., Kwon, I.C. et al., Effects of the controlled-release TGF-beta 1 from chitosan microspheres on chondrocytes cultured in a collagen/chitosan/glycosaminoglycan scaffold, *Biomaterials* 25, 4163–4173, 2004.
96. Wu, X., Black, L., Santacana-Laffitte, G., and Patrick, C.W., Jr., Preparation and assessment of glutaraldehyde-crosslinked collagen chitosan hydrogels for adipose tissue engineering, *J. Biomed. Mater. Res. A* 81, 59–65, 2007.
97. Ma, L., Gao, C., Mao, Z. et al., Thermal dehydration treatment and glutaraldehyde cross-linking to increase the biostability of collagen-chitosan porous scaffold used as dermal equivalent, *J. Bimater. Sci. Polym. Ed.* 14, 961–974, 2003.
98. Putnam, F.W., Protein denaturation, in *The Proteins*, Vol. 1, Pt. G, Eds. H. Neurath and K. Bailey, Academic Press, New York, 1953.
99. Lumry, R. and Eyring, H., Conformational changes of proteins, *J. Phys. Chem.* 58, 110–120, 1954.
100. Misawa, S. and Kumagai, I., Refolding of therapeutic proteins produced in *Escherichia coli* as inclusion bodies, *Biopolymers* 51, 297–307, 1999.
101. Singh, S.M. and Panda, A.K., Solubilization and refolding of bacterial inclusion body proteins, *J. Biosci. Bioeng.* 99, 303–310, 2005.
102. Cromwell, M.E.M., Hilario, E., and Jacobsen, F., Protein aggregation and bioprocessing, *Am. Assoc. Pharm. Sci. J.* 8, E572–E579, 2006.
103. Murphy, R.M. and Kendrick, B.S., Protein misfolding and aggregation, *Biotechnol. Prog.* 23, 548–552, 2007.
104. Glaser, C.B., Busby, T.F., Ingham, K.C., and Childs, A., Thermal denaturation of $alpha_1$-protease inhibitor. Stabilization by neutral salts and sugars, *Am. Rev. Respir. Dis.* 128, 77–81, 1983.
105. Raman, B., Ramakrishana, T., and Rao, C.M., Refolding of denatured and denatured/reduced lysozyme at high concentrations, *J. Biol. Chem.* 271, 17067–17072, 1996.
106. Hammarström, P., Persson, M., and Freskgård, P.O., Structural mapping of an aggregation nucleation site in a molten globule intermediate, *J. Biol. Chem.* 274, 32897–32903, 1999.
107. Militella, V., Vetri, V., and Leone, M., Conformational changes involved in thermal aggregation processes of bovine serum albumin, *Biophys. Chem.* 105, 133–141, 2003.
108. Tsai, C.-J., Lin, S.L., Wolfson, H.J., and Nussinov, R., Studies of protein-protein interfaces: A statistical analysis of the hydrophobic effect, *Protein Sci.* 6, 53–64, 1997.
109. Kundu, B. and Guptasarma, P., Hydrophobic dye inhibits aggregation of molten carbonic anhydrase during thermal unfolding and refolding, *Proteins* 37, 321–324, 1999.
110. Kundu, B. and Guptasarma, P., Use of a hydrophobic dye to indirectly probe the structural organization and conformational plasticity of molecules in amorphous aggregates of carbonic anhydrase, *Biochem. Biophys. Res. Commun.* 293, 572–577, 2002.
111. Cheung, J.K., Raverker, P.S., and Truskett, T.M., Analytical model for studying how environmental factors influences protein conformational stability in solution, *J. Chem. Phys.* 125:234903, 2006.

112. Rezaei-Ghaleh, N., Ramshini, H., Ebrahim-Habibi, A. et al., Thermal aggregation of α-chymotrypsin: Role of hydrophobic and electrostatic interactions, *Biophys. Chem.* 132, 23–32, 2008.
113. McNally, K.M., Sorg, B.S., Chan, E.K. et al., Optimal parameters for laser tissue soldering, Part I: Tensile strength and scanning electron microscopy analysis, *Lasers Surg. Med.* 24, 319–331, 1999.
114. Tanaka, A., Differential scanning calorimetric studies on thermal unfolding of *Pseudomonas cepcia* lipase in the absence and presence of alcohols, *J. Biochem.* 123, 289–293, 1998.
115. Kurganov, B.I., Lyoborev, A.E., Sanchez-Ruiz, J.M., and Shriynov, V.L., Analysis of differential scanning calorimetry data for proteins. Criteria of validity of a one-step mechanism of irreversible protein denaturation, *Biophys. Chem.* 69, 125–135, 1997.
116. Catanzano, F., Giancola, C., Graziano, G., and Barone, G., Temperature-induced denaturation of ribonuclease S: A thermodynamic study, *Biochemistry* 35, 13378–13884, 1996.
117. Waldner, J.C., Lahr, S.J., Egdell, M.H., and Pielak, G.J., Nonideality and protein thermal denaturation, *Biopolymers* 49, 471–479, 1999.
118. Relkin, P., Thermal unfolding of β-lactoglobulin, α-lactalbumin and bovine serum albumin. A thermodynamic approach, *Crit. Rev. Food Sci. Nutr.* 36, 565–606, 1996.
119. Weijers, M., Barnneveld, P.A., Cohen Stuart, M.A., and Visecher, R.W., Heat-induced denaturation and aggregation of ovalbumin at neutral pH described by irreversible first-order kinetics, *Protein Sci.* 12, 2693–2703, 2003.
120. Bleustein, C.B., Sennett, M., King, R.T. et al., Differential scanning calorimetry of albumin solders: Interspecies differences and fatty acid binding effect on protein denaturation, *Lasers Surg. Med.* 27, 465–470, 2000.
121. Matthews, S., Friess, W., and Mahlen, H.-C., FTIR and nDSC as analytical tools for high-concentration protein formulations, *Pharm. Res.* 23, 1350–1363, 2006.
122. Watson, E.B., O'Neill, M.J., Justin, J., and Brenner, N., A differential scanning calorimeter for quantitative differential thermal analysis, *Anal. Chem.* 36, 1233–1236, 1964.
123. O'Neill, M.J., Measurement of exothermic reactions by differential scanning calorimetry, *Anal. Chem.* 47, 630–637, 1975.
124. Riga, A. and Collins, R., Differential scanning calorimetry and differential thermal analysis, in *Encyclopedia of Analytical Chemistry*, John Wiley & Sons, Chichester, U.K., Volume 15, pp. 13147–13179, 2000.
125. Groeneewand, W., *Characterisation of Polymers by Thermal Analysis*, Elsevier, Amsterdam, Netherlands, 2001.
126. Streicher, W.W., and Makhaladze, G.I., Advances in the analysis of conformational transitions in peptide using differential scanning calorimetry, in *Protein Folding Pathways*, Eds. Y. Bai and R. Nussinov (*Methods in Molecular Biology*, Volume 350), Humana Press, Totowa, NJ, Chapter 7, pp. 105–113, 2007.
127. Biltonen, R.L. and Freire, E., Thermodynamic characterization of conformational states of biological macromolecules using differential scanning calorimetry, *CRC Crit. Rev. Biochem.* 5, 85–124, 1978.
128. Plum, G.R. and Breslaver, K.J., Calorimetry of proteins and nucleic acids, *Curr. Opin. Struct. Biol.* 5, 582–590, 1995.
129. Hendrix, T., Griko, Y.V., and Privalov, P.L., A calorimetric study of the influence of calcium on the stability of bovine α-lactalbumin, *Biophys. Chem.* 84, 27–34, 2000.
130. Bruylants, G., Woulters, J., and Miceaux, C., Differential scanning calorimetry in life sciences: Thermodynamics, stability, molecular recognition, and applications in drug design, *Curr. Med. Chem.* 12, 2011–2020, 2005.
131. Mihalyi, E., Review of the some unusual aspects of calcium binding to fibrinogen, *Biophys. Chem.* 112, 131–140, 2004.

132. Privalov, P.L. and Dragon, A.I., Microcalorimetry of biological macromolecule, *Biophys. Chem.* 126, 16–24, 2007.
133. Ahmad, A., Akhter, M.S., and Bhakuni, V., Monovalent cation-induced conformation change in glucose oxidase leading to stabilization of the enzyme, *Biochemistry* 40, 1945–1955, 2001.
134. Potter, S.Z., Zhu, H., Shaw, B.F. et al., Binding of a single zinc ion in the one subunit of copper-zinc superoxide dismutase apoprotein substantially influences the structure and stability of the entire homodimeric protein, *J. Am. Chem. Soc.* 129, 4575–4583, 2007.
135. Casares, S., López-Mayurga, O., Vega, M.C. et al., Cooperative propagation of local stability changes from low-stability and high-stability regions in a SH3 domain, *Proteins* 67, 531–547, 2007.
136. Foglia, F., Mandrich, L., and Pezzollo, M., Role of the N-terminal region for conformational stability of esterase 2 from *Alicyclobacillus acidocaldarius, Biophys. Chem.* 127, 113–122, 2007.
137. Freudenberg, U., Behrens, S.H., Welzel, P.B. et al., Electrostatic interactions modulate the conformation of collagen 1, *Biophys. J.* 92, 2108–2119, 2007.
138. Muthsamy, R., Gromino, M.M., and Ponnuswamy, P.K., On the thermal unfolding character of globular proteins, *J. Prot. Chem.* 19, 1–10, 2000.
139. Febo-Ayala, W., Morera-Felix, S.L., Hrycana, C.A., and Thompson, D.H., Functional reconstitution of the integral membrane enzyme, isoprenylcysteine carboxyl methyltransferase, in synthetic bolalipid membrane vesicles, *Biochemistry* 45, 14683–14694, 2006.
140. de Groot, J., Kosters, H.A., and de Jongh, H.H., Deglycosylation of ovalabumin prohibits the formation of a heat-stable conformer, *Biotechnol. Bioeng.* 97, 735–741, 2007.
141. Tzannis, S.T. and Prestrelski, S.J., Moisture effects on protein-excipient interactions in spray-dried powders. Nature of destabilizing effects of sucrose, *J. Pharm. Sci.* 88, 360–370, 1999.
142. Ghirlando, R., Lund, J., Goodall, M., and Jefferis, R., Glycosylation of human IgG-Fc: Influences on structure revealed by differential scanning micro-calorimetry, *Immunol. Lett.* 68, 47–52, 1999.
143. Hinrichs, W.L., Prinsen, M.G., and Frijlink H.W., Inulin glasses for the stabilization of therapeutic proteins, *Int. J. Pharm.* 215, 163–174, 2001.
144. Davidson, P. and Sun, W.Q., Effect of sucrose/raffinose mass ratios on the stability of co-lyophilized protein during storage above the Tg, *Pharm. Res.* 18, 474–479, 2001.
145. Tang, H.R., Covington, A.D., and Hancock, R.A., Use of DSC to detect the heterogeneity of hydrothermal stability in the polyphenol-treated collagen matrix, *J. Agric. Food Chem.* 51, 6652–6656, 2003.
146. Chen, B., Bautista, R., Yu, K. et al., Influence of histidine on the stability and physical properties of a fully human antibody in aqueous and solid forms, *Pharm. Res.* 20, 1952–1960, 2003.
147. Liao, Y.H., Brown, M.B., and Martin, G.P., Investigation of the stabilization of freeze-dried lysozyme and the physical properties of the formulations, *Eur. J. Pharm. Biopharm.* 58, 15–24, 2004.
148. Passot, S., Fonseca, F., Alarcon-Lorca, M. et al., Physical characterization of formulations for the development of two stable freeze-dried proteins during both dried and liquid storage, *Eur. J. Pharm. Biopharm.* 60, 335–348, 2005.
149. Chang, L.L., Shepherd, D., Sun, J. et al., Mechanism of protein stabilization by sugars during freeze-drying and storage: Native structure preservation, specific interaction, and/or immobilization in a glassy matrix?, *J. Pharm. Sci.* 94, 1427–1444, 2005.
150. Ihnat, P.M., Vellekamp, G., Obenauer-Kutner, L.J. et al., Comparative thermal stabilities of recombinant adenoviruses and hexon protein, *Biochim. Biophys. Acta* 1726, 138–151, 2005.

151. Tang, X.C. and Pikal, M.J. The effect of stabilizers and denaturants on the cold denaturation temperatures of proteins and implications for freeze-drying, *Pharm. Res.* 22, 1167–1175, 2005.
152. Tang, X.C. and Pikal, M.J., Measurement of the kinetics of protein unfolding in viscous systems and implications for protein stability in freeze-drying, *Pharm. Res.* 22, 1176–1185, 2005.
153. Noorjahan, S.E. and Sastry, T.P., Physiologically clotted fibrin-calcined bone composite—a possible bone graft substitute, *J. Biomed. Mater. Res. B. Appl. Biomater.* 75, 343–350, 2005.
154. Jovanović, N., Bouchard, A., Hofland, G.W. et al., Distinct effects of sucrose and trehalose on protein stability during supercritical fluid drying and freeze-drying, *Eur. J. Pharm. Sci.* 27, 336–345, 2006.
155. Hawe, A. and Friess, W., Physico-chemical lyophilization behavior of mannitol, human serum albumin formulations, *Eur. J. Pharm. Sci.* 28, 224–232, 2006.
156. Matheus, S., Friess, W. and Mahler, H.C., FTIR and nDSC as analytical tools for high-concentration protein formulations, *Pharm. Res.* 23, 1350–1363, 2006.
157. Harn, N., Allan, C., Oliver, C., and Middaugh, C.R., Highly concentrated monoclonal antibody solutions: Direct analysis of physical structure and thermal stability, *J. Pharm. Sci.* 96, 532–546, 2007.
158. Hédoux, A., Willart, J.F., Ionov, R. et al., Analysis of sugar bioprotective mechanisms on the thermal denaturation of lysozyme from Raman scattering and differential scanning calorimetry investigations, *J. Phys. Chem. B* 110, 22886–22893, 2006.
159. Hawe, A. and Friess, W., Physicochemical characterization of the freezing behavior of mannitol-human serum albumin formulations, *AAPS PharmSciTech.* 7, 94, 2006.
160. Garber, E. and Demarest, S.J., A broad range of Fab stabilities within a host of therapeutic IgGs, *Biochem. Biophys. Res. Commun.* 355, 751–757, 2007.
161. Yoshioka, S., Miyazaki, T., Aso, Y., and Kawanishi, T., Significance of local mobility in aggregation of β-galactosidase lyophilized with trehalose, sucrose or stachyose, *Pharm. Res.* 24, 1660–1667, 2007.
162. Bhatnagar, B.S., Pikal, M.J., and Bogner, R.H., Study of the individual contributions of ice formation and freeze-concentration on isothermal stability of lactate dehydrogenase during freezing, *J. Pharm. Sci.* 97, 798–814, 2008.
163. Conesa, C., Sánchez, L., Pérez, M.D., and Calvo, M., A calorimetric study of thermal denaturation of recombinant human lactoferrin from rice, *J. Agric. Food Chem.* 55, 4848–4853, 2007.
164. Nesarikar, V.V. and Nassar, M.N., Effect of cations and anions on glass transition temperatures in excipient solutions, *Pharm. Dev. Technol.* 12, 259–264, 2007.
165. Liu, H., Bulseco, G.G., and Sun, J., Effect of posttranslational modifications on the thermal stability of a recombinant monoclonal antibody, *Immunol. Lett.* 106, 144–153, 2006.
166. Broering, J.M. and Bommarius, A.S., Evaluation of Hofmeister effects on the kinetic stability of proteins, *J. Phys. Chem. B. Condens. Matter Mater. Surf. Interface Biophys.* 109, 20612–20619, 2005.
167. Tang, S.Y., Le, Q.T., Shim, J.H. et al., Enhancing thermostability of maltogenic amylase from *Bacillus thermoalkophilus* ET2 by DNA shuffling, *FEBS J.* 273, 3335–3345, 2006.
168. Hédoux, A., Willart, J.F., Ionov, R. et al., Analysis of sugar bioprotective mechanisms on the thermal denaturation of lysozyme from Raman scattering and differential scanning calorimetry investigations, *J. Phys. Chem. B.* 110, 22886–22893, 2006.
169. Efimova, Y.M., Haemers, S., Wierczinski, B. et al., Stability of globular proteins in H_2O and D_2O, *Biopolymers* 85, 264–273, 2007.
170. Tadeo, X., Pons, M., and Millet, O., Influence of the Hofmeister anions on protein stability as studied by thermal denaturation and chemical shift perturbation, *Biochemistry* 46, 917–923, 2007.

171. Woodhead–Galloway, J., *Collagen: The Anatomy of a Protein,* E. Arnold, London, 1980.
172. *Collagen*, Eds. B.R. Olsen and M.E. Nimni, CRC Press, Boca Raton, FL, 1989.
173. Brinkermann, J. and Notbohm, H., *Collagen: Primer in Structure, Processing and Assembly*, Springer-Verlag, Berlin, Germany, 2005.
174. Fratzl, P., *Collagen: Structure and Mechanics*, Springer-Verlag, Berlin, Germany, 2008.
175. Kivirikko, K.I. and Pihlajaniemi, T., Collagen hydroxylase and the protein disulfide isomerase subunit of prolyl 4-hydroxylase, *Adv. Enzymol. Relat. Areas Mol. Biol.* 72, 325–398, 1998.
176. Kefalides, N.A, Basement membranes: Current concepts of structure and synthesis, *Dermatologica* 150, 4–15, 1975.
177. Reiser, K., McCormick, R.J., and Rucker, R.S., Enzymatic and nonenzymatic cross-linking of collage and elastin, *FASEB J.* 6, 2439–2449, 1992.
178. Robins, S.P., Analysis of the crosslinking components in collagen and elastin, *Methods Biochem. Anal.* 28, 329–379, 1982.
179. Anttinen, H., Intracellular enzymes of collagen biosynthesis in human skin and serum as a function of age, *Scand. J. Soc. Med. Suppl.* 14, 69–74, 1977.
180. Lucero, H.A. and Kagan, H.M., Lysyl oxidase: An oxidative enzyme and effector of cell function, *Cell Mol. Life Sci.* 63, 2304–2316, 2006.
181. Guzman, N.A., Prolyl-4-hydroxylase: An overview, in *Prolyl hydroxylase, Protein Disulfide Isomerase, and Other Structurally Related Proteins,* Ed. N.A. Guzman, Marcel Dekker, New York, 1998.
182. Monnier, V.M., Mustata, G.T., Biemel, K.L. et al., Cross-linking of the extracellular matrix by the Maillard reaction in aging and diabetes: An update on "a puzzle nearing resolution," *Ann. N. Y. Acad. Sci.* 1043, 533–544, 2005.
183. Avery, N.C. and Bailey, A.J., The effects of the Maillard reaction on the physical properties and cell interactions of collagen, *Pathol. Biol.* (Paris), 54, 387–395, 2006.
184. Greenwald, S.E., Aging of the conduit arteries, *J. Pathol.* 211, 157–172, 2007.
185. Nimni, M.E. and Harkness, R.D., Molecular structures and function of collagen, in *Collagen,* Vol. 1, Ed. M.E. Nimni, CRC Press, Boca Raton, FL, Chapter 1, pp. 1–77, 1988.
186. Kuznetsova, N. and Leikin, S., Does the triple helical domain of type 1 collagen encode molecular recognition and fiber assembly while telopeptides serve as catalytic domains? Effect of proteolytic cleavage on fibrillogenesis and on collagen-collagen interaction in fibers, *J. Biol. Chem.* 274, 36083–36088, 1999.
187. Gray, R.E., Seng, N., Mackay, L.R., and Rowley, M.J., Measurement of antibodies to collagen II by inhibition of collagen fibril formation *in vitro*, *J. Immunol. Methods.* 285, 55–61, 2004.
188. Tenni, R., Sonaggere, M., Viola, M. et al., Self-aggregation of fibrillar collagens I and II involves lysine side chains, *Micron* 37, 640–647, 2006.
189. Fullerton, G.D. and Rahal, A., Collagen structure: The molecular source of the tendon magic angle effect, *J. Magnetic Resonance Imaging* 25, 345–361, 2007.
190. Gobeaux, F., Belamie, E., Mosser, G. et al., Cooperative ordering of collagen triple helices in the dense state, *Langmuir* 23, 6411–6417, 2007.
191. Timpl, R., Fujiwara, S., Dziadek, M. et al., Laminin, proteoglycan, nidogen and collagen IV: Structural models and molecular interactions, *Ciba Found. Symp.* 108, 25–43, 1984.
192. Scott, J.E., Proteoglycan-fibrillar collagen interactions, *Biochem. J.* 252, 313–323, 1988.
193. Scott, J.E., Proteoglycan: Collagen interactions in connective tissues. Ultrastructural, biochemical, functional and evolutionary aspects, *Int. J. Biol. Macromol.* 13, 157–161, 1991.
194. Babu, P.B. and Sudhakaran, P.R., Isolation of heparin sulfate proteoglycan from beneath the monolayers of rat hepatocytes and its binding to type IV collagen, *J. Cell. Biochem.* 46, 48–53, 1991.
195. Scott, J.E., Proteoglycan: Collagen interactions and corneal ultrastructure, *Biochem. Soc. Trans.* 19, 877–881, 1991.

196. Satoh, H., Kawasaki, K., Khara, I., and Nakano, Y., Importance of type IV collagen, laminin, and heparin sulfate proteoglycan in the regulation of labyrinthine fluid in the rat cochlear duct, *Eur. Arch. Otorhinolaryngol.* 255, 285–288, 1998.
197. Svensson, L, Närlid, I., and Oldberg, Å., Fibromodulin and lumican bind to the same region on collagen type I fibrils, *FEBS Lett.* 470, 178–182, 2000.
198. Bartolini, B., Sonaggere, M., Giudici, C. et al., Fibromodulin interactions with type I and type II collagens, *Connect. Tissue. Res.* 48, 141–148, 2007.
199. Goldberg, A., Kalamajski, S., Salnikov, A.V. et al. Collagen-binding proteoglycan fibromodulin can determine stroma matrix structure and fluid balance in experimental carcinoma, *Proc. Natl. Acad. Sci. USA* 104, 13966–13971, 2007.
200. Brass, L.F. and Bensusan, H.B., The role of collagen quaternary structure in the platelet: Collagen interaction, *J. Clin. Invest.* 54, 1480–1487, 1974.
201. Chesney, C.M. Pifer, D.D., Crofford, L.J., and Huch, K.M., Reevaluation of the role of the polar groups of collagen in the platelet-collagen interaction, *Am. J. Pathol.* 112, 200–206, 1983.
202. Probstmeier, R., Kühn, K., and Schachner, M., Binding properties of the neural cell adhesion molecule to different components of the extracellular matrix, *J. Neurochem.* 53, 1784–1801, 1989.
203. Renner, C., Saccà, B., and Moroder, L., Synthetic heterotrimeric collagen peptides as mimics of cell adhesion sites of the basement membrane, *Biopolymers* 76, 34–47, 2004.
204. Yeh, A.T. and Hirshburg, J., Molecular interactions of exogenous chemical agents with collagen—implications for tissue optical clearing, *J. Biomed. Opt.* 11: 014003, 2006.
205. Leitinger, B. and Hohenester, E., Mammalian collagen receptors, *Matrix Biol.* 26, 146–155, 2007.
206. Zeebregts, C.J., Heijmen, R.H., van den Dungen, J.J., and van Schilfgaade, R. Non-suture method of vascular anastomosis, *Brit. J. Surg.* 90, 261–271, 2003.
207. Rabau, M.Y., Wasserman, I., and Shashan, S., Healing process of laser-welded intestinal anastomosis, *Lasers Surg. Med.* 14, 13–17, 1994.
208. Bass, L.S. and Treat, M.R., Laser tissue welding: A comprehensive review of current and future clinical applications, in *Laser Surgery and Medicine*, Ed. A. Puliafito, Wiley-Liss, New York, Chapter 12, pp. 381–415, 1996.
209. Tang, J., Godleioski, G., Ray, S., and Delacrétaz, G., Morphologic changes in collagen fibers after 830 nm diode laser welding, *Lasers Surg. Med.* 21, 436–443, 1997.
210. Tang, J., O'Callaghan, D., Rovy, S., and Godliewski, G., Quantitative changes in collagen levels following 830 nm diode laser welding, *Lasers Surg. Med.* 20, 207–211, 1998.
211. Hasegawa, M., Sakurai, T., Matsushita, M. et al., Comparison of argon-laser welded and sutured repair of inferior vena cava in a canine model, *Lasers Surg. Med.* 29, 62–69, 2001.
212. Rossi, F., Pini, R., Menabuoni, L. et al., Experimental study on the healing process following laser welding of the cornea, *J. Biomed. Opt.* 10:024004, 2005.
213. Matteini, P., Rossi, F., Menabuoni, L., and Pina, R., Microscopic characterization of collagen modifications induced by low-temperature diode-laser welding of corneal tissue, *Lasers Surg. Med.* 39, 597–604, 2007.
214. Amiel, D., Frank, C.B., Harwood, F.L. et al., Collagen alteration in medial collateral ligament healing in a rabbit model, *Connect. Tissue Res.* 16, 357–366, 1987.
215. Capon, A. and Mardon, S., Can thermal laser promote skin wound healing?, *Am. J. Clin. Dermatol.* 4–12, 2003.
216. Posten, W., Wrone, D.A., Dover, J.S. et al., Low-level laser therapy for wound healing: Mechanisms and efficacy, *Dermatol. Surg.* 31, 334–340, 2005.
217. Demidova-Rice, T.N., Salomatina, E.V., Yaroslavsky, A.N. et al., Low-level light stimulates excisional wound healing in mice, *Laser Surg. Med.* 39, 706–715, 2007.

218. Al-Watban, F.A.R., Zhang, X.Y., and Andres, B.L., Low-level laser therapy enhanced wound healing in diabetic rats: A comparison of different lasers, *Photomed. Laser Surg.* 25, 72–77, 2007.
219. Hawkins, D. and Abrahamse, H., Influence of broad-spectrum and infrared light in combination with laser irradiation on the proliferation of wounded skin fibroblasts, *Photomed. Laser Surg.* 25, 159–169, 2007.
220. Hwang, K., Kim, S.G., Kim, D.J., and Lee, C.H., Laser welding of rat's facial nerve, *J. Craniofacial Surg.* 16, 1102–1106, 2005.
221. Bhatta, K., Laser in urology, *Lasers Surg. Med.* 16, 312–320, 1995.
222. Pappas, D.P. and Scheer, D.S., Laser tissue welding: A urological surgeon's perspective, *Haemophilia* 4, 456–462, 1998.
223. Savage, H.E., Halder, R.K., Kartazayeu, U. et al., NIR laser tissue welding of *in vitro* porcine cornea and sclera tissue, *Lasers Surg. Med.* 35, 292–303, 2004.
224. Pen, Z., Xie, H., Lagerquist, K.A. et al., Optimal dye concentration and irradiance for laser-assisted vascular anastomosis, *J. Clin. Laser Med. Surg.* 22, 81–86, 2004.
225. Chuck, R.S., Oz, M.C., Delohery, T.M. et al., Dye-enhanced laser tissue welding, *Lasers Surg. Med.* 9, 471–477, 1989.
226. Oz, M.C., Johnson, J.P., Parangi, S., Tissue soldering by use of indocyanine green dye-enhanced fibrinogen with the near infrared diode laser, *J. Vasc. Surg.* 11, 716–725, 1990.
227. Hoffman, G.T., Byrd, B.D., Soller, E.C. et al., Effect of varying chromophores used in light-activated protein solders on tensile strength and thermal damage profile of repairs, *Biomed. Sci. Instrum.* 39, 12–17, 2003.
228. Talmor, M., Bleustein, C.B., and Poppas, D.P., Laser tissue welding: A biotechnological advance for the future, *Arch. Facial Plast. Surg.* 3, 207–213, 2001.
229. Murray, L.W., Su, L,. Kopchok, G.E., and White, R.A., Crosslinking of extracellular matrix proteins: A preliminary report on a possible mechanism of argon laser welding, *Lasers Surg. Med.* 9, 490–496, 1989.
230. Fecko, C.J., Munson, K.M., Saunders, A. et al., Comparison of femtosecond laser and continuous wave UV sources for protein-nucleic acid cross-linking, *Photochem. Photobiol.* 83, 1394–1404, 2007.
231. Constantinescu, M.A., Alfieri, A., Mihalache, G. et al., Effect of laser soldering irradiation on covalent bonds of pure collagen, *Lasers Med. Sci.* 22, 10–14, 2007.
232. Gayen, T.K., Katz, A., Savage, H.E. et al., Aorta and skin tissue welded by near-infrared Cr^{4+}:YAG laser, *J. Clin. Laser Med. Surg.* 21, 259–269, 2003.
233. Bass, L.S., Moazami, N., Pocsido, J. et al., Changes in type I collagen following laser welding, *Laser Surg. Med.* 12, 500–505, 1992.
234. Raman, B., Ramakrishna, T., and Rao, C.M., Refolding of denatured and denatured/reduced lysozyme at high concentrations, *J. Biol. Chem.* 271, 17067–17072, 1996.
235. Hammerström, P., Persson, M., Freskgård, P.O. et al., Structural mapping of an aggregation nucleation site in a molten globule, *J. Biol. Chem.* 274, 32897–32903, 1999.
236. Militeilo, V., Vetri, V., and Leone, M., Conformational changes involved in thermal aggregation processes of bovine serum albumin, *Biophys. Chem.* 105, 133–141, 2003.
237. Flock, S.T. and Machitto, K.S., Progress toward seamless tissue fusion for wound closure, *Otolaryngol. Clin. North Am.* 38, 295–305, 2005.
238. Starling, E.H., On the absorption of fluids from the connective tissue spaces, *J. Physiol.* 19, 312–326, 1896.
239. Foster, J.F., Plasma albumin, in *The Plasma Proteins*, Volume 1, Ed. F.W. Putnam, Academic Press, New York, Chapter 6, pp. 179–239, 1960.
240. Quinlan, G.J., Martin, G.S., and Evans, T.W., Albumin: Biochemical properties and therapeutic potential, *Hepatology* 41, 1211–1219, 2005.

241. Newhauser, L.R. and Loznen, E.L., Studies on human albumin in military medicine: The standard Army-Navy package of serum albumin (concentrated), *U.S. Navy Med. Bull.* 40, 796–799, 1942.
242. Heyl, J.T., Gibson, J.G., 2nd, and Janeway, C.W., Studies on the plasma proteins. V. The effect of concentrated solutions of human and bovine serum albumin in man, *J. Clin. Invest.* 22, 763–773, 1943.
243. *Albumin and the Systemic Circulation*, Eds. B. Blauhut and P. Lundsgaard-Hansen, Karger, Berlin, 1986.
244. Sen, S. and Williams, R., New liver support devices in acute liver failure: A critical evaluation, *Semin. Liver Dis.* 23, 283–294, 2003.
245. Barshes, N.R., Gay, A.N., Williams, B. et al., Support for the acutely failing liver: A comprehensive review of historic and contemporary strategies, *J. Am. Coll. Surg.* 201, 458–476, 2005.
246. Berchane, N.S., Andrews, M.J., Kerr, S. et al., On the mechanical properties of bovine serum albumin (BSA) adhesives, *J. Mater. Sci. Mater. Med.* 19, 1831–1838, 2008.
247. Duly, E.B., Grimason, S., Grimaon, P. et al., Measurement of serum albumin by capillary zone electrophoresis, bromocresol green, bromocresol purple, and immunoassay methods, *J. Clin. Pathol.* 56, 780–781, 2003.
248. Taylor, J.F., The isolation of proteins, in *The Proteins. Chemistry, Biological Activity and Methods*, Volume I. Pt. A, Eds. H. Neurath and K. Bailey, Chapter 1, pp. 1–85, 1953.
249. Cooper, G.R., Electrophoretic and ultracentrifugal analysis of normal human serum, in *The Plasma Proteins*, Ed. F.W. Putman, Academic Press, New York, 1960, Chapter 3, pp. 51–103, 1960.
250. Ott, B., Zuger, B.J., Erni, D. et al., Comparative *in vitro* study of tissue welding using a 808 nm diode laser and Ho:YAG laser, *Lasers Med. Sci.* 16, 260–266, 2001.
251. Shrake, A., Finlayson, J.S., and Ross, P.D., Thermal stability of human albumin measured by differential scanning calorimetry. I. Effects of caprylate and *N*-acetyltryptophanate, *Vox Sang.* 47, 7–18, 1984.
252. Michnik, A., Thermal stability of bovine serum albumin DSC studies, *J. Thermal Anal. Calorimetry* 71, 509–519, 2003.
253. Shrake, A., Frazier, D., and Schwarz, F.P., Thermal stabilization of human albumin by medium and short chain *n*-alkyl fatty acid anions, *Biopolymers* 81, 235–248, 2006.
254. Shrake, A. and Ross, P.D., Origins and consequences of ligand-induced multiphasic thermal protein denaturation, *Biopolymers* 32, 925–940, 1992.
255. D'Auria, S., Rossi, M., Barone, G. et al., Temperature-induced denaturation of β-glycosidase from the archaeon *Sulfolobus solfatoricus*, *J. Biochem.* 120, 292–300, 1996.
256. Catabzano, F., Giancola, C., Graziano, G., and Barone, G., Temperature-induced denaturation of ribonuclease S: A thermodynamic study, *Biochemistry* 35, 13378–13385, 1996.
257. Ulrih, N.P., Anderluh, G., Maček, P., and Chalikian, I.V., Salt-induced oligomerization of partially folded intermediates of equinatoxin II, *Biochemistry* 43, 9536–9545, 2004.
258. Ren, Z., Zie, H., Lagerquist, K.A. et al., Optimal dye concentration and irradiation for laser-assisted vascular anastomosis, *J. Clin. Laser Med. Surg.* 22, 81–86, 2004.
259. McNally, K.M., Sorg, B.S., and Welch, A.J., Novel solid protein solder designs for laser-assisted tissue repair, *Lasers Surg. Med.* 27, 147–157, 2000.
260. Kirsch, A.J., Cooper, C.S., Gath, J. et al., Laser tissue soldering for hypospadias repair: Results of a controlled prospective clinical trial, *J. Urol.* 165, 574–577, 2001.
261. Xie, H., Shaffer, B.S., Prahl, S.A., and Gregory, K.W., Intraluminal albumin stent assisted laser welding for urethral anastomosis, *Lasers Surg. Med.* 31, 225–229, 2002.
262. Xie, H., Bendre, S.C., Burke, A.P. et al., Laser-assisted vascular and end to end anastomosis of elastic heterograft to carotid artery with an albumin stent: A preliminary *in vivo* study, *Lasers Surg. Med.* 35, 201–205, 2004.

263. Maitz, P.K.M., Trickett, R.J., Dekker, P. et al., Sutureless microvascular anastomoses by a biodegradable laser-activated solid protein solder, *Plast. Reconstruct. Surg.* 104, 1726–1731, 1999.
264. Wright, B., Vicaretti, M., Schwaiger, N. et al., Laser-assisted end-to-end Bioweld® anastomosis in an ovine model, *Lasers Surg. Med.* 39, 667–673, 2007.
265. Lord, S.T., Fibrinogen and fibrin: Scaffold proteins in hemostasis, *Curr. Opin. Hematol.* 14, 236–241, 2007.
266. McKee, P.A., Mattock, P., and Hill, R.L., Subunit structure of human fibrinogen, soluble fibrin, and cross-linked insoluble fibrin, *Proc. Natl. Acad. Sci. USA* 66, 738–744, 1970.
267. Marx, G., Mou., X., Hotovely-Salomon, A. et al., Heat denaturation of fibrinogen to develop a biomedical matrix, *J. Biomed. Res. Part B* 84B, 49–57, 2008.
268. Oz, M.C., Johnson, J.P., Parangi, S. et al., Tissue soldering by use of indocyanine green dye-enhanced fibrinogen the near infrared diode laser, *J. Vasc. Surg.* 11, 718–725, 1990.
269. Wider, T.M., Libutti, S.K., Greenwald, D.P. et al., Skin closure with dye-enhanced laser welding and fibrinogen, *Plastic Reconstruct. Surg.* 88, 1018–1025, 1991.
270. Mueller, M.P., Kopchok, G.E., Tabbara, M.R. et al., Argon laser-welded bovine heterograft anastomoses, *J. Clin. Laser Med. Surg.* 11, 1–5, 1993.
271. Shohet, J.A., Reinisch, L., and Ossoff, R.H., Prevention of pharyngocutaneous fistulas by means of laser-weld techniques, *Laryngoscope* 105, 717–722, 1995.
272. Khadem, J., Truong, T., and Ernest, J.T., Photodynamic biological tissue glue, *Cornea* 13, 406–410, 1994.
273. Edwards, A.M. and Silva, A., Effect of visible light on selected enzymes, vitamins and amino acids, *J. Photochem. Photobiol. B* 63, 126–131, 2002.
274. Spoerl, E., Mrochen, M., Sliney, D. et al., Safety of UVA-riboflavin cross-linking of the cornea, *Cornea* 26, 385–389, 2007.
275. Mirelman, D. and Siegel, R.C., Oxidative deamination of epsilon-aminolysine residues and formation of Schiff base cross-linkages in cell envelopes of *Escherichia coli*, *J. Biol. Chem.* 254, 571–574, 1979.
276. Lundblad, R.L, Bradshaw, R.A., Gabriel, D. et al., A review of the therapeutic uses of thrombin, *Thromb. Haemostas.* 91, 851–860, 2004.
277. Wider, T.M., Libutti, S.K., Greenwald, D.P. et al., Skin closure with dye enhanced laser welding and fibrinogen, *Plast. Reconstruct. Surg.* 88, 1018–1025, 1991.
278. Marx, G., Mou, X., Hotovely-Salomon, A. et al., Heat denaturation of fibrinogen to develop a biomedical matrix, *J. Biomed. Mater. Res. Part B: Appl. Biomater.* 84B, 49–57, 2008.
279. Zamarron, C., Ginsberg, M.H., and Plow, E.F., Monoclonal antibodies specific for a conformationally altered state of fibrinogen, *Thromb. Haemostas.* 64, 41–60, 1990.
280. Tang, L., Wu, Y., and Timmons, R.B., Fibrinogen adsorption and host tissue responses to plasma functionalized surfaces, *J. Biomed. Mater. Res.* 42, 156–163, 1998.
281. Hu, W.J., Eaton, J.W., Ugarova, T.P., and Yang, L., Molecular basis of biomaterial-mediated foreign body reactions, *Blood* 98, 1231–1238, 2001.
282. Tassman, I.J., Experimental studies with physiological glue (autologous plasma plus thrombin) for use in the eyes, *Am. J. Ophthalmol.* 33, 870–878, 1950.
283. Stepanov, V.K., Lyophilized plasma as a biological glue for laminar keratoplasty, *Vestn. Optalmol.* 7, 6–8, 1972.
284. Pearl, R.M., Wustrack, K.O., Harburg, C. et al., Microvascular anastomosis using a blood product sealant-adhesive, *Surg. Obstet. Gynecol.* 144, 227–231, 1977.
285. Dees, J.E. and Fox, H., The properties of human fibrinogen coagulum—preliminary report, *J. Urology* 49, 503–511, 1943.
286. Hartman, A.R., Galanakis, D.K., Honig, M.P. et al., Autologous whole plasma fibrin gel. Intraoperative procurement, *Arch. Surg.* 127, 357–359, 1992.

287. Roukis, T.S., Zgonis, T., and Tiernan, B., Autologous platelet-rich plasma for wound and osseous healing: A review of the literature and commercially available products, *Adv. Ther.* 23, 218–237, 2006.
288. Evarts, P.A., Knape, J.T., Weibrich, G. et al., Platelet-rich plasma and platelet gel: A review, *J. Extra-Corpor. Technol.* 38, 174–187, 2006.
289. Evarts, P.A., Brown-Mahoney, C., Hoffman, J.J. et al., Platelet-rich plasma preparation using three devices: Implications for platelet activation and platelet growth factor release, *Growth Factors* 24, 165–171, 2006.
290. Rulla, R., Feste, V.M., Guida, L., and Laino, G., Bone grafting with platelet-rich plasma in alveolar cleft. Case report, *Minerva Stomatol.* 56, 63–71, 2007.
291. Han, J., Meng, H.X., Tang, J.M. et al., The effect of different platelet-rich plasma concentrates on proliferation and differentiation of human periodontal ligament cells *in vitro, Cell Prolif.* 40, 241–252, 2007.
292. Roussy, Y., Bertrand-Duchesne, M.P., and Gagnon, G., Activation of human platelet-rich plasma: Effect on growth factor release, cell division, and *in vitro* bone formation, *Clin. Oral. Implants Res.* 18, 639–648, 2007.
293. Wu, W., Chen, F., Liu, Y. et al., Autologous injectable tissue-engineered cartilage by using platelet-rich plasma: Experimental study in a rabbit model, *J. Oral Maxillofac. Surg.* 65, 1951–1957, 2007.
294. Sánchez, A.R., Sheridan, P.J., and Kup, L.T., Is platelet-rich plasma the perfect enhancement factor?, *Int. J. Oral Maxillofac. Implants* 18, 93–103, 2003.
295. Rock, G., Neurath, D., Lu, M. et al., The contribution of platelets in the production of cyroprecipitates for use in a fibrin glue, *Vox Sang.* 91, 252–255, 2006.
296. De Somer, F., DeBrauwer, V., Vanderkerckhaue, M. et al., Can autologous thrombin with a rest fraction of ethanol be used safely on nerves?, *Eur. Spine. J.* 15, 501–505, 2006.
297. Man, D., Plasker, H., and Winland-Brown, J.E., The use of autologous platelet-rich plasma (platelet gel) and autologous platelet-poor plasma (fibrin sealant) in cosmetic surgery, *Plast. Reconstruct. Surg.* 107, 229–237, 2001.
298. Altmeppen, J., Hansen, E., Bonnländer, G.L. et al., Composition and characteristics of an autologous thrombocyte gel, *J. Surg. Res.* 117, 202–207, 2004.
299. Yazawa, M., Ogata, M., Nakajima, T. et al., Basic studies on the clinical applications of platelet-rich plasma, *Cell Transplant.* 12, 509–518, 2003.
300. Oz, M.C., Jeevanadam, V., Smith, C.A. et al., Autologous fibrin glue from intraoperatively collected platelet-rich plasma, *Ann. Thorac. Surg.* 53, 530–531, 1992.
301. Kjaergard, H.K. and Weis-Fogh, U.S., Important factors influencing the strength of autologous fibrin glue: The fibrin concentration and reaction time—comparison of strength with commercial fibrin sealant, *Eur. Surg. Res.* 26. 273–276, 1994.
302. Kjaergard, H.K. and Weis-Fohg, U.S., Autologous fibrin glue for sealing vascular prosthesis of high porosity, *Cardiovasc. Surg.* 2, 45–47, 1994.
303. Isaacson, G. and Herman, J.H., Autologous plasma fibrin glue: Rapid preparation and selective use, *Am. J. Otolaryngol.* 17, 92–94, 1998.
304. Gammon, R.R., Prum, B.E., Jr., Avery, M., and Mintz, P.D., Rapid preparation of small-volume autologous fibrinogen concentrates and its same day use in bleb leaks after glaucoma filtration surgery, *Ophthalmol. Surg. Lasers* 29, 1010–1012, 1998.
305. Yoshide, H., Hirozane, K., and Kamiya, A., Comparative study of autologous fibrin glues prepared by cryo-centrifugation, cryo-filtration, and ethanol precipitation methods, *Biol. Pharm. Bull.* 22, 1222–1225, 1999.
306. Yoshide, H., Hirozane, K., and Kamiya, A., Adhesive strength of autologous fibrin glue, *Biol. Pharm. Bull.* 23, 313–317, 2000.
307. Blumenkranz, M.S., Ohana, E., Shaikh, S. et al., Adjuvant methods in macular hole surgery: Intraoperative plasma-thrombin mixture and postoperative fluid-gas exchange, *Ophthalmic Surg. Lasers* 32, 198–207, 2001.

308. Thorn, J.J., Sørensen, H., Weis-Fogh, U., and Andersen, M., Autologous fibrin glue with growth factors in reconstructive maxillofacial surgery, *Int. J. Oral Maxillofac. Surg.* 33, 95–100, 2004.
309. Christenson,. J.J. and Kalangos, A., Autologous fibrin glue reinforced by platelets in surgery of ascending aorta, *Thorac. Cardiovasc. Surg.* 52, 225–229, 2004.
310. Alston, S.M., Solen, K.A., Broderick, A.H. et al., New method to prepare autologous fibrin glue on demand, *Transl. Res.* 149, 187–195, 2007.
311. Hofmann, M. and Jenner, P., Variability in the fibrinogen and von Willebrand factor content of cryoprecipitate, *Am. J. Clin. Pathol.* 93, 694–697, 1990.
312. Lindner, A., Elliott, M., and Holzer, F., Optimizing the fibrinogen-thrombin adhesive system, *Wien. Klin. Wochenschr. Suppl.* 109, 1–9, 1980.
313. Wolberg, A.S., Gabriel, D.A,. and Hoffman, M., Analyzing fibrin clot structure using a microplate reader, *Blood Coagul. Fibrinol.* 13, 533–539, 2002.
314. Budzynski, A., Dependence of fibrin clot formation on the expression of polymerization sites, in *Fibrinogen 3*, Eds. M.W. Mosesson, D.L. Amrani, K.R. Siebenlist, and J.D. DiOrio, Excerpta Medica, Amsterdam, Netherlands, pp. 71–79, 1988.
315. Weisel, J.W., Veklich, Y., and Gorkun, O., The sequence of cleavage of fibrinopeptides from fibrinogen is important for protofibril formation and enhancement of lateral aggregation in fibrin clots, *J. Mol. Biol.* 232, 285–297, 1993.
316. Kita, R. Takahashi, A., Kaibara, M., and Kubota, K., Formation of fibrin gel in fibrinogen-thrombin system: Statics and dynamic light scattering study, *Biomacromolecules* 3, 1013–1020, 2002.
317. Weisel, J.W., Fibrinogen and Fibrin, *Adv. Prot. Chem.* 70, 247–299, 2005.
318. Miloszewski, K.J.A. and Lasowsky, M.S., Fibrin stabilization and factor XIII deficiency, in *Fibrinogen, Fibrin Stabilization, and Fibrinolysis*, Ed. J.L. Francis, VCH/Ellis Horwood, Weinheim, Germany/London, Chapter 6, pp. 175–202, 1988.
319. Marsh, N.A., The fibrinolytic enzyme system, in *Fibrinogen, Fibrin Stabilization, and Fibrinolysis*, Ed. J.L. Francis, VCH/Ellis Horwood, Weinheim, Germany/London, Chapter 8, pp. 203–263, 1988.
320. Blombäck, B., Fibrnogen and fibrin formation and its role in fibrinolysis, *Biotechnology* 19, 225–279, 1991.
321. Collen, D. and Lijnen, H.B., Fibrin-specific fibrinolysis, *Ann. N. Y. Acad. Sci.* 667, 259–271, 1992.
322. Gabriel, D.A., Muga, K., and Boothroyd, E.M., The effect of fibrin structure on fibrinolysis, *J. Biol. Chem.* 267, 24259–24263, 1992.
323. Sidelmann, J.J., Gram, J., Jespersen, J., and Kluft, C., Fibrin clot formation and lysis: Basic mechanisms, *Semin. Thromb. Haemost.* 26, 605–618, 2000.
324. Medved, L. and Nieuwenhuizen, W., Molecular mechanisms of initiation of fibrinolysis by fibrin, *Thromb. Haemost.* 89, 409–419, 2003.
325. Weisel, J.W., Structure of fibrin: Impact on clot stability, *J. Thromb. Haemost.* 5(Suppl. 1), 116–124, 2007.
326. Carr, M.E. and Gabriel, D.A., Dextran-induced changes in fibrin fiber size and density based on wavelength dependence of gel turbidity, *Macromolecules* 13, 1473–1477, 1980.
327. Marguerie, G., The binding of calcium to fibrinogen: Some structural features, *Biochim. Biophys. Acta* 494, 172–181, 1977.
328. Procyk, R. and Blombäck, B., Disulfide bond reduction in fibrinogen: Calcium protection and effect on clottability, *Biochemistry* 29, 1501–1507, 1990.
329. Ohta, N. and Yotsuyanagi, T., Alteration of fibrinogen secondary structures by *cis*-diaminedichloroplatinum(II) and calcium protection, *Biol. Pharm. Bull.* 16, 631–634, 1993.
330. Takebe, M., Soe, G., Kohno, I. et al., Calcium ion-dependent monoclonal antibody against human fibrinogen: Preparation, characterization, and application to fibrinogen purification, *Thromb. Haemost.* 73, 662–667, 1995.

331. Odrljin, T.M., Rybarczyk, B.J., Francis, C.W. et al., Calcium modulates plasmin cleavage of the fibrinogen D fragment gamma chain *N*-terminus mapping of monoclonal antibody J88B to a plasmin sensitive domain of the gamma chain, *Biochim. Biophys. Acta* 1298, 69–77, 1996.
332. Mihalyi, E., Review of some unusual effects of calcium binding to fibrinogen, *Biophys. Chem.* 112, 131–140, 2004.
333. Okada, M. and Blombäck, B., Calcium and fibrin gel structure, *Thromb. Res.* 29, 269–280, 1983.
334. Perizzolo, K.E., Sullivan, S., and Waugh, D.F., Effects of calcium binding and of EDTA and CaEDTA on the clotting of bovine fibrinogen by thrombin, *Arch. Biochem. Biophys.* 237, 520–534, 1985.
335. Milhalyi, E., Clotting of bovine fibrinogen, calcium binding to fibrin during clotting and its dependence on release of fibrinopeptide B, *Biochemistry* 27, 967–976, 1988.
336. Profumo, A., Turci, M., Damonte, G. et al., Kinetics of fibrinopeptide release by thrombin as a function of $CaCl_2$ concentration: Different susceptibility of FPA and FPB and evidence for a fibrinogen isoform specific effect at physiological Ca^{2+} concentration, *Biochemistry* 42, 12335–12348, 2003.
337. Godal, H.C., Delayed fibrin polymerization due to removal of calcium ions, *Scand. J. Clin. Lab. Invest.* 24, 29–33, 1969.
338. Boyer, M.H., Shainoff, J.R., and Ratnoff, O.D., Acceleration of fibrin polymerization by calcium ions, *Blood* 39, 382–387, 1972.
339. Endres, G.F. and Scheraga, H.A., Equilibria in the fibrinogen-fibrin conversion. IX. Effects of calcium ions on the reversible polymerization of fibrin monomer, *Arch. Biochem. Biophys.* 153, 266–278, 1972.
340. Shen, L., McDonagh, R.F., McDonagh, J., and Hermans, J., Jr., Fibrin gel structure: Influence of calcium and covalent cross-linking on the elasticity, *Biochem. Biophys. Res. Commun.* 56, 793–798, 1974.
341. Brass, E.P., Forman, W.B., Edwards, R.V., and Linden, O., Fibrin formation: Effect of calcium ions, *Blood* 52, 654–658, 1978.
342. Furlan, M., Rupp, C., Beck, E.A., and Svendsen, L., Effect of calcium and synthetic peptides on fibrin polymerization, *Thromb. Haemost.* 47, 118–121, 1982.
343. Váradi, A. and Scheraga, H.A., Localization of segments essential for polymerization and for calcium binding in the gamma-chain of human fibrinogen, *Biochemistry* 25, 519–528, 1986.
344. Marx, G., Divalent cations induce protofibril gelation, *Am. J. Hematol.* 27, 104–109, 1988.
345. Kostelansky, M.S., Lounes, K.C., Ping, L.F. et al., Calcium-binding site β-2, adjacent to the "b" polymerization site, modulates lateral aggregation of protofibrils during fibrin polymerization, *Biochemistry* 43, 2475–2483, 2004.
346. Carr, M.E. and Powers, P.L., Differential effects of divalent cations on fibrin structure, *Blood Coagul. Fibrinol.* 2, 741–747, 1991.
347. Siedentop, K.H., Harris, D.M., and Sanchez, B., Autologous fibrin tissue adhesive: Factors influencing bonding power, *Laryngoscope* 98, 731–733, 1988.
348. Byrne, D.J., Hardy, J., Wood, R.A. et al., Effect of fibrin glues on the mechanical properties of healing wounds, *Br. J. Surg.* 78, 841–843, 1991.
349. Sierra, D.M., Feldman, D.S., Saltz, R., and Huang, S., A method to determine shear adhesive strength of fibrin sealants, *J. Appl. Biomater.* 3, 147–151, 1992.
350. Park, M.S., Autologous fibrin glue for tympanoplasty, *Am. J. Otol.* 15, 687–689, 1994.
351. Velada, J.L., Hollingsbee, D.A., Menzies, A.R. et al., Reproducibility of the mechanical properties of Vivostat system patient-derived fibrin sealant, *Biomaterials* 23, 2249–2254, 2002.
352. Dickneite, G., Metzner, R., Pfeifer, T. et al., A comparison of fibrin sealants in relation to their *in vitro* and *in vivo* properties, *Thromb. Res.* 112, 73–82, 2003.

353. Janus, T.J., Lewis, S.D., Lorand, L., and Shafer, J.A., Promotion of thrombin-catalyzed activation of factor XIII by fibrinogen, *Biochemistry* 22, 6269–6272, 1983.
354. Greenberg, C.S., Achyuthan, K.E., Rajagopalan, S., and Pizzo, S.V., Characterization of the fibrin polymer structure that accelerates thrombic cleavage of plasma factor XIII, *Arch. Biochem. Biophys.* 262, 142–146, 1988.
355. Hornyak, T.J. and Shafter, J.A., Interactions of factor XIII with fibrin as substrate and cofactor, *Biochemistry* 31, 423–429, 1992.
356. Procyk, R., Bishop, P.D., and Kudryk, B., Fibrin—recombinant factor XIII A-subunit association, *Thromb. Res.* 71, 127–138, 1993.
357. Moaddel, M., Falls, l.A., and Farrell, D.N., The role of γA/γ' fibrinogen in plasma factor XIII activation, *J. Biol. Chem.* 275, 32125–32140, 2000.
358. Maurer, M.C., Trumbo, T.A., Isetti, G., and Turner, B.T., Jr., Probing interactions between the coagulants thrombin, factor XIII, and fibrinogen, *Arch. Biochem. Biophys.* 445, 36–45, 2006.
359. Dickneite, G., Metzner, H.J., Kroez, M. et al., The importance of factor XIII as a component of fibrin sealants, *J. Surg. Res.* 107, 186–195, 2002.
360. Phillips, M., Dickneite, G., and Metzner, H., Fibrin sealants in supporting surgical techniques: Strength in factor XIII, *Cardiovasc. Surg.* 11(Suppl. 1), 13–15, 2003.
361. Wozniak, G., Fibrin sealants in supporting surgical techniques: The importance of individual components, *Cardiovasc. Surg.* 11(Suppl. 1), 17–23, 2003.
362. Marx, G. and Mou, X., Characterizing fibrin glue performance as modulated by heparin, aprotinin, and factor XIII, *J. Lab. Clin. Med.* 140, 152–160, 2002.
363. Glidden, P.F., Malaska, C., and Herring, S.W., Thromboelastograph assay for measuring the mechanical strength of fibrin sealant, *Clin. Appl. Thromb. Hemost.* 6, 226–233, 2000.
364. Bar, L., Malka, O., Nabolchenko, E., and Nur, I., The binding of fibrin sealant to collagen is influenced by the method of purification and the cross-linked fibrinogen-fibronectin (heteronectin) content of the "fibrinogen" component, *Blood Coagul. Fibrinol.* 16, 111–117, 2005.
365. Carr, M.E., Jr., Gabriel, D.A., and McDonagh, J., Influence of factor XIII and fibronectin on fiber size and density in thrombin-induced fibrin gels, *J. Lab. Clin. Med.* 110, 747–752, 1987.
366. Litinov, R.I. and Zabairov, D.M., Influence of fibronectin on the conversion of fibrinogen to fibrin, *Biochemistry (Moscow)* 53, 1046–1055, 1988.
367. Okada, M., Blombäck, M., Chang, M.D., and Horowitz, B., Fibronectin and fibrin gel structure, *J. Biol. Chem.* 260, 1811–1820, 1985.
368. Procyk, R., Adamson, L., Block, M., and Blombäck, B., Factor XIII catalyzed formation of fibrinogen-fibronectin oligomers—a thiol enhanced process, *Thromb. Res.* 40, 833–852, 1985.
369. Wilson, C.L. and Schwarzbauer, J.E., The alternatively spliced-V region contributes to the differential incorporation of plasma and cellular fibronectin into fibrin clots, *J. Cell Biol.* 119, 923–933, 1992.
370. Corbett, S.A., Wilson, C.L., and Schawarzbauer, J.E., Changes in cell-spreading and cytoskeleton organization are induced by adhesion to a fibronectin-fibrin-matrix, *Blood* 88, 158–166, 1996.
371. Corbett, S.A. and Schwarzbauer, J.E., Fibronectin-fibrin cross-linking: A regulator of cell behavior, *Trends Cardiovascul. Med.* 8, 357–362, 1998.
372. Eissner, A., Mazur, G., and Vogelmeier, C., Inhibition of factor XIIIa mediated incorporation of fibronectin into fibrin by pulmonary surfactant, *Am. J. Physiol.—Lung Cell. Molec. Physiol.* 276, L625–L630, 1999.
373. Corbett, S.A. and Schwarzbauer, J.E., Requirements for $\alpha_5\beta_1$ integrin-mediated retraction of fibronectin-fibrin matrices, *J. Biol. Chem.* 274, 20943–20948, 1999.

374. Ikari, Y., Yee, K.O., and Schwartz, S.M., Role of $\alpha_5\beta_1$ and $\alpha_v\beta_3$ integrins on smooth muscle cell spreading and migration in fibrin gels, *Thromb. Haemost.* 84, 701–705, 2000.
375. Midwood, K.S., Valenick, L.V., Hsin, H.C. et al., Coregulation of fibronectin signaling and matrix contraction by tenascin-C and syndecan-4, *Mol. Biol. Cell.* 15, 5670–5677, 2004.
376. Nenci, G.G., Parise, P., Morini, A. et al., Fibrin clots obtained from plasma containing heparin show a higher sensitivity to t-PA-induced lysis, *Blood Coagul. Fibrinol.* 3, 279–285, 1992.
377. Parise, P., Morini, M., Agnelli, G. et al., Effects of low molecular weight heparins on fibrin polymerization and clot sensitivity to t-PA-induced lysis, *Blood Coagul. Fibrinol.* 4, 721–727, 1993.
378. Collen, A., Smorenburg, S.M., Peters, E. et al., Unfractionated and low molecular weight heparin affect fibrin structure and angiogenesis *in vitro*, *Cancer Res.* 60, 6196–6200, 2000.
379. Van Hinsbergh, V.W.M., Collen, A., and Koowijk, P., Role of fibrin matrix in angiogenesis, *Ann. N. Y. Acad. Sci.* 936, 426–437, 2001.
380. Ku, C.S. and Fiedel, B.A., Modulation of fibrin clot formation by human serum amyloid P component (SAP) and heparin, *J. Exp. Med.* 158, 767–780, 1983.
381. Meyer, K., Smith, R., and Williams, R.C., Inhibition of fibrin polymerization by serum amyloid P component and heparin, *Thromb. Haemost.* 57, 345–348, 1987.
382. Zarge, J.I., Huang, P., Husak, V. et al., Fibrin glue containing fibroblast growth factor type 1 and heparin with autologous endothelial cells reduces intimal hyperplasia in a canine carotid artery ballon injury model, *J. Vasc. Surg.* 25, 840–848, 1997.
383. Zarge, J.I., Gosselin, C., Huang, P. et al., Platelet deposition on ePTFE grafts coated with fibrin glue with or without FGF-1 and heparin, *J. Surg. Res.* 75, 4–8, 1998.
384. Zarge, J.I., Husik, V., Huang, P., and Greisler, H.P., Fibrin glue containing fibroblast growth factor type 1 and heparin decreases platelet deposition, *Am. J. Surg.* 174, 188–192, 1997.
385. Moon, M.C., Molnar, A., Yau, L., and Zahradka, P., Perivascular delivery of losartan with surgical fibrin glue prevents neointimal hyperplasia after arterial injury, *J. Vasc. Surg.* 40, 130–137, 2004.
386. Pardue, M.T., Hejny, C., Gilbert, J.A. et al., Retinal function after subconjunctival injection of carboplatin in fibrin sealant, *Retina* 24, 776–782, 2004.
387. Han, D.K., Kim, C.S., Jung, U.W. et al., Effect of a fibrin-fibronectin sealing system as a carrier for recombinant human bone morphogenetic protein-4 on bone formation in rat calvarial defects, *J. Periodontol.* 76, 2216–2222, 2005.
388. Tredwell, S., Jackson, J.K., Hamilton, D. et al., Use of fibrin sealant for the localized, controlled release of cefazolin, *Can. J. Surg.* 49, 347–352, 2006.
389. Ishil, I., Mizura, H., Sei, A. et al., Healing of full-thickness defects of the articular cartilage in rabbits using fibroblast growth factor-2 an a fibrin sealant, *J. Bone Joint Surg. Br.* 89, 693–700, 2007.
390. Brodsky, B. and Persikov, A.V. Molecular structure of the collagen triple helix, *Adv. Prot. Chem.* 70, 302–339, 2005.
391. Wess, T.J., Collagen fibril form and function, *Adv. Prot. Chem.* 70, 341–374, 2005.
392. Knupp, C. and Squire, J.M., Molecular parking in network-forming collagens, *Adv. Prot. Chem.* 70, 375–403, 2005.
393. Teti, A., Regulation of cellular functions by extracellular matrix, *J. Am. Soc. Nephrol.* 2, S83–S87, 1992.
394. Brodsky, B. and Shah, N.K., The triple-helix motif in proteins, *FASEB J.* 9, 1537–1546, 1995.
395. Heino, J., The collagen receptor integrins have distinct ligand recognition and signaling functions, *Matrix Biol.* 19, 319–323, 2000.
396. Eble, J.A., Collagen-binding integrins as pharmaceutical targets, *Curr. Pharm. Design* 11, 867–880, 2005.

397. Jones, M. and Gabriel, D.A., Influence of the subendothelial basement membrane components on fibrin assembly, *J. Biol. Chem.* 263, 7043–7048, 1988.
398. Mosesson, M.W., Collagen and the development of resistance to fibrin clot lysis, *J. Lab. Clin. Med.* 117, 265, 1991.
399. Mirshahi, M. Azzarone, B., Soria, J. et al., The role of fibroblasts in organization and degradation of a fibrin clot, *J. Lab. Clin. Med.* 117, 274–281, 1991.
400. Koken Company, Ltd. http://www.kokenmpc.co.jp/english/products/medical_plastics/surgery/implant/index.html.
401. Nomori, H., Horio, H., and Suemasu, K., Mixing collagen with fibrin glue to strengthen the sealing effect for pulmonary air leakage, *Ann. Thorac. Surg.* 70, 1666–1701, 2000.
402. Bannister, D.W. and Burns, A.B., Pepsin treatment of avian skin collagen. Effect on solubility, subunit composition and aggregation properties, *Biochem. J.* 129, 677–681, 1972.
403. Epstein, E.H. Jr., $[_{\alpha}1(\mathrm{III})]_3$ Human skin collagen. Release by pepsin digestion and preponderance in fetal life, *J. Biol. Chem.* 249, 3225–3331, 1974.
404. Lenaers, A. and Lapiere, C.M., Type III procollagen and collagen in skin, *Biochim. Biophys. Acta* 400, 121–131, 1975.
405. Condell, R.A., Hanko, V.P., Larenas, E.A. et al., Analysis of native collagen monomers and oligomers by size-exclusion high-performance liquid chromatography and its application, *Anal. Biochem.* 212, 436–445, 1993.
406. Jakob, H. Campbell, C.D., Stemberger, A. et al., Combined application of heterologous collagen and fibrin sealant, *J. Surg. Res.* 36, 572–577, 1984.
407. Andreason, T.T. and Jorgenson, P.H., Biomechanical properties and collagen formation in subcutaneously implanted cellulose sponges treated with fibrin sealant, *Eur. Surg. Res.* 17, 264–268, 1985.
408. Izbicki, J.R, Kreusser, T., Meier, M. et al., Fibrin-glue-coated collagen fleece in lung surgery—experimental comparison with infrared coagulation and clinical experiences, *Thorac. Cardiovasc. Surg.* 42, 306–309, 1994.
409. Prior, J.J, Wallace, D.G., Harner, A., and Powers, N., A sprayable hemostat containing fibrillar collagen, bovine thrombin, and autologous plasma, *Ann. Thorac. Surg.* 68, 479–485, 1999.
410. Czerny, M., Verrel, F., Weber, H. et al., Collagen patch coated with fibrin glue components. Treatment of suture hole bleedings in vascular reconstruction, *J. Cardiovasc. Surg.* 41, 553–557, 2000.
411. Petter-Puchner, A.H., Froetscher, W., Krametter-Froetscher, R. et al., The long-term neurocompatibility of human fibrin sealant and equine collagen as biomatrices in experimental spinal cord injury, *Exp. Toxicol. Pathol.* 58, 237–245, 2007.
412. Gabriel, D.A., Smith, L.A., Folds, J.D. et al., The influence of immunoglobulin (IgG) on the assembly of fibrin gels, *J. Lab. Clin. Med.* 101, 545–552, 1983.
413. Jones, M., McDonagh, J., Johnson, C.S., and Gabriel, D.A., Paraprotein modified fibrin gels, in *Fibrinogen 3*, Eds. M.W. Mosesson et al., Elsevier-North Holland, Amsterdam, Netherlands, 1988.
414. Coleman, M., Vigliano, E.M., Weksler, M.E, and Nachman, R.L., Inhibition of fibrin monomer polymerization by lambda myeloma globulins, *Blood* 39, 210–223, 1972.
415. Davey, R.F., Gordon, G.B., Borsl, L.I., and Gottlieb, A.J., Gamma globulin inhibition of fibrin clot formation, *Ann. Clin. Lab. Sci.* 6, 72–77, 1976.
416. Klingemann, H.G., Egbring, R., and Havemann, K., Incomplete fibrin formation and highly elevated Factor XIII activity in multiple myeloma, *Scand. J. Haematol.* 27, 253–262, 1981.
417. Wisloff, F., Michaelsen, T.E., and Godal, H.C., Inhibition of acceleration of fibrin polymerization by monoclonal immunoglobulins and immunoglobulin fragments, *Thromb. Res.* 35, 81–90, 1984.

418. Panzer, S. and Thaler, E. An acquired cryoglobulinemia which inhibits fibrin polymerization in a patient with IgG kappa myeloma, *Haemostasis* 23, 69–76, 1993.
419. Handin, R.I. and Cohen, H.J., Purification and binding properties of human platelet factor four, *J. Biol. Chem.* 251, 4273–4282, 1976.
420. Chesterman, C.N., McGready, J.R, Doyle, D.J., and Morgan, F.J., Plasma levels of platelet factor 4 measured by radioimmunoassay, *Br. J. Haematol.* 40, 489–500, 1978.
421. Kaplan, K.L. and Owen, J., Plasma levels of β-thromboglobulin and platelet factor 4 as indices of platelet activation *in vivo*, *Blood* 57, 199–202, 1981.
422. Carr, M.E., White, G.C., 2nd, and Gabriel, D.A., Platelet factor 4 enhances fibrin fiber polymerization, *Thromb. Res.* 45, 539–543, 1987.
423. Weisel, J.W. and Nagaswami, C., Computer modeling of fibrin polymerization kinetics correlated with electron microscope and turbidity observations: Clot structure and assembly are kinetically controlled, *Biophys. J.* 63, 111–128, 1992.
424. Amelot, A.A., Tagzirt, M., Ducouret, G. et al., Platelet factor 4 (CXCL4) seals blood clots by altering the structure of fibrin, *J. Biol. Chem.* 282, 710–720, 2007.
425. Carney, S.L. and Muir, H., The structure and function of cartilage proteoglycans, *Physiol. Rev.* 68, 858–910, 1988.
426. Scott, J.E., The chemical morphology of the vitreous, *Eye* 6, 553–555, 1992.
427. Scott, J.E., Extracellular matrix, *J. Anat.* 187, 259–269, 1995.
428. Liao, Y.H., Jones, S.A., Forbes, B. et al., Hyaluronan: Pharmaceutical characterization and drug delivery, *Drug Deliv.* 12, 327–342, 2005.
429. Jiang, D., Liang, J., and Noble, J.W., Hyaluronan in tissue injury and repair, *Annu. Rev. Cell Dev. Biol.* 23, 435–461, 2007.
430. LaBoeuf, R.D., Raja, R.H., Fuller, G.M., and Weigel, P.H., Human fibrinogen specifically binds hyaluronic acid, *J. Biol. Chem.* 261, 12586–12592, 1986.
431. Frost, S.J. and Weigel, P.H., Binding of hyaluronic acid to mammalian fibrinogens, *Biochim. Biophys. Acta* 1034, 39–45, 1990.
432. LaBoeuf, R.D., Gregg, R.R., Weigel, P.H., and Fuller, G.M., Effects of hyaluronic acid and other glycosaminoglycans on fibrin polymer formation, *Biochemistry* 26, 6052–6057, 1987.
433. Fournier, N. and Doillon, C.J., *In vitro* angiogenesis in fibrin matrices containing fibronectin or hyaluronic acid, *Cell Biol. Int. Rep.* 16, 12510–1263, 1992.
434. Hayen, W., Goebeler, M., Kumar, S. et al., Hyaluronan stimulates tumor cell migration by modulating the fibrin fiber architecture, *J. Cell Sci.* 112, 2241–2245, 1999.
435. Nehls, V. and Hayen, W., Are hyaluronan receptors involved in three-dimensional cell migration? *Histol. Histopathol.* 15, 629–636, 2000.
436. Hubbell, J.A., Materials as morphogenetic guides in tissue engineering, *Curr. Opin. Biotechnol.* 14, 551–558, 2003.
437. Chou, C.H., Cheng, W.T., Kuo, T.F. et al., Fibrin glue mixed with gelatin/hyaluronic acid/chondroitin-6-sulfate tri-copolymer for articular cartilage tissue engineering: The results of real-time polymerase chain reaction, *J. Biomed. Mater. Res. A* 82, 757–767, 2007.
438. Torbet, J., Fibrin assembly in human plasma and fibrinogen/albumin mixtures, *Biochemistry* 25, 5309–5314, 1986.
439. Galanakis, D.K., Anticoagulant albumin fragments that bind to fibrinogen/fibrin: Possible implications, *Semin. Thromb. Hemost.* 18, 44–52, 1992.
440. Carr, M.E., Jr., Turbidimetric evaluation of the impact of albumin on the structure of thrombin-mediated fibrin gelation, *Haemostasis* 17, 189–194, 1987.
441. Carr, M.E., Fibrin formed in plasma is composed of fibers more massive that those formed from purified fibrinogen, *Thromb. Haemost.* 59, 535–539, 1988.
442. Nair, C.B., Ashar, A., and Dhall, D.P., The effects of some plasma proteins on fibrin network structure, *Blood Coagul. Fibrinol.* 1, 469–473, 1990.

443. Nair, C.H. and Dhall, D.F., Studies on fibrin network structure: The effect of some plasma proteins, *Thromb. Res.* 61, 315–325, 1991.
444. Veis, A., *The Macromolecular Chemistry of Gelatin*, Academic Press, New York, 1964.
445. *The Science and Technology of Gelatin*, Eds. A.G. Ward and A. Courts, Academic Press, London, 1977.
446. Buenel, J.P., Morris, E.R., and Ross-Murphy, S.B., Interpretation of the renaturation kinetics of gelatin solutions, *Int. J. Biol. Macromol.* 11, 119–125, 1989.
447. Gekko, K., and Fukamizu, M., Effect of pressure on the sol-gel transition of gelatin, *Int. J. Biol. Macromol.* 13, 295–300, 1991.
448. Giraudier, S., Hellio, D., Djabourov, M., and Larreta-Garde, V., Influence of weak and covalent bonds on formation and hydrolysis of gelatins, *Biomacromolecules* 5, 1662–1666, 2004.
449. Gornall, J.L. and Terentjev, E.M., Concentration-temperature superposition of helix folding rates in gelatin, *Phys. Rev. Lett.* 99, 028304, 2007.
450. Kushibiki, T. and Tabata, Y., A new gene delivery system based on controlled releases technology, *Curr. Drug Deliv.* 1, 153–163. 2004.
451. Sutter, M., Siepmann, J., Henink, W.E., and Jiskoot, W., Recombinant gelatin hydrogels for the sustained release of proteins, *J. Control. Release* 22, 301–312, 2007.
452. Strickland, W.A., Jr. and Moss, M., Water vapor sorption and diffusion through hard gelatin capsules, *J. Pharm. Sci.* 51, 1002–1005, 1962.
453. Digenis, G.A., Gold, T.B., and Sheh, V.P., Cross-linking of gelatin capsules and its relevance to their *in vitro* and *in vivo* performance, *J. Pharm. Sci.* 83. 915–921, 1994.
454. Bigi, A., Bracci, B., Cojazzi, G. et al., Drawn gelatin films with improved mechanical properties, *Biomaterials* 19, 2335–2340, 1998.
455. Farrugia, C.A. and Groves, M.J., Gelatin behavior in dilute aqueous solution: Designing a nanoparticulate formulation, *J. Pharm. Pharmacol.* 51, 643–649, 1999.
456. Aguilar-Mendez, M.A., Martni-Martinez, E.S., Morales, J.E. et al., Photothermal techniques applied to the determination of the water vapor diffusion coefficient and thermal diffusivity of edible films, *Anal. Sci.* 23, 457–461, 2007.
457. MacNight, C., Clinical implications of bovine spongiform encephalopathy, *Clin. Infect. Dis.* 32, 1726–1731, 2001.
458. Grobben, A.H., Steele, P.J., Somerville, R.A. et al., Inactivation of the BSE agent by the heat and pressure process for manufacturing gelatine, *Vet. Rec.* 157, 277–281, 2005.
459. Alexander, J., *Glue and Gelatin*, The Chemical Catalog Company, New York, 1923.
460. Mo, X., Iwata, H., Matsuda, S., and Ikada, Y., Soft tissue adhesive compound of modified gelatin and polysaccharides, *J. Biomater. Sci. Polym. Ed.* 11, 341–351, 2000.
461. Chen, T., Janjua, R., McDermott, M.K. et al., Gelatin-based biomimetic tissue adhesive. Potential for retinal reattachment, *J. Biomed. Mater. Res. B. Appl. Biomater.* 77, 416–422, 2006.
462. Baekeland, L.H., The chemical constitution of resinous phenolic condensation product, *J. Industr. Engineer. Chem.* 5, 506–521, 1913.
463. Ellis, C., *The Chemistry of Synthetic Resins*, Reinhold Publishing, New York, 1935.
464. Megson, N.J.L., *Phenolic Resin Chemistry*, Academic Press, London, 1958.
465. Santana, M.A.E., Baumann, M.G.D., and Conner, A.H., Resol resins prepared with tannin liquefied in phenol, *Holzforschung* 49, 146–152, 1995.
466. Goldsmith, B.B., Composition of Matter and Process of Making the Same, U.S. Patent 1,076,417, 1913.
467. Morris, P.R., Laminating Glass and Process of Making the Same, U.S. Patent 1,974,624, 1934.
468. Dressler, H., *Resorcinol Its Uses and Derivatives*, Plenum Press, New York, NY, 1994.
469. Durairag, R.B., *Resorcinol Chemistry, Technology and Applications,* Springer, Berlin, Germany, 2005.

470. Donawa, M.E., Meeting European and U.S. requirements for design and development documentation. Part II, *Med. Device Technol.* 7, 10–15, 1996.
471. Lasky, F.D. and Boser, R.B., Deigning in quality through design control: A manufacturer's perspective, *Clin. Chem.* 43, 866–872, 1997.
472. Pharmaceutical Development, Annex to Q8, http://www.ich.org.
473. Laurian, C., Gigou, F., and Guilmet, D., Gelatin resorcin formaldehyde glue in vascular surgery, *Nouv. Presse Med.* 6, 3221–3223, 1977.
474. Guilmet, D., Bachet, J., Goudot, B. et al., Use of biological glue in acute aortic dissection. Preliminary clinical results with a new surgical technique, *J. Thorac. Cardiovasc. Surg.* 77, 516–521, 1979.
475. Bachet, J., Gigou, F., Laurian, C. et al., Four year clinical experience with the gelatin-resorcin-formal biological glue in acure aortic dissection, *J. Thorac. Cardiovasc. Surg* 82, 212–217, 1982.
476. Droegemueller, W., Greer, B.E., Davie, J.R. et al., Cryocoagulation of the endometrium at the uterine cornua, *Am. J. Obstet. Gynecol.* 131, 1–9, 1978.
477. Hake, U. and Uelert, H., Surgical therapy of thoracic aortic dissection, *Herz* 17, 357–376, 1992.
478. Basu, S., Marini, C.P., Bauman, G. et al., Composition study of biological glues: Cryoprecipitate glue, two-component fibrin sealant and "French" glue, *Ann. Thorac. Surg.* 60, 1255–1262, 1995.
479. Chao, H.H. and Torchiana, D.F., BioGlue: Albumin/glutaraldehyde sealant in cardiac surgery, *J. Card. Surg.* 18, 500–503, 2003.
480. Braunwald, N.J., Gay, W., and Tatooles, C.J., Evaluation of crosslinked gelatin as a tissue adhesive and hemostatic agent: An experimental study, *Surgery* 59, 1024–1030, 1966.
481. Tatooles, C.J. and Braunwald, N.J., The use of crosslinked gelatin as a tissue adhesive to control hemorrhage from liver and kidney, *Surgery* 60, 857–861, 1966.
482. Banchek, L.I. and Braunwald, N.J., Experimental evaluation of a cross-linked gelatin adhesive in gastrointestinal surgery, *Ann. Surg.* 165, 420–424, 1966.
483. Cooper, C.-W. and Falb, R.D., Surgical adhesives, *Ann. N. Y. Acad. Sci.* 146, 214–224, 1968.
484. Grode, G.A., Pavkov, K.L., and Falb, R.D., Feasibility study on the use of a tissue adhesive for the nonsurgical blocking of fallopian tubes. Phage I: Evaluation of a tissue adhesive, *Fert. Steril.* 22, 552–555, 1971.
485. Jenkins, H.P. and Clarke, J.S., Gelatin sponge, a new hemostatic substance, *Arch. Surg.* 51, 253–261, 1945.
486. Correll, J.T. and Vanderpoel, J.C., Biologic absorption of insolubilized gelatin films, *Proc. Soc. Exp. Med. Biol.* 71, 134–136, 1949.
487. Albes, J.M., Krettek, C, Hausen, B. et al., Biological properties of the gelatin-resorcin-formaldehyde/glutaraldehyde adhesive, *Ann. Thorac. Surg.* 56, 910–915, 1993.
488. Bachet, J., Goudot, B., Dreyfus, G. et al., The proper use of glue: A 20-year experience with the GRF glue in acute aortic dissection, *J. Card. Surg.* 12(Suppl. 2), 243–253, 1997.
489. Fukunaka, S., Karck, M., Harringer, W. et al., The use of gelatin-resorcin-formalin glue in acute aortic dissection, type A, *Europ. J. Card-Thorac. Surg.* 15, 564–570, 1999.
490. Bachet, J. and Guilmet, D., The use of biological glue in aortic surgery, *Cardiol. Clin.* 17, 779–796, 1999.
491. Hata, H., Takano, H., Matsumiya, G. et al., Late complications of gelatin-resorcinol-formalin glue in the repair of acute type A aortic dissection, *Ann. Thorac. Surg.* 83, 1621–1627, 2007.
492. Nishimura, H., Hata, A., and Sasaguri, S., Optimal application of gelatin-resorcin-formaldehyde glue with special reference to the quality of mixing, *Ann. Thorac. Surg.* 69, 1299, 2000.

493. Gerhart, T.N., Hayes, W.C., and Stern, S.H., Biomechanical optimization of a model particulate composite for orthopaedic applications, *J. Ortho. Res.* 4, 76–85, 1986.
494. Gerhart, T.N., Miller, R.L., Kleshinski, S.J., and Hayes, W.C., *In vitro* characterization and biomechanical optimization of a biodegradable particulate composite bone cement, *J. Biomed. Mater. Res.* 22, 1071–1082, 1988.
495. Witschger, P.M., Gerhart, T.N., Goldman, J.B. et al., Biomechanical evaluation of a biodegradable composite as an adjunct to internal fixation of proximal femur fractures, *J. Ortho. Res.* 9, 48–53, 1991.
496. Ennker, I.C., Ennker, J., Schoo, D. et al., Formaldehyde-free collagen glue in experimental lung gluing, *Ann. Thorac. Surg.* 57, 1622–1627, 1994.
497. Sung, H.-W., Huang, D.-M., Chang, W.-H. et al., Gelatin-derived bioadhesives for closing skin wounds: An *in vivo* study, *J. Biomater. Sci. Polymer Edn.* 10, 754–771, 1999.
498. Kücükaksu, D.S., Akgül, A., Cağli, K., and Taşdemir, O., Beneficial effect of Bioglue surgical adhesive in repair of iatrogenic aortic dissection, *Tex. Heart Inst. J.* 27, 307–308, 2000.
499. Passage, J., Jalali, H., Tom, R.K. et al., BioGlue® surgical adhesive—An appraisal of its indications in cardiac surgery, *Ann. Thorac. Surg.* 74, 432–437, 2002.
500. Frehrenacher, J.W. and Siderys, H., Use of BioGlue® in aortic surgery: Proper application techniques and results in 92 patients, *Heart Surg. Forum* 9, E794–E799, 2006.
501. Zehr, K.J., Use of bovine albumin-glutaraldehyde in cardiovascular surgery, *Ann. Thorac. Surg.* 84, 1048–1052, 2007.
502. De Somer, F., Delanghe, J., Somers, P. et al., Mechanical and chemical characteristics of an autologous glue, *J. Biomed. Mater. Res. A*, 86, 1106–1112, 2008.
503. Metzner, M., Horber, J., Rademacher, G., and Klee, W., Application of the glutaraldehyde test in cattle, *J. Vet. Med. A* 54, 449–454, 2007.
504. Cheung, D.T. and Nimni, M.E., Mechanisms of crosslinking of proteins by glutaraldehyde II. Reaction with monomeric and polymeric collagen, *Connect. Tissue Res.* 10, 210–216, 1982.
505. Sung, H.W., Huang, D.M., Chang, W.H. et al., Evaluation of gelatin hydrogel crosslinked with various crosslinking agents as bioadhesives: *In vitro* study, *J. Biomed. Mater. Res.* 46, 520–530, 1999.
506. Richards, F.M. and Knowles, J.R. Glutaraldehyde as a protein cross-linking reagent, *J. Mol. Biol.* 37, 231–233, 1968.
507. Wine, Y., Cohen-Hadar, N., Freeman, A., and Frolow, F., Elucidation of the mechanism and end products of glutaraldehyde crosslinking reaction by X-ray structure analysis, *Biotechnol. Bioengineer.* 98, 711–718, 2007.
508. Overberger, C.G., Ishida, S., and Ringsdorf, H., Intra-intermolecular polymerization of glutaraldehyde, *J. Polymer Sci.* 62, S1–S2, 1962.
509. Moyer, W.W., Jr. and Grev, D.A., Linear polyglutaraldehyde, *Polymer Lett.* 1, 29–32, 1963.
510. Anderson, P.J., Purification and quantitation of glutaraldehyde and its effect on several enzyme activities in skeletal muscle, *J. Histochem. Cytochem.* 15, 652–660, 1967.
511. Goff, C.W. and Oster, M.O., Formation of 235-nanometer absorbing substance during glutaraldehyde fixation, *J. Histochem. Cytochem.* 22, 913–915, 1974.
512. Walt, D.R. and Agayn, V.I., The chemistry of enzyme and protein immobilization with glutaraldehyde, *Trends Anal. Chem. TRAC* 13, 425–530, 1994.
513. Fadouloglou, V.E., Kokkinidis, M., and Glykos, N.M., Determination of protein oligomerization states: Two approaches based on glutaraldehyde crosslinking, *Anal. Biochem.* 373, 404–406, 2008.
514. Speer, D.P., Chvapil, M., Eskelson, C.D., and Ulreich, J., Biological effects of residual glutaraldehyde in glutaraldehyde-tanned collagen biomaterials, *J. Biomed. Mater. Res.* 14, 753–764, 1980.

515. Huanglee, H.H.L., Cheung, D.T., and Nimni, M.E., Biochemical changes and cytotoxicity associated with the degradation of polymeric glutaraldehyde derived cross-links, *J. Biomed. Mater. Res.* 24, 1185–1201, 1990.
516. Simionescu, A., Simionescu, D., and Deac, R., Lysine-enhanced glutaraldehyde cross-linking of collagenous biomaterials, *J. Biomed. Mater. Res.* 25, 1495–1505, 1991.
517. Gratzer, P.E., Pereira, C.A., and Lee, J.M., Solvent environment modulates effects of glutaraldehyde crosslinking on tissue-derived biomaterials, *J. Biomed. Mater. Res.* 31, 533–543, 1996.
518. Salles, C.A., Buffolo, E., Andrade, J.C. et al., Mitral vale replacement with glutaraldehyde preserverd aortic allografts, *Eur. J. Cardiothorac. Surg.* 13, 135–143, 1998.
519. Malashenkov, A.I., Rusanov, N.I., Muratov, R.M. et al., Eight years clinical experience with the replacement of the ascending aorta using composite xenopericardial conduit, *Eur. J. Cardiothorac. Surg.* 18, 168–173, 2000.
520. Isenburg, J.C., Simionescu, D.T., and Vyavahare, N.R., Elastin stabilization in cardiovascular implants: Improved resistance to enzymatic degradation by treatment with tannic acid, *Biomaterials* 25, 3293–3302, 2004.
521. Nuefand, A., Espinola-Klein, C., Dorweiler, B. et al., Femoropopliteal prosthetic bypass with glutaraldehyde stabilized human umbilical vein (HUV), *J. Vasc. Surg.* 46, 280–288, 2007.
522. Waite, J.H. and Tanzer, M.L., The bioadhesive of *Mytilus bysssus*: A protein containing L-DOPA, *Biochem. Biophys. Res. Commun.* 1554–1561, 1980.
523. Waite, J.H., Evidence for a repeating 3,4-dihydroxyphenylalanine- and hydroxyproline-containing decapeptide in the adhesive protein of the mussel, *Mytilus edulis L.*, *J. Biol. Chem.* 258, 2911–2915, 1083.
524. Waite, J.H., Mussel glue from *Mytilus californianus* Conrad: A comparative study, *J. Comp. Physiol.* 156, 491–496, 1986.
525. Zhao, H., Robertson, N.B., Jewhurst, S.A., and Waite, J.H., Probing the adhesive footprint of *Mytilus californianus* byssus, *J. Biol. Chem.* 281, 11090–11096, 2006.
526. van der Leeden, M.C., Are conformational changes induced by osmotic variations the underlying mechanism of controlling the adhesive activity of mussel adhesive protein? *Langmuir* 21, 11373–11379, 2005.
527. Hemler, M.E. and Mihich, E., *Cell Adhesion Molecules: Cellular Recognition Mechanisms,* Plenum Press, New York, NY, 1993.
528. Roberts, D.D. and Mecham, R.P., *Cell Surface and Extracellular Glycoconjugate: Structure and Function,* Academic Press, San Diego, CA, 1993.
529. Adams, J.C., *Methods in Cell-Matrix Adhesion*, Academic Press, San Diego, CA, 2002.
530. Phillips, D.R., Fitzgerald, L.A, Charo, I.E., and Parise, L.V., The platelet membrane glycoprotein IIb/IIIa complex. Structure, function, and relationship to adhesive protein receptors in nucleated cells, *Ann. N. Y. Acad. Sci.* 509, 177–187, 1987.
531. Hemler, M.W., Adhesive protein receptors on hematopoietic cells, *Immunol. Today* 9, 109–113, 1988.
532. Uede, T., Katagiri, Y., Iizuka, J., and Murakami, M., Osteopontin, a coordinator of host defense system: A cytokine or an extracellular adhesive protein?, *Microbiol. Immunol.* 41, 641–648, 1997.
533. Dole, A., Modiano, D., Doumbo, O., Bosman, A. et al., Thrombospondin related adhesive protein (TSAP), a potential malaria vaccine candidate, *Parassitologia* 41, 425–428, 1999.
534. Lin, Q., Gourdon, D., Sun, C. et al., Adhesion mechanisms of the mussel foot proteins mfp-1 and mfp-3, *Proc. Natl. Acad. Sci. USA* 104, 3782–3785, 2007.
535. Loizou, E, Weissen, J.T., Dundigallo, A. et al., Structural effects of crosslinking a biopolymer hydrogel derived from marine mussel adhesive protein, *Macromol. Biosci.* 6, 711–718, 2006.

536. Martine, T.J., Farris, D.B., and Graham, D.G., Covalent crosslinking of neurofilament protein by oxidized catechols as a potential mechanism of Lewy body formation, *J. Neuropathol. Exp. Neurol.* 54, 311–319, 1995.
537. Höök, F., Kasema, B., Nylander, T. et al., Variations in coupled water, viscoelastic properties, and film thickness of a Mepf-1 protein film during adsorption and crosslinkage: A quartz crystal microbalance with dissipation monitoring ellipsometry, and surface Plasmon resonance study, *Anal. Chem.* 73, 5796–5804, 2001.
538. Holl, S.M., Hansen, D., Waite, J.H., and Schaefer, J., Solid-state NMR analysis of crosslinking in mussel protein glue, *Arch. Biochem. Biophys.* 302, 255–258, 1993.
539. Stewart, R.J., Weaver, J.C., Morse, D.E., and Waite, J.H., The tube cement of *Phragmatopoma californica*: A solid foam, *J. Exp. Biol.* 207, 4727–4734, 2004.
540. Green, K., Berdecia, R., and Cheeks, L., Mussel adhesive protein: Permeability characteristics when used as a basement membrane, *Curr. Eye Res.* 8, 835–838, 1987.
541. Strausberg, R.L. and Link, R.P., Protein-based medical adhesives, *Trends Biotechnol.* 8, 53–57, 1990.
542. Tay, F.R. and Pashley, D.H., Dental adhesives of the future, *J. Adhes. Dent.* 4, 91–103, 2002.
543. Ninan, L., Strashine, R.L., Wilker, J.J., and Shi, R., Adhesive strength and curing rate of marine mussel protein extracts on porcine intestinal submucosa, *Acta Biomater.* 3, 687–694, 2007.
544. Notter, M.F., Selective attachment of neural cells to specific substrates including Cell-Tek®, a new cellular adhesive, *Expt. Cell Res.* 177, 237–246, 1988.
545. Olivieri, M.P., Rittle, K.H., Tweden, K.S., and Loomis, R.E., Comparative biophysical study of adsorbed calf serum, fetal bovine serum and mussel adhesive protein films, *Biomaterials* 13, 201–208, 1992.
546. Burzio, V.A., Silva, T., Pardo, J., and Burzio, L.O., Mussel adhesive enhances the immobilization of human chorionic gonadotrophin to a solid support, *Anal. Biochem.* 241, 190–194, 1996.
547. Zahn, M., Renken, J., and Seeger, S., Fluorimetric multiparameter cell assay at the single cell level fabricated by optical tweezers, *FEBS Lett.* 443, 337–340, 1999.
548. Williams, J.L., Stimulation of *Plasmodium falciparum* gametogenesis by conditioned medium from parasite cultures, *Am. J. Trop. Med. Hyg.* 60, 7–13, 1999.
549. Statz, A.R., Meagher, R.J., Barron, A.E., and Messersmith, P.B., New peptidomimetic polymers for antifouling surfaces, *J. Am. Chem. Soc.* 127, 7972–7973, 2005.
550. Lee, H., Dellatore, S.M., Miller, W.M., and Messersmith, P.B., Mussel-inspired surface chemistry for multifunctional coating, *Science* 318, 426–430, 2007.
551. Hwang, D., Gim, Y., and Cho, H.J., Expression of functional recombinant mussel adhesive protein type 3A in *Escherichia coli*, *Biotechnol. Prog.* 21, 965–970, 2005.
552. Hwang, D.S., Sim, S.B., and Cho, H.J., Cell adhesion biomaterial based on mussel adhesive protein fused with RGD peptide, *Biomaterials* 28, 4039–4046, 2007.

7 Protein Drug Delivery

Protein drug products* can be administered topically such as with thrombin in topical hemostasis[1] or with peptide growth factors as cosmeceuticals or in wound healing.[3,4] With topical administration, the drug product is being supplied at the site of action. Topical (mucosal) administration has been used for viral vector in gene therapy for immunization purposes.[5–7] Local administration of viral vectors for gene therapy is also a subject of considerable study.[8–14] Local administration of protein therapeutics such as collagen[15,16] is well accepted in practice and bone morphogenic protein[17,18] in concept. Local administration of botulinium toxin (Botox®) is well known for cosmetic purposes.[19]

Most protein drugs need to be administered in a manner that will eventually get them into the circulatory system. Parenteral administration has been the method of choice for drugs based on oligosaccharides, oligonucleotides, or proteins. However, there has been considerable interest in the oral administration of protein drugs dating back to early treatment for bleeding disorders.[20,21] Oral glandular kallikrein has been evaluated for treatment of male infertility[22–26] with more positive than negative results. There also have been studies on the use of glandular kallikrein for the treatment of hypertension, but the results were problematic.[27–29] Bläckberg and Ohlsson[30] studied the distribution of ^{125}I-radiolabeled glandular kallikrein after intraduodenal administration in three normal subjects. The radiolabel, mostly as ^{125}I, was excreted in 72 h. No intact ^{125}I-labeled glandular kallikrein was found in plasma, and it is suggested that the administration of glandular kallikrein is not useful. Oral thrombin has been studied for the treatment of gastrointestinal bleeding.[31–33] This was the topical administration of thrombin; there was no evidence of transport into the circulatory system. There were positive and negative results, and it was concluded that it was unlikely that the thrombin is effective in the control of gastrointestinal bleeding but is likely useful for the

* This chapter will focus on protein drug products, but much of the material has application to oliogonucleotide and oligosaccharide drug products. Some of the material is also applicable to products such as viruses that are used in gene therapy. See Ludwig, C. and Wagner, R., Virus-like particles-universal toolboxes, *Curr. Opin. Biotechnol.* 18, 537–545, 2007; Jager, L. and Ehrhardt, A., Emerging adenoviral vectors for stable correction of genetic disorders, *Curr. Gene Ther.* 7, 272–283, 2007; Power, A.T. and Bell, J.C., Taming the Trojan horse: Optimizing dynamic carrier cell/oncolytic virus systems for cancer biotherapy, *Gene Therapy* 15, 772–779, 2008; Hu, Y.C., Baculovirus vectors for gene therapy: A review, *Curr. Gene Therapy* 8, 54–65, 2008 ; Coura Rdos, S. and Nardi, N.B., The state of the art of adeno-associated virus-based vectors is gene therapy, *Virol. J.* 4:99, 2007. Delivery of cells is beyond the scope of the current work. See Paintaud, G., Tonelli, D., and Postaire, E., Biotherapies: Are they just like any other drugs?, *Therapie* 62, 229–239, 2007; Khurdayan, V.K., Stem cells: Therapeutic present and future, *Timely Top. Med. Cardiovasc. Dis.* 11:E14, 2007; Laurencin, C.T. and El-Amin, S.F., Xenotransplantation in orthopaedic surgery, *J. Am. Acad. Orthop. Surg.* 16, 4–8, 2008; Prieto, J., Fernandez-Ruiz, V., Kawa, M.P. et al., Cells as vehicles for therapeutic genes to treat liver diseases, *Gene Ther.* 15, 765–771, 2008; Power, A.T. and Bell, J.C., Taming the Trojan horse: Optimizing dynamic carrier cell/oncolytic virus systems for cancer biotherapy, *Gene Ther.* 15, 772–779, 2008.

control of esophageal bleeding. Orally administered bovine immunoglobulin retains function.[34–36] The oral administration of insulin has been a goal for many years[37] but has not yet achieved success.[38] The oral administration of insulin is still attracting considerable interest for reasons other than direct glucose control.[39–42]

This limited data suggests that a protein of average size and shape (60–80 kDa; 83 nm^3; 9.5 × 5 × 5 nm)[43] can pass from the gastrointestinal tract to the circulation; however, the efficiency of the process is quite low, precluding implementation without changes in formulation and possible structure. For example, neocarzinostatin is an acidic single-chain polypeptide with a mass of 12 kDa[44] and having poor oral availabilty.[45] A conjugate of neocarzinostatin with poly(styrene-maleic acid) SMANCS was shown to be effective with oral administration in formulations containing phosphatidyl choline and polyglycerine dioleate.[46–48] A combination of structural alterations and formulation was required to provide oral bioavailability.

Alternate routes for protein/oligonucleotide/oligosaccharide drug delivery include oral (buccal and gastrointestinal), nasal, pulmonary, rectal, vaginal, and transdermal. There are advantages and disadvantages to each of these approaches. Our discussion will now focus on transport of macromolecules across a biological membrane, the mucosal membrane; these considerations will be applicable to other surface areas[49–51] such as skin.[52–54] Transport of a macromolecule across the mucosal membrane is (1) retrograde to normal; and (2) assuming successful traverse of the mucosal membrane, there is a substantial osmotic gradient against protein movement from the extravascular space to the intravascular space. Subcutaneous or intramuscular delivery of macromolecules faces similar issues; uptake of proteins is thought to occur via the lymphatic system.[55–60] Route of administration does make a difference with interferon,[61] and TNF-α[62] with intramuscular injection gives lower[62] but more sustained response.[61]

The major challenge for the alternative delivery of macromolecules is size.[55,63–66] Although the lipid solubility (log P) and charge/hydrophilicity characteristics suggest that proteins/oligosaccharides/oligonucleotides will be poor drugs for nonparenteral administration, these properties can be altered; however, size is a factor that cannot be modified. As noted earlier for carzinostatin, solubility characteristics can be changed and, as discussed in the previous chapter, macromolecules can be cationized to promote cellular absorption.

The alimentary tract is lined with epithelial cells[67] that are part of the mucosa. The epithelial cells are polarized with the apical end abutting the lumen and the distal end abutting the basement membrane. Tight junctions and adherens junctions join the cells in the pore region. Alimentary epithelial cells are coated with mucus consisting of aqueous "solution" of polydisperse glycoproteins. The glycoproteins contain up to 80% carbohydrate with considerable asymmetry.[68,69] The mucins contribute to the formation of the unstirred layer, which protects the mucosa. The unstirred layer presents a challenge to the delivery of material to the mucosa.

Assuming that the macromolecule in question has traversed the mucus or unstirred layer, the options for further movement from the lumen to the circulation through the epithelia are either transcellular transport or paracellular transport. For all practical purposes, transcellular transport requires the presence of a specific receptor and processing pathway such as that for the processing of IgG.[70,71] Such a process does occur with Peyer's patches[65] which are involved in the process of mucosal immunity.[72]

There is a mucosal transport system for specific peptides.[73,74] The endocytotic process is a complex process in which different receptors utilize highly individualized endocytotic mechanisms.[75] There are also intestinal epithelial exosomes that function as a link between lumen antigens and the local immune system.[76–78]

Paracellular transport is a passive process which can also function in peptide drug absorption.[79] Transport via the paracellular pathway is limited by size. Hisada and coworkers[80] observed that the transport of hyaluronan in a Caco-2 monolayer system.[81–83] Hess and coworkers[84] observed that the transport of the members of a family of hexapeptides were transported by a paracellular process regardless of the structural (N-methylation or C-methylation) or conformational (cyclization) modifications. Transport velocity (Papp) was increased by cyclization due to decrease in molecular volume.

Peptide drug applications have included peptidomimetics (Figure 7.1)[79] and peptide-based prodrugs (Figure 7.2 and 7.3).[85,86] Regardless of whether there is a transcellular or paracellular pathway for absorption in the intestine, challenges remain[87,88] including exopeptidase and endopeptidase activity in the brush border membrane.[89,90] Pulmonary[91–93] and nasal[94,95] delivery of peptide drugs have also been areas of active study. Desmopressin (DDAVP, 1-deamino-8-arginine-vasopressin is a small (Mr 1069) peptide drug used for the treatment of mild von Willebrand disease[96] in adults and nocturanal enuresis in a pediatric population[97] that uses a nasal administration route. DDAVP is also used as a marker protein for epithelial transport.[98,99] Folkesson and colleagues[100] observed that both bovine serum albumin and DDAVP pass from the lung into the circulation in a pig model system. The amount of protein or peptide transports is inversely proportional to size. It is noteworthy that attempts to administer insulin via the pulmonary route have been unsuccessful.[93] Recent interest in peptide drug delivery is focused on the use of nanocarriers.[101–106]

The previous text shows that even small peptides have difficulty being absorbed from the mucosal to the serosal side of the epithelial cell layer. There are materials that can enhance absorption of peptides[107] but appear to be nonspecific,[108] in that the mechanism is the enhancement of paracellular absorption.[109–116] These absorption enhancers also function with proteins somewhat less effectively. A fragment, designated delta G, derived from zonula occuldens toxin (ZOT), is being developed to enhance the absorption of macromolecules[117–121] Zonula occludens toxin is described[119] as an effective absorption enhancer that reversibly opens tight junctions, thereby increasing intestinal permeability to hydrophilic and hydrophobic materials.

Morishita and Peppas[66] discussed the issues in oral protein drug delivery and suggested the following options: (1) modification of physicochemical properties; (2) addition of a novel function; and (3) improved carriers. Missing from this list is size, which would appear to be the major determinant,[122] and was also discussed earlier. Modification of physicochemical properties is discussed in Chapter 9, and addition of a novel function (formation of a bioconjugate) is discussed in Chapter 4. Neither of these approaches reduces protein size, and it is unlikely that the known specialized transport pathways for immunoglobulin[123] can be used for the transport of therapeutic macromolecules in the gastrointestinal tract. There is also the binding of antigen of M cells discussed earlier, but this does not appear to be useful for transport. It is possible that nanotechnology might be useful.[124,125] There has been recent

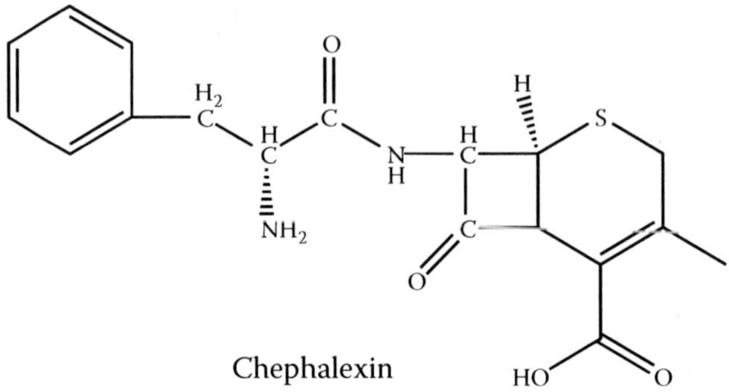

Chephalexin

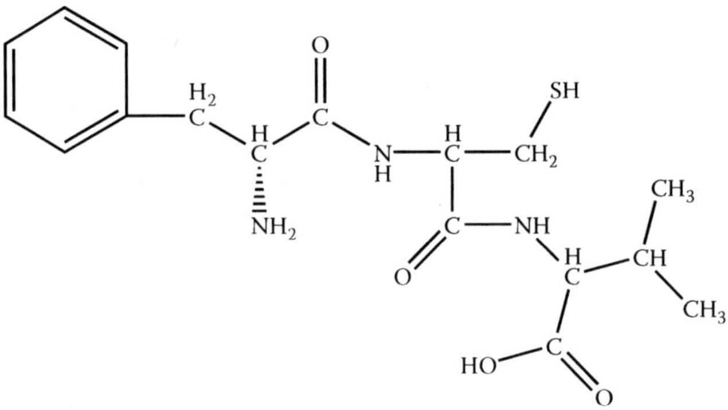

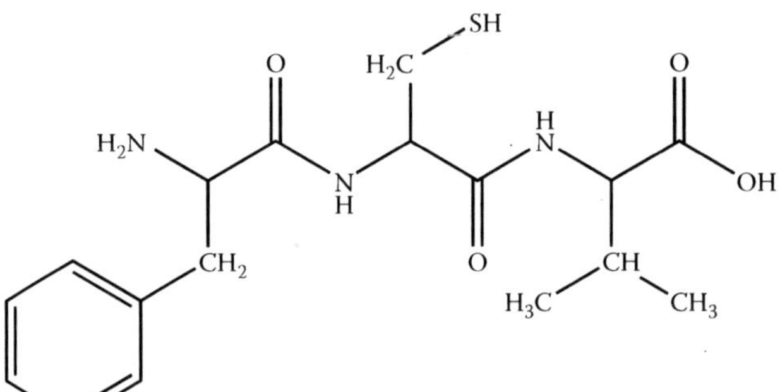

Phe-Cy-Val

FIGURE 7.1 An Illustration of Peptidomimetics. Peptidomimetics are derived from peptide structures where the peptide bond is replaced by an analogous linkage not susceptible to hydrolysis by peptidases or proteases. (See Yang, C.Y., Dantzig, A.H., and Pidgeon, C., Intestinal peptide transport systems and oral drug availability, *Pharm.Res.* 16, 1331–1343, 1999; Majumdar D., Alexander, M.D., and Coward, J.K., Synthesis of isopeptide epoxide peptidomimetrics, *J. Org. Chem.* 74, 617–627, 2009.)

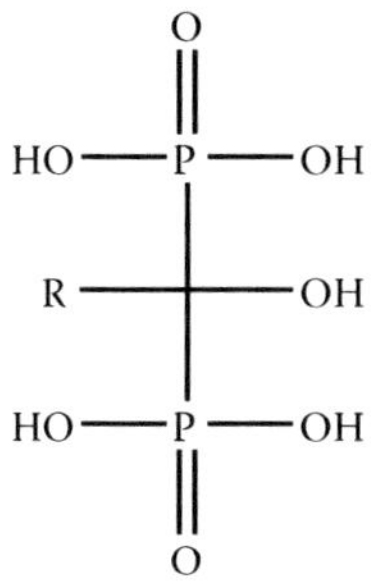

Parent bisphosphonate

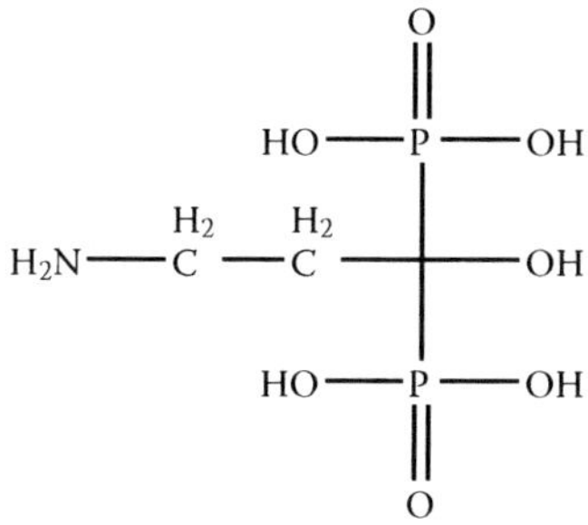

Pamidronate; panidronic acid, ADP

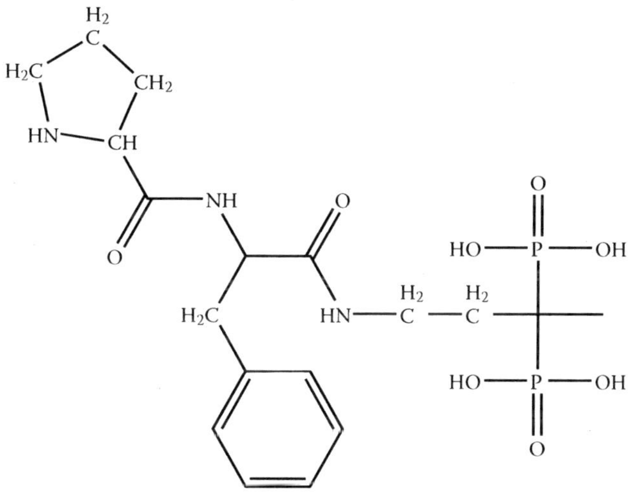

Pro-Phe-pamidronate

FIGURE 7.2 A Peptidyl Derivative of a Bisphosphonate Drug. The structure is based on Pamidronate (See Reitsma, P.H., Bijvoet, O.L., Frijlink, W.B., et al., Pharmacology of disodium (3-amino-1-hydroxypropylidene)-1,1-bisphosphonate, *Adv. Exp. Med. Biol.* 128, 219–227, 1980). (The peptidyl structure shown is derived from Ezra, A., Hoffman, A., Breuer, E., et al., A peptide prodrug approach for improving bisphosphonate oral absorption, *J. Med. Chem.* 43, 3641–3652, 2000.)

interest in the use of chitosan particles,[126–129] because chitosan has enhancing effects on mucosal absorption as well as mucoadhesive properties. There is also interest in the colon as a site of absorption.[130–136]

Our discussion suggests that, without the development of new paradigms, there is not much promise for the nonparenteral delivery of macromolecules.[137] That is not to say that work should cease on this problem. Rather, the results to date suggest that the most significant issue is size. If the size problems were solved with a mimetic,

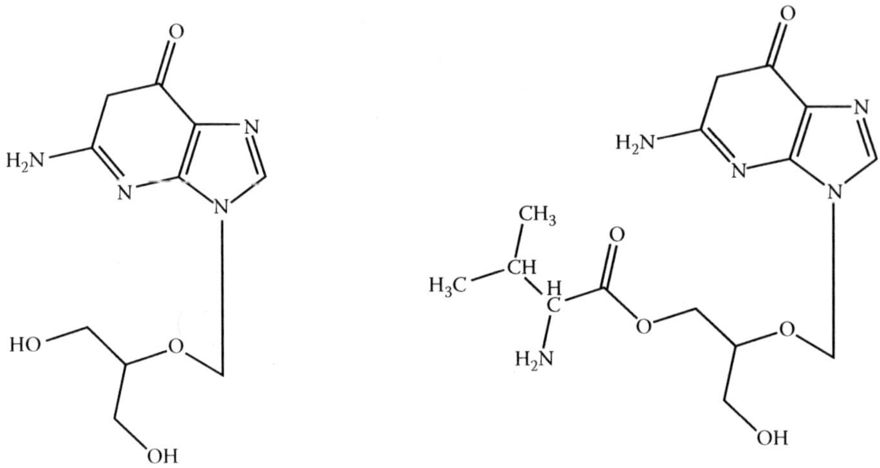

FIGURE 7.3 Structure of a drug (Ganciclover) and a Prodrug Derivative (Valganociclovir) (See Li, F., Maag, H., and Alfredson, T., Prodrugs of nucleoside analogues for improved oral absorption and tissue targeting, *J. Pharm. Sci.* 97, 1009–1134, 2008.)

then sufficient information would be available to design a vehicle that could deliver the drug in the colon. The use of an enhancer is a bit more problematic in that one does not wish to make the barrier permeable to all large molecules.

REFERENCES

Chapter References

1. Lundblad, R.L., Bradshaw, R.A., and Gabriel, D., A review of the therapeutic use of thrombin, *Thromb. Haemost.* 91, 851–860, 2004.
2. Mehta, R.C. and Fitzpatrick, R.E., Endogenous growth factors as cosmeceuticals, *Dermatol. Ther.* 20, 350–359, 2007.
3. Ma, Y., Zhao, H., and Zhou, X., Topical treatment with growth factors for tympanic membrane perforations: Progress towards clinical applications, *Acta Otolaryngnol.* 122, 586–599. 2002.
4. Braund, R., Hook, S, and Medlicott, N.J., The role of topical growth factors in chronic wounds, *Curr. Drug. Deliv.* 4, 195–204, 2007.
5. Morrow, C.D., Novak, M.J., and Ansardi, D.C. et al., Recombinant viruses as vectors for mucosal immunity, *Curr. Top. Microbiol. Immunol.* 236, 255–273. 1999.
6. Crotty, S. and Andino, R., Poliovirus vaccine strains as mucosal vaccine vectors and their potential use to develop an AIDS vaccine, *Adv. Drug Deliv. Rev.* 56, 835–852, 2004.
7. Vajdy, M. and Singh, M. Intranasal delivery of vaccines against HIV, *Expert Opin. Drug Deliv.* 3, 247–259, 2006.
8. Harrington, K.J., Spitzweg, C., Bateman, A.R. et al., Gene therapy for prostate cancer: Current status and future prospects, *J. Urol.* 166, 1220–1233, 2001.
9. Akporiaye, E.T. and Hersh, E., Clinical aspects of intratumoral gene therapy, *Curr. Opin. Mol. Ther.* 1, 443–453, 1999.

10. Harrington, K., Alvarez-Vallina, L., Crittenden, M. et al., Cells as vehicles for cancer gene therapy: The missing link between targeted vectors and systemic delivery, *Hum. Gene Ther.* 13, 1263–1280, 2002.
11. Tenenbaum, L., Chtarto, A., Lehtonen, E. et al., Neuroprotective gene therapy for Parkinson's disease, *Curr. Gene Ther.* 2, 451–483, 2002.
12. Schlachetzki, F., Zhang, Y., Boado, R.J., and Pardridge, W.M., Gene therapy of the brain: The transvascular-approach, *Neurology* 62, 1275–1281, 2004.
13. Wang, Y. and Yuan, F., Delivery of viral vectors to tumor cells: Extracellular transport, systemic distribution, and strategies for improvement, *Ann. Biomed. Eng.* 34, 114–127. 2006.
14. Ochiya, T., Nagahara, S., Sano, A. et al., Biomaterials for gene therapy: Atelocollagen-mediated controlled release of molecular medicines, *Curr. Gene Ther.* 1, 31–52, 1991.
15. Kuniyasu, H., Hirose, Y., Ochi, M. et al., Bone augmentation using rhGDF-5-collagen composites, *Clin. Oral. Implants Res.* 14, 490–499, 2003.
16. Matarasso, S.L., The use of injectable collagens for aesthetic rejuventation, *Semin. Cutan. Med. Surg.* 25, 151–157, 2006.
17. Poynton, A.R. and Lane, J.M., Safety profile for the clinical use of bone morphogenetic proteins in the spine, *Spine* 27(16 suppl 1), S40–S48, 2002.
18. Kim, J., Kim, I.S., Cho, T.H. et al., Bone regeneration using hyaluronic acid-based hydrogel with bone morphogenic protein-2 and human mesenchymal stem cells, *Biomaterials* 28, 1830, 2007.
19. Kostrzewa, R.M. and Segura-Aguilar, J., Botulinum neurotoxin: Evolution from poison, to research tool—onto medical therapeutic and future pharmaceutical panacea, *Neurotox. Res.* 12, 275–290, 2007.
20. Eley, R.C., Green, A.A., and McKhann, C.F., The use of a blood coagulant extract from the human placenta in the treatment of hemophilia, *J. Pediat.* 8, 135–147, 1936.
21. Bendien, W.M. and van Crevald, S., Investigations on hemophilia, *J. Dis. Children* 54, 713–725, 1937.
22. Schill, W.B., Improvement of sperm motility in patients with asthenozoospermia by kallikrein treatment, *Int. J. Fertil.* 20, 61–63, 1975.
23. Schill, W.B., Krizic, A., and Rjork, H., Determination of various serum parameters and sex hormone levels in subfertile men during kallikrein therapy, *Adv. Exp. Med. Biol.* 120A, 537–546, 1979.
24. Mittelbach, E. and Nümberger, F., Peroral kallikrein therapy for male infertility—comparison between smokers and non-smokers, *Andrologia* 15, Special. No. 515–522, 1983.
25. Saiteh, S., Kunamoto, Y., Shimamoto, K., and Iimura, O., Kallikrein in the male reproductive system, *Arch. Androl.* 19, 133–147, 1987.
26. Glezerman, M., Lunenfeld, E., Potashnik, G. et al., Efficacy of kallikrein in the treatment of oligozoospermia and asthenozoospermia, *Fertil. Steril.* 60, 1052–1056, 1993.
27. Overlack, A. Stumpe, K.O., Ressel, C., and Krück, F., Low urinary kallikrein excretion and elevated blood pressure normalized by orally kallikrein in essential hypertension, *Clin. Sci.* 57(suppl 5), 263s–265s, 1979.
28. Overlook, A., Stumpe, K.O., Ressel, C. et al., Decreased urinary kallikrein activity and elevated blood pressure normalized by orally applied kallikrein in essential hypertension, *Klin. Wochenschr.* 58, 37–42, 1980.
29. Bönner, G., Toussaint, C., and Claus, M. et al., Lack of oral kallikrein in lowering systemic blood pressure in primary hypertension, *Agents Action Suppl.* 38, 294–303, 1992.
30. Bläckberg, M. and Ohlsson, K., Turnover of ^{125}I-labelled tissue kallikrein following intraduodenal or intravenous administration, *Scand. J. Clin. Lab. Invest.* 61, 57–67, 2001.
31. Rogers, T.M., Management of gastric hemorrhage using topical thrombin, *J. Am. Med. Assoc.* 137, 1035–1036, 1948.
32. Daly, B.M., Use of buffer thrombin in the treatment of gastric hemorrhage, *Arch. Surg.* 55, 208–212, 1947.

33. Edmunds, V., Oral thrombosis in the treatment of haematemesis, *Brit. Med. J.* 1(4824), 1371–1372, 1953.
34. Kelly, C.P., Chetham, S., Keates, S. et al., Survival of anti-*Clostridium difficile* bovine immunoglobulin concentrate in the human gastrointestinal tract, *Antimicrob. Agents Chemother.* 41, 236–241, 1997.
35. Warny, M., Fatimi, A., Bostwick, E.F. et al., Bovine immunoglobulin concentrate *Clostridium difficile* retains *C. difficile* toxin neutralizing activity after passage through the human stomach and small intestine, *Gut* 44, 212–217, 1999.
36. Rodriquez, C., Blanch, F., Romano, V. et al., Porcine immunoglobulins survival in the intestinal tract of adult dogs and cats fed dry food kibbles containing spray-dried porcine plasma (SDPP) or porcine immunoglobulin concentrate (PIC), *Animal Feed Sci. Technol.* 139, 201–211, 2007.
37. Murlin, J.R., Gibbs, C.B., Romansky, M.J. et al., Effectiveness of pre-oral insulin in human diabetes, *J. Clin. Invest.* 19, 709–722, 1940.
38. Arbit, E., The physiologic rationale for oral insulin administration, *Diabetes Technol. Ther.* 6, 510–517, 2004.
39. Sukhotnik, I., Shehadeh, N., Mogilner, J. et al., Beneficial effects of oral insulin on intestinal recovery following ischemia-reperfusion injury in rat, *J. Surg. Res.* 128, 108–113. 2205.
40. Skyler, J.S., Krischer, J.P., Wolfsdorf, J. et al., Effect of oral insulin in relatives of patients with type 1 diabetes,: The Diabetes Prevention Trial—Type 1, *Diabetes Care* 28, 1068–1076, 2005.
41. Babu, V.R., Patel, P., Mundargi, R.C. et al., Developments in polymeric devices for oral insulin delivery, *Expert Opin. Drug. Deliv.* 5, 403–415, 2008.
42. Butty, V., Campbell, C., Mathis, D. et al., Impact of diabetes susceptibility loci on progression from pre-diabetes to diabetes in at-risk individuals of the DPT1 trial, *Diabetes*, 57(9): 2348–2359, Sept., 2008.
43. Jachimska, B., Wasilewska, M., and Adamczyk, Z., Characterization of globular protein solutions by dynamic light scattering, electrophoretic mobility, and viscosity measurements, *Langmuir* 24, 6866–6872, 2008.
44. Maeda, H., Kumagi, K., and Ishida, N., Characterization of neocarzinostatin, *J. Antibiot.* 19, 253–259, 1966.
45. Toriyama, K., Fujita, H., and Ishida, N., Absorption, distribution, and excretion of neocarzinostatin (NCS) in mice after oral administration, *J. Antibiot.* 28, 611–621, 1972.
46. Oka, K., Miyamoto, Y., Matsumura, Y. et al., Enhanced intestinal absorption of a hydrophobic polymer-conjugated protein drug, SMANCS, in an oily formulations, *Pharm. Res.* 7, 852–855, 1990.
47. Schmidtt, D.A., Kisanuki, K., Kimura, S. et al., Antitumor activity of orally administered SMANCS, a polymer-conjugated protein drug in mice bearing various murine tumors, *Anticancer Res.* 12, 2219–2224, 1992.
48. Suzuki, F., Matsumoto, K., Schmidtt, D.A. et al., Immunomodulating activity of orally administered SMANCS, a polymer-conjugated derivatives of the proteinaceous antibiotic neocarzinostatin in an oily formulation, *Int. J. Immunopharmacol.* 15, 175–183, 1996.
49. *Topical Drug Bioavailability, Bioequivalence, and Penetration*, Eds. V.P. Shah and H.I. Maibach, Plenum Press, New York, 1993.
50. *Aulton's Pharmaceutics. The Design and Manufacture of Medicines*, Ed. M.E. Aulton, Churchill Livingstone/Elsevier, Edinburgh, Scotland, U.K., 2007.
51. *Theory and Practice of Contemporary Pharmaceutics*, Eds. T.K. Ghosh and B.P. Jasti, CRC Press, Boca Raton, FL, 2006.
52. *Skin Delivery Systems. Transdermal, Dermatological and Cosmetic Actives,* Ed. J. Wille, Blackwell, Ames, IA, 2006.

53. Banga, A., *Electrically Assisted Transdermal and Topical Drug Delivery,* Taylor and Francis, London, 1998.
54. *Mechanisms of Transdermal Drug Delivery*, Eds. R.O. Potts and R.H. Guy, Marcel Dekker, New York, 1997.
55. Crommelin, D.J.A., van Winden, E., and Mekking, A., Delivery of pharmaceutical proteins, in *Aulton's Pharmaceutics*, Ed. M.E. Aulton, Churchill Livingstone/Elsevier, Edinburgh, Scotland, UK, 2007.
56. Rivière, G., Choumet, V., Saliou, B. et al., Absorption and elimination of viper venom after antivenom administration, *J. Pharmacol. Exp. Ther.* 285, 490–495, 1998.
57. Bocci, V., Pessina, G.P., Nicoletti, C., and Paulesu, L., The lymphatic route. VII. Distribution of recombinant human interleukin-2 in rabbit plasma and lymph, *J. Biol. Regul. Homeost. Agents* 4, 25–29, 1990.
58. Porter, C.J., Edwards, G.A., and Charman, S.A., Lymphatic transport of proteins after s.c injection: Implications of animal model selection, *Adv. Drug Deliv. Rev.* 50, 157–171, 2001.
59. Kagan, L., Gershkovich, P., Mendelman, A. et al., The role of the lymphatic system in subcutaneous absorption of macromolecules in the rat model, *Eur. J. Pharm. Biopharm.* 67, 759–765, 2007.
60. Kota, J., Machavaram, K.K., McLennan, D.N. et al., Lymphatic absorption of subcutaneously administered proteins: Influence of different injection sites on the adsorption of darbepoetin alpha using a sheep model, *Drug. Metab. Dispos.* 35, 2211–2217, 2007.
61. Anderson, P.M. and Sorenson, M.A., Effects of route and formulation on clinical pharmacokinetics of interleukin-2, *Clin. Pharmacokinet.* 27, 19–31, 1994.
62. Saks, S. and Rosenblum, M., Recombinant human TNF-α: Preclinical studies and results from early clinical trials, *Immunol. Ser.* 56, 567–587, 1992.
63. Fix, J.A., Oral controlled release technology for peptides: Status and future prospects, *Pharm. Res.* 13, 1760–1764, 1991.
64. Laine, M.E. and Corrigan, O.I., Paracellular and transcellular pathways facilitate insulin permeability in rat gut, *J. Pharm. Pharamcol.* 58, 271–275, 2006.
65. Liang, E., Kabcenell, A.K., Coleman, J.R. et al., Permeability measurement of macromolecules and assessment of mucosal antigen sampling using *in vitro* converted M cells, *J. Pharmacol. Toxicol. Methods* 46, 93–101, 2002.
66. Morishita, M. and Peppas, N.A., Is the oral route possible for peptide and protein drug delivery?, *Drug Discov. Today* 11, 905–911, 2006.
67. Madara, J.L. and Anderson, J.M., Epithelia: Biologic principles of organization, in *Textbook of Gastroenterology*, 4th edn., Eds. T. Yamada, D.H. Alpers, N. Kaplowitz, L. Laine, C. Owyang, and D.W. Powell, Chapter 8, pp. 151–165, Lippincott, Philadelphia, PA, 2005.
68. Sheehan, J.K. and Carlstedt, I., Hydrodynamic properties of human cervical-mucus glycoproteins in 6M-guanidinium chloride, *Biochem. J.* 217, 93–101, 1984.
69. Sheehan, J.K. and Carlstedt, I., The effect of guanidinium chloride on the behaviour of human cervical-mucus glycoproteins. Evidence for unfolding regions of ordered structure in 6M-guanidinium chloride, *Biochem. J.* 221, 499–504, 1984.
70. Ghetie, V. and Ward, E.S., Transcytosis and catabolism of antibody, *Immunol. Res.* 25, 97–113, 2002.
71. Lencer, W.I. and Blumberg, R.S., A passionate kiss, then run: Exocytosis and recycling of IgG by FcR_n, *Trends Cell Biol.* 15, 5–9, 2005.
72. *Mucosal Vaccines*, Ed. H, Kiyono, P.L. Ogra, and J.R. McGhee, Academic Press, San Diego, CA, 1996.
73. Bai, J.P.F. and Amidon, G.L., Structural specificity of mucosal-cell transport and metabolism of peptide drugs. Implications for oral peptide drug delivery, *Pharm. Res.* 9 969–978, 1992.

74. Walter, E., Kissel, T., and Amidon, G.L., The intestinal peptide carrier: A potential transport system for small peptide-derived drugs, *Adv. Drug Deliv. Rev.* 20, 33–58, 1996.
75. Perret, E., Lakkaraju, A., Deborde, S. et al, Evolving endosomes: How many varieties and why?, *Curr. Opin. Cell Biol.* 17, 423–434, 2005.
76. Mallegol, J., van Niel, G., and Heyman, M., Phenotypic and functional characterization of intestinal epithelial exosomes, *Blood Cells Mol. Dis.* 35, 11–16, 2005.
77. Lin, X.P., Almqvist, N., and Telemo, E., Human small intestinal epithelial cells constitutively express the key elements for antigen processing and the production of exosomes, *Blood Cells Mol. Dis.* 35, 122–128, 2005.
78. Hunderfean, G., Zimmer, K.P., Strobel, S. et al., Luminal antigens access late endosomes of intestinal epithelial cells enriched in MHC I and MHC II molecules: *In vivo* study in Crohn's ileitis, *Am. J. Physiol. Gastrointest. Liver Physiol.* 293, 798–808, 2007.
79. Yang, C.Y., Dentzig, A.H., and Pidgeon, C., Intestinal peptide transport systems and oral drug availability, *Pharm. Res.* 16, 1331–1343, 1999.
80. Hisada, N., Satsu, H., Mori, A. et al., Low-molecular-weight hyaluronan permeates through human intestinal Caco-2 cell monolayers via the paracellular pathway, *Biosci. Biotechnol. Biochem.* 72, 1111–1114, 2008.
81. Ingels, F.M. and Augustijns, P.F., Biological, pharmaceutical, and analytical considerations with respect to the transport media used in the absorption screening system, Caco-2, *J. Pharm. Sci.* 92, 1545–1558, 2003.
82. van Breeman, R.B. and Li, Y., Caco-2 cell permeability assays to measure drug absorption, *Expert Opin. Drug Metab. Toxicol.* 1, 175–185, 2005.
83. Sambuy, Y., De Angelis, I., Ranaldi, G. et al., The Caco-2 cell line as a model of the intestinal barrier: Influence of cell and culture-related factors on Caco-2 cell functional characteristics, *Cell. Biol. Toxicol.* 21, 1–26, 2005.
84. Hess, S., Ovadia, O., Shaley, D.E. et al., Effect of structural and conformational modifications including backbone cyclization, of hydrophobic hexapeptides on their intestinal permeability and enzymatic stability, *J. Med. Chem.* 50, 6201–6211, 2007.
85. Ezra, A., Hoffman, A., Breuer, E. et al., A peptide prodrug approach for improving bisophosphonate oral absorption, *J. Med. Chem.* 43, 3641–3652, 2000.
86. Li, F., Maag, H., and Alfredson, I., Prodrugs of nucleotide analogues for improved oral absorption and tissue targeting, *J. Pharm. Sci.* 97, 1109–1124, 2008.
87. Soltero, R. and Erwurikbe, N., The oral delivery of protein and peptide drugs, *Innov. Pharmacol. Technol.* 01, 106–110, 2001.
88. Hamman, J.H., Enstin, G.M., and Kotze, A.F., Oral delivery of peptide drugs: Barriers and developments, *Biodrugs* 19, 166–177, 2005.
89. Sterchi, E.E. and Woodley, J.F., Peptide hydrolases of the human small intestinal mucosa: Distributions of activities between brush border membranes and cytosol, *Clin. Chim. Acta* 102, 49–56, 1980.
90. Tobey, N., Herzer, W., Yeh, R. et al, Human intestinal brush border peptidases, *Gastroenterology* 88, 913–926, 1985.
91. Yu, J. and Chien, Y.W., Pulmonary drug delivery: Physiologic and mechanistic aspsects, *Crit.* 68. Thwaites, D.T., Hirst, B.H., and Simmons, N.L., Passive transepithelial absorption of thyrotropin-releasing hormone (TRH) via a paracellular route in cultured intestinal and renal epithelial cell lines, *Pharm. Res.* 10, 674–681, 1993.
92. Strack, T., Pulmonary administration of peptide and protein drugs, *Am. Pharm. Res.* 10, 129–133, 2007.
93. Levy, W., Peptide progress, *Pharm. Formulation Quality* 10(3), 28–32, June, 2008.
94. Hussain, M.A., Ashok, B., and Rove, S.M., The use of α-aminoboronic acid derivatives to stabilize peptide drugs during their intranasal absorption, *Pharm. Res.* 6, 186–189, 1989.
95. Morita, T. and Yamahara, H., Nasal delivery systems for peptide drugs: Market trends and technology development, *Drug. Deliv. Systems* 21, 425–434, 2006.

96. Wilde, J.T., Von Willebrand disease, *Clin. Med.* 7, 629–632, 2007.
97. Weaver, A. and Dobson, F., Nocturnal enuresis in children, *J. Fam. Health Care* 17, 159–161, 2007.
98. Pantzar, N., Lundin, S., Wester, L., and Westroem, B.R., Bidirectional small-intestinal permeability in the rat to some common marker molecules *in vitro*, *Scand. J. Gastroenterol.* 29, 703–709, 1994.
99. Li,L., Mathias, N.R., Heran, C. et al., Carbopol-mediated paracellular transport enhancement in Calv-3 cell layers, *J. Pharm. Sci.* 95, 326–335, 2005.
100. Folkesson, H.G., Westroem, B.R., Pierzynowski, S.G. et al., Lung to blood passage of albumin and a nonapeptide after intratracheal instillation in the young developing pig, *Acta Physiol. Scand.* 147, 173–178, 1993.
101. Prego, C., Garcia, M., Torres, D., and Alonso, M.J., Transmucosal macromolecular drug delivery, *J. Conrol. Release* 101, 151–162, 2005.
102. Martins, S., Sarmento, B., Ferriera, D.C., and Souto, E.B., Lipid-based colloidal carriers for peptide and protein delivery—liposomes versus lipid nanoparticles, *Int. J. Nanomedicine* 2, 595–607, 2007.
103. Damgé, C., Reis, C.P., and Maincent, F., Nanoparticle strategies for the oral delivery of insulin, *Expert Opin. Drug Deliv.* 5, 45–68, 2008.
104. Shayele, S.A., Engineering protein particles for pulmonary drug delivery, *Methods Mol. Biol.* 437, 149–160, 2008.
105. Rytting, E., Nguyen, J., Wang, X., and Kissel, T., Biodegradable polymeric nanoparticles for pulmonary drug delivery, *Expert. Opin. Drug. Deliv.* 5, 629–639, 2008.
106. Qian, F., Cui, F., Ding, J. et al., Chitosan graft copolymer nanoparticles for oral protein drug delivery: Preparation and characterization, *Biomacromolecules* 7, 2722–2727, 2006.
107. Chao, A.C., Nguyen, J.V., Broughall, M. et al., *In vitro* and *in vivo* evaluation of sodium caprate on enteral peptide absorption and on mucosal morphology, *Int. J. Pharm.* 191, 15–24, 1999.
108. Kondohn, M. and Yagi, K., Progress in absorption enhancers based on tight junction, *Expert Opin. Drug Deliv.* 4, 275–286, 2007.
109. Lane, M.E. and Corrigan, O.I., Paracellular and transcellular pathways facilitate insulin permeability in rat gut, *J. Pharm. Pharmacol.* 58, 271–275, 2006.
110. Salama, N.N., Eddington, A., Fuhii, M. et al., Tight junction modulation and its relationship to drug delivery, *Adv. Drug Deliv. Rev.* 58, 15–28, 2006.
111. Kondoh, M., Takahashi, A., Fujii, M. et al., A novel strategy for a drug delivery system using a claudin modulator, *Biol. Pharm. Bull.* 29, 1783–1789, 2006.
112. Hayashi, M. and Tomita, M., Mechanistic analysis for drug permeation through intestinal membrane, *Drug Metab. Pharmacokinet.* 22, 67–77, 2007.
113. Zhang, Z.N., Xu, J., Tang, L.H. et al., Influence on intestinal mucous permeation of paclitaxel of absorption enhancers and dosage forms based on electron spin resonance spectroscopy, *Pharmaie* 62, 368–371, 2007.
114. Loftsson, T., Vogensen, S.B., Brewster, M.E., and Konrádsdóttir, F., Effects of cyclodextrins on drug delivery through biological membranes, *J. Pharm. Sci.* 96, 2532–2546, 2007.
115. Shah, F., Jogani, V., Mishra, P. et al., Modulation of ganciclovir intestinal absorption in presence of absorption enhancers, *J. Pharm. Sci.* 96, 2710–2722, 2007.
116. Gao, Y., He, L., Katsumi, H. et al., Improvement of intestinal absorption of water-soluble macromolecules by various polyamines: Intestinal mucosal toxicity and absorption-enhancing mechanism of spermine, *Int. J. Pharm.* 354, 126–134, 2008.
117. Salama, N.N., Fasano, A., Lu. R., and Eddington, N.D., Effect of the biologically active fragment of *Zonula occludens* toxin, delta G, on the intestinal paracellular transport and oral absorption of mannitol, *Int. J. Pharm.* 251, 113–121, 2003.
118. Salama, N.N., Fasano, A., Thakar, M., and Eddington, N.D., The effect of delta G on the transport and oral absorption of macromolecules, *J. Pharm. Sci.* 93, 1310–1319, 2004.

119. Salama, N.N., Fasano, A., Thakar, M., and Eddington, N.D., The impact of Δ G on the oral bioavailability of low bioavailable therapeutic agents, *J. Pharmacol. Exp. Ther.* 312, 199–205, 2005.
120. Song, K.H., Fasano, A., and Eddington, N.D., Enhanced nasal absorption of hydrophilic markers after dosing with AT1002, a tight junction modulator, *Eur. J. Pharm. Biopharm.* 69, 231–237. 2008.
121. Song, K.H., Fasano, A., and Eddington, N.D., Effect of the six-mer synthetic peptide (AT1002) fragment of *Zonula occludens* toxin on the intestinal absorption of cyclosporin A, *Int. J. Pharm.* 351, 8–14, 2008.
122. Lennernäs, H., Intestinal permeability and its relevance for absorption and elimination, *Xenobiotica* 37, 1015–1051, 2002.
123. Yoshida, M., Claypool, S.M., Wagner, J.S. et al., Human neonatal Fc mediates transport of IgG into luminal secretions for delivery of antigens in mucosal dendritic cells, *Immunity* 20, 769–783, 2004.
124. Behrens, I., Pena, A.I., Alonso, M.J., and Kissel, T., Comparative uptake studies of bioadhesive and non-bioadhesive nanoparticles in human intestinal cell lines and rats: The effect of mucus on particle adsorption and transport, *Pharm. Res.* 19, 1185–1193, 2002.
125. Pinto-Alphandary, H., Aboubakar, M., Jaillard, D. et al., Visualization of insulin-loaded nanocapsules: *In vitro* and *in vivo* studies after oral administration to cats, *Pharm. Res.* 20, 1071–1084, 2003.
126. Prego, C., Torres, D., and Alonso, M.J., Chitosan nanocapsules as carriers for oral peptide delivery: Effect of chitosan molecular weight and type of salt of the *in vitro* behaviour and *in vivo* effectiveness, *J. Nanosci. Nanotechnol.* 6, 2921–2928, 2006.
127. Sandri, C., Bonferoni, M.C., Rossi, S. et al., Nanoparticles based on *N*-trimethylchitosan: Evaluation of absorption properties using *in vitro* (Caco-2 cells) and ex vivo (excised rat jejunum) models, *Eur. J. Pharm. Biopharm.* 65, 68–77, 2007.
128. Bonferoni, M.C., Sandri, G., Rossi, S. et al., Chitosan citrate as multifunctional polymer for vaginal delivery. Evaluation of penetration enhancement and peptide inhibition properties, *Eur. J. Pharm. Sci.* 33, 166–176., 2008.
129. Sadeghi, A.M., Dorkoosh, F.A., Avadi, M.R. et al., Permeation enhancer effect of chitosan and chitosan derivatives: Comparison of formulations as soluble polymers and nanoparticulate systems on insulin absorption in Caco-2 cells, *Eur. J. Pharm. Biopharm.* in press, 2008.
130. Hu, Z., Mawatari, S., Shimokawa, T. et al., Colon delivery efficiencies of intestinal pressure-controlled colon delivery capsules prepared by a coating machine in human subjects, *J. Pharm. Pharmacol.* 52, 1187–1193, 2000.
131. Yang, L., Chu, J.S., and Fix, J.A., Colon-specific drug delivery: New approaches and *in vitro/in vivo* evaluation, *Int. J. Pharm.* 235, 1–15, 2002.
132. Haupt, S. and Rubinstein, A., The colon as a possible target for orally administered peptide and protein drugs, *Crit. Rev. Ther. Drug Carrier Syst.* 19, 499–551, 2002.
133. Basit, A.W., Advances in colonic drug delivery, *Drugs* 65, 1991–2007, 2005.
134. Kosarju, S., Colon targeted delivery systems: Review of polysaccharides for encapsulation and delivery, *Crit. Rev. Food Sci. Nutr.* 45, 251–258, 2005.
135. Patel, M., Shah, T., and Amin, A., Therapeutic opportunities in colon-specific drug-delivery systems, *Crit. Rev. Ther. Drug Carrier Syst.* 24, 147–202, 2007.
136. Malik, D.K, Baboota, S., Ahuja, A. et al., Recent advances in protein and peptide drug delivery systems, *Curr. Drug Deliv.* 4, 141–151, 2007.
137. Brown, L.R., Commercial challenges of protein drug delivery, *Expert Opin. Drug Deliv.* 2, 29–42, 2007.

8 Application of Solution Protein Chemistry to Proteomics

The purpose of this chapter is to address issues of the applications of solution protein chemistry to proteomics that are not discussed in detail by others and to avoid, when possible, redundance in the coverage of information that is discussed in considerable detail in other sources.[1–7]

Traditional solution protein chemistry,[8–12] including spectroscopy,[13,14] ultracentrifugation,[15–17] and chemical modification, is the basis for proteomics. The chemical modification of proteins has extensive use in proteomics in several different categories. Chemical proteomics could be described as the application of "classical" solution chemistry, including the use of alkylating agents such as isotope-coded affinity tags and fluorescent probes in quantitative proteomics.[18–25]

Chemiproteomics, which on the surface would appear to be a closely related concept to proteomics, has been defined as an approach using small molecules as affinity materials for the discovery of small-molecule binding proteins.[26] This approach is best characterized by the use of affinity chromatography.[27–30] There has been another publication[31] on chemiproteomics suggesting that it has not gathered traction as a descriptor.

Functional proteomics can be defined as changes in protein expression during challenge or differentiation as measured by changes in activity, concentration, or interaction.[32–39] Techniques include modification of enzymes *in situ* and measurement of interaction by affinity methods.[40–42]

Quantitative proteomics refers to the use of the aforementioned alkylating agents and fluorescent dyes for the labeling of proteins and peptides prior to analysis,[43,44] permitting the measurement of differences in protein expression in response to system challenges.[45,46] Quantitative proteomics has also been described as comparative proteomics.[47]

Activity-based proteomics is an approach within functional proteomics that has been used to describe methods for the identification and measurement of enzymes *in situ* in cell and biological fluids.[48–51] Activity-based proteomics is based on the modification of proteins (usually enzymes) at specific binding sites (usually enzyme-active sites). Modification of proteins at specific binding sites is mostly driven by selective binding as opposed to functional group reactivity, although functional groups at enzyme-active site are usually more reactive than the same group outside the enzyme active site.

Activity-based proteomics uses the various technologies developed for affinity labeling of enzymes,[52–57] suicide enzyme inhibitors,[58–62] and enzyme histochemistry.[63–66]

Photoaffinity reagents (Figure 8.1) have been quite useful in the study of sites in proteins where binding of reagent is dominant over site reactivity.[67–73]

Several approaches developed for the study of serine proteases have been subsequently modified for use in proteomics. One approach has promoted derivatives of diisopropylphosphorofluoridate (DFP) that is one of the earliest described inhibitors of serine proteases.[74–76] Alkylphosphorofluoridates react with the serine residues at the active site of enzymes such as trypsin and chymotrypsin resulting in phosphorylation and inactivation of the enzyme (Figure 8.2). There are some mechanistic similarities to active-site titration[77] and the reaction of proteases with protease inhibitors.[78] Reaction also occurs with enzymes other than proteases, including esterases such as acetylcholine esterase.[79]

Cravatt and colleagues[80] have prepared several fluorophosphate derivatives (Figure 8.3) for use in active-based proteomics. One derivative [fluorophosphonyl-biotin; 10-(fluoroethoxyphosphinyl)-*N*-(biotinamidopentyl) decanamide] has been used to isolate modified proteins, whereas another derivative, labeling with fluorescein instead of biotin, has been used for intracellular localization of serine proteases. The FP-biotin derivative [10-(fluoroethoxyphosphinyl)-*N*-(biotinamidopentyl) decanamide] was reacted with either purified proteins or tissue homogenates in 50 mM Tris-0.32 M sucrose, pH 8.0. Samples were subjected to electrophoresis. Detection of the modified proteins was accomplished by a Western blot technique using avidin-horseradish peroxidase with chemiluminescence reagents. Nonspecific binding was determined with heat-treated samples. Under these conditions it is reasonable to assume that the fluorophosphorate derivatives were reacting with serine residues at enzyme active sites. Examination of different tissues suggested that this technique could evaluate the tissue-dependent expression of serine proteases. In a subsequent study,[81] this group developed a FP-biotin derivative containing a more hydrophilic linker region (FP-Peg-Biotin) (Figure 8.4). A comparison of the two derivatives was obtained with rat testis, and a similar pattern of reaction was observed; however, there were some differences in the rates of reaction with the two derivatives, and it is suggested that the multiplexing with the two FP derivatives would be the most useful approach. By the combined use of the two reagents, it was possible to identify a number of serine peptidases, esterase, and lipases, including fatty acid amide hydrolase. A fluorophosphate probe [fluorophosphonate-poly(ethylene)glycol-(6-carboxyltetramethylrhodamine)][82] was used to demonstrate that Orlistat® was an inhibitor of fatty acid synthase and had antitumor activity (Figure 8.5). Orlistat is a β-lactone developed as an antiobesity drug. This study reported that Orlistat blocked the reaction of the fluorophosphate probe with fatty synthase in prostate tumor cell extracts. Subsequent studies with a cell-permeable fluorophosphate demonstrated that Orlistat also blocked the reaction with fatty acid synthase in intact cells. Although alkylphosphorofluoridate derivatives are reasonably specific for the modification of serine at enzyme active sites, reaction at other residues has been reported. There are also previous studies on the use of radiolabeled DFP for the identification of serine esterases in tissue homogenates.[83] The derivative formed is reasonably stable but as it is analogous to an acyl-enzyme intermediate,[84–87] the inactivated enzymes can undergo a slow reactivation with loss of the label (Figure 8.2). The reactivation occurs more rapidly at acid pH (phthalate buffer,

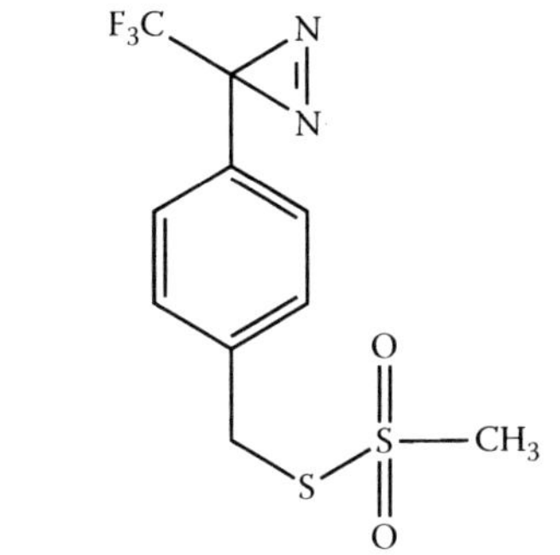

Diazirine methylthiosulfate derivative (1)

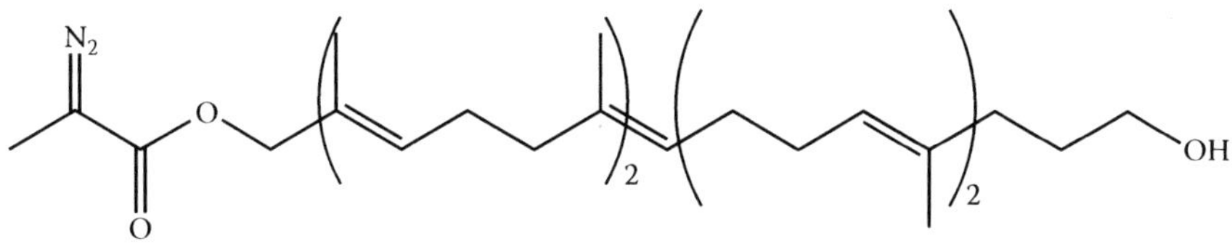

Trinorsqualine alcohol, diazo derivative (2)

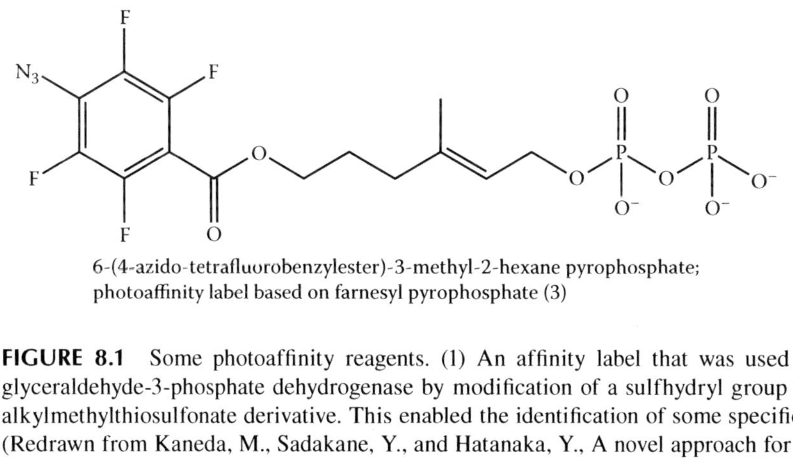

6-(4-azido-tetrafluorobenzylester)-3-methyl-2-hexane pyrophosphate; photoaffinity label based on farnesyl pyrophosphate (3)

FIGURE 8.1 Some photoaffinity reagents. (1) An affinity label that was used to label glyceraldehyde-3-phosphate dehydrogenase by modification of a sulfhydryl group with the alkylmethylthiosulfonate derivative. This enabled the identification of some specific ligand. (Redrawn from Kaneda, M., Sadakane, Y., and Hatanaka, Y., A novel approach for affinity-based screening of target specific ligands: Application of photoreactive D-glyceraldehyde-3-phosphate dehydrogenase, *Bioconjug. Chem.* 14, 849–852, 2003.) (2) An alkyl diazo derivative for labeling the substrate binding site of squalene epoxidase. (Redrawn from Lee, H.-K., Zheng, Y.F., Ziao, X.-Y. et al., Photoaffinity labeling identifies the substrate-binding site of mammalian squalene epoxidase, *Biochem. Biophys. Res. Commun.* 315, 1–9, 2004.) (3) A photoaffinity analog of farnesyl pyrophosphate. (Redrawn from Chehade, K.A.H., Kiegiel, K., Isaacs, R.J. et al., Photoaffinity analogues of farnesyl pyrophosphate transferable by protein farnesyl transferase, *J. Am. Chem. Soc.* 124, 8206–8219, 2002.)

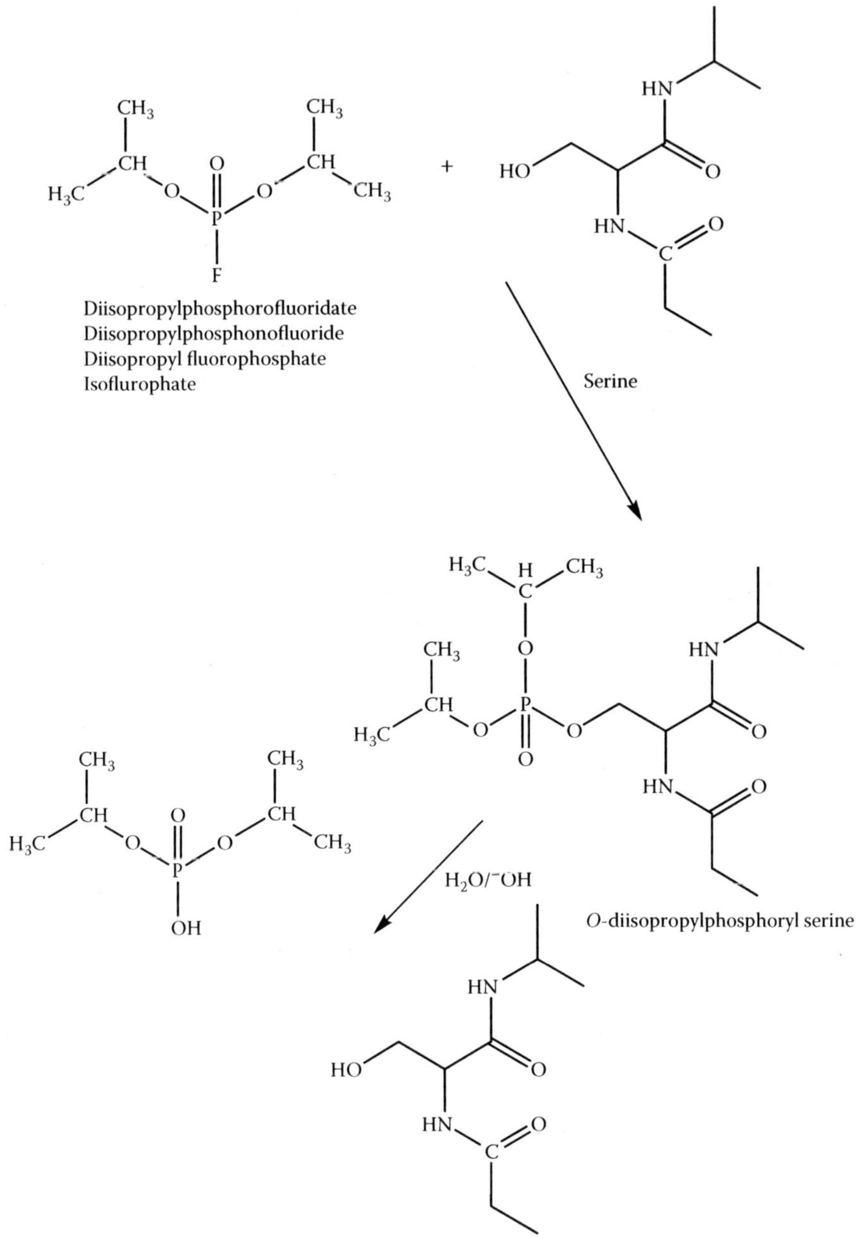

FIGURE 8.2 Reaction of diisopropylphosphorofluoridate (diisopropylfluorophosphate, DFP) with serine residues in enzymes. Also shown is the dephosphorylation reaction that occurs more rapidly at alkaline pH; the presence of a nucleophile such as hydroxylamine will accelerate the dephosphorylation reaction. The modified serine is also somewhat unstable under conditions of the Edman degradation. (See Manco, G., Camardella, L., Febbraio, F. et al., Homology modeling and identification of serine 160 as nucleophile of the active site in a thermostable carboxylesterase from the archeon *Archaeoglobus fulgidus*, *Protein Eng.* 13, 197–200, 2000.) Hydrolysis of the phosphoroflouridate is also an issue and is enhanced at alkaline pH.

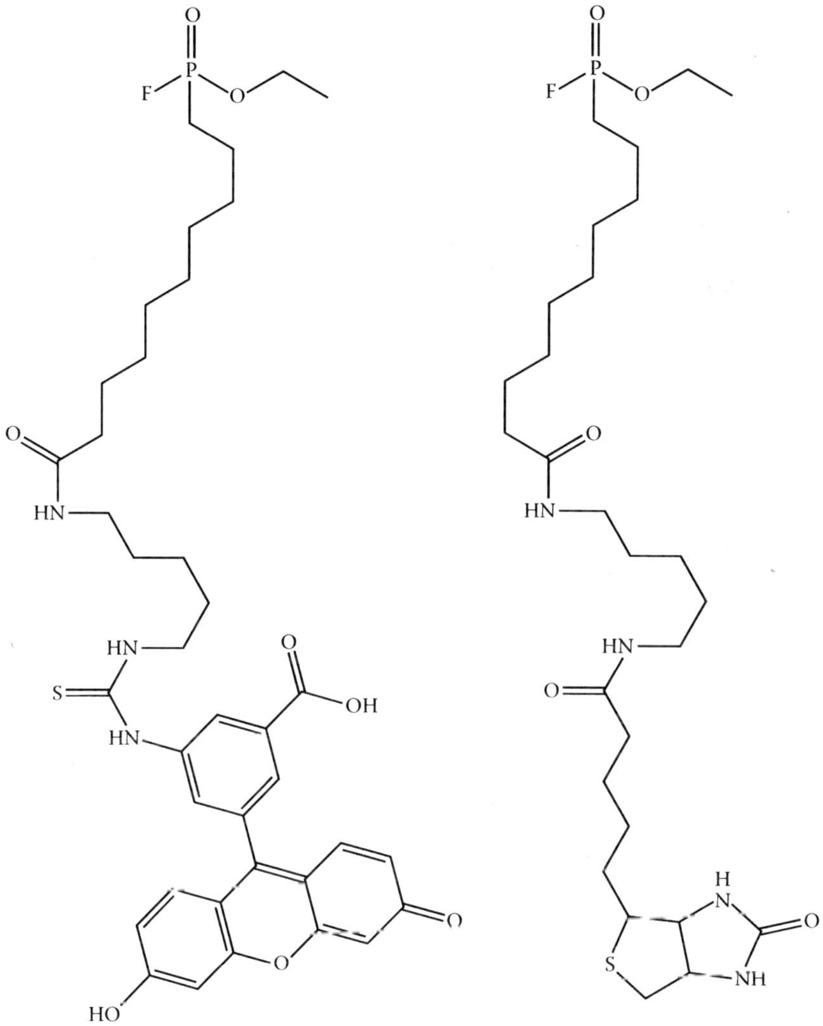

FIGURE 8.3 Some examples of activity-based fluorophosphate affinity probes for use in proteomics. One derivative contains a "tag," in this example, biotin, which can be used for the isolation of enzymes. The other derivative contains a fluorescent label. These derivatives can be used for the identification of a class of serine enzymes, most notably proteases but also esterases.

pH 5.1) than at basic pH (borate, pH 8.7) but is still on the order of days. Reactivation does occur more rapidly with addition of nucleophiles[84,66–88] such as oximes and hydroxamic acids. Similar reactivation occurs with choline esterases inactivated with organophosphorous compounds.[89] Those investigators who have worked with DFP may remember having 2-PAM (2-pyridinealdoxime) close by in case of contact with the reagent. There are more complex reactions involving rearrangement

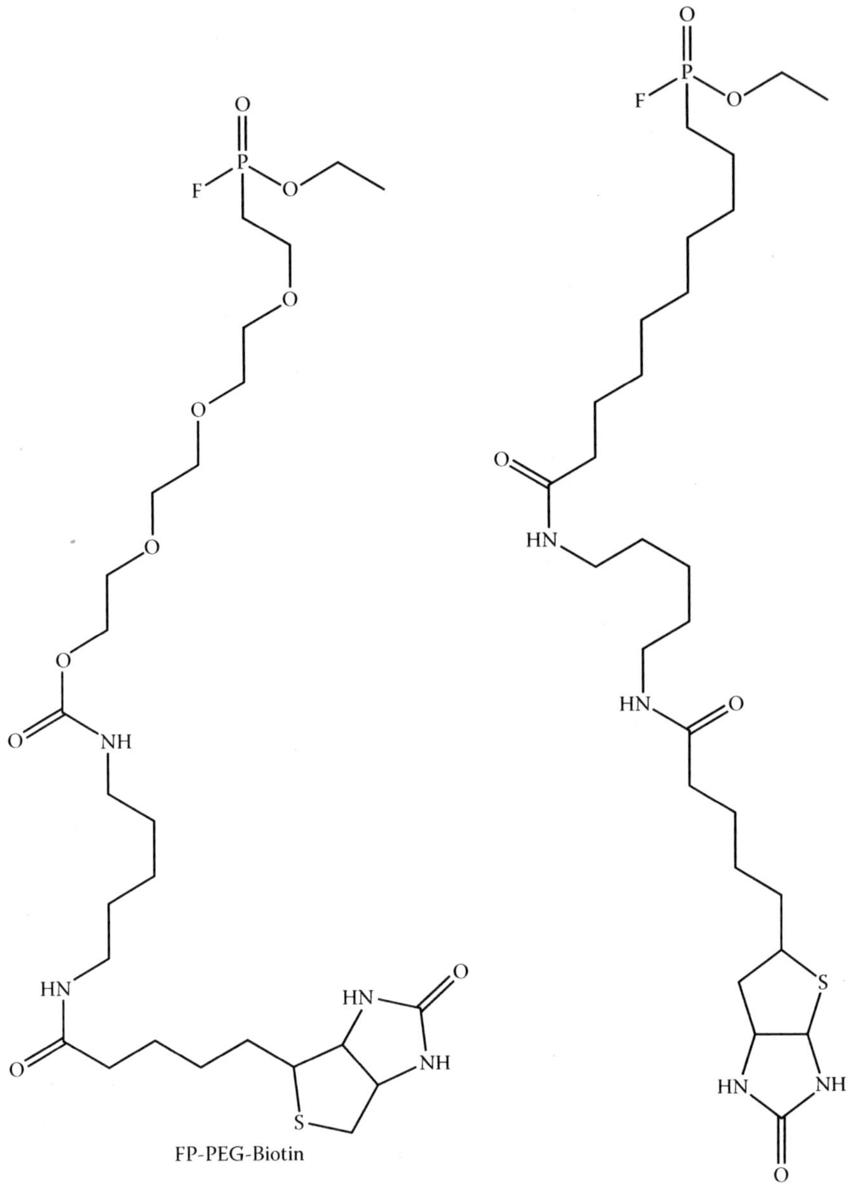

FP-PEG-Biotin

FP-Biotin

FIGURE 8.4 Some examples of activity-based affinity probes for labeling serine enzymes. The poly(ethylene) glycol derivative is designed to have a more hydrophilic quality than the parent DFP-based affinity probe shown on the right. (See Kidd, D., Liu, Y., and Cravatt, B.F., Profiling serine hydrolase activities in complex proteomics, *Biochemistry* 40, 4005–4015, 2001.)

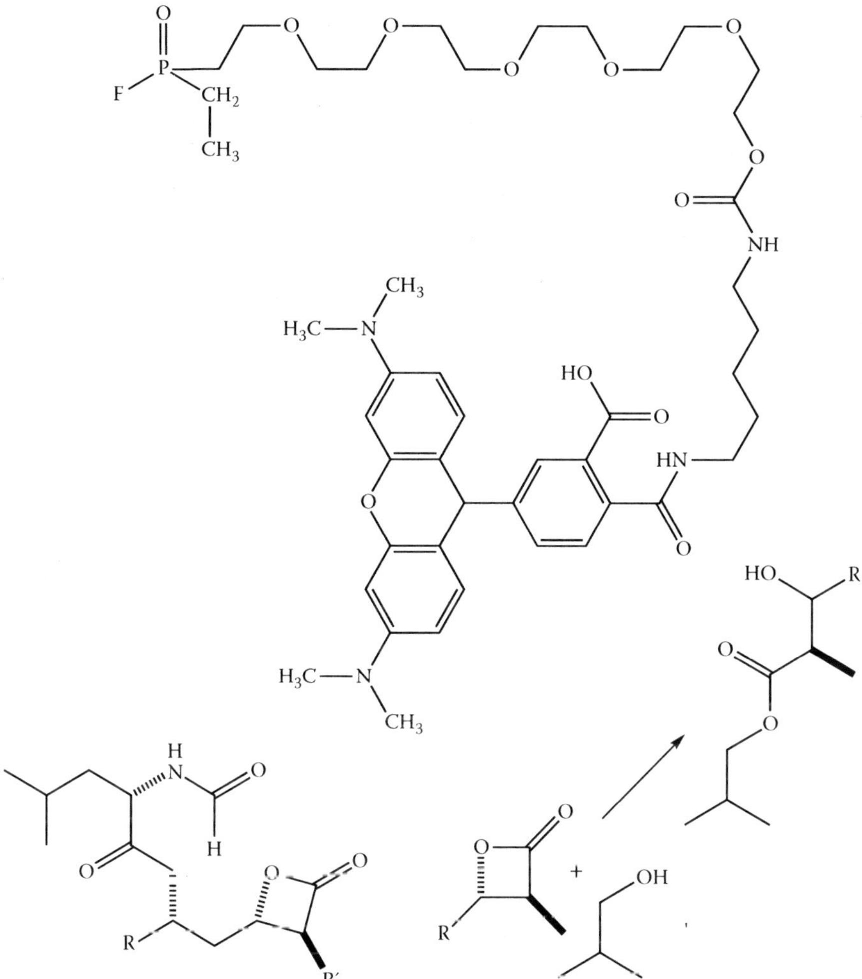

Serine

Tetrahydrolipstatin, Orlistat
$R = nC_{11}H_{23}$; $R' = nC_6H_{13}$

FIGURE 8.5 A fluorescent fluorophosphate probe. Fluorophosphonate-poly(ethylene)glycol-(6-carboxyltetramethylrhodamine) was demonstrated to shown that Orlistat® is a covalent inhibitor of fatty acid synthase reacting via a lactone mechanism. (See Mancini, M.C. and Halpern, A. , Orlistat in the prevention of diabetes in the obese patient, *Vasc. Health Risk Manag.* 4, 325–336, 2008.)

of the alkylphosphoryl label[90] that tend to retard reactivation of the enzyme. These "aging" reactions are thought to involve the hydrolytic removal of one of the *O*-alkyl substituent groups.[91] Reactivation of human neuropathy target esterase with *N,N'*-diisopropyl-phosphorodiamidic fluoride (Mipafox, Pestox) (Figure 8.6) occurs by the alkyl substituent loss as well as another pathway involving deprotonation.[92] The potential complexity of the labeling of cellular contents with "specific" reagents is

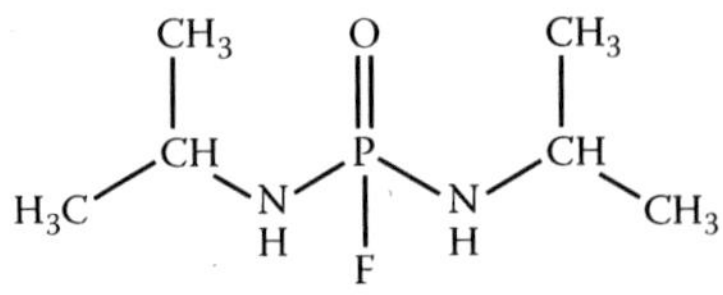

N,N'-Diisopropylphosphoramidic fluoride (Mipafox)

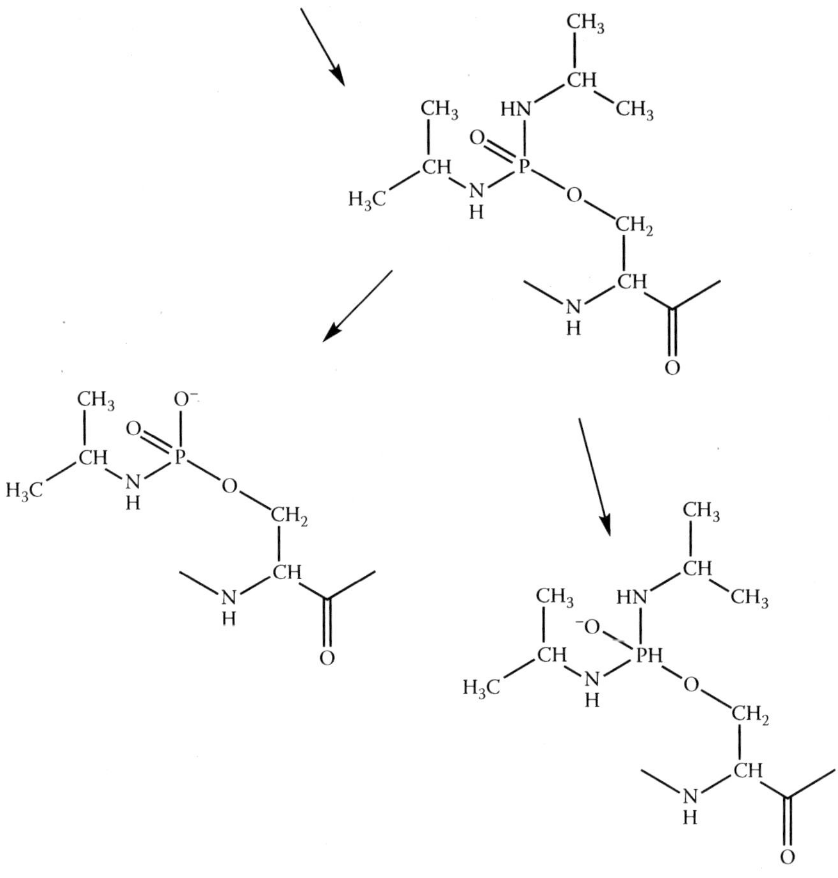

FIGURE 8.6 The reaction of serine enzymes with *N,N'*-diisopropylphosphorodiamidic fluoride (Mipafox, Pestox). Shown are two pathways for the "decomposition of the reaction product. These chemical changes are described as an aging response of the reaction product. (See Clothier, B. and Johnson, M.K., Rapid aging of neurotoxic esterase after inhibition by di-isopropyl phosphorofluoridate, *Biochem. J.* 177, 549–559, 1979.)

illustrated by the studies of the neuropathy target esterase (neurotoxic esterase).[92–94] There are several studies on the modification of this protein in tissue extracts and the ability of pretreatment of tissues with Paraoxon (diethyl-*p*-nitrophenyl phosphate) to block the reaction of DFP and related compounds with cholinesterase, which permitted more specific modification and identification of the neuropathy esterase.[94–97] IgM serine protease activity has been modified by reaction with an amidino phosphodiester.[98] A representative structure for these "covalently reactive analogs" is shown in Figure 8.7. More complex derivatives incorporate peptide segments to mimic EGF antigen and GP120 antigen.[99]

Peptide halomethyl ketones, most notably tosyl-phenylalanyl chloromethyl ketone (TPCK) and tosyl-lysylchloromethyl ketone, react with classical serine protease by alkylation of a histidine residue (Figure 8.8).[100,101] Tosyl-phenylalanyl chloromethyl ketone (TPCK) reacts preferentially with chymotrypsin-like enzymes, whereas TLCK reacts preferentially with trypsin-like enzymes. The chloro function was selected because of the low reactivity of chloroacetate/chloroacetamide with histidine residues in proteins as compared to the bromo- or iodo-derivatives. 5-(6-Carboxyfluoresceinyl)-L-phenylalanyl-chloromethyl ketone has been used as a specific inhibitor for chymotrypsin-like enzymes[102] (Figure 8.9). This compound and a Texas Red derivative of TPCK[154] are available from Imgenex under the name of Serpas™. There are other interesting derivatives of peptide chloromethyl ketones that might be useful. N-terminal biotin-labeled peptide chloromethyl ketones (Figure 8.10) developed by Williams, Mann, and coworkers[103] have been used to measure various activated coagulation factors in complex mixtures.[104] Biotin-labeled[105] and fluorescein-labeled[106] peptide chloromethyl ketones have been used to identify caspases in

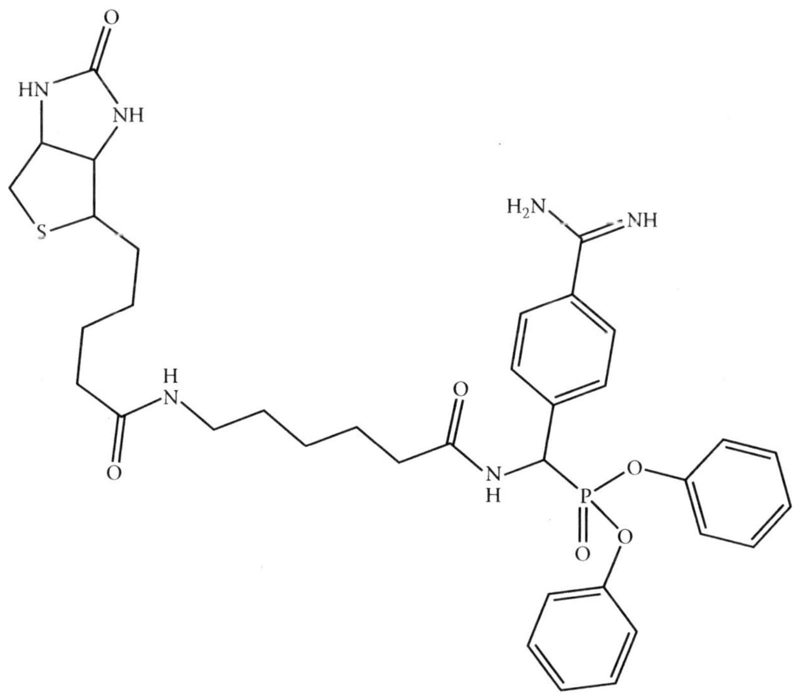

FIGURE 8.7 Phosphonate ester probes for proteolytic antibodies (hapten covalently reactive analogs). (See Paul, S. et al., Phosphonate ester probes for proteolytic antibodies, *J. Biol. Chem.* 276, 28314–28320, 2001.)

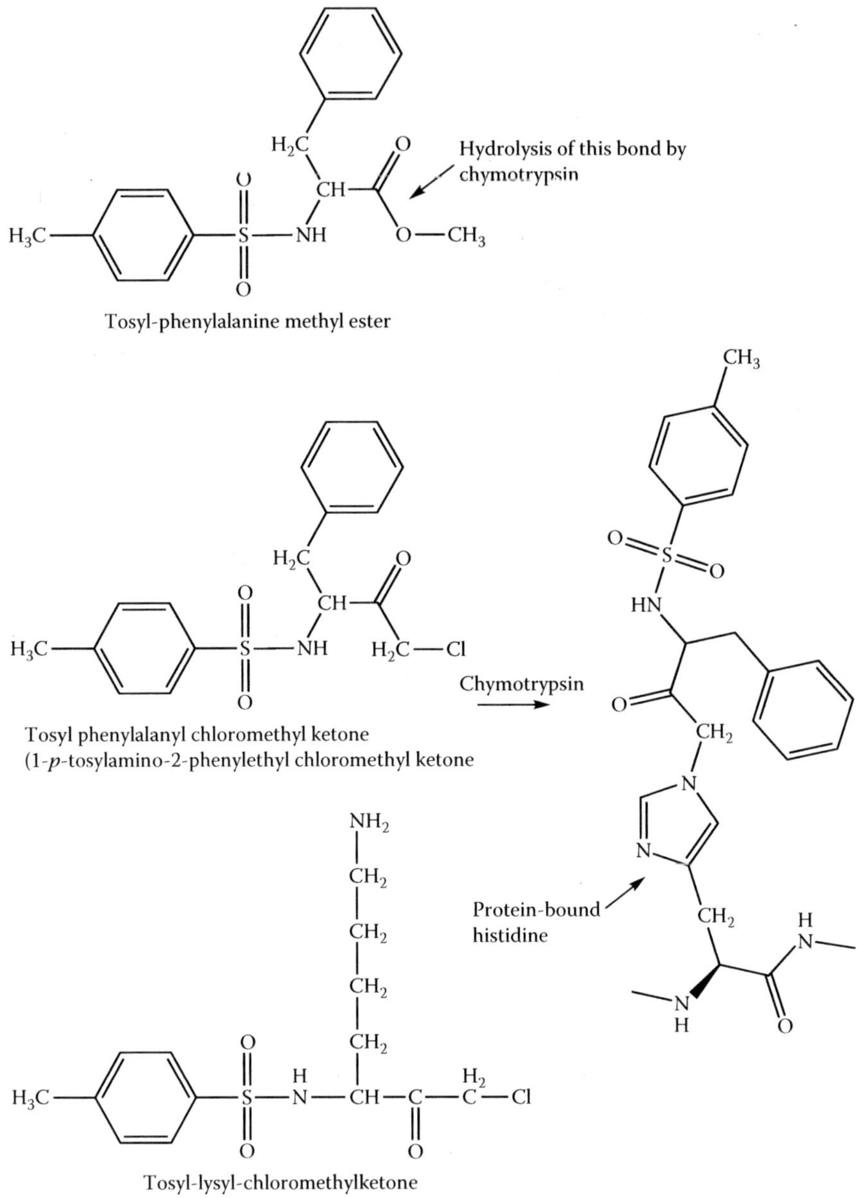

FIGURE 8.8 The design of affinity labels for enzymes. Shown is the basis for the design of tosyl-phenylalanine chloromethyl ketone. This is an affinity label designed to modify the enzyme active site of chymotrypsin. Shown is the structure of an ester substrate for chymotrypsin, tosyl-phenylalanine methyl ester; also shown is the structure of tosyl-phenylalanine chloromethyl ketone (TPCK) and the reaction with the active site histidine. (Redrawn from Shoellman, G. and Shaw, E., Direct evidence for histidine at the active center of chymotrypsin, *Biochemistry* 2, 252–255, 1963.)

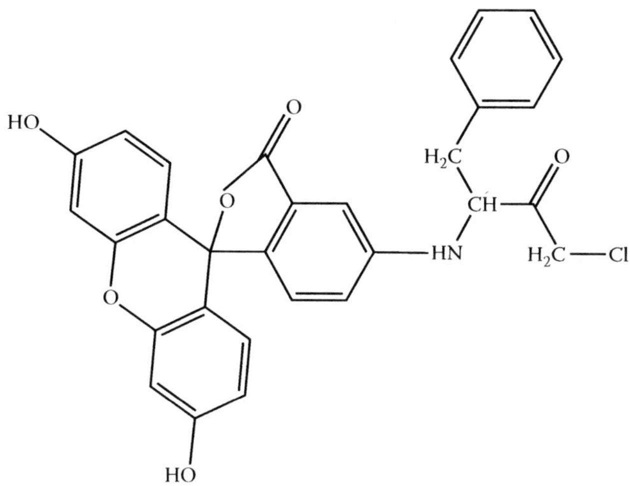

Carboxyfluoresceinyl-phenylalanine chloromethyl ketone

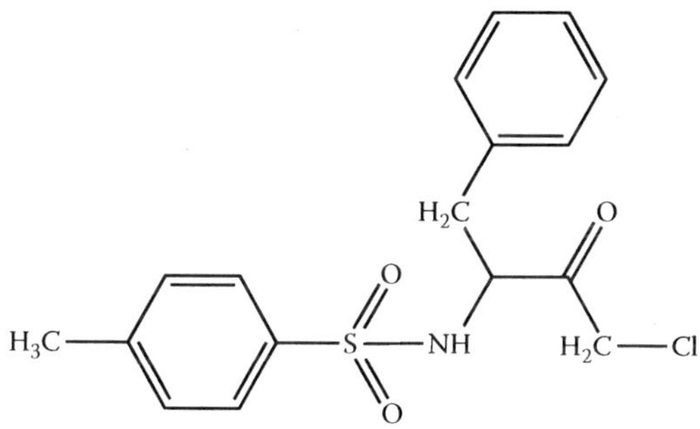

Tosylphenylalanine chloromethyl ketone

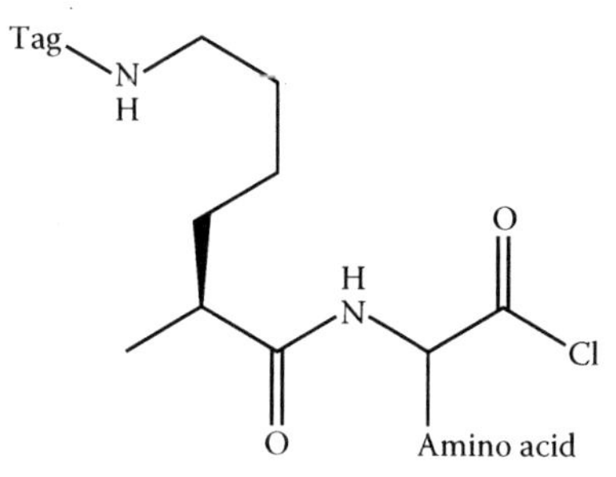

FIGURE 8.9 Carboxyfluoresceinyl phenylalanine chloromethyl ketone. This is a fluorescent analog of TPCK that can be used to modify a class of proteolytic enzymes. For a general consideration of affinity tags, see Campbell, D.A. and Szardenings, A.K., Functional profiling of the proteome with affinity labels, *Curr. Opin. Chem. Biol.* 7, 296–303, 2003; Schmidinger, H., Hermetter, A., and Birner-Gruenberger, R., Activity-based proteomics: Enzymatic activity profiling in complex proteomes, *Amino Acids* 30, 333–350, 2006.

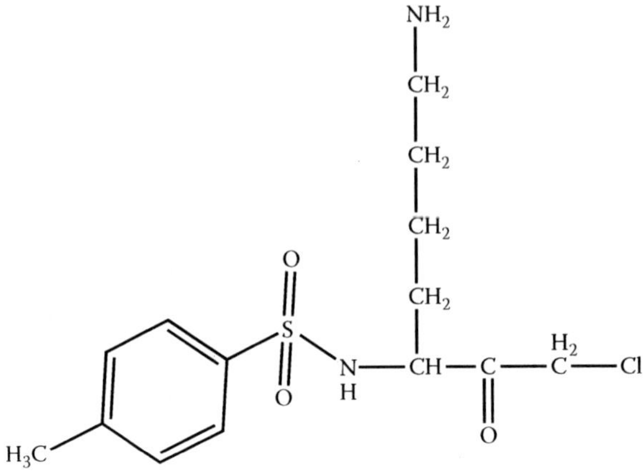

Tosyl-lysine-chloromethylketone, TLCK (1)

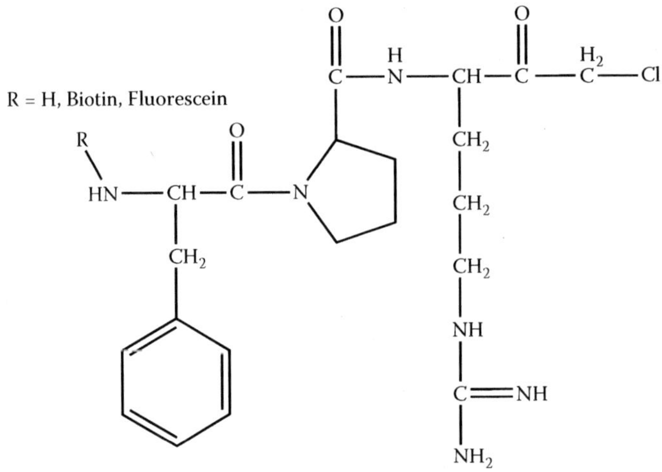

Phenylalanylprolylarginine chloromethyl ketone, PPACK (2)

FIGURE 8.10 Peptide chloromethyl ketones. Shown is a simple peptide chloromethyl ketone, tosyl-lysine chloromethyl ketone, and a more complex peptide chloromethyl ketone, phenylalanine arginine chloromethyl ketone. Relationships between simple and complex peptidyl diazomethyl ketones as inhibitors of thiol proteases are presented by Green, D.D.J. and Shaw, E., Peptidyl diazomethyl ketones are specific inactivators of thiol proteinases, *J. Biol. Chem*. 256, 1923–1928, 1981.

complex mixtures. As a precautionary note, peptide chloromethyl ketones do react with sulfhydryl groups in a variety of proteins.[107,108] Shaw and coworkers[109] have observed that peptidyl fluoromethyl ketones are less reactive than the chloromethyl derivatives and can be useful for the study of thiol proteases. Peptide fluoromethyl ketones have been used as relatively specific inhibitors of caspase activity.[110] Bergen and coworkers[111] have observed "nonspecific" modification of protein with the components of protease inhibitor "cocktails."

Zhang and coworkers[112] have developed affinity probes for the modification of the active site of protein tyrosine phosphatases (Figure 8.11). These inhibitors are based on earlier work developing α-bromobenzylphosphonate.[113] These reagents demonstrated high specificity in the modification of protein tyrosine phosphatases in *Escherichia coli* cell lysates.

Another approach has been developed by Ovaa and colleagues[114] to detect ubiquitin proteases with vinyl sulfone derivatives and related compounds. These are fairly complex reagents (Figure 8.12) that react with cysteine proteases to form stable derivatives, which can be detected. Bogyo and coworkers[115] have developed epoxide probes (Figure 8.13) for the modification of cysteine proteases. Epoxides inactivate

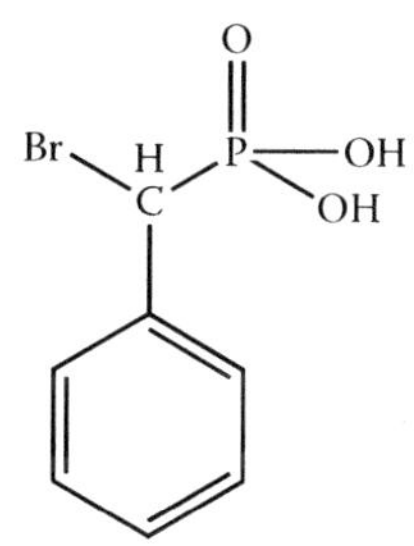

Alpha-bromobenzylphosphonate[1]

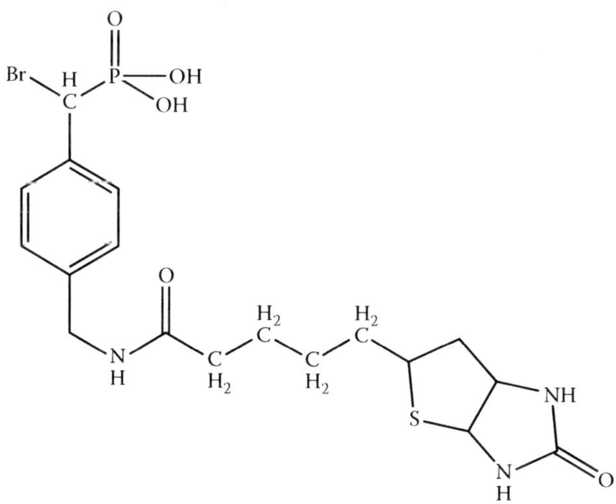

Protein Tyrosine Phosphatase Activity-Based Probe[2]

FIGURE 8.11 Protein tyrosine phosphatase activity-based probes for use in proteomic research. (Redrawn from Taylor, W.P., Zhang, Z.-Y., and Widlanski, T.S., Quantative affinity inactivators of protein tyrosine phosphatases, *Bioorg. Medicinal Chem.* 4, 1515, 1996; Kuman, S. et al., Activity-based probes for protein tyrosine phosphatases, *Proc. Natl. Acad. Sci. USA* 101, 7943, 2004.)

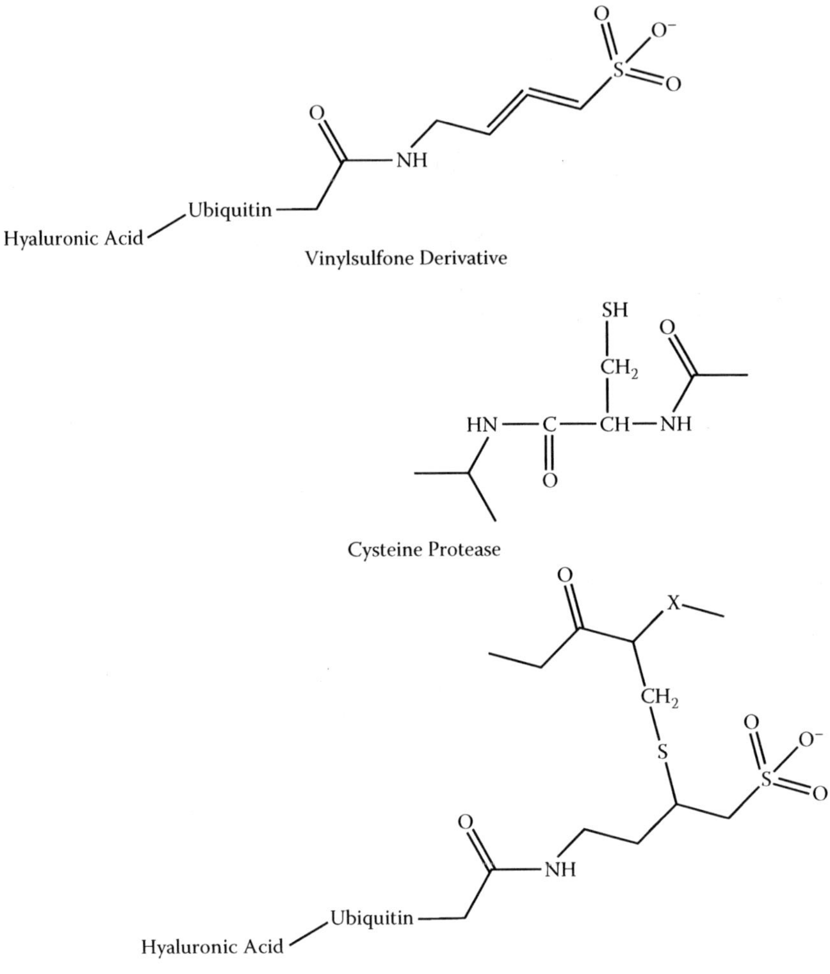

FIGURE 8.12 The reaction of a complex vinyl sulfone derivative with a cysteine proteinase. (Adapted from Hemelaar, J., Galardy, P.J., Borodovsky, A. et al., Chemistry-based functional proteomics: Mechanism-based activity profiling tools for ubiquitin and ubiquitin-like specific proteases, *J. Proteome. Res.* 3, 268, 2004. See also Reddick, J.J., Cheng, J., and Roush, W.R., Relative rates of Michael reactions of 2′-(phenethyl)thiol with vinyl sulfones, vinyl sulfonate esters, and vinyl sulfonamides relevant to vinyl sulfonyl cysteine protease inhibitors, *Org. Lett.* 5, 1967–1970, 2003; Uttamchandani, M., Liu, M., Panicker, R.D., and Yan, S.Q., Activity-based fingerprinting and inhibitor discovery of cysteine proteases in a microarray, *Chem. Commun.* 1518–1520, 2007.)

cysteine proteases by alkylation of the active site cysteine residue via a mechanism-based reaction.[116]

A novel approach developed by Dupont and colleagues[117] uses an antibody directed at the cleavage site region in a specific substrate; after cleavage of the substrate protein, the antibody no longer reacts with the protein. This approach has been used

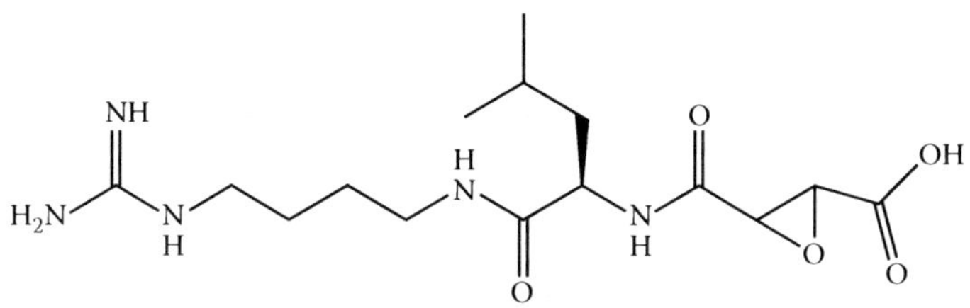

E-64 from *Aspergillus japonicus*

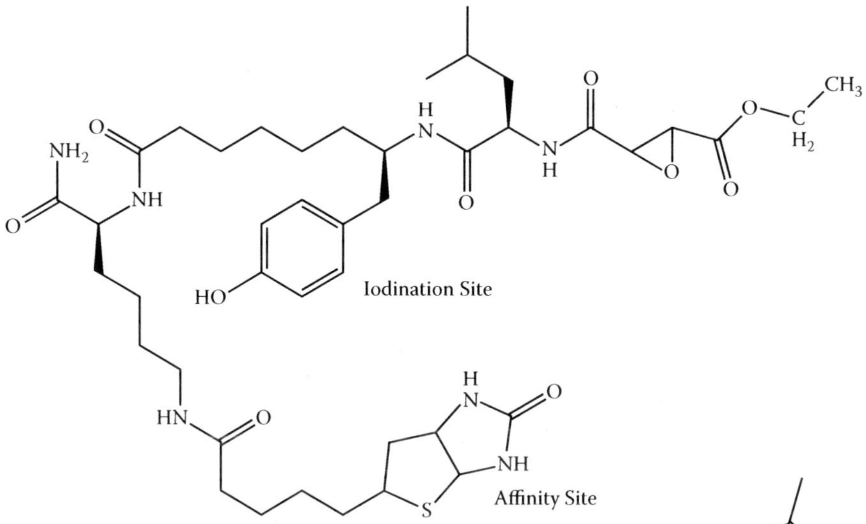

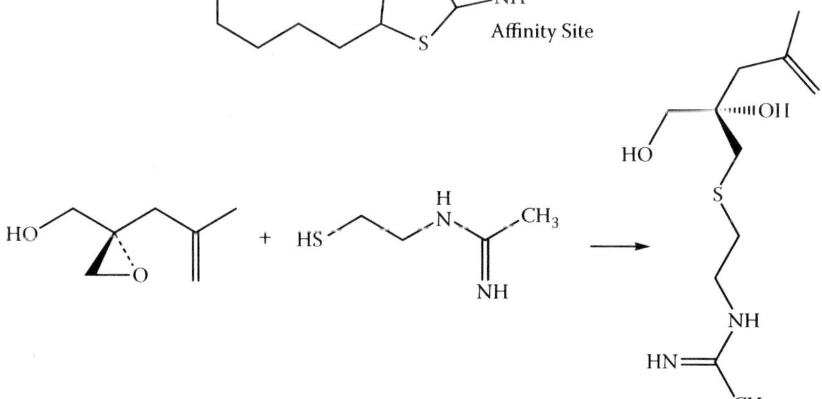

FIGURE 8.13 Epoxide-based inhibitors of cysteine proteinases. Shown is E-64 [*trans*-epoxysuccinyl-leucylamido(4-guanidino)butane] (See Barrett, A.J., Kembhavi, A.A., Brown, M.A. et al., L-trans-epoxysuccinyl-leucylamido(4-guanidino)butane) (E-64) and its analogues as inhibitors of cysteine proteinases including cathepsins B, H, and L. *Biochem. J.* 201(1): 189–198, January 1, 1982 and a more complex derivative; Greenbaum, D., Medzihradszky, K.F., Burlingame, A.L. et al., Epoxide electrophiles as activity-dependent cysteine protease profiling and discovery tools, *Chem. Biol.* 7, 569–581, 2000). The presence of a phenolic functional group (tyrosine analog) permits the preparation of a radiolabeled form of the inhibitor.

to evaluate casein hydrolysis[118] and neutrotoxin hydrolysis.[119] Another proteomic approach to the study of intracellular proteases used 2D differential gel electrophoresis (2D DIGE)[120] to identify substrates for granzyme A and granzyme B in YAC-1 (mouse lymphoma) cell lysates.

Reagents have been developed that are not designed to be active-site directed[121] but may react with functional groups on proteins (see Chapter 1), which can then be captured for proteomic analysis. One example is provided by sulfonate esters (Figure 8.14). There has not been much research on the reaction of these compounds

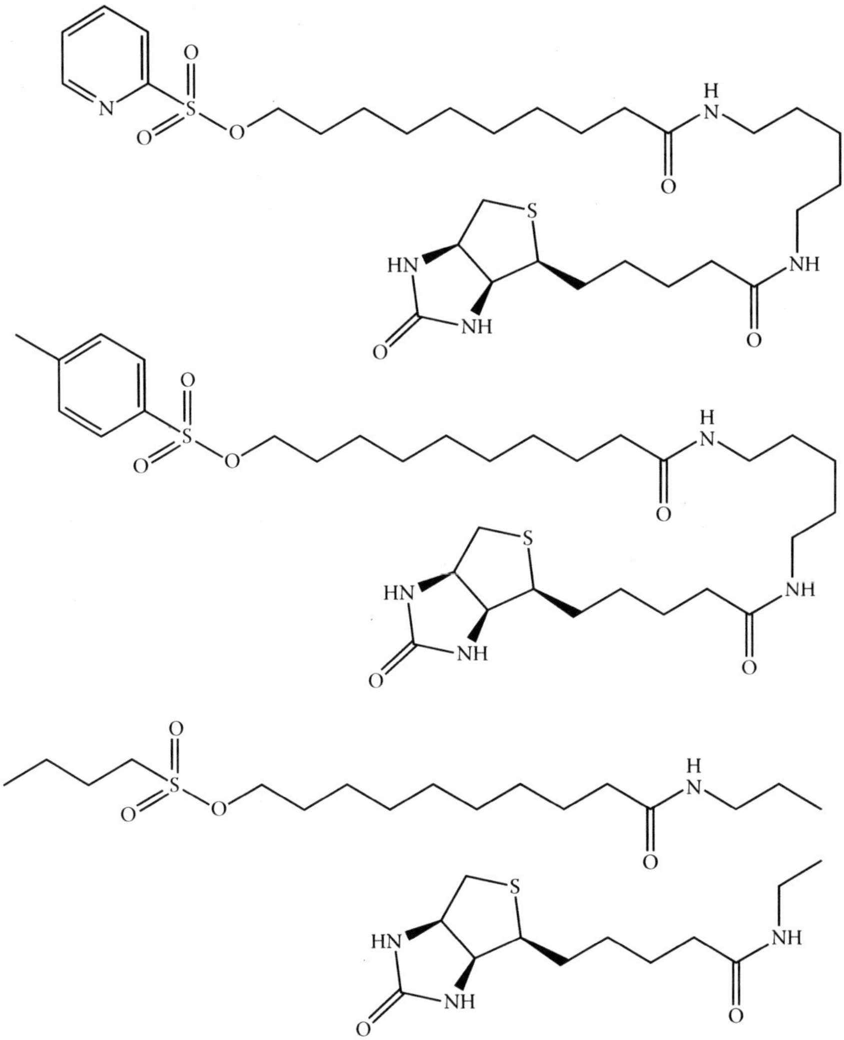

FIGURE 8.14 Some complex alkyl sulfonate esters used for proteomic profiling. (See Adam, G.C., Cravatt, B.F., and Sorensen, E.J., Profiling the specific reactivity of the proteome with non-directed activity-based probes, *Chem. Biol.* 9, 81, 2001.)

with proteins. Although there is extensive information on reaction with nucleic acids, there is a limited amount of information on the reaction of simple alkyl sulfonate esters such as methylmethanesulfonate with proteins.[122,123] In principle, these reagents would have the potential to react with a variety of amino functional groups in proteins such as the amino-terminal α-amino group, the ε-amino group of lysine, imidazole nitrogens of histidine, the mercaptide function of cysteine, and the hydroxyl group of tyrosine. In fact, the potential reactive sites are probably similar to those subject to alkylation by α-halo acid and would then also include glutamic acid. Reaction at a given residue should be a function of (1) nucleophilicity of the protein bound functional group, (2) the electronegativity of the alkyl/aryl sulfenyl leaving group, and (3) steric accessibility (see Chapter 1).

A series of biotinylated alkylsulfonate esters (Figure 8.15) that have been developed by Adam and coworkers[121] were used to screen tissue homogenates for uniquely reactive proteins. Reaction was assessed after electrophoretic separation on the basis of reaction with an avidin-based detection system. Specificity for reaction with native proteins was demonstrated by the lack of reaction with heat-treated (80°C/5 min) preparations. An aldehyde dehydrogenase was preferentially modified with the pyridyl derivative. In subsequent work,[124] a combined chemical modification approach was used. In order to improve the sensitivity, a rhodamine-labeled probes was used in initial studies and a biotin-labeled probe was used to isolate the modified proteins. Activity-based protein labeling occurred with protein extracts that had not been heat-treated. A variety of probes were prepared with different alkyl or aryl sulfenyl leaving groups. The highest extent of labeling of tissue homogenates was obtained with the phenyl derivative and the least with the octyl derivative. The phenyl derivative identified a variety of different enzymes, including aldehyde dehydrogenase, acetyl CoA acetyltransferase, epoxide hydrolase, and NAD/NADP-dependent oxidoreductase. This same group also developed a sulfonate ester probe that contained both a rhodamine function and a biotin function.[125] In a more recent work,[126] this group further characterized the reaction of the phenyl sulfonate derivative with enzymes. In this study, a rhodamine tag was inserted via the alkylation reaction. The proteins were subjected to proteolytic digestion, and the rhodamine-labeled peptides were isolated by immunoaffinity chromatography on a rhodamine monoclonal affinity column. Mass spectrometric analysis combined with oligonucleotide-directed mutagenesis established the sites of modification in various enzymes. Reaction occurred at a cysteine residue in a glutathione *S*-transferase, at aspartic acid residues in enoyl CoA transferase, and at a glutamic residue in aldehyde dehydrogenase-1. Other studies[127,128] have used alkyl sulfonate esters with terminal azide or alkyne functions, for reaction with either alkyne or azide functions, respectively[129,130] (see Figure 8.16).

Reaction with 2,4-dinitrophenylhydrazine, fluorescein hydrazide, and thiosemicarbizide is used for the detection of carbonyl products occurring as a result of oxidation.[131,132] Nitrotyrosine (3-nitrotyrosine) is formed as a result of peroxynitrite action.[133,134] A method for detecting 3-nitrotyrosine in the proteome has been described by Tannenbaum and coworkers.[135] These investigators reduced the 3-nitrotyrosine residues to the corresponding 3-aminotyrosine with sodium dithionite. The 3-aminotyrosine was modified with sulfosuccinimidyl-2-(biotinamio)ethyl-1,3-dithiopropionate

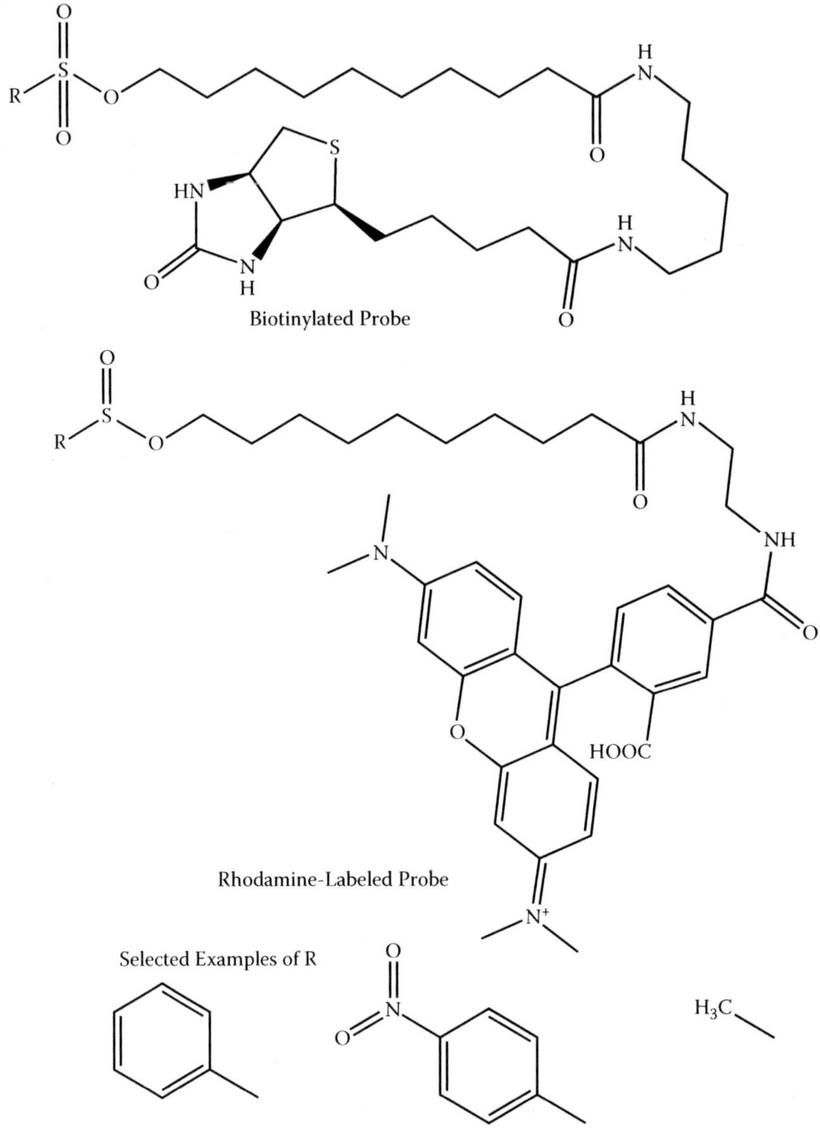

FIGURE 8.15 Alkyl and aryl sulfonate esters used for non-activity directed probes in proteomic research. (See Evans, M.J. and Cravatt, B.F., Mechanism-based profiling of enzyme families, *Chem. Rev.* 106, 3279–3301, 2006.)

(Figure 8.17). Selective alkylation of the 3-aminotyrosine is obtained at pH 5.0, as the p*K*a of the aromatic amino group is 4.75.[136] After tryptic cleavage, the biotin-containing peptides are obtained by adsorption to a streptavidin matrix. Reduction of the disulfide linkage releases the peptide for subsequent analysis by mass spectrometry. Two somewhat more complex approaches to the identification of phosphorylated

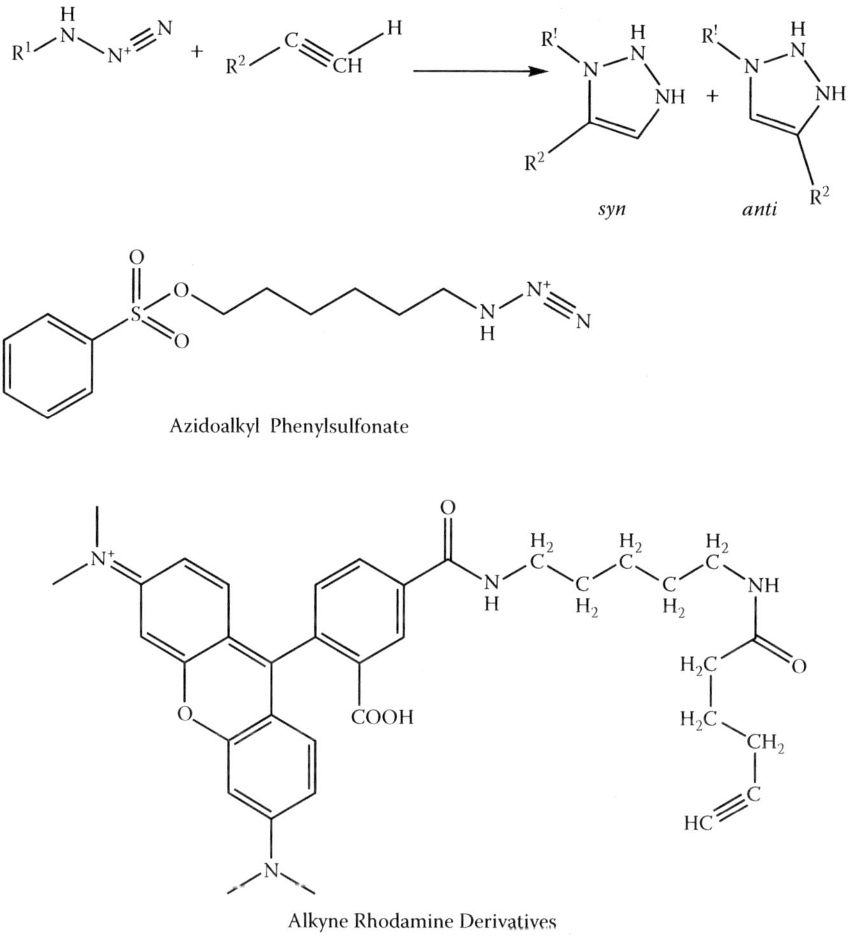

FIGURE 8.16 Azide- and alkyne-based chemical probes for proteomic profiling. Note that these reagents can be bifunctional and useful for the study of protein–protein interaction (cf. Breinbauer, R. and Kohn, M., Azide-alkyne coupling: A powerful reaction for bioconjugate chemistry, *ChemBioChem* 4, 1147–1149, 2003).

peptides have been developed.[137–140] One approach is based on the β-elimination of phosphoserine or phosphothreonine under alkaline conditions and elevated temperature followed by Michael addition of a dithiol and subsequent modification of the thiol group with a biotin-containing alkylating reagent (see Figure 8.18). The other approach[138] involves a series of chemical reactions resulting in the selective modification of the phosphoryl group and subsequent affinity isolation (Figure 8.19). As a word of caution, β-elimination has been observed for nonphosphorylated residues under the preceding reaction conditions.[141] Thaler and colleagues have extended some of this earlier work on β-elimination.[142] First, solvent conditions for the β-elimination reaction have been optimized by increasing the amount of dimethyl sulfoxide. Second, an alterative approach to the isolation of the thiol derivative has been developed using

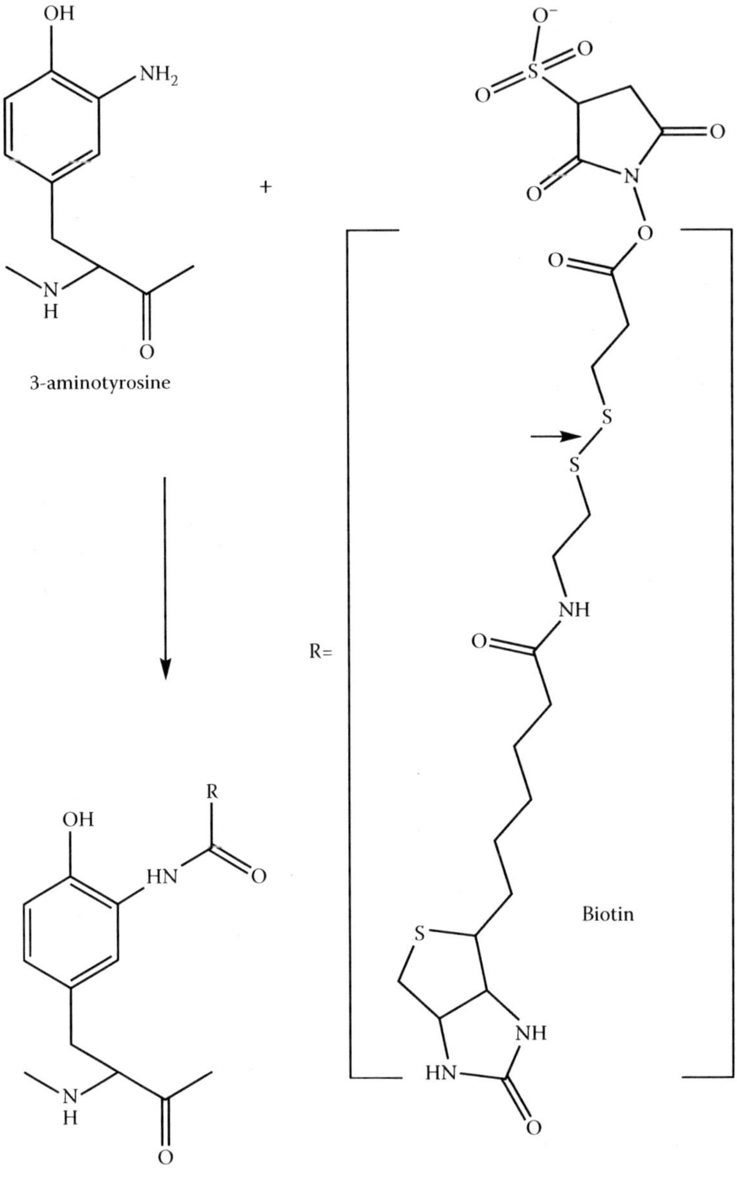

FIGURE 8.17 The reduction of nitrotyrosine to aminotyrosine and subsequent modification with an affinity label. This shows a method for the isolation of nitrotyrosine peptides. Nitrotyrosine, either endogenous as a result of reaction with peroxynitrite or exogenous by modification with tetranitromethane, is reduced to aminotyrosine. As a result of the lower p*K*a, this amino group can be selectively modified with the *N*-hydroxysuccinimide derivative containing a biotin tag. The biotin tag permits the isolation of the modified protein or peptide on a streptavidin matrix; the disulfide bond can be reduced with the release of the isolated material.

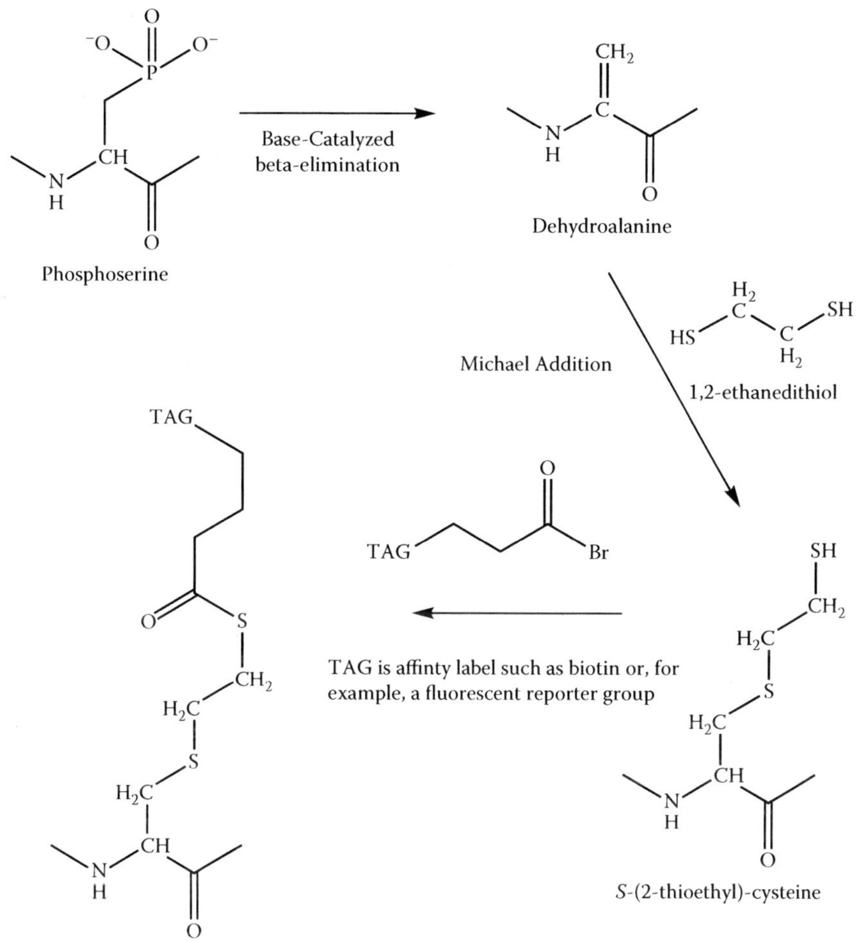

FIGURE 8.18 Modification and isolation of phosphoserine peptides via beta-elimination followed by a Michael addition. (See Tintette, S., Feyereisen, R., and Robichon, A., Approach to systematic analysis of serine/threonine phosphoproteomic using β-elimination and subsequent side effects: Intramolecular linkage and/or racemization, *J. Cell. Biochem.* 100, 875–882, 2007.)

disulfide exchange on a dithiopyridine matrix. Affinity labeling of integral membrane proteins has been obtained by using (+)biotinyl-iodoacetamidyl-3,6-dioxaoctanediamine (Figure 8.20) as an alkylating agent for cysteine residues.[143]

The use of stable isotope labeling for quantitative proteomics was introduced by Aebersold and coworkers.[144] This is a clever application of the concept of isotope dilution, which is a well-accepted technique in analytical chemistry.[145,146] Specifically, Aebersold and coworkers developed a somewhat complex reagent composed of a "tag" that binds to an affinity matrix facilitating purification of a labeled peptide or protein, a "linker" region that can be labeled with a "heavy" isotope such as deuterium in the place of hydrogen, and a reactive probe such as an α-ketohalo function

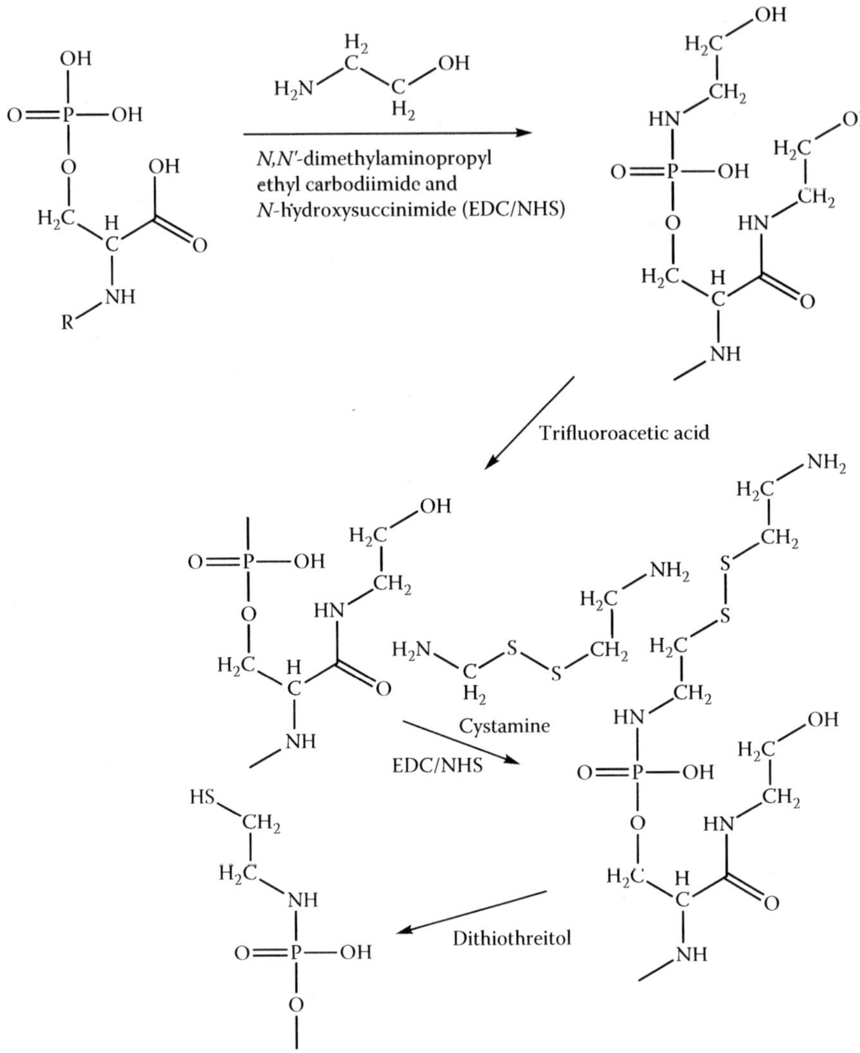

FIGURE 8.19 The differential modification of phosphoserine with carbodiimide in phosphoproteins. (See Tao, W.A., Wollscheid, B., O'Brien, R. et al., Quantitative phosphoproteome analysis using a dendrimer conjugation chemistry and tandem mass spectrometry, *Nat. Methods* 2, 591–598, 2005.)

analogous to iodoacetamide (Figure 8.21). These reagents were described as isotope-coded affinity tags (ICAT).[147–149] ICAT reagents enable the relatively specific introduction of a deuterium-labeled moiety on the sulfhydryl groups in a protein mixture. The use of a chemically identical modifying reagent not containing deuterium on a different protein allows the comparison of protein expression.[150] As it was presumed that, with the exception of the isotope, the derived peptides are chemically identical, a difference in molecular weight on mass spectrometric analysis would differentiate

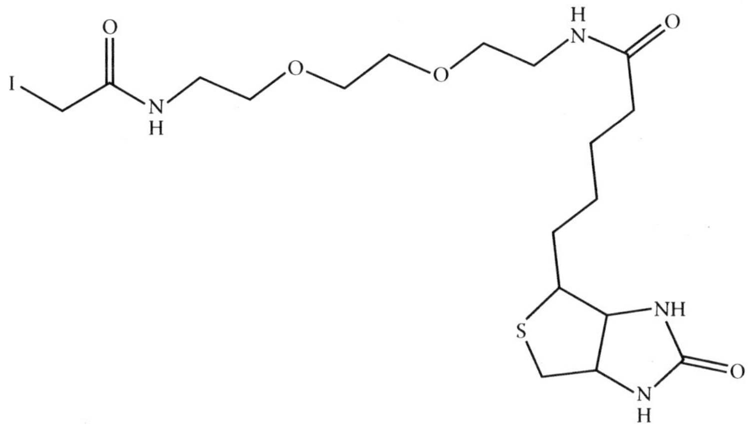

1-iodoacetamidyl,8-biotinyl, 3,5-dioxaoctanediamine (idoacetyl-PEO-biotin)

FIGURE 8.20 A complex probe for the modification and isolation of cysteine-containing membrane proteins. (See Goshe, M.B., Blonder, J., and Smith, R.D., Affinity labeling of highly hydrophobic integral membrane proteins for proteome-wide analysis, *J. Proteome Res.* 2, 153, 2003; Blonder, J., Goshe, M.B., Xiao, W. et al., Global analysis of the membrane subproteome of *Pseudomonas aeruginosa* using liquid chromatography-tandem mass spectrometry, *J. Proteome Res.* 3, 434–444, 2004; Ramus, C., Gonzalez de Peredo, C., Gallaher, M. et al., An optimized strategy for ICAT quantification of membrane proteins, *Mol. Cell. Proteomics* 5, 68–78, 2006.)

between the peptides in the two samples. Consider an experiment: two cell cultures, one of which is challenged, are compared by labeling the experimental sample with a deuterated reagent. Proteins from the "naïve" culture are reduced and modified with a reagent consisting of an iodoalkyl function with a linker containing a terminal biotin moiety. Proteins from the "challenged" culture are labeled with the same reagent except that it contains deuterium. Proteolysis of the combined alkylated protein peptides yields sulfhydryl peptides that have been modified with biotinylated reagents, which can then be isolated by affinity chromatography on streptavidin/avidin matrices. The isolated peptides are then analyzed by mass spectrometry. The ratio of deuterated peptide to unlabeled peptide is an indication of a change induced by the change. Subsequent work has refined this technique,[151–154] and reagents that target residues other than cysteine have been developed.[155–157] Maier and colleagues[158] developed a hydrazide-based ICAT reagent for the measurement of protein–lipid oxidation products derived from acrolein or 4-hydroxy-2-hexanal.

The development of a reagent with an acid labile link to a resin permits the facile purification of peptides.[159] More recently, visible ICAT reagents (VICAT reagents) have been developed (Figure 8.22).[160,161] Regnier and coworkers demonstrated that the deuterated derivatives were, in fact, chemically different and could be separated on HPLC.[162] Regnier and colleagues then introduced a different type of ICAT reagents (Figure 8.23) that did not have this problem because the reagents were labeled with ^{13}C.[163] This reagent modified amino groups with a ^{13}C-labeled acetyl

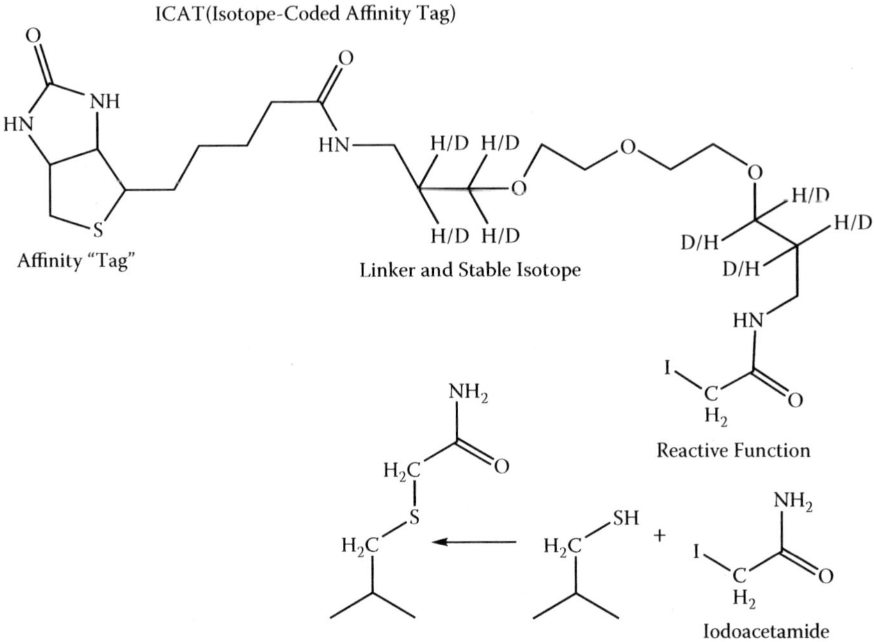

FIGURE 8.21 Labeling of proteins with stable isotope chemical reagents for proteomic research.

group via an *N*-hydroxysuccinimide reagent, and is referred to as Global Internal Standard Technology (GIST).

The use of isobaric reagents in proteomics provides an increase in the development of quantitative measurement. Pappin and coworkers[164] introduced isobaric reagents that modified amino groups. Collision-induced dissociation (CID) allows identification of an individual member of a multiplexed group. This approach has been used as a technique described as iTRAQ (isobaric tags for relative and absolute quantification).[165–169]

There are a variety of applications that stem directly from earlier research in protein chemistry as discussed in Chapter 1. One particular example is provided from the work of Vanderkerckhove and coworkers.[170] These investigators used the change in chromatographic mobility of methionine-containing peptides after oxidation with hydrogen peroxide (H_2O_2) to form the sulfoxide.[171] This is a useful approach that is based on earlier work by Hartley and colleagues on diagonal electrophoresis.[172] Another approach to the selective isolation of methionine-containing peptides is derived from the selectivity of alkylation at low pH.[173] In this approach, methionine-containing peptides are isolated by reaction with solid-phase bromoacetyl derivative[174,175] and subsequently eluted with a mild reducing agent such as 2-mercaptoethanol. This approach has been used to differentiate methionine peptides from methionine sulfoxide peptides.[176]

Hsieh-Wilson and colleagues have reported an extremely novel approach to the proteomic analysis of *O*-linked β-*N*-acetylglucosamine-modified proteins.[177] An

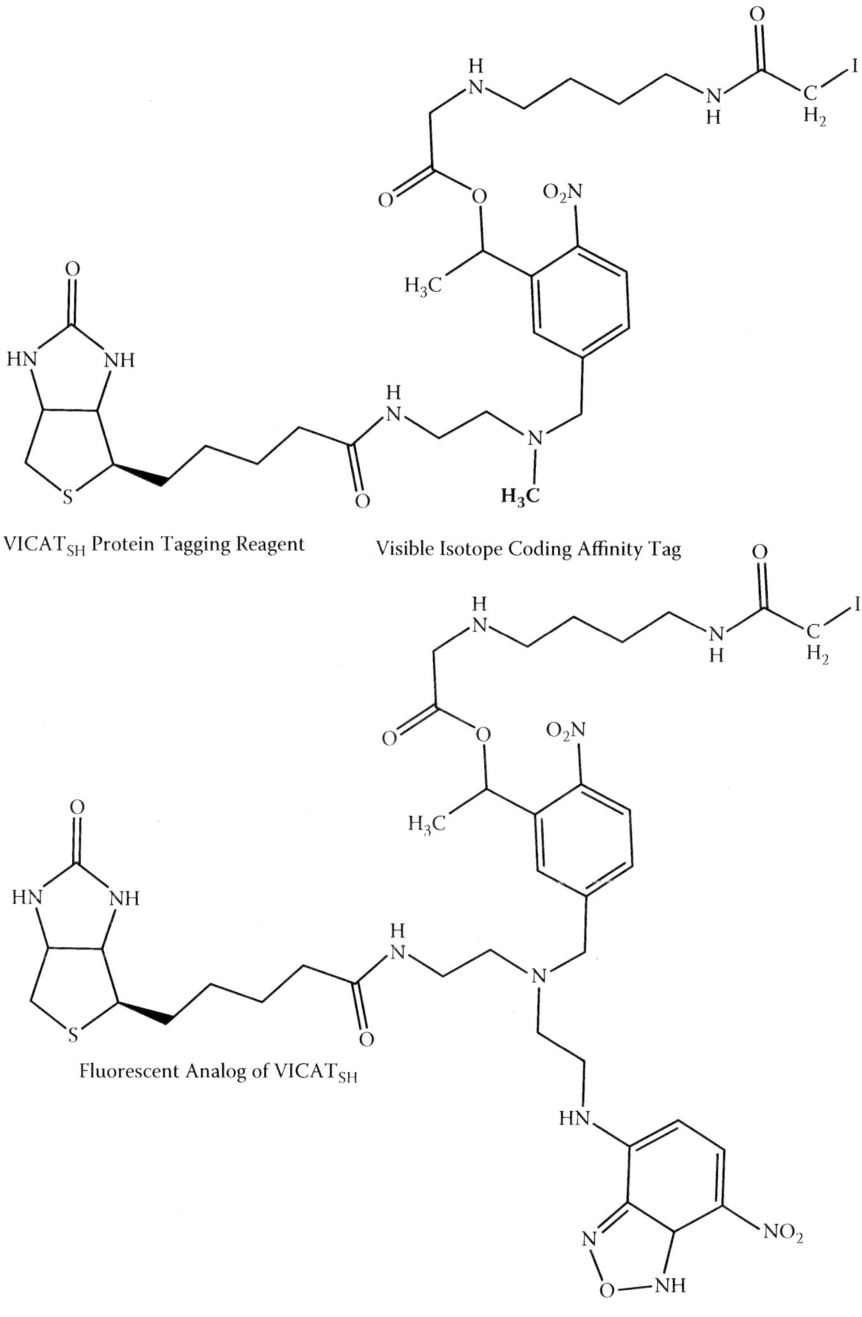

FIGURE 8.22 Visible isotope coding affinity tag.

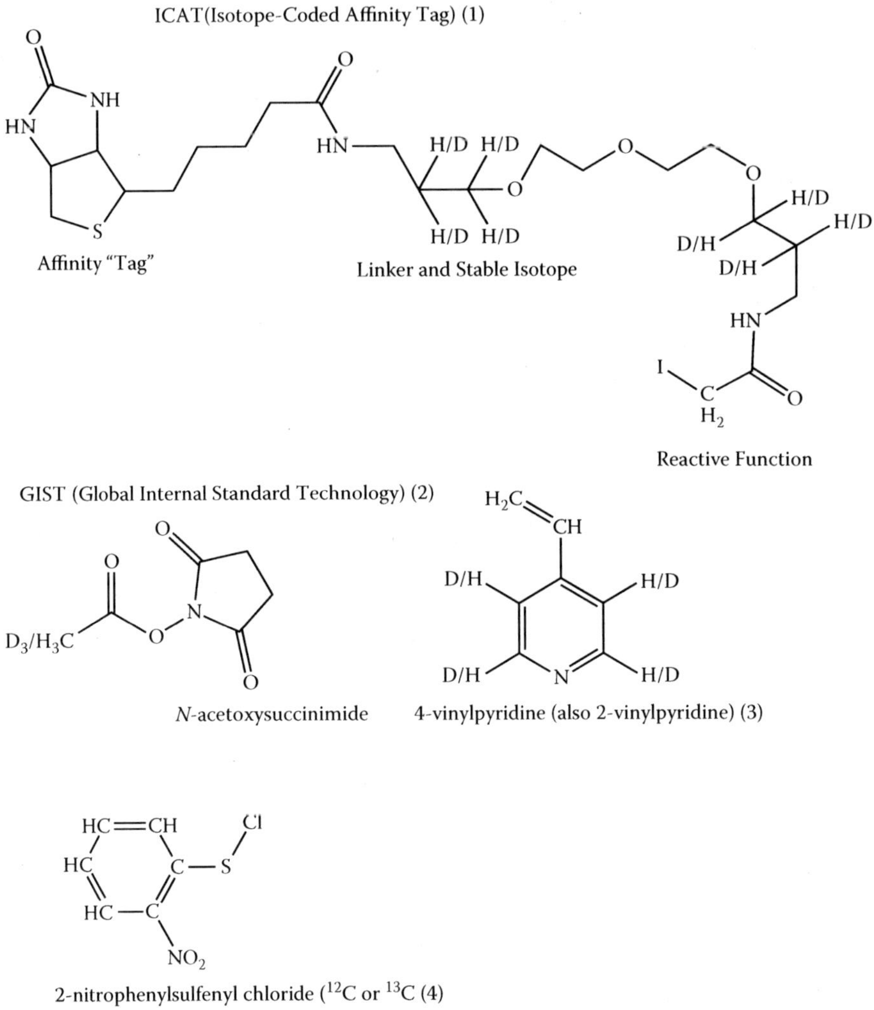

FIGURE 8.23 A selection of stable isotope chemical reagents. The use of these reagents for the determination of differential protein expression is based on the inclusion of "light" or "heavy" isotopes. (See Gygi, S.P., Rist, B., Gerber, S.A., Turecek, F., Gels, M.H., and Aebersold, R., Quantitative analysis of complex protein mixtures using isotope-coded affinity tags, *Nat. Biotechnol.* 17, 994–999, 1999; Chakraborty, A. and Regnier, F.E., Global internal standard technology for comparative proteomics, *J. Chromatog. A.* 949, 173–184, 2002; Sebastino, R., Cirreria, A., Lapadula, M., and Righetti. P.G., A new deuterated alkylating agents for quantitative proteomics, *Rapid Commun. Mass Spectrom.* 17, 2380–2386 , 2003; Kuyama, H., Watanabe, M., Todo, C., Ando, E., Tanaka, K., and Nishimura, O., An approach to quantitative proteome analysis by labeling tryptophan residues, *Rapid Commun. Mass Spectrom.* 17, 1642–1650, 2003.)

engineered galactosyltransferase is used to label *O*-linked β-*N*-acetylglucosamine proteins with a ketone-biotin tag [*N*-(aminooxyacetyl)-*N*′-(D-biotinoyl) hydrazine] (Figure 8.24). A related approach to the identification of *O*-linked β-*N*-acetylglucosamine has been developed by Bertozzi and colleagues.[180,181] In this approach, *N*-azidoacetyl-glucosamine is incorporated into proteins and derivatized at the azido function via Staudinger ligation (Figure 8.25). This same group has used this chemistry for the incorporation of a fluorogenic dye.[182]

A combination of an enzymatic approach and a chemical approach has been developed for the identification of phosphorylation sites in proteins. A clever approach to this problem has been developed by Zhao and coworkers.[183] The use

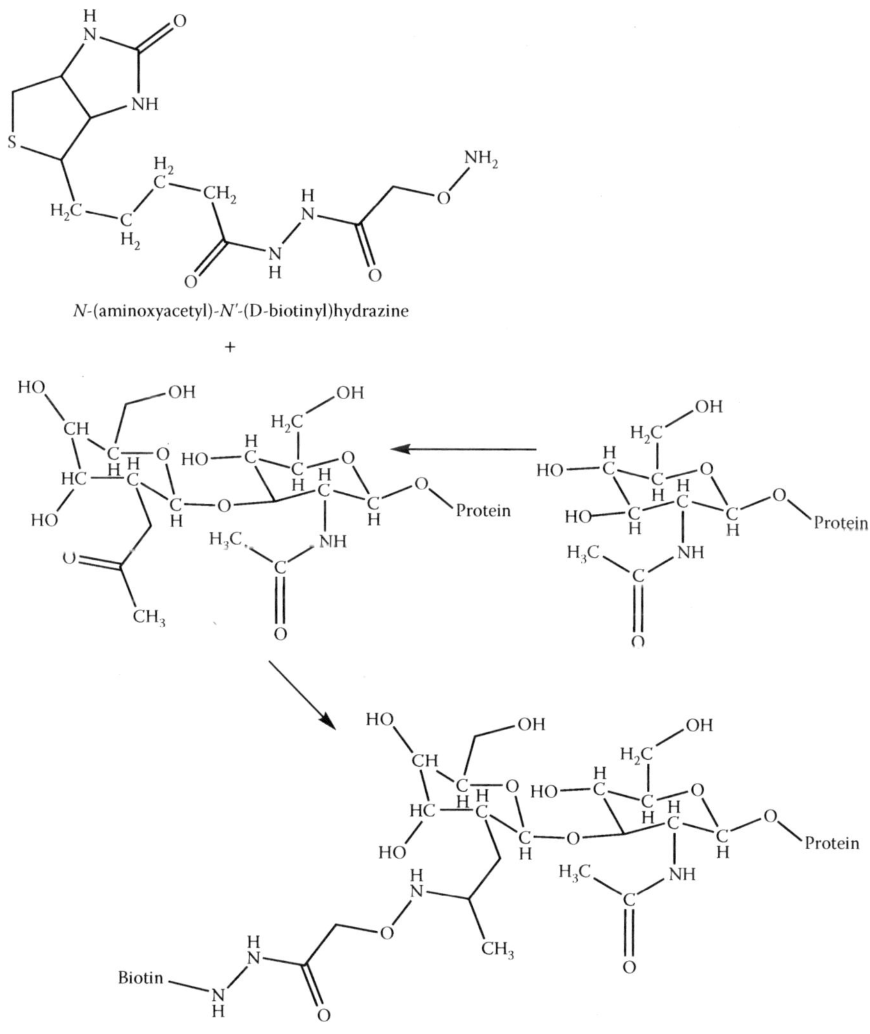

FIGURE 8.24 The combination of a chemical and enzymatic approach to the detection of *O*-acetylglucosamine in proteins.

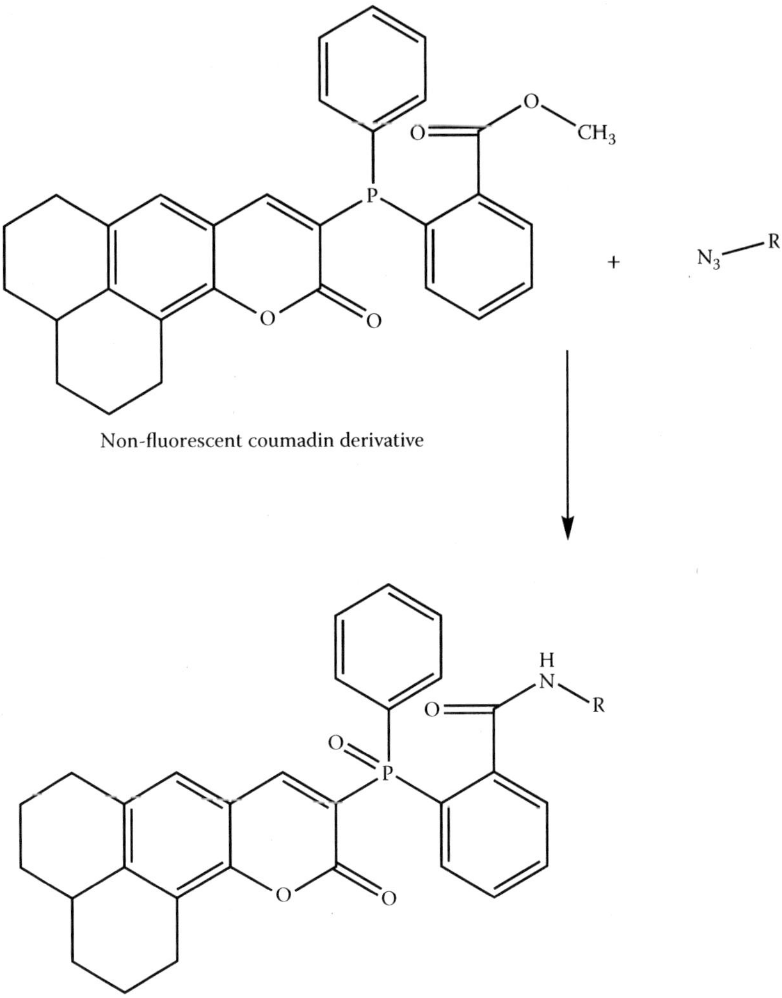

FIGURE 8.25 An example of a fluorogenic dye activated by the Staudinger ligation. (See Lemieux, G.A., De Graffenried, C.L., and Bertozzi, C.R., A fluorogenic dye activated by the Staudinger ligation, *J. Am. Chem. Soc.* 125, 470804709, 2003.)

of adenosine 5′-*O*-(thiotriphosphate) resulted in the labeling of sites of phosphorylation with a thiophosphoryl function. The difference in the acid dissociation constant for the thiophosphoryl group and protein sulfhydryl groups permitted the selective alkylation of the thiophosphoryl function with a tagged reagent allowing isolation and identification. This reaction is shown in Figure 8.26. This group[184]

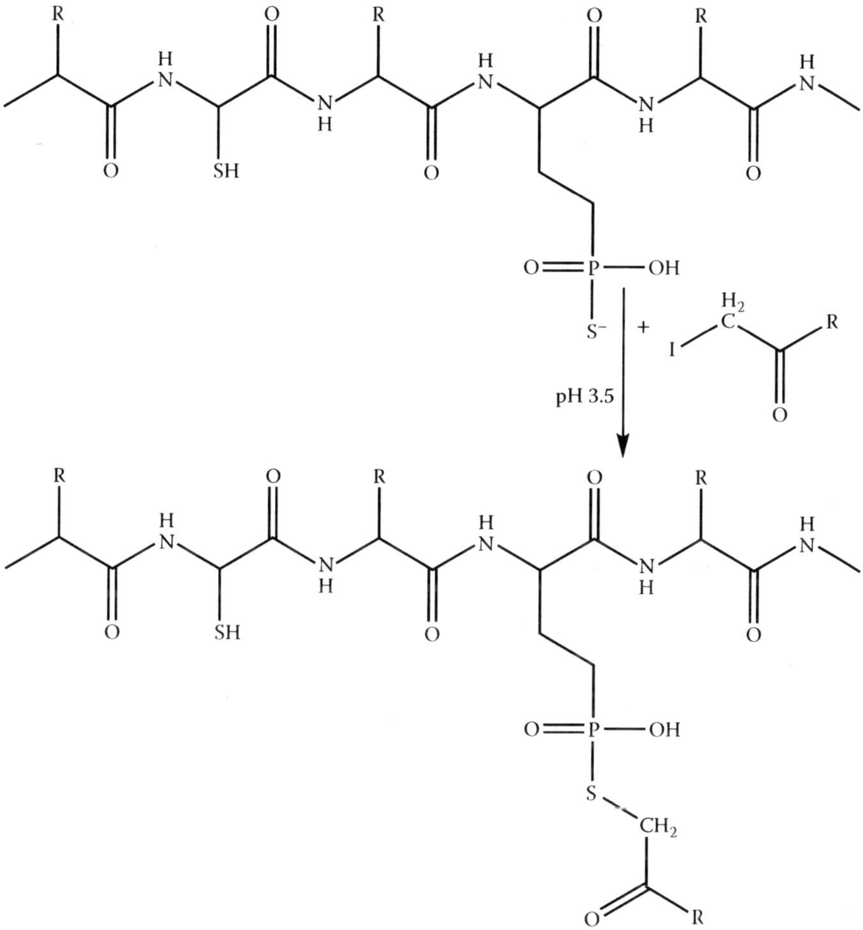

FIGURE 8.26 Modification of a thiophosphorylated serine residues in proteins. Selectively of alkylation is obtained by reaction at low pH.

has also used the Staudinger reaction advanced by Bertozzi and coworkers to characterize farnesyl-modified proteins. An azide–farnesyl intermediate is coupled to the recipient protein. Subsequent condensation with a phosphine derivative allows the placement of a probe such as biotin. The chemistry for this process is outlined in Figure 8.27.

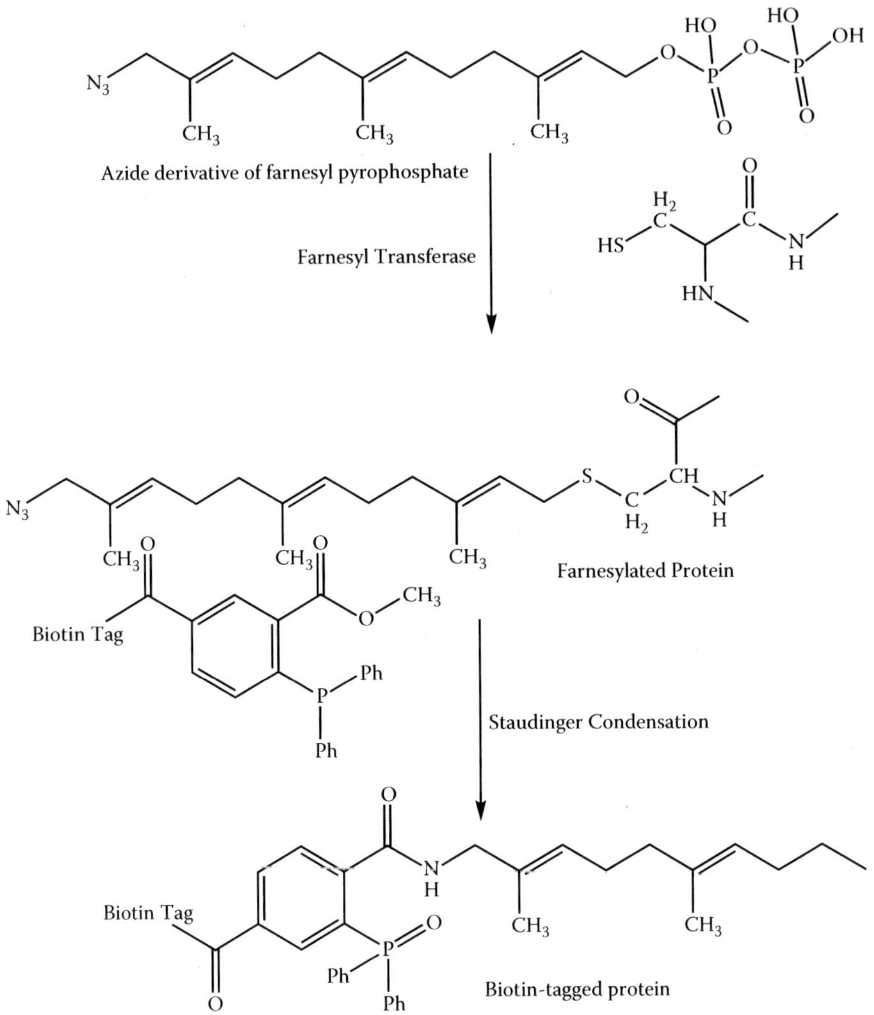

FIGURE 8.27 A process for the detection and isolation of farnesylated proteins. (See Kuo, Y., Kim, S.C., Jiang, C. et al., A tagging-via-substrate technology for detection and proteomics of farnesylated proteins, *Proc. Natl. Acad. Sci. USA* 101, 12479–12484, 2004.)

REFERENCES

Chapter References

1. *Proteins and Proteomics: A Laboratory Manual,* Ed. R.J., Simpson, Cold Spring Harbor Laboratory Press, Cold Spring Harbor, New York, 2003.
2. *Purifying Proteins for Proteomics: A Laboratory Manual*, Ed. R.J. Simpson, Cold Spring Harbor Laboratory Press, Cold Spring Harbor, New York, 2004.
3. *Metabolic Profiling: Its Role in Biomarker Discovery and Gene Function Analysis*, Eds. G.G. Arrigan and R. Goodacre, Kluwer, Boston, 2003.
4. Cullis, C.A., *Plant Genomics and Proteomics*, John Wiley and Sons, Hoboken, NJ, 2004.

5. *Toxicogenomics: Principles and Applications*, Eds. H.K. Hamadeh and C.A. Asfari, Wiley-Liss, Hoboken, NJ, 2004.
6. *Capillary Electrophoresis of Proteins and Peptides*, Eds. M.A. Strege and A.L. Lagu, Humana Press, Towata, NJ, 2004.
7. Pechkova, E., *Proteomics and Nanocrystallography*, Kluwer Academic/Plenum Publishers, 2003.
8. Mayya, V. and Hen, D., Proteomic applications of protein quantification by isotope-dilution mass spectrometry, *Exp. Rev. Proteomics* 3, 561–610, 2006.
9. Haselberg, R., de Jong, G.L., and Somsen, G.W., Capillary electrophoresis–mass spectrometry for the analysis of intact proteins, *J. Chromatog. A.* 1159, 181–209, 2007.
10. Godin, J.P., Fay, L.B., and Hopfgartner, G., Liquid chromatography combined with mass spectrometry for ^{13}C isotope analysis in life sciences research, *Mass Spectrom. Rev.* 26, 751–774, 2007.
11. Dominguez, D.C., Lopex, R., and Torres, M.L., Proteomics technologies, *Clin. Lab. Sci.* 20, 239–244, 2007.
12. Shen, H., Li, X., Biebenoh, C.J., and Frey, D.D., Reducing sample complexity in proteomics by chromatofocusing with simple buffer mixtures, *Methods Mol. Biol.* 424, 187–203, 2008.
13. Geerlof, A., Brown, J., Coutard, B. et al., The impact of protein characterization in structural proteomics, *Acta Crystallog. D Biol. Crystallog.* 62, 1125–1136, 2006.
14. Ho, H.A., Najan, A., and Lecterc, M., Optical detection of DNA and protein with cationic polythiophenes, *Acc. Chem. Res.* 41, 168–178, 2008.
15. Huber, C.G., Walcher, W., Timperio, A.M. et al., Multidimensional proteomic analysis of photosynthetic membrane proteins by liquid extraction-ultracentrifugation-liquid chromatography-mass spectrometry, *Proteomics* 4, 3989–3920, 2004.
16. Oliva, A., Habrés., and Fariña, J.B., Application of multi-cycle laser light-scattering detection in the analysis of human plasma lipoproteins, *Proteomics* 5, 2619–2630, 2005.
17. Arai, Y., Hayashi, M., and Nishimrua, M., Proteomic analysis of highly purified peroxisomes from etiolated soybean cotyledons, *Plant Cell Physiol.* 49, 526–539, 20084.
18. Smolka, M.B, Zhou, H., Purkayastha, S., and Aebersold, R., Optimization of the isotope-coded affinity tag-labeling procedure for quantitative proteome analysis, *Anal. Biochem.* 297, 25, 2001.
19. Watt, S.A., Patschkowski, T., Kalinowski, J., and Niehaus, K., Qualitative and quantitative proteomics by two-dimensional gel electrophoresis peptide mass fingerprinting and a chemical coded affinity tag (CCAT), *J. Biotechnol.* 106, 287, 2003.
20. Parker, K.C., Patterson, D., Williamson, B. et al., Depth of proteome issues: A yeast isotope-coded affinity tag reagent study, *Mol. Cell. Proteomics* 3, 625, 2004.
21. Ünlü, M., Morgan, M.E., and Minden, J.S., Difference gel electrophoresis: A single gel method for detecting changes in protein extracts, *Electrophoresis* 18, 2071, 1997.
22. Jeffrey, D.A. and Bogyo, M., Chemical proteomics and its application to drug discovery, *Curr. Opin. Biotechnol.* 14, 87, 2003.
23. Kocks, C., Maehr, R., Overlkeeft, H.S. et al., Functional proteomics of the active cysteine protease content in Drosophila S2 cells, *Mol. Cell. Proteomics* 2, 1188, 2003.
24. Hemelaar, J., Galardy, P.J., Borodovsky, A. et al., Chemistry-based functional proteomics: Mechanism-based activity-profiling tools for ubiquitin and ubiquitin-like specific proteases, *J. Proteome Res.* 3, 268, 2004.
25. Spears, A.E. and Cravatt, B.F., Chemical strategies for activity-based proteomics, *ChemBioChem.* 5, 41, 2004.
26. Figeys, D., Novel approaches to map protein interactions, *Curr. Opin. Biotechnol.* 14, 119, 2003.
27. Cuatrecasas, P., Wilchek, M., and Anfinsen, C.B., Selective enzyme purification by affinity chromatography, *Proc. Natl. Acad. Sci. USA* 61, 636, 1968.

28. Porath, J. and Kristiansen, T., Biospecific affinity chromatography and related methods, in *The Proteins*, 3rd ed., Eds. H. Neurath and R.L. Hill, Academic Press, New York, Vol. 1, Chapter 2, pp. 95, 1975.
29. Lolli, G., Thaler, F., Valsanina, B. et al., Inhibitor affinity chromatography: Profiling the specific reactivity of the proteome with immobilized molecules, *Proteomics* 3, 1287, 2003.
30. Lee, W.-C. and Lee, K.H., Applications of affinity chromatography in proteomics, *Anal. Biochem.* 324, 1, 2004.
31. Tien, R., Jiang, X., Li, X. et al., Biological fingerprinting analysis of the interactome of a kinase inhibitor in human plasma by a chemiproteomic approach, *J. Chromatog. A* 1134, 134–142, 2006.
32. Lefkovits, I., Functional and structural proteomics: A critical appraisal, *J. Chromatog. B* 787, 1, 2003.
33. Yanagadi, M., Functional proteomics: Current achievements, *J. Chromatog.* 771, 89, 2002.
34. Hemelaar, J., Galardy, P.J., Borodovsky, A. et al., Chemistry-based functional proteomics: Mechanism-based activity-profiling tools for ubiquitin and ubiquitin-like specific proteases, *J. Proteome Res.* 3, 268–273, 2004.
35. Monti, M., Orrù, S., Pagnozzi, D., and Pucci, P., Functional proteomics, *Clin. Chim. Acta* 357, 140–150, 2005.
36. Uttamchandani, M., Li, J., Sun, H., and Yao, S.Q., Activity-based protein profiling: New developments and directions in functional proteomics, *Chembiochem.* 9, 667–675, 2008.
37. Prinz, A., Reither, G., Diskar, M., and Schultz, C., Fluorescence and bioluminescence procedures for functional proteomics, *Proteomics* 8, 1179–1196, 2008.
38. Köcher, T. and Superti-Furga, G., Mass spectrometry-based functional proteomics: From molecular machines to protein networks, *Nat. Methods* 4, 807–815, 2007.
39. Yan, G.R. and He, Q.Y., Functional proteomics to identify critical proteins in signal transduction pathways, *Amino Acids*, 35, 267–274, 2008.
40. Berggård, T., Linse, S., and James, P., Methods for the detection and analysis of protein-protein interactions, *Proteomics* 7, 2833–2842, 2007.
41. Gingras, A.C., Gstaiger, M., Raught, B., and Aebersold, R., Analysis of protein complexes using mass spectrometry, *Nat. Rev. Mol. Cell. Biol.* 8, 645–654, 2007.
42. Kiernan, U.A., Quantitation of target proteins and post-translational modifications in affinity-based proteomics approaches, *Expert. Rev. Proteomics* 4, 421–428, 2007.
43. Righetti, P.G., Campostrini, N., Pascali, J., Herndon, M., and Astner, H., Quantitative proteomics: A review of the different methodologies, *Eur. J. Mass. Spectrom.* 10, 335, 2004.
44. Gygli, S.P., Rist, B., Gerber, S.A. et al., Quantitative analysis of complex protein mixtures using isotope-coded affinity tags, *Nat. Biotechnol.* 17, 994, 1999.
45. Kubota, K., Kosaka, T., and Ichikawa, K., Combination of two-dimensional electrophoresis and shotgun peptide sequences in comparative proteomics, *J. Chromatog. B.* 815, 2, 2005.
46. Whiteleggee, J.P., Mass spectrometry for high throughput quantitative proteomics in plant research: Lessons from thylakoid membranes, *Plant Physiol. Biochem.* 42, 919, 2004.
47. Masselson, C., Paša-Tolić, L., Tolić, N. et al., Targeted comparative proteomics by liquid chromatography-tandem Fourier ion cyclotron resonance mass spectrometry, *Anal. Chem.* 77, 400, 2005.
48. Adam, G.C., Sorensen, E.J., and Cravatt, B.F., Proteomic profiling of mechanistically distinct enzyme classes using a common chemotype, *Nat. Biotechnol.* 20, 805, 2002.
49. Kozarich, J.W., Activity-based proteomics: Enzyme chemistry redux, *Curr. Opin. Chem. Biol.* 7, 78, 2003.
50. Speers, A.E. and Cravatt, B.F., Chemical strategies for activity-based proteomics, *ChemBioChem.* 5, 41, 2004.
51. Hagenstein, M.C. and Sewald, N., Chemical tools for activity-based proteomics, *J. Biotechnol.* 124, 56–73, 2006.

52. Plapp, B.V., Application of affinity labeling for studying structure and function of enzymes, *Methods Enzymol.* 87, 469, 1982.
53. Colman, R.F., Affinity labeling of purine nucleotide sites in proteins, *Annu. Rev. Biochem.* 52, 67, 1983.
54. Sweet, F. and Murdock, G.L., Affinity labeling of hormone-specific proteins, *Endocrinol. Rev.* 8, 154, 1987.
55. Cooperman, B.S., Affinity labeling of ribosomes, *Methods Enzymol.* 164, 341, 1988.
56. Ji, T.H., Nishimura, R., and Ji, I., Affinity labeling of binding proteins for the study of endocytic pathways, *Methods Cell Biol.* 32, 277, 1989.
57. Hohenegger, M., Freissmuth, M., and Nanoff, C., Covalent modification of G proteins by affinity labeling, *Methods Mol. Biol.* 83, 179, 1997.
58. Marcotte, P. and Walsh, C., Vinylglycine and proparglyglycine—complementary suicide substrates for L-amino acid oxidase and D-amino acid oxidase, *Biochemistry* 15, 3070, 1976.
59. Hanzlik, R.P., Kishore, V., and Tullman, R., Cyclopropylamides as suicide substrates for cytochromes P-450, *J. Med. Chem.* 22, 759, 1979.
60. Alston, T.A., Suicide substrates for mitochondrial enzymes, *Pharmcol. Ther.* 12, 1,1981.
61. Walsh, C.T., Suicide substrates, mechanism-based enzyme inactivations: Recent developments, *Annu. Rev. Biochem.* 53, 493, 1984.
62. Decodts, G. and Wakselman, M., Suicide inhibitors of proteases. Lack of activity of halomethyl derivatives of some aromatic lactams, *Eur. J. Med. Chem.* 18, 107, 1983.
63. Asen, A., Progress in focus: Recent advances in histochemistry and cell biology, *Histochem. Cell. Biol.* 118, 507, 2002.
64. Brown, R.E., and Boyle, J.L., Mesenchymal chondrosarcoma: Molecular characterization by a proteomic approach with morphogenic and therapeutic implications, *Ann. Clin. Lab. Sci.* 33, 131, 2003.
65. Kiernan, J.A., Indigogenic substrates for detection and localization of enzymes, *Biotech. Histochem.* 82, 73–103, 2007.
66. Meier-Ruge, W.A. and Bruder, E., Current concepts of enzyme histochemistry in modern pathology, *Pathobiology* 75, 233–243, 2008.
67. Chowdhry, V. and Westheimer, F.H., Photoaffinity labeling of biological systems, *Annu. Rev. Biochem.* 48, 293, 1979.
68. Dorman, G. and Prestwich, G.D., Using photolabile ligands in drug discovery and development, *Trends Biotechnol.* 18, 64, 2000.
69. Evans, S.J. and Moore, F.L., Nonradioactive photoaffinity labeling of steroid receptors using western blot detection systems, *Methods Mol. Biol.* 76, 261, 2001.
70. Hatanaka, Y. and Sadkane, Y., Photoaffinity labeling in drug discovery and developments: Chemical gateway for entering proteomic frontier, *Curr. Top. Med. Chem.* 2, 271, 2002.
71. Gartner, C.A., Photoaffinity ligands in the study of cytochrome p450 active site structure, *Curr. Med. Chem.* 10, 671, 2003.
72. Knorre, D.G. and Godovikova, T.S., Photoaffinity labeling as an approach to study supramolecular structure, *FEBS Lett.* 433, 9, 1998.
73. Robinette, D., Namati, N., Tomer, K.B., and Borchers, C.H., Photoaffinity labeling combined with mass spectrometric approaches as a tool for structural proteomics, *Expert Rev. Proteomics* 3, 399–408, 2006.
74. Jansen, E.P., Nutting, M.D.F. Jong, R., and Balls, A.K., Inhibition of the proteinase and esterase activities of trypsin and chymotrypsin by diisopropyl fluorophosphate: Crystallization of the inhibited chymotrypsin, *J. Biol. Chem.* 179. 189, 1949.
75. Oosterbaan, R.A., Kunst, P., and Cohen, J.A., The nature of the reaction between diisopropylfluorophosphate and chymotrypsin, *Biochim. Biophys. Acta* 16, 299–300, 1955.
76. Jansen, E.P., Nutting, M.-D., F., and Balls, A.K., Mode of inhibition of chymotrypsin by diisopropyl fluorophosphate, I. Introduction of phosphorous, *J. Biol. Chem.* 179, 201, 1949.

77. Cook, R.R. and Powers, J.C., Benzyl-*p*-guanidinothiobenzoate hydrochloride, a new active-site titrant for trypsin and trypsin-like enzymes, *Biochem. J.* 215, 287, 1983.
78. Radisky, E.S. and Koshland, D.E., Jr., A clogged gutter mechanism for serine proteases, *Proc. Nat. Acad. Sci. USA* 99, 10316, 2002.
79. Mazur, A. and Bodensky, O., The mechanism of *in vitro* and *in vivo* inhibition of cholinesterase activity by diisopropyl fluorophosphate, *J. Biol. Chem.* 163, 261, 1945.
80. Liu, Y., Patricelli, M.P., and Cravatt, B.F., Activity-based protein profiling: The serine hydrolases, *Proc. Natl. Acad. Sci. USA* 96, 14694, 1999.
81. Kidd, D., Liu, Y., and Cravatt, B.F., Profiling serine hydrolase activity in complex proteomes, *Biochemistry* 40, 4005, 2001.
82. Kridel, S.J., Axelrod, F., Rozenkrantz, N., and Smith, J.W., Orlistat is a novel inhibitor of fatty acid synthase with antitumor activity, *Cancer Res.* 64, 2070, 2004.
83. Keshavarz-Shokri, A., Suntornwat, O., and Kitos, P.A., Identification of serine esterases in tissue homogenates, *Anal. Biochem.* 267, 406, 1999.
84. Green, A.L. and Nichols, J.D., The reactivation of phosphorylated chymotrypsin, *Biochem. J.* 72, 70, 1959.
85. Cohen, W. and Erlanger, B.F., Studies on the reactivation of diethylphosphoryl-chymotrypsin, *J. Am. Chem. Soc.* 82, 3928, 1960.
86. Cohen, W., Lache, M., and Erlanger, B.F., The reactivation of diethylphosphoryltrypsin, *Biochemistry* 1, 686, 1962.
87. Caplow, M. and Jencks, W.P., The effect of substituents on the deacylation of benzoylchymotrypsin, *Biochemistry* 1, 883, 1962.
88. Moutier, L.A., Reactivation of organophosphate-inhibited trypsin by hydroxylamine, *Biochim. Biophys. Acta* 77, 301, 1963.
89. Gray, A.P., Design and structure-activity relationships of antidotes to organophosphorous anticholinesterase agents, *Drug Metab. Rev.* 15, 557, 1984.
90. Clothier, B. and Johnson, M.K., Rapid aging of neurotoxic esterase after inhibition by di-isopropyl phosphorofluoridate, *Biochem. J.* 177, 549, 1979.
91. Johnson, M.K., Organophosphorous esters causing delayed neurotoxic effects: Mechanism of action and structure activity studies, *Arch. Toxicol.* 34, 259, 1975.
92. Kropp, T.J., Glynn, P., and Richardson, R.J., The Mipafox-inhibited catalytic domain of human neuropathy target esterase ages by reversible proton loss, *Biochemistry* 43, 3716, 2004.
93. Li, Y., Dinsdale, D., and Glynn, P., Protein domains, catalytic activity, and subcellular distribution of neuropathy target esterase in mammalian cells, *J. Biol. Chem.* 278, 8820, 2003.
94. Zaccheo, O., Dinsdale, D., Meacock, P.A., and Glynn, P., Neuropathy target esterase and its yeast homologue degrade phosphatidylcholine to glycerophosphocholine in living cells, *J. Biol. Chem.* 279, 24024, 2004.
95. Estevez, J. Garcia-Perez, A.G., Barril, J., Pellin, M., and Vilanova, E., The inhibition of the high sensitive peripheral nerve soluble esterases by mipafox. A new mathematical processing for the kinetics of inhibition of esterases by organophosphorous compounds, *Toxicol. Lett.* 151, 171, 2004.
96. Williams, D.G. and Johnson, M.K., Gel-electrophoretic identification of hen brain neurotoxic esterase, labelled with tritiated di-isopropyl phosphorofluoridate, *Biochem. J.* 199, 323, 1981.
97. Williams, D.G., Intramolecular group transfer is a characteristic of neurotoxic esterase and is independent of the tissue source of the enzyme, *Biochem. J.* 209, 817, 1983.
98. Planque, S., Bangale, Y., Song, X.-T. et al., Ontogeny of proteolytic immunity. IgM serine protease, *J. Biol. Chem.* 279, 14024, 2004.
99. Planque, S., Taguchi, H., Burr, G. et al., Broadly distributed chemical reactivity of natural antibodies expressed in coordination with specific antigen binding activities, *J. Biol. Chem.* 278, 20436–20443, 2003.

100. Shaw, E. and Glover, G., Further observations on substrate-derived chloromethyl ketones that inactivate trypsin, *Arch. Biochem. Biophys.* 139, 298, 1970.
101. Schoellmann, G. and Shaw, E., Direct evidence for the presence of histidine in the active center of chymotrypsin, *Biochemistry* 2, 252, 1963.
102. Grabarek, J., Dragan, M., Lee, M.W. et al., Activation of chymotrypsin-like serine protease(s) during apoptosis detected by affinity-labeling of the enzymatic center with fluoresceinated inhibitor, *Int. J. Oncol.* 20, 225, 2002.
103. Williams, E.B. and Mann, K.G., Peptide chloromethyl ketones as labelling reagents, *Methods Enzymol.* 222, 503, 1993.
104. Lundblad, R.L., Bergstrom, J., De Vreker, R. et al., Measurement of active coagulation factors in Autoplex®-T with colorimetric active site-specific assay technology, *Thromb. Haemost.* 80, 811, 1998.
105. Mack, A., Furmann, C., and Hacker, G., Detection of capase-activation in intact lymphoid cells using standard capase substrates and inhibitors, *J. Immunol. Methods* 241, 19, 2000.
106. Jayaraman, S., Intracellular determination of activated capases (IDAC) by flow cytometry using a pancapase inhibitor labeled with FITC. *Cytometry* 56A, 104, 2003.
107. Wolthers, B.C., Kinetics of inhibition of papain by TLCK and TPCK in the presence of BAEE as substrate, *FEBS Lett.* 2, 143, 1969.
108. Jonák, J., Rychlík, I., Smrt, J., and Holý, A., The binding site for the 3′-terminus of aminoacyl-tRNA in the molecule of elongation factor Tu for *Escherichia coli. FEBS Lett.* 98, 329, 1979.
109. Shaw, E., Angliker, H., Rauber, P. et al., Peptidyl fluoromethyl ketones as thiol protease inhibitors, *Biomed. Biochem. Acta* 45, 1397, 1986.
110. Falk, M., Ussat, S., Reiling, N. et al., Capase inhibition blocks human T cell proliferation by suppressing appropriate regulation of IL-2, CD25, and cell-cycle associated proteins, *J. Immunol.* 173, 5077, 2004.
111. Bergen, H.R, III, Klug, M.G., Bolander, M.E., and Muddiman, D.C., Informed use of proteolytic inhibitors in biomarker discovery, *Rapid Commun. Mass Spectrometry* 18, 1001, 2004.
112. Kumar, S., Zhou, B., Liang, F. et al., Activity based probes for protein tyrosine phophatases, *Proc. Natl. Acad. Sci. USA* 101, 7942, 2004.
113. Taylor, W.P., Zhang, Z.-Y., and Widlanski, T.S., Quiescent affinity inactivators of protein tyrosine phosphatases, *Bioorgan. Med. Chem.* 4, 1515, 1996.
114. Hemelaar, J., Galardy, P.J., Borodovsky, A. et al., Chemistry-based functional proteomics: Mechanism-based activity-profiling tools for ubiquitin and ubiquitin-like specific proteases, *J. Proteome Res.* 3, 268, 2004.
115. Greenbaum, D., Medzihradszky, K.F., Burlingame, A., and Bogyo, M., Epoxide electrophiles as activity-dependent cysteine protease profiling and discovery tools, *Chem. Biol.* 7, 569, 2000.
116. Albeck, A. and Kliper, S., Inactivation of cysteine proteases by peptidyl epoxides: Characterization of the alkylation sites on the enzyme and the inactivator, *Biochem. J.* 346, 71, 2000.
117. Dupont, D., Rolet-Repecaud, O., and Senocq, D., A new approach to monitoring proteolysis phenomena using antibodies specifically directed against the enzyme cleavage site on its substrate. *Anal. Biochem.* 317, 240, 2003.
118. Dupont, D., Lugand, D., Rolet-Repecaud, O., and Degelaen, J., ELISA to detect proteolysis of ultrahigh-temperature milk upon storage, *J. Agric. Food Chem.* 55, 6857–6862, 2007.
119. Kegel, B., Behrensdorf, H.A., Bonitas, H.A.J. et al., An *in vitro* assay for detection of tetanus neurotoxin activity: Using antibodies for recognizing the proteolytically generated cleavage products, *Toxicol. In Vitro* 21, 1641–1649, 2007.

120. Bredemeyer, A.J., Lewis, R.M., Malone, J.P., Davis, A.E., Gross, J., Townsend, R.R., and Ley, T.J., A proteomic approach for the discovery of proteases substrates, *Proc. Natl. Acad. Sci. USA* 101, 11785, 2004.
121. Adam, G.C., Cravatt, B.F., and Sorensen, E.J., Profiling the specific reactivity of the proteome with non-directed activity-based probes, *Chem. Biol.* 8, 81, 2001.
122. Chuang, L.S.-H., Tan, E.H.-H., Oh, H.-K., and Li, B.F.-L., Selective depletion of human DNA-methyltransferase DNMT1 proteins by sulfonate-derived methylating agents, *Cancer Res.* 62, 1592, 2002.
123. Cloutier, J.-F., Castonguay, A., O'Connor, T.R., and Drouin, R., Alkylating agent and chromatin structure determine sequence context-dependent formation of alkylpurines, *J. Mol. Biol.* 306, 169, 2001.
124. Adam, G.C., Sorensen, E.J., and Cravatt, B.F., Proteomic profiling of mechanistically distinct enzyme classes using a common chemotype, *Nat. Biotechnol.* 20, 805, 2002.
125. Adam, G.C., Sorensen, E.J., and Cravatt, B.F., Trifunctional chemical probes for the consolidated detection and identification of enzyme activities from complex proteomes, *Mol. Cellular Proteomics* 1, 828, 2002.
126. Adam, G.C., Burbaum, J., Kozarich, J.W. et al. Mapping enzyme active sites in complex proteomes, *J. Am. Chem. Soc.* 126, 1363, 2004.
127. Speers, A.E., Adam, G.C., and Cravatt, B.F., Activity-based protein profiling *in vivo* using a copper(I)-catalyzed azide-alkyne [3 + 2] cycloaddition, *J. Am. Chem. Soc.* 125, 4686, 2003.
128. Speers, A.E. and Cravatt, B.F., Profiling enzyme activities *in vivo* using click chemistry methods, *Chem. Biol.* 11, 535, 2004.
129. Breinbauer, R. and Köhn, M., Azide-alkyne coupling: A powerful reaction for bioconjugate chemistry, *ChemBioChem.* 4, 1147, 2003.
130. Wang, Q., Chan, T.R., Hilgraf, R. Fokin, V.V., Sharpless, K.B., Finn, M.G. et al., Bioconjugation by copper(I)-catalyzed azide-alkyne [3 + 2] cycloaddition, *J. Am. Chem. Soc.* 125, 3192, 2003.
131. Levine, R.L., Williams, J.A., Stadtman, E.R., and Shacter, E., Carbonyl assays for determination of oxidatively modified proteins, *Methods Enzymol.* 233, 346, 1994.
132. Ghezzi, P. and Bonetto, V., Redox proteomics: Identification of oxidatively modified proteins, *Proteomics* 3, 1145, 2003.
133. Pietraforte, D., Salzano, A.M., Marino, G., and Minetti, M., Peroxynitrite-dependent modifications of tyrosine residues in hemoglobin. Formation of tyrosyl radical(s) and 3-nitrotyrosine, *Amino Acids* 25, 341–350, 2003.
134. Uppu, R.M., Nossaman, B.D., Greco, A.J. et al., Cardiovascular effects of peroxynitrite, *Clin. Exp. Pharmacol. Physiol.* 34, 933–977, 2007.
135. Nikov, G., Bhat, V., Wishnok, J.S., and Tannenbaum, S.R., Analysis of nitrated proteins by nitrotyrosine-specific affinity probes and mass spectrometry, *Anal. Biochem.* 320, 214, 2003.
136. Sokovlovsky, M., Riordan, J.F., and Vallee, B.L., Conversion of 3-nitrotyrosine to 3-aminotyrosine in peptides and proteins, *Biochem. Biophys. Res. Commun.* 27, 20–25, 1967.
137. Goshe, M.B., Conrads, T.P., Panisko, E.A. et al., Phosphoprotein isotope coded affinity tag approach for isolating and quantitating phosphopeptides in proteome-wide analyses, *Anal. Chem.* 73, 2578, 2001.
138. Zhou, H., Watts, J.D., and Aebersold, R., A systematic approach to the analysis of protein phosphorylation, *Nat. Biotechnol.* 19, 375, 2001.
139. Oda, Y., Nagasu, T., and Chait, B.T., Enrichment analysis of phosphorylated proteins as a tool for probing the phosphoproteome, *Nat. Biotechnol.* 19, 379, 2001.
140. Conrads, T.P., Issaq, H.J., and Veenstra, T.D., New tools for quantitative phosphoproteome analysis, *Biochem. Biophys. Res. Commun.* 290, 885, 2002.

141. Li, W., Backlund, P.S., Boykins, R.A. et al., Susceptibility of the hydroxyl groups in serine and threonine to β-elimination/Michael addition under commonly used conditions moderately high-temperature conditions, *Anal. Biochem.* 323, 94, 2003.
142. Thaler, F., Valsasina, B., Baldi, R. et al., A new approach to phosphoserine and phosphothreonine analysis in peptides and proteins: Chemical modification, enrichment via solid-phase reversible binding, and analysis by mass spectrometry, *Anal. Bioanal. Chem.* 376, 366, 2003.
143. Goshe, M.B., Blonder, J., and Smith, R.D., Affinity labeling of highly hydrophobic integral membrane proteins for proteome-wide analysis, *J. Proteome Res.* 2, 153, 2003.
144. Zhang, H., Yan, W., and Aebersold, R., Chemical probes and tandem mass spectrometry: A strategy for the quantitative analysis of proteomes and subproteomes, *Curr. Opin. Chem. Biol.* 8, 66, 2004.
145. Whitehead, J.K. and Dean, H.G., The isotope derivative method in biochemical analysis, *Methods Biochem. Anal.* 16, 1, 1968.
146. Szpunar, J., Advances in analytical methodology for bioinorganic speciation analysis: Metallomics, metalloproteomics and heteroatom-tagged proteomics and metabolomics, *Analyst* 130, 442, 2005.
147. Gygi, S.P., Rist, B., Gerber, S.A. et al., Quantitative analysis of complex protein mixtures using isotope-coded affinity tags. *Nat. Biotechnol.* 17, 994, 1999.
148. Griffin, T.J. and Aebersold, R., Advances in proteome analysis by mass spectrometry. *J. Biol. Chem.* 276, 45479, 2001.
149. Smolka, M.B., Zhou, H., Purkayastha, S., and Aebersold, R., Optimization of the isotope-coded affinity tag-labeling procedure for quantitative proteome analysis. *Anal. Biochem.* 297, 25, 2001.
150. Smolka, M., Zhou, H., and Aebersold, R., Quantitative protein profiling using two-dimensional gel electrophoresis, isotope-coded affinity tag labeling, and mass spectrometry. *Mol. Cell. Proteomics* 1, 19, 2002.
151. Tao, W.A. and Aebersold, R., Advances in quantitative proteomics via stable isotope tagging and mass spectrometry. *Curr. Opin. Biotechnol.* 14, 110, 2003.
152. Zhang, Z., Edwards, P.J., Roeske, R.W., and Guo, L., Synthesis and self-alkylation of isotope-coded affinity tag reagents, *Bioconjug. Chem.* 16, 458, 2005.
153. Kim, Y.J.K., Zhan, P., Field, B. et al., Reproducibility assessment of relative quantization strategies for LC-MS based proteomics, *Anal. Chem.* 79, 5651–5658, 2007.
154. Guaragna, A., Amoresano, A., Pinto, V. et al., Synthesis and proteomic activity evaluation of a new isotope-coded affinity tagging (ICAT) reagent, *Bioconjug. Chem.* 19, 1095–1101, 2008.
155. Goshe, M.B., Conrads, T.P., Panisko, E.A. et al., Phosphoprotein isotope-coded affinity tag approach for isolating and quantitating phosphopeptides in proteome-wide analyses. *Anal. Chem.* 73, 2578, 2001.
156. Kuyama, H., Watanabe, M., Toda, C. et al., An approach to quantitative proteome analysis by labeling tryptophan residues. *Rapid Commun. Mass. Spectrom.* 17, 1642, 2003.
157. Goshe, M.B. and Smith, R.D., Stable isotope-coded proteomic mass spectrometry. *Curr. Opin. Biotechnol.* 14, 101, 2003.
158. Han, B., Stevens, J.F., and Maier, C.S., Design, synthesis, and application of hydrazide-functionalized isotope-coded affinity tags for the quantization of oxylipid-protein conjugates, *Anal. Chem.* 79, 3342–3354, 2007.
159. Qiu, Y., Sousa, E.A., Hewick, R.M., and Wang, J.H., Acid-labile isotope-coded extractants: A class of reagents for quantitative mass spectrometric analysis of complex protein mixtures. *Anal. Chem.* 74, 4969, 2002.
160. Lu, Y., Bottari, P., Turecek, F. et al., Absolute quantification of specific proteins in complex mixtures using visible isotope-coded affinity tags, *Anal. Chem.* 76,, 4104, 2004.

161. Lu, A., Bottari, P., Aebersold, R. et al., Absolute quantification of specific proteins in complex mixtures using visible isotope-coded affinity tags, *Methods Mol. Biol.* 359, 159–176, 2007.
162. Zhang, R., Sioma, C.S., Wang, S., and Regnier, F.E., Fractionation of isotopically labeled peptides in quantitative proteomics. *Anal. Chem.* 73, 5142, 2001.
163. Chakraborty, A. and Regnier, F.E., Global internal standard technology for comparative proteomics, *J. Chromatog. A.* 949, 173, 2002.
164. Ross, P.L., Huang, Y.N., Marchese, J.N. et al., Multiplexed protein quantization in *Saccharomyces cerevisiae* using amine-reactive isobaric tagging reagents, *Mol. Cell. Proteomics* 3, 1154–1169, 2004.
165. Choe, L.H., Aggarwal, K., Franck, AZ., and Lee, K.H., A comparison of the consistency of proteome quantitation using two-dimensional electrophoresis and shotgun isobaric tagging in *Escherichia coli* cells, *Electrophoresis* 26, 2437–2449, 2005.
166. Hardt, M., Witkowska, H.E., Webb. S. et al., Assessing the effects of diurnal variation on the composition of human parotid saliva: Quantitative analysis of native peptides using iTRAQ reagents, *Anal. Chem.* 77, 4947–4954, 2005.
167. Sachon, E., Mohammed, S., Bache, N., and Jensen, O.N., Phosphopeptide quantization using amine-reactive isobaric tagging reagents and tandem mass spectrometry: Application to proteins isolated by gel electrophoresis, *Rapid Commun. Mass. Spectrom.* 20, 1127–1134, 2006.
168. Bantscheff, M., Boesche, M., Eberhard, D. et al., Robust and sensitive iTRAQ quantification on an LTQ Orbitrap mass spectrometer, *Mol. Cell. Proteomics*, 7, 1702–1713, 2008.
169. Quaglia, M., Pritchard, C., Hall, E., and O'Connor, G., Amine-reactive isobaric tagging reagents: Requirements for absolute quantification of proteins and peptides, *Anal. Biochem.* 379, 164–169, 2008.
170. Gevaert, K., Van Damme, J., Goethals, M. et al., Chromatographic isolation of methionine-containing peptides for gel-free proteome analysis, *Mol. Cell. Proteomics* 1, 896, 2002.
171. Savige, W.E. and Fontana, A., Interconversion of methionine and methionine sulfoxide, *Methods Enzymol.* 47, 453, 1977.
172. Tang, J. and Hartley, B.S., A diagonal electrophoretic method for the purification of methionine peptides, *Biochem. J.* 102, 593, 1967.
173. Gundlach, G., Moore, S., and Stein, W.H., The reaction of iodoacetate with methionine, *J. Biol. Chem.* 234, 1761, 1959.
174. Weinberger, S.R., Viner, R.I., and Ho, P., Tagless extraction-retentate chromatography: A new global protein digestion strategy for monitoring differential protein expression, *Electrophoresis* 23, 3182, 2002.
175. Shen, M., Guo, L., Wallace, A. et al., Isolation and isotope labeling of cysteine- and methionine-containing tryptic peptides. Application to the study of cell surface proteolysis, *Mol. Cell. Proteomics* 2, 315, 2003.
176. Grunert, T., Pock, K., Buchacher, A., and Allmaier, G., Selective solid-phase isolation of methionine-containing peptides and subsequent matrix-assisted laser desorption mass spectrometric detection of methionine- and methionine-sulfoxide-containing peptides, *Rapid Commun. Mass Spectrom.* 17, 1815, 2003.
177. Khidekel, N., Ficarro, S.B., Peters, E.C., and Hsieh-Wilson, L.C., Exploring the *O*-GlaNac-modified proteins from the brain, *Proc. Natl. Acad. Sci.* 101, 13132, 2004.
178. Datta, D., Wang, P., Carrico, I.S., Mayo, S.L., and Tirrell, D.A., A designed phenylalanyl-tRNA synthetase variant allows efficient *in vivo* incorporation of aryl ketone functionality into proteins, *J. Am. Chem. Soc.* 124, 5652, 2002.

179. Khidekel, N., Arndt, S., Lamarre-Vincent, N. et al., A chemoenzymatic approach toward the rapid and sensitive detection of *O*-GlcNAc posttranslational modifications, *J. Am. Chem. Soc.* 125, 16162, 2003.
180. Saxon, E. and Bertozzi, C.R., Cell surface engineering by a modified Staudinger reaction, *Science* 287, 2007, 2000.
181. Vacadlo, D.J., Hang, H.C., Kim, E.-J. et al., A chemical approach for identifying *O*-GlcNAc-modified proteins in cells, *Proc. Natl. Acad. Sci. USA* 100, 9116, 2003.
182. Lemieux, G.A., de Graffenried, C.L., and Bertozzi, C.R., A fluorogenic dye activated by the Staudinger ligation, *J. Am. Chem. Soc.* 125, 4708, 2003.
183. Kwon, S.W., Kim, S.C., Jaunbergs, J. et al., Selective enrichment of thiophosphorylated polypeptides as a tool for the analysis of protein phosphorylation, *Mol. Cell. Proteomics* 2, 242, 2003.
184. Kho, Y., Kim, S.C., Jiang, C. et al., A tagging-via-substrate technology for detection and proteomics of farnesylated proteins, *Proc. Natl. Acad. Sci. USA* 101, 12479, 2004.

9 Use of Chemical Modification to Produce Biopharmaceutical Products

It can be argued that the term *biopharmaceutical* is more of a marketing term than a technical term and has been hijacked to include a variety of drugs which are not biological polymers[1,2]; for the purpose of the current discussion, the term *biopharmaceutical* includes materials such as peptides/proteins, oligonucleotides/polynucleotides or oligosaccharides/polysaccharides. Albumin was the first approved protein therapeutic; other early approved therapeutics include plasma protein fraction, thrombin, and intravenous immunoglobulin (IVIG).[3–5] More recently there have been a number of therapeutic protein and peptide products.[6–9] Therapeutic preparations of carbohydrate such as dextran and hydroxyethyl starch are colloids used for plasma expanders.[10–12] Antisense nucleotides and small, interfering RNA molecules (siRNA) are being developed as therapeutics.[13–16] The reader is directed to excellent recent reviews by Walsh, which provide global coverage of various biological molecules.[17–19]

Therapeutic preparations (active pharmaceutical ingredients; final drug products) of protein, carbohydrate, lipid, or nucleic acids, or combinations thereof are usually prepared either by purification from natural sources such as blood or derived from fermentation or cell culture. There are some examples of chemical synthesis in the preparation of lipid-derived therapeutics such as liposomes,[20,21] nucleic acids,[22,23] and in the synthesis of peptides.[24,25] While it has been possible to achieve the "semisynthesis" of a protein by chemical ligation[26] (see Chapter 4), the total chemical synthesis of a protein has been achieved only for pancreatic ribonuclease.[27] Other proteins[28–31] have been altered only in part.

Since the goal of these exercises is the production of biopharmaceuticals, there are at least three issues to keep in mind when approaching the problem in order to reach the goal of a licensed product. While modification of nucleic acids and polysaccharides may well yield new products, modification of proteins will yield preparations with the same indications as the "original" material but enhanced characteristics such as storage stability or increased circulatory half-life. Thus, to some extent, the modified proteins will be "follow-on" biologicals (biosimilars; general biologics).[32,33]

It is useful to consider briefly some aspect of regulatory guidelines that have impact on the use of chemical medication to manufacture biopharmaceuticals.

1. The modification reaction must be robust and, as such, able to be validated.[34–46] This enables compliance with CGMP manufacturing regulations.
2. Reagents that are used in the modification/manufacturing process must be removed during the process OR shown to be innocuous. Removal is far superior. As a corollary, you must have an assay to measure the reagent/products as well as proof of toxic levels. The reader is directed to ICH Documents[37,38] entitled "Impurities in New Drug Substances" and "Impurities in Drug Products" for a discussion of these issues.
3. It is necessary to demonstrate rational development[39] and control of the overall manufacturing process[40] and have excellent characterization of the active pharmaceutical ingredient.[41–43] This characterization is assumed to include mass spectrometry[44–46] as well and NMR and capillary electrophoresis.[47,48]

While the total chemical synthesis of a large polysaccharide, nucleic acid, or intact protein is a formidable challenge, the chemical modification of such a macromolecule derived from natural or recombinant sources would appear to be a more reasonable proposition. However, there are still both technical and regulatory challenges.

Currently approved drugs/biologicals obtained by chemical modification are, with the exception of some oligonucleotides, limited to products obtained by derivatization of existing products such as asparaginase, interferon, or erythropoietin. The "approved" protein or polysaccharide is then the starting material for the process; in other words, the active pharmaceutical ingredient that is formulated into the final drug product in one process is now the biological source material[49] for another process. This would require documentation of such as Drug Master File.[50] It will likely simplify matters if the specifications for this material are the same as those for the intermediate step in the manufacturing process prior to formulation of the final drug product. The management of this material is critical as storage conditions may influence reactivity in the subsequent manufacturing process.

The development of a product should include consideration of Quality by Design.[51] This concept was recently adopted by the biotechnology community, and it is of longstanding in other manufacturing processes.[52–55] Fundamentally, this requires that one understand the manufacturing and formulation process that results in the final drug product. The use of chemical modification requires a fundamental understanding of the basic chemical reaction and environmental factors such as solvent effects and conformation issues which influence the chemical reaction (see Chapter 1).

CHEMICAL MODIFICATION OF OLIGOSACCHARIDES/POLYSACCHARIDES TO PRODUCE THERAPEUTIC PRODUCTS

Hydroxyethyl starch (HES; Hetastarch) has a long history of use as a volume expander.[56–61] Hydroxyethyl starch is prepared by the esterification of starch with ethylene chlorohydrin in pyridine (Figure 9.1).[62] Figure 9.1 shows modification at the 6 position; substitution can also occur at the 2 position.[60,61] The primary use of HES is as a plasma volume expander; there has been use in other areas. Whistler and Belfort[56] showed that hydroxyethyl starch produced the same weight gain in laboratory

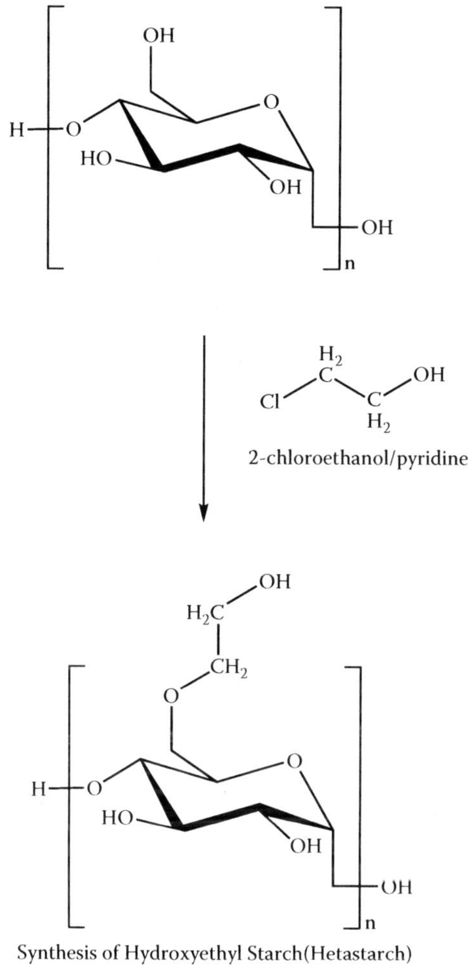

FIGURE 9.1 Preparation of hydroxyethyl starch.

animals as unmodified corn starch, whereas oxidized corn starch produced a lower weight gain. Hydroxyethyl starch has also been used as cyropreservative.[57,58]

Cellulose is subjected to chemical modification to produce a variety of products including membranes for hemodialys.[63–67] Dextran sulfate (Figure 9.2) was developed as a synthetic analog of heparin.[68] Dextran sulfate is used as matrix in apheresis for removal of low-density lipoproteins.[69–71]

There is more information on the modification of carbohydrate in Chapter 10 on food chemistry and Chapter 4 on bioconjugates.

CHEMICAL MODIFICATION OF NUCLEIC ACIDS

The chemical modification of nucleic acids is not as complex as that of proteins since there are fewer monomer units and, for all practical purposes, only nitrogen

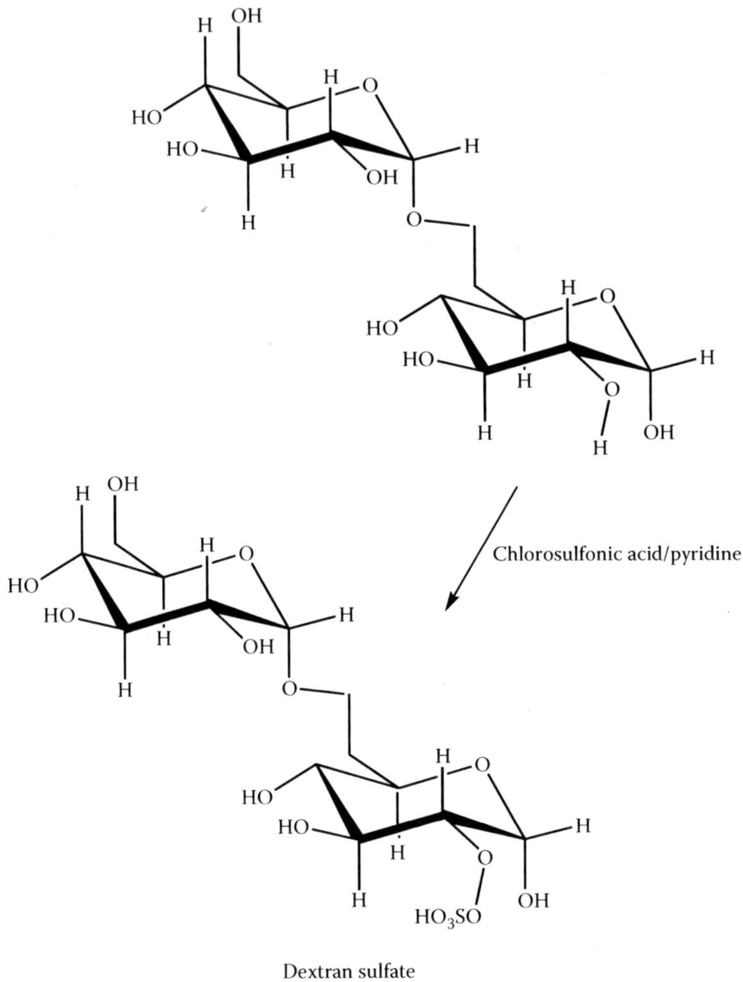

FIGURE 9.2 Dextran sulfate.

as a nucleophilic reactive group; the nitrogen is reactive as a primary and secondary amine. Reaction at the primary amine groups of, for example, adenine, is referred to as *exocyclic modification* whereas reaction at the imine nitrogens of pyrimidines and purine rings is referred to as an *endocyclic modification*. There are also ring-opening reactions and cross-linking reactions. These are reactions which occur with oligonucleotides and polynucleotides; the reactions are not as specific as the modification of amino acid residues and, to that extent, it is difficult to obtain "site-specific" modification of a nucleic acid. Reagents such as diethylpyrocarbonate (Figure 9.3),[72] dimethyl sulfate (Figure 9.4),[73] or nitrous acid (Figure 9.5)[74] provide a more general modification of the nucleic acid molecule. The use of chemical modification in the nucleic acids described below occurs with the precursor nucleotides or nucleosides

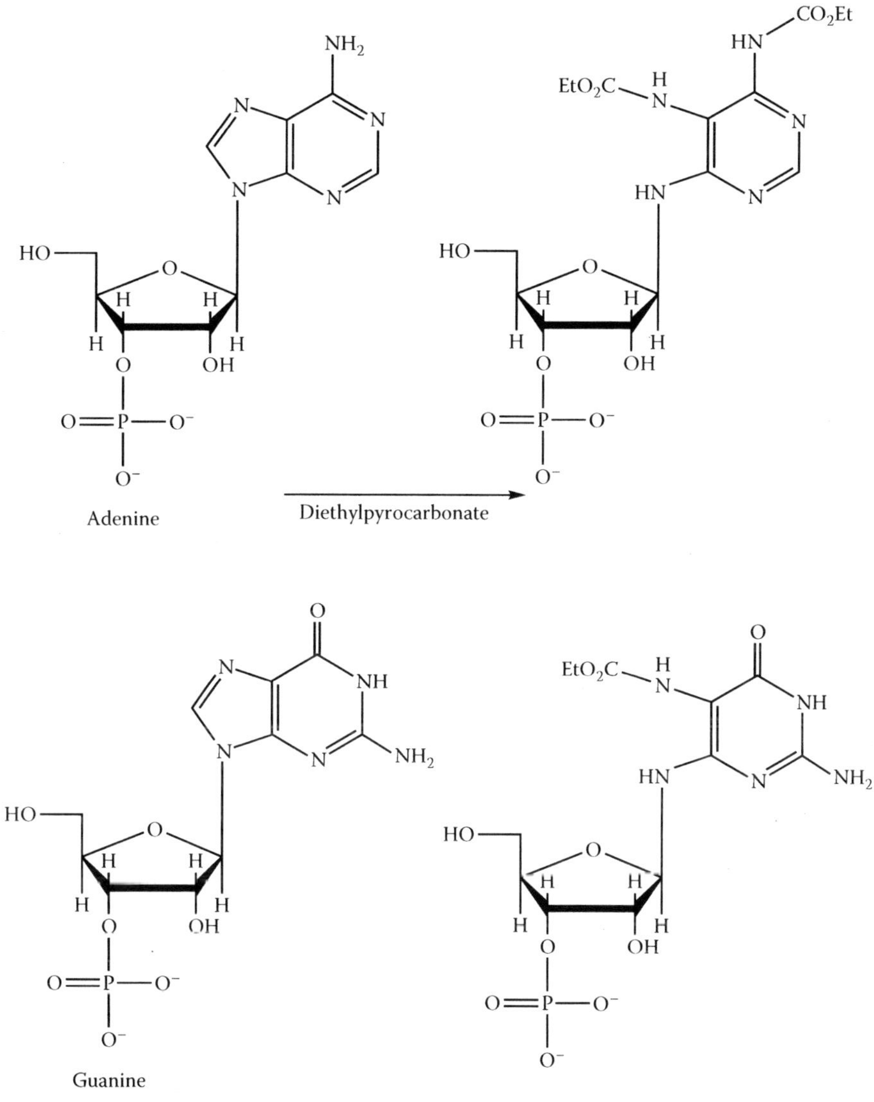

FIGURE 9.3 Diethylpyrocarbonate and nucleic acids.

rather than with the final oligonucleotide product. Thus, it is similar to the use of amino acid analogues[75–79] which can be incorporated into proteins and peptides.

The discovery of RNA interference has stimulated the chemical modification of nucleic acid-based drug products.[80] Small interfering RNA (siRNA) are approximately 20 base pairs in length and the "active" strand, which is complementary to an mRNA sequence, is in a double-strand complex with an inactive strand that is lost in the formation of the RNA-induced silencing complex (RISC). siRNA are closely related to antisense oligonucleotides (ASO), and technologies developed with

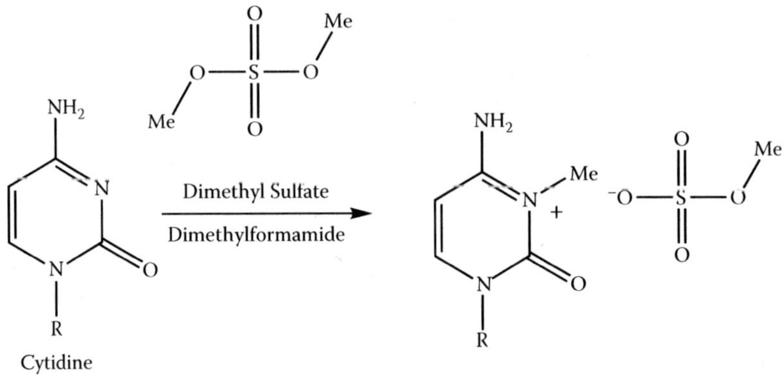

FIGURE 9.4 Dimethyl sulfate and nucleic acids.

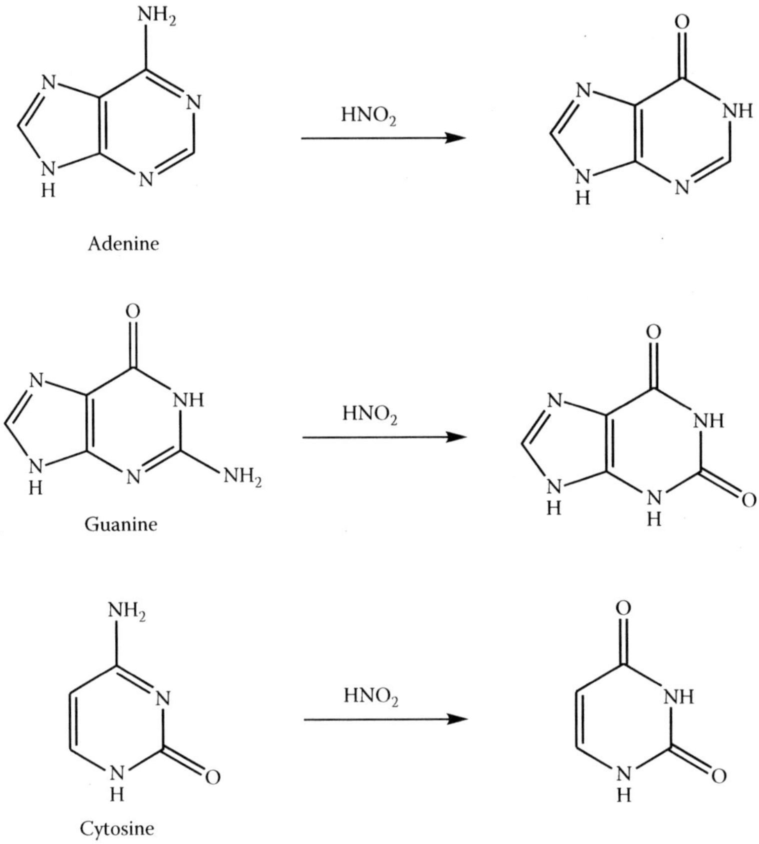

FIGURE 9.5 Nitrous acid and nucleic acids.

antisense oligonucleotides have proved valuable in the development of siRNA.[81–89] Antisense oligonucleotides inhibit expression of a specific gene product by binding to a complementary RNA sequence; this technology has resulted in drug products.[90,91]

Aptamers are another group of nucleic-acid drugs[92–98] that has been demonstrated to have specific protein-binding properties. Aptamers have been used as probes in microarrays, so there has been interest in chemistry associated with binding to matrices.[99] The chemistry for the attachment is described in Chapter 3.

Unlike aptamers, ASO drug products and siRNA drug products function inside the cell; therefore, there are issues of stability to nuclease degradation (also an issue with aptamers), target cell specificity, and cell penetration to consider in the development of a drug product.[80] Modifications have focused on improving stability by modification of the sugar moiety (Figure 9.6),[100–104] and modification of the phosphodiester backbone (Figure 9.7).[105–108] It has been possible to obtain some specificity by the modification of protein–nucleic acid complexes (hydroxyl radicals) and by using results to design a specific siRNA for the protein.[109]

CHEMICAL MODIFICATION AND THE MANUFACTURE OF THERAPEUTIC PROTEINS

Chemical modification has been used to produce novel biotherapeutics.[110–114] Recombinant DNA technology combined with protein engineering has permitted the development of novel proteins resulting from the ability to either insert or remove reactive amino acid residues. The insertion of cysteine residues as discussed earlier provide a reactive nucleophilic site available for modification,[115–119] whereas the removal of lysine residues permits specific modification at the amino-terminal residues[120]; modification at the amino-terminal residue can also be accomplished by performing the reaction at slightly acidic pH.[121] Protein design is still somewhat limited by the biosynthetic machinery of either the prokaryotic or eukaryotic cells, although some novel strategies are being developed for the incorporation of non-naturally occurring amino acids into proteins.[75–79,122–127] Some of these, such as the incorporation of *m*-acetyl L-phenylalanine,[123] provide a novel functional group for subsequent reaction. Protein ligation as discussed in Chapter 4 on bioconjugates also provides a mechanism for the insertion of unique reactive sites in a protein.

CHEMICAL GLYCOSYLATION

The coupling of proteins and nucleic acids to carbohydrate via periodate oxidation or hydrazide coupling has been discussed in Chapter 4 (bioconjugation). Glycation is a reaction that occurs between reducing sugars or compounds such as methyl glyoxal and amino groups and is a well-described phenomena.[128,129] Advanced glycation end products (AGE) are biomarkers of diseases such as diabetes. Chemical glycosylation, is the site-specific chemical modification of a protein or nucleic acid (lipids and other compounds are not excluded but are somewhat rare; the enabling chemistry would be the same) with the goal of improving product quality.[130,131] The approach to chemical glycosylation builds upon the experience developed with reagents used

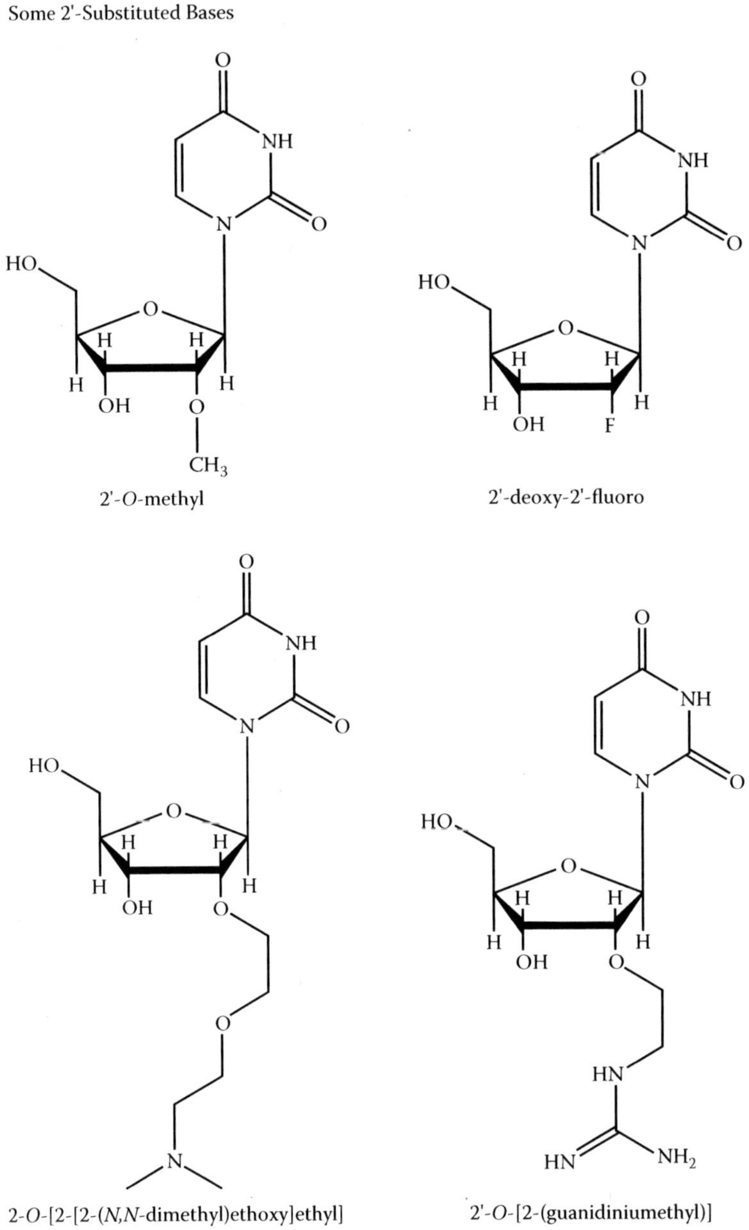

FIGURE 9.6 Some 2′-substituted nucleosides used in the chemical modification of therapeutic oligonucleotides. (Adapted from Prakash, T.P. and Bhat, B., 2′-modified oligonucleotides for antisense therapeutics, *Curr. Top. Med. Chem.* 7, 641–649, 2007.) Uridine is used for an example in this figure.

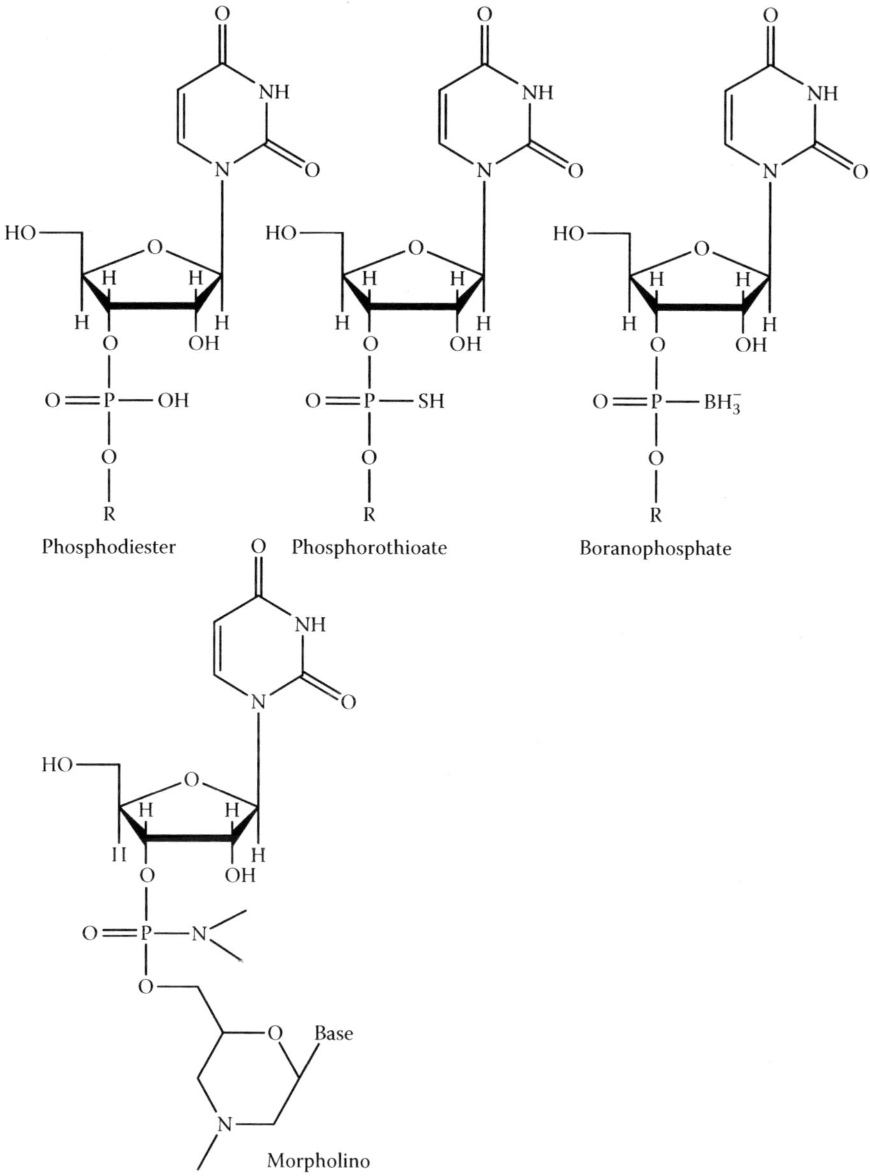

FIGURE 9.7 Some backbones used for chemically modified oligonucleotides. (Taken in part from Taylor, M.F., Paulaukis, J.D., Weller, D.D., and Kobzik, L., *In vitro* efficacy of morpholino modified antisense oligomers directed against tumor necrosis factor-α mRNA, *J. Biol. Chem.* 271, 17445–17452, 1996; and Zhang, H.-Y., Du, Q., Wahlestedt, C., and Liang, Z., RNA interference with chemically modified SiRNA, *Curr. Top. Med. Chem.* 6, 893–900, 2006.)

for structure-function studies in proteins as described in Chapter 1. Sabesan and Linna[132] described the use of acyl azides (Figure 9.8) as developed by Lemieux and coworkers[133] for chemical glycosylation. This approach has been used recently for the chemical glycosylation of tumor necrosis factor α with *N*-acetylneuraminic acid.[134] Acyl azide chemistry has also been used for the immobilization of proteins on agarose matrices.[135] MacMillan and others[136] used the cysteine mutagenesis strategy described above to replace asparagine residues with cysteine residues in human erythropoietin; these residues were then modified with β-*N*-glycosyl iodoacetamides (Figure 9.9) to obtain a homogeneous glycoform. Other approaches have been used[137]

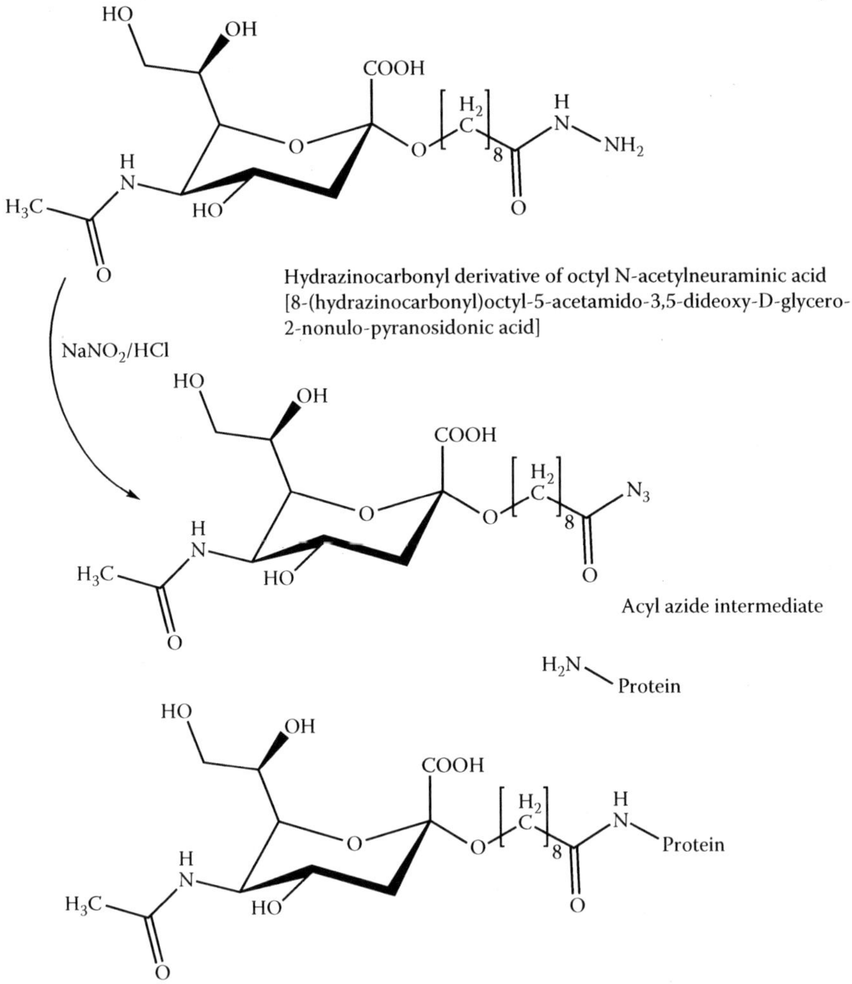

FIGURE 9.8 The use of acyl azide chemistry for chemical glycosylation. (Adapted from Hayashi, A., Chiba, T., Hayashi, H., et al., Synthesis of glycosylated human tumor necrosis factor α coupled with N-A/acetylneuraminic acid, *Cancer Immunol. Immunother.* 56, 545–553, 2007.)

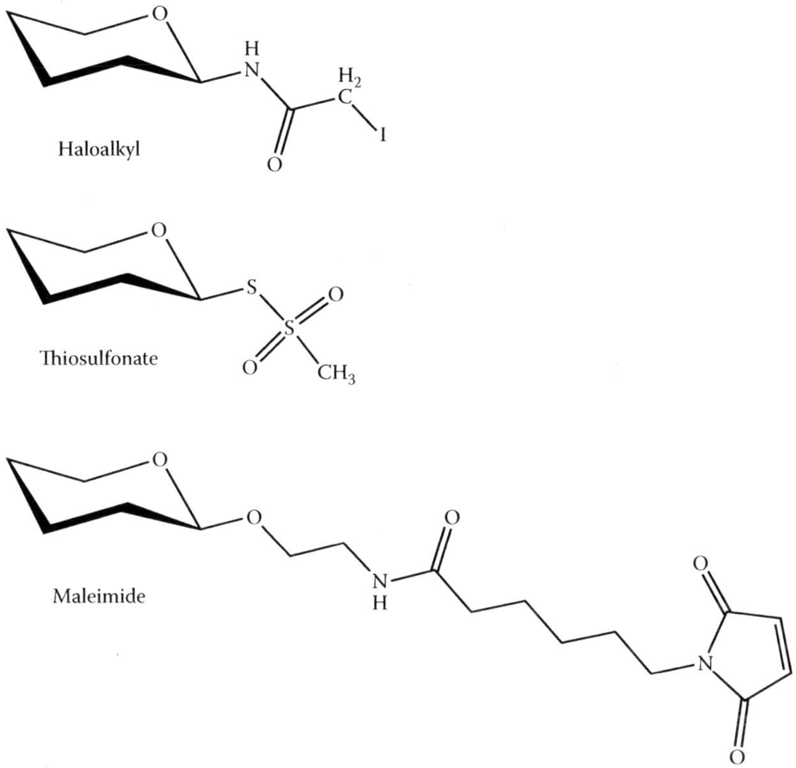

FIGURE 9.9 Some reagents used to introduce carbohydrate by the modification of cysteine. (Adapted from Macmillan, D., Bill, R.M., Sage, K.A. et al., Selective *in vitro* glycosylation of recombinant proteins: semi synthesis of novel homogeneous glycoforms of human erythropoietin, *Chem. Biol.* 8, 133–143, 2001; Ni, J.H., Singh, S., and Wang, L.X., Synthesis of maleimole-activated carbohydrate as chemoselective tags for site-specific glycosylation of peptides and proteins, *Bioconjug.Chem.* 14, 232–238, 2003; and Swanwick, R.S., Daines, A.M., Flitsch, S.I., and Allemann, R.K., Synthesis of homogeneous site-selectively glycosylated proteins, *Org. Biomol. Chem.* 3, 572–574, 2005.)

including maleimides[138] and thiosulfonates.[139] Galonić and Gn[140] used a variety of approaches to the synthesis of glycoconjugate antitumor vaccines. Langenhan and coworkers[141] prepared a "library" of neoglycosideds by the reaction of reducing sugars and aglycon forms (alkoxyamide) for cardiac glycosides (Figure 9.10).

ALLERGOIDS

Allergoids are chemically modified allergens (allergenic components)[142] that are used as a vaccine for type 1 allergies, which affect approximately 25% of the population.[143] Allergens (usually proteins) are components of pollens, which cause a hypersensitivity defined as an allergy. This hypersensitivity (allergic response) is an inflammatory response to the interaction of allergens with the IgG component,

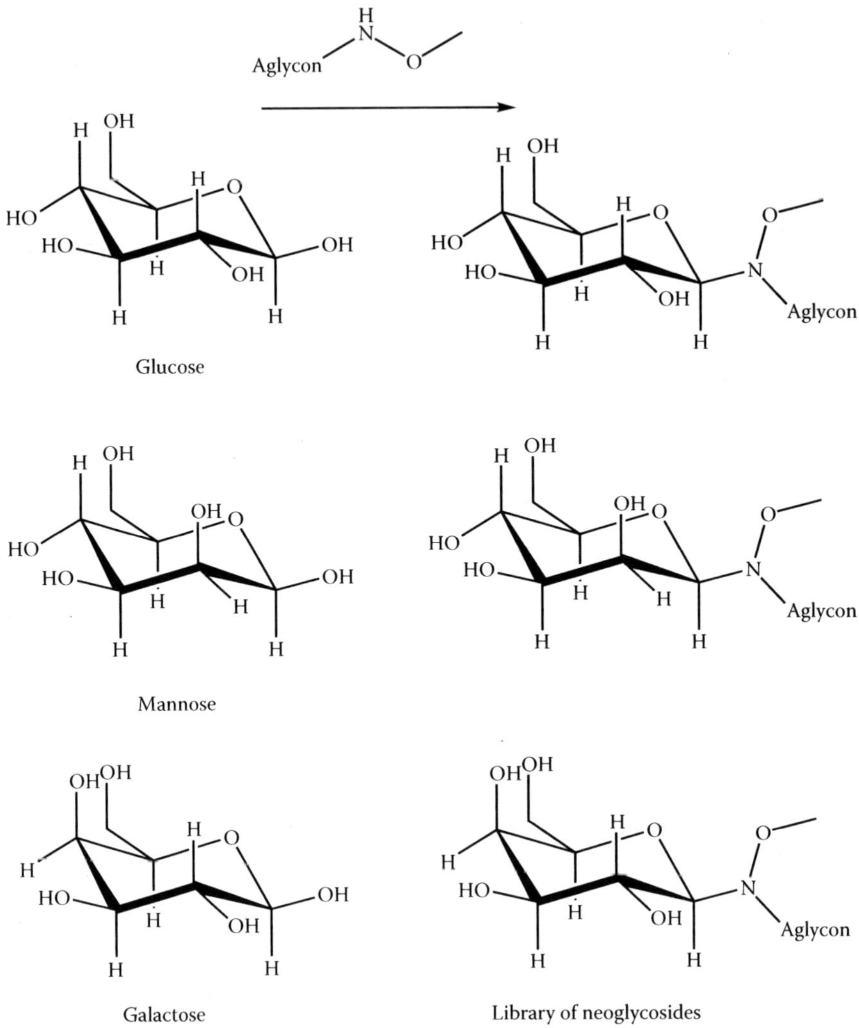

FIGURE 9.10 Formation of a library of neoglycosides by modification of aglycons by reaction with reducing sugars. (Adapted from Langenhan, J.M., Peters, N.R., Guzel, I.A. et al., Enhancing the anticancer properties of cardiac glycosides by neoglycorandomization, *Proc. Natl. Acad. Sci. USA* 102, 12305–12310, 2005.)

which causes the release of histamine and leukotrienes from mast cells; activation of specific T-cells may also be included in this pathophysiology. The consequences of an allergic response range from mild irritation to life-threatening.

The seminal paper by Marsh and colleagues in 1970[142] extended earlier observations[144] which showed that formalin (formaldehyde) could reduce the activity of toxins by created nontoxic derivatives that could be used as a vaccine. Marsh and colleagues[142] showed that treatment of rye grass pollen (2.0 mg/mL in 0.1 M sodium phosphate, pH 7.5 containing 1 part per 10,000 merthiolate) with 60 mM formaldehyde for 32 days at 32°C resulted in a loss of allergenicity (*in vitro* histamine release) but with substantial retention of immunogenicity; the inclusion of ornithine, for example, protected against some loss of allergenicity with an increase in immunogenicity. These modified allergens (allergoids) did elicit the formation of antibody, which blocked the allergic response.[143] It is assumed that the reaction of the allergen with formaldehyde "inactivated" the epitopes reacting with IgG while retaining determinants which would elicit the formation of blocking antibodies.[143]

The reaction of formaldehyde with biological materials occurs at primary amine (see Chapter 1) and can be a complex process (Figure 9.11). The chemistry of formaldehyde relative to biopharmaceuticals is discussed in greater detail below. Despite the scientific and regulatory complexity of formaldehyde, it continues to be used for the preparation of allergoids,[145–149] and at least one commercial product is available.[150] Salgado and colleagues[149] explored the reaction of formaldehyde with ovalbumin in detail as a model system for allergoid synthesis. Modification of ovalbumin (2 mg/mL) with formaldehyde (pH 7.5/10°C/0.5 M formaldehyde) resulted in an allergoid with 50% reduced binding to IgE, a 1000-fold reduction in the histamine release assay but induced a higher production of IgG. It is suggested that assays in this study provide a framework for the preclinical testing allergoids.

Since formaldehyde cross-links proteins, glutaraldehyde (Figure 9.12) has been evaluated as a reagent for the preparation of allergoids.[151–153] Polymerization is not necessary for the preparation of allergoids as carbamylation (Figure 9.13) has been sucessful.[154,155] Modification of allergens with acid anhydrides such as maleic anhydride and succinic anhydride (Figure 9.12) has also been useful in producing allergoids.[156] In subsequent work, Ćirković, and coworkers[157] also showed that modification with succinic anhydride was most effective in reducing the allergenicity of *Artemisia vulgaris* pollen, while preservation of antigenicity was greatest with the carbamyl derivative.

Recombinant DNA technology has also been used to prepare allergoids.[158–161]

CROSS-LINKAGE

The cross-linking of macromolecules has proved useful in the preparation of biopharmaceuticals; use in the manufacture of hydrogels is discussed in Chapter 5. As is noted above in the discussion of chemical modification (Chapter 1) and in the chapter on protein hydrogels (Chapter 5), there are a variety of chemistries available for the cross-linking of proteins. It is therefore of some interest that one of the oldest cross-linking reagents, glutaraldehyde,[162] is widely used for the preparation of protein-based hydrogels (see Chapter 5). Glutaraldehyde is widely used for a variety of purposes[163–169] from the fixation of animal heart valves for use in cardiovascular surgery to the preparation of tissues for electron microscopy, and thus is "understood" in the medical device and diagnostic community. Glutaraldehyde treatment reduces or eliminates the antigenicity of cardiovascular tissue permitting its use as implant material. It is used as histological preservative where the masking of antigens can be reversed by heat treatment.[169–171] Glutaraldehyde is used to produce allergens (see above) but the immunological response of proteins to modification can be variable.[172–174]

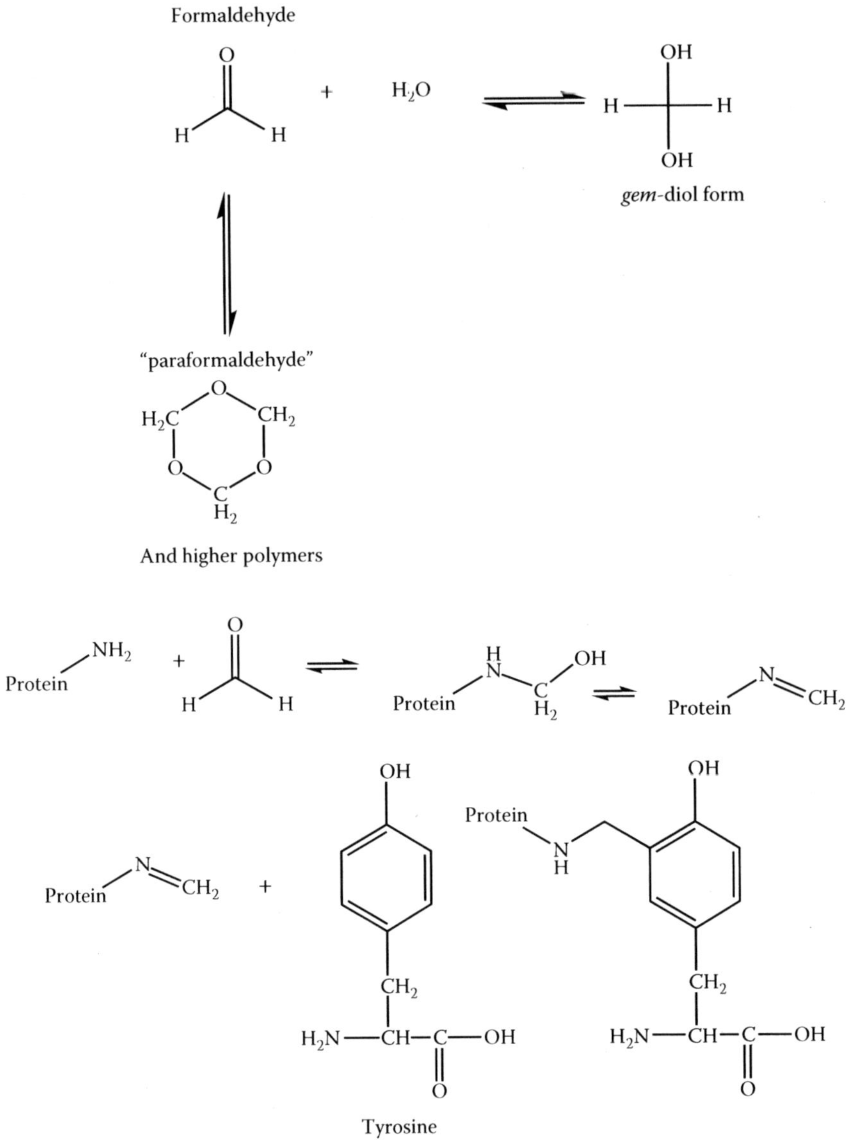

FIGURE 9.11 The structure of formaldehyde and a scheme for the reaction with amino groups in proteins. (Adapted from Walker, J.F., *Formaldehyde* 3rd ed., ACS Monograph, Reinhold, New York, 1964; and Metz, B., Kersten, G.F.A., Hoogerhout, P. et al., Identification of formaldehyde-induced modifications in proteins. Reactions with model peptides, *J. Biol. Chem.* 279, 6225–6243, 2004.)

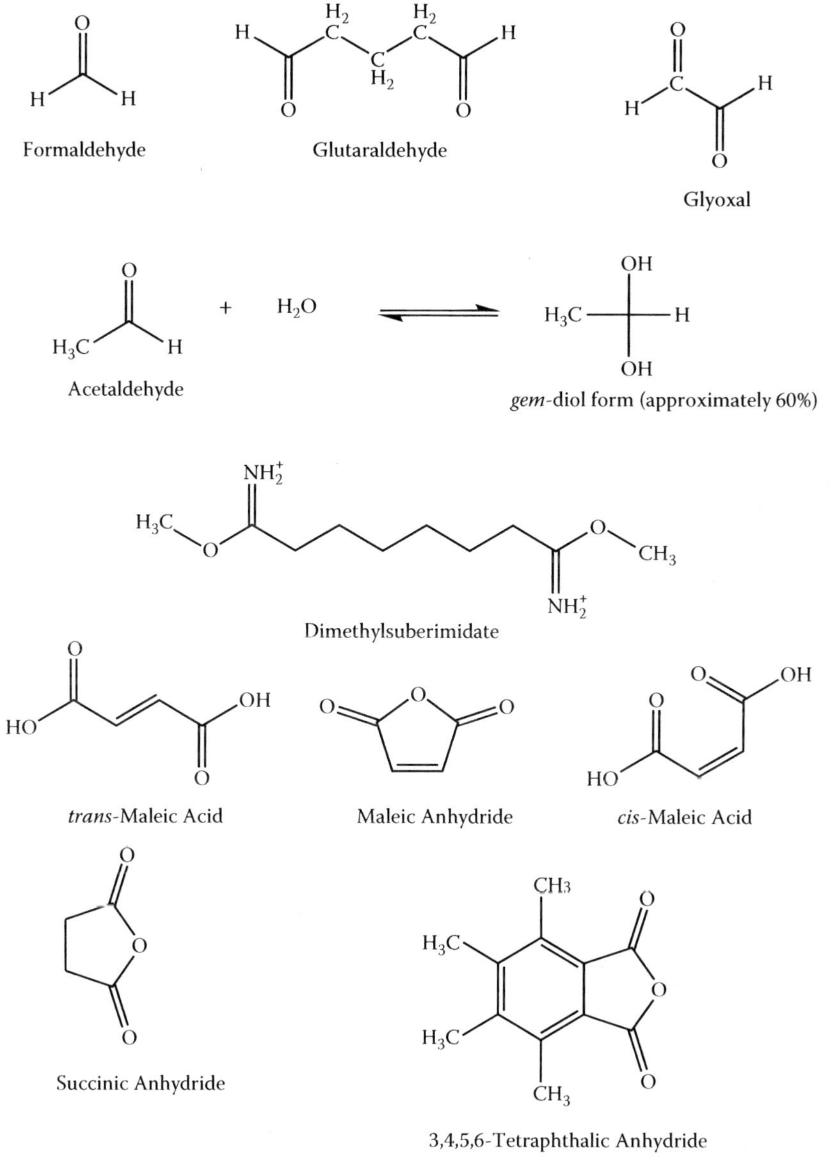

FIGURE 9.12 Glutaraldehyde and other protein cross-linking reagents.

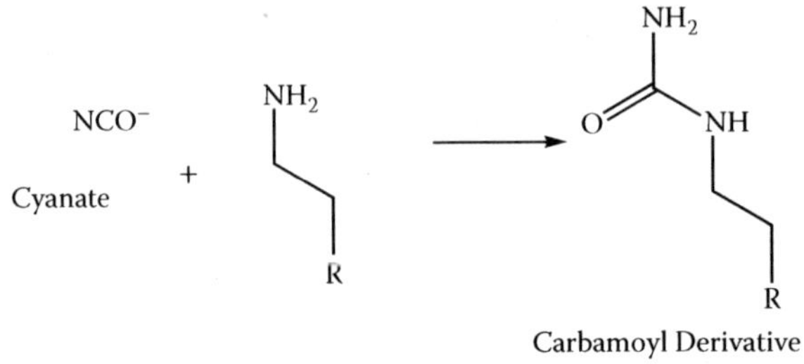

FIGURE 9.13 The reaction of cyanate with amino groups.

It is of equal interest that although glutaraldehyde is widely used, the chemistry is still poorly understood. First, glutaraldehyde exists largely as a polymer in solution or as a hydrate (Figure 9.14).[175,176] Second, although glutaraldehyde has been used extensively for the cross-linking of proteins, the chemistry of the reaction is only poorly understood. The presence of terminal aldehyde functional groups (α, ω) would suggest that cross-linking occurs via reaction with primary amino groups, most likely the ε-amino groups of lysine residues. The initial product of the reaction should be Schiff bases. This reaction should be reversible and it is not, suggesting that the cross-linking reaction occurs via a different mechanism(s) (Figures 9.15, 9.16).[176–179]

FORMALDEHYDE

The use of formaldehyde for the manufacture of allergoids has been described above and suggests that the modification of a "toxin" reduces "toxicity" with retention of sufficient native structure to generate neutralizing antibodies against the "toxin." The use of formaldehyde to produce attenuated pathogens for use as vaccines has been known for some time[180–185] and continues in use today.[186–193] In one of the earlier studies, Schultz and Gebhardt[180] showed that while formalin inactivated bacteriophage (0.018% formaldehyde for 24 hours at 37°C), activity could be regained by dilution in H_2O and storage for 10–15 days. Formaldehyde treatment (with heating) formed the basis for the Salk vaccine for polio.[182]

The use of formaldehyde to "fix" tissues for histology has a considerable history.[194] Recent interest in formaldehyde fixation has focused on antigen retrieval after tissue fixation.[195–199] Antigen retrieval is a term describing a process by which immunological reactivity is recovered from formaldehyde-treated tissue samples, permitting the subsequent application of immunochemistry.[200] The reaction of formaldehyde with proteins is a complex process resulting in a range of products from the simple to the complex. Some of these appear to be reversed by heating or chaotropic agents. There is an example where chemical modification with citraconic anhydride results in antigen.[195] It would seem as if denaturation of the formaldehyde-fixed protein results in exposure of epitopes constrained by formaldehyde cross-linking; this would be consistent with the importance of linear epitopes.[197]

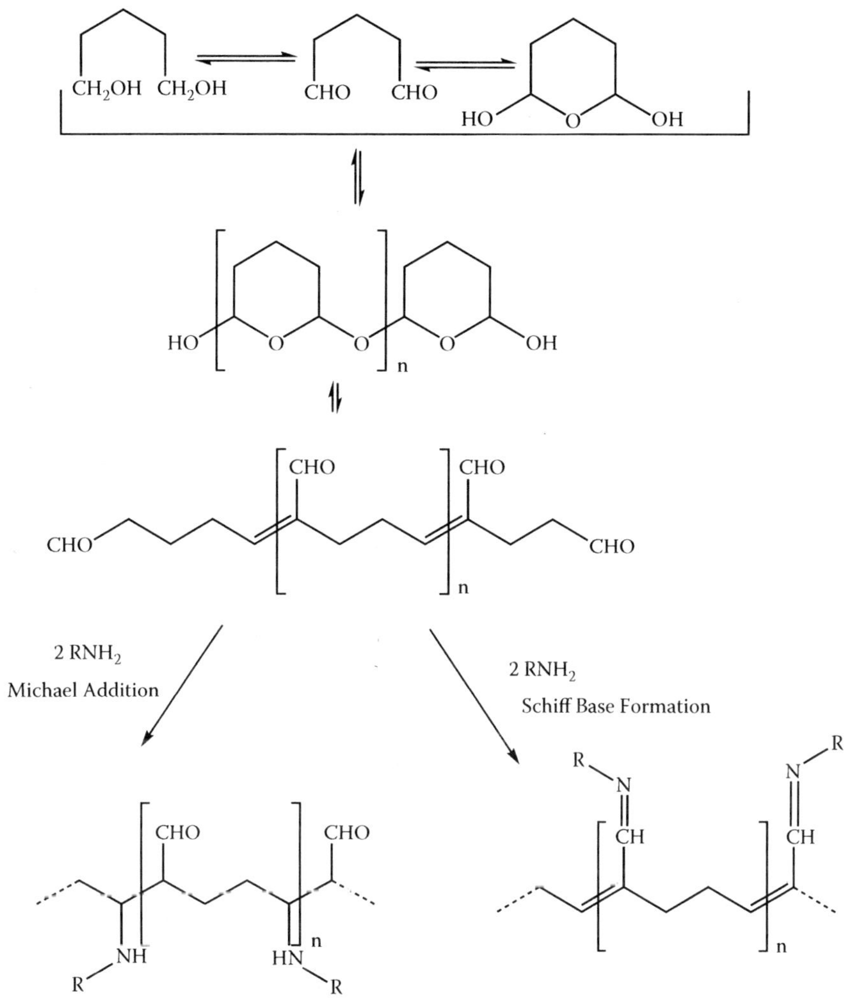

FIGURE 9.14 The reactions of glutaraldehyde in aqueous solution. (Adapted from Migneault, I., Dartiguernave, C., Bertrand, M.J., and Waldron, K.C., Glutaraldehyde: behavior in aqueous solution, rection with proteins, and application to enzyme cross-linking, *BioTechniques* 37, 790–802, 2004.)

ACTIVE-SITE BLOCKED ENZYMES AS COMPETITIVE INHIBITORS

Certain enzymes, usually hydrolases, which act on large substrates and thus have extended binding sites have been inactivated by modification at the active site and subsequently demonstrated to act as inhibitors of the native enzyme. *In vitro* lipolysis by human pancreatic lipase is inhibited by lipase modified at the active site serine with octyl-undecyl-phosphonate.[201] Lipolysis is also blocked by a mutant enzyme where the active site serine is replaced with a glycine residue. Blood coagulation

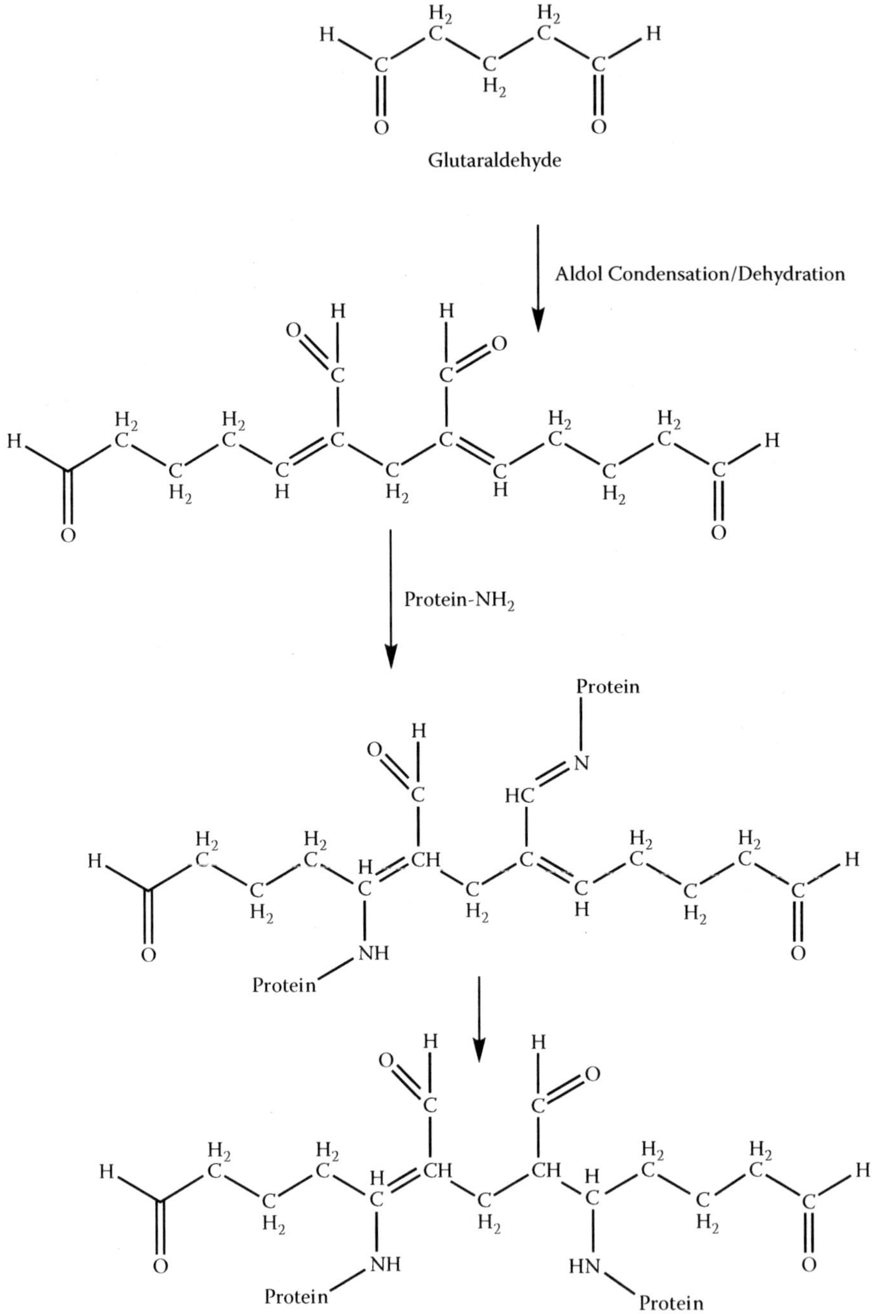

FIGURE 9.15 The reaction of glutaraldehyde with proteins. (See Richards, F.H. and Knowles, J.R., Glutaraldehyde as a protein cross-linking reagent, *J. Mol. Biol.* 37, 231–233, 1968.)

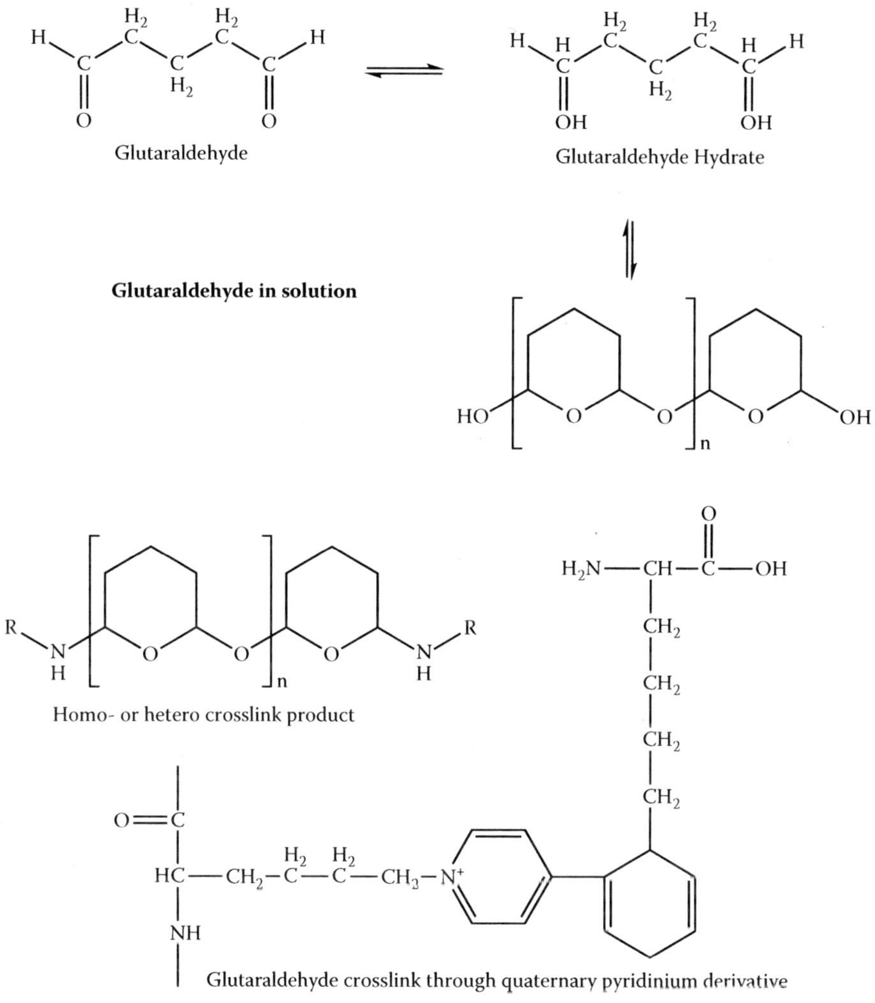

FIGURE 9.16 The reaction of glutaraldehyde with proteins. (See Walt, D.R. and Agayn, V.I., The chemistry of enzyme and protein immobilization with glutaraldehyde, *TRAC* 13, 425–430, 1994.)

factor IXa has been modified with a peptide chloromethyl ketone that reacts with the active site histidine resulting in the loss of activity. The inactivated factor IXa (Factor IXai) inhibits the biological action of factor IXa and is a potentially useful highly specific anticoagulant.[202,203] A similar effect is seen with Factor Xa.[204] It is important to note that the effectiveness of these derivatives depends on the existence of important binding sites distant from the enzyme active site.

MISCELLANEOUS CHEMICAL MODIFICATION OF PROTEINS HAVING THERAPEUTIC VALUE

There are a number of useful modifications of proteins producing therapeutic value that do not fit conveniently into the categories above. Transglutaminases are enzymes that catalyze acyl transfer reactions between the γ-carboxamide group of glutamine (receptor function) and a primary amine (donor function). Sato and colleagues have used microbial transglutaminases to modify recombinant interleukin-2 with alkylamine derivatives of poly (ethylene) glycol or a galactose-terminated triantennary glycoside.[205–207] While the majority of this work was focused on the coupling poly(ethylene)glycol derivatives, transglutaminases can be used with other donor molecules.[208,209] Jordan and colleagues have shown that the chemical modification of murine red blood cells by oxidation (ascorbate/Fe^{++}) or a band-3 cross-linking [3, 3′-dithiobis(sulfosuccinimidyl)propionate enhances phagocytosis by macrophages, providing an approach to the delivery of drugs to macrophages.[210] Insulin has been modified with alkyl acyl functions (e.g., octanoic acid, palmitic acid) to extend circulatory half-life.[211–213] Uludag and Yang[214] have coupled a bisphosphonate, 1-amino-1′,1″- diphosphonate methane, to recombinant bovine serum albumin and demonstrated enhanced uptake of the protein into bone. There was, however, substantial uptake of the conjugate into other organs. More recent work[215] from this group coupled the bisphosphonate to an oxidized carbohydrate (sodium periodate) in fetuin using 4- (maleimidomethyl)cyclohexane-1-carboxyl hydrazide. The amino bisphosphonate was converted to a thiol derivative for this reaction by modification with 1-iminothiolane (Traut's reagent).[216–218] The properties of this derivative of fetuin were compared to a bisphosphonate derivative obtained by coupling to a lysine residue with succinimidyl-4-(*N*-maleimidomethyl)-cyclohexane-1-carboxylate. The carbohydrate-linked bisphosphonate bound more efficiently to bone matrices *in vivo* than the lysine-coupled biphosphonate.

Backer and colleagues have developed an interesting concept for the targeted delivery of both therapeutics and diagnostics.[219–221] This approach involves the expression of a protein/peptide probe as a fusion protein with a "tag" sequence. The "tag" sequence is selected on the basis of its ability to bind tightly with another protein or peptide which is chemically linked to a diagnostic or a therapeutic material. An example is provided with vascular endothelial growth factor (VEGF) expressed as a fusion protein with bovine pancreatic ribonuclease S-peptide. A conjugate of ribonuclease S-protein and polyethyleneimine, polyethyleneimine, was modified with 2-iminothiolane to generate sulfydryl groups and coupled to the S-protein using the heterobifunctional cross-linking agent, 4-maleimido-benzoyl-*N*-hydroxysuccinimide. The noncovalent complex formed between the VEGF conjugate and the polyethyleneimine conjugate was used to target DNA (bound to the polyethyleneimine) to cells over-overexpressing VEGF receptor 2.

Futami and colleagues[222,223] created a cytotoxic derivative of ribonuclease A or ribonuclease 1 (human counterpart of ribonuclease A) by modification of the carboxyl groups. Modification with ethylene diamine resulted in a cationic derivative that was more efficient adsorbed to the negatively charged cell. Chemical cationization of proteins has generated considerable interest with studies on cationized ovalbumin,[224] albumin,[225] carrier proteins,[226] catalase,[227] and p53.[228] The pharmacokinetics of monoclonal antibodies have been improved by changing the net charge to negative, making nonspecific binding to negative cell surfaces less likely and providing more rapid clearance of radiolabeled materials.[229]

REFERENCES

Chapter References

1. Rader, R.E., What is a biopharmaceutical? Part 1 (bio) technology-based definitions, *Bioexecutive Intenational*, March, 2005, pp. 60–65.
2. Rader, R.E., What is a biopharmaceutical? Part 2 Company and industrial definitions, *Bioexecutive International*, May, 2005, pp. 43–49.
3. Finlayson, J.S., Therapeutic plasma fractions and plasma fractionation, *Sem. Throm. Hemost.* 6, 1–11, 1979.
4. Swisher, S.N. and Petz, L.D., Plasma protein and blood substitutes, in *Clinical Practice of Transfusion Medicine*, 2nd ed., Eds. K.D. Petz and S.N. Swisher, Churchill Livingstone, New York, 1989.
5. Farrugia, A. and Robert, P., Plasma protein therapies: Current and future perspectives, *Best Pract. Res. Clin. Haematol.* 19, 243–258, 2006.
6. Liu, T.-Y., Natural and biotech-derived therapeutic proteins. What is the future?, *Electrophoresis* 21, 1914–1917, 2000.
7. Evens, R.P., Biotechnology and biological preparations, in *Encyclopedia Biotechnology*, Marcel Dekker, New York, 20002.
8. Birch, J.R. and Onakunle, Y., Biopharmaceutical proteins. Opportunities and challenges, *Methods Mol. Biol.* 308, 1–15, 2005.
9. Reichert, J.M., Trends in U.S. approvals: New biopharmaceuticals and vaccines, *Trends Biotechnol.* 24, 293–298, 2006.
10. Barron, M.E., Wilkes, M.M., and Navickis, R.J., A systematic review of the comparative safety of colloids, *Arch. Surg.* 139, 552–563, 2004.
11. Hemington-Gorse, S.J., Colloid or crystalloid for resuscitation of major burns, *J. Wound Care* 14, 256–258, 2005.
12. Bunn, F., Trivedi, D., and Ashraf, S., Colloid solutions for fluid resuscitation, *Cochrane Database Sys. Rev.* 23: CD001319, 2008.
13. Esau, C.C. and Monia, B.P., Therapeutic potential for microRNAs, *Adv. Drug Deliv. Rev.* 59, 101–114, 2007.
14. Gjertsen, B.T., Bredholt, T., Anensen., N., and Vintermyr, O.K., Bcl-2 antisense in the treatment of human malignancies: A delusion in targeted therapy, *Curr. Pharm. Biotechnol.* 8, 373–381, 2007.
15. Singh, S.K., RNA interference and its therapeutic potential against HIV infection, *Expert Opin. Biol. Ther.* 8, 449–461, 2008.
16. Shrivastava, N. and Srivastava, A., RNA interference: An emerging generation of biologicals, *Biotechnol. J.* 3, 339–353, 2008.
17. Walsh, G., *Proteins Biochemistry and Biotechnology*, John Wiley and Sons, Ltd., Chichester, U.K., 20002.
18. Walsh, G., Pharmaceutical biotechnology products approved within the European Union, *Eur. J. Pharm. Biopharm.* 55, 3–10, 2003.
19. Walsh, G., Biopharmaceuticals: Recent approvals and likely directions, *Trends Biotechnol.* 23, 553–558, 2003.

20. Shilpi, S., Jain, A., Gupta, Y., and Jain, S.K., Colloidosomes: An emerging vesicular system in drug delivery, *Crit. Rev. Ther. Drug Carrier Syst.* 24, 361–391, 2007.
21. Dass, C.R., Drug delivery in cancer using liposomes, *Methods Mol. Biol.* 437, 177–182, 2008.
22. Kurreck, J., Antisense technologies. Improvement through novel chemical modification, *Eur. J. Biochem.* 270, 1628–1644, 2003.
23. Beaucage, S.L., Solid-phase synthesis of siRNA oligonucleotides, *Curr. Opin. Drug Discov. Dev.* 11, 203–216, 2008.
24. Niederhavner, P., Sebestik, J., and Jezek, J., Peptide dedrimers, *J. Pept. Sci.* 11, 757–788, 2005.
25. Bode, J.W., Emerging methods in amide- and peptide-bond formation, *Curr. Opin. Drug Discov. Dev.* 9, 765–775, 2006.
26. Boerema, D.J., Tereshko, V.A., and Kent, S.B., Total synthesis by modern chemical ligation methods and high X-ray structure of ribonuclease A, *Biopolymers* 90, 278–286, 2008 .
27. Gutte, B. and Merrifield, R.B., The total synthesis of an enzyme with ribonuclease A activity, *J. Am. Chem. Soc.* 91, 501–502, 1969.
28. Ontjes, D.A. and Anfinsen, C.B., Solid phase synthesis of a 42-residue fragment of staphylococcal nuclease: Properties of a semisynthetic enzyme, *Proc. Nat. Acad. Sci. USA* 64, 428–435, 1969.
29. Gutte, B., A synthetic 70-amino acid residue analog of ribonuclease S-protein with enzymic activity, *J. Biol. Chem.* 250, 889–904, 1975.
30. Gutte, B., Study of RNase mechanism and folding by means of synthetic 63-residue analogs, *J. Biol. Chem.* 252, 663–700, 1977.
31. Cerovským V. and Scheraga, H.A., Combined solid-phase/solution synthesis of large ribonuclease A C-terminal peptides containing a non-natural proline analog, *J. Pept. Res.* 65, 518–528, 2005.
32. Di Girolamo, G., Kaufman, M.A., Gonzalez, E. et al., Bioequivalence of two subcutaneous pharmaceutical products of interferon β1a, *Arnzneimittelforschung* 58, 193–198, 2008.
33. Pavlovic, M., Girardin, E., Kapetanovic, L. et al., Similar biological medicinal products containing recombinant human growth hormone: European regulations, *Horm. Res.* 69, 14–21, 2008.
34. Guideline of General Principles of Process Validation, USA Food and Drug Administration, Rockville, Maryland, USA, May, 1987; http://www.fda.gov/cder/guidance/pv.htm.
35. Nash, R.A. (1999), The validation of pharmaceutical processes, in *Preparing for FDA Pre-Approval Inspections*, Ed. M.D. Hynes, III, Marcel Dekker, New York, Chapter 7, pp. 161–185.
36. ICH Harmonised Tripartite Guideline Q7, Good Manufacturing Practice Guide for Active Pharmaceutical Ingredients, http://www.ich.org.
37. ICH Harmonised Tripartite Guideline Q3A, Impurities in New Drug Substances, http://www.ich.org.
38. ICH Harmonised Tripartite Guideline Q3B, Impurities in New Drug Products, http://www.ich.org.
39. ICH Harmonised Tripartite Guideline, Q8 Pharmaceutical Development, http://www.ich.org.
40. ICH Harmonised Tripartite Guideline Q7. Good Manufacturing Practice Guide for Active Pharmaceutical Ingredients, http://www.ich.org.
41. ICH Harmonised Tripartite Guideline Q6A, Specifications: Test Procedures and Acceptance Criteria for New Drug Substances and New Drug Products: Chemical Substances, http://www.ich.org.

42. ICH Harmonised Tripartite Guideline Q6A, Specifications: Test Procedures and Acceptance Criteria for Biotechnological/Biological Products, http://www.ich.org.
43. ICH Harmonised Tripartite Guideline Q2)R1, Validation of Analytical Procedures, http://www.ich.org.
44. Rademacher, T.W. (1998), Recent advances in glycoconjugate analysis and glycobiology, *Curr. Opin. Biotechnol.* 9, 74–79.
45. Reubsaet, J.L.E., Beijnen, J.H., Bult, A., van Maanen, R.J., Marchal, J.A.D., and Underberg, W.J.M. (1998), Analytical techniques used to study the degradation of protein and peptides: Chemical instability, *J. Pharmaceut. Biomed. Anal.* 17, 955–978.
46. Glish, G.L. and Vachet, R.W. (2003), The basics of mass spectrometry in the twenty-first century, *Nat. Rev. Drug Discovery* 2, 140–150.
47. Guerrini, M., Beccati, D., Shriver, Z. et al., Oversulfated chondroitin sulfate is a contaminant in heparin associated with adverse clinical events, *Nat. Biotechnol.* 26, 669–675, 2008.
48. Kisimoto, T.K., Viswanathan, K., and Ganguly, T., Contaminated heparin associated with adverse clinical events and activation of the contact system, *N. Engl. J. Med.* 358, 2457–2467, 2008.
49. Robertson, J.S., Changes in biological source material, *Biologics* 34, 61–63, 2006.
50. Guideline for Drug Master Files, http://www.fda.gov/cder/guidance/dmf/htm.
51. Yu, L.X., Pharmaceutical quality by design: Product and process development, understanding and control, *Pharm. Res.* 25, 781–791, 2008.
52. Anonymous, What's in Quality by Design, *Engineering* 217. 217–219, 1977.
53. Deming, S.N., Quality by Design. 1., *Chemtech* 18, 560–566, 1988.
54. Clausing, D. and Simpson, B.H., Quality by Design, *Quality Progress* 23, 41–44, 1990.
55. Bisgaard, S., Quality management and Juran's legacy, *Qual. Reliability Engineer. Int.* 23, 665–677. 2007.
56. Whistler, R.L. and Belfort, A.M., Nutritional value of chemically modified corn starches, *Science* 133, 1599–1600, 1961.
57. Thompson, W.L. and Walton, R.P., Circulatory responses to intravenous infusions of hydroxyethyl starch solutions, *J. Pharmacol. Exp. Ther.* 146, 359–364, 1964.
58. Sillett, H.K., Whicher, J.T., and Trejdosiewicz, L.K., Effects of resuscitation fluids on nonadaptive immune responses, *Transfusion* 37, 953–959, 1997.
59. Persidsky, M.D. and Ellett, M.H., Hydroxyethyl starch as a cryopresevative for nucleated mammalian cells, *Cryobiology* 8, 586–588, 1971.
60. Waitzinger, J., Bepperling, F., Pabst, G. et al., Pharmacokinetics and tolerability of a new hydroxyethyl starch (HES) specification [HEW 130/0.4] after single-dose infusion of 6% or 10% solutions in healthy volunteers, *Clin. Drug. Investig.* 16, 151–160, 1998.
61. Blasco, V., Leone, M., Antonini, F. et al., Comparison of the novel hydroxyethyl starch 200/0.6 in brain-dead donor resuscitation on renal function after transplantation, *Br. J. Anaesth.* 100, 504–508, 2008.
62. Treib, J., Baron, J.-F., Grauer, M.T., and Strauss, R.G., An international view of hydroxyethyl starches, *Intensive Care Med.* 25, 258–268, 1999 .
63. Bowry, S.K. and Rintelen, T.H., Synthetically modified cellulose (SMC): A cellulosic hemodialysis membrane with minimized complement activation, *ASAIO J.* 44, M579–M583, 1998 .
64. Stevens, C.V., Meriggi, A., and Booten, K., Chemical modification of inulin, valuable renewable resource, and its industrial applications, *Biomacromolecules* 2, 1–16, 2001.
65. Diamantoglou, M., Platz, J., and Vienken, J., Cellulose carbamates and derivatives as hemocompatible materials for hemodialysis, *Artif. Organs* 23, 15–22, 1999.

66. Ye, S.H., Watanabe, J., Takai, M. et al., Design of functional hollow fiber membranes modified with phospholipid polymers for application in total hemopurification system, *Biomaterials* 26, 5032–5041, 2005 .
67. Stulzer, H.K., Silva, M.A., Fernandes, D., and Assreuy, J., Development of controlled release captopril granules coated with ethylcelluose and methylcellulose by fluid bed dryer, *Drug Deliv.* 15, 11–18, 2008 .
68. Ricketts, C.R., Dextran sulphate—A synthetic analogue of heparin, *Biochem. J.* 51, 129–133, 1951.
69. Kojima, S., Low-density lipoprotein apheresis and changes in plasma components, *Ther. Apher.* 5, 232–238, 2001.
70. Bambauer, R., Schiel, R., and Latza, R., Low-density lipoprotein apheresis: An overview, *Ther. Apher. Dial.* 7, 382–390, 2003.
71. Bosch, T., Recent advances in therapeutic apheresis, *J. Artif. Organs* 6, 1–8, 2003.
72. Herr, W., Diethyl pyrocarbonate: A chemical probe for secondary structure in negatively supercoiled DNA, *Proc. Nat. Acad. Sci. USA* 82, 8009–8013, 1985.
73. Kirkegaard, K., Buc, H., Spassky, A., and Wang, J.C., Mapping of single-stranded regions in duplex DNA at the sequence level: Single-strand-specific cytosine methylation in RNA polymerase-promoter complexes, *Proc. Nat. Acad. Sci. USA* 80, 2544–2548, 1983.
74. Shapiro, R. and Yamaguchi, H., Nucleic acid reactivity and conformation I. Deamination of cytosine by nitrous acid, *Biochim. Biophys. Acta.* 281, 501–506, 1972.
75. Cardillo, G., Gentilucci, L., and Tolomelli, A., Unusual amino acids: Synthesis and introduction into naturally occurring peptides and biologically active analogues, *Mini Rev. Med. Chem.* 6, 293–304, 2006.
76. Hino, N., Hayashi, A., Sakamoto, K., and Yokoyama, S., Site-specific incorporation of non-natural amino acids into proteins in mammalian cells with an expanded genetic code, *Nat. Protoc.* 1, 2957–2962, 2006.
77. Liu, W., Brock, A., Chen, S. et al., Genetic incorporation of unnatural amino acids into proteins in mammalian cells, *Nat. Methods* 4, 239–244, 2007.
78. Chen, S., Schultz, P.G., and Brock, A., An improved system for the generation and analysis of mutant proteins containing unnatural amino acids in *Saccharomyces cerevisiae*, *J. Mol. Biol.* 371, 112–122, 2007.
79. Xie, J., Supekova, L., and Schultz, P.G., A genetically encoded metabolically stable analogue of phosphotyrosine in *Escherichia coli*, *ACS Chem. Biol.* 2, 474–478, 2007.
80. Corey, D.R., Chemical modification: The key to clinical application of RNA interference, *J. Clin. Invest.* 117, 3615–3622, 2007.
81. Rayburn, E.R. and Zhang, R., Antisense, RNAi, and gene silencing strategies for therapy: Mission possible or impossible?, *Drug Disc. Today* 13, 513–521, 2008.
82. Ross, D.W., Sense and antisense oligonucleotides, *Arch. Pathol. Lab. Med.* 114, 1296, 1990.
83. Zon, G., Innovations in the use of antisense oligonucleotides, *Ann. N. Y. Acad. Sci.* 616, 161–172, 1990.
84. Eck, S.L. and Nabel, G.J., Antisense oligonucleotides for therapeutic intervention, *Curr. Opin. Biotechnol.* 1, 897–904, 1991.
85. Lebdeva, I. and Stein, C.A., Antisense oligonucleotides: Promise and reality, *Annu. Rev. Pharmacol. Toxicol.* 41, 403–419, 2001.
86. Urban, E. and Noe, C.R., Structural modifications of antisense oligonucleotides, *Farmaco* 58, 243–258, 2003.
87. Sazani, P. and Kole, R., Therapeutic potential of antisense oligonucleotides as modulators of alternative splicing, *J. Clin. Invest.* 112, 481–486, 2003.
88. Chan, J.H., Lim, S., and Wong, W.S., Antisense oligonucleotides: From design to therapeutic application, *Clin. Exp. Pharmacol. Physiol.* 33, 533–540, 2006.

89. Moizumi, M., True antisense oligonucleotides with modified nucleotides restricted in the N-conformation, *Curr. Top. Med. Chem.* 7, 661–665, 2007.
90. Anonymous, Oblimersen: Augmerosen, BCL-2 antisense oligonucleotides—Genta, G 3139, GC 3129, *Drugs R D* 8, 321–334, 2007.
91. Athyros, V.G., Kakafika, A.I., Tziomalos, K. et al., Antisense technology for the prevention or the treatment of cardiovascular disease: The next blockbuster? *Expert Opin. Invest. Drugs* 17, 969–972, 2008.
92. Osborne, S.E., Matsumara, I., and Ellington, A.D., Aptamers as therapeutic and diagnostic reagents: Problems and prospects, *Curr. Opin. Chem. Biol.* 1, 5–9, 1997.
93. Patel, D.J., Structural analysis of nucleic acid aptamers, *Curr. Opin. Chem. Biol.* 1, 32–46, 1997.
94. Patel, D.J., Suri, A.K., Jiang, F. et al., Structure, recognition and adaptive binding in RNA aptamers complexes, *J. Mol. Biol.* 272, 645–664, 1997.
95. Younes, C.K., Boisgard, R., and Tavitian, B., Labelled oligonucleotides as radiopharmaceuticals: Pitfalls, problems and perspectives, *Curr. Pharm. Des.* 8, 1451–1466, 2002.
96. Kulbachinsky, A.V., Methods for selection of aptamers to protein targets, *Biochemistry (Moscow)* 72, 1505–1518, 2007.
97. Kaur, G. and Roy, I., Therapeutic applications of aptamers, *Expert Opin. Investig. Drug.* 17, 43–60, 2008.
98. Eckstein, F., The versatility of oligonucleotides as potential therapeutics, *Expert. Opin. Biol. Ther.* 7, 1021–1034, 2007.
99. Balamurugan, S., Obubuafo, A., Soper, S.A., and Spivak, D.A., Surface immobilization methods for aptamers diagnostic applications, *Anal. Bioanal. Chem.* 390, 1009–1021, 2008.
100. Zhang, H.-Y., Du, Q., Wahlestedt, C., and Liang, Z., RNA interference with chemical modified siRNA, *Curr. Top. Med. Chem.* 6, 893–900, 20068.
101. Choung, S., Kim, Y.J., Kim, S. et al., Chemical modification of siRNAs to improve serum stability without loss of efficacy, *Biochem. Biophys. Res. Commun.* 342, 919–927. 2006.
102. Hoerter, J.A.H. and Walter, N.G., Chemical modification resolves the asymmetry of siRNA strand degradation in human blood serum, *RNA* 13, 1887–1892, 2007.
103. Prakash, T.P. and Bhat, B., 2'-Modified oligonucleotides for antisense therapeutics, *Curr. Top. Med. Chem.* 7, 641–649, 2007.
104. Faria, M. and Ulrich, H., Sugar boost: When ribose modifications improve oligonucleotide performance, *Curr. Opin. Mol. Ther.* 10, 168–175, 2008.
105. De Mesmacker, A., Altmann, K.-H., Waldner, A., and Wendeborn, S., Backbone modifications in oligonucleotides and peptide nucleic acid systems, *Curr. Opin. Struct. Biol.* 5, 343–355, 1995.
106. Taylor, M.F., Paulauskis, J.D., Weller, D.D., and Kobzik, L., *In vitro* efficacy of morpholino-modified antisense oligomers directed against tumor necrosis factor-α mRNA, *J. Biol. Chem.* 271, 17445–177452, 1996.
107. Gaynor, J.W., Brazier, J., and Cosstick, R., Synthesis of 3'-*S*-phosphorothiolate oligonucleotides for their potential use in RNA interference, *Nucleos. Nucleot. Nucl. Acids* 26, 709–712, 2007.
108. Wang, Z., Olsen, P., and Ravikumar, V.T., A novel universal linker for efficient synthesis of phosphorothioate oligonucleotides, *Nucleos. Nucleot. Nucl. Acids* 26, 259–269, 2007.
109. Puthnveetil, S., Whitby, L., Ren, J. et al., Controlling activation of the RNA-dependent protein kinase by siRNAs using site-specific chemical modification, *Nucl. Acids Res.* 34, 4900–4911, 2006.
110. Smith, R.A. et al., Chemical derivatization of therapeutic proteins, *Trends Biotechnol.* 11, 297, 1993.

111. Pozansky, M.J., Soluble enzyme-albumin conjugates: New possibilities for enzyme replacement therapy, *Methods Enzymol.* 137, 566, 1988.
112. Brader, M.L. et al., Hybrid insulin cocrystals for controlled release delivery, *Nat. Biotechnol.* 20, 800, 2002.
113. Lundblad, R.L. and Bradshaw, R.A., Application of site-specific chemical modification in the manufacture of biopharmaceuticals: I. An overview, *Biotechnol. Appl. Biochem.* 26, 143, 1997.
114. Davis, B.G., Chemical modification of biocatalysts, *Curr. Opin. Biotechnol.* 14, 379–286, 2003.
115. Tsutsumi, Y., Onda, M., Nagata, S. et al., Site-specific chemical modification with polyethylene glycol of recombinant immunotoxin anti-Tac(Fv)-PE38 (LMB-2) improves antitumor activity and reduces animal toxicity and immunogenicity, *Proc. Natl. Acad. Sci. USA* 97, 8548–8553, 2000.
116. Albrecht, H., Burke, P.A., Natarajan, A. et al., Production of soluble ScFv with C-terminal free thiol of site-specific conjugation or stable dimeric ScFv on demand, *Bioconjug. Chem.* 15, 16–26, 2004.
117. Long, D.L., Doherty, D.H., Eisenberg, S.P. et al., Design of homogeneous, monopegylated erythropoietin analogs with preserved *in vitro* bioactivity, *Exp. Hematol.* 34, 697–704, 2006.
118. Kim, Y., Ho, S.O., Gassman, N.R. et al., Efficient site-specific labeling of proteins via cysteines, *Bioconjug. Chem.* 19, 786–791, 2008.
119. Hegazy, U.M., Tars, K., Hellman, U., and Mannervik, B., Modulating catalytic activity by unnatural amino acid residues in a GSH-binding loop of GST P1-1, *J. Mol. Biol.* 376, 811–826, 2008.
120. Yamamoto, Y., Tsutsumi, Y., Yoshinoka, Y., Nichibato, T., Kobayashi, K., Oakamoto T., Mukai, Y., Shimizu, T., Nakagama, S., Nagoto, S., and Najumi, T., Site-specific pegylation of a lysine-deficient TNF-α with full bioactivity, *Nat. Biotechnol.* 211, 546–552, 2003.
121. Lee, H. Jang, I.H, Ryo, S.H., and Park, T.G., N-Terminal site-specific mono-PEGylation of epidermal growth factor, *Pharmaceut. Res.* 20, 818–825, 2003.
122. Eisenhauer, B.M. and Hecht, S.M. (2002), Site-specific incorporation of (aminooxy) acetic acid into proteins. *Biochemistry* 41, 11472–11478, 2002.
123. Cornish, V.W., Hahn, K.M., and Schultz, P.G., Site-specific protein modification using a ketone handle, *J. Am. Chem. Soc.* 118, 5150–5151, 1996.
124. Zhang, Z., Smith, B.A., Wang, L., Brock, A., Cho, C., and Schultz, P.G., A new strategy for the site-specific modification of proteins *in vivo*, *Biochemistry* 42, 6735–6746, 2003.
125. Ikeda, Y., Kawahara, S.-i., Taki, M., Kumo, A., Hasegawa, T., and Taira, K., Synthesis of a novel histidine analogue and its efficient incorporation into a protein *in vivo. Protein Eng.* 16, 699–706, 2003.
126. Kretsinger, J.K. and Schneider, J.P., Design and application of basic amino acids displaying enhanced hydrophobicity, *J. Am. Chem. Soc.* 125, 7907–7913, 2003.
127. Broos, J., Gabellieri, E., Biemans-Oldehinkel, E., and Strambini, G.B., Efficient biosynthetic incorporation of tryptophan and indole analogs in an integral membrane protein, *Protein Sci.* 12, 1991–2000, 2003.
128. Horvat, S. and Jakas, A., Peptide and amino acid glycation: New insights into the Maillard reaction, *J. Pept. Sci.* 10, 119–137, 2004.
129. Bidmon, C., Frischmann, M., and Pischetsrieder, M., Analysis of DNA-bound advanced glycation end-products by LC and mass spectrometry, *J. Chromatog. B. Anal. Technol. Biomed. Life Sci.* 855, 51–58, 2006.
130. Davis, B.G. and Robinson, M.A. (2002), Drug-delivery systems based on sugar-macromolecule conjugates, *Curr. Opin. Drug. Discovery Dev.* 5, 279–288.

131. Sola, R.J., Rodríguez-Martinez, J.A., and Griebenow, K., Modulation of protein biophysical properties by chemical glycosylation: Biochemical insights and biomedical implications, *Cell. Mol. Life. Sci.* 64, 2133–2152, 2007.
132. Sabesan, S. and Linna, T.J., Chemical glycosylation of recombinant interleukin 2, *Methods Enzymol.* 242, 46–46. 1994.
133. Lemieux, R.U., Baker, D.A., and Bundle, D.R., A methodology for the production of carbohydrate-specific antibody, *Can. J. Biochem.* 55, 507–512, 1997.
134. Hayashi, A., Chiba, T., Hayashi, H. et al., Synthesis of glycosylated human tumor necrosis factor α coupled with *N*-acetylneuraminic acid, *Cancer Immunol. Immunother.* 56, 545–553, 2007.
135. Loeffler, L.J. and Pierce, J.V., Acyl-azide derivatives in affinity chromatography. Immobilization of enzymatically active trypsin on beaded agarose and porous glass, *Biochim. Biophys. Acta* 327, 20–27, 1973.
136. Macmillan, D., Bill, R.M., Sage, K.A. et al., Selective *in vitro* glycosylation of recombinant proteins: Semi-synthesis of novel homogeneous glycoforms of human erythropoietin, *Chem. Biol.* 8, 133–143, 2001.
137. Nicotra, F., Cipolla, L., Peri, F. et al., Chemoselective neoglycosylation, *Adv. Carbohy. Chem. Biochem.* 61, 353–398, 2008.
138. Ni, J.H., Singh, S., and Wang, L.X., Synthesis of maleimde-activated carbohydrate as chemoselective tags for site-specific glycosylation of peptides and proteins, *Bioconjug. Chem.* 14, 232–238, 2003.
139. Swanwick, R.S., Daines, A.M., Flitsch, S.I., and Allemann, R.K., Synthesis of homogeneous site-selectively glycosylated proteins, *Org. Biomol. Chem.* 3, 572–574, 2005.
140. Galonić, D.P. and Gin, D.Y., Chemical glycosylation in the synthesis of glycoconjugate antitumour vaccines, *Nature* 446, 1000–1007, 2001.
141. Langenhan, J.M., Peters, N.R., Guzei, I.A. et al., Enhancing the anticancer properties of cardiac glycosides by neoglycorandomization, *Proc. Natl. Acad. Sci. USA* 102, 12305–12310, 2005.
142. Marsh, D.G., Lichtenstein, L.M., and Campbell, D.H., Studies on allergoids prepared from naturally occurring allergens. I. Assay of allergenicity and antigenicity of formalized rye group component, *Immunology* 18, 702–722, 1970.
143. Flicker, S. and Valenta, R., Renaissance of the blocking antibody concept in type 1 allergy, *Int. Arch. Allergy Immunol.* 132, 13–24, 2003.
144. Linggood, F.V., Stevens, M.F., Fulthorpe, A.J. et al., The toxoiding of purified diphtheria toxin, *Brit. J. Exp. Pathol.* 44, 177–188, 1963.
145. Mueller, U., Reisman, R., Elliott, W., Steger, R., Walsh, S., Wypych, J., and Arbesman, C. (1982), Studies of chemically modified honeybee venom. I. Biochemical, toxicologic and immunologic characterization, *Int. Arch. Allergy Appl. Immun.* 68, 312–319.
146. Puttonen, E., Massch, H.J., and Pilström, L. (1982), Studies on allergen and allergoid preparations from purified Timothy (*Phleum pretense*) pollen extracts. I. Physicochemical characteristics and binding to allergen-specific human IgE, *Int. Arch. Appl. Immun.* 68, 1–6.
147. Puttonen, E., Pilström, L., Wahn, U., and Maasch, H.J. (1982), Studies on allergen and allergoid preparations from purified Timothy (*Phleum pretense*) pollen extracts. II. Anaphylaxis studies in rats and histamine release from human leukocytes, *Int. Arch. Allergy Appl. Immun.* 68, 7–12.
148. Dormann, D., Ebner, C., Jarman, E.P. et al., Responses of human birch pollen allergen-reactive T cells to chemically modified allergen (allergoids), *Clin. Exp. Allergy* 28, 1374–1383, 1998.
149. Salgado, J., Casadevall, G., Puigneró, V., and Queralt, J., Characterization of allergoids from ovalabumin *in vitro* and *in vivo*, *Immunobiology* 196, 375–386, 1996.

150. Lund, L., Henmar, H., Würtzen, P.A. et al., Comparison of allergenicity and immunogenicity of an allergen vaccine and commercially available products for birch pollen immunotherapy, *Clin. Exp. Allergy* 37, 564–571, 2007.
151. Ibarrola, I., Sanz, M.L., Gamboa, P.M. et al., Biological characterization of glutaraldehyde-modified *Parietaria judaica* pollen extracts, *Clin. Exp. Allergy* 34, 303–309, 2004.
152. Hopkins, M., Lees, B.G., Richardson, D.G. et al., Standardisation of glutaraldehyde-modified tyrosine-adsorbed tree pollen vaccines containing the Th1-inducing adjuvant, monophosphoryl lipid A (MPL), *Allergol. Immunopathol.* (Madr) 29, 245–254, 2001.
153. Ibarrola, I, Sanz, M.I., Gamboa, P.M. et al., Biological characterization of glutaraldehyde-modified *Parietoria judaica* pollen extracts, *Clin. Exp. Allergy* 34, 303–309, 2004.
154. Mistrello, G. et al., Monomeric chemically modified allergens: Immunologic and physiochemical characterization, *Allergy* 51, 8, 1999.
155. Bagnasco, M. et al., Pharmacokinetics of an allergen and a monomeric allergoid for oromuscosal immunotherapy in allergic volunteers, *Clin. Exp. Allergy* 31, 54, 2001.
156. Ćirković, T.D., Bukilica, M.N., Gavrović, Vujčić, Petrović, S., and Jankov, R.M. (1999), Physicochemical and immunologic characterization of low-molecular weight alleregoids of *Dactylis glomerata* pollen proteins, *Allergy* 54, 128–134.
157. Ćirković, T.D, Gavrović-Jankulović, M., Prišić, S. et al., The influence of a residual group in low-molecular-weight allergoids of *Artemia vulgaris* pollen on their allergenicity, IgE- and IgG-binding properties, *Allergy* 57, 1013–1020, 2002.
158. Ferreira, F., Briza, P., Inführ, D. et al., Modified recombinant allergens for safer immunotherapy, *Inflamm. Allergy Drug Targets* 5, 5–14, 2006.
159. Vrtala, S., Focke-Tejkl, M., Swoboda, I. et al., Strategies for converting allergens into hypoallergenic vaccine candidates, *Methods* 32, 313–320, 2005.
160. Akdis, C.A. and Blaser, K., Regulation of specific immune responses by chemical and structural modifications of allergens, *Int. Arch. Allergy Immunol.* 121, 261, 2000.
161. Müller, U.R. (2001), New developments in the diagnosis and treatment of hymenoptera venom allergy, *Int. Arch. Allergy Immunol.* 124, 447–453.
162. Sabatini, D.D., Bensch, K., and Barnett, R.J., Cytochemistry and electron microscopy. The preservation of cellular ultrastructure and enzymatic activity by aldehyde fixation, *J. Cell. Biol.* 17, 19–58, 1963.
163. Goetz, W.A., Lim, H.S., Lansac, E. et al., A temporarily stented, autologous pericardial aortic valve prosthesis, *J. Heart Valve. Dis.* 11, 696–702, 2002.
164. Schussler, O., Shen, M., Shen, L. et al., Effect of human immunoglobulins on the immunogenicity of porcine bioprotheses, *Ann. Thorac. Surg.* 71(Suppl. 5), S396–S400, 2001.
165. Neethling, W.M.L., Hodge, A.J., and Glancy, R., Glutaraldehyde-fixed kangaroo aortic wall tissue: Histology, crosslink stability and calcification potential, *J. Biomed. Mater. Res.* 66B, 356–363, 2003.
166. Wengerter, K. and Dardik, H., Biological vascular grafts, *Semin. Vasc. Surg.* 12, 46–51, 1999.
167. Chen, R.H. and Adams, D.H., Decreased porcine valve antigenicity with *in vitro* culture, *Ann. Thorac. Surg.* 71(Suppl. 5), S393–S395, 2001.
168. Hopwood, D., Fixation and fixatives, in *Theory and Practice of Histological Techniques*, Eds. J.D. Bancroft and A. Stevens, 4th ed., Churchill Livingstone, Chapter 2, pp. 23–45, 1996.
169. Jamur, M.C., Faraco, C.D., Lunardi, L.O. et al., Microwave fixation improves antigenicity of glutaraldehyde-sensitive antigens while preserving ultrastructural detail, *J. Histochem. Cytochem.* 43, 307–311, 1995.
170. Brorson, S.H., Fixative-dependent increase in immunogold labeling following antigen retrieval on acrylic and epoxy sections, *Biotech. Histochem.* 74, 248–260, 1999.

171. Yamashita, S. and Okada, Y., Application of heat-induced antigen retrieval to aldehyde-fixed fresh frozen sections, *J. Histochem. Cytochem.* 53, 1421–1432, 2005.
172. Zipeto, D., Matucci, A., Ripamonte, C. et al., Induction of human immunodeficiency virus neutralizing antibodies using fusion complexes, *Microbes Infect.* 8, 1424–1433, 2006.
173. Zhu, X., Chu, W., Wang, T. et al., Variations in dominant antigen determinants of glutaraldehyde polymerized human, bovine, and porcine hemoglobin, *Artif. Cells Blood Substit. Immobil. Biotechnol.* 35, 518–532, 2007.
174. Wu,K.J., Wang, C.Y., and Lu, H.K., Effect of glutaraldehyde on the humoral immunogenicity and structure of porcine dermal collagen membranes, *Arch. Oral. Biol.* 49, 305–311, 2004.
175. Kawahara, J.-i., Ohmori, T., Ohkubo, T., Hattori, S., and Kawamura, M., The structure of glutaraldehyde in aqueous solution determined by ultraviolet adsorption and light scattering. *Anal. Biochem.* 201, 94–98, 1992.
176. MIgneault, I., Dartiguenave, C., Bertrand, M.J., and Waldron, K.C., Glutaraldehyde: Behavior in aqueous solution, reaction with proteins, and application to enzyme cross-linking, *BioTechniques* 37, 790–802, 2004.
177. Richards, F.H. and Knowles, J.R., Glutaraldehyde as a protein cross-linking reagent. *J. Mol. Biol.* 37, 231–233, 1968.
178. Walt, D.R. and Agayn, V.I., The chemistry of enzyme and protein immobilization with glutaraldehyde, *TRAC* 13, 425–430, 1994.
179. Wine, Y., Cohen-Hadar, N., Freeman, A., and Frolow, F., Elucidation of the mechanism and end products of glutaraldehyde crosslinking reaction by X-ray structure analysis, *Biotechnol. Bioengineer.* 98, 711–718, 2007.
180. Schultz, E.W. and Gebhardt, L.P., Nature of formalin inactivation of bacteriophage, *Proc. Soc. Exp. Biol. Med.* 32, 1111–1112, 1934/35.
181. Salk, J.E., Krech, U., Youngner, J.S. et al., Formaldehyde treatment and safety testing of experimental poliomyelitis vaccines, *Am. J. Pub. Health Nations Health* 44, 563–570, 1954.
182. Salk, J.E., Considerations in the preparation and use of poliomyelitis vaccine, *J. Am. Med. Assoc.* 158, 1239–1248, 1955.
183. Salk, J.E. and Gori, J.B., A review of theoretical experimental, and practical considerations in the use of formaldehyde for the inactivation of poliovirus, *Ann. N.Y. Acad. Sci.* 83, 609–637, 1960.
184. Laufs, R. Immunization of marmoset monkeys with killed oncogenic Herpesvirus, *Nature* 249, 571–572, 1974.
185. Ablashi, D.V. and Eaton, J.M., Preventive vaccination against Herpesvirus salmini-induced neoplasia, *Cancer Res.* 36, 701–703, 1976.
186. Ohno, T., Autologous formalin-fixed tumor vaccine, *Curr. Pharm. Des.* 11, 1181–1188, 2005.
187. Poon, B., Hsu, J.F., Gudeman, V. et al., Formaldehyde-treated, heat-inactivated virions with increased human immunodeficiency virus type 1 env can be used to induce high-titer neutralizing antibody responses, *J. Virol.* 79, 10210–10217, 2005.
188. Rezapkin, G., Martin, J., and Chumakov, K., Analysis of antigenic profiles of inactivated poliovirus vaccine and vaccine-derived poliovirus by block-ELISA method, *Biologicals* 33, 29–39, 2006.
189. Thaysen-Andersen, M., Jørgensen, S.B., Wilhelmsen, E.S. et al., Investigation of the detoxification mechanism of formaldehyde-treated tetanus toxin, *Vaccine* 25, 2213–2227, 2007.
190. Lagergård, T., Lundqvist, A., Wising, C. et al., Formaldehyde treatment increases the immunogenicity and decreases the toxicity of *Haemophilus ducreyi* cytolethal distending toxin, *Vaccine* 25, 3606–3614, 2007.

191. Tano, Y., Shimizu, H., Martin, J. et al., Antigenic characterization of a formalin-inactivated poliovirus vaccine derived from live-attenuated Sabin strain, *Vaccine* 25, 7041–7046, 2007.
192. Ruat, C., Caillet, C., Bidaut, A. et al., Vaccination of macques with adjuvanted formalin-inactivated influenza A virus (H5N1) vaccines: Protection against 55N1 challenge without disease enhancement, *J. Virol.* 82, 2565–2569, 2008.
193. Salnikova, M.S., Joshi, S.B., Rytting, J.H. et al., Physical characterization of *Clostridium difficile* toxins and toxoids: Effect of the formaldehyde crosslinking on thermal stability, *J. Pharm. Sci.* 97, 3735–3752, 2008.
194. Fox, C.H., Johnson, F.B., Whiting, J., and Roller, P.P., Formaldehyde fixation, *J. Histochem. Cytochem.* 33, 845–853, 1985.
195. Namimatsu, S., Ghazizadeh, M., and Sugisaki, Y., Reversing the effects of formalin fixation with citraconic anhydride and heat: A universal antigen retrieval method, *J. Histochem. Cytochem.* 53, 3–11, 2005.
196. Yamashita, S. and Okada, Y., Application of heat-induced antigen retrieval to aldehyde-fixed fresh frozen sections, *J. Histochem. Cytochem.* 53, 1421–1432, 2005.
197. Sompuran, S.R., Vani, K., and Bogen, S.A., A molecular model of antigen retrieval using a peptide array, *Am. J. Clin. Pathol.* 125, 91–98, 2006.
198. Xu, H., Yang, L., Wang, W. et al., Antigen retrieval for proteomic characterization of formalin-fixed and paraffin embedded tissues, *J. Proteome Res.* 7, 1098–1108, 2008.
199. Fowler, C.B., O'Leary, T.J., and Mason, J.T., Modeling formalin fixation and histological processing with ribonuclease A: Effects of ethanol dehydration on reversal of formaldehyde cross-links, *Lab. Invest.* 88, 185–195, 2008.
200. Leong, T. Y.-M. and Leong, A.S.-Y., How does antigen retrieval work?, *Adv. Anal. Pathol.* 14, 129–131, 2007.
201. Miled, N., Berti-Dupuis, L., Riviere, M., Carrière, F., and Verger, R., *In vitro* lipolysis by human pancreatic lipase is specifically abolished by its inactive forms, *Biochim. Biophys. Acta* 1645, 241–246, 2003.
202. Benedict, C.R., Ryan, J., Wolitzky, B., Ramos, R., Gerlach, M., Tijburg, P., and Stern, D., Active site-blocked factor XIa prevents intravascular thrombus formation in the coronary vasculature without inhibiting extravascular coagulation in a canine thrombosis model, *J. Clin. Invest.* 88, 1760–1765, 1991.
203. Choudri, T.F., Hoh, B.L., Prestigiacomo, C.J., Huang, J., Kim, L.J., Schmidt, A.M., Kisiel, W., Connolly, E.S., Jr., and Pinsky, D.J., Targeted inhibition of intrinsic coagulation limits cerebral injury in stroke without increasing intracerebral hemorrhage, *J. Exp. Med.* 180, 91–99, 1999.
204. Himber, J., Refino, C.J., Burckln, L., Roux, S., and Kirchhofer, D., Inhibition of arterial thrombosis by a soluble tissue factor mutant and active site-blocked factors IXa and Xa in the guinea pig, *Thromb. Haemost.* 85, 475–481, 2001.
205. Sato, H., Ikeda, M., Suzuki, K., and Hirayama, K., Site-specific modification of interleukin-2 by the combined use of genetic engineering techniques and transglutaminases, *Biochemistry* 35, 13072–13080. 1996.
206. Sato, H., Hayashi, E., Yamada, N., Yatagai, M., and Takahara, Y., Further studies on the site-specific protein modification by microbial transglutaminases, *Bioconjug. Chem,* 12, 701–710, 2001.
207. Sato, H., Enzymatic procedure for site-specific pegylation of proteins, *Adv. Drug Delivery Rev.* 54, 487–504, 2002.
208. Taki, M., Shiota, M., and Taira, K., Transglutaminase-mediated *N*- and C-terminal fluorescein labeling of a protein can support the native activity of the modified protein, *Protein Eng. Des. Sel.* 17, 119–126. 2004.
209. Meusel, M., Synthesis of hapten-protein conjugates using microbial transglutaminases, *Methods Mol. Biol.* 283, 109–124, 2004 .

210. Jordan, J.A,, Alvarez, F.J., Latero, L.A., Herraez, A., Diaz, J.C., and Trejedon, M.C., *In vitro* phagocytosis of carrier mouse red blood cells is increased by band 3 cross-linking or diamide treatment, *Biotechnol. Appl. Biochem.* 34, 143–149, 2001.
211. Myers, S.R., Yakubu-Madus, F.E., Johnston, V.T., Baker, J.E., Cusick, T.S., Williams, V.K., Tinsley, F.C., Kriauciunas, A., Manetta, J., and Chen, V.J., Acylation of human insulin with palmitic acid extends the time action of human insulin in diabetic dogs, *Diabetes* 46, 637–642, 1997.
212. Broder, M.L., Sukuman, M., Pekar, A.H., McClellan, D.S., Chone, R.F., Flora, D.B., Cox, A.L., Irwin, L., and Myers, S.R., Hybrid insulin cocrystals for controlled release delivery, *Nat. Biotechnol.* 20, 800–804, 2002.
213. Bhatnagar, S., Srivastava, D., Jayadev, M.S., and Dubey, A.K., Molecular variants and derivatives of insulin for improved glycemic control in diabetes, *Prog. Biophys. Mol. Biol.* 91, 199–228, 2006.
214. Uludag, H. and Yang, J., Targeting systemically administered proteins to bone by bisphosphonate conjugation, *Biotechnol. Prog.* 18, 604–616. 2002.
215. Gittens, S.A., Matyas, J.R., Zennicke, R.F., and Uludag, H., Imparting bone affinity to glycoproteins through conjugation of bisphosphonates, *Pharmaceut. Res.* 20, 978–987, 2003.
216. Jue, R., Lambert, J.M., Pierce, L.R., and Traut, R.R., Addition of sulfydryl groups to *Escherichia coli* ribosomes by protein modification with 2-iminothiolane (methyl-4-mercaptoburyimidate), *Biochemistry* 17, 5399–5406, 1978.
217. Mokotoff, M., Mocarski, Y.M., Gentsch, B.L., Miller, M.R., Zhou, J.-H., Chen, J., and Ball, E.D., Caution in the use of 2-iminothiolane (Traut's reagent) as a cross-linking agent for peptides. The formation of *N*-peptidyl-2-iminothiolanes with bombesin (BN) antagonist (D-Trp^{6}-Leu^{13}-ψ[CH_2NH]-Phe^{14})$BN_{6\text{-}14}$ and D-Trp-Gln-Trp-NH_2, *J. Peptide Res.* 57, 383–389, 2001.
218. Bernkop-Schnürch, A., Guggi, D., and Pinter, Y. (2004), Thiolated chitosans: Development and *in vitro* evaluation of a mucoadhesive, permeation enhancing oral drug delivery system, *J. Control. Release* 94, 177–186, 2004.
219. Gaidamakova, E.K., Backer, M.V., and Backer, J.M., Molecular vehicles for target-mediated delivery of therapeutics and diagnostics, *J. Control Release* 74, 341–347, 2001.
220. Backer, M.V., Budker, V.G., and Backer, J.M., Shiga-like toxin-VEGF fusion proteins are selectively cytotoxic to endothelial cells overexpressing VEGR-2, *J. Control Release* 74, 249–355, 2001.
221. Backer, M.V. , Aloise, R., Przekop, K., Stoletov, K., and Backer, J.M., Molecular vehicles for targeted drug delivery, *Bioconjug. Chem.* 13, 462–467, 2002.
222. Futami, J., Maeda, T., Kitazoe, M. et al., Preparation of potent cytotoxic ribonucleases by cationization: Enhanced cellular uptake and decreased interaction with ribonuclease inhibitor by chemical modification of carboxyl groups, *Biochemistry* 40, 7518–7524, 2001.
223. Futami, J., Nukui, E., Maeda, T. et al., Optimum modification for the highest cytotoxicity of cationized ribonuclease, *J. Biochem.* 132, 223–228, 2002.
224. Ikenaga, T., Yamasaki, Y., Shakushiro, K. et al., Induction of cytotoxic T lymphocytes following immunization with cationized soluble antigen, *Vaccine* 22, 2609–2616, 2004.
225. Ma, S.F., Nichikawa, M., Katsumi, H. et al., Cationic charge-dependent hepatic delivery of amidated serum albumin, *J. Control. Release* 102, 583–594, 2005.
226. Kitazoe, M., Murata, H. Futami, J. et al., Protein transduction assisted polyethyleneimine-cationized carrier proteins, *J. Biochem.* 137, 693–701, 2005.
227. Ma. S.F., Nishikawa, M., Katsumi, H. et al., Liver targeting of catalase by cationization for prevention of acute liver failure in mice, *J. Control. Release* 110 273–282, 2006.

228. Murata, H., Sakaguchi, M., Futami, J. et al., Denatured and reversibly cationized p53 readily enters cells and simultaneously folds to the functional protein in the cells, *Biochemistry* 45, 6124–6132, 2006.
229. Sharifi, J., Khawali, L.A., Hornick, J.L., and Epstein, A.L., Improving monoclonal antibody pharmacokinetics via chemical modification, *Quart. J. Nucl. Med.* 42, 242–249, 1009.

10 Food and Agricultural Chemistry

Food protein has a functional/structural role in food as well as value in nutrition. Legume proteins, milk proteins, and egg proteins have functional roles in gelling, viscosity, forming fiber, forming and stabilizing foam, and in forming dough. Edible films may be formed from protein.[1–6] The reader is also directed to a discussion of protein films with protein plastics in Chapter 6.

Chemical modification is an important part of food processing.* Such chemical modification includes the Maillard reaction, which is important for the color and taste of cooked foods,[7–19] and the Strecker reaction (Figure 10.1) in the production of aldehydes important for aroma.[21–26] These processes are distinct for the complex chemistry that forms heterocyclic amines during the cooking of meat.[26–29] These heterocyclic amines are carcinogenic and form adducts with proteins[27] and DNA.[30]

Chemical modification of macromolecules has extensive use in the processing of food. The best understood (from a technical perspective) is the hydrolysis of macromolecules. One example is the enzymatic or chemical hydrolysis of food protein. Soy protein hydrolysate is used in nutrition[31–33] and as supplement to animal product-free cell culture.[34–36] The effectiveness of soy protein hydrolysate is a function of the method of preparation; material prepared by enzymatic hydrolysis was more effective than that of a chemical hydrolysis product in stabilizing a varicella viral vaccine.[36] It is assumed that chemical hydrolysis involves either hydrochloride acid or sulfuric acid which would destroy tryptophan and oxidize amino acids such as methionine and cystine/cysteine. Enzymatic hydrolysis usually involves pronase[37] or other proteolytic enzymes.[38,39]

Chemical modification of food protein has been used as a way of improving functional properties.[40] Functional properties, as defined above, may not be of nutritional importance, but they are of importance in food qualities such as mouthfeel.[41–43] Food proteins, particularly those derived from plant sources, required modification to improve physical properties such as solubility and emulsification ability.[40] Functional properties of food components such as proteins are separate from the concept of functional foods,[44] which have a positive effect on health separate from their direct nutritional value. An example of the modification of the functional properties of food proteins is provided by the modification of whey protein by acidification, heating (80°C for 90 minutes), gelation, and spray drying.[45] This treatment markedly alters the effect of calcium ions on the viscosity properties of modified whey protein. This

* The term *processing* is used for any step between raw food product and final food product that would include pasteurization, filtration, separation, freezing, and cooking. This list is not meant to be inclusive.

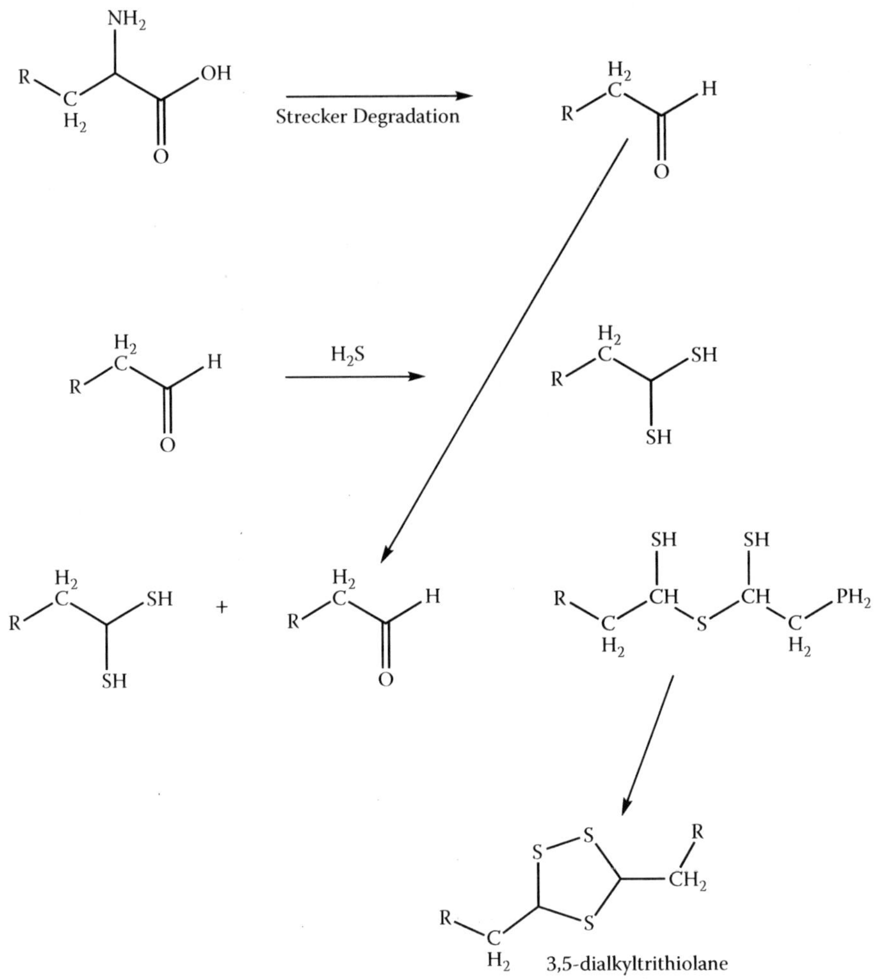

FIGURE 10.1 The Strecker reaction.

procedure is usually associated with deamidation, oxidation, and perhaps isolated peptide bond cleavage, relatively gross changes compared to the selective chemical modification used in protein chemistry and biopharmaceutical processing. Such results, however, are reproducible in commercial food processing as a result of guidelines such as HACCP (Hazard Analysis and Critical Process Control).[46,47]

Acylation of protein amino groups and hydroxyl groups has been used to modify food proteins. Sung and colleagues[48] used papain to acylate the amino groups of soy protein with *N*-acetyl-L-homocysteine thiolactone, which modifies the amino group and introduces a sulfydryl group. Solubility and the ability of foam to form an emulsion was increased by the modification. A literature search suggests that this is the only use of this novel reaction. Papain has been used for the synthesis of peptide bonds,[49] and the reader is directed to an excellent paper by Hindle and

Kirsch.[50] Traut's reagent, 2-iminothiolane, is the reagent of choice for introducing sulfydryl groups via modification of lysine residues[51–53] (see also Chapter 1). Pea protein was acylated with acetic anhydride or succinic anhydride.[54] As with the acylation of soy protein with *N*-acetyl-l-homocysteine thiolactone, emulsification ability and foam capacity/stability was increased by acylation with these organic carboxylic acid anhydrides. It is noted that modification with acetic anhydride provides charge neutralization while reaction with succinic anhydride provides charge reversal from moderate positive charge to negative charge. Modification of ovalbumin with succinic anhydride resulted in considerable conformational change (change in Stoke's radius, susceptibility to tryptic digestion, intrinsic viscosity, UV difference spectra, and intrinsic fluorescence) in the protein, which appeared proportional to the extent of modification.[55] Kosters and coworkers[56] observed that succinylation decreased thermal stability of ovalbumin, while modification of carboxyl groups with methylamine with a carbodiimide resulted in a more marked decrease in thermal stability. Zhao and colleagues used Ramen spectroscopy to study the conformational change in food proteins on succinylation.[57] New stretching vibrations were detected at 1420 and 1737 cm^{-1}. Ramen spectroscopy also demonstrated a change from an ordered conformation to a disordered conformation. Succinylation may well improve the functional properties of food proteins, but there might be a decrease in the nutrition value of such modified proteins.[58] Ali and Younus[59] reported that succinylation of antibodies resulted in a loss of activity and substantial conformational change (intrinsic fluorescence).

Earlier studies on the acylation of casein and vegetable globulins reported changes in the relative nutritive value associated with the modification of lysine.[60] Subsequent studies by Schwende on the chemical modification (acylation with succinic anhydride) of storage protein from rapeseed stabilized the protein with respect to thermal activation.[61] This group extended their studies to pea protein.[62] More recently, Gerbanowski and coworkers[63] studied the effect of modification of rapeseed protein with several aliphatic and aromatic acid anhydrides; sulfamidation was also explored in this study. Evaluation of the modified proteins with CD spectroscopy showed that the extent of conformation change increased with the increased hydrophobicity of the substituent group, while the extent of modification decreased; 95% of the amino groups in the 2S protein were modified with acetic anhydride with a decrease in α-helix from 41% to 30%, while 43% of the amino groups were modified with tosyl chloride with a decrease in α-helix to 9%. There was an increase in surface hydrophobicity as measured by the binding of 8-anilino-1-naphthalenesulfonic acid. These workers discuss the possibility of such modification providing more useful materials for preparing protein plastics. Swart and coworkers[64] observed that charge modification of plasma or milk proteins with succinic anhydride resulted in antiviral activity. A similar effect was observed with the succinylation of human serum albumin.[65] These investigators also noted normal degradation of the modified albumin to constituent amino acids include ε-*N*-succinyllysine, which was not further degraded but excreted in the urine.

Hen egg white lysozyme is a sweet-tasting protein.[66] Lysozymes from other species such as turtle and turkey are also sweet, whereas human protein is not a sweet-tasting protein. Reduction of the disulfide bonds as well as heating at

95°C for 18 hours eliminated the sweet-taste activity.[67] Modification of carboxyl groups by carbodiimide-mediated coupling of glycine inactivated enzyme activity but had no effect on the sweet taste. Modification of lysine residues with acetylation or phosphopyridoxylation increased sweetness while guanidation had no effect.[68]

Aldehydes react with the protein amino groups, resulting in a complex group of products including cross-linked protein products[69] (see also Chapter 1). Glycation is also a reaction of an aldehyde with a protein but is considered elsewhere with the discussion of Maillard reaction. Reactive aldehydes can be derived from the oxidation of lipids[70–72] with 4-hydroxy-2-nonenal or methylglyoxal derived from advanced glycation end products.[73–75]

Phosphorylation of food proteins has proved to be a useful approach to improving functional properties. The various approaches described below used an "activated" form of orthophosphate, resulting in the nonspecific phosphorylation of various amino acid residues. This is another instance of the importance of a process which can be validated through a concept such as HACCP or FMEA (Failure Mode and Effect Analysis).

Sodium trimetaphosphate (sodium cyclotriphosphate, $Na_3P_3O_9$, FW 305.89), prepared by thermal treatment using sodium dihydrogen phosphate, is available from commercial sources, and has been known to obtain useful derivatives.[76–78] This reaction occurs at pH 11.5 (maintained by addition of 40% (w/v) NaOH) at 35°C for three hours.[76] The modified soy protein had improved functional properties (solubility, emulsification ability, and foam characteristics [whippability]). Application of this technique to a soy protein hydrolysate[78] yielded a product with potential as a calcium carrier. Sodium trimetaphosphate has subsequently been used to phosphorylate various food proteins including brewer's yeast protein,[79,80] soy protein,[81,82] and vicilin.[83] There has also been extensive use of sodium trimetaphosphate for cross-linking carbohydrates.[84–90] This reaction has been useful in developing hydrogels for drug delivery. This subject is considered in more detail in Chapter 5.

Protein phosphorylation is a subject of considerable interest in the regulation of biological activity.[91–96] It is somewhat remarkable that the enzymatic phosphorylation of proteins had not been demonstrated in 1953.[97] Casein had been demonstrated to be a phosphoprotein for some time[98] but the phosphorylation was associated more with nutrition than with regulation.[99]

Phosphorylation of proteins usually occurs in the hydroxyl group of serine but can occur with threonine, hydroxyproline, tyrosine, histidine, arginine, and lysine residues.[100–118] The primary sites of biological phosphorylation are serine and tyrosine. Phosphorylation of free histidine using phosphoramidate can occur at either the 1-position or 3-position[108]; the diphosphoderivative appears over a longer period of time. The *O*-phosphoryl derivatives of serine and threonine are relatively stable in acid but rapidly hydrolyzed in base.[110,112] Tyrosine *O*-phosphate is somewhat more stable in base.[112] The *N*-phosphoryl derivative (phosphohistidine, phospholysine, and phosphoarginine) are relatively stable in base but rapidly hydrolyzed in acid.[82g,82n] The phosphoryl derivates of cysteine in thioredoxin are stable below pH 2 and between 6 and 8 and are least stable between pH 2.5 and pH 4.0.[114] While acyl derivatives[105] are possible, such compounds would be unstable and have not been demonstrated in phosphorylated proteins.

Early efforts to phosphorylate proteins by chemical means used somewhat harsh conditions. Based on earlier studies[119] on the sulfation of proteins with sulfuric acid (concentration sulfuric acid/<0°C; warmed to 23°C and poured over ice; dialyzed and lyophilized), Fraenkel-Conrat and coworkers[120] phosphorylated proteins with phosphorous pentoxide in phosphoric acid (23°C/3 days/dessicator). There was significant hydrolysis and denaturation of proteins. A gel was observed on the modification of wheat gluten but not as strong as the gel obtained with sulfation. The majority of the phosphate was labile and removed easily while the remainder of the protein-bound phosphate was quite stable. In subsequent work Fraenkel–Conrat and Fraenkel–Conrat[121] modified bovine insulin with phosphorous pentoxide in phosphoric acid under the above conditions and retained 38% of the activity (*in vivo* in mice). It was not possible to find later studies using phosphorous pentoxide for protein modification.

The chemical phosphorylation of proteins continues to be of interest in the food industry. Most studies use phorphorous oxychloride for the chemical phosphorylation of proteins. This technique was proposed by Rimington[122] in 1927 who added phosphorous oxychloride (dissolved in CCl_4) dropwise to a protein solution at 0°C. The reaction was allowed to proceed for 6–8 hours and the pH maintained between 6 and 8 by the addition of sodium hydroxide. Phosphorous oxychloride has been used for the synthesis of phosphoamino acids.[109,111,113,115] There has been continued use of phorphorous oxychloride to present for the phosphorylation of food proteins.[123–128] There is a recent report on the use of pyrophosphate in dry heat for the phosphorylation of proteins.[129]

Protein cross-linking is a reaction associated with the Maillard reaction and is important for food protein function.[130–132] The specific aspects of the Maillard reaction will be discussed below. There is also cross-linking which occurs during processing[133] as well as the *in vivo* cross linking of collagen contributing to meat quality.[134–136]

Cross-linking of food protein such as soy protein is important in the development of protein-based adhesives[137] and in the formation of a soy protein-PEG hydrogel[138] useful for wound dressing. In the former study,[137] the water resistance of soy protein as an adhesive was enhanced by cross-linking with glutaraldehyde. This work is part of an effort to improve the utility of proteins as adhesive materials (see Chapter 6). In the later study, PEG was modified with *p*-nitrophenyl chloroformate[139] to yield the *p*-nitrophenylcarbonate derivative, which was then coupled with soy protein to yield the hydrogel. Improvement of water resistance of gliadin films was also obtained by cross-linkage.[140] Formaldehyde treatment yielded stronger films that did either glyoxal or glutaraldehyde.

Sodium bisulfite is added as a preservative to food and beverage products. Sodium bisulfite also converts cytosine to uracil[141] and catalyzed transamination reactions between cytosine and amine donors[142–144] including proteins.[145] Cross-linking of proteins in milk is observed on storage[146,147] and is likely a result of glycation reactions. This cross-linkage process is thought to be involved in the gelation of milk,[114] important for the production of yogurt.[148] Transglutaminase (discussed below) has also been used to cross-link milk proteins.[149–152]

The majority of food cross-linking reactions involve the use of transglutaminase, which catalyzes a transamination reaction between lysine and glutamine. Aboumahmoud and Savello[153] used guinea pig liver transglutaminase to cross-link whey protein fractions. The cross-linking reaction required calcium ion and the presence of dithiothreitol. Wilcox and Swaisgood[154] used an immobilized (controlled-pore glass) microbial transglutaminase (calcium independent)[155] to cross-link whey proteins. The cross-linking process results in a protein product with increased intrinsic viscosity and decreased gelation temperature, which formed a more brittle gel on heating. Cozzolino and coworkers[156] incorporated whey into cheese curd using microbial (*Streptoverticillium mobaraense*) transglutaminase, producing a novel dairy product. There are other studies on the use of microbial transglutaminase to convert whey protein into a more useful function product.[157–159]

Microbial transglutaminase has also been used to prepare products from soy protein.[160–163] Mizuno and coworkers[160,161] observed that cross-linking of soy protein results in a product with decreased glass transition temperature. Zhang and coworkers[162] reported differences between products obtained with microbial transglutaminase and tissue transglutaminase. Tang and coworkers[163] observed modification of the properties of soy protein films with treatment of precursor soy protein isolate with microbial transglutaminase. High-pressure treatment of protein prior to modification with transglutaminase increased the extent of cross-linking through the exposure of more glutamine residues.[164,165] There were some deleterious effects of high pressure destabilized fish protein.[164] The prior infusion of carbohydrate provided both baroprotection and cryoprotection. Transglutaminase has also been used for the polymerization of peanut protein.[166]

Other examples of the use of transglutaminase include oat globulin.[167,168] The product had improved water-binding and fat-binding properties, as well as higher solubility at lower pH. The gel obtained with transglutaminase had characteristics similar to that of a classical polymer gel with permanent cross-links. Treatment of wheat dough with microbial transglutaminase created a product with less rigidity and extensibility.[169] There appeared to be a greater effect with organic flour than conventional flour. Treatment with transglutaminase improves the baking quality of weak flours.

Transglutaminase has been used to obtain hybrid products including gelatin–chitosan,[170] chitosan–whey protein edible film,[171] and fennel–phaseolin.[172]

Limited proteolysis is of considerable benefit in food processing. The use of papain in tenderizing meat is well known.[173,174] Proteolysis also has a role in flavor development[175,176] and changes in functional processes such as tack which would be beneficial in the development of pressure-sensitive adhesives.[176]

The Maillard reaction is a complex reaction between carbohydrates and amino groups in amino acids, peptides, and proteins critical in various aspects of food preparation and storage.[177–196] It is referred to as the nonenzymatic browning reaction[197–204] to differentiate from the browning reaction catalyzed by the action of polyphenol oxidase (PPO) on phenolic compounds in fruits and vegetables.[205–209] The nonenzymatic browning refers to the color of complex pyroles, furans, imidazolones, and related compounds resulting from the reaction of carbohydrates with carbonyl compounds[201,210–212] (Figure 10.2). The Maillard reaction is a glycation reaction. Glycation is the term used to identify the reaction of reducing sugars with proteins. This involves the initial formation of a Schiff base followed by rearrangement[213] to form the Amadori product (Figure 10.3), eventually resulting in advanced glycation end (AGE) products.[214] Reaction can occur at lysine and arginine residues with resulting cross-link formation.[215–219] The Amadori products can rearrange to form a variety of products including dicarbonyl derivatives.[220–224] A wide variety of products are formed via the Maillard reaction pathway, including the formation of acrylamide from asparagine (Figure 10.4).[225–235] Even though asparagine has the internal structure sufficient to generate acrylamide, acrylamide does not form in the absence of added carbohydrate.[225]

Pressure and temperature are the two primary variables in the food processing industry. While the functional consequences of these variables are known in the quality of the food product, there is a lack of study of the effect of pressure on isolated biochemical reactions. Most biochemical reactions are performed under what are described as ambient conditions: 23°C and 1 atm pressure (760 mm Hg [29.92 in], 0.101 MPa). IUPAC defines standard atmospheric pressure as 101,325 Pa and uses the symbol atm.[236–238] High-pressure is used extensively in the food processing industry[239–244] for issues other than pathogen inactivation.[245–250] Le Chatelier's principle governs the effect of pressure on chemical reactions; reactions where the products occupy a small volume (synthetic reactions) are accelerated by pressure while reactions where products occupy a larger volume such as hydrolysis of an ester are inhibited by pressure.

Le Chatelier's principle states that when an exterior force such as pressure or temperature is applied to a system in equilibrium, the equilibrium will shift to compensate for the force.[251–254] When there is a decrease in volume ($\Delta < 0$), an increase in pressure shifts the reaction in that direction; conversely, when there is an increase in volume ($\Delta V < 0$) such as with hemolytic bond cleavage, pressure will inhibit that reaction.[255] The solubility of gases in liquids is usually increased by pressure, while the solubility of solids in liquids is usually decreased by pressure. It is noted the compression of liquids is relatively small compared to the compression of gases. Standard high pressure is 500 MPa (1 kbar = 100 MPa = 0.1 GPa = 14503.8 psi = 982s.92 atm), and it is technically difficult to attain pressures greater than 500 MPa.[256]

High pressure has also proved useful in the study of basic protein–protein interactions and enzymology.[257] Masson has studied the effect of high pressure on the electrophoresis of proteins.[258] At moderate pressure ($P < 1$ kbar), there are alterations of binding properties and catalytic activity but no gross structural changes. Pressures greater that 1.5 kbar causes conformational change, which could be considered denaturation, while there is significant unfolding of the protein only above 5 kbar. It is suggested that ionic interactions and hydrophobic interactions are disrupted by pressure while stacking interactions between aromatic rings and charge-transfer interactions are increased; it is thought that hydrogen bonds are insensitive to pressure. It is important to realize that this is occurring in an aqueous environment where there would be an increase in temperature and a decrease in pH (increased ionization of water). Denaturation and unfolding of a protein is consistent with the increased availability of glutamine residues to participate as substrate in transamination.[164,165]

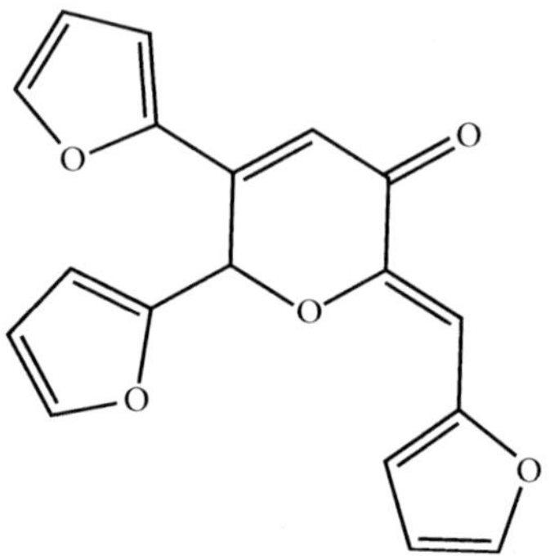

(Z)-2-[(2-furyl)methylidene]-5,6-(2-furyl)-6*H*-pyran-3-one

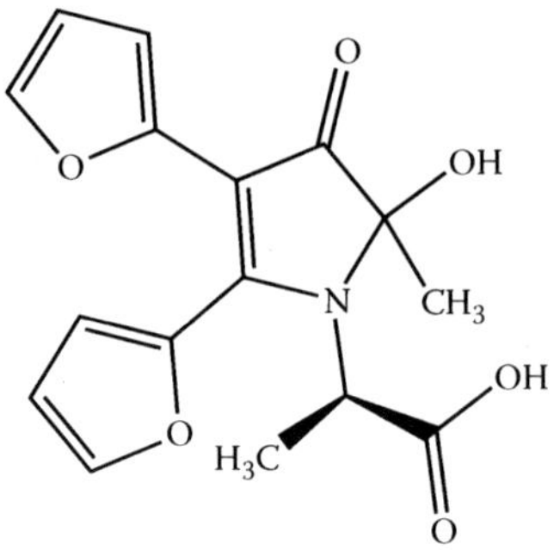

(*S,S*)/(*S,R*)-2=[4,5-bis(2-furyl)-2-hydroxy-2-methyl-3(2*H*)-pyrrol-1-yl]propionic acid

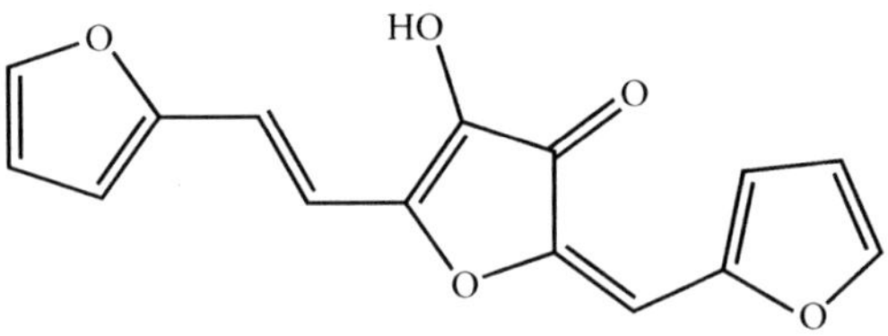

2-[2-furylmethylidene]=4-hydroxy-[(*E*)-2-furylmethylidene-2*H*-furan-3-one]

FIGURE 10.2 Some products from the nonenzymatic browning reaction with carbohydrate and protein (Maillard reaction).

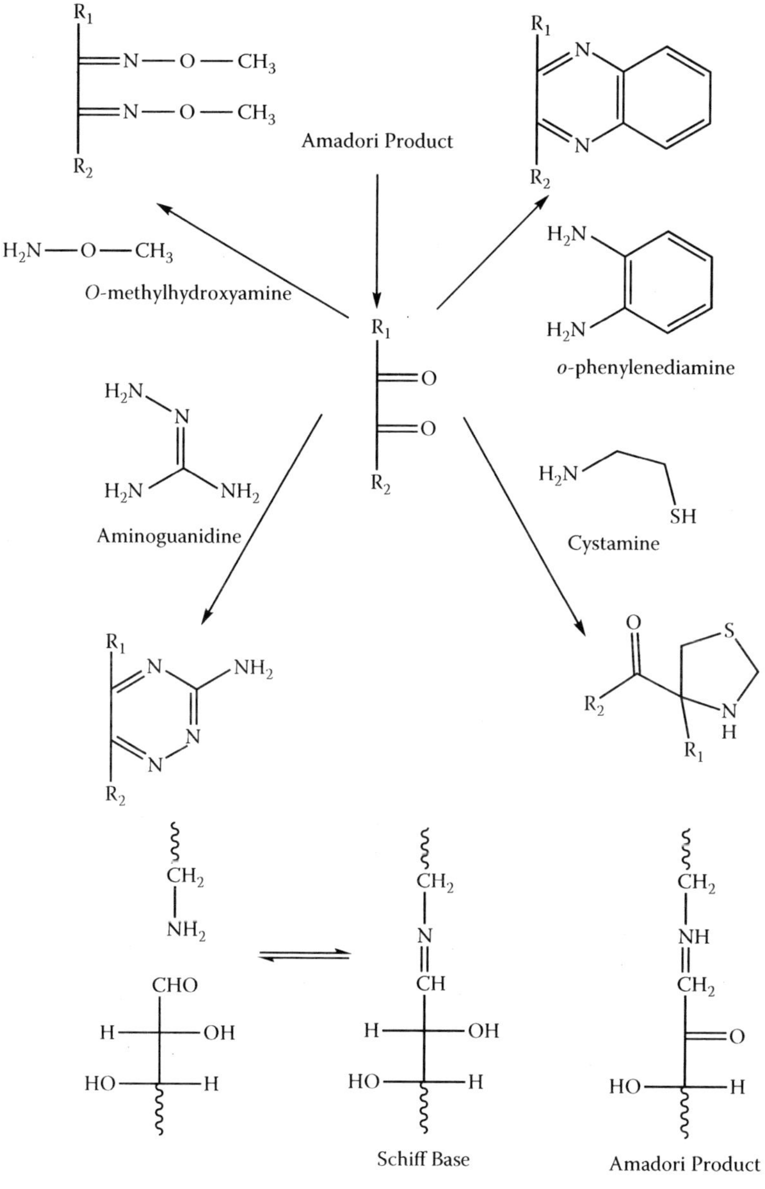

FIGURE 10.3 The various pathways in the formation of the Amadori product on the reaction of carbohydrate with protein.

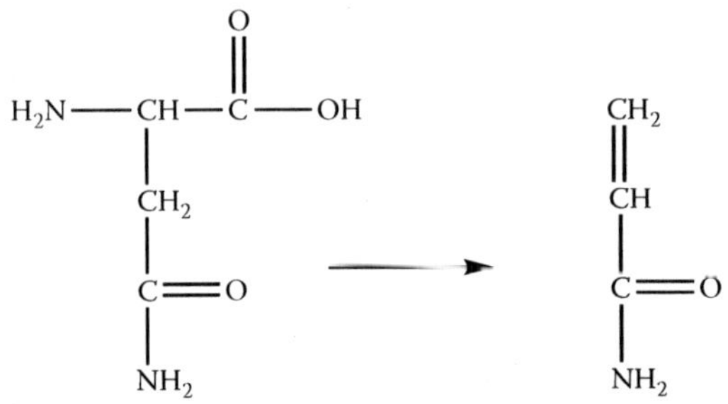

FIGURE 10.4 The formation of acrylamide from asparagine.

REFERENCES

Chapter References

1. Brandenburg, B.H., Weller, C.L., and Testin, R.F., Edible films and coatings from soy proteins, *J. Food Sci.* 58, 1086–1089, 1993.
2. Chen, H., Functional properties and applications of edible films made of milk proteins, *J. Dairy Sci.* 78, 2563–2583, 1995.
3. McHugh, T.H., Protein-lipid interactions in edible films and coatings, *Nahrung* 44, 148–151, 2000.
4. Khwaldia, K., Perez, C., Boron, S. et al., Milk protein for edible films and coatings, *Crit. Rev. Food Sci. Nutr.* 44, 239–251, 2004.
5. Di Pierro, P., Chico, P., Villalonga, R. et al., Chitosan-whey protein edible films produced in the absence and presence of transglutaminase: Analysis of their mechanical and barrier properties, *Biomacromolecules* 7, 744–749, 2006.
6. Hernandez-Izquierdo, V.M. and Krochta, J.M., Thermoplastic processing of protein for film formation—a review, *J. Food Sci.* 73, R30–R39, 2008.
7. *Chemical Markers for Processed and Stored Foods*, Eds. T.C. Lee and H.J. Kim, American Chemical Society, Washington, DC, 1996.
8. Feather, M.S., Massire, V., and Hirsch, J., The use of aminoguanidine to trap and measure dicarbonyl intermediates produced during the Maillard reaction, in *Chemical Markers for Processed and Stored Foods*, Eds. T.C. Lee and H.J. Kim, American Chemical Society, Washington, DC, Chapter 3, pp. 224–31, 1996.
9. Matsuda, J. and Kato, Y., Haptenic sugar antigens as immunochemical markers for the Maillard reaction of processed and stored milk products, in *Chemical Markers for Processed and Stored Foods*, Eds. T.C. Lee and H.J. Kim, American Chemical Society, Washington, DC, Chapter 19, pp. 221–226, 1996.
10. Kato, Y. and Matsuda, J., Inactivation of egg trypsin inhibitors by the Maillard reaction, in *Chemical Markers for Processed and Stored Foods*, Eds. T.C. Lee and H.J. Kim, American Chemical Society, Washington, DC, Chapter 20, pp. 227–231, 1996.
11. Reynald, T.H., Chemistry of nonenzymatic browning. I. The reaction between aldoses and amines, *Adv. Food Res.* 12, 1–52, 1963.
12. Didzbalis, J. and Ho, C.-T., Analysis of low molecular weight aldehydes formed during the Maillard reaction, in *Aroma Active Compounds in Foods Chemistry and Sensory Properties*, Eds. G.R. Takeoka, M. Günert, and K.H. Engel, American Chemical Society, Washington, DC, 2001.

13. *Bioactive Compounds in Food Effect of Processing and Storage*, Eds. T.-C Lee and C.-T Ho, American Chemical Society, Washington, DC, 2002.
14. Nursten, H., The *Maillard Reaction Chemistry, Biochemistry and Implications,* Royal Society of Chemistry, Cambridge, 2005.
15. Ellis, G.P., The Maillard reaction, *Adv. Carbohydrate Chem.* 14, 63–134, 1959.
16. Fayle, S.E. and Gerrard, J.A., *The Maillard Reaction*, Royal Society of Chemistry, Cambridge, 2002.
17. Dai, Z., Nemet, I., Shen, W., and Monnier, V.M., Isolation, purification and characterization of histidine-threosidine, a novel Maillard reaction product crosslink from threose, lysine, and histidine, *Arch. Biochem. Biophys.* 463, 78–88, 2007.
18. Wedzicha, B.L. and Leong, L.P., Rate constants for individual steps in the reaction, in *The Maillard Reaction in Food and Medicine*, Ed. J. O'Brien et al., Royal Society of Chemistry, Cambridge, 1998.
19. Mundt, S. and Wedzicha, B.L., A kinetic model for the glucose–fructose–glycine browning reaction, *J. Agric. Food Chem.* 51, 3651–3655, 2003.
20. Saraiva, M.A., Borges, C.M., and Florêncio, M.H., Reactions of a modified lysine with aldehydes and diketonic dicarbonyl compounds: An electrospray mass spectrometry structure/activity study, *J. Mass Spectrom.* 41, 216–228, 2006.
21. Hofmann, T. and Schieberle, P., Formation of aroma-active Strecker aldehydes by a direct oxidative degradation of Amadori compounds, *J. Agric. Food Chem.* 48, 4301–4305, 2000.
22. Weenen, H. and van der Ven, J.G.M., The function of Strecker aldehydes, in *Aroma Active Compounds in Foods Chemistry and Sensory Properties*, Eds. G.R. Takeoka, M. Günert, and K.H. Engel, American Chemical Society, Washington, DC, 2001.
23. Hildago, F.J. and Zamora, R., Strecker-type degradation produced by the lipid oxidation products, 4,5-epoxy-2-alkenals, *J. Agric. Food Chem.* 52, 7126–7131, 2004.
24. Bittner, J., When quinones meet amino acids; chemical, physical, and biological consequences, *Amino Acids* 30, 205–224, 2006.
25. Rotsatchakul, P., Chaiseri, S., and Cadwallader, K.R., Identification of characteristic aroma compounds of Thai chili paste, *J. Agric. Food Chem.* 56, 528–536, 2008.
26. Song, H. and Cadwallader, K.R., Aroma compounds of American country ham, *J. Food Sci.* 73, C29–C35, 2008.
27. Sabbioni, G. and Schülze, D., Hemoglobin binding of bicyclic aromatic amines, *Chem. Res. Toxicol.* 11, 471–483, 1998.
28. Sinha, R., Rothman, N., Salmon, C.P. et al., Heterocyclic amine content in beef cooked by different methods to varying degrees of doneness and gravy made from meat drippings, *Food Chem. Technol.* 36, 279–287, 1998.
28a. Knize, M.G. and Felton, J.S., Formation and human risk of carcinogenic heterocyclic amines formed from natural precursors in mean, *Nutr. Rev.* 63, 158–165, 2005.
29. Murkovic, M., Analysis of heterocyclic aromatic amines, *Anal. Bioanal. Chem.* 389, 139–146, 2007.
30. Turesky, R.J., Formation and biochemistry of carcinogenic heterocyclic aromatic amines formed from natural precursors in meat, *Nutr. Rev.* 63, 158–163, 2005.
31. Nishiwaki, I., Yoshimizu, S., Furuta, M., and Hayashi, K., Debittering of enzymatic hydrolysate using an aminopeptidases from the edible basidiomyocete *Grifola frandosa*, *J. Biosci. Engineer.* 93, 60–63, 2002.
32. Fitzgerald, R.J. and O'Cuinn, G., Enzymatic debittering of food protein hydrolysates, *Biotechnol. Adv.* 24, 234–237, 2006.
33. Cho, S.J., Juilkerat, M.A., and Lee, C.H., Cholesterol lowering mechanism of soy bean protein hydrolysate, *J. Agric. Food Chem.* 55, 10599–10604, 2007.

34. Sung, Y.H., Lin, S.W., Chung, J.Y., and Lee, G.M., Yeast hydrolysate as a low cost additive to serum-free medium for the production of human erythropoietin in suspension cultures of Chinese hamster ovary cells, *Appl. Microbiol. Biotechnol.* 63, 527–536, 2004.
35. Chun, B.H., Kim, J.H., Lee, H.J., and Chung, N., Usability of size-excluded fractions of soy protein hydrolysates for growth and viability of Chinese hamster ovary cells in protein-free suspension culture, *Bioresourc. Technol.* 98, 1000–1005, 2007.
36. Chun, B.H., Lee, Y.K., Lee, B.C., and Chung, N., Development of a varicella virus vaccine containing no animal derived components, *Biotechnol. Lett.* 26, 807–812, 2004.
37. Franek, F., Hohenwarter, O., and Katinger, H., Plant protein hydrolysates: Preparation of defined peptide fractions promoting growth and production in animal cells culture, *Biotechnol. Progress* 16, 688–692, 2000.
38. Liu, F. and Yasuda, M., Debittering effect of *Monascus* carboxypeptidase during the hydrolysis of soybean protein, *J. Ind. Microbiol. Biotechnol.* 32, 487–489, 2005.
39. Ringseis, R., Matthes, B., Lehmann, V. et al., Peptides and hydrolysates from casein and soy protein modulate the release of vasoactive substances from human aortic endothelial cells, *Biochim. Biophys. Acta* 1721, 89–97, 2005.
40. Shih, F.F., Modification of food proteins by non-enzymatic methods, in *Biochemistry of Food Proteins*, Ed. B.J.F. Hudson, Elsevier, London, 1992.
41. Chen, J., Surface tension of foods: Perception and characterization, *Crit. Rev. Food Sci. Nutr.* 47, 583–598, 2007.
42. Setser, C.S. and Racette, W.L., Macromolecule replacers in food products, *Crit. Rev. Food Sci. Nutr.* 32, 275–297, 1992.
43. Lal, S.N., O'Connor, C.J., and Eyres, L, Application of emulsifiers/stabilizers in dairy products of high rheology, *Adv. Colloid Interface Sci.* 16, 123–126, 2006.
44. Roberfroid, M.B., Global view on functional foods: European perspectives, *Brit. J. Nutrition* 88(Suppl. 2), S133–S138, 2002.
45. Clare, D.A., Lillard, S.J., Ramsey, S.R. et al., Calcium effects on the functionality of a modified whey protein ingredient, *J. Agric. Food Chem.* 55, 10932–10940, 2007.
46. Arvanitoyannis, I.S. and Traikou, A., A comprehensive review of the implementation of hazard analysis critical control, *Crit. Rev. Food Sci. Nutr.* 45, 327–370, 2005.
47. Varzakas, T.H. and Arvanitoyannis, I.S., Application of Failure Mode and Effect Analysis (FEMA), cause and effect analysis, and Pareto diagram in conjunction with HACCP to a corn curl manufacturing plant, *Crit. Rev. Food Sci. Nutr.* 47, 363–387, 2007.
48. Sung, H.-Y., Chen, H.-J., Liu, T.-Y., and Su, J.-C., Improvement of the functionality of soy protein by introduction of new thiol groups through papain-catalyzed acylation, *J. Food Sci.* 48, 722–725, 1983.
49. Schuster, M., Jakubke, H.D., and Kasche, V., Nucleophile specificity in papain-catalyzed acyl transfer reactions, *Biomed. Biochim. Acta* 50, S122–126, 1991.
50. Hindle, P.M. and Kirsch, J.F., Kinetics and thermodynamics of the reactions of acyl-papains. Effects of pH, temperature, solvents, ionic strength, and added nucleophiles, *Biochemistry* 10, 3700–3707, 1971.
51. Lasch, J., Niedermann, G., Bogdanov, A.A., and Torchilin, V.P., Thiolation of preformed liposomes with iminothiolane, *FEBS Lett.* 214, 113–116, 1987.
52. Miyata, K., Kakizawa, Y., Nishiyama, N. et al., Block catiomer polyplexes with regulated densities of charge and disulfide cross-linking directed to enhance gene expression, *J. Amer. Chem. Soc.* 126, 2355–2361, 2004.
53. Hahn, S.K., Park, J.K., Tomimatsu, T., and Shimoboji, T., Synthesis and degradation test of hyaluronic acid hydrogels, *Int. J. Biol. Macromol.* 40, 374–380, 1007.
54. Johnson, E.A. and Brekke, C.J., Functional properties of acylated pea protein isolates, *J. Food Sci.* 48, 722–725, 1983.

55. Kidwai, S.A., Ansari, A.A., and Salahuddin, A., Effect of succinylation (3-carboxypropionylation) on the conformation and immunological properties of ovalbumin, *Biochem. J.* 155, 171–180, 1976.
56. Kosters, H.A., Broersen, K., de Groot, J. et al., Chemical processing as a tool to generate ovalbumin variants with changed stability, *Biotechnol. Bioengineer.* 84, 61–70, 2003.
57. Zhao, Y., Ma, C.Y., Yuan, S.N., and Phillips, D.L., Study of succinylated food proteins by Raman spectroscopy, *J. Agric. Food Chem.* 52, 1815–1823, 2004.
58. Wallace, R.J., Min, W.K., and Witt, M.W., Uptake of acetylated peptides from the small intestine in sheep and their nutritive value in rats, *Br. J. Nutr.* 80, 101–108, 1998.
59. Ali, J. and Younus, H., Effect of succinylation of antibodies on their conformation and interaction with the allergen, *Biochemistry* (Moscow) 71, 1336–1340, 2006.
60. Schwende, K.D., Prahl, L., and Enders, B., Modification of proteins by reaction with carbonyl compounds, *Nahrung* 19, 921–927, 1975.
61. Schwende, K.D., Structural studies on native and chemically modified storage proteins from rapeseed (*Brassica rapis* L) and related plant proteins, *Nahrung* 34, 225–240, 1990.
62. Schwenke, K.D., Kim, Y.H., Kroll, J. et al., Selected physicochemical properties of succinylated legumin from pea (*Pisum sativum* L), *Nahrung* 37, 519–527, 1993.
63. Gerbanowski, A., Malabat, C., Rabiller, C., and Guéguen, J., Grafting of aliphatic and aromatic probes on rapeseed 2S and 12S proteins: Influence on their structural and physicochemical properties, *J. Agric. Food Chem.* 47, 5218–5226, 1999.
64. Swart, P.J., Harmsen, M.C., Kuipers, M.E. et al., Charge modification of plasma and milk proteins results in antiviral active compounds, *J. Pept. Sci.* 5, 563–576, 1999.
65. Swart, P.J., Kuipers, M.E., Smit, C. et al., The metabolic fate of the anti-HIV active drug carrier succinylated human serum albumin after intravenous administration in rats, *J. Drug. Target.* 9, 95–109, 2001.
66. Maehashi, K. and Udaka, S., Sweetness of lysozymes, *Biosci. Biotechnol. Biochem.* 62, 605–606, 1998.
67. Matsuda, T., Ueno, Y., and Kitabatake, N., Sweetness and enzymatic activity of lysozyme, *J. Agric. Food Chem.* 49, 4937–4941, 2001.
68. Matsuda, T., Ide, N., and Kitabatake, N., Effects of chemical modification of lysine residues on the sweetness of lysozyme, *Chem. Senses* 30, 253–264, 2005.
69. Deyl, Z. and Miksik, I., Post-translational non-enzymatic modification of proteins. I. Chromatography of marker adducts with special emphasis to glycation reactions, *J. Chromatog. B. Biomed. Sci. Appl.* 699, 287–309, 1997.
70. Kjærgard, I.V.H., Nørrelykke, M.E., Baron, C.P., and Jessen, F., Identification of carbonylated protein in frozen rainbow trout (*Oncorhynchus mykissi*) fillets and development of protein oxidation during storage, *J. Agric. Food Chem.* 54, 9437–9446, 2006.
71. Chaijan. M., Benjakul, S., Visessanguan, W. et al., The effect of freezing and aldehydes on the interaction between myoglobin and myofibrillar proteins, *J. Agric. Food Chem.* 55, 4562–4568, 2007.
72. Labuza, T.P. and Bergquist, S., Kinetics of oxidation of potato chips under constant temperature and sine wave temperature conditions, *J. Food Sci.* 48, 712–715, 1983.
73. Chung, S.Y. and Champagne, E.T., Association of end-product with increased IgE binding of roasted peanuts, *J. Agric. Food Chem.* 49, 3911–3916, 2001.
74. Nemet, I., Varga-Deftedarović, L., and Turk, Z., Methylglyoxal in food and living organisms, *Mol. Nutr. Food Res.* 50, 1105–1117, 2006.
75. Ahmad, M.X., Pischetrieder, M., and Ahmed, N., Aged garlic extract and *S*-allyl cysteine prevent formation of advanced glycation end products, *Eur. J. Pharmacol.* 561, 32–36, 2007.
76. Sung, H.Y., Chen, H.-J., Liu, T.-Y., and Su, J.-C., Improvement of the functionalities of soy protein isolate through chemical phosphorylation, *J. Food Sci.* 48, 716–721, 1983.

77. Chen, H.-J., Lin, S.-J., and Sung, H.-Y., enzymatic preparation of seasoning 5'-nucleotides from baker's yeast, *Proc. Natl. Sci. Counc. Repub. China B* 8, 124–128, 1984.
78. Lee, S.H., Yang, J.I., and Hong, S.M., Phosphorylation of peptides derived from isolated soybean protein: Effects on calcium binding, solubility and influx into Caco-2 cells, *Biofactors* 23, 121–128, 2005.
79. Giec, A., Stasinska, B., and Skupin, J., A protein isolate for food by phosphorylation of yeast homogenates, *Food Chem.* 31, 279–288, 1989.
80. Pacheco, M.T.B. and Sgarbieri, V.C., Hydrophilic and rheological properties of brewer's yeast protein concentrates, *J. Food Sci.* 63, 238–243, 1998.
81. Kim, N.S., Kwon, D.Y., and Nam, Y.J., Effects of phosphorylation and acetylation on functional properties of soy protein, *Han'guk Sikp'um Kwadhakhoechi* 20, 625–630, 1988.
82. Liu, T. and Chi, Y., Studies on phosphorylated modification of soybean protein isolate and its functional properties, *Shipin Yu Fajiao Gongye* 30, 118–121, 2004.
83. Casanova, F., Nakamura, A., Masuda, H. et al., Functionality of phosphorylated vicilin exposed to chemical and physical agents, *Food Chem.* 107, 1138–1143, 2008.
84. Gliko-Kabir, I., Yagen, B., Penhasi, A., and Rubenstein, A., Phosphated crosslinked guar for colon-specific drug delivery. I. Preparation and physicochemical characterization, *J. Control. Release* 63, 121–127, 2000.
85. Gliko-Kabir, I., Yagen, B., Baluom, M., and Rubenstein, A., Phosphated crosslinked guar for colon-specific drug delivery. II. *In vitro* and *in vivo* evaluation in the rat, *J. Control. Release* 63, 129–134, 2000.
86. Machy, D., Cateron, P., and Jozefonvicz, J., A new vascular prosthesis impregnated with cross-linked dextran, *J. Biomater. Sci. Polym. Ed*, 963–975, 2002.
87. Autissier, A., Letourneur, D., and Le Visage, C., Pullulan-based hydrogel for smooth muscle cell culture, *J. Biomed. Mater. Res. A.* 82, 336–342, 2007.
88. Lack, S., Dulong, V., Picton, L. et al., High-resolution nuclear magnetic resonance spectroscopy studies of polysaccharides crosslinked by sodium trimetaphosphate: A proposal for the reaction mechanism, *Carbohydr. Res.* 342, 943–953, 2007.
89. Thébaud, N.B., Pierron, D., Bareille, R. et al., Human endothelial progenitor cell attachment to polysaccharide-base hydrogels: A pre-requisite for vascular tissue engineering, *J. Mater. Sci. Mater. Med.* 18, 339–345, 2007.
90. Liu, M., Fan, J., Wang, K., and He, Z., Synthesis, characterization, and evaluation of phosphated cross-linked konjac glucomannan hydrogels for colon-targeted drug delivery, *Drug. Deliv.* 14, 397–407, 2007.
91. Rubin, C.S. and Rosen, O.M., Protein phosphorylation, *Annu. Rev. Biochem.* 44, 831–887, 1975.
92. Nimmo, H.G. and Cohen, P., Hormonal control of protein phosphorylation, *Adv. Cyclic Nucleotide Res.* 8, 145–266, 1977.
93. Krebs, E.G., Historical perspectives on protein phosphorylation and a classification system for protein kinases, *Philos. Trans. R. Soc. Lond. B. Biol. Sci.* 302, 3–11, 1983.
94. Ptacek, J. and Snyder, M., Charging it up: Global analysis of protein phosphorylation, *Trends Genet.* 22, 545–554, 2006.
95. Ubersax, J.A. and Ferrell, J.E., Jr., Mechanisms of specificity in protein phosphorylation, *Nat. Rev. Mol. Cell Biol.* 8, 530–541, 2007.
96. Stamm, S., Regulation of alternative splicing by reversible protein phosphorylation, *J. Biol. Chem.* 283, 1223–1227, 2008.
97. Putman, F.W., The chemical modification of proteins, in *The Proteins*, Eds. H. Neurath and K. Bailey, Academic Press, New York, 1953.
98. Taborsky, G., Phosphoproteins, *Adv. Protein Chem.* 28, 1–210, 1974.
99. McKenzie, H.A., Milk proteins, *Adv. Protein Chem.* 22, 55–234, 1967.
100. Corbridge, D.E.C., *Phosphorous: An Outline of Its Chemistry, Biochemistry and Technology*, 4th ed., Elsevier, Amsterdam, Netherlands, 1990.

101. Durocher, Y. and Chevalier, S., Detection of phosphotyrosine in glutaraldehyde-cross-linked and alkali-treated phosphoproteins following their partial hydrolysis in gels, *J. Biochem. Biophys. Methods* 28, 101–113, 1994.
102. Hiraishi, H., Yokoi, F., and Kumon, A., 3-Phosphohistidine and 6-phospholysine are substrates of a 56kDa inorganic pyrophosphatase from bovine liver, *Arch. Biochem. Biophys* 349, 381–387, 1988.
103. McTique, J.J. and Van Etten, R.L., An essential active-site histidine residue human prostatic acid phosphatase. Ethoxyformylation by diethyl pyrocarbonate and phosphorylation by a substrate, *Biochim. Biophys. Acta* 523, 407–421, 1978.
104. Pas, H.H. and Roubillard, G.T., *S*-Phosphocysteine and phosphohistidine are intermediates in the phosphoenolpyruvate-dependent mannitol transport catalyzed by *Escherichia coli* EIIM+1, *Biochemistry* 27, 5835–5839, 1988.
105. Koshland, D.E., Jr., Effect of catalysis on the hydrolysis of acetyl phosphate. Nucleophilic displacement mechanisms in enzymatic reactions, *J. Am. Chem. Soc.* 74, 2286–2292, 1952.
106. Ostrowski, W., Isolation of tau-phosphohistidine from a phosphoryl enzyme intermediate of human prostatic acid phosphatase, *Biochim. Biophys. Acta* 526, 147–153, 1978.
107. Congote, L.F., Dansylation and high-performance liquid chromatographic separation of phosphoserine, phosphothreonine and phosphortyrosine from ^{32}P-labeled protein hydrolyzate. *J. Chromatog.* 253, 270–282, 1982.
108. Hultqvist, D.E., Mayer, R.W., and Boyer, P.D., The preparation and characterization of 1-phosphohistine and 3-phosphohistidine, *Biochemistry* 5, 322–3332, 1966.
109. Severin, S.E., Discovery of carnosine and anserine. Some of their properties, *Biokhimiia* 57, 1285–1295, 1992.
110. Bionidi, R.M., Walz, K., and Issinger, O.G. et al., Discrimination between acid and alkali-labile phosphorylated residues in Immobilon: Phosphorylation studies of nucleoside diphosphate kinase, *Anal. Biochem.* 260,165–171, 1996.
111. Attwood, P.V., Piggett, M.J., Xu, Z.L., and Besart, P.G., Focus on phosphohistidine, *Amino Acids* 32, 145–156, 2007.
112. Martensen, T.M., Chemical properties, isolation and analysis of *O*-phosphates in proteins, *Methods Enzymol.* 107, 3 22, 1984.
113. Fujitaki, J.M. and Smith, R.A., Techniques in the detection and characterization of phosphoramidate-containing proteins, *Methods Enzymol.* 107, 23–36, 1984.
114. Pigiet, V. and Conley, R.P., Isolation and characterization of phosphoredoxin from *Escherichia coli*, *J. Biol. Chem.* 253, 1910–1920, 1978.
115. Zetterqvist, O. and Engström, L., Isolation of *N*-ε-[^{32}P]phosphoryl lysine from rat liver cell sap after incubation with [^{32}P]adenosine triphosphate, *Biochim. Biophys. Acta* 141, 523–532, 1967.
116. Kamps, M.P., Determination of phosphoramino acid composition by acid hydrolysis of protein blotted to Immobilon, *Methods Enzymol.* 201, 21–27, 1991.
117. Plimmer, R.H.A., Esters of phosphoric acid. 4. Phosphoryl hydroxyamino-acids, *Biochem. J.* 35, 464–469, 1941.
118. Ringler, D.P., Separation of phosphotyrosine, phosphoserine and phosphothreonine by high-performance liquid chromatography, *Methods Enzymol.* 201, 3–21, 1991.
119. Reitz, H.C., Ferrel, R.E., Fraenkel-Conrat, H., and Olcott, H.S., Action of sulfating agents on proteins and model substances. I. Concentrated sulfuric acid, *J. Am. Chem. Soc.* 68, 1024–1031, 1946.
120. Ferrel, R.E., Olcott, H.S., and Fraenkel-Conrat, H., Phosphorylation of proteins with phosphoric acid containing excess phosphorous pentoxide, *J. Am. Chem. Soc.* 70, 2101–2102, 1948.
121. Fraenkel-Conrat, J. and Fraenkel-Conrat, H., The essential groups of insulin, *Biochim. Biophys. Acta* 5, 89–97, 1950.

122. Rimington, C.L., Phosphorylation of proteins, *Biochem. J.* 21, 272–281, 1927.
123. Matheis, G. and Whittaker, J.R., Chemical phosphorylation of food proteins: An overview and prospectus, *J. Agric. Food Chem.* 32, 699–704, 1984.
124. Matheis, G., Phosphorylation of food proteins with phosphorous oxychloride—improvement of functional and nutritional properties: A review, *Food Chem.* 39, 19–26, 1991.
125. Medina, A., Colas, B., Le Meste, M. et al., Physicochemical and dynamic properties of caseins modified by chemical phosphorylation, *J. Food Sci.* 57,617–621, 1992.
126. Medina, A.L., Mesnier, D., Tainturier, G. et al., Chemical phosphorylation of bovine casein: Relationship between the reacting mixture and the binding sites of the phosphoryl moiety, *Food Chem.* 57, 261–265, 1996.
127. Schwenke, K.D., Mothes, R., Dudek, S., and Görnitz, E., Phosphorylation of the 12S globulin from rapeseed (*Brassica napus* L.) by phosphorous oxychloride: Chemical and conformational aspects, *J. Agric. Food Chem.* 48, 708–714, 2000.
128. Nayak, S.K., Arora, J., Sindu, J.S., and Sangwan, R.B., Effect of chemical phosphorylation on the solubility of buffalo milk protein, *Int. Dairy J.* 16, 261–265, 1996.
129. Enomoto, H., Li, C.P., Morizane, K. et al., Improvement of functional properties of bovine serum albumin through phosphorylation by dry-heating in the presence of pyrophorphate, *J. Food Sci.* 73, C84–C91, 2008.
130. Lotan, R. and Sharon, N., Modification of the biological properties of plant lectins by chemical crosslinking, *Adv. Expt. Med. Biol.* 86A, 149–168, 1977.
131. Hoseney, R.C. and Roger, D.E., The formation and properties of wheat flour doughs, *Crit. Rev. Food Sci. Nutr.* 29, 73–93, 1990.
132. Gerrard, J.A., Meade, S.J., Miller, A.G. et al., Protein cross-linking in food, *Ann. N. Y. Acad. Sci.* 1043, 97–103, 2005.
133. Whitaker, J.R. and Feeney, R.E., Chemical and physical modification of proteins by the hydroxide ion, *Crit. Rev. Food Sci. Nutr.* 19, 173–212, 1983.
134. Skiorski, Z.E., Scott, D.N., and Buisson, D.H., The role of collagen in the quality and processing of fish, *Crit. Rev. Food Sci. Nutr.* 20, 301–343, 1984.
135. Rath, N.C., Huff, G.R., Huff, W.E., and Balog, J.M., Factors regulating bone maturity and strength in poultry, *Poult. Sci.* 79, 1024–1032, 2000.
136. McCormick, R.J., Extracellular modifications to muscle collagen: Implications for meat quality, *Poult. Sci.* 78, 785–791, 1999.
137. Wang, Y., Mo, X., Sun, X.S., and Wang, D., Soy protein adhesion enhanced by glutaraldehyde crosslink, *J. Appl. Polym. Sci.* 104, 130–136, 2007.
138. Snyders, R., Shingel, K.I., Zabeida, O. et al., Mechanical and microstructural properties of hybrid poly(ethylene glycol)-soy protein hydrogels for wound dressing applications, *J. Biomed. Mater. Res.* 83A, 88–97, 2007.
139. Ernst-Cabrera, K. and Wilchek, M., Coupling of ligands to primary hydroxyl-containing silica for high-performance affinity chromatography. Optimization of conditions, *J. Chromatog.* 397, 187–196, 1985.
140. Hernández-Muñoz, P., Kanavouras, A., Lagaron, J.M., and Gavara, R., Development and characterization of films based on chemically cross-linked gliadins, *J. Agric. Food Chem.* 53, 8216–8223, 2005.
141. Hayatsu, H., Wataya, Y., Kai, K., and Iida, S., Reaction of sodium bisulfite with uracil, cytosine, and their derivatives, *Biochemistry* 9, 2858–2865, 1970.
142. Draper, D.E., Attachment of reporter groups to specific, selected cytidine residues in RNA using a bisulfite-catalyzed transamination reaction, *Nucleic Acids Res.* 12, 989–1002, 1984.
143. Avignolo, C., Valente, F., Cai, S. et al., Biotinylation of double stranded DNA after transamination, *Biochem. Biophys. Res. Commun.* 170, 243–250, 1990.
144. Miller, P.S. and Cushman, C.D., Selective modification of cytosines in oligodeoxyribonucleotides, *Bioconjug. Chem.* 3, 74–79, 1992.

145. Shapiro, R. and Gazit, A., Crosslinking of nucleic acids and proteins by bisulfite, *Adv. Exp. Med. Biol.* 86A, 633–640, 1977.
146. Klostermeyer, H. and Reimerdes, E.H., Heat induced crosslinks in milk proteins and consequences for the milk system, *Adv. Exp. Med. Biol.* 86B, 263–275, 1977.
147. Venkatachalam, N., McMahon, D.J., and Savello, P.A., Role of protein and lactose interactions in the age gelation of ultra-high temperature processed concentrated skim milk, *J. Dairy Sci.* 76, 1882–1894, 1993.
148. Lauber, S., Klostermeyer, H., and Henle, T., On the influence of non-enzymatic crosslinking of caseins on the gel strength of yogurt, *Nahrung* 45, 215–217, 2001.
149. O'Sullivan, M.M., Kelly, A.L., and Fox, P.F., Effects of transglutaminase on the heat stability of milk: A possible mechanism, *J. Dairy Sci.* 85, 1–7, 2002.
150. Vasbinder, A.J., Rollema, H.S., Bot, A., and de Kruif, C.G., Gelation mechanism of milk as influenced by temperature and pH; studied by the use of transglutaminase cross-linked casein micelles, *J. Dairy Sci.* 86, 1556–1563, 2003.
151. Menendez. P. Schwarzenbolz, U., Rohm, H., and Henle, T., Casein gelation under simultaneous action of transglutaminase and glucono-Δ-lactone, *Nahrung* 48, 165–168, 2004.
152. Partschefeld, C., Schwarzenbolz, U., Richter, S., and Henle, T., Crosslinking of casein by microbial transglutaminase and its resulting influence on the stability of micelle structure, *Biotechnol. J.* 2, 456–461, 2007.
153. Aboumahmoud, R. and Savello, P., Crosslinking of whey protein by transglutaminase, *J. Dairy Sci.* 73, 256–263, 1990.
154. Wilcox, C.P. and Swaisgood, H.P., Modification of the rheological properties of whey protein isolate through the use of an immobilized microbial transglutaminase, *J. Agric. Food Chem.* 50, 5546–5551, 2002.
155. Yokoyama, K., Nio, N., and Kituchi, Y., Properties and applications of microbial transglutaminase, *Appl. Microbial. Biotechnol.* 64, 447–454, 2004.
156. Cozzolino, A., Di Pierro, P., Marieniello, L. et al., Incorporation of whey proteins into cheese curd by using transglutaminase, *Biotechnol. Appl. Biochem.* 38, 285–295, 2003.
157. Truong, V.D., Clare, D.A., Catignani, G.L., and Swaisgood, H.E., Cross-linking and rheological changes of whey proteins treated with microbial transglutaminase, *J. Agric. Food Chem.* 52, 1170–1176, 2004.
158. Eissa, A.S., Bisram, S., and Khan, S.A., Polymerization and gelation of whey protein isolates at low pH using transglutaminase enzyme, *J. Agric. Food Chem.* 52, 4456–4464, 2004.
159. Eissa, A.S. and Khan, S.A., Acid-induced gelation of enzymatically modified, pretreated whey proteins, *J. Agric. Food Chem.* 53, 5010–5017, 2005.
160. Mizuno, A., Mitsuiki, M., Motoki, M. et al., Relationship between the glass transition of soy protein and molecular structure, *J. Agric. Food Chem.* 48, 3292–3297, 2000.
161. Mizuno, A., Mitsuiki, M., and Motoki, M., Effect of transglutaminase treatment on glass transition of soy proteins, *J. Agric. Food Chem.* 48, 3286–3291, 2000.
162. Zhang, G., Matsumara, Y., Matsumoto, S. et al., Effects of Ca^{2+} and sulfydryl reductant on the polymerization of soybean glycinin catalyzed by mammalian and microbial transglutaminase, *J. Agric. Food Chem.* 51, 236–243, 2003.
163. Tang, C.H., Jiang, Y., Wen, Q.B., and Yang, X.Q., Effect of transglutaminase treatment on the properties of cast films of soy protein isolates, *J. Biotechnol.* 120, 296–307, 2005.
164. Lanier, T.C., High pressure processing effects on fish proteins, *Adv. Exp. Med. Biol.* 434, 45–55, 1998.
165. Partschelfeld, C., Richter, S., Schwarzenbolz, U., and Henle, T., Modification of β-lactoglobulin by microbial transglutaminase under high hydrostatic pressure: Localization of reactive glutamine residues, *Biotechnol. J.* 2, 462–468, 2007.
166. Clare, D.A, Gharst, G., and Sanders, T.H., Transglutaminase polymerization of peanut proteins, *J. Agric. Food Chem.* 55, 432–438, 2007.

167. Siu, N.C., Ma, C.Y., and Mine, Y., Physicochemical and structural properties of oat globulin polymers formed by a microbial transglutaminase, *J. Agric. Food Chem.* 50, 2660–2665, 2002.
168. Siu, N.C., Ma, C.Y., Mock, W.Y., and Mine, Y., Functional properties of oat globulin modified by a calcium-independent microbial transglutaminase, *J. Agric. Food Chem.* 50, 2666–2672, 2002.
169. Autio, K., Kruus, K., Knaapila, A. et al., Kinetics of transglutaminase-induced cross-linking of wheat proteins in dough, *J. Agric. Food Chem.* 53, 1039–1045, 2005.
170. Chan, T., Embree, M.D., Brown, E.M. et al., Enzyme-catalyzed gel formation of gelatin and chitosan: Potential for *in situ* applications, *Biomaterials* 24, 2831–2841, 2003.
171. Di Pierro, P., Chico, B., Villalonga, R. et al., Chitosan-whey protein edible films produced in the absence or presence of transglutaminase: Analysis of their mechanical and barrier properties, *Biomacromolecules* 7, 744–749, 2006.
172. Marinello, L., Giosafatto, C.V., Moschetti, G. et al., Fennel waste-based films suitable for protecting cultivations, *Biomacromolecules* 8, 3008–3014, 2007.
173. Katzuda, S., Ito, M., and Waseda, T., Papain-hydrolyzed port meat reduces serum cholesterol level and premature atherosclerosis in dietary-induced hypercholesterolemic rabbits, *J. Nutr. Sci. Vitaminol.* 46, 180–187, 2000.
174. Kaul, P., Sathish, H.A., and Prakash, V., Effect of metal ions on structure and activity of papain from *Carios papaya*, *Nahrung* 2002, 46, 2–6, 2002.
175. Toldrá, F. and Flores, M., The role of muscle proteases and lipases in flavor development during the processing of dry-cured ham, *Crit. Rev. Food Sci. Nutr.* 38, 331–352, 1998.
176. Pommet, M., Redi, A., Guilbert, S. et al., Impact of protein-size distribution on gluten thermal reactivity and functional properties, *J. Agric. Food Chem.* 53, 3943–3949, 2005.
177. Maillard, L.C., Action des acides amines sur les sucres: Formation des melanoidines par voie methodique. *C. R. Hebd. Seances Acad. Sci.* 154, 66, 1912.
178. Tanaka, M., Kimiager, M., and Lee, T.C., Effect of Maillard browning reaction on nutritional quality of protein, *Adv. Exp. Med. Biol.* 86B, 321–341, 1977.
179. Waller, G.R. and Feather, M.S., *The Maillard Reaction in Foods and Nutrition*, American Chemical Society, Washington, DC, 1983.
180. O'Brien, J. and Morrissey, P.A., Nutritional and toxicological aspects of the Maillard browning reaction in foods, *Crit. Rev. Food Sci. Nutr.* 28, 211–248, 1989.
181. Kaanane, A. and Labuza, T.R., The Maillard reaction in foods, *Prog. Clin. Biol. Res.* 204, 301–327, 1989.
182. Finot, P.A., *The Maillard Reaction in Food Processing, Human Nutrition, and Physiology*, Birkhäuser Verlag, Basel, Switzerland, 1990.
183. Wedzicha, B.L., Bellion, I., and Goddard, S.J., Inhibition of browning by sulfites, *Adv. Exp. Med. Biol.* 289, 217–236, 1991.
184. Labuza, T.P., *Maillard Reaction in Chemistry, Food, and Health,* Royal Society of Chemistry, Cambridge, 1994.
185. Shahadi, F. and Ho, C.-T., *Process-Induced Changes in Foods*, Plenum Press, New York, 1994.
186. Ikan, R., *The Maillard Reaction: Consequences for the Chemical and Life Sciences,* Wiley, Chichester, U.K., 1996.
187. O'Brien, J., *The Maillard Reaction in Foods and Medicine,* Royal Society of Chemistry, Cambridge, 1998.
188. Hofmann, T., Bors, W., and Stettmaier, K., Studies on radical intermediates in the early stages of the nonenzymatic browning reaction of carbohydrates and amino acids, *J. Agric. Food Chem.* 47, 379–390, 1999.
189. Ajandouz, E.W. and Puigserver, A., Nonenzymatic reaction of essential amino acids: Effect of pH on carmelization and Maillard reaction kinetics, *J. Agric. Food Chem.* 47, 1786–1793, 1999.

190. Ide, N., Lau, B.H., Ryu, K. et al., Antioxidant effects of fructosyl arginine, a Maillard reaction product in aged garlic extracts, *J. Nutr. Biochem.* 10, 372–376, 1989.
191. Fayle, S.E. and Gerrard, J.A., *The Maillard Reaction*, Royal Society of Chemistry, Cambridge, United Kingdom, 2002.
192. Mundt, S. and Wedzicha, B.L., A kinetic model for the glucose-fructose-glycine browning reaction, *J. Agric. Food Chem.* 51, 3651–3655, 2003.
193. Nursten, H., *The Maillard Reaction: Chemistry, Biochemistry, and Implications,* Royal Society of Chemistry, Cambridge, 2005.
194. Durkham, S.C., Dodson, A. T., Bakker, J., and Ames, J.M., Volatile flavor components of baked potato flesh. A comparison of eleven potato cultivars, *Nahrung*, 45, 317–323, 2001.
195. Frazier, R.A., Ames, J.M., and Nursten, H.E., The development and application of capillary electrophoresis methods for food analysis, *Electrophoresis*, 20, 3156–3180, 1999.
196. Finot, P.-A., Historical perspective of the Maillard reaction in food science, *Ann. N. Y. Acad. Sci.* 1043, 1–8, 2005.
197. Danehy, J.P., Maillard reactions: Nonenzymatic browning in food systems with special reference to the development of flavor, *Adv. Food Res.* 30, 77–138, 1986.
198. Dyer, D.G., Blackledge, J.A, Thorpe, S.R., and Baynes, J.W., Formation of pentosidine during nonenzymatic browning of proteins by glucose. Identification of glucose and other carbohydrates as possible precursors of pentosidine *in vivo*, *J. Biol. Chem.* 266, 11654–11660, 1991.
199. Ajandouz, E.H. and Puigserver, A., Nonenzymatic browning reaction of essential amino acids: Effect of pH on carmelization and Maillard reaction kinetics, *J. Agric. Food Chem.* 47, 1786–1793, 1999.
200. Zamora, R., Alaiz, M., and Hidalgo, F.J., Contribution of pyrrole formation produced by amino-carbonyl reactions, *J. Agric. Food Chem.* 48, 3152–3158, 2000.
201. Frank, O. and Hofmann, T., Characterization of key chromophores formed by nonenymatic browning of hexoses and L-alanine by using the color activity concept, *J. Agric. Food Chem.* 48, 6303–6311, 2000.
202. Miao, S. and Roos, Y.H., Nonenzymatic browning kinetics of a carbohydrate-based low-moisture food system at temperatures applicable to spray drying, *J. Agric. Food Chem.* 52, 5250–5257, 2004.
203. Ge, P.G., Oliver, C.M., and Melton, L.D., α-Dicarbonyl compounds formed by nonenzymatic browning during the dry heating of caseinate and lactose, *J. Agric. Food Chem.* 54, 6852–6857, 2006.
204. Pan, G.G. and Melton, L.D. Nonenzymatic browning of lactose and caseinate during dry heating at different relative humidities, *J. Agric. Food Chem.* 55, 10036–10042, 2007.
205. Chisari, M,. Barbagallo, R.N., and Spagna, G., Characterization of polyphenol oxidase and peroxidase and influence on browning of cold stored strawberry fruit, *J. Agric. Food Chem.* 55, 3469–3476, 2007.
206. Segovia-Bravo, K.A., Jarén-Galan, M., García-García, P., and Garrido-Fernandez, A., Characterization of polyphenol oxidase from the Manzanilla cultivar (*Olea europaea pomiformis*) and prevention of browning reactions in bruised olive fruits, *J. Agric. Food Chem.* 55, 6515–6520, 2007.
207. Arias, E., González, J., Oria, R., and Lopez-Buesa, P., Ascorbic acid and 4-hexylresorcinol effects on pear PPO and PPO catalyzed browning reaction, *J. Food Sci.* 208, C422–C429, 2007.
208. Arias, J., González, J., Peiró, J.M. et al., Browning prevention by ascorbic acid and 4-hexylresorcinol: Different mechanisms of action on polyphenol oxidase in the presence and absence of substrates, *J. Food Sci.* 72, C464–C470, 2007.

209. Chisari, M., Barbagallo, R.N., and Spagna, G., Characterization and role of polyphenol oxidase and peroxidase in browning of fresh-cut melon, *J. Agric. Food Chem.* 56, 132–138, 2008.
210. Frank, O., Heuberger, S., and Hofmann, T., Structure determination of a novel 3(6H)-pyranone chromophore and clarification of its formation from carbohydrates and primary amino acids, *J. Agric. Food Chem.* 49, 1595–1600, 2001.
211. Herderich, M., Mass spectrometry techniques in food analysis, *Nachrichten aus der Chemie* 49, 378–381, 2001.
212. Mancilla-Margalli, N.A. and Lopez, M.G., Generation of Maillard compounds from inulin during thermal processing of *Agave tequilana Weber* var. azul, *J. Agric. Food Chem.* 50, 806–812, 2002.
213. Ulrich, P. and Cerami, A., Protein glycation, diabetes, and aging, *Recent Prog. Horm. Res.* 56, 1, 2001.
214. Biemel, K.M. and Lederer, M.O., Site-specific quantitative evaluation of the protein glycation product N(6) 0 (2,3-Dihydroxy-5,6-dioxohexyl-1-lysinate by LC-(ESI) MS peptide mapping: Evidence for its key role in AGE formation, *Bioconjug. Chem.* 14, 619, 2003.
215. Tessier, F.J., Monnier, V.M., Sayre, L.M., and Kornfield, J.A., Triosidines: Novel Maillard reaction products and cross-links from the reaction of triose sugars with lysine and arginine residues, *Biochem. J.* 369, 705, 2003.
216. Thornally, P.J., Battah, S., Ahmed, N. et al., Quantitative screening of advanced glycation end products in cellular and extracellular proteins by tandem mass spectrometry, *Biochem. J.* 375, 581, 2003.
217. Voziyan, P.A., Khalifah, R.G., and Thibaudeau, C., Modification of proteins *in vitro* by physiological levels of glucose. Pyridoxamine inhibits conversion of amadori intermediate to advanced glycation end-products through binding of redox metal ions, *J. Biol. Chem.* 278, 46616, 2003.
218. Argirov, O.K., Lin, B., Olesen, P., and Ortwerth, B.J., Isolation and characterization of a new advanced glycation endproduct of dehydroascorbic acid and lysine, *Biochim. Biophys. Acta* 1620, 235, 2003.
219. Miller, A.G., Meade, S.J., and Gerrard, J.A., New insights into protein crosslinking via the Maillard reaction: Structural requirements, the effects on enzyme function, and predicted efficacy of crosslinking inhibitors as anti-aging therapeutics, *Bioorg. Med. Chem.* 11, 843, 2003.
220. Glomb, M.A. and Tschirnich, R., Detection of α-dicarbonyl compounds in Maillard reaction systems and *in vivo*, *J. Agric. Food Chem.* 49, 5543–5550, 2001.
221. Humeny, A., Kislinger, T., Becker, C.H. et al., Qualitative determination of specific protein glycation products by matrix-assisted laser desorption mass spectrometry peptide mapping, *J. Agric. Food Chem.* 50, 2153–2160, 2002.
222. Hollnagel, A. and Kroh, L.W., 3-Deoxypentosulose: An α-dicarbonyl compound predominating in nonenzymatic browning of oligosaccharides in aqueous solution, *J. Agric. Food Chem.* 50, 1659–1664, 2002.
223. Sarvaiva. M.A., Borges, C.M., and Florêncio, M.H., Reactions of a modified lysine with aldehydic and diketonic dicarbonyl compounds: An electrospray mass spectrometry structure/activity study, *J. Mass Spectrom.* 41, 216–228, 2006.
224. Munanairi, A., O'Banion, S.K., Gamble, R. et al., The multiple Maillard reactions of ribose and deoxyribose sugars and sugar phosphates, *Carbohydr. Res.* 342, 2575–2592, 2007.
225. Yaylayan, V.A., Wnorowski, A., and Perez Locas, C., Why asparagine needs carbohydrates to generate acrylamide, *J. Agric. Food Chem.* 51, 1753–1757, 2003.

226. Stadler, R.H., Robert, E., Riediker, S. et al., In-depth mechanistic study on the formation of acrylamide and other vinylogous compounds by the Maillard reaction, *J. Agric. Food Chem.* 52, 5550–5558, 2004.
227. Yaylayan, V.A. and Stadler, R.H., Acrylamide formation in foods: A mechanistic perspective, *J. AOAC Int.* 88, 262–267, 2005.
228. Blank, I., Robert, F., Goldman, T. et al., Mechanism of acrylamide formation: Maillard-induced transformation of asparagine, *Adv. Exp. Med. Biol.* 561, 171–189, 2005.
229. Vauthey, S., Milo, C., Frossard, P. et al., Structured fluids as microreactors for flavor formation by the Maillard reaction, *J. Agric. Food Chem.* 48, 4808–4816, 2000.
230. Ottinger, H., Bareth, A., and Hofmann, T., Characterization of natural "cooling" compounds formed from glucose and l-proline in dark malt by application of taste dilution analysis, *J. Agric. Food Chem.* 49, 1336–1344, 2001.
231. Spanier, A.M., Flores, M., Toldrá, F. et al., Meat flavor: Contribution of proteins and peptides to the flavor of beef, *Adv. Exp. Med. Biol.* 542, 33–49, 2004.
232. Adams, A., Tehrani, K.A., Kersiene, M., and De Kimpe, N., Detailed investigation of the production of the bread flavor component 6-acetyl-1,2,3,4-tetrahydropyridine in proline/1,3-dihydroxyacetone model systems, *J. Agric. Food Chem.* 52, 5685–5693, 2004.
233. Aliani, M. and Farmer, L.J., Precursors of chicken flavor. I. Determination of some flavor precursors in chicken muscle, *J. Agric. Food Chem.* 53, 6067–6072, 2005.
234. Hofmann, T., Taste-active Maillard reaction products: The "tasty" world of nonvolatile Maillard reaction products, *Ann. N. Y. Acad. Sci.* 1043, 20–29, 2005.
235. Adams, A. and De Kimpe, N., Chemistry of 2-acetyl-1-pyrroline, 6-acetyl-1,2,3,4-tetrahydropyridine, 2-acetyl-2-thiazoline, and 5-acetyl-2,3-dihydro-4H-thiazine: Extraordinary Maillard flavor compounds, *Chem. Rev.* 106, 2299–2319, 2006.
236. CODATA Bulletin 63, 1, 1986.
237. CODATA Bulletin 68, 993, 1996.
238. *Compendium of Chemical Terminology IUPAC Recommendations*, 2nd ed., McNaught, A.D. and Wilkinson, A., Blackwell, Oxford, UK, 1997.
239. San Martín, M.F., Barbosa-Cánovas, G.V., and Swanson, B.G., Food processing by high hydrostatic pressure, *Crit. Rev. Food Sci. Nutr.* 42, 627–645, 2003.
240. Ludikhuyre, L, Van Loey, A., Indrawati, S.C., and Hendrickx, M., Effects of combined pressure and temperature on enzyme related to quality of fruits and vegetables: From kinetic information to process engineering aspects, *Crit. Rev. Food Sci. Nutr.* 43, 527–586, 2003.
241. Dumay, E., Picart, L, Regnault, S., and Thiebaud, M., High pressure-low temperature processing of food proteins, *Biochem. Biophys. Acta* 1764, 599–618, 2006.
242. Knorr, D., Heinz, V., and Buckow, R., High pressure application for food biopolymers, *Biochim. Biophys. Acta* 1764, 619–631, 2006.
243. Rastogi, N.K., Rafhavarao, K.S., Balasubramaniam, V.M. et al., Opportunities and challenges in the high pressure processing of foods, *Crit. Rev. Food Sci. Nutr.* 47, 69–112, 2007.
244. Augustin, M.A. and Udabage, P., Influence of processing on functionality of milk and dairy proteins, *Adv. Food Nutr. Res.* 53, 1–38, 2007.
245. Smelt, J.P., Hellemons, J.C., Wouters, P.C., and van Gerwen, S.J., Physiological and mathematical aspects in setting criteria for decontamination of foods by physical means, *Int. J. Food Microbiol.* 78, 57–77, 2002.
246. Ross, A.I., Griffiths, M.W., Mittal, G.S., and Deeth, H.C., Combining nonthermal technologies to control foodborne microorganisms, *Int. J. Food Microbiol.* 89, 1250138, 2003.

247. Li, S.Q., Zhang, H.Q., Balasubramaniam, V.M. et al., Comparison of effects of high-pressure and heat treatment on immunoactivity of bovine milk immunoglobulin G in enriched soymilk under equivalent microbial inactivation levels, *J. Agric. Food Chem.* 54, 739–746, 2006.
248. Picart, L., Thiebaud, M., and René, M., Effects of high pressure homogenization of raw bovine milk on alkaline phosphatase and microbial inactivation. A comparison with continuous short-time thermal treatments, *J. Dairy Res.* 73, 454–463, 2006.
249. Muñoz, M., De Ancos, B., Sánchez-Moreno, C., and Pilar Cano M., Evaluation of chemical and physical (high-pressure and temperature) treatments to improve the safety of minimally processed mung bean sprouts during refrigerated storage, *J. Food Prot.* 69, 2395–402, 2006.
250. Koseki S, and Yamamoto K., A novel approach to predicting microbial inactivation kinetics during high pressure processing, *Int. J. Food Microbiol.* 116, 275–82, 2007.
251. Nosaka, S., Kamaya, H., and Ueda, I., High pressure and anesthesia: Pressure stimulates or inhibits bacterial bioluminescence depending upon temperature, *Anesth. Analg.* 67, 988–992, 1988.
252. Tauscher, B., Pasteurization of food by hydrostatic high pressure: Chemical aspects, *Z. Lebensm. Unters Forsch.* 200, 3–13, 1995.
253. Mentré, F. and Hui Bon Hoa, G., Effects of high hydrostatic pressure of living cells: A consequence of the properties of macromolecules and macromolecule-associated water, *Int. Rev. Cytol.* 201, 1–84, 2001.
254. Dmitreiv, L.F., A new approach to the problem of enzymatic catalysis based on excited state of proteins and Le Chatelier's principle, *Arch. Biochem. Biophys.* 464, 12–18, 2007.
255. *High-Pressure Chemistry Synthetic Mechanistic and Supercritical Applications*, Eds. R. van Eldik, and F.G. Klärner, Wiley-VCH, Weinheim, Germany, 2002.
256. *High-Pressure Techniques in Chemistry and Physics*, Eds. W.B. Holzapfel and N.S. Issacs, Oxford University Press, Oxford, 1992.
257. Balny, C., What lies in the future of high-pressure bioscience?, *Biochim. Biophys. Acta* 1764, 632–639, 2006.
258. Masson, P., Electrophoresis of proteins under high hydrostatic pressures, in *High-Pressure Techniques in Chemistry and Physics*, Eds. W.B. Holzapfel and N.S. Issacs, Oxford University Press, Oxford, Chapter 10, pp. 353–373, 1992.

Index

A

D

E

F

G

H

I

O

P

Q

R

S